Wolf-Heidegger's

Atlas of Human Anatomy

Volume 2

Wolf-Heideggers

Atlas der Anatomie des Menschen

Band 2

Petra Köpf-Maier, Berlin

Wolf-Heidegger's

Atlas of Human Anatomy

Volume 2

Head and Neck, Thorax, Abdomen, Pelvis, CNS, Eye, Ear

5th, completely revised and supplemented edition, 2000

866 figures of which 677 are in color

KARGER

Petra Köpf-Maier, Berlin

Wolf-Heideggers

Atlas der Anatomie des Menschen

Band 2

Kopf und Hals, Brust, Bauch, Becken, ZNS, Auge, Ohr

5., völlig neu bearbeitete und ergänzte Auflage, 2000

866 Abbildungen, davon 677 in Farbe

KARGER

Editor
Univ.-Prof. Dr. med. Petra Köpf-Maier
Professor of Anatomy
Freie Universität Berlin
Königin-Luise-Strasse 15
D–14195 Berlin (Germany)

Herausgeberin
Univ.-Prof. Dr. med. Petra Köpf-Maier
Professorin für Anatomie
Freie Universität Berlin
Königin-Luise-Strasse 15
D–14195 Berlin (Deutschland)

This Atlas is published in two volumes:
Volume 1: Systemic Anatomy, Body Wall, Upper and Lower Limbs
Volume 2: Head and Neck, Thorax, Abdomen, Pelvis, CNS, Eye, Ear

Dieser Atlas besteht aus zwei Bänden:
Band 1: Allgemeine Anatomie, Rumpfwand, obere und untere Extremität
Band 2: Kopf und Hals, Brust, Bauch, Becken, ZNS, Auge, Ohr

Until 1989 the Atlas was published as
'Atlas of Systematic Human Anatomy', vol. I–III
1st edition 1954
2nd edition 1960
3rd edition 1972
Spanish translation: Salvat Editores S.A., Barcelona
Portuguese translation: Editors Guanabara Koogan S.A., Rio de Janeiro

Der Atlas wurde bis 1989 unter dem Titel
«Atlas der systematischen Anatomie des Menschen», Bd. I–III, publiziert
1. Auflage 1954
2. Auflage 1960
3. Auflage 1972
Spanische Übersetzung: Salvat Editores S.A., Barcelona
Portugiesische Übersetzung: Editors Guanabara Koogan S.A., Rio de Janeiro

4th edition 1990
Published as 'Wolf-Heidegger's Atlas of Human Anatomy'
Japanese translation: Nishimura Co., Ltd., Tokyo
Indonesian translation: Penerbit Widya Indonesia, Jakarta

4. Auflage 1990
publiziert unter dem Titel «Wolf-Heideggers Atlas der HumanAnatomie»
Japanische Übersetzung: Nishimura Co., Ltd., Tokyo
Indonesische Übersetzung: Penerbit Widya Indonesia, Jakarta

Library of Congress Cataloging-in-Publication Data

Wolf-Heidegger's atlas of human anatomy. – 5th, completely rev. and supplemented ed. / Petra Köpf-Maier.
Includes bibliographical references and index.
Contents: v. 1. Systemic anatomy, body wall, upper and lower limbs –
v. 2. Head and neck, thorax, abdomen, pelvis, CNS, eye, ear.
English and German.
Added title page title: Wolf-Heideggers Atlas der Anatomie des Menschen.
Other title: Atlas der Anatomie des Menschen.
ISBN 3–8055–6754–5 (v. 1: hardcover). – ISBN 3–8055–6755–3 (v. 2: hardcover) – ISBN 3–8055–5442–7 (complete set)
1. Human anatomy–Atlases. I. Wolf-Heidegger, G. (Gerhard)
II. Köpf-Maier, P. (Petra) III. Title: Wolf-Heideggers Atlas der Anatomie des Menschen
[DNLM: 1. Anatomy atlases. QS 17W859 2000]
QM25.W633 2000 611'.022'2–dc21

Die Deutsche Bibliothek – CIP Einheitsaufnahme

Wolf-Heideggers Atlas der Anatomie des Menschen / Petra Köpf-Maier. – Basel; Freiburg [Breisgau]; Paris; London; New York; New Delhi; Bangkok; Singapore; Tokyo; Sydney: Karger
Parallelt.: Wolf-Heidegger's atlas of human anatomy
4. Aufl. u.d.T.: Wolf-Heideggers Atlas der HumanAnatomie
ISBN 3–8055–5442–7 (beide Bände zusammen)
Bd. 1. Allgemeine Anatomie, Rumpfwand, obere und untere Extremität. –
5., völlig neu bearb. und erg. Aufl. – 2000
ISBN 3–8055–6754–5
Bd. 2. Kopf und Hals, Brust, Bauch, Becken, ZNS, Auge, Ohr. –
5., völlig neu bearb. und erg. Aufl. – 2000
ISBN 3–8055–6755–3

An English nomenclature version, prepared in cooperation with Arthur W. English, Emory University, Atlanta, Georgia, is also available under the title 'Wolf-Heidegger's Atlas of Human Anatomy'.
Vol. 1: Systemic Anatomy, Body Wall, Upper and Lower Limbs: ISBN 3–8055–6852–5
Vol. 2: Head and Neck, Thorax, Abdomen, Pelvis, CNS, Eye, Ear: ISBN 3–8055–6853–3
Complete set: ISBN 3–8055–6854–1

KARGER
Basel · Freiburg · Paris · London · New York · New Delhi · Bangkok · Singapore · Tokyo · Sydney

Printed in Switzerland on acid-free paper by Reinhardt Druck, Basel
ISBN 3–8055–6755–3

The editor dedicates
this book to her grandson

Leander Leonin

Die Herausgeberin widmet
dieses Buch ihrem Enkel

Leander Leonin

Homage to Those Who Bequeathed Their Bodies to Science

Den Körperspendern zum Gedenken

'Hic locus est ubi mors gaudet succurrere vitae'

'This is the place where death delights in helping life'
(Inscription above the Anatomical Theatre of Bologna)

The present atlas of human anatomy shall not begin without paying due homage and returning thanks to those who freely bequeath their bodies to anatomy. Such donations testify to an admirable, unselfish, and idealistic sense of sacrifice and nothing can compensate for the invaluable service rendered to science and society. Anatomy and medicine owe these individuals a tremendous debt of gratitude. By bequeathing their bodies, they enable medical students to learn through real observation and direct 'grasping', and even now, at the end of the twentieth century, there is no alternative to this. Thus, even beyond death, these altruistic people help the living – medical students, physicians, and their patients alike. This is how the above inscription should be interpreted. Students should make every endeavour to be worthy of these voluntary and generous body donations by respecting and honoring the dead as well as by working hard and learning eagerly.

«Hier ist der Ort, wo der Tod sich freut, dem Leben zu helfen»
(Inschrift über dem Theatrum anatomicum von Bologna)

Der vorliegende Atlas der Anatomie des menschlichen Körpers soll nicht begonnen werden, ohne denjenigen Menschen Ehrerbietung entgegenzubringen und Dank abzustatten, die sich der Anatomie vermachen und ihren Leichnam durch eine freiwillige testamentarische Willenserklärung für Lehr- und Forschungszwecke zur Verfügung stellen. Diese Körperspende setzt eine bewundernswerte, selbstlose, idealistische Hingabebereitschaft voraus und stellt einen unbezahlbaren, nicht vergeltbaren Dienst an der Wissenschaft und Gesellschaft dar. Diesen hingabebereiten Menschen verdankt die Anatomie, und auch die Medizin, unermesslich viel. Durch ihre Körperspende ermöglichen sie den Studenten der Medizin ein Lernen in konkreter Anschauung und in «direktem Begreifen», was – auch am Ende des zwanzigsten Jahrhunderts – durch nichts anderes ersetzt werden kann. Hierdurch helfen diese hingabebereiten Menschen über ihren Tod hinaus den noch Lebenden, und zwar sowohl den Studenten der Medizin wie auch der Ärzteschaft und ihren Patienten. In diesem Sinne ist auch die oben zitierte Inschrift zu verstehen. Die Studenten sollten sich dieser freiwilligen und grossherzigen Körperspende durch Achtung und Ehrerbietung den Toten gegenüber sowie durch Fleiss und eifriges Lernen würdig erweisen.

Contents

XII Preface to the 5th Edition
XIV Preface to the 1st Edition
XVI Concept of the New Version of the Atlas and Illustration Credits
XX Information for Users

1–102 Head and Neck

2–17 Skull
18–25 Bones of the skull
26–27 Temporomandibular joint
28–32 Superficial and deep muscles of the scalp and face
33 Cervical fascia
34–37 Muscles of the neck
38–43 Oral cavity and teeth
44–45 Salivary glands and tonsils
46–47 Tongue
48–53 Pharynx
54–57 External nose and nasal cavity
58–63 Sections and tomograms of the face
64–69 Larynx
70–71 Thyroid gland
72–75 Sections and tomograms of the neck
76–80 Blood vessels of the head and neck
81 Lymph vessels and lymph nodes of the head and neck
82–83 Cutaneous innervation of the head and neck
84–90 Nerves and blood vessels of the face
91–101 Nerves and blood vessels of the neck

103–168 Thoracic Viscera

104 Mediastinum and thymus
105–107 Esophagus
108–110 Trachea and bronchi
111–117 Lungs, bronchial tree and bronchopulmonary segments
118 Aortic arch and its branches
119–121 Pulmonary arteries and bronchial tree
122–123 Pericardial sac
124–127 Heart and large blood vessels
128–131 Sections and tomograms of the heart
132–139 Inner cavities and valves of the heart
140–141 Conducting system of the heart and myocardium
142–148 Arteries and veins of the heart
149–153 Thoracic organs
154–161 Sections and tomograms of the thoracic organs
162–165 Large blood vessels and lymph vessels in the mediastinum
166–167 Autonomic nervous system in the thorax and neck

169–262 Abdominal and Pelvic Viscera

170–171 Peritoneal cavity
172 Surface projections of abdominal viscera
173–179 Stomach, small intestine, colon and rectum
180–187 Liver, segmentation of the liver, gall bladder and bile duct system
188–189 Pancreas and spleen
190–195 Viscera of the upper and lower abdomen
196–203 Blood vessels and lymph nodes in the abdominal region
204–207 Large intestine and posterior abdominal wall
208–217 Kidneys, suprarenal glands, renal pelvis and ureter
218–219 Great blood vessels in the retroperitoneal space
220–225 Sections and tomograms of the abdominal viscera
226–227 Lumbosacral plexus
228–229 Great blood vessels in the lower retroperitoneal space and pelvis
230–231 Blood vessels and nerves of the rectum
232–245 Male and female pelvis with urogenital organs and blood vessels
246–247 Connective tissue of the lesser pelvis
248–249 Autonomic nervous system in the retroperitoneal space
250–261 Sections and tomograms of the female and male pelvis

263–282 Pelvic Floor and External Genitalia

264–267 Pelvic floor
268–269 External genitals of a female
270–271 Testis and epididymis
272–273 Spermatic cord and inguinal region of a male
274–278 External genitals of a male
279–281 Blood vessels and nerves of the perineal region

283–362 Central Nervous System

284–288 Spinal cord, spinal meninges and blood vessels of the vertebral canal
289–302 Cranial meninges and blood vessels of the cranial meninges and the brain
303–306 Brain
307–311 Cerebellum and brain stem
312–319 Cerebrum
320–323 Ventricles of the brain and subarachnoidal space
324–325 Fornix, hippocampus and limbic system
326 Choroid plexuses
327–328 Thalamus, hypothalamus and mid brain
329–333 Nuclei of the forebrain and internal capsule
334 Cerebellorubral connections
335 Visual pathway
336–362 Sections and tomograms of the brain

363–388 Visual Organ and Orbital Cavity

364–366 Eye, lacrimal apparatus and eyelids
367–369 Orbital cavity
370–371 Extraocular muscles
372–377 Eyeball
378–383 Contents and topography of the orbital cavity
384–387 Sections and tomograms of the orbital cavity

389–410 Vestibulocochlear Organ

390–392 External ear and tympanic membrane
393–399 Middle ear with mastoid air cells, tympanic cavity, facial nerve canal, auditory tube and auditory ossicles
400–405 Bony and membranous labyrinth with the inner ear
406–409 Sections and tomograms of the petrous part of the temporal bone

411 Subject Index

Contents of Volume 1:
Systemic Anatomy, Body Wall, Upper and Lower Limbs

Inhalt

XIII Vorwort zur 5. Auflage
XV Vorwort zur 1. Auflage
XVIII Konzept der Atlasneugestaltung und Abbildungsnachweise
XXII Hinweise für die Benutzer

1–102 **Kopf und Hals**

2–17 Knöcherner Schädel
18–25 Knochen des Schädels
26–27 Kiefergelenk
28–32 Oberflächliche und tiefe Gesichtsmuskeln
33 Halsfaszie
34–37 Halsmuskeln
38–43 Mundhöhle und Zähne
44–45 Speicheldrüsen und Tonsillen
46–47 Zunge
48–53 Rachen
54–57 Äussere Nase und Nasenhöhle
58–63 Schnitte und Tomogramme des Gesichts
64–69 Kehlkopf
70–71 Schilddrüse
72–75 Schnitte und Tomogramme des Halses
76–80 Blutgefässe von Kopf und Hals
81 Lymphgefässe und Lymphknoten von Kopf und Hals
82–83 Hautinnervation von Kopf und Hals
84–90 Nerven und Blutgefässe des Gesichts
91–101 Nerven und Blutgefässe des Halses

103–168 **Brusteingeweide**

104 Mediastinum und Thymus
105–107 Speiseröhre
108–110 Luftröhre und Bronchien
111–117 Lungen, Bronchialbaum und Lungensegmente
118 Aortenbogen mit Ästen
119–121 Lungenarterien und Bronchialbaum
122–123 Herzbeutel
124–127 Herz und herznahe Gefässe
128–131 Schnitte und Tomogramme des Herzens
132–139 Innenräume und Klappen des Herzens
140–141 Erregungsleitungssystem und Muskulatur des Herzens
142–148 Arterien und Venen des Herzens
149–153 Brustorgane
154–161 Schnitte und Tomogramme der Brustorgane
162–165 Grosse Blutgefässe und Lymphgefässe im Mediastinum
166–167 Autonomes Nervensystem im Brustraum und Hals

169–262 **Bauch- und Beckeneingeweide**

170–171 Peritonealhöhle
172 Projektion der Bauchorgane
173–179 Magen, Dünndarm, Dickdarm und Mastdarm
180–187 Leber, Lebersegmente, Gallenblase und Gallenwege
188–189 Bauchspeicheldrüse und Milz
190–195 Oberbauch- und Unterbauchorgane
196–203 Blutgefässe und Lymphknoten im Bauchraum
204–207 Dickdarm und hintere Bauchwand
208–217 Nieren, Nebennieren, Nierenbecken und Harnleiter
218–219 Grosse Blutgefässe im Retroperitonealraum
220–225 Schnitte und Tomogramme der Bauchorgane
226–227 Plexus lumbosacralis
228–229 Grosse Blutgefässe im unteren Retroperitonealraum und Becken
230–231 Blutgefässe und Nerven des Mastdarms
232–245 Männliches und weibliches Becken mit Harn- und Geschlechtsorganen und Blutgefässen
246–247 Bindegewebe des kleinen Beckens
248–249 Autonomes Nervensystem im Retroperitonealraum
250–261 Schnitte und Tomogramme des weiblichen und männlichen Beckens

263–282 Beckenboden und äussere Geschlechtsorgane

264–267 Beckenboden
268–269 Äussere weibliche Geschlechtsorgane
270–271 Hoden und Nebenhoden
272–273 Samenstrang und Leistenregion eines Mannes
274–278 Äussere männliche Geschlechtsorgane
279–281 Blutgefässe und Nerven der Dammregion

283–362 Zentralnervensystem

284–288 Rückenmark, Rückenmarkhäute und Blutgefässe des Wirbelkanals
289–302 Hirnhäute und Blutgefässe der Hirnhäute und des Gehirns
303–306 Gehirn
307–311 Kleinhirn und Hirnstamm
312–319 Grosshirn
320–323 Ventrikel des Gehirns und Subarachnoidalraum
324–325 Fornix, Hippocampus und limbisches System
326 Adergeflechte
327–328 Thalamus, Hypothalamus und Mittelhirn
329–333 Kerne des Vorderhirns und innere Kapsel
334 Zerebellorubrale Verbindungen
335 Sehbahn
336–362 Schnitte und Tomogramme des Gehirns

363–388 Sehorgan und Augenhöhle

364–366 Auge, Tränenapparat und Lider
367–369 Augenhöhle
370–371 Äussere Augenmuskeln
372–377 Augapfel
378–383 Inhalt und Topographie der Augenhöhle
384–387 Schnitte und Tomogramme der Augenhöhle

389–410 Gehör- und Gleichgewichtsorgan

390–392 Äusseres Ohr und Trommelfell
393–399 Mittelohr mit Mastoidzellen, Paukenhöhle, Fazialiskanal, Ohrtrompete und Gehörknöchelchen
400–405 Knöchernes und häutiges Labyrinth mit Innenohr
406–409 Schnitte und Tomogramme des Felsenbeins

411 Sachwortverzeichnis

Inhalt von Band 1:

Allgemeine Anatomie, Rumpfwand, obere und untere Extremität

Preface to the 5th Edition

Macroscopic anatomy is a fundamental branch of medicine without which clinical facts cannot be understood.

Throughout history, the importance of anatomy for medicine – and thus for medical studies – has fluctuated considerably. Five hundred years ago, at the end of the Renaissance, Leonardo da Vinci and Andreas Vesal laid the foundation stones of modern anatomy and modern medicine. In those days, anatomy – then exclusively macroscopic – was the only fundamental speciality medical students were confronted with during their studies, along with the clinical subjects internal medicine, surgery and botany (in the meaning of use of herbal drugs).

The first half of the twentieth century saw the development of microscopic anatomy besides macroscopic anatomy; physiology became an independent speciality and physiological chemistry and biochemistry made huge progress. Research in these fields provided new knowledge on functional and molecular interactions in the mammalian organism which fundamentally altered our understanding of diseases and opened new perspectives in clinical diagnosis and therapy. As a consequence of these developments, macroscopic anatomy was somehow relegated to the background during the 1960s and 1970s, and seemed to have retained its essential importance only for surgical specialities.

Apart from these developments, new diagnostic imaging technologies have become clinically established in the second half of the twentieth century: computed tomography, magnetic resonance imaging and ultrasonography. These imaging techniques opened up new visions of the morphology of the living organism, enabled a very detailed identification of structures and thus laid the foundation stone of rapid and unexpected progress in clinical diagnosis. However, the interpretation of normal and pathologically altered structures in two-dimensional images of the human body with all these techniques demands extremely precise anatomical knowledge. In recent years, this has led to the revival and to a considerable increase in the significance of macroscopic anatomy both for clinical medicine and the education of medical students.

Successful clinical work without well-founded knowledge in topographical and sectional anatomy is thus no longer possible. This is the reason why the editor urges present and future medical students to study macroscopic anatomy intensively.

As a matter of fact, it is the establishment of the new imaging techniques in clinical medicine that prompted this new revised version of Professor Wolf-Heidegger's Atlas of Human Anatomy and the supplementation of its 5th edition with numerous anatomical sections, computed and magnetic resonance imaging tomograms and ultrasonograms. Such a new design of an atlas of the anatomy of the whole human body is only possible with the collaboration of many enthusiastic forces. Thus I am deeply indebted to Dr. R. Andresen and Priv.-Doz. Dr. D. Banzer (Berlin) for most of the new radiographs as well as the computed and magnetic resonance tomograms included in this atlas. Prof. Dr. G. Bogusch (Berlin), Prof. Dr. E. Fleck (Berlin), Dr. M. Jäckel (Göttingen), Dr. H. Kellner (Munich), Priv.-Doz. Dr. T. Riebel (Berlin), Priv.-Doz. Dr. C. Sohn (Heidelberg), Dr. D. Zeidler (Berlin) and Prof. Dr. W.G. Zoller (Munich) contributed some further radiographs, tomograms and ultrasonograms for which I would like to thank them.

I am moreover deeply indebted to Prof. Dr. M. Herrmann (Ulm) who provided the anatomical sections for most of the computed and magnetic resonance tomograms of the present atlas and thus considerably enriched it. The sections on which these illustrations are based were prepared and photographed by Mr. E. Voigt (Ulm), whom I would like to thank as well.

I also express my thanks to Mrs. G. Heymann-Monhof, Mr. H. Jonas, Mrs. H. Heinen, Mrs. I. Tripke, Mrs. C. Naujok and Mr. F. Geisler who prepared about 230 new anatomical drawings for the present edition.

My special thanks go to Dr. h.c. Th. Karger for his constructive collaboration during the past years. Dr. Karger always lent an understanding ear to my concepts, which were often difficult and expensive to realize, and was a partner whose expert advice and understanding always helped me in my work with the atlas. Many thanks in particular to Mr. B. Pfäffli as well as to all the personnel of S. Karger Publishers and Neue Schwitter AG who helped in the production of Wolf-Heidegger's atlas.

Mrs. M. Risch, my secretary, has been a great and dependable help over the past years, which has eased my work in many respects. I would like to thank her as well.

This new edition of Wolf-Heidegger's *Atlas of Systematic Human Anatomy* has been supplemented by numerous new anatomical drawings, radiographs, tomograms, ultrasonograms and anatomical sections. As the editor, I am confident that this new edition will indeed 'help one to see' – one of the most difficult things, according to the quote from Goethe, which Wolf-Heidegger chose as the motto for the first edition of his atlas – and that it will give medical students better access to anatomy and clinical medicine:

> 'What is the most difficult of all? What seems the easiest:
> To see with one's eyes what is in front of one's eyes.'
>
> Johann Wolfgang von Goethe,
> Distichon 155 of the *'Xenien'* (translated)

Berlin, Spring 1999 — Petra Köpf-Maier

Vorwort zur 5. Auflage

Die makroskopische Anatomie ist ein grundlegendes Fach der Medizin, ohne das auch heute ein Verständnis für klinische Zusammenhänge nicht möglich ist.

In der Geschichte unterlag die Bedeutung der Anatomie für die Medizin und damit auch für das Studium der Medizin erheblichen Schwankungen. Vor 500 Jahren, am Ende der Renaissance, wurde durch Leonardo da Vinci und Andreas Vesal der Grundstein für eine neuzeitliche Betrachtung der Anatomie und der Medizin gelegt. Zu dieser Zeit war die Anatomie, und zwar damals ausschliesslich repräsentiert durch die makroskopische Anatomie, das einzige Grundlagenfach, mit dem sich – in heutiger Terminologie gesprochen – ein Student der Medizin neben den klinischen Fächern Innere Medizin, Chirurgie und Botanik (im Sinne der Heilkräuteranwendung) während seines Studiums zu beschäftigen hatte.

In der ersten Hälfte des zwanzigsten Jahrhunderts blühte dann die mikroskopische neben der makroskopischen Anatomie auf, es löste sich die Physiologie als eigenständiges Fach ab, und die physiologische Chemie und Biochemie entwickelten sich mächtig. Durch die Forschungsaktivität auf diesen Gebieten wurden neue Erkenntnisse über das funktionelle und molekulare Zusammenspiel im Säugetierorganismus erschlossen, die das Verständnis von Krankheitsbildern grundlegend veränderten und zu neuen Wegen der klinischen Diagnostik und Therapie führten. Aufgrund dieser Entwicklung trat die makroskopische Anatomie in den sechziger und siebziger Jahren dieses Jahrhunderts etwas in den Hintergrund und schien nur noch für klinisch-operative Fächer von wesentlicher Bedeutung zu sein.

Abseits dieser Entwicklung wurden in der zweiten Hälfte des zwanzigsten Jahrhunderts neue Technologien der bildgebenden Diagnostik klinisch etabliert. Hierzu gehören die Röntgen-Computertomographie, die Kernspin-Magnetresonanztomographie und die Ultraschall-Sonographie. Diese bildgebenden Verfahren eröffneten neue Einblicke in die Morphologie des lebenden Organismus, ermöglichten eine sehr detaillierte Strukturerkennung und legten damit den Grundstein für rasante und zuvor ungeahnte Fortschritte der klinischen Diagnostik. Die Interpretation der normalen anatomischen und pathologisch veränderten Strukturen in den durch alle drei Techniken dargestellten zweidimensionalen Schnittbildern des menschlichen Körpers setzen allerdings ein hohes Mass an genauen Anatomiekenntnissen voraus. Dies wiederum führte in den vergangenen Jahren zu einer Wiederbelebung und einem enormen Bedeutungszuwachs der makroskopischen Anatomie sowohl für die klinische Medizin als auch die Ausbildung von Medizinstudenten.

Eine erfolgreiche klinische Tätigkeit ohne fundierte Kenntnisse der topographischen Anatomie und der Schnittanatomie ist daher heute nicht mehr möglich. Aus diesem Grund ist es der Herausgeberin ein Anliegen, den jetzigen und künftigen Studenten der Medizin das Studium der makroskopischen Anatomie intensiv ans Herz zu legen.

Die klinische Etablierung der neuen bildgebenden Verfahren war auch der Anstoss für die neuerliche Überarbeitung des von Professor Wolf-Heidegger begründeten und den Professoren Frick, Kummer und Putz fortgeführten Atlasses der menschlichen Anatomie und seine Erweiterung durch zahlreiche anatomische Schnittbilder, Computertomogramme, Kernspintomogramme und Ultraschallbilder in der jetzt vorliegenden 5. Auflage. Eine solche Neugestaltung eines Atlasses der Anatomie des gesamten menschlichen Körpers ist nur unter Aufbietung vieler engagierter Kräfte möglich. So verdanke ich den Herren Dr. R. Andresen und Priv.-Doz. Dr. D. Banzer (Berlin) den Grossteil der in den vorliegenden Atlas neu aufgenommenen Röntgenbilder, Computertomogramme und Kernspintomogramme. Ausserdem stellten dankenswerterweise die Herren Prof. Dr. G. Bogusch (Berlin), Prof. Dr. E. Fleck (Berlin), Dr. M. Jäckel (Göttingen), Dr. H. Kellner (München), Priv.-Doz. Dr. T. Riebel (Berlin), Priv.-Doz. Dr. C. Sohn (Heidelberg), Dr. D. Zeidler (Berlin) und Prof. Dr. W.G. Zoller (München) weitere Röntgenbilder, Tomogramme und Ultraschallbilder zur Verfügung.

Zu herzlichem Dank verpflichtet bin ich darüber hinaus Herrn Prof. Dr. M. Herrmann (Ulm), der zu den meisten Computer- und Kernspintomogrammen des vorliegenden Atlasses die entsprechenden anatomischen Schnittbilder zur Verfügung stellte und den Atlas dadurch enorm bereicherte. Die diesen Abbildungen zugrundeliegenden anatomischen Schnitte wurden von Herrn Präparator E. Voigt (Ulm) angefertigt und photographiert. Ihm möchte ich an dieser Stelle meinen Dank aussprechen.

Weiterhin danke ich Frau G. Heymann-Monhof, Herrn H. Jonas, Frau H. Heinen, Frau I. Tripke, Frau C. Naujok und Herrn F. Geisler für die Erstellung der insgesamt etwa 230 neu gezeichneten anatomischen Abbildungen der vorliegenden Auflage.

Mein besonderer Dank gilt Herrn Dr. h.c. Th. Karger für die konstruktive Zusammenarbeit während der vergangenen Jahre. Herr Dr. Karger war aufgrund seines stets offenen Ohres, seines sachkundigen Rates und seines Verständnisses für meine manchmal nur schwierig und kostspielig zu realisierenden Vorstellungen ein Verhandlungspartner, der mir die Arbeit an dem Atlas stets erleichterte. Insbesondere danke ich Herrn B. Pfäffli sowie allen anderen Mitarbeitern des Karger-Verlags und der Neuen Schwitter AG, die die vorliegende Neuauflage des Wolf-Heidegger-Atlasses mitgestalteten.

Mein Dank gebührt weiterhin Frau M. Risch, die mir während der vergangenen Jahre als Sekretärin verlässlich zur Seite stand und mich in vieler Hinsicht entlastete.

Als Herausgeberin hoffe ich, dass die vorliegende Neugestaltung des von Herrn Professor Wolf-Heidegger begründeten *Atlas der Systematischen Anatomie des Menschen* aufgrund der darin realisierten Erweiterung durch zahlreiche neue anatomische Bilder, Röntgenbilder, Tomogramme, Ultraschallbilder und anatomische Schnittbilder die Umsetzung des von Wolf-Heidegger der 1. Auflage vorangestellten Goethe-Zitats erleichtern und möglichst vielen Studenten der Human- und Zahnmedizin den Zugang zur Anatomie und klinischen Medizin eröffnen wird:

> «Was ist das Schwerste von allem? Was dir das Leichteste dünket:
> Mit den Augen zu sehn, was vor den Augen dir liegt.»
>
> Johann Wolfgang von Goethe, Distichon 155 aus den *«Xenien»*

Berlin, im Frühjahr 1999 — Petra Köpf-Maier

Preface to the 1st Edition

«Was ist das Schwerste von allem? Was dir das Leichteste dünket:
Mit den Augen zu sehn, was vor den Augen dir liegt.»*

Accustomed during his school years to place greater trust in the written word than in his own senses, the young medical student in his first pre-clinical term is faced with a problem which Goethe aptly describes as 'hardest of all': He has to learn how to see. To teach him to do so, by the aid of anatomical preparations and plates as the most effective means at his disposal, is the foremost task of the pre-clinical instructor. The aim of the present Atlas is to give to the medical student and to the physician wishing to revise his anatomical knowledge a picture, as true and exact as possible, of the organs of our human body. The drawings were made partly from specimens preserved in the large collection of the Basle Anatomical Institute, partly from special preparations. Nearly all the plates in the section on muscles were drawn from fresh preparations in order to exclude the deformities caused by preservation. Our aim was always to avoid individual peculiarities and, by using a larger number of similar preparations, to produce as general and universal a picture as possible. With a few exceptions, noted in the legends, the right side of the body was always chosen in all bilaterally symmetrical organs or parts.

We were for a long while undecided whether or not the illustrative material should be accompanied by a short written text. As stated above, we are of the opinion that the Atlas is the primary aid in anatomical instruction, but it neither can nor should be a substitute for the detailed textbook and the spoken word; these are indispensable in preparing the student for what he is to see and in fixing what he has seen firmly in his mind. Students tend to regard a short Atlas text as a source of information sufficient for their needs, but it can never deal exhaustively with all noteworthy and necessary aspects of the subject; we therefore decided finally to publish the present volume without text, but to pay great attention to the labelling of the separate illustrations. The Atlas can thus be used in combination with any textbook of anatomy. On the other hand, for the sake of clarity, care was taken not to overload the separate plates with too many pointers; thus, parts and details which have already been shown are not re-labelled in plates in which they are not important for purposes of instruction. Sketches of the body surface, copied partly from well-known sculptures, have been inserted beside the plates showing the superficial muscular layers; it is hoped that these will help the student and qualified doctor to fit the muscle relief into the body of the patient. X-ray photographs of all important skeletal parts and junctures have been included with the intention of preparing the student for a form of examination which is of vital importance in clinical medicine and only possible on the basis of a sound knowledge of the normal anatomical picture. We had also planned, and partly completed, some treatment of general morphology, constitutional types, evolution, and the mechanics of joints, also a summarising survey in tabular form of the musculature; but all this had to be omitted in order to keep the volume of handy size and accessible price for the student.

Pending the establishment – we hope at a not too distant date – of a standard anatomical nomenclature, internationally recognised and scientifically and linguistically acceptable, we have made use in the present work of the Jena nomina anatomica; this is the terminology most widely used in the German-speaking countries. The Basle nomenclature has, however, been substituted for a few linguistically incorrect or in our opinion inappropriate terms.

I wish to take this oppportunity of expressing once again may sincere thanks to the publisher, Dr Heinz Karger, who by his energy and expert knowledge, his optimism and kindly, confident encouragement has made possible the wearisome and costly realisation of this work. I wish to thank further my faithful artistic collaborators: Mr Adolph Dressler (junctures), Mr Rolf Muspach (osteology), and above all Mr Robert Schlumpf (myology), who as sculptor with many years' dissecting room experience has in the course of our prolonged collaboration far surpassed his original function as artist and become a knowledgeable and indispensable scientific colleague, invaluable at every stage of the work from the preparation of muscle specimens to the typographical composition of the plates and the correction of proofs. For untiring and invaluable help I owe sincere thanks to my former Viennese assistant, Dr Arthur von Hochstetter (now Fribourg, Switzerland). Nearly all the X-ray photographs I owe to the kindness of Dr Emil A. Zimmer (Basle/Berne). For important suggestions and active help I wish to thank in particular my kind and highly esteemed anatomy instructor, Professor Eugen Ludwig, M.D. (Basle), and also Dr Walter Bejdl (Vienna/Basle), Dr Leopold Drexler (Vienna), Mr Willy Jäggi of S. Karger Ltd., Dr Walter Krause (Vienna), Dr Kurt S. Ludwig (Basle), Dr Carl Rudolf Pfaltz (Basle), Professsor Joseph Tomasch, M.D. (Kingston, Canada; formerly Vienna/Basle), Mr Armin Wolf, dissector (Basle), and Dr Wolfgang Zürcher (Basle).

In deep gratitude I wish finally to pay tribute to the memory of my mother who by her devoted and untiring energy made it possible for me, after the early death of my father, to follow the profession of my choice and thus to bring this Atlas into being.

Basle, Autumn 1953 Gerhard Wolf-Heidegger

* J.W. Goethe: 'Xenien.' From the posthumous papers.
Weimar Edition, Vol. 5, part 1, p. 275, No. 45, 1893.

Vorwort zur 1. Auflage

«Was ist das Schwerste von allem? Was dir das Leichteste dünket:
Mit den Augen zu sehn, was vor den Augen dir liegt.»*

Dieses Goethe-Wort umschreibt die Hauptaufgabe des vorklinischen Lehrers: den jungen Mediziner, der es sich in seiner Schulzeit meist angewöhnt hat, dem dogmatischen Buchstaben mehr denn seinen Sinnesorganen zu trauen, wieder sehen zu lehren. Anatomische Präparate und Abbildungen sind hierzu seit jeher die besten Lehrmittel. Der vorliegende Atlas soll dem Medizinstudenten sowie dem auf Repetition seines anatomischen Wissens bedachten Arzt einen möglichst naturgetreuen Aspekt der Organsysteme unseres Körpers vermitteln. Die Abbildungen wurden zum Teil nach dem überaus reichhaltigen Material der Basler anatomischen Sammlung, zum Teil nach besonders zu diesem Zweck hergestellten Präparaten gezeichnet. Als Grundlage für die Muskelbilder wurden fast ausschliesslich Präparationen an frischen Leichen verwandt, um Konservierungsdeformitäten auszuschalten. Stets wurde darauf geachtet, nicht einen Individualfall wiederzugeben, sondern auf Grund einer grösseren Zahl gleichartiger Präparate ein möglichst allgemeingültiges Bild zu vermitteln. Bei allen bilateral-symmetrischen Organen oder Regionen wurde – von wenigen, in der entsprechenden Bildunterschrift besonders vermerkten Ausnahmen abgesehen – die rechte Körperseite dargestellt.

Wir haben uns lange Zeit überlegt, ob dem Bildmaterial ein kurzer Text beigegeben werden sollte oder nicht. Wohl stellt nach unserer oben zum Ausdruck gebrachten Überzeugung der Atlas das primäre anatomische Lehrmittel dar; doch kann und soll er weder ein ausführliches Lehrbuch noch das lebendige Unterrichtswort ersetzen; beide sind unentbehrlich, um das Verstehen des zu Sehenden anzubahnen und des Gesehenen zu festigen. Da ein kurzer Atlastext dem Studenten häufig als Wissensquelle ausreichend erscheint, das Wissenswerte und -notwendige des Stoffes jedoch nie erschöpfend behandeln kann, wurde von der Beifügung eines Textes abgesehen, der Bildbeschriftung jedoch grösste Beachtung geschenkt. Der Atlas kann somit zusammen mit jedem Lehrbuch der Anatomie benutzt werden. Andererseits wurde im Interesse der Übersicht und Klarheit darauf geachtet, die Beschriftung des einzelnen Bildes nicht zu überladen; so haben wir auf die nochmalige Nennung von Begriffen häufig verzichtet, wenn sie für die pädagogische Hauptaufgabe der Darstellung nicht von Bedeutung sind und in der Beschriftung vorhergehender Abbildungen bereits vorkommen. Neben den Bildern der oberflächlichen Muskelschichten wurden jeweils Skizzen der Körperoberfläche, z.T. unter Benützung bekannter Plastiken, wiedergegeben, um dem werdenden und fertigen Arzt die Übertragung des anatomischen Muskelreliefs in den Körper des Patienten zu erleichtern. Röntgenbilder aller wichtigen Skeletstücke und Juncturen sollen das Verständnis für die in der klinischen Medizin unentbehrliche, nur bei völliger Kenntnis der normal-anatomischen Grundlage deutbare Röntgenuntersuchung anbahnen. Auf die Wiedergabe verschiedener z.T. bereits ausgeführter, z.T. geplanter Darstellungen zur allgemeinen Morphologie, zur Lehre von den Körperbautypen, zur Entwicklungsgeschichte und zur Gelenkmechanik sowie auf eine tabellarische Zusammenstellung der Muskulatur musste zunächst verzichtet werden, um den Umfang des Bandes und damit seinen Preis für den Studenten tragbar zu halten.

Bis zur hoffentlich baldigen Realisierung einer international anerkannten, sprachlich und sachlich annehmbaren anatomischen Nomenklatur haben wir im vorliegenden Atlas die jetzt im deutschen Sprachgebiet fast allgemein gebräuchlichen Jenaer Nomina anatomica angewandt; allerdings wurden sprachlich falsche sowie einige uns sinnlos erscheinende Begriffe durch solche der Basler Nomenklatur ersetzt.

Auch an dieser Stelle möchte ich vor allem meinem Verleger, Herrn Dr. Heinz Karger, danken, der durch seine fachmännische Tatkraft, seinen Optimismus und sein wohlwollend-freundschaftliches Vertrauen die langwierige und kostspielige Realisierung meines Atlas-Planes ermöglichte. Ich danke ferner meinen getreuen künstlerischen Mitarbeitern, Herrn Universitätszeichner Adolph Dressler (Juncturen), Herrn Universitätszeichner Rolf Muspach (Osteologie) sowie vor allem Herrn Robert Schlumpf (Myologie), der – als Bildhauer mit mehrjähriger Seziersaal-Erfahrung – im Laufe der Jahre über die Erfüllung seiner ursprünglichen, zeichnerischen Aufgabe hinaus zu einem unersetzlichen, sachkundigen Mitarbeiter geworden ist und sich von der Herstellung der Muskelpräparate bis zur typographischen Komposition des Werkes und zur Korrektur der Druckbogen in hervorragendem Masse bewährt hat. Für unermüdliche und wertvollste Mitarbeit bin ich meinem früheren Wiener Assistenten, Herrn Dr. Arthur v. Hochstetter (jetzt Fribourg/Schweiz) zu herzlichem Dank verpflichtet. Fast alle Röntgenbilder verdanke ich der Liebenswürdigkeit von Herrn Priv. Doz. Dr. Emil A. Zimmer (Basel/Bern). Für wichtige Anregungen, viele Hinweise und rege Hilfe danke ich insbesondere meinem hochverehrten, lieben Lehrer, Herrn Prof. Dr. Eugen Ludwig (Basel) sowie den Herren Priv. Doz. Dr. Walter Bejdl (Wien/Basel), Dr. Leopold Drexler (Wien), Verlagslektor Willy Jäggi (Basel), Priv. Doz. Dr. Walter Krause (Wien), Priv. Doz. Dr. Kurt S. Ludwig (Basel), Dr. Carl Rudolf Pfaltz (Basel), Prof. Dr. Joseph Tomasch (Kingston, Canada; früher Wien/Basel), Präparator Armin Wolf (Basel) und Dr. Wolfgang Zürcher (Basel).

In tiefster Dankbarkeit gedenke ich schliesslich meiner verstorbenen Mutter, die durch ihre rastlos-gütige Energie es dem früh Vaterlosen ermöglichte, den ersehnten Lebensberuf zu ergreifen und damit auch diesen Atlas zu schaffen.

Basel, im Herbst 1953 Gerhard Wolf-Heidegger

* J.W. Goethe: 'Xenien'. Aus dem Nachlass.
Weimarer Ausgabe, Bd. 5, 1. Abt., p. 275, No. 45, 1893.

Concept of the New Version of the Atlas and Illustration Credits

The present 5th edition of the *Atlas of Human Anatomy,* published in 1954, 1960, and 1972 by Professor Dr. Gerhard Wolf-Heidegger and edited in 1990 by Prof. Dr. H. Frick, Prof. Dr. B. Kummer, and Prof. Dr. R. Putz, has been thoroughly revised in several aspects and supplemented in comparison to the previous four editions.

1. Retained Anatomical Drawings

The classical drawings of the three previous editions prepared by Wolf-Heidegger and his illustrators have been retained, recolored and – in the case of black-and-white drawings – colored didactically in order to make them clear also for beginners. Moreover, most of the figures prepared by Frick, Kummer, and Putz for the 4th edition, were revised and incorporated into the current 5th edition.

2. New Anatomical Pictures

The original illustrations of the previous four editions have been supplemented with about 230 new, mostly topographical drawings. These drawings were realized by six illustrators from Berlin, whom I would like to thank here for their enthusiasm. First of all, Mrs. Gertrud Heymann-Monhof, who possesses the talent to represent anatomical situations both true to detail and in an aesthetically convincing fashion. Mr. Hendrik Jonas drew most of the new illustrations of the locomotor apparatus and the head and succeeded very well in maintaining them in the style of the earlier editions. Mrs. Hildegard Heinen prepared numerous schematized drawings in a didactically clear manner. Other, mostly smaller drawings were done by Mrs. Ilona Tripke; three illustrations whose originals had been lost where painted in water colors by Mrs. Corinna Naujok. Mr. Frank Geisler prepared some new pictures and revised several others of the last edition.

Mrs. Gertrud Heymann-Monhof
Volume 1: Cover picture; Figs. 18, 19, 20, 21, 48b,c, 54b, 59, 60c, 75, 77a,b, 79a, 82, 83, 87a, 134a, 170, 171, 244a, 246a–c, 253, 254
Volume 2: Cover picture; Figs. 79, 95a, 107a, 113, 115b, 118a, 160c, 166, 172, 209a–c, 211b, 227, 243a, 250a,b, 251a,b, 256a–c, 285a, 336a, 342, 352, 382a, 395c

Mr. Hendrik Jonas
Volume 1: Figs. 90a,b, 99a, 103a, 123b, 125a–c, 127a–c, 131a–d, 137b,c, 138a,b, 155a, 163a, 167c–e, 175b, 176c, 177b, 181e, 189a–c, 191a,b, 193c, 196a,b, 200b–e, 201b–e, 209b, 213b, 220b, 221a,b, 225a,b, 226b, 249b,c, 250a,b, 257a–c, 280b
Volume 2: Figs. 264a,b, 265a, 276b, 277b, 281a,b, 290a,b, 292a, 321a,b, 324b, 325b, 329a,b, 332a,b, 368a, 369a,b, 394a,b, 395a, 399a

Mrs. Hildegard Heinen
Volume 1: Figs. 40b,c, 41b,c, 67a,b, 69b,c, 71c, 73a
Volume 2: Figs. 95b,c, 106b, 114c, 122b, 123b, 137a–c, 144a–c, 145a–c, 146a,c, 147a,c, 178b, 212a, 235b,c, 237b, 246b,c, 247a,b, 272a–e, 273a,b, 392a,b,d

Mrs. Ilona Tripke
Volume 1: Figs. 5c, 23b, 24, 63b, 79, 134b, 244b, 248b, 249a
Volume 2: Figs. 45b, 49b, 60a,b, 80a, 81a, 84a,b, 86b, 87b, 88b, 89b, 90b,c, 183c, 184a,b, 199b, 214b–e, 231b, 249, 293a–c, 371d, 379a, 382b, 383b

Mrs. Corinna Naujok
Volume 1: Fig. 252
Volume 2: Figs. 320a,b

Mr. Frank Geisler
Volume 1: Figs. 6a–c, 18, 19, 20, 21
Volume 2: Figs. 27b, 53a,b, 405a

Other anatomical illustrations, that is 3D reconstructions of the coronary arteries (Vol. 2, Figs. 148a–d), were contributed by Prof. Dr. Eckart Fleck and Dr. Helmut Oswald, Deutsches Herzzentrum Berlin. I acknowledge Dr. Martin Jäckel, Universitätsklinik Göttingen, for the laryngoscopic pictures in Volume 2 (Figs. 69a–d) of the present atlas, and Prof. Dr. Dieter Sasse, Universität Basel, for giving us access to the Anatomical Collection of the University of Basel and allowing Hansjörg Stöcklin to photograph the corrosion casts of the pulmonary, hepatic and renal vessels for the present atlas (Vol. 2, Figs. 115a, 119a, 183a, 213c, 214a).

3. Presentation of Imaging Techniques

The present atlas also aimed at giving imaging techniques due attention. Most radiographs, CT[1] and MRI[2] images published in the previous edition were technically superseded and, thus, replaced by new pictures. Besides conventional radiographs, the editor was anxious to incorporate computed and MRI tomograms of the whole human body and to represent ultrasonography by some selected images. For an anatomist, this could only be achieved by the close collaboration with enthusiastic radiologists: two radiologists from Berlin, Dr. Reimer Andresen and Priv.-Doz. Dr. Dietrich Banzer, Krankenhaus Zehlendorf, Behring-Krankenhaus Berlin. They have both untiringly searched for 'normal' anatomical images, which proved much more difficult and time-consuming than expected. Most MRI tomograms were done by use of a Philips Gyroscan ACS-NT MRI tomograph which had fortunately been installed a few years ago in the Zehlendorf Hospital.

[1] CT = Computed tomography, computed tomogram
[2] MRI = Magnetic resonance imaging, magnetic resonance image

Nearly 200 radiographs, CT and MRI tomograms as well as ultrasonograms were taken from Dr. Reimer Andresen's and Priv.-Doz. Dr. Dietrich Banzer's 'treasury':

Volume 1: Figs. 3a,b, 7b, 8d, 33b, 36a,b, 37b, 38a,b, 39b, 43, 45d, 51a,b, 55c, 71a,b, 73b, 85b, 87b, 91a,b, 93a,b, 97a,b, 100a,b, 101a,c, 103c, 106b, 120, 121, 123a, 129b, 137a, 139, 144b, 145b, 146b, 147b, 149a–c, 151b, 155b, 156b, 157b, 158b, 159b, 163b, 164b, 166a, 167a, 173a,b, 181f, 185, 188c, 194a,b, 195a–c, 197a,b, 202a–c, 203a–c, 204a,c,e, 205, 208b, 209a, 210a,b, 247a,b, 251, 258c,d, 259b, 260b, 261b, 262b, 263b, 264b, 265b, 269, 272b, 273b, 274b, 275b, 276a–c, 281a,b

Volume 2: Figs. 5, 7, 12b, 26c,d, 41b, 43c, 53c,d, 59a, 61a,b, 71a–c, 72a, 75a, 78, 110a, 118b, 119b, 126, 127, 128a, 129a, 130a, 131a, 156a, 157a, 159a,b, 175a–c, 182a,b, 186b,d, 188b, 189c,d, 194b, 197a,b, 198b, 201a,b, 203b, 208c, 215a–e, 217a,b, 218c, 219, 221, 223b, 224b, 229, 244b, 253a,b, 255a,b, 257a–c, 259a–c, 261a,b, 271b, 278a,b, 287a–c, 305, 336b, 339, 341, 343, 347, 349, 367b, 378b, 385a,b, 387a,b

Moreover, I gratefully acknowledge the following colleagues for other radiographs and ultrasonograms:

Prof. Dr. Eckart Fleck and Dr. Helmut Oswald,
Deutsches Herzzentrum Berlin:
Volume 2: Figs. 136a–c, 138a–h, 146b,d, 147b,d

Dr. Martin Jäckel, HNO-Klinik, Universität Göttingen:
Volume 2: Figs. 9, 11, 395b, 406a–c, 407a–c

Dr. Herbert Kellner, Medizinische Poliklinik,
Klinikum Innenstadt, Universität München:
Volume 1: Fig. 173c

Priv.-Doz. Dr. Thomas Riebel, Strahlenklinik,
Virchow-Klinikum, Humboldt-Universität Berlin:
Volume 1: Fig. 177a

Priv.-Doz. Dr. Christof Sohn, Frauenklinik,
Universität Heidelberg:
Volume 2: Figs. 74a, 238b, 242a–d, 245b,c

Dr. Diethmar Zeidler, dentist, Berlin:
Volume 2: Figs. 40e,f

Prof. Dr. Wolfram Zoller, Medizinische Poliklinik,
Klinikum Innenstadt, Universität München:
Volume 2: Figs. 185a, 186c, 199c, 210e, 218b, 236b,c

I thank Prof. Dr. Gottfried Bogusch, formerly Institut für Anatomie der Freien Universität Berlin, now Humboldt-Universität Berlin, for giving me the permission to inspect the collection of radiographs and tomograms set up by him and to use the following pictures for the atlas:

Volume 1: Figs. 194a, 199a,b, 214b, 222b

Volume 2: Figs. 23c, 59b, 73a, 74a, 107a, 158a,b, 160a,b 161a,b, 176b, 213a, 222b 225b, 278c,d, 301b, 302a,b

4. Anatomical Sections

In order to facilitate the understanding of the radiological sections (CTs and MRIs), the present atlas was designed to enable direct comparison with anatomical sections. While many institutes of anatomy pay attention to sectional anatomy, only few institutes possess complete series of sections of the whole human body. I would like to thank very much Prof. Dr. Martin Herrmann and Mr. Ernst Voigt, Abteilung Anatomie der Universität Ulm, for these illustrations; Mr. Voigt prepared and photographed all the sections.

A total of 90 anatomical sections contributed by Prof. Dr. M. Herrmann, Universität Ulm, have been included in the present new edition of Wolf-Heidegger's atlas:

Volume 1: Figs. 37a, 39a, 60b, 101b,d, 103b, 129a, 144a, 145a, 146a, 147a, 148a,b, 156a, 157a, 158a, 159a, 166b, 167b, 204b,d,f, 208a, 211, 258a,b, 259a, 260a, 261a, 262a, 263a, 264a, 265a, 272a, 273a, 274a, 275a, 277a–c

Volume 2: Figs. 62a,b, 63a,b, 68c, 72b, 73b, 74b, 75b, 128b, 129b, 130b, 131b, 156b, 157b, 158c, 159c, 164, 165, 213b, 220, 222a, 223a, 224a, 225a, 252a,b, 254a,b, 258a–c, 260a,b, 355a,b, 356a,b, 357a,b, 358a,b, 359a,b, 360a,b, 361a,b, 362a,b, 386a,b, 400b,c, 408a,b

5. Organization

The first three editions of Wolf-Heidegger's *Atlas of Systematic Human Anatomy* were published in three volumes organized in a somewhat modified manner according to the classical division into three parts:

Volume 1: Bones, joints, muscles
Volume 2: Viscera, skin, sensory organs
Volume 3: Nervous system, vascular system

In the 4th edition, prepared by H. Frick, B. Kummer, and R. Putz, the main contents of the previous three volumes were concentrated into one volume and reorganized according to topographical aspects.

The present 5th edition is divided into two volumes; the topographical arrangement of the illustrations has been taken over from the 4th edition, but the chapters have been arranged as follows:

Volume 1: Systemic anatomy, body wall, upper and lower limbs
Volume 2: Head and neck, thorax, abdomen, pelvis, central nervous system, eye, and ear.

This arrangement is based on the organization of the dissection courses into two main parts in many institutes of anatomy:

Part 1 – dissection of the locomotor apparatus: Ventral and dorsal body wall, upper and lower limbs

Part 2 – dissection of the viscera: Thorax, abdomen, pelvis, neck, head, brain

This division of the atlas into two volumes should make it easier for students to carry the atlas around during the dissection course and should moreover facilitate future supplementations.

Konzept der Atlasneugestaltung und Abbildungsnachweise

Die vorliegende 5. Auflage des 1954, 1960 und 1972 von Professor Dr. Gerhard Wolf-Heidegger veröffentlichten und 1990 von den Professoren Dr. H. Frick, Dr. B. Kummer und Dr. R. Putz neu bearbeiteten Atlasses der Anatomie des Menschen wurde im Vergleich zu den vorherigen vier Auflagen in mehrfacher Hinsicht grundlegend überarbeitet und erweitert.

1. Übernommene anatomische Abbildungen

Die von Wolf-Heidegger und seinen Zeichnern erstellten klassischen Abbildungen der ersten drei Auflagen wurden übernommen, farblich überarbeitet und – soweit sie als Schwarzweiss-Zeichnungen vorlagen – unter Verwendung von didaktischen Farben koloriert, um sie so verständlicher und auch dem Anfänger zugänglich zu machen. Überdies wurden auch zahlreiche der für die von Frick, Kummer und Putz bearbeiteten 4. Auflage neugezeichneten Abbildungen farblich überarbeitet und eingebaut.

2. Neue anatomische Abbildungen

Der übernommene Bilderschatz der vorherigen vier Auflagen wurde durch etwa 230 neu angefertigte, zumeist topographisch ausgerichtete Abbildungen ergänzt. An der Erstellung dieser neuen Bilder waren insgesamt sechs Graphiker aus Berlin beteiligt, für deren Engagement ich mich an dieser Stelle herzlich bedanken möchte. Allen voran Frau Gertrud Heymann-Monhof, die mit künstlerischem Feingefühl anatomische Situationen detailgetreu und ästhetisch überzeugend zu gestalten weiss. Herr Hendrik Jonas zeichnete die meisten neuen Abbildungen des Bewegungsapparates und des Kopfes mit grossem Einfühlungsvermögen in den Stil der ersten Auflagen des Wolf-Heidegger-Atlasses. Frau Hildegard Heinen erstellte zahlreiche schematisierende anatomische Zeichnungen in didaktisch klarer Wiedergabe. Weitere, zumeist kleinere Abbildungen fertigte Frau Ilona Tripke an, Frau Corinna Naujok aquarellierte drei Abbildungen, deren Originale verschollen waren. Herr Frank Geisler trug einige neue Bilder bei und überarbeitete mehrere Abbildungen der letzten Auflage.

Frau Gertrud Heymann-Monhof
Band 1: Umschlagbild; Abb. 18, 19, 20, 21, 48b,c, 54b, 59, 60c, 75, 77a,b, 79a, 82, 83, 87a, 134a, 170, 171, 244a, 246a–c, 253, 254
Band 2: Umschlagbild; Abb. 79, 95a, 107a, 113, 115b, 118a, 160c, 166, 172, 209a–c, 211b, 227, 243a, 250a,b, 251a,b, 256a–c, 285a, 336a, 342, 352, 382a, 395c

Herr Hendrik Jonas
Band 1: Abb. 90a,b, 99a, 103a, 123b, 125a–c, 127a–c, 131a–d, 137b,c, 138a,b, 155a, 163a, 167c–e, 175b, 176c, 177b, 181e, 189a–c, 191a,b, 193c, 196a,b, 200b–e, 201b–e, 209b, 213b, 220b, 221a,b, 225a,b, 226b, 249b,c, 250a,b, 257a–c, 280b
Band 2: Abb. 264a,b, 265a, 276b, 277b, 281a,b, 290a,b, 292a, 321a,b, 324b, 325b, 329a,b, 332a,b, 368a, 369a,b 394a,b, 395a, 399a

Frau Hildegard Heinen
Band 1: Abb. 40b,c, 41b,c, 67a,b, 69b,c, 71c, 73a
Band 2: Abb. 95b,c, 106b, 114c, 122b, 123b, 137a–c, 144a–c, 145a–c, 146a,c, 147a,c, 178b, 212a, 235b,c, 237b, 246b,c, 247a,b, 272a–e, 273a,b, 392a,b,d

Frau Ilona Tripke
Band 1: Abb. 5c, 23b, 24, 63b, 79b, 134b, 244b, 248b, 249a
Band 2: Abb. 45b, 49b, 60a,b, 80a, 81a, 84a,b, 86b, 87b, 88b, 89b, 90b,c, 183c, 184a,b, 199b, 214b–e, 231b, 249, 293a–c, 371d, 379a, 382b, 383b

Frau Corinna Naujok
Band 1: Abb. 252
Band 2: Abb. 320a,b

Herr Frank Geisler
Band 1: Abb. 6a–c, 18, 19, 20, 21
Band 2: Abb. 27b, 53a,b, 405a

Anatomische Abbildungen stellten weiterhin die Herren Prof. Dr. Eckart Fleck und Dr. Helmut Oswald, Deutsches Herzzentrum Berlin, in Form von 3D-Rekonstruktionen der Herzkranzarterien (Band 2, Abb. 148a–d) zur Verfügung. Herrn Dr. Martin Jäckel, Universitätsklinik Göttingen, verdanke ich die Kehlkopfspiegelbilder im 2. Band (S. 69), und Herrn Prof. Dr. Dieter Sasse, Universität Basel, danke ich für den Zugang zur Anatomischen Sammlung der Universität Basel und die Möglichkeit, die Korrosionspräparate der Lungen-, Leber- und Nierengefässe (Band 2, Abb. 115a, 119a, 183a, 213c, 214a) durch Herrn Hansjörg Stöcklin photographieren zu lassen.

3. Darstellungen bildgebender Verfahren

Ein weiterer Schwerpunkt des vorliegenden Atlasses war es, die bildgebenden Verfahren in gebührendem Rahmen zu berücksichtigen. Die meisten in den vorherigen Auflagen vorhandenen Röntgen-, CT[1]- und MRI[2]-Bilder waren technisch überholt. Sie wurden herausgenommen und durch neue Abbildungen ersetzt. Neben der konventionellen Röntgendarstellung war die Herausgeberin bestrebt, Computer- und Kernspintomogramme des gesamten menschlichen Körpers einzubauen sowie die Sonographie (Ultraschalldarstellung) durch einige ausgewählte Bilder zu repräsentieren. Dieses Anliegen ist für einen Anatomen nur in enger Zusammenarbeit mit engagierten Radiologen möglich, die die Herausgeberin in den beiden Berliner Radiologen Herrn Dr. Reimer Andresen und Herrn Priv.-Doz. Dr. Dietrich Banzer, Krankenhaus Zehlendorf, Behring-Krankenhaus Berlin, fand. Beide suchten in unermüdlichem Einsatz «normale anatomische Bilder», was viel schwieriger und zeitaufwendiger ist, als man es erwartet. Vorteilhaft war in diesem Zusammenhang, dass vor wenigen Jahren im Krankenhaus Zehlendorf ein neues MRI-Gerät vom Typ Gyroscan ACS-NT der Firma Philips installiert worden war, mit dem die meisten Kernspinaufnahmen des vorliegenden Atlasses gefertigt wurden.

[1] CT = Computertomographie, Computertomogramm
[2] MRI = «Magnetic resonance imaging», Kernspintomographie, Kernspintomogramm

Nahezu 200 Röntgen-, CT-, MRI- und Ultraschallbilder der vorliegenden beiden Atlasbände entstammen der «Schatztruhe» von Dr. Reimer Andresen und Priv.-Doz. Dr. Dietrich Banzer:

Band 1: Abb. 3a,b, 7b, 8d, 33b, 36a,b, 37b, 38a,b, 39b, 43, 45d, 51a,b, 55c, 71a,b, 73b, 85b, 87b, 91a,b, 93a,b, 97a,b, 100a,b, 101a,c, 103c, 106b, 120, 121, 123a, 129b, 137a, 139, 144b, 145b, 146b, 147b, 149a–c, 151b, 155b, 156b, 157b, 158b, 159b, 163b, 164b, 166a, 167a, 173a,b, 181f, 185, 188c, 194a,b, 195a–c, 197a,b, 202a–c, 203a–c, 204a,c,e, 205, 208b, 209a, 210a,b, 247a,b, 251, 258c,d, 259b, 260b, 261b, 262b, 263b, 264b, 265b, 269, 272b, 273b, 274b, 275b, 276a–c, 281a,b

Band 2: Abb. 5, 7, 12b, 26c,d, 41b, 43c, 53c,d, 59a, 61a,b, 71a–c, 72a, 75a, 78, 110a, 118b, 119b, 126, 127, 128a, 129a, 130a, 131a, 156a, 157a, 159a,b, 175a–c, 182a,b, 186b,d, 188b, 189c,d, 194b, 197a,b, 198b, 201a,b, 203b, 208c, 215a–e, 217a,b, 218c, 219, 221, 223b, 224b, 229, 244b, 253a,b, 255a,b, 257a–c, 259a–c, 261a,b, 271b, 278a,b, 287a–c, 305, 336b, 339, 341, 343, 347, 349, 367b, 378b, 385a,b, 387a,b

Weitere röntgenologische und sonographische Abbildungen wurden dankenswerterweise von folgenden Kollegen zur Verfügung gestellt:

Prof. Dr. Eckart Fleck und Dr. Helmut Oswald,
Deutsches Herzzentrum Berlin:
Band 2: Abb. 136a–c, 138a–h, 146b,d, 147b,d

Dr. Martin Jäckel, HNO-Klinik, Universität Göttingen:
Band 2: Abb. 9, 11, 395b, 406a–c, 407a–c

Dr. Herbert Kellner, Medizinische Poliklinik,
Klinikum Innenstadt, Universität München:
Band 1: Abb. 173c

Priv.-Doz. Dr. Thomas Riebel, Strahlenklinik,
Virchow-Klinikum, Humboldt-Universität Berlin:
Band 1: Abb. 177a

Priv.-Doz. Dr. Christof Sohn, Frauenklinik,
Universität Heidelberg:
Band 2: Abb. 74a, 238b, 242a–d, 245b,c

Dr. Diethmar Zeidler, Zahnarzt, Berlin:
Band 2: Abb. 40e,f

Prof. Dr. Wolfram Zoller, Medizinische Poliklinik,
Klinikum Innenstadt, Universität München:
Band 2: Abb. 185a, 186c, 199c, 210e, 218b, 236b,c

Herrn Prof. Dr. Gottfried Bogusch, früher Institut für Anatomie der Freien Universität Berlin, jetzt Humboldt-Universität Berlin, danke ich für seine Erlaubnis, die von ihm eingerichtete Sammlung von Röntgenbildern und Tomogrammen zu sichten und folgende Bilder in den Atlas zu übernehmen:

Band 1: Abb. 194a, 199a,b, 214b, 222b

Band 2: Abb. 23c, 59b, 73a, 74a, 107a, 158a,b, 160a,b, 161a,b, 176b, 213a, 222b, 225b, 278c,d, 301b, 302a,b

4. Anatomische Schnitte

Um die radiologischen Schichtbilder (CTs und MRIs) verständlicher zu machen, war es das Konzept des vorliegenden Atlasses, ihnen anatomische Schnittbilder gegenüberzustellen. Zwar wird die Schnittanatomie in vielen Anatomieinstituten gepflegt, komplette Schnittserien des gesamten menschlichen Körpers sind jedoch nur in wenigen Instituten verfügbar. Diese Abbildungen verdanke ich Herrn Prof. Dr. Martin Herrmann und Herrn Präparator Ernst Voigt, Abteilung Anatomie der Universität Ulm, wobei Herr Voigt alle vorliegenden Schnitte anfertigte und photographierte.

Insgesamt wurden über 90 anatomische Schnittbilder aus der Ulmer Anatomie in die vorliegende Neuauflage eingebaut:

Band 1: Abb. 37a, 39a, 60b, 101b,d, 103b, 129a, 144a, 145a, 146a, 147a, 148a,b, 156a, 157a, 158a, 159a, 166b, 167b, 204b,d,f, 208a, 211, 258a,b, 259a, 260a, 261a, 262a, 263a, 264a, 265a, 272a, 273a, 274a, 275a, 277a–c

Band 2: Abb. 62a,b, 63a,b, 68c, 72b, 73b, 74b, 75b, 128b, 129b, 130b, 131b, 156b, 157b, 158c, 159c, 164, 165, 213b, 220, 222a, 223a, 224a, 225a, 252a,b, 254a,b, 258a–c, 260a,b, 355a,b, 356a,b, 357a,b, 358a,b, 359a,b, 360a,b, 361a,b, 362a,b, 386a,b, 400b,c, 408a,b

5. Gliederung

Der Wolf-Heidegger-Atlas der menschlichen Anatomie erschien in seinen ersten drei Auflagen als dreibändiges Werk, gegliedert in etwas modifizierter Form gemäss der klassischen Dreiteilung:

Band I: Knochen, Gelenke, Muskeln
Band II: Eingeweide, Haut, Sinnesorgane
Band III: Nervensystem, Gefässe

In der 4. Auflage, bearbeitet von H. Frick, B. Kummer und R. Putz, wurde der wesentliche Inhalt der drei vorherigen Bände in einem Band verdichtet und unter topographischen Gesichtspunkten neu geordnet.

In der vorliegenden 5. Auflage wurde ein zweibändiges Konzept realisiert, wobei die topographische Anordnung der Bilder aus der 4. Auflage übernommen, die Kapitel jedoch wie folgt aufgeteilt wurden:

Band 1: Allgemeine Anatomie, Rumpfwand, obere und untere Extremität

Band 2: Kopf und Hals, Brust, Bauch, Becken, ZNS, Auge, Ohr

Diese Aufteilung lehnt sich an die in vielen Anatomieinstituten durchgeführte Gliederung des Präparierkurses in zwei grosse Abschnitte an:

Abschnitt 1 – Präparation des Bewegungsapparates: Vordere und hintere Leibeswand, Arm, Bein

Abschnitt 2 – Präparation der Eingeweide: Brust, Bauch, Becken, Hals, Kopf, Gehirn

Durch diese Zweibändigkeit soll den Studenten die «Tragearbeit» während des Präparierkurses erleichtert werden. Für künftige Auflagen eröffnet sie die Möglichkeit, den Atlas auch weiterhin durch neue Abbildungen zu ergänzen.

Information for Users

1. Anatomical Nomenclature

In the present atlas, the designation of anatomical structures follows the official international anatomical nomenclature, the *Nomina Anatomica* (NA), in its latest edition (6th edition, 1989) available upon completion of the manuscript. However, compound words (such as *articulatio sacro-iliaca*) were spelt without hyphens and words with diphthongs (e.g., glutaeus instead of gluteus) were used.

2. Abbreviations

The following abbreviations have been used throughout:

Singular			Plural		
A.	=	*Arteria*	*Aa.*	=	*Arteriae*
Lig.	=	*Ligamentum*	*Ligg.*	=	*Ligamenta*
M.	=	*Musculus*	*Mm.*	=	*Musculi*
N.	=	*Nervus*	*Nn.*	=	*Nervi*
Proc.	=	*Processus*	*Procc.*	=	*Processus*
R.	=	*Ramus*	*Rr.*	=	*Rami*
V.	=	*Vena*	*Vv.*	=	*Venae*

3. Brackets

The synonyms of the 6th edition of the NA are placed in square brackets [ ], e.g., *Gaster [Ventriculus]* or *Splen [Lien]*. These equivalent terms also appear in the index; an arrow refers to the main term outside the square brackets.

Parentheses ( ) are used to note terms also shown in parentheses in the NA, and for designating varieties, additional information and explanations. In the legends, the relative size of the figure is given as percentage in parentheses.

Commonly used, but not official NA terms are noted in pointed parentheses 〈 〉.

4. Formation of Genitives

Unless they are official designations of the NA, complicated Latin genitives have been avoided, since these days only few students have learned Latin sufficiently. Instead, we have first indicated the generic term (bone, artery or nerve) and then, on the following line, without separation by a comma, the specific term:

Examples:	*Os occipitale* *Pars basilaris*	or	*Glandula thyroidea* *Lobus sinister*

5. Dashes

A dash *following* (left column) or *preceding* (right column) an entry indicates that one or several specific entries for the same body part will follow. The generic term is shown above it – usually without a pointer:

Examples:	*Corpus fibulae*	or	*Plexus brachialis*
	Facies lateralis –		*Pars infraclavicularis*
	Margo anterior –		*– Fasciculus lateralis*
	Facies medialis –		*– Fasciculus medialis*
	Margo interosseus –		*– Fasciculus posterior*

6. Pointers and Dots

If dots on a pointer identify two or more anatomical structures or if several dots appear on the pointer, the various designations are separated by a comma; their order follows that of the arrangement of the anatomical structures in the figure. In both columns, the labellings are arranged according to the following principle: left first, then right; in the case of branched pointers, above first, then below.

7. Notation of Sizes

Unless otherwise indicated in the legends, the anatomical drawings in the present atlas always represent the situation in adults; the percentages given in parentheses in the legends denote the relative size of the image referred to the original. With a view to the considerable biological variations in body size, the percentages have been rounded off and should only be considered as indicative.

8. MRI Tomograms

Enhancement of the tissue-specific relaxation parameters T_1 and T_2 in MRI tomograms is noted in the legends as T_1 or T_2 weighting. T_1- and T_2-weighted tomograms represent the various structures of the human body in different brightnesses and different contrasts. Thus, in T_1-weighted tomograms, liquid-filled spaces are shown black, muscles dark and the bone marrow white. In T_2-weighted tomograms, liquid-filled spaces appear white, bones dark and muscles light grey.

9. Tomograms and Anatomical Sections

In current clinical practice, transverse computed and magnetic resonance imaging tomograms of the human body are always viewed from caudal, that is from below and looking up. This is the reason why, in the present atlas, the anatomical sections – with the few exceptions noted in the legends – are also viewed from caudal, that is from the feet of the patient. While this view of the tomograms and sections is doubtless difficult for beginners, it does correspond to the physician's perspective when he approaches the supine patient from the foot end of the patient's bed. The accompanying figure illustrates this view from caudal (bottom) to cranial (top) and makes clear that in this perspective the organs located on the patient's right (R) side appear on the left in the figure and the organs located on the left (L) side appear on the right side of the figure.

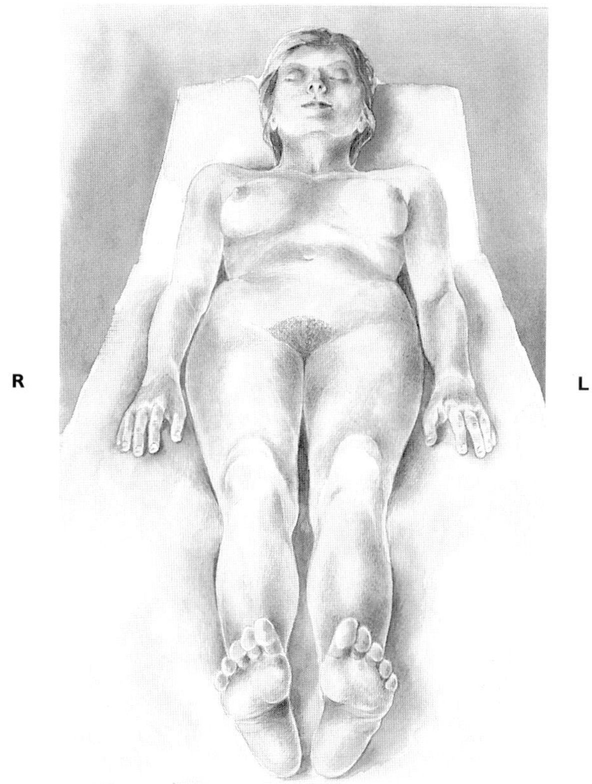

(Painted by G. Heymann-Monhof, Berlin)

Hinweise für die Benutzer

1. Anatomische Nomenklatur

Die Benennung der anatomischen Strukturen erfolgt im vorliegenden Atlas durch die offiziellen lateinischen Begriffe der internationalen anatomischen Nomenklatur, die «Nomina anatomica» (NA), in der bei Abschluss des Manuskripts letzten Fassung (6. Auflage, 1989). Bindestriche in zusammengesetzten Wörtern (z.B. in *Articulatio sacro-iliaca*) wurden jedoch weggelassen, auch kamen Begriffe mit Umlaut (z.B. *glutaeus* statt *gluteus*) zur Anwendung.

2. Abkürzungen

Folgende Abkürzungen wurden durchgängig verwendet:

Singular			Plural		
A.	=	*Arteria*	*Aa.*	=	*Arteriae*
Lig.	=	*Ligamentum*	*Ligg.*	=	*Ligamenta*
M.	=	*Musculus*	*Mm.*	=	*Musculi*
N.	=	*Nervus*	*Nn.*	=	*Nervi*
Proc.	=	*Processus*	*Procc.*	=	*Processus*
R.	=	*Ramus*	*Rr.*	=	*Rami*
V.	=	*Vena*	*Vv.*	=	*Venae*

3. Klammern

In eckigen Klammern [ ] stehen Synonyma der 6. Auflage der NA, z.B. *Gaster [Ventriculus]* oder *Splen [Lien]*. Diese bedeutungsgleichen Begriffe sind auch im Register erfasst, wobei hier ein Pfeil auf den ausserhalb der eckigen Klammer stehenden Hauptbegriff verweist.

Runde Klammern ( ) enthalten Begriffe, die auch in den NA in runden Klammern stehen, Bezeichnungen von Varietäten, ergänzende Hinweise und Erläuterungen. In den Legenden wird die relative Grösse der Abbildung als Prozentzahl in runden Klammern angegeben.

In spitze Klammern 〈 〉 sind Begriffe gesetzt, die zwar gebräuchlich sind, aber keine offiziellen Nomina der NA darstellen.

4. Genitivbildung

Sprachlich umständliche lateinische Genitive werden vermieden, sofern sie nicht offizielle Termini der NA sind, da heute nur noch wenige Studenten die lateinische Sprache ausreichend beherrschen. Statt der Genitivbildung wird im allgemeinen zunächst der Oberbegriff (Knochen, Arterie oder Nerv) genannt und – ohne Trennung durch ein Komma – in der nächsten Zeile die Detailbezeichnung angegeben.

Beispiele:	*Os occipitale*	oder	*Glandula thyroidea*
	Pars basilaris		*Lobus sinister*

5. Spiegelstriche

Spiegelstriche, die in der linken Spalte der Beschriftung *hinter*, in der rechten Spalte *vor* dem zugehörigen Terminus stehen, weisen darauf hin, dass hier eine oder mehrere Detailangaben für den gleichen Körperteil folgen. Der Oberbegriff ist – zumeist ohne Hinweislinie – darüber gesetzt.

Beispiele:	*Corpus fibulae*	oder	*Plexus brachialis*
	Facies lateralis –		*Pars infraclavicularis*
	Margo anterior –		– *Fasciculus lateralis*
	Facies medialis –		– *Fasciculus medialis*
	Margo interosseus –		– *Fasciculus posterior*

6. Hinweispunkte und -linien

Bezeichnet ein Hinweispunkt zwei oder mehrere anatomische Strukturen oder sind auf einer Hinweislinie mehrere Hinweispunkte eingetragen, so werden die verschiedenen Termini jeweils durch ein Komma getrennt. Die Reihenfolge der Bezeichnungen ergibt sich aus der Anordnung der anatomischen Strukturen in der Abbildung. In der linken wie auch der rechten Beschriftungsspalte gilt als Ordnungsprinzip: links vor rechts; bei gegabelten Hinweislinien: oben vor unten.

7. Grössenangaben

Die anatomischen Zeichnungen des vorliegenden Atlasses geben, sofern die Legende nichts anderes aussagt, die Verhältnisse beim Erwachsenen wieder. Die in Klammern gesetzten Prozentzahlen in den Legenden orientieren über die relative Grösse des Abbildes bezogen auf das Original. Die Zahlen sind gerundet und im Hinblick auf die erhebliche biologische Variabilität der Körpergrösse nur als grobe Richtwerte zu verstehen.

8. Kernspintomogramme

Die in den Kernspintomogrammen (MRIs) vorliegende Betonung der gewebespezifischen Relaxationsparameter T_1 und T_2 ist als T_1- bzw. T_2-Gewichtung in den Legenden vermerkt. T_1- und T_2-gewichtete Bilder stellen die verschiedenen Strukturen des menschlichen Körpers in jeweils unterschiedlicher Helligkeit und in unterschiedlichem Kontrast dar. So werden in T_1-gewichteten Aufnahmen flüssigkeitsgefüllte Räume schwarz, Muskeln dunkel und Knochenmark weisslich dargestellt. T_2-betonte Bilder hingegen lassen flüssigkeitsgefüllte Räume weiss, Knochen dunkel und Muskeln hellgrau erscheinen.

9. Tomogramme und anatomische Schnitte

Transversale Computer- und Kernspintomogramme des menschlichen Körpers werden heute in der Klinik stets in der Ansicht von kaudal, d.h. von unten nach oben blickend, dargestellt. Aus diesem Grund werden in dem vorliegenden Atlas auch die anatomischen Körperquerschnitte – von wenigen, in den Legenden vermerkten Ausnahmen abgesehen – von kaudal, d.h. von den Füssen des Patienten aus betrachtet. Diese Sicht der Tomogramme und Schnitte ist für den Anfänger sicherlich schwierig. Sie entspricht der Perspektive des Arztes, der sich dem liegenden Patienten vom Fussende des Patientenbettes her nähert. Die nebenstehende Abbildung soll diese Ansicht von kaudal (unten) nach kranial (oben) verdeutlichen und klarlegen, dass bei dieser Betrachtungsweise die rechts (R) gelegenen Organe des Patienten im Bild links, die links (L) gelegenen Organe des Patienten im Bild rechts erscheinen.

R

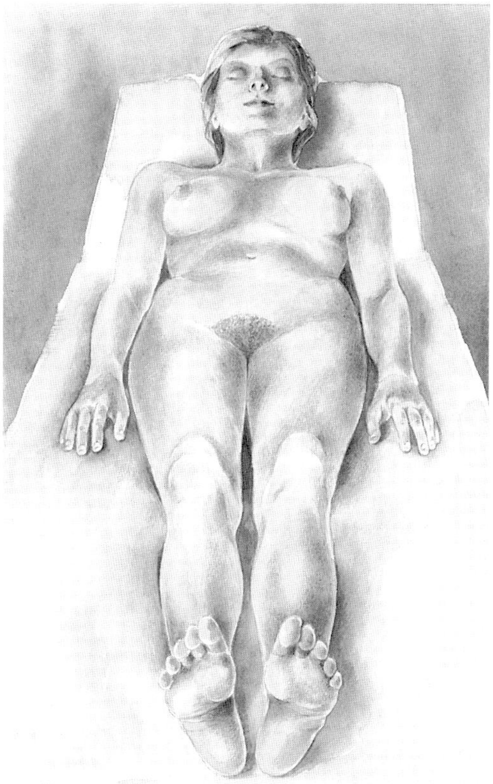

L

(Gemalt von G. Heymann-Monhof, Berlin)

Caput et Collum

Head and Neck

Kopf und Hals

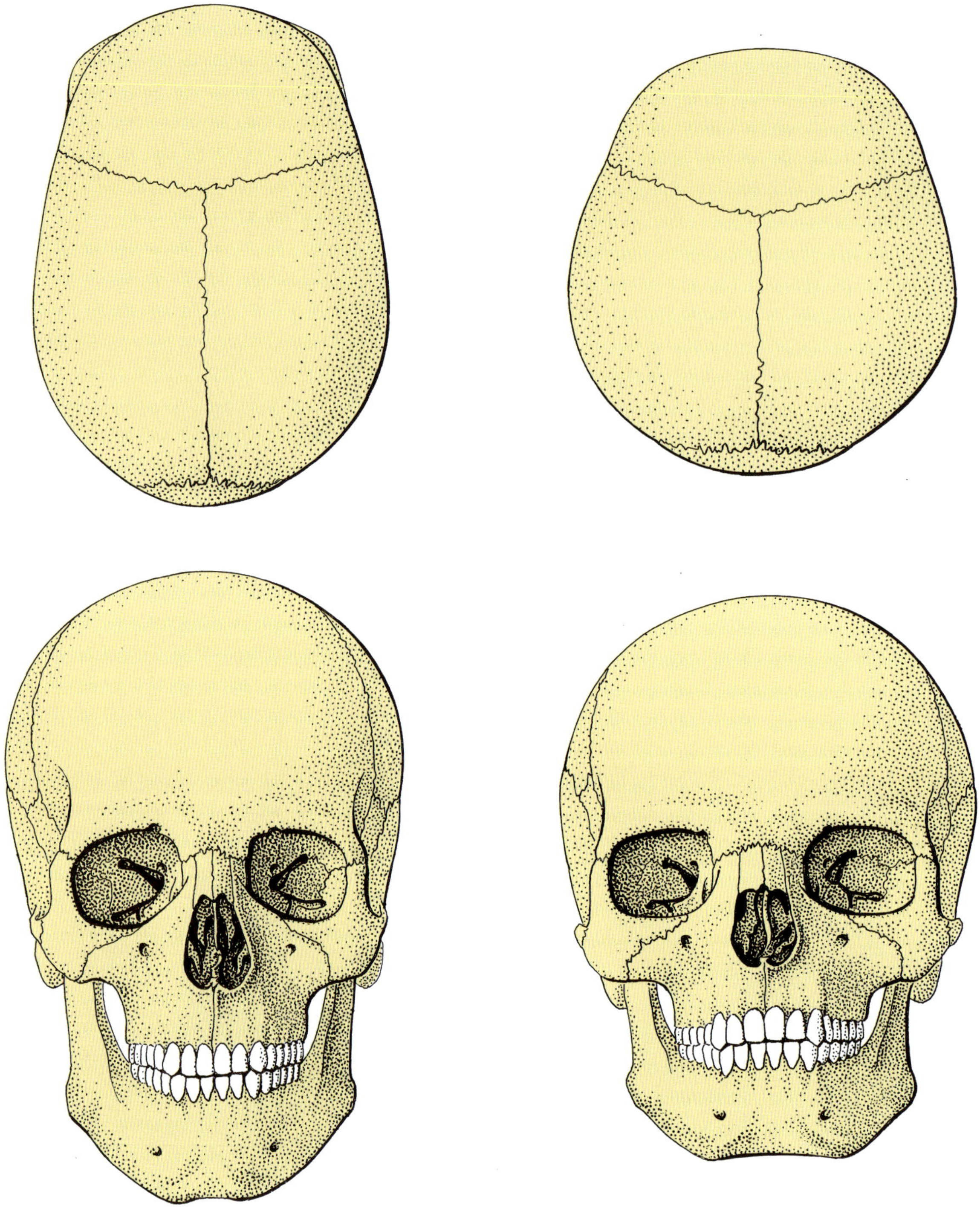

2 Cranium

2 Shape differences in the skull

- a, b Skull cap, norma verticalis (superior aspect)
- c, d Skull, norma frontalis (anterior aspect)
- a Long (dolichocephalic) skull
- b Broad (brachycephalic) skull
- c Long (dolichocephalic) skull with a narrow face (leptoprosopia)
- d Broad (brachycephalic) skull with a broad face (euryprosopia)

2 Unterschiede in der Schädelform

- a, b Schädeldach, Norma verticalis (Ansicht von oben)
- c, d Schädel, Norma frontalis (Ansicht von vorne)
- a Langschädel (Dolichozephalie)
- b Breitschädel (Brachyzephalie)
- c Langschädel (Dolichozephalie) mit Schmalgesicht (Leptoprosopie)
- d Breitschädel (Brachyzephalie) mit Breitgesicht (Euryprosopie)

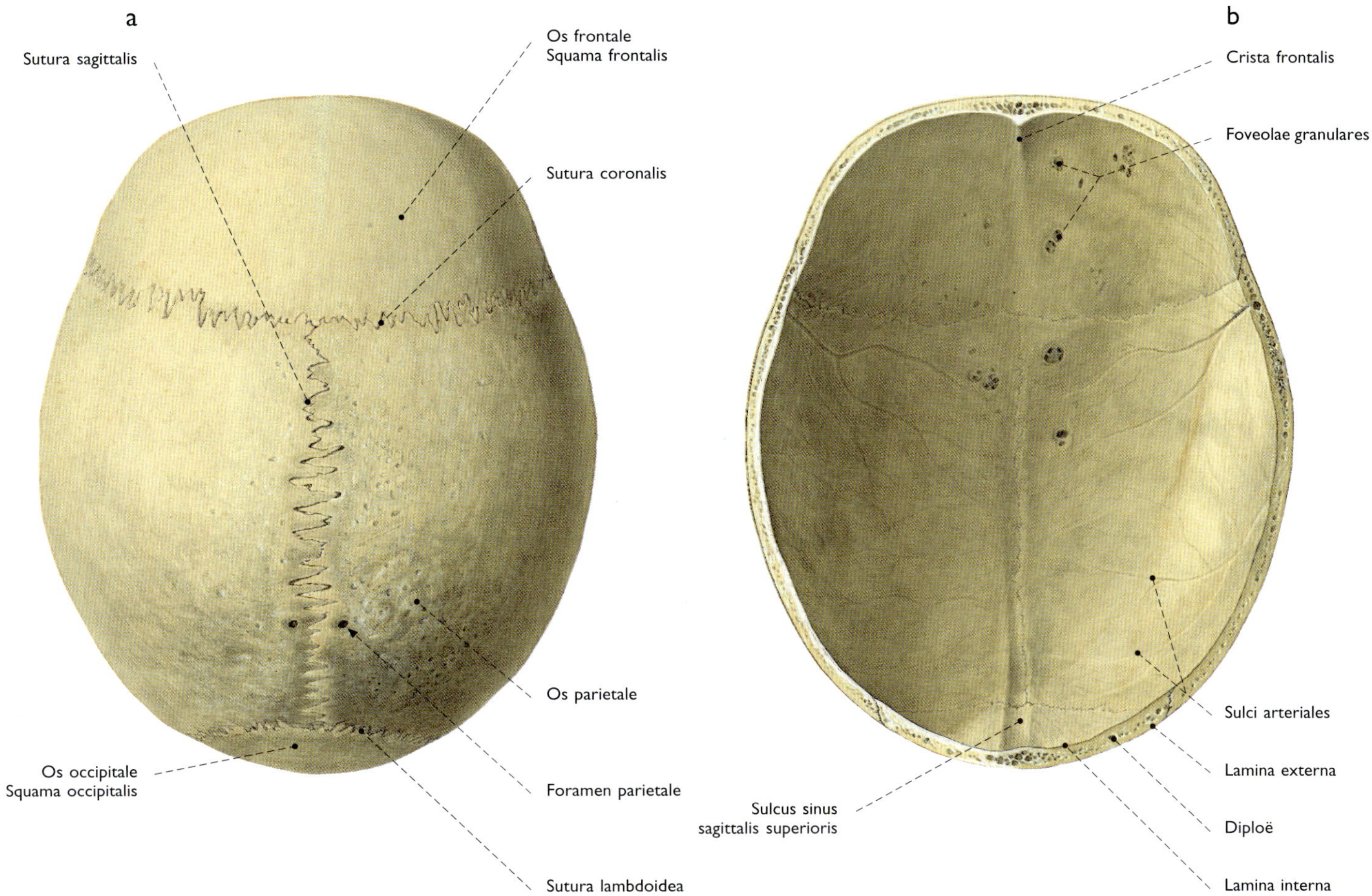

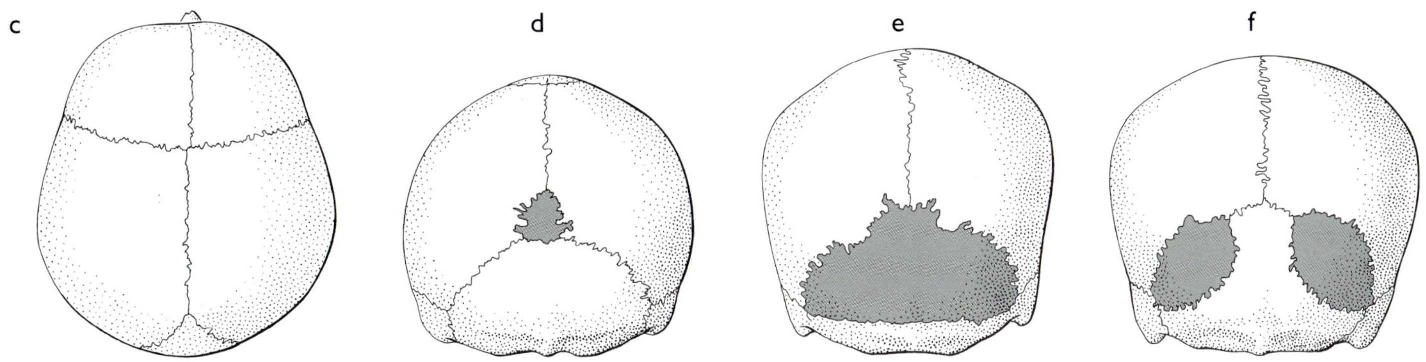

3 Calvaria et Ossa suturalia

3 Skull cap and sutural bones

a Norma verticalis (50%)
b Internal aspect (50%)
c Persisting frontal suture
d Sutural bone (grey) in the sagittal suture
e Inca bone (interparietal bone) (grey)
f Sutural bones (grey) in the lambdoid suture

3 Schädeldach und Nahtknochen

a Norma verticalis (50%)
b Innenansicht (50%)
c Persistierende Stirnnaht
d Nahtknochen (grau) in der Sutura sagittalis
e Inkabein (Os interparietale) (grau)
f Nahtknochen (grau) in der Sutura lambdoidea

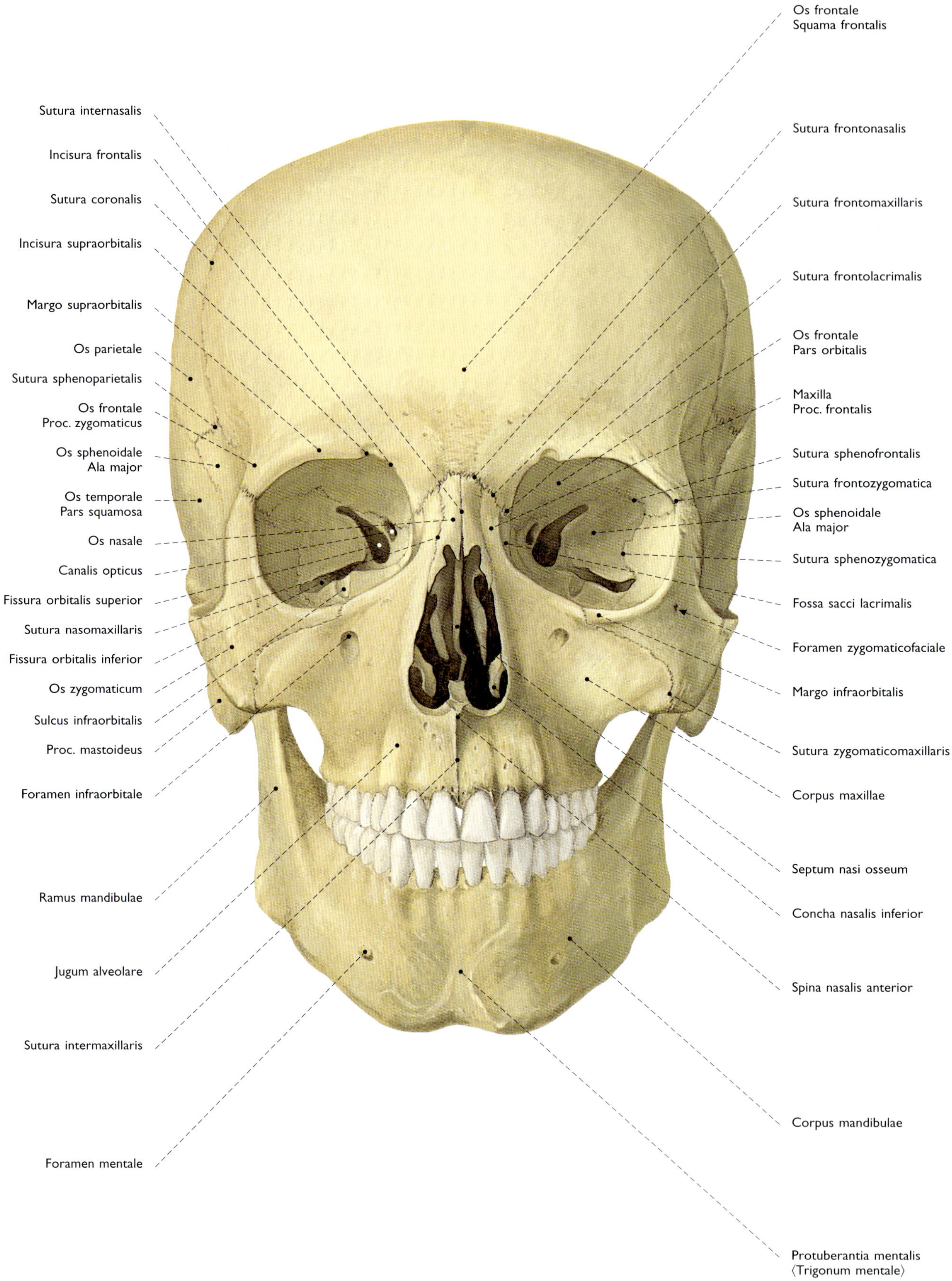

4 Cranium

4 Skull (75%)
Norma frontalis (anterior aspect)

4 Schädel (75%)
Norma frontalis (Vorderansicht)

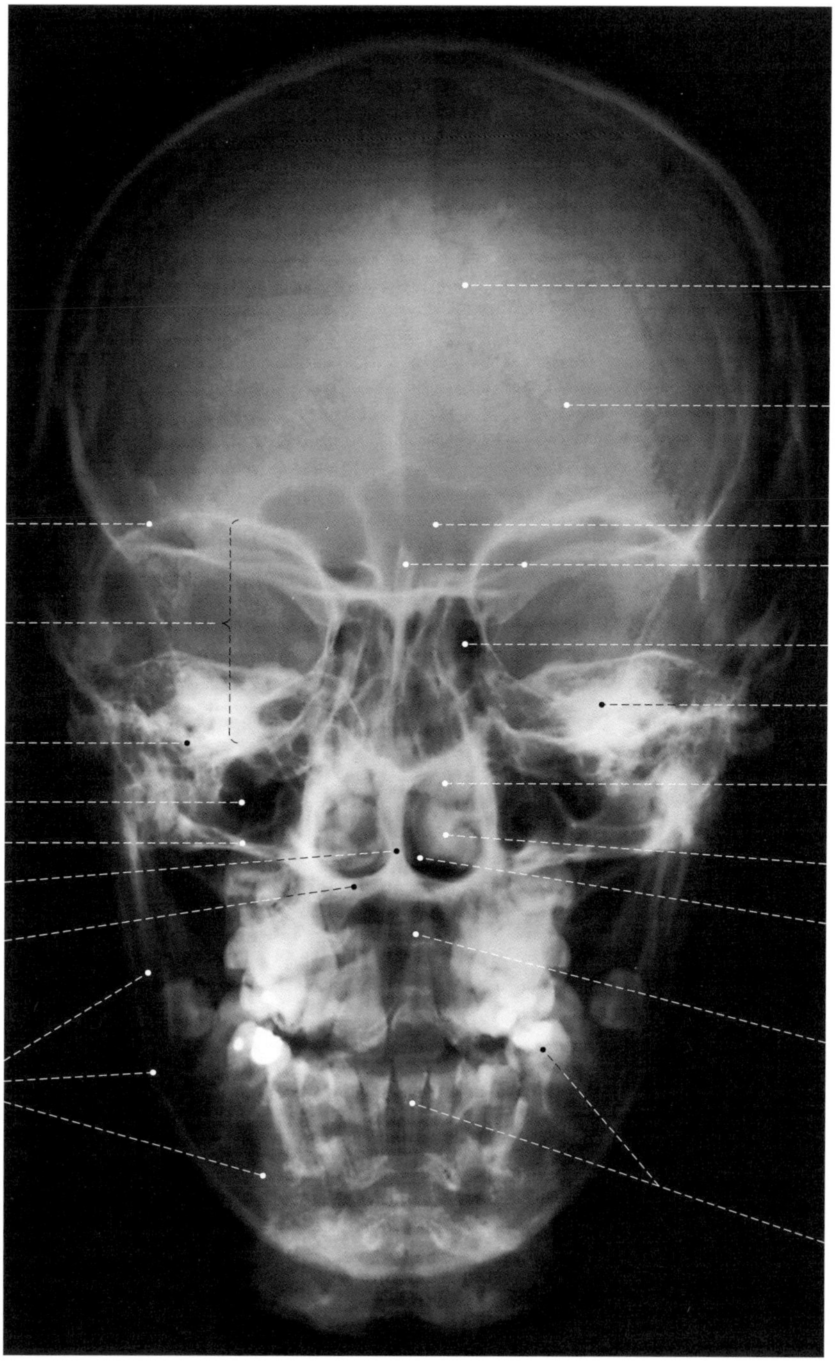

5 Cranium

5 Skull (75%)

Posteroanterior radiograph of the skull of a young female

5 Schädel (75%)

Posteroanteriore Röntgenaufnahme des Schädels einer jungen Frau

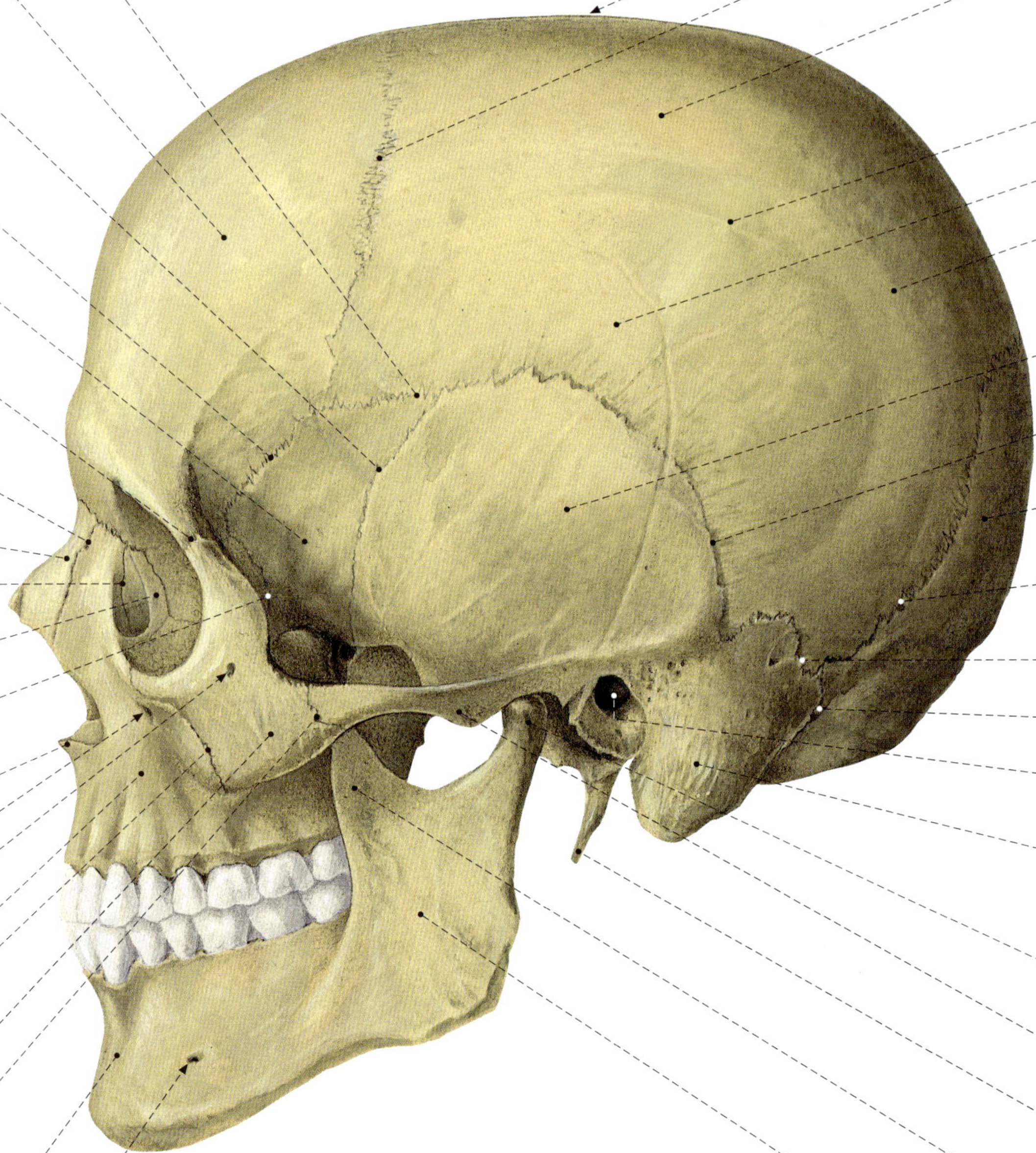

6 Cranium

6 Skull (75%)
Left norma lateralis (left lateral aspect)

6 Schädel (75%)
Linke Norma lateralis (linke Seitenansicht)

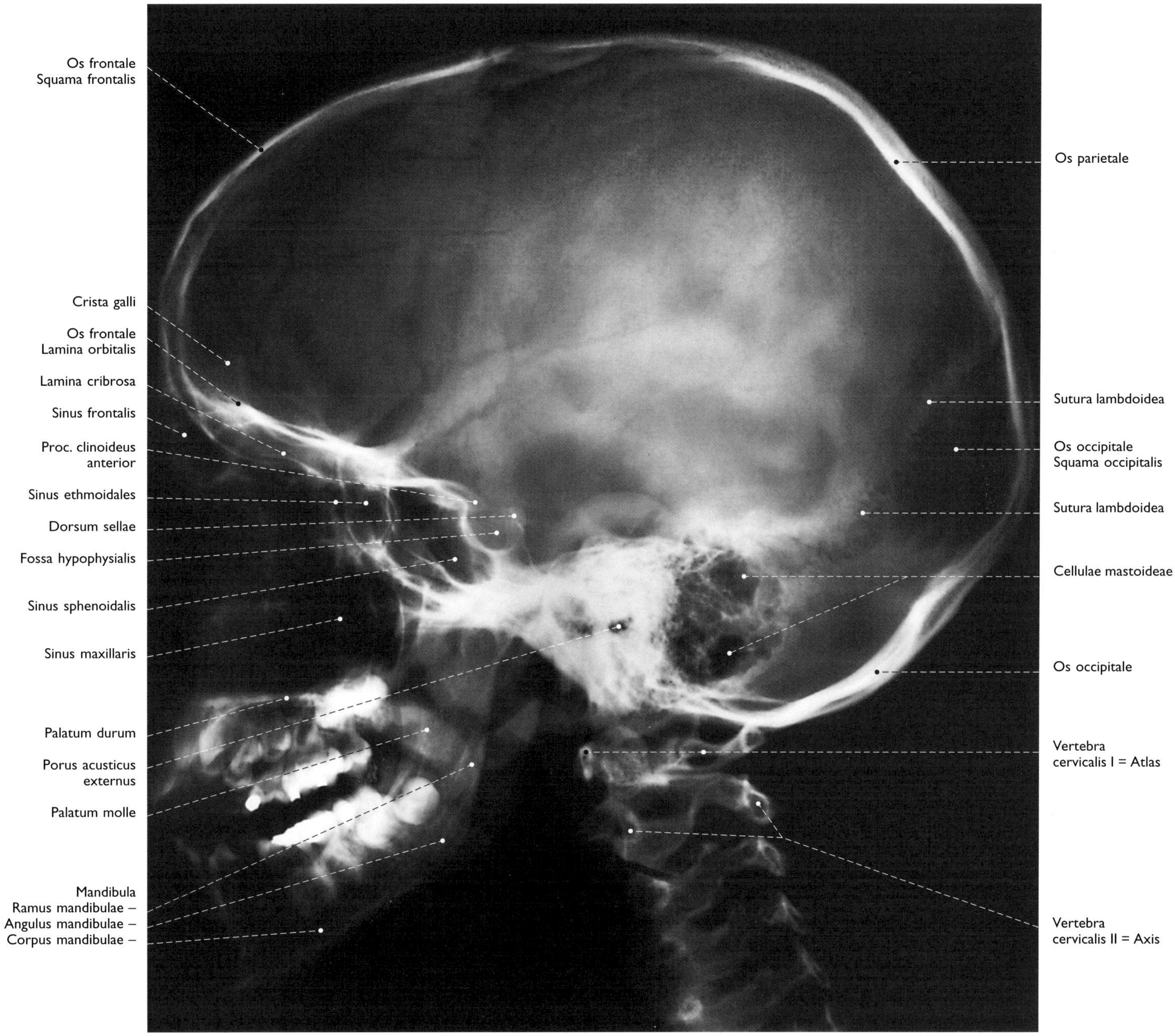

7 Cranium

7 Skull (75%)

Lateral radiograph of the skull of a young female

7 Schädel (75%)

Seitliche Röntgenaufnahme des Schädels einer jungen Frau

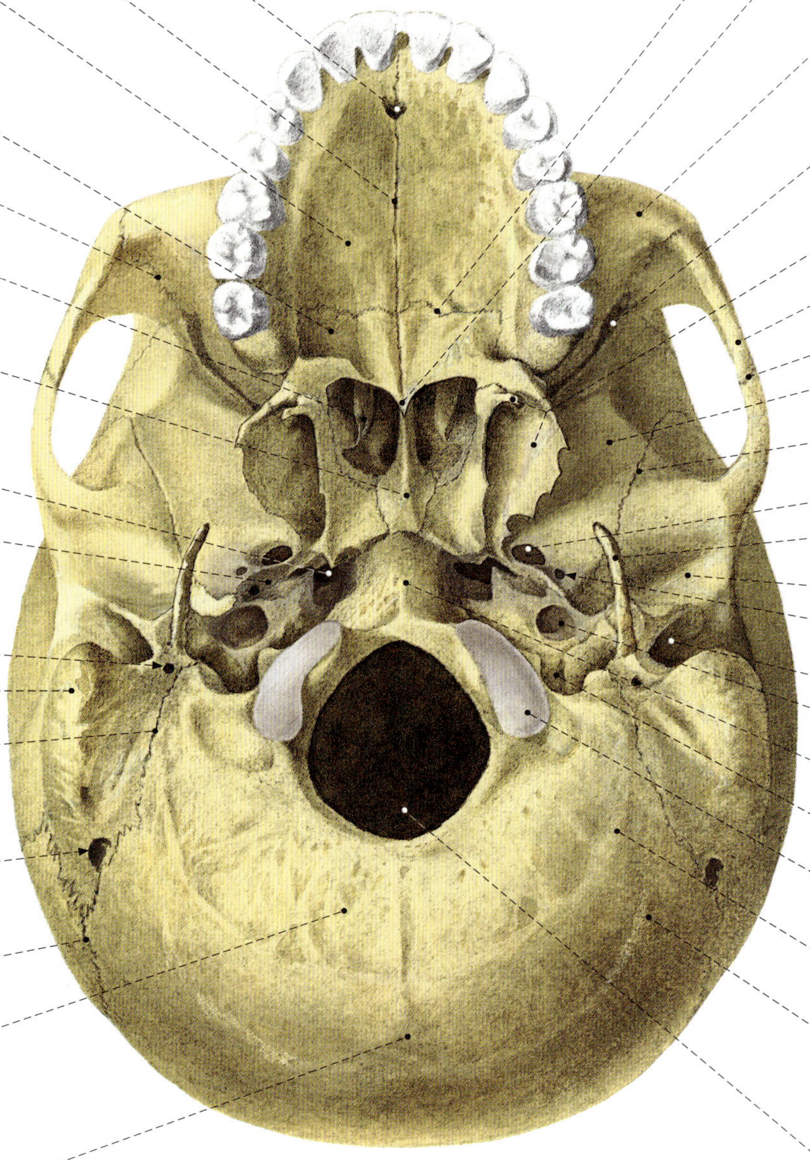

8 Basis cranii externa

8 Base of the skull (75%)
Norma basalis (inferior aspect)

8 Äussere Schädelbasis (75%)
Norma basalis (Ansicht von unten)

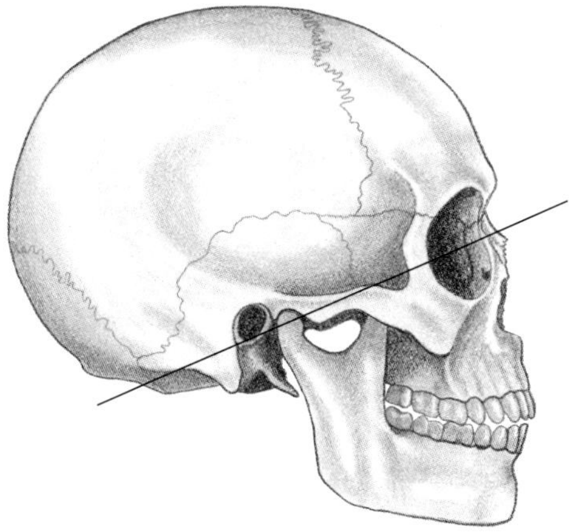

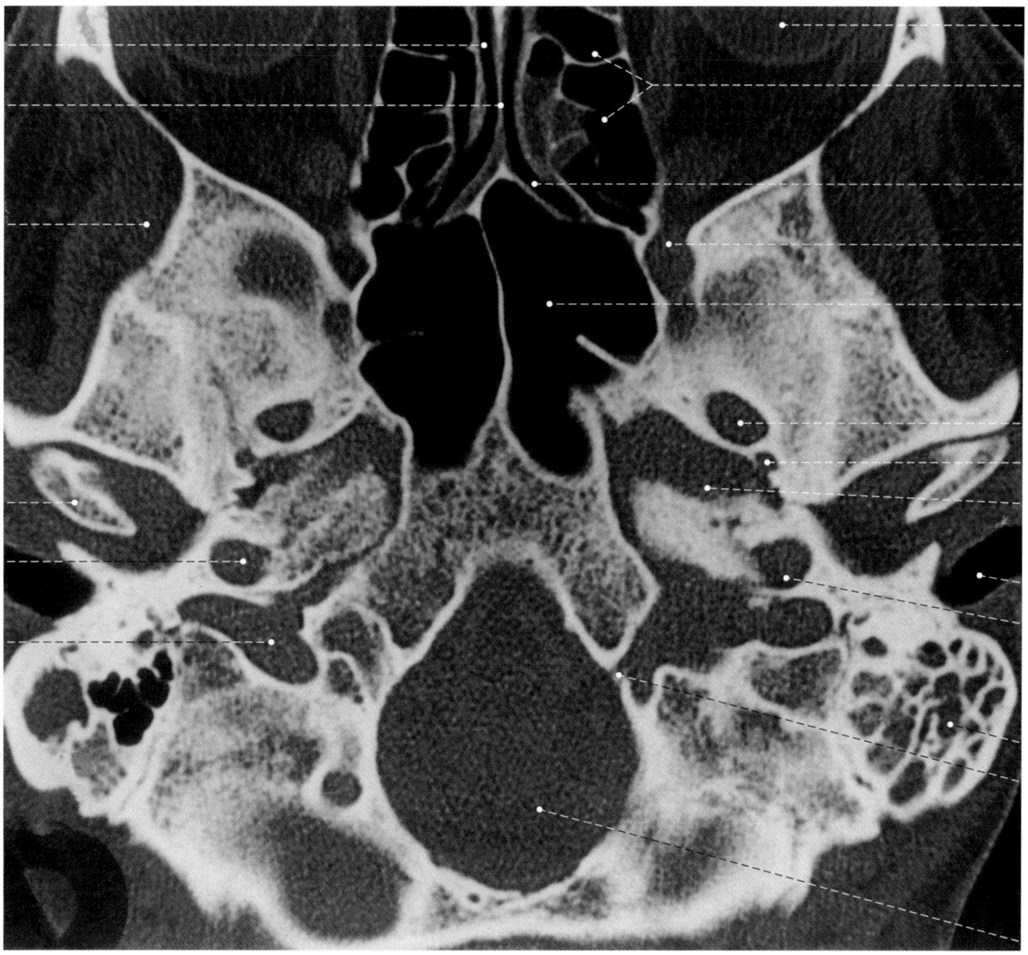

9 Basis cranii externa

9 Base of the skull (100%)

Transverse computed tomogram (CT), bone window, plane as indicated in the sketch above

9 Äussere Schädelbasis (100%)

Transversales Computertomogramm (CT), Knochenfenster, Ebene wie in Skizze angegeben

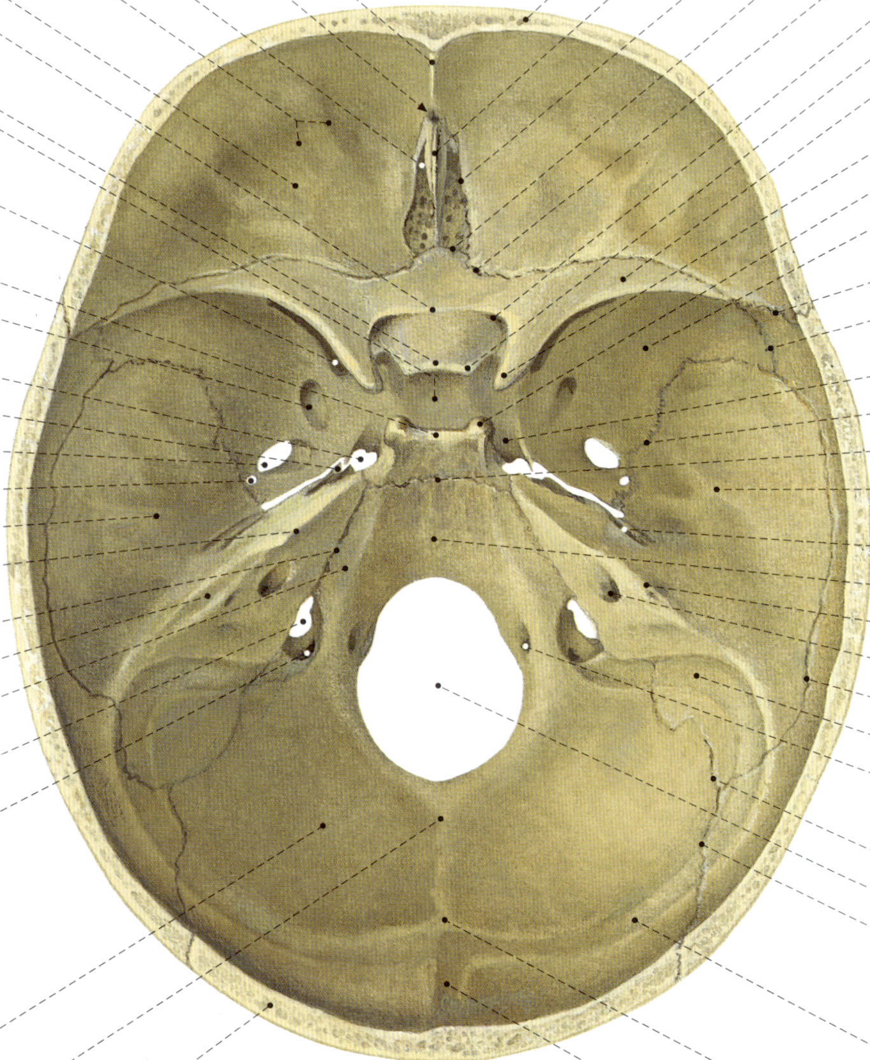

10 Basis cranii interna

10 Base of the skull (75%)
Internal surface, superior aspect

10 Innere Schädelbasis (75%)
Ansicht von oben

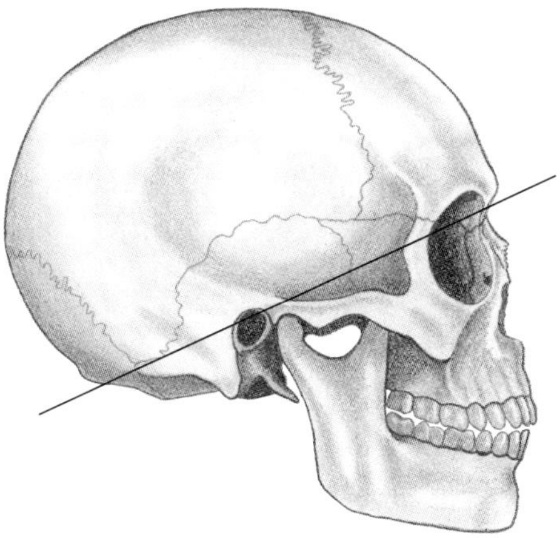

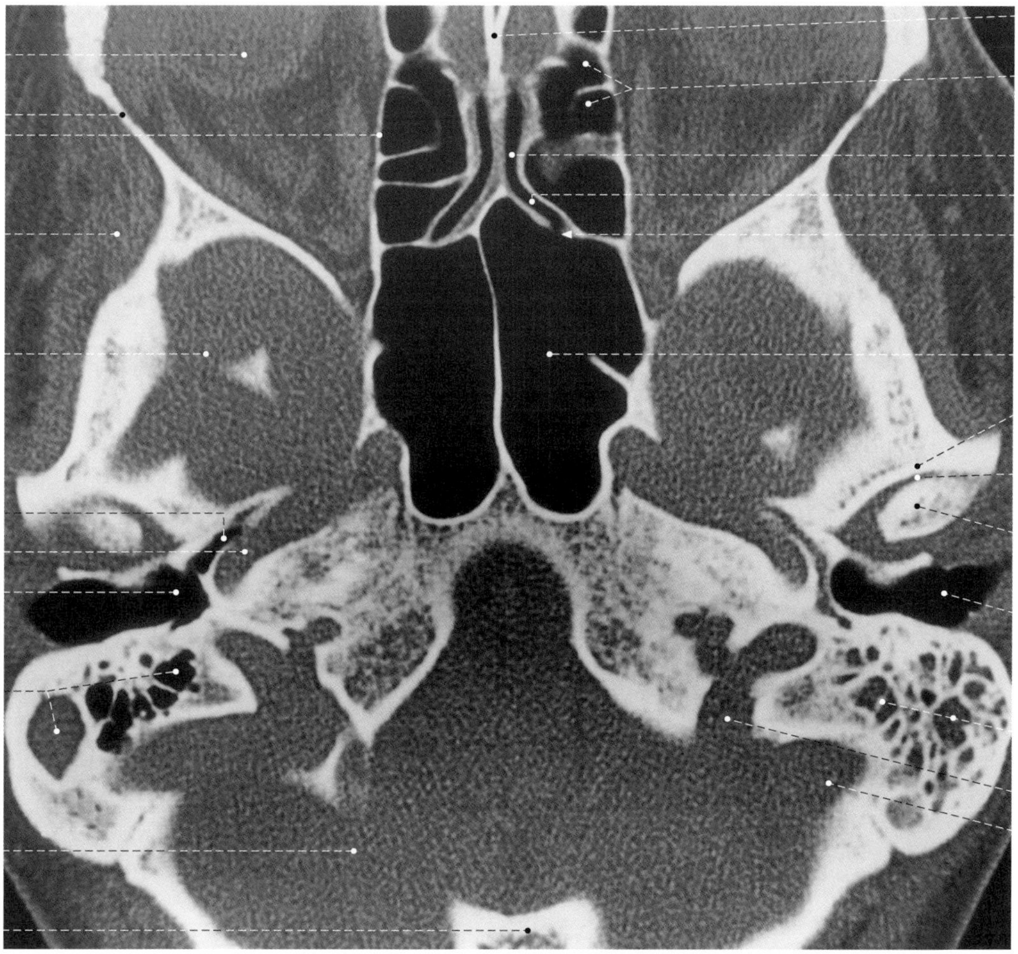

11 Basis cranii interna

11 Base of the skull (100%)

Transverse computed tomogram (CT), bone window, plane as indicated in the sketch above

11 Innere Schädelbasis (100%)

Transversales Computertomogramm (CT), Knochenfenster, Ebene wie in Skizze angegeben

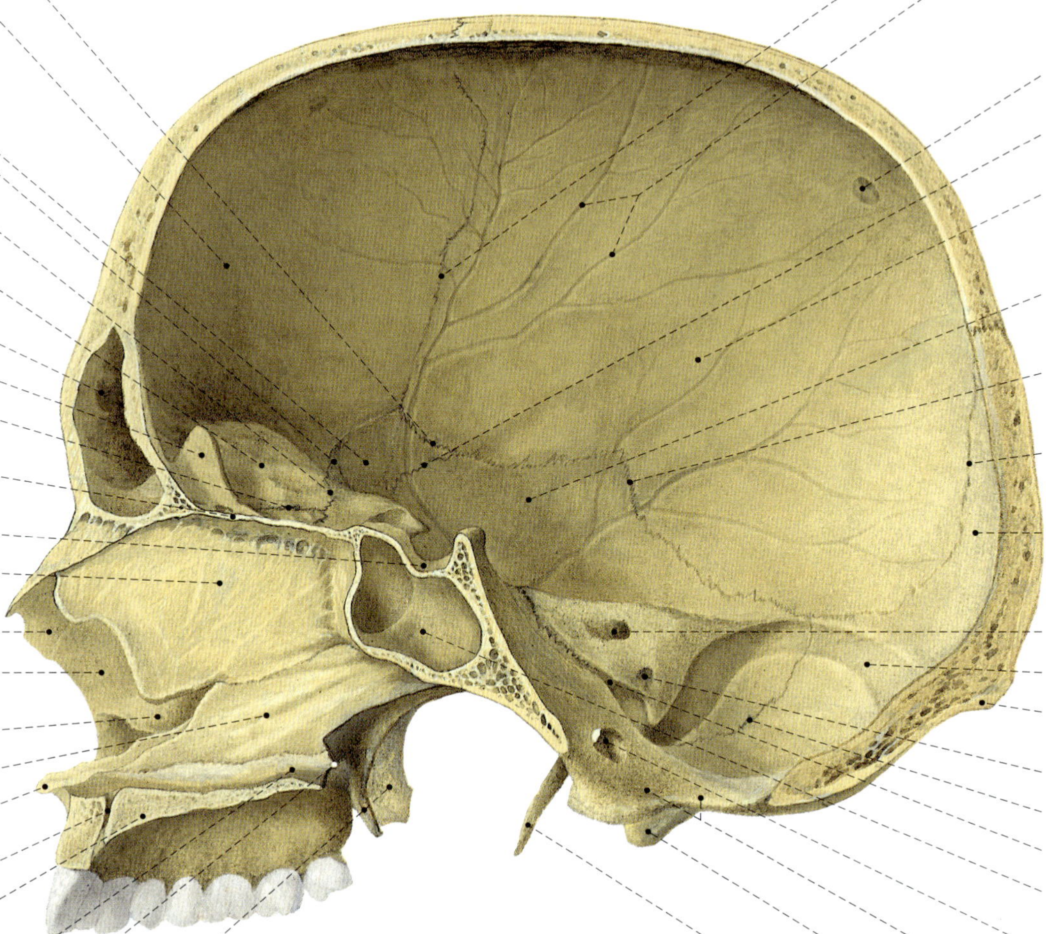

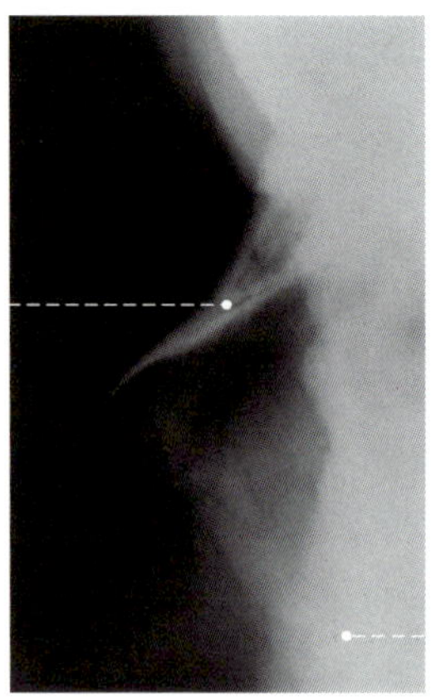

12 Cranium

12 Skull (75%)

a Sagittal section slightly to the left of the median plane, medial aspect
b Lateral radiograph of the nasal bone

12 Schädel (75%)

a Sagittalschnitt etwas links der Medianebene, Medialansicht
b Seitliche Röntgenaufnahme des Nasenbeins

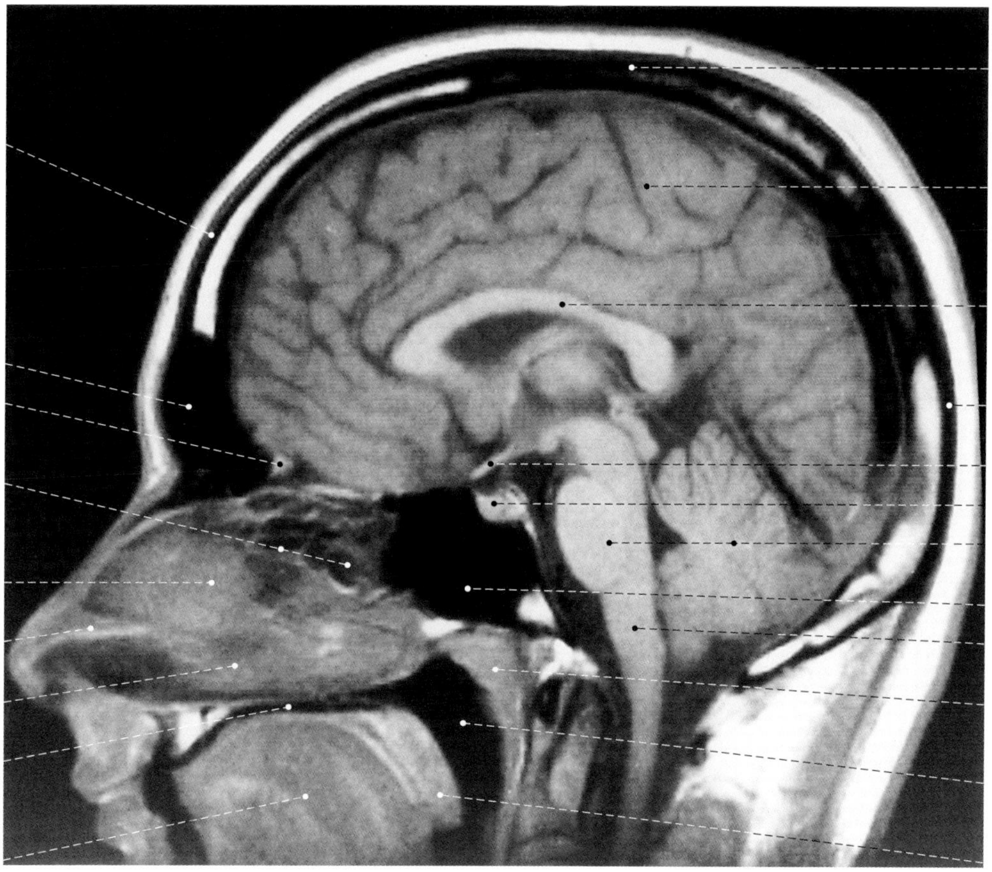

13 Caput

13 Head (55%)

Sagittal, paramedian magnetic resonance image (MRI, T_1-weighted)

13 Kopf (55%)

Sagittales, paramedianes Kernspintomogramm (MRI, T_1-gewichtet)

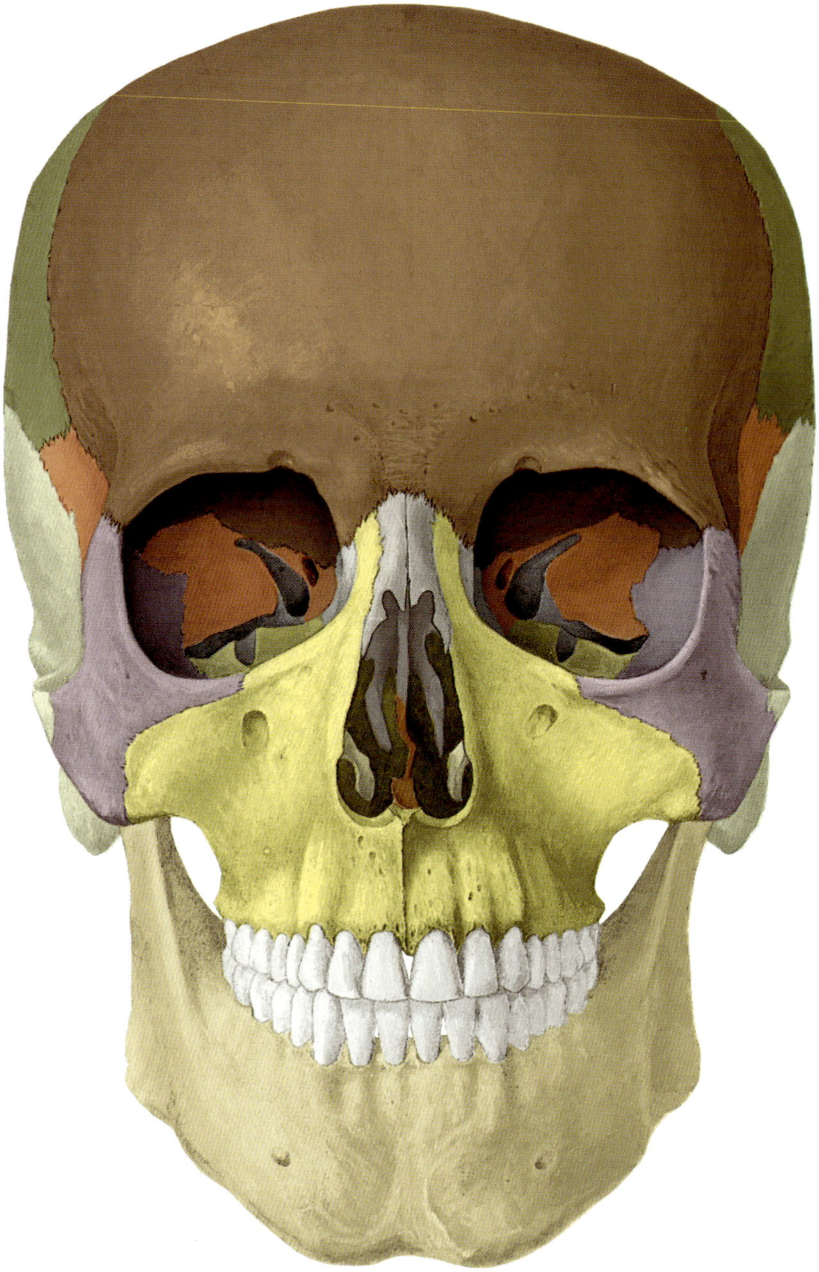

- Os frontale
- Os parietale
- Os sphenoidale
- Os temporale
- Os nasale
- Os lacrimale
- Os ethmoidale
- Os zygomaticum
- Maxilla
- Concha nasalis inferior
- Mandibula
- Vomer

14 Cranium

14 Skull (80%)

The skull bones are marked by different colors. Norma frontalis (anterior aspect)

14 Schädel (80%)

Schädelknochen in verschiedenen Farben, Norma frontalis (Vorderansicht)

a

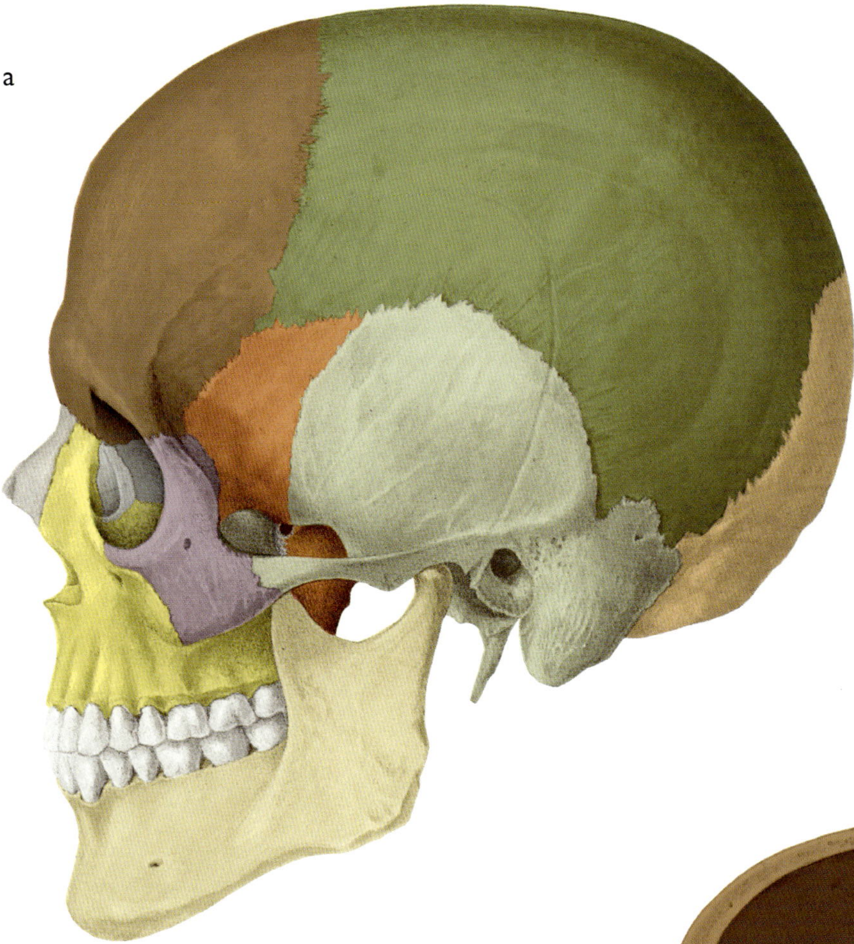

b

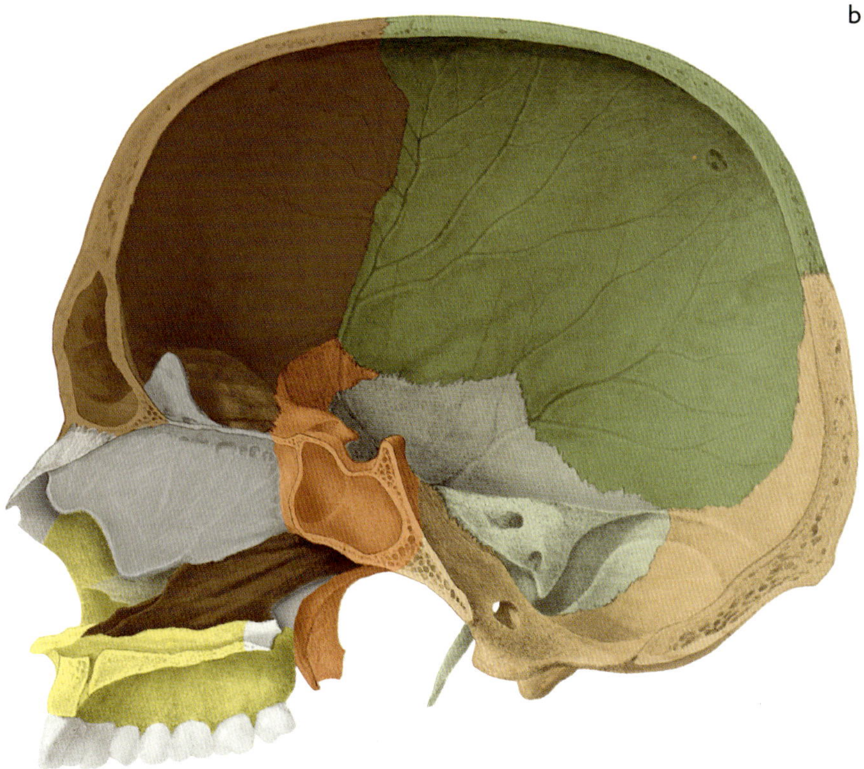

- Os frontale
- Os parietale
- Os sphenoidale
- Os occipitale
- Os temporale
- Os nasale
- Os lacrimale
- Os ethmoidale
- Os zygomaticum
- Maxilla
- Mandibula
- Os palatinum
- Vomer

15 Cranium

15 Skull

The skull bones are marked by different colors.

a Left norma lateralis (lateral aspect)

b Medial aspect

15 Schädel

Schädelknochen in verschiedenen Farben

a Linke Norma lateralis (Seitenansicht)

b Medialansicht

a

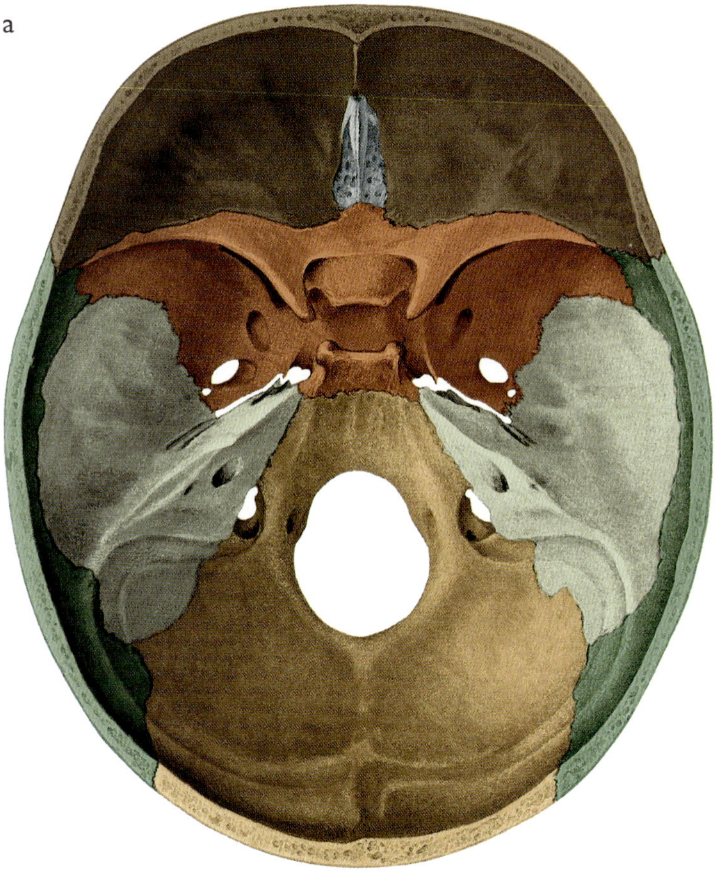

b

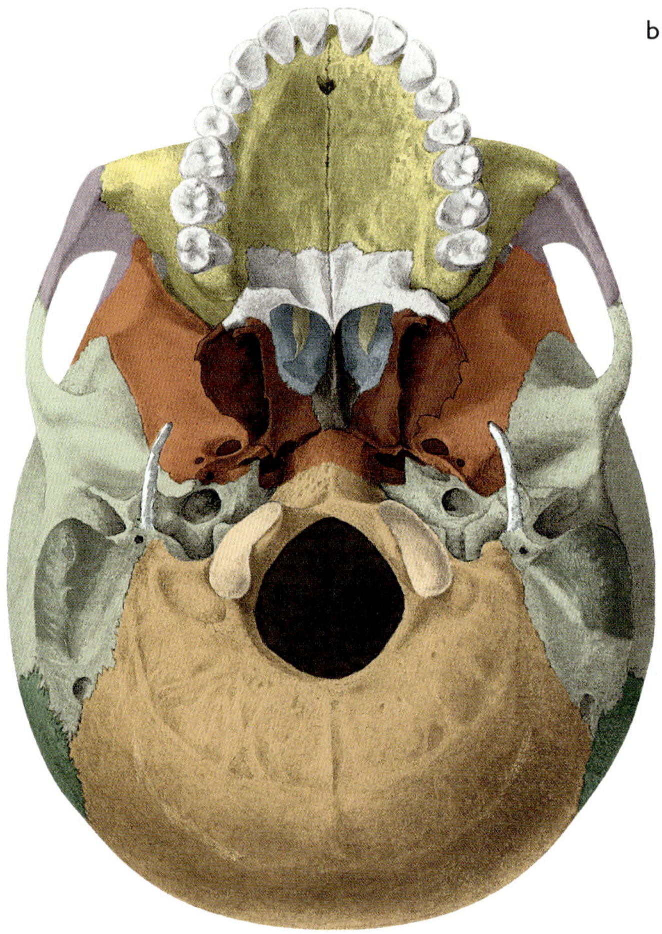

16 Basis cranii

16 Base of the skull (70%)

The skull bones are marked by different colors.

a Internal surface
b Outer surface

16 Schädelbasis (70%)

Schädelknochen in verschiedenen Farben

a Innenansicht
b Aussenansicht

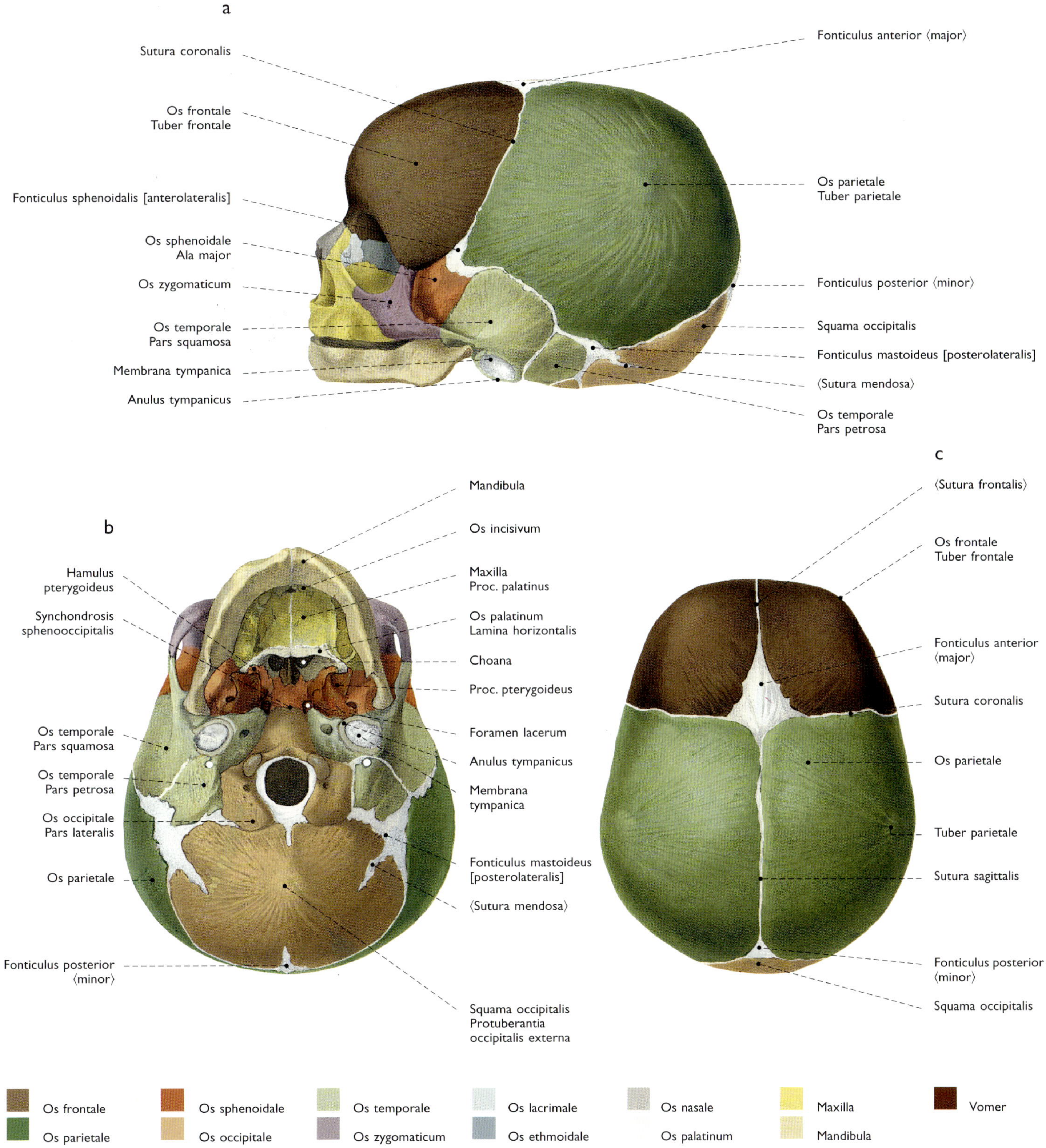

17 Cranium neonati

17 Skull of a newborn child (80%)

The skull bones are marked by different colors.

a Left norma lateralis (left lateral aspect)
b Base of the skull, norma basalis
c Skull cap, norma verticalis

17 Schädel eines Neugeborenen (80%)

Schädelknochen in verschiedenen Farben

a Linke Norma lateralis (linke Seitenansicht)
b Äussere Schädelbasis, Norma basalis
c Schädeldach, Norma verticalis

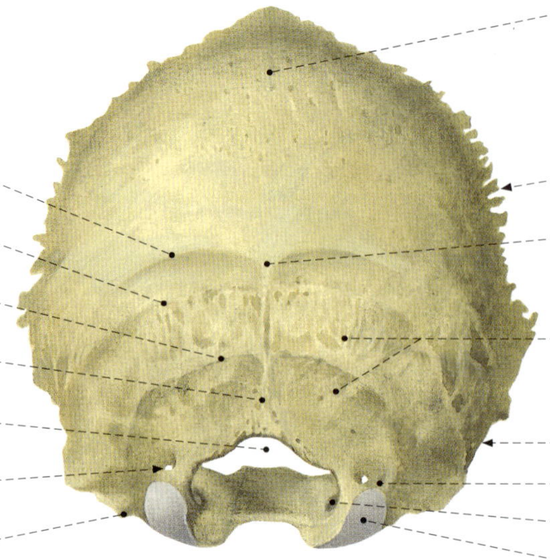

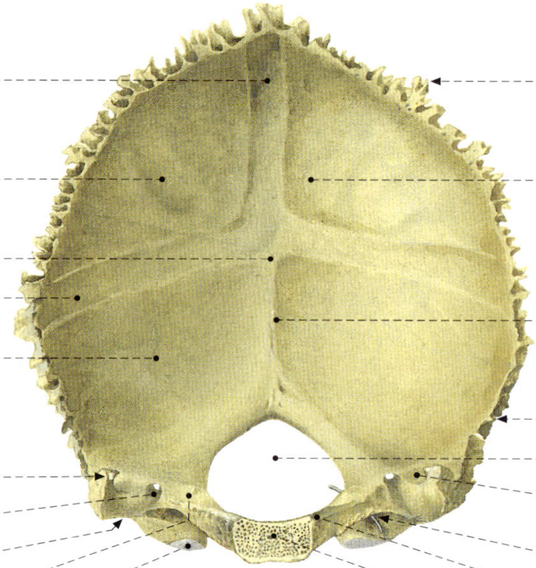

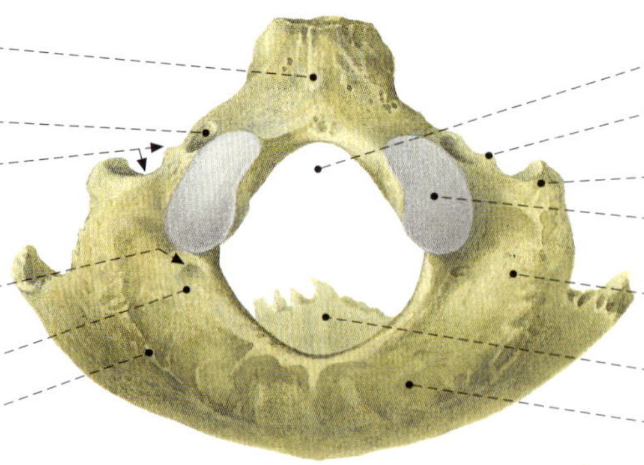

18 Os occipitale

18 Occipital bone (70%)

a External surface, occipital aspect
b Internal surface, frontal aspect
c External surface, inferior aspect

18 Hinterhauptbein (70%)

a Facies externa, Okzipitalansicht
b Facies interna, Frontalansicht
c Facies externa, Basalansicht

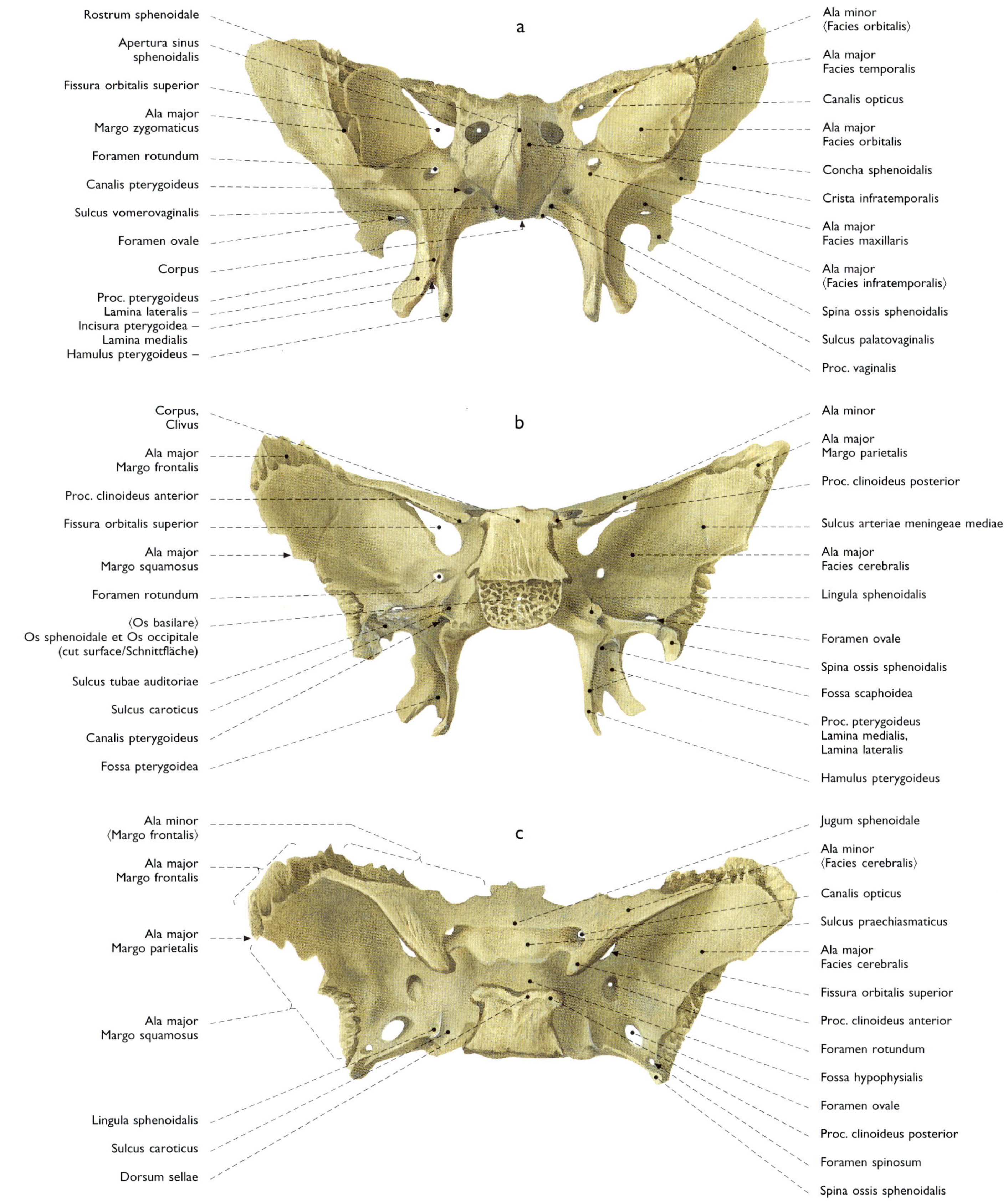

19 Os sphenoidale

19 Sphenoid bone (70%)

a Frontal aspect
b Occipital aspect
c Parietal aspect

19 Keilbein (70%)

a Frontalansicht
b Okzipitalansicht
c Parietalansicht

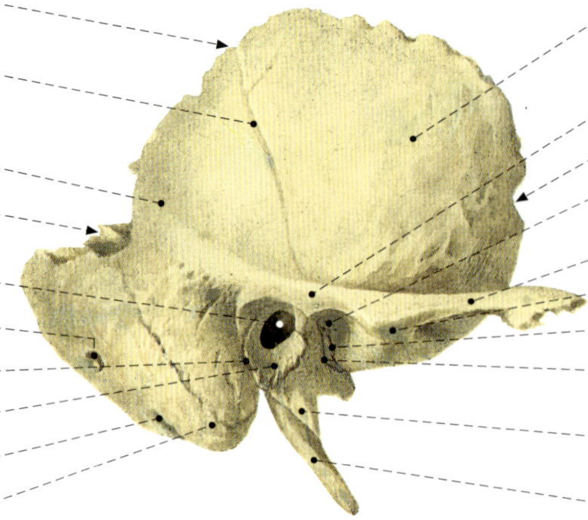

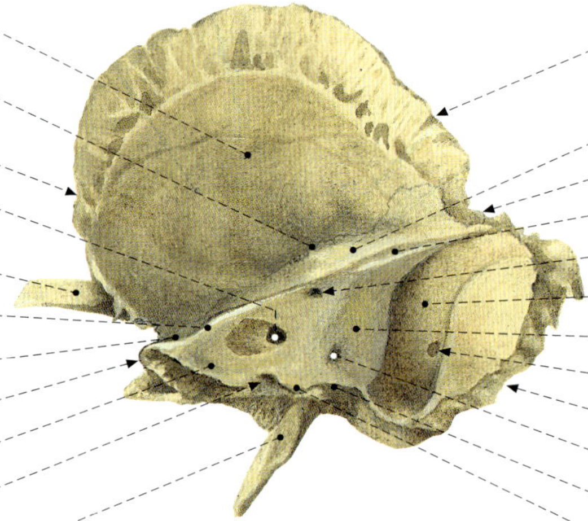

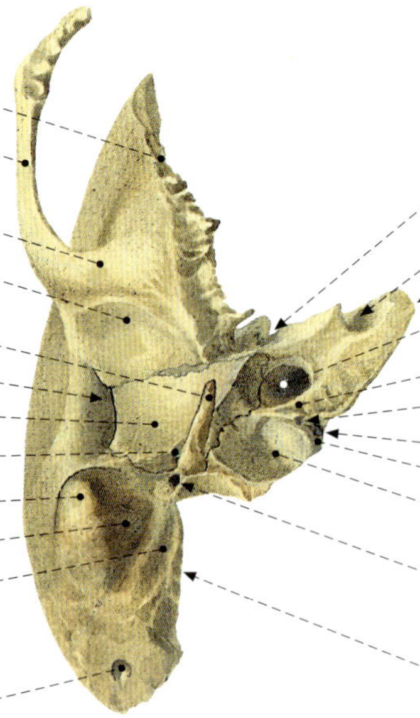

20 Os temporale

20 Right temporal bone (70%)

a Lateral aspect
b Medial aspect
c Inferior aspect

20 Rechtes Schläfenbein (70%)

a Seitenansicht
b Medialansicht
c Basalansicht

a

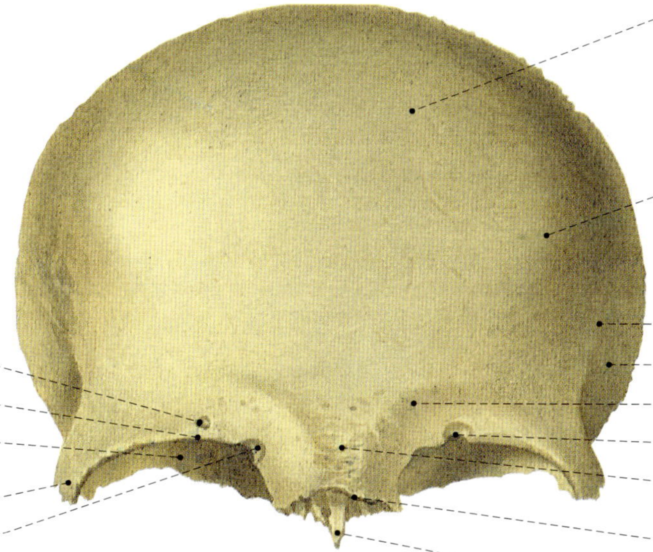

b

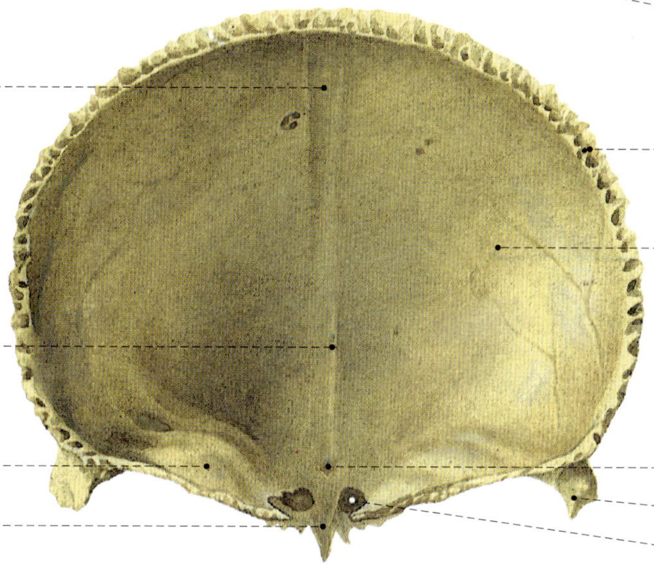

c

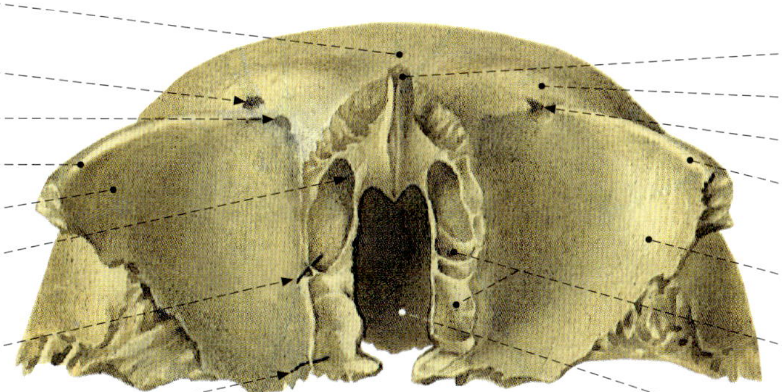

21 Os frontale

21 Frontal bone (90%)

a External surface, frontal aspect
b Internal surface, occipital aspect
c Orbital part, inferior aspect

21 Stirnbein (90%)

a Facies externa, Frontalansicht
b Facies interna, Okzipitalansicht
c Pars orbitalis, Basalansicht

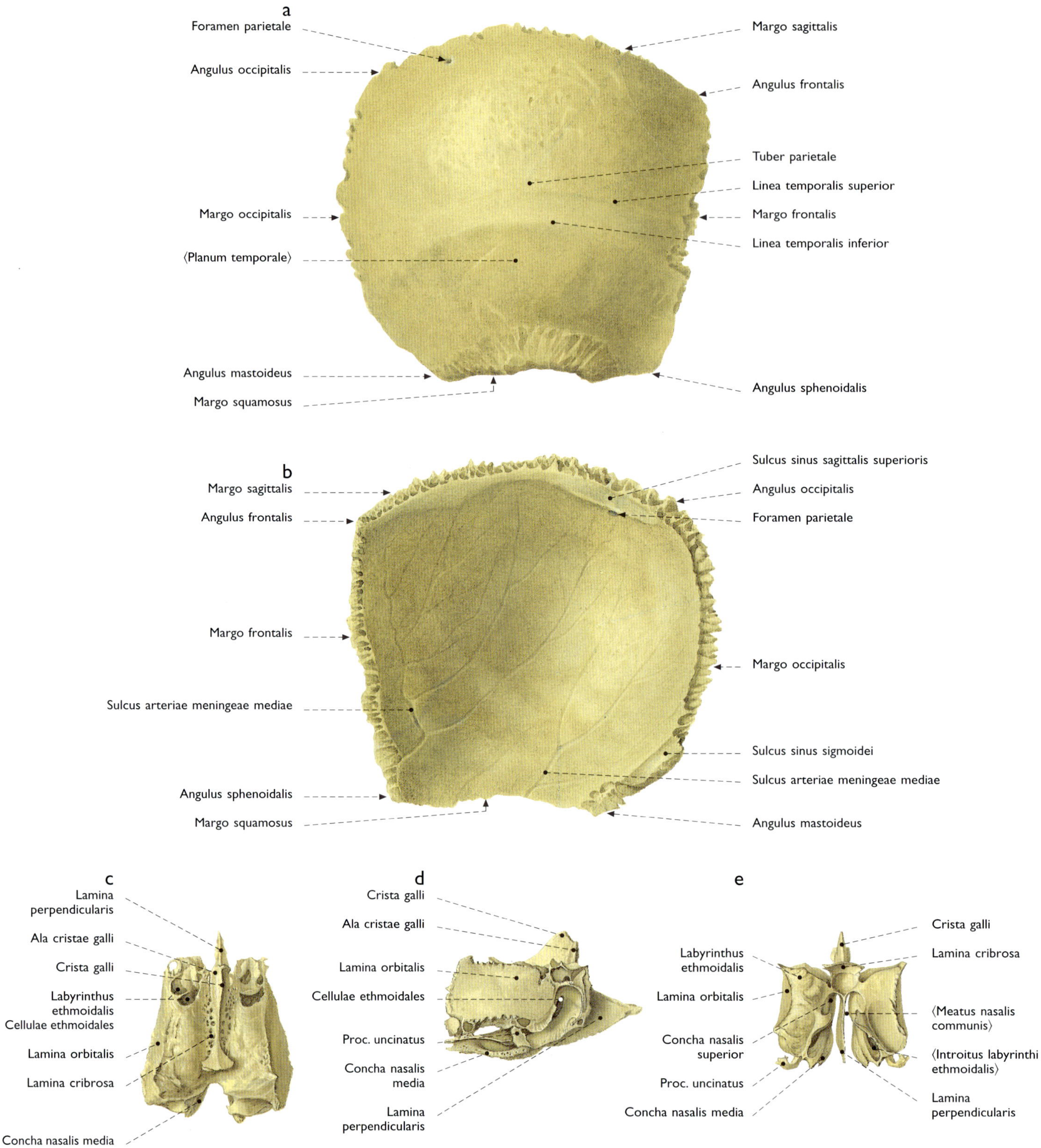

22 Ossa parietale et ethmoidale

22 Parietal and ethmoid bones (70%)

a, b Parietal bone
- a External surface, lateral aspect
- b Internal surface, medial aspect

c–e Ethmoid bone
- c Parietal aspect
- d Lateral aspect
- e Occipital aspect

22 Scheitel- und Siebbein (70%)

a, b Scheitelbein
- a Facies externa, Seitenansicht
- b Facies interna, Medialansicht

c–e Siebbein
- c Parietalansicht
- d Seitenansicht
- e Okzipitalansicht

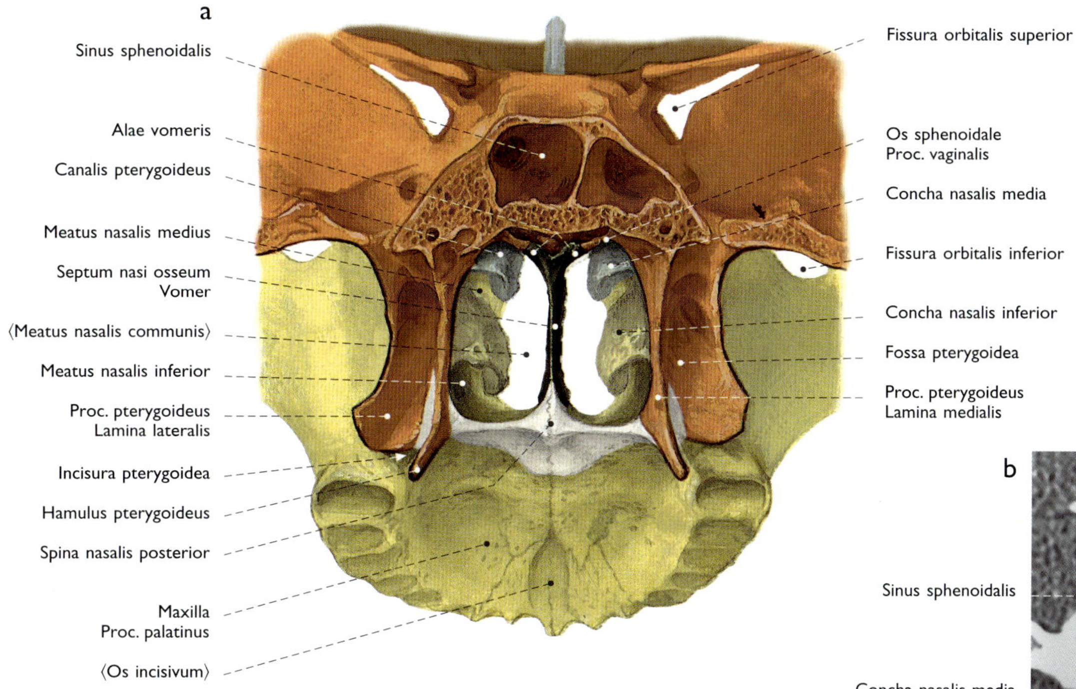

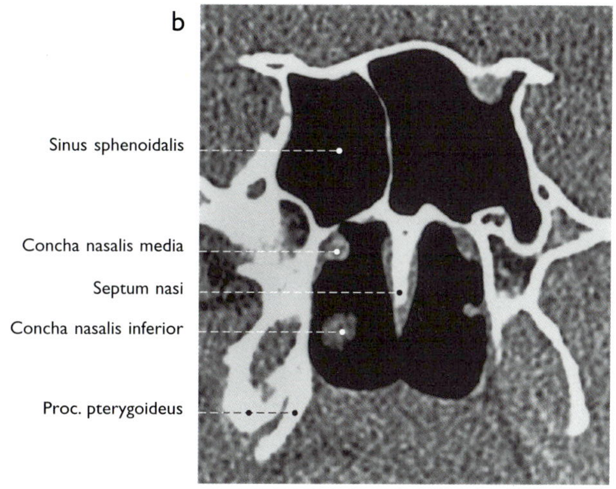

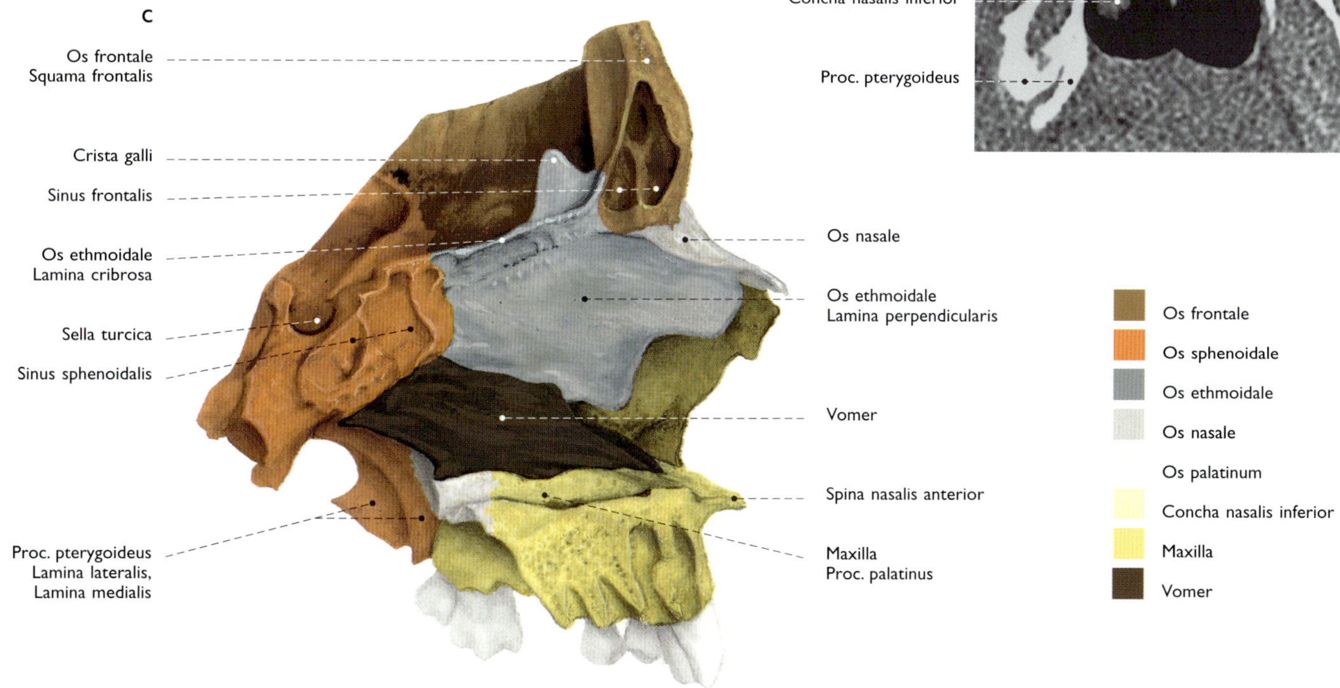

23 Choanae et Septum nasi osseum

23 Skeleton of the nose (100%)

a, c The different bones are marked by various colors.

a Posterior nasal apertures (choanae), occipital aspect

b Coronal computed tomogram (CT) through the choanae and the sphenoidal sinuses

c Bony nasal septum. Medial aspect of the left part of the skull, sagittal section slightly to the right of the median plane

23 Nasenskelett (100%)

a, c Die verschiedenen Knochen sind durch unterschiedliche Farben dargestellt.

a Hintere Nasenöffnungen (Choanen), Okzipitalansicht

b Frontales Computertomogramm (CT) durch die Choanen und die Keilbeinhöhlen

c Knöcherne Nasenscheidewand. Medialansicht des linken Schädelteils. Sagittalschnitt etwas rechts der Medianebene

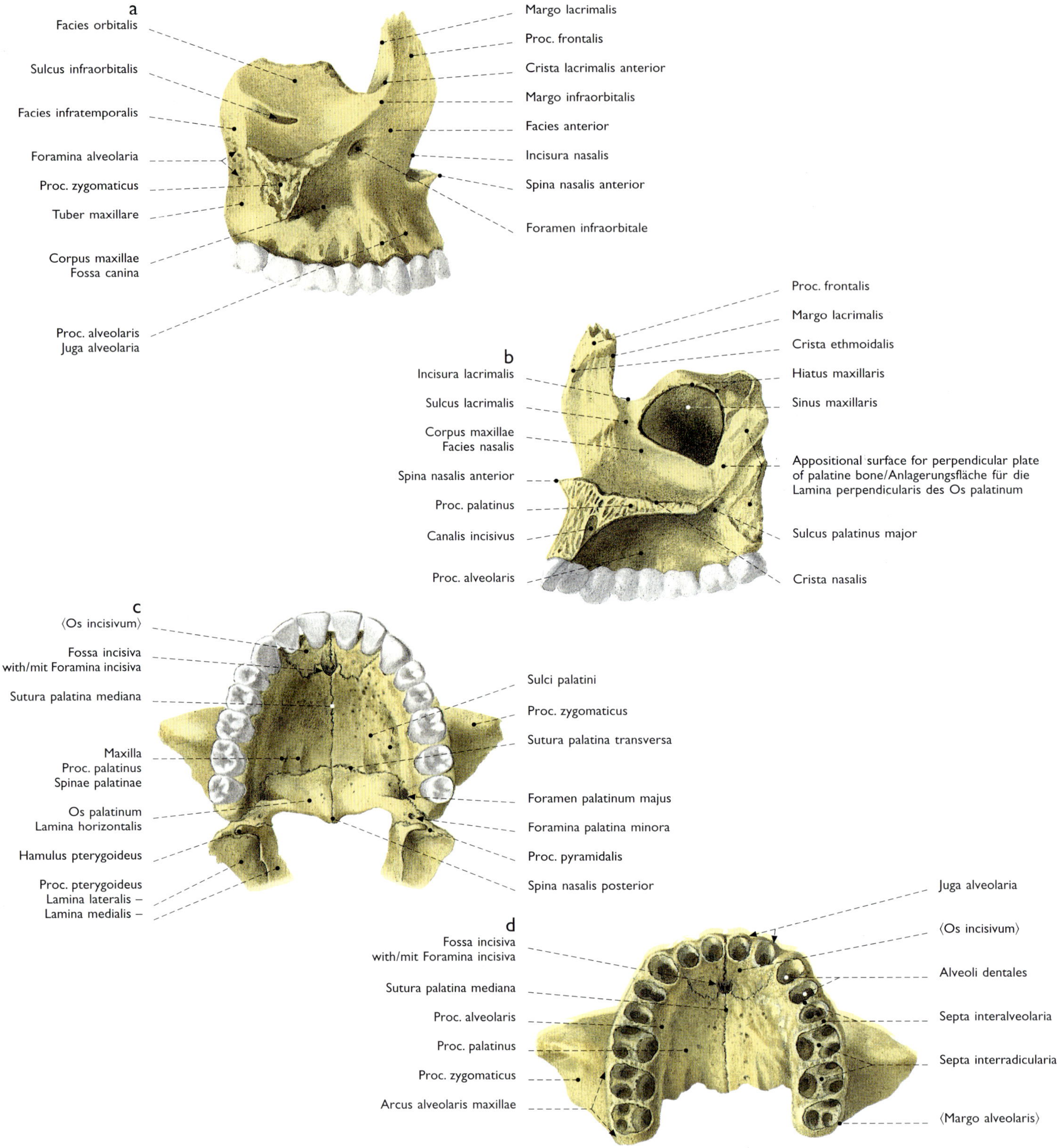

24 Maxilla

24 Maxilla (70%)

a Lateral aspect
b Medial aspect
c, d Inferior aspect
c Alveolar arch with the teeth
d Alveolar arch after extraction of the teeth

24 Oberkiefer (70%)

a Seitenansicht
b Medialansicht
c, d Basalansicht
c Zahnbogen mit Zähnen
d Zahnbogen nach Extraktion der Zähne

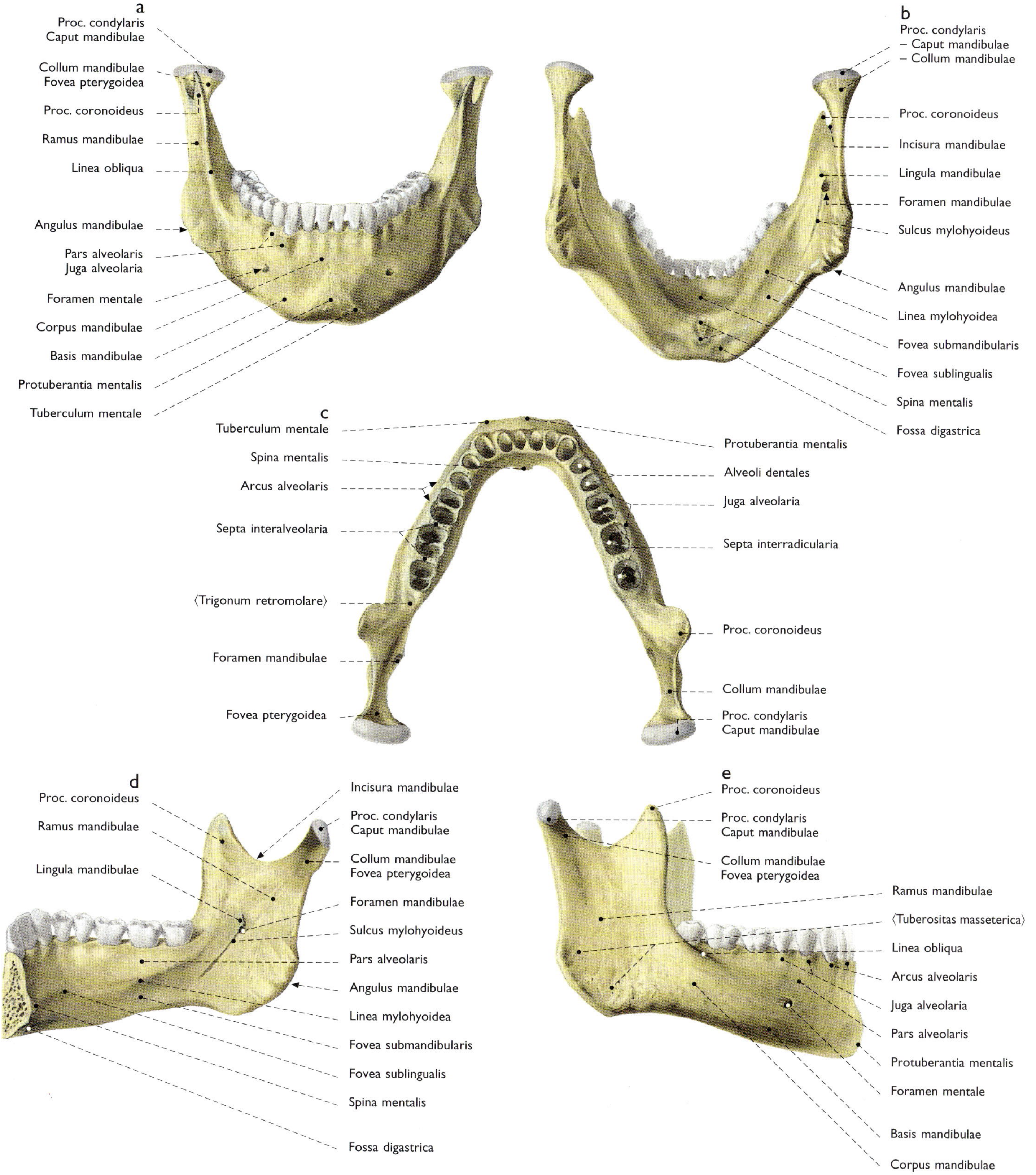

25 Mandibula

25 Mandible (55%)

a Frontal aspect
b Occipital aspect
c Parietal aspect, alveolar arch after extraction of the teeth
d Medial aspect of the right half of the mandible
e Right lateral aspect

25 Unterkiefer (55%)

a Frontalansicht
b Okzipitalansicht
c Parietalansicht, Zahnbogen nach Extraktion der Zähne
d Medialansicht der rechten Hälfte des Unterkiefers
e Rechte Seitenansicht

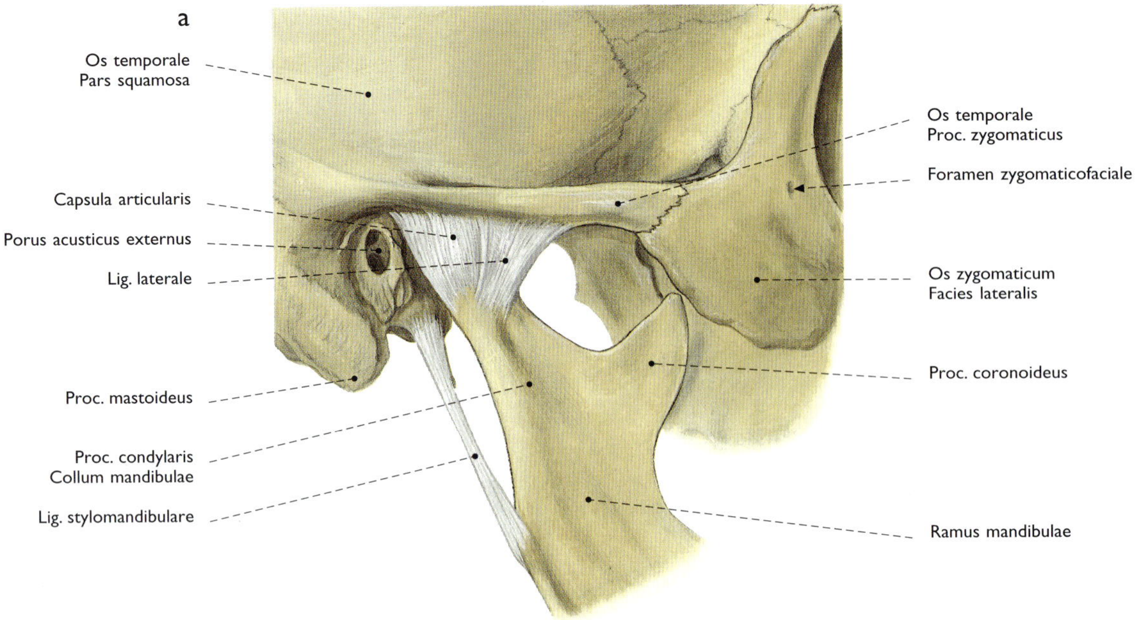

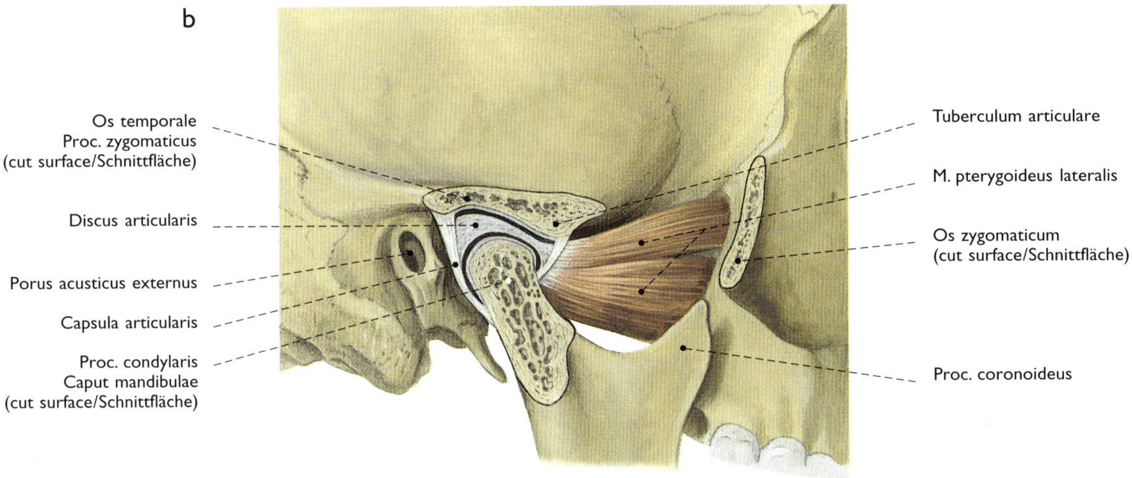

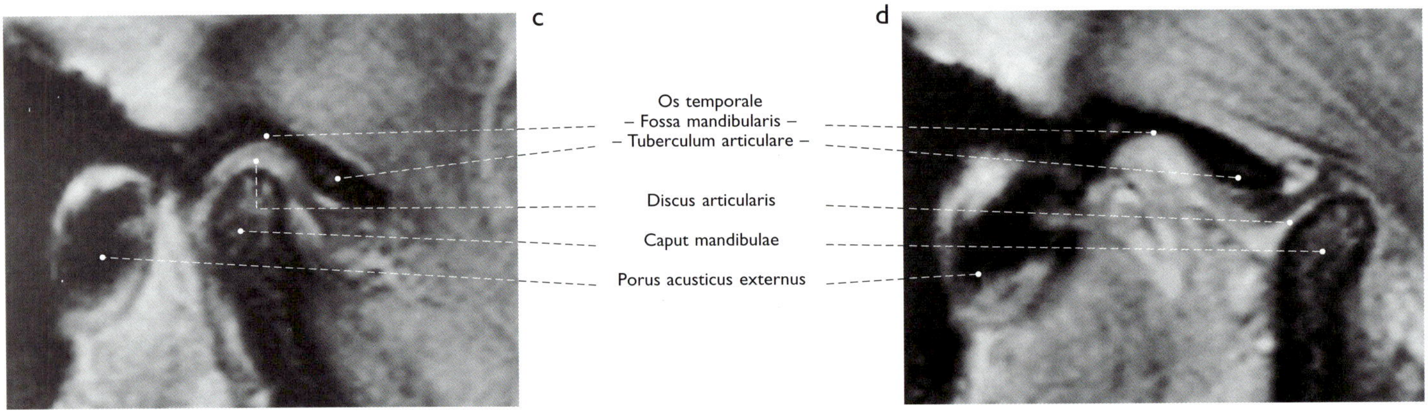

26 Articulatio temporomandibularis

26 Temporomandibular joint

a, b Right lateral aspect (80%)
- a Capsule and ligaments
- b Sagittal section, the lateral pterygoid muscle is left in position

c, d Sagittal magnetic resonance image (MRI, T_2-weighted) through the temporomandibular joint
- c when mouth closed
- d when mouth opened

26 Kiefergelenk

a, b Rechte Seitenansicht (80%)
- a Kapsel und Bänder
- b Sagittalschnitt, Musculus pterygoideus lateralis in situ

c, d Sagittales Kernspintomogramm (MRI, T_2-gewichtet) durch das Kiefergelenk
- c bei geschlossenem Mund
- d bei geöffnetem Mund

a

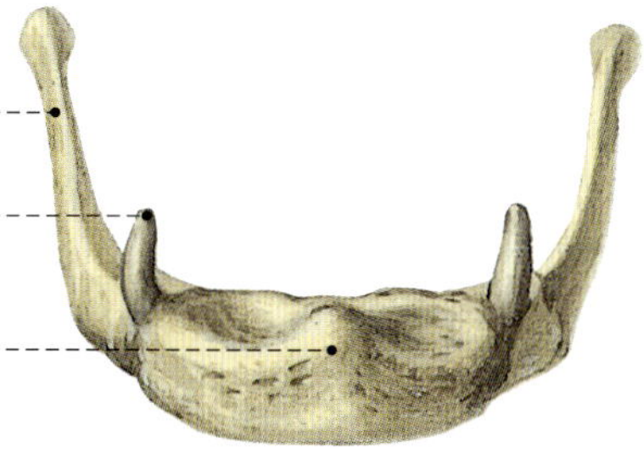

b

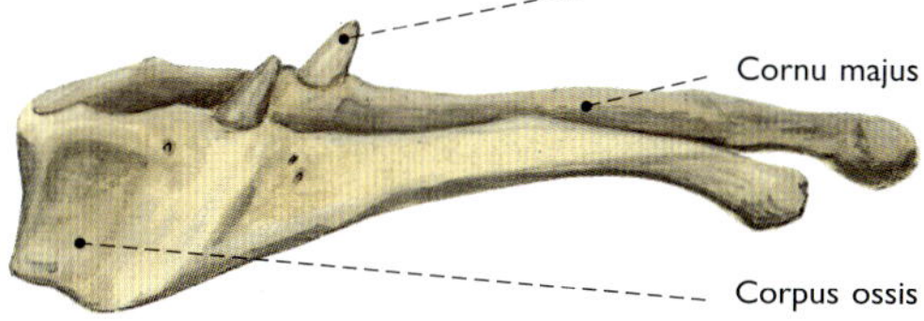

c

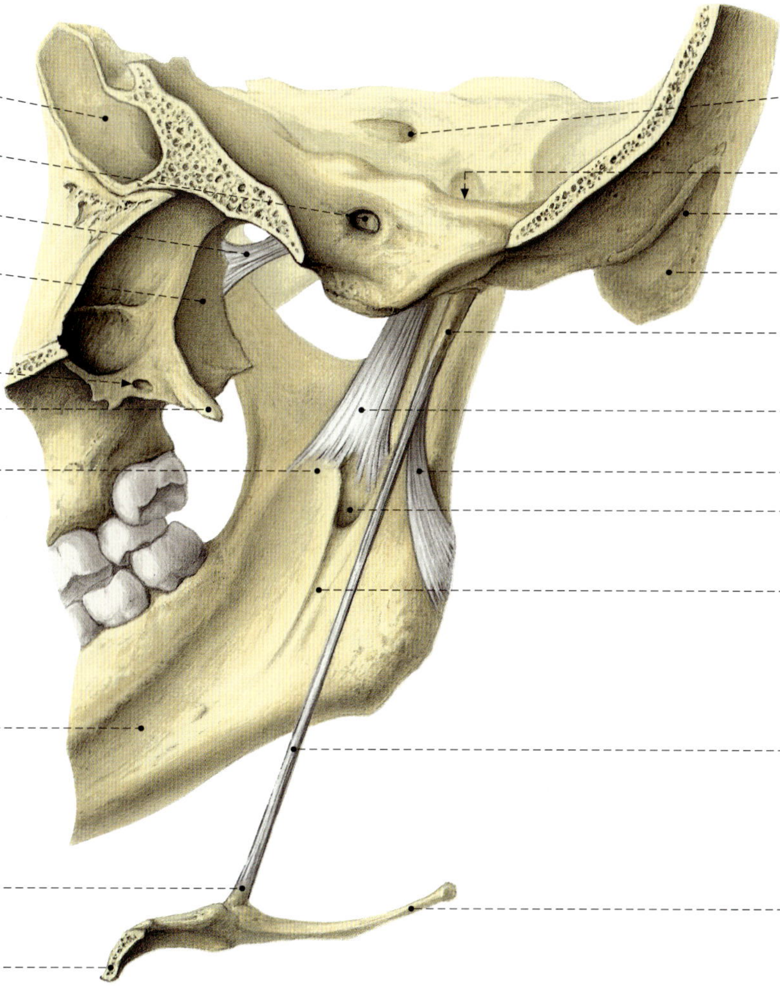

27 Os hyoideum et Articulatio temporomandibularis

27 Hyoid bone and ligaments of the temporomandibular joint

a, b Hyoid bone (80%)
a Ventral aspect
b Left lateral aspect
c Ligaments of the right temporomandibular joint (100%), medial aspect

27 Zungenbein und Bänder des Kiefergelenks

a, b Zungenbein (80%)
a Ventralansicht
b Linke Seitenansicht
c Bänder des rechten Kiefergelenks (100%), Medialansicht

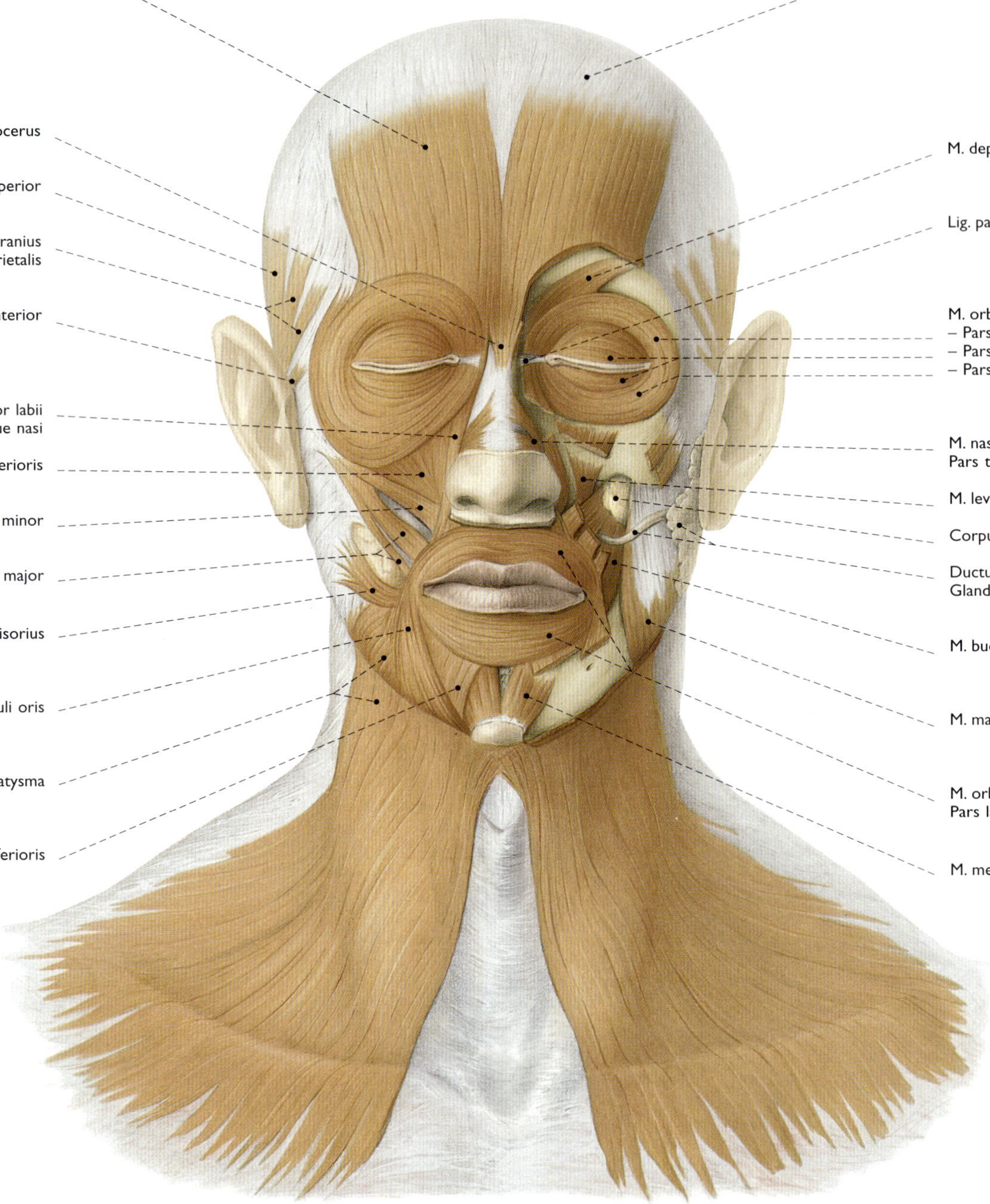

28 Musculi faciales

28 Muscles of the scalp and face (50%)

On the right side of the face, the superficial layer of the facial musculature is demonstrated, the deep layer and the masseter muscle are shown on the left side. Frontal aspect

28 Gesichtsmuskeln (mimische Muskulatur) (50%)

Auf der rechten Seite sind die oberflächlichen, auf der linken die tiefen Gesichtsmuskeln und der Musculus masseter dargestellt. Frontalansicht

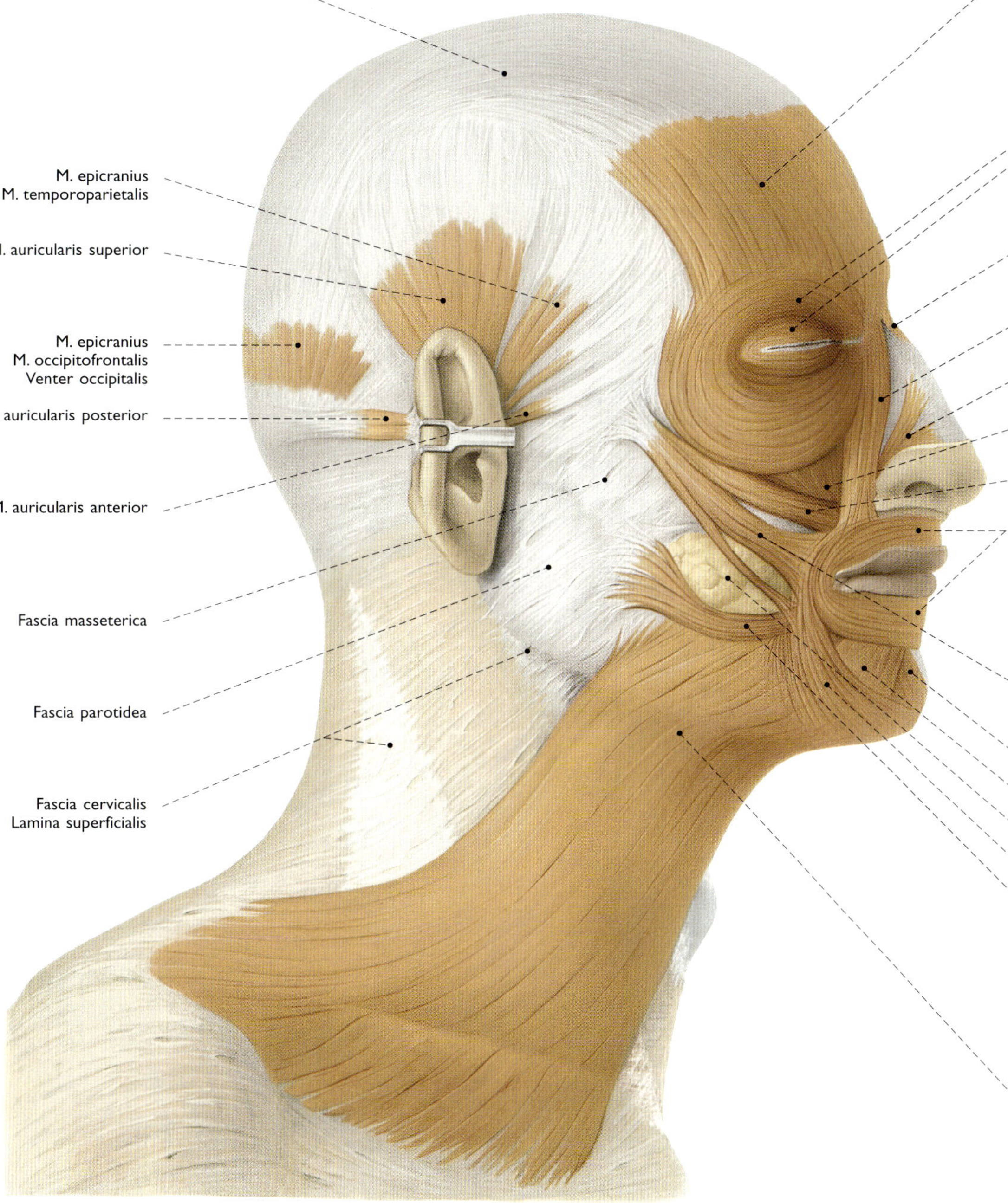

29 Musculi faciales

29 Muscles of the scalp and face (50%)
Superficial layer, right lateral aspect

29 Gesichtsmuskeln (mimische Muskulatur) (50%)
Oberflächliche Schicht, rechte Seitenansicht

a

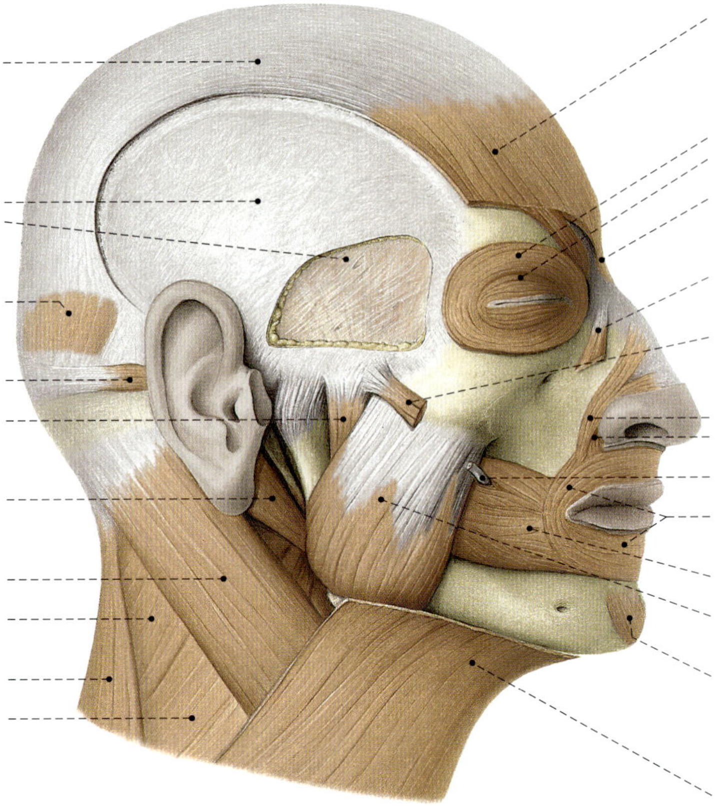

b

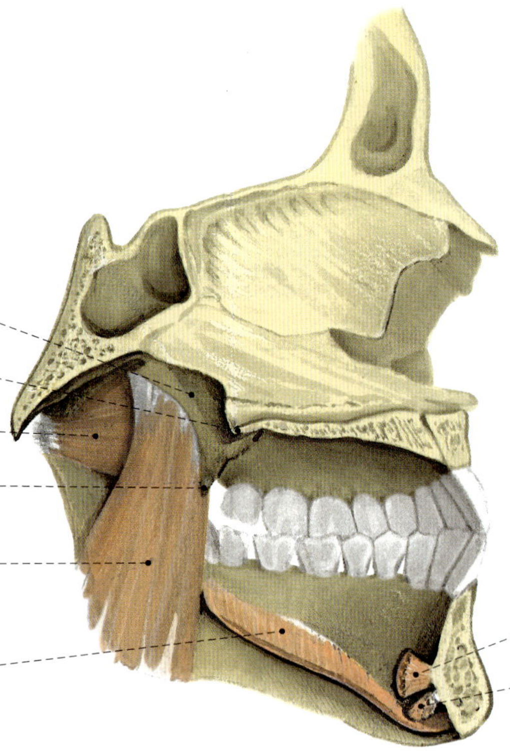

30 Musculi faciales, masseter et pterygoidei

30 Deep muscles of the face, masseter and pterygoid muscles (50%)

a Deep muscles of the face and masseter muscle after removal of the parotid gland and the superficial muscles of the face, right lateral aspect

b Pterygoid muscles, medial aspect

30 Tiefe Gesichtsmuskeln, Musculus masseter und Musculi pterygoidei (50%)

a Tiefe Gesichtsmuskeln und Musculus masseter nach Entfernung der Ohrspeicheldrüse und der oberflächlichen Gesichtsmuskeln, rechte Seitenansicht

b Musculi pterygoidei, Medialansicht

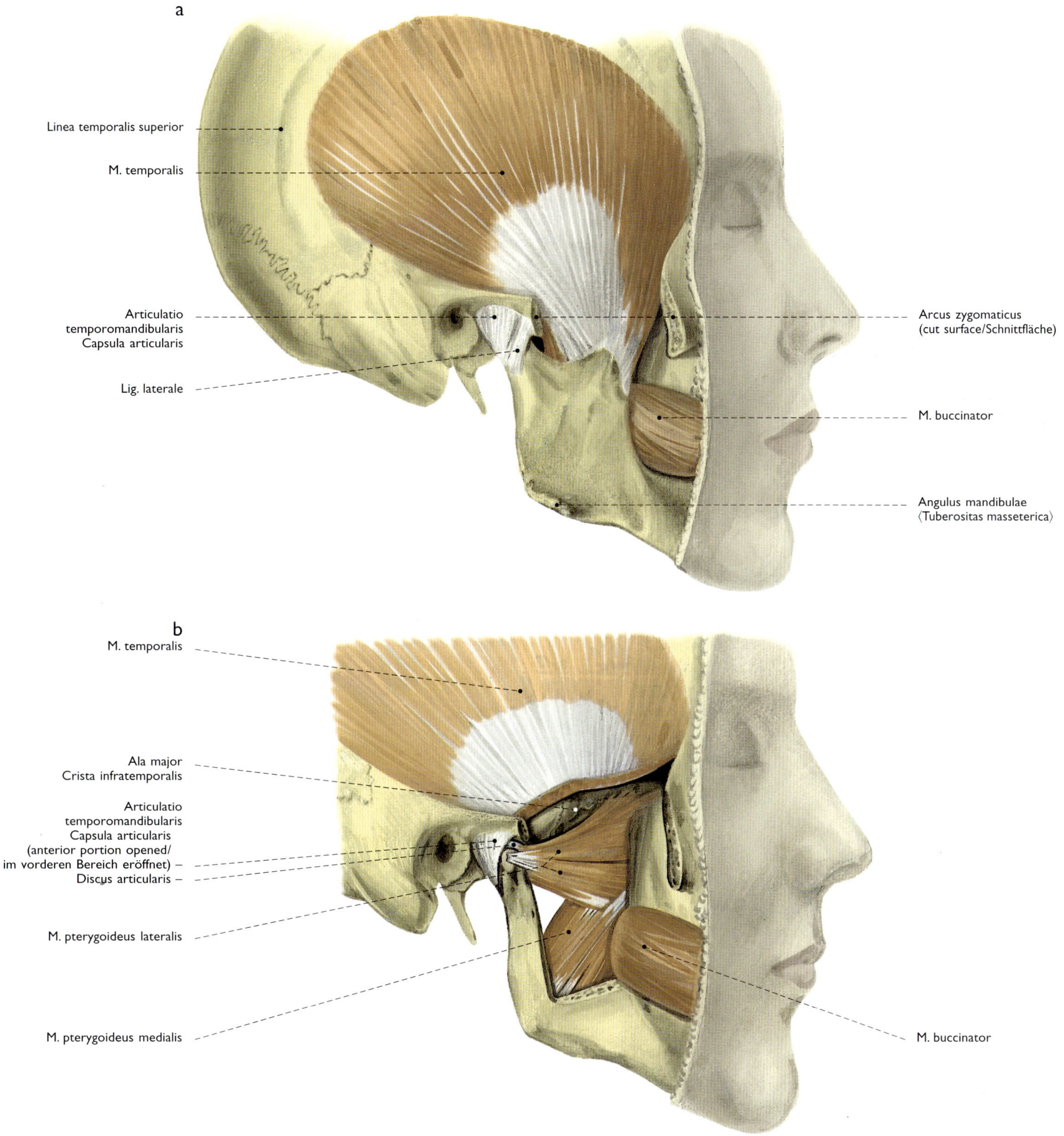

31 Musculi buccinator, temporalis et pterygoidei

31 Buccinator, temporalis and pterygoid muscles (60%)

Right lateral aspect

- a The zygomatic arch and the masseter muscle were removed.
- b The ramus of the mandible and the temporalis muscle were additionally excised partially.

31 Musculi buccinator, temporalis und pterygoidei (60%)

Rechte Seitenansicht

- a Jochbogen und Musculus masseter sind entfernt.
- b Unterkieferast und Musculus temporalis sind zusätzlich teilweise abgetragen.

a

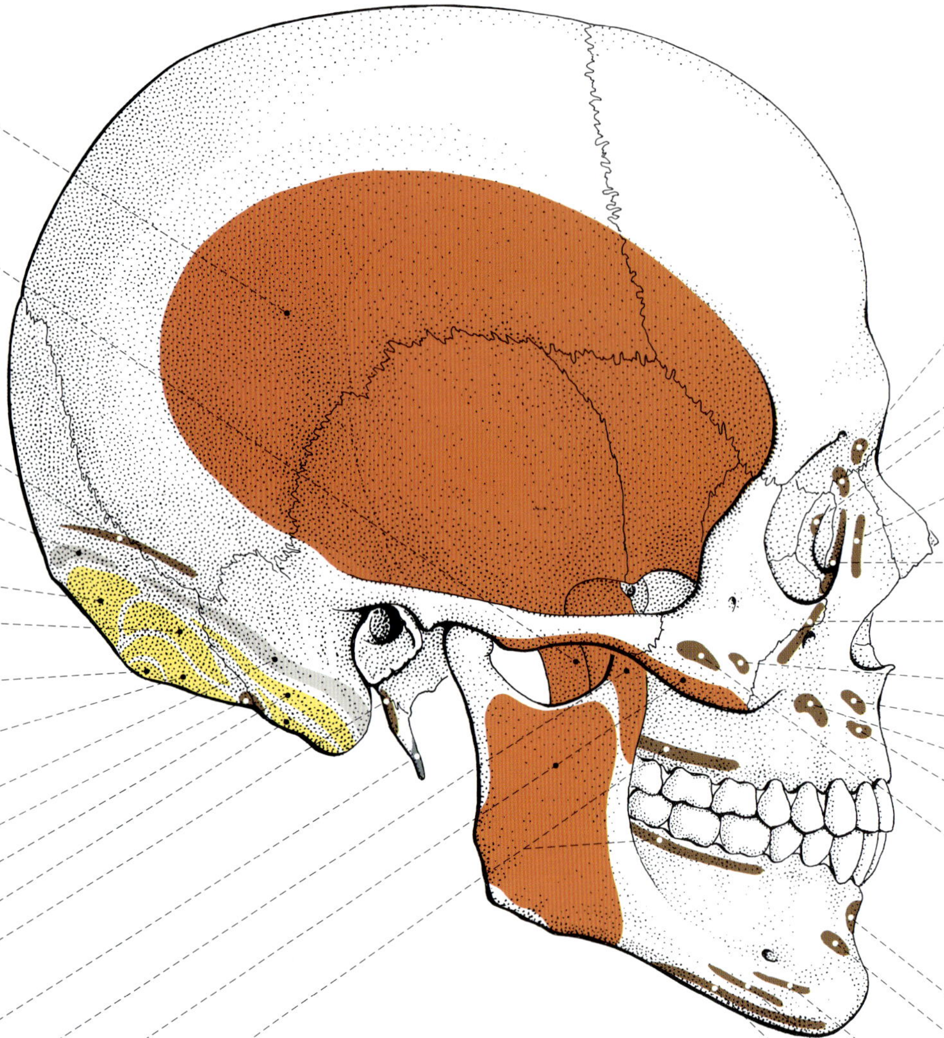

b

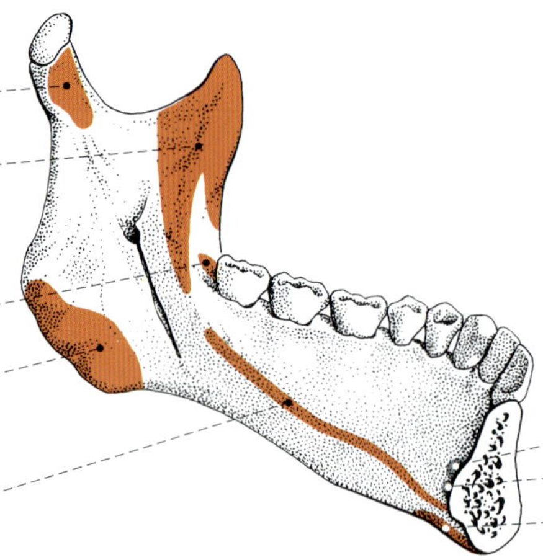

32 Musculi capitis, colli et dorsi

32 Muscle attachments on the skull

Attachments on the
a right surface of the skull
b inner side of the mandible

32 Muskelanheftungen am Schädel

Anheftungen an der
a rechten Seite des Schädels
b Innenseite des Unterkiefers

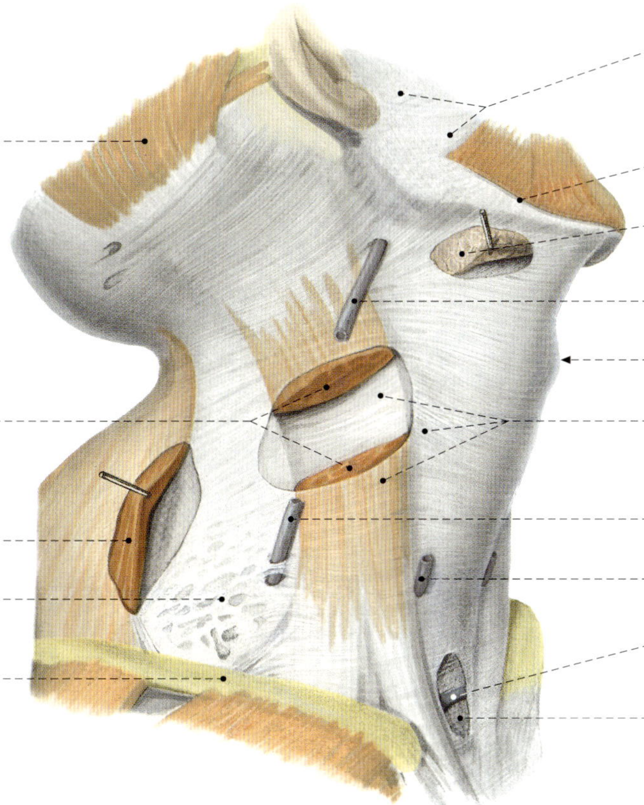

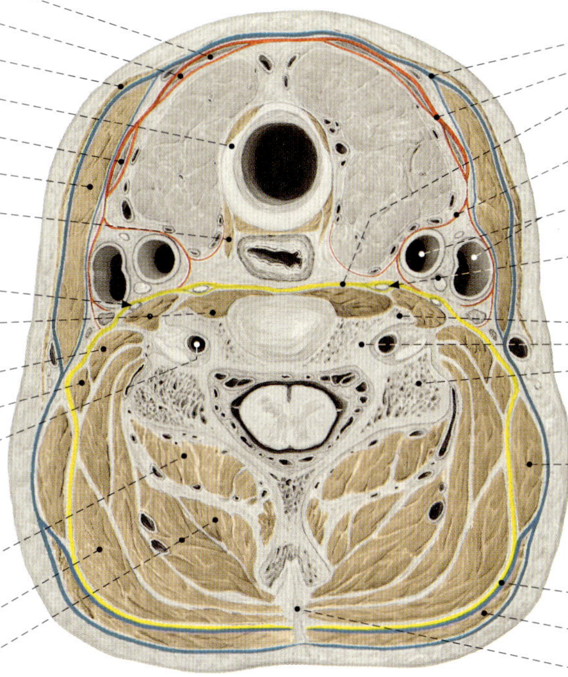

33 Fascia cervicalis

33 Cervical fascia

a Superficial layer of the deep cervical fascia (45%), lateral aspect. Several fascial compartments were opened.

b Transverse section through the neck at the level of the sixth cervical vertebra (70%), schematic representation of the deep cervical fascia. The diverse laminae are indicated by different colors.

33 Halsfaszie

a Oberflächliches Blatt der Halsfaszie (45%), Seitenansicht. Einige Faszienlogen sind eröffnet.

b Transversalschnitt durch den Hals in Höhe des sechsten Halswirbels (70%), schematische Darstellung der Schichten der Halsfaszie.
Die verschiedenen Blätter der Halsfaszie sind in unterschiedlichen Farben dargestellt.

a

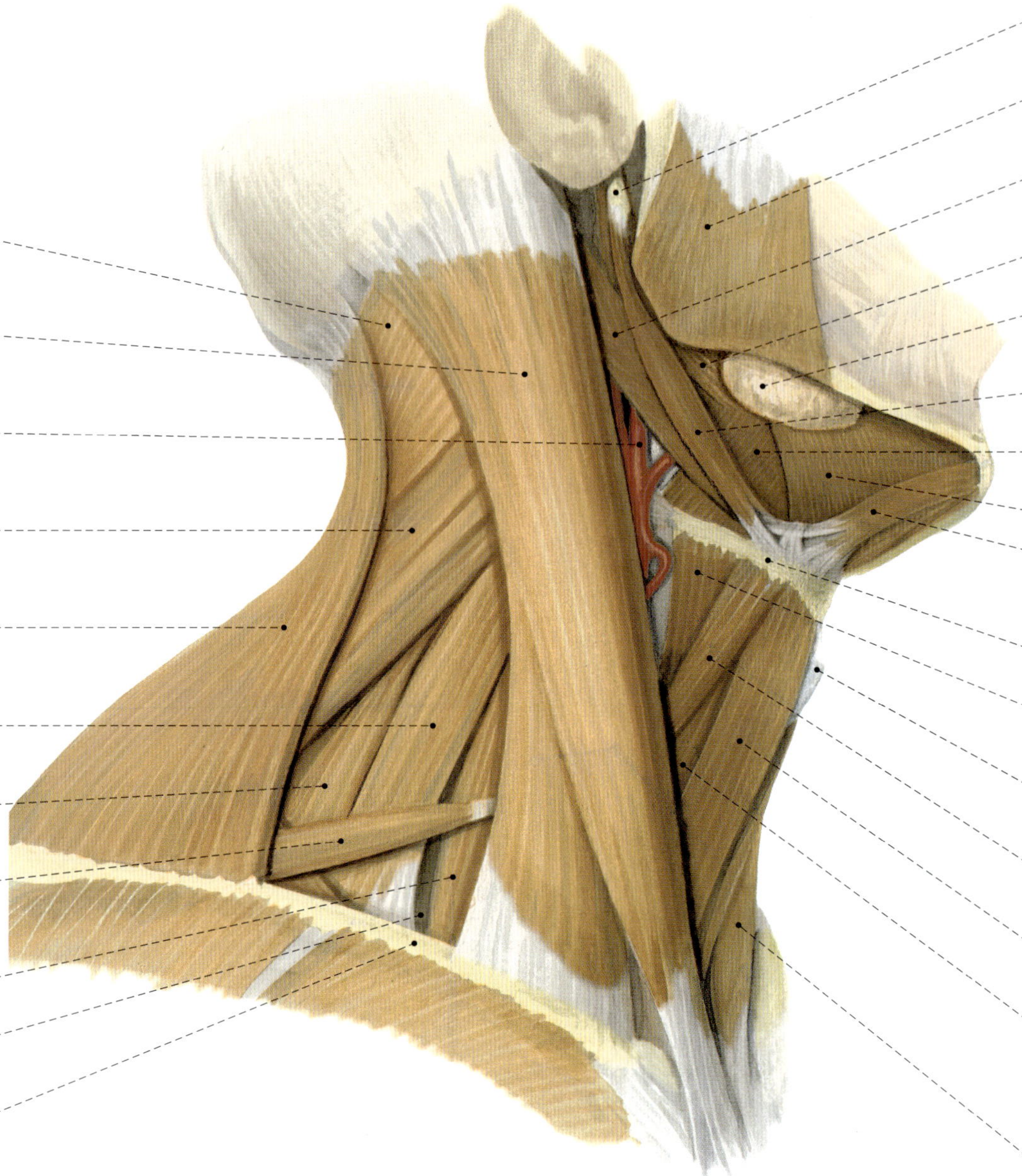

b

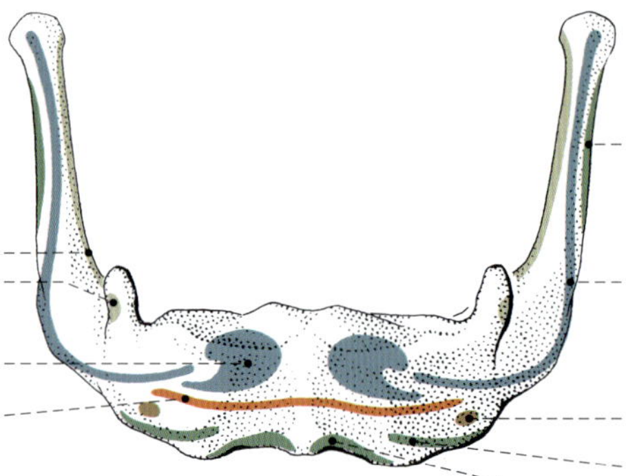

34 Musculi colli

34 Muscles of the neck

a The platysma and the cervical fascia were removed (60%). Right lateral aspect
b Muscle attachments on the outer and superior surface of the hyoid bone

34 Halsmuskeln

a Das Platysma und die Halsfaszie sind entfernt (60%). Rechte Seitenansicht
b Muskelanheftungen an der Aussen- und Oberseite des Zungenbeins

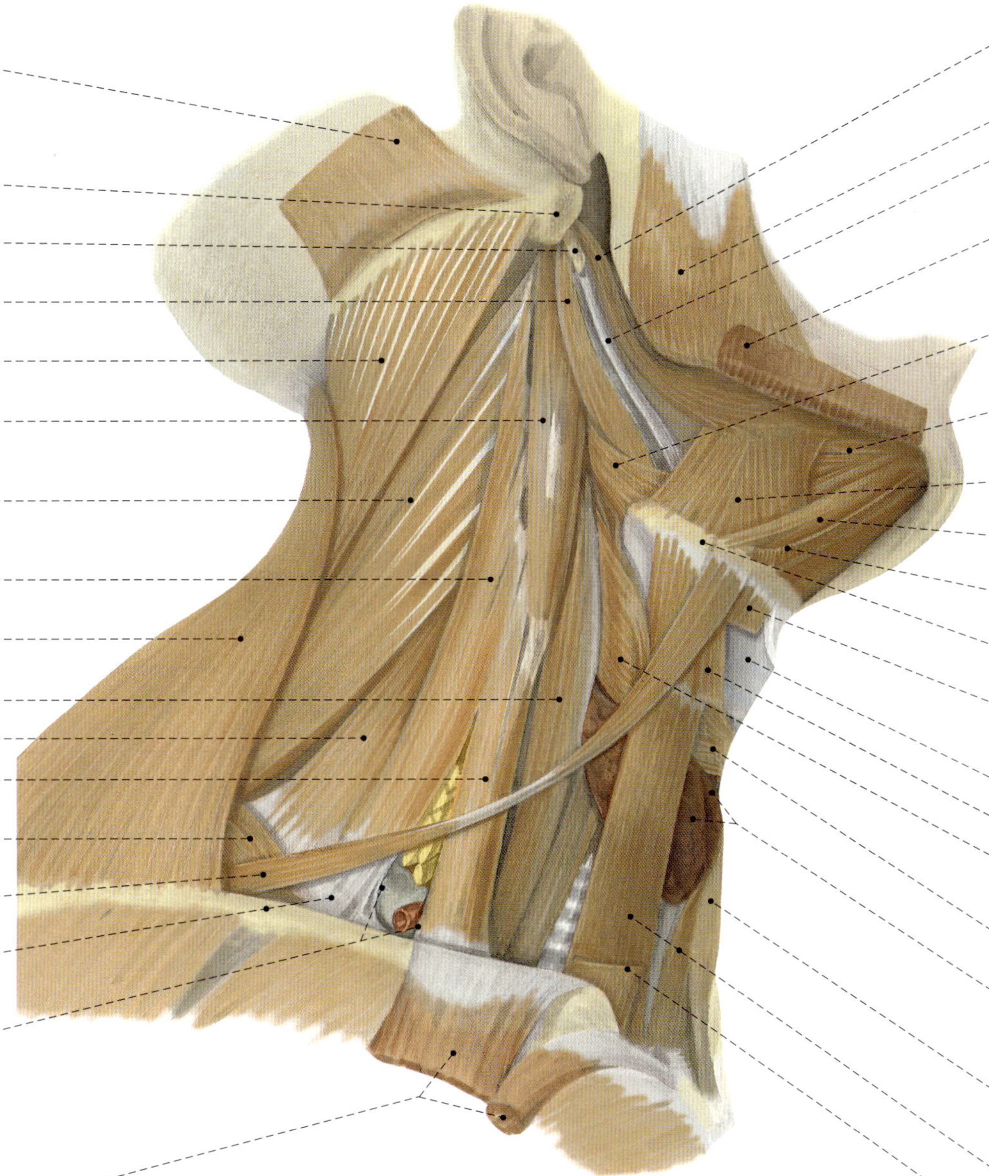

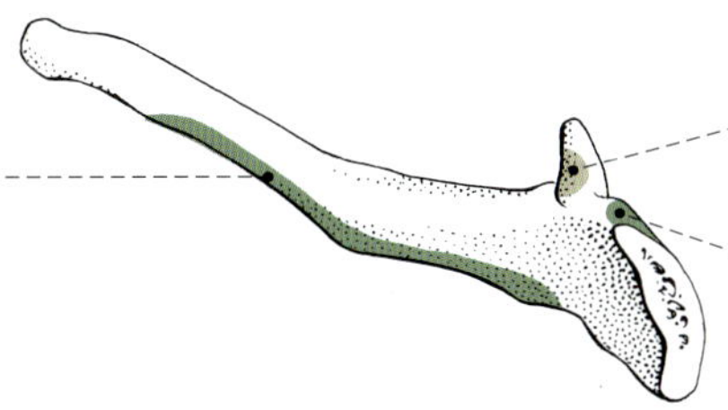

35 Musculi colli

35 Muscles of the neck

a In addition to the platysma and the cervical fascia, the sternocleidomastoid muscle was severed and partially removed (60%). Right lateral aspect
b Muscle attachments on the inner side of the hyoid bone

35 Halsmuskeln

a Zusätzlich zum Platysma und zur Halsfaszie ist der Musculus sternocleidomastoideus teilweise entfernt (Muskelenden umgeklappt) (60%). Rechte Seitenansicht
b Muskelanheftungen an der Innenseite des Zungenbeins

a

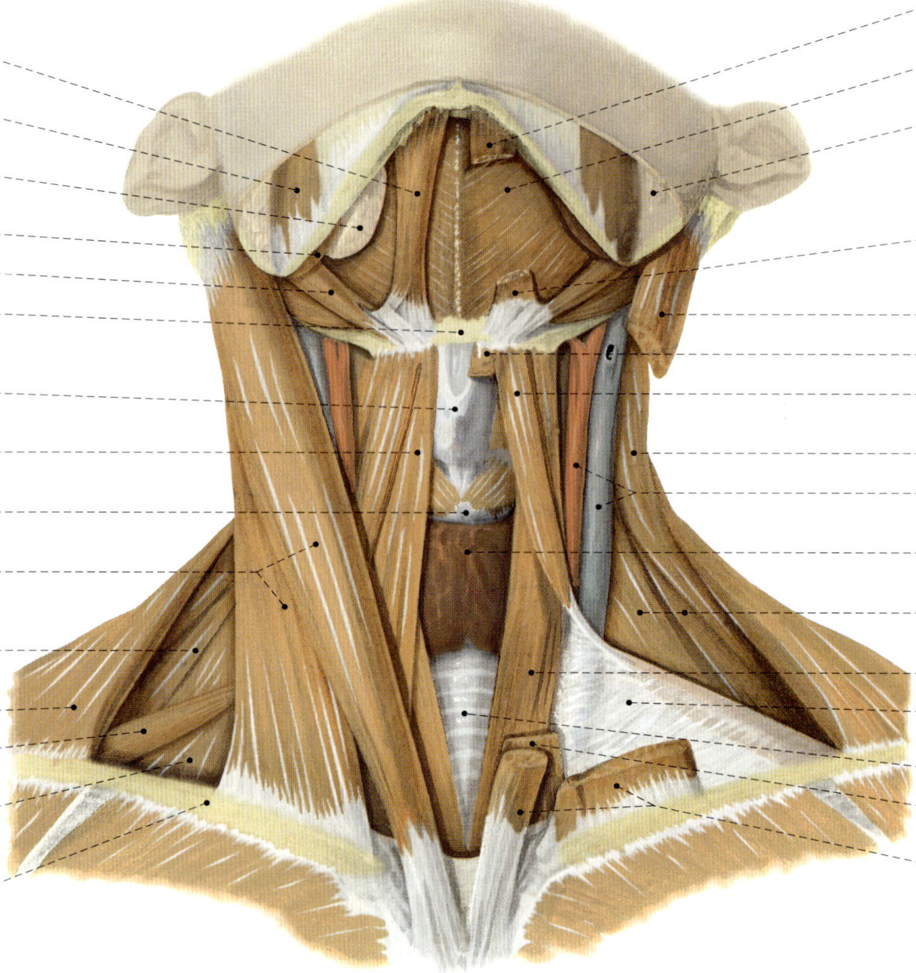

b

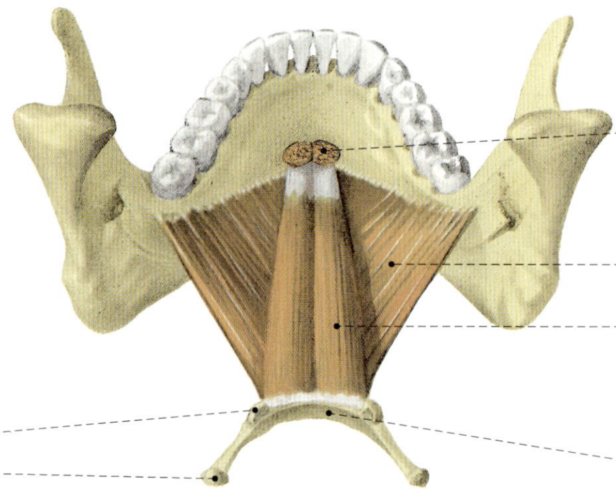

36 Musculi colli et suprahyoidei

36 Neck and suprahyoid muscles (60%)

a On the right side, the platysma was removed. On the left side, parts of the sternocleidomastoid and sternohyoid muscles were excised additionally. Ventral aspect

b Muscles of the floor of the oral cavity, occipital aspect

36 Hals- und obere Zungenbeinmuskeln (60%)

a Rechts ist das Platysma entfernt. Links sind die Musculi sternocleidomastoideus et sternohyoideus zusätzlich teilweise abgetragen. Ventralansicht

b Mundbodenmuskeln, Okzipitalansicht

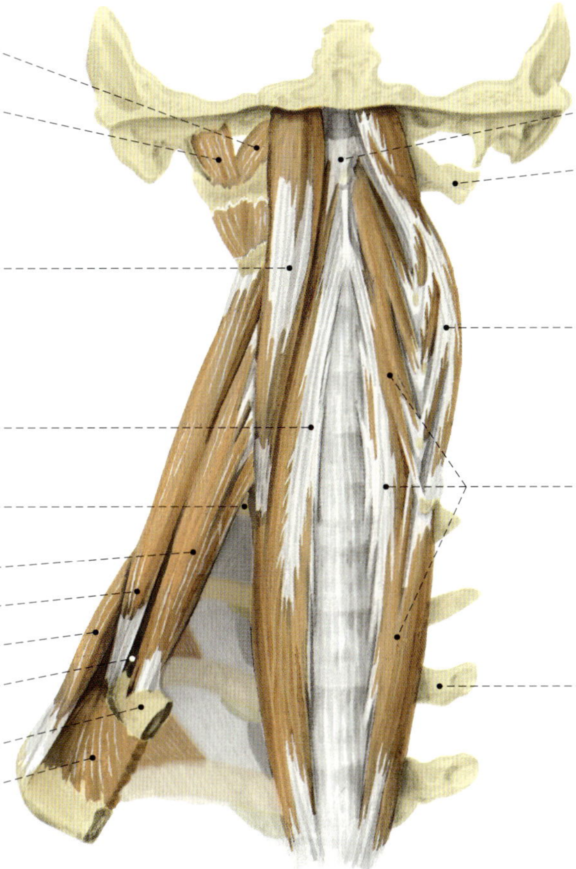

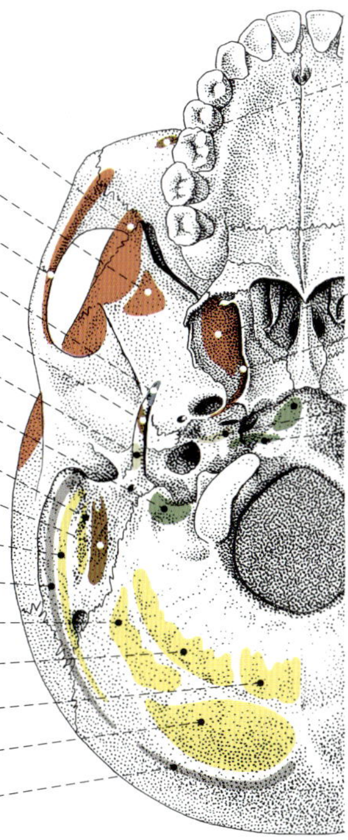

37 Musculi colli, capitis et dorsi

37 Muscles of the neck, the face and the back

a Anterior and lateral vertebral muscles (55%), ventral aspect
b Muscle attachments on the base of the skull

37 Hals-, Kopf- und Rückenmuskeln

a Tiefe Halsmuskeln (55%), Ventralansicht
b Muskelanheftungen an der Schädelbasis

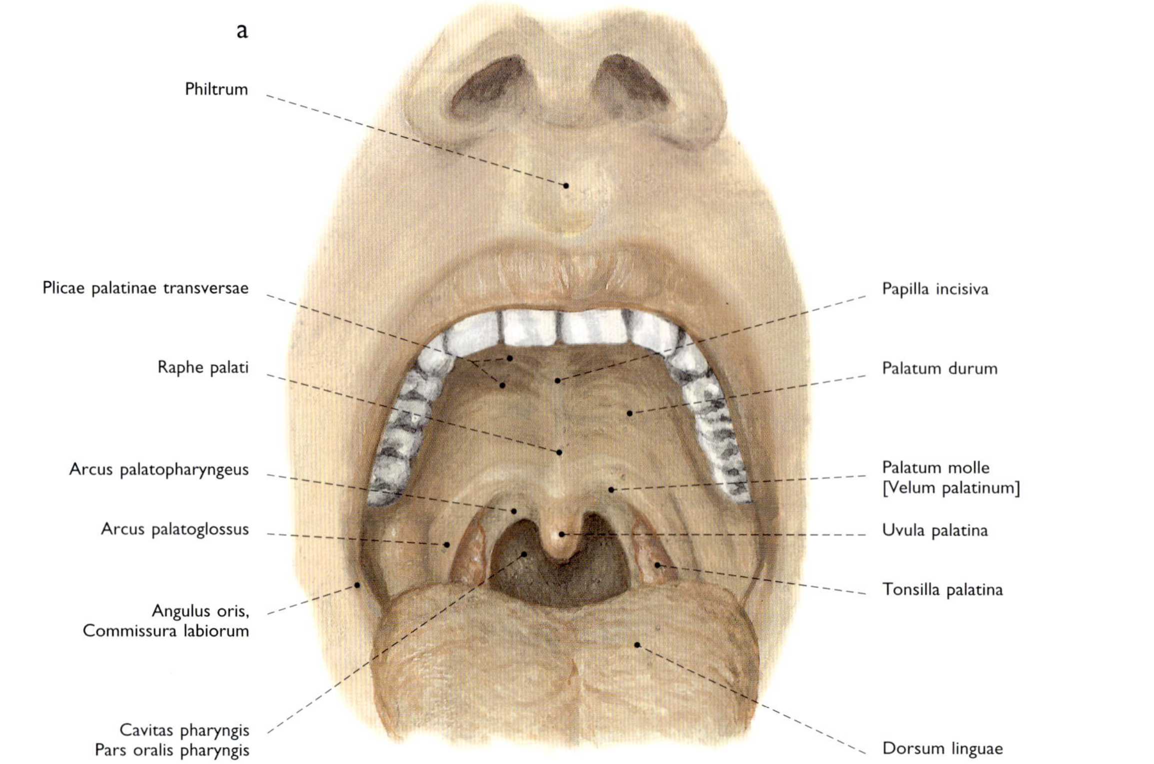

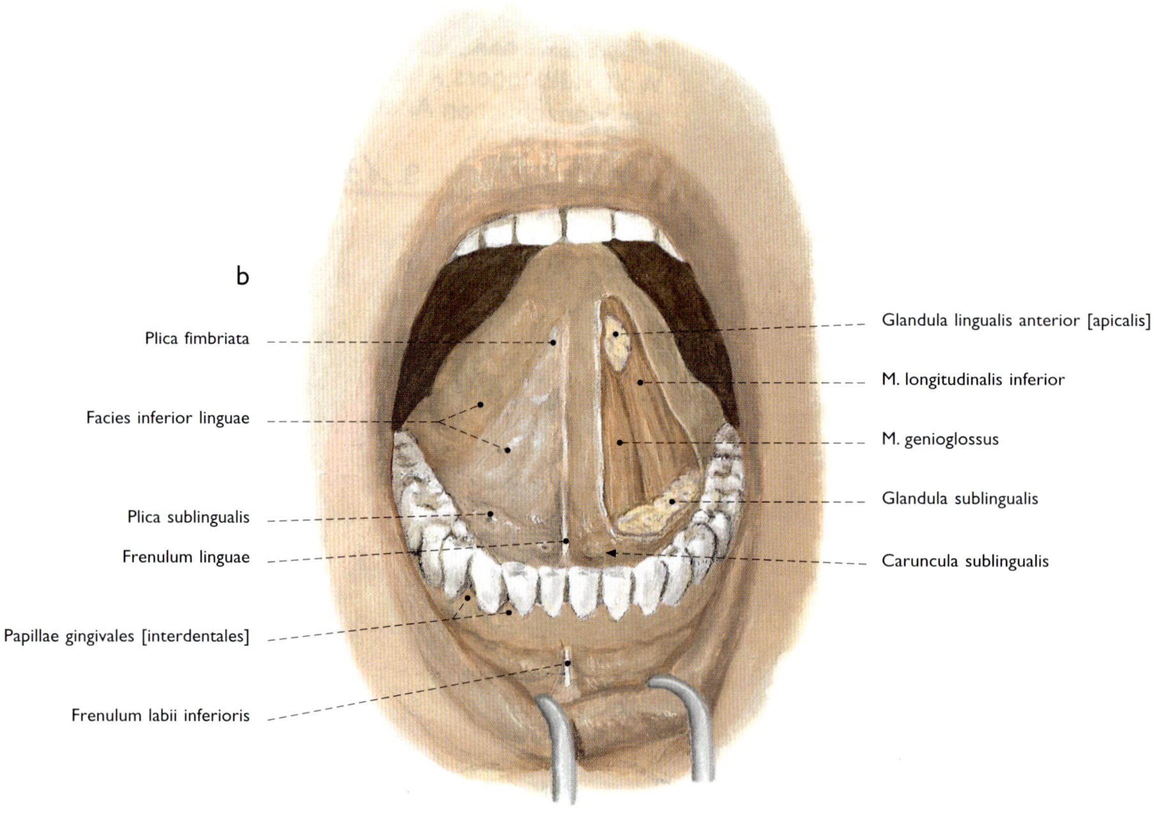

38 Cavitas oris

38 Oral cavity (100%)

Frontal aspect

a The mouth is wide open and the tongue extended.

b The tip of the tongue is turned upwards. Fenestration of the mucosa on the left side exposes the tongue muscles and the glands.

38 Mundhöhle (100%)

Frontalansicht

a Weit geöffnete Mundspalte, Zunge herausgestreckt

b Zungenspitze gaumenwärts gerichtet. Auf der linken Seite im Schleimhautfenster sind die Zungenmuskeln und Drüsen dargestellt.

a

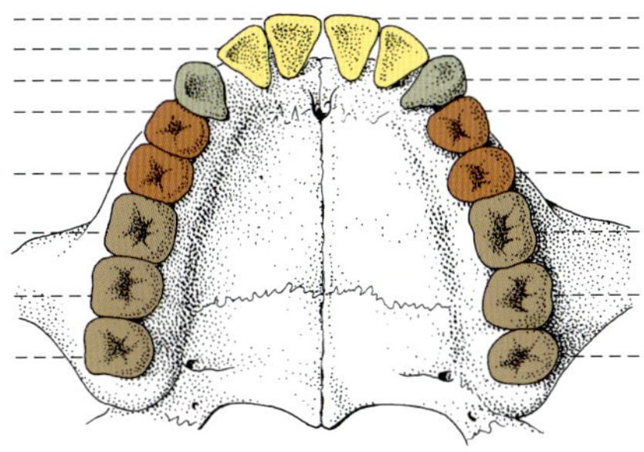

b

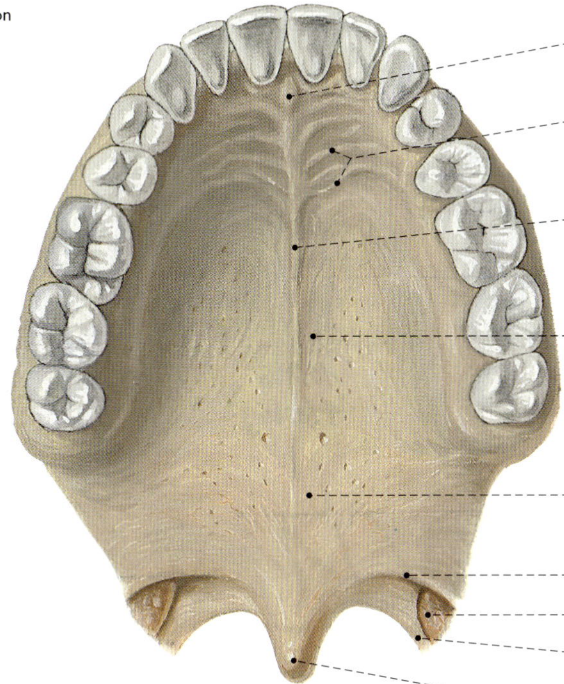

c

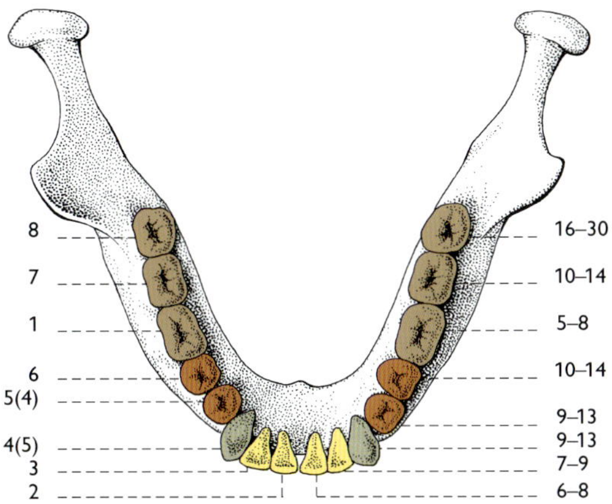

Dentes incisivi
Dentes canini
Dentes praemolares
Dentes molares

d

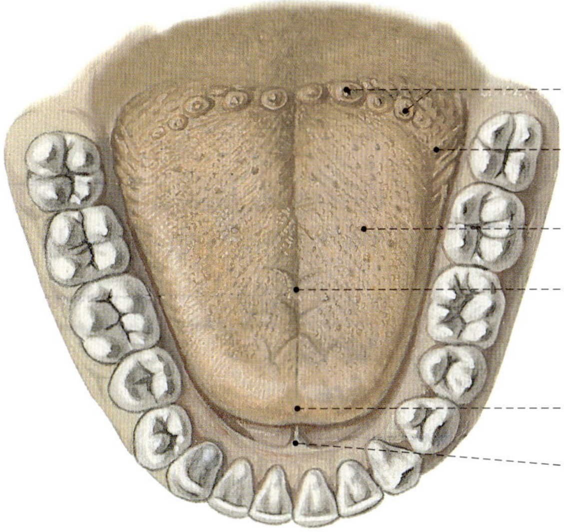

e

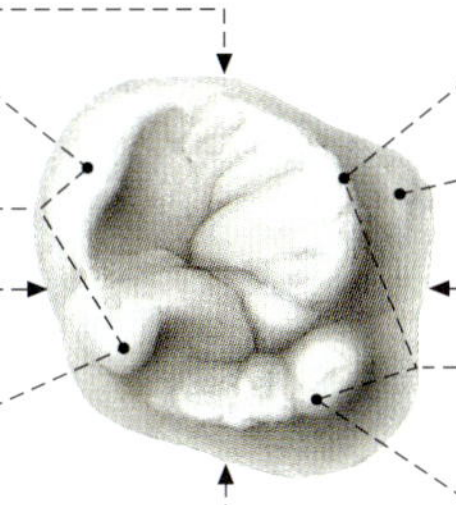

39 Dentes permanentes et Cavitas oris

39 Permanent dentition and oral cavity

a, c Permanent dentition
 a Upper jaw, inferior aspect
 c Lower jaw, superior aspect
b, d Oral cavity (100%)
 b Roof of the oral cavity, inferior aspect
 d Floor of the oral cavity, superior aspect
 e Occlusal surface of the first upper molar (400%)

39 Definitives Gebiss und Mundhöhle

a, c Definitives Gebiss
 a Oberkiefer, Ansicht von unten
 c Unterkiefer, Ansicht von oben
b, d Mundhöhle (100%)
 b Mundhöhlendach, Ansicht von unten
 d Mundboden, Ansicht von oben
 e Kaufläche (Facies occlusalis) des ersten oberen Molaren (400%)

a

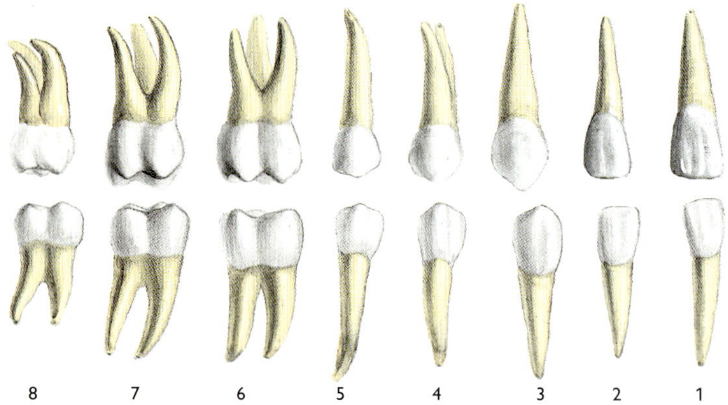

b

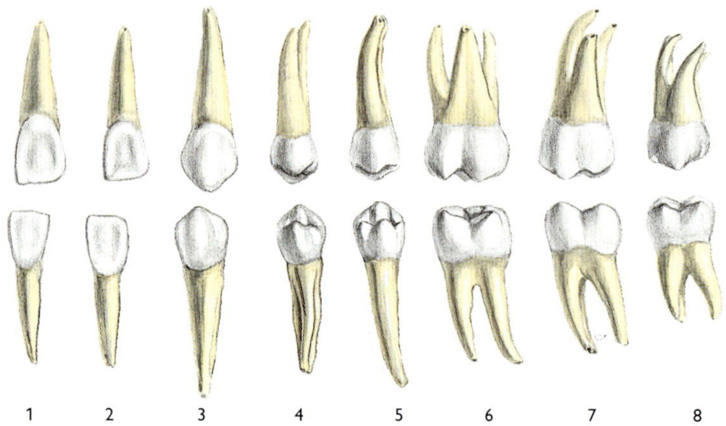

c

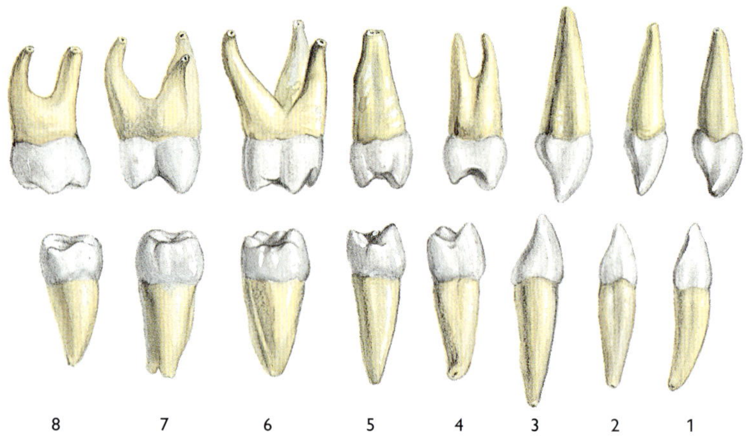

d

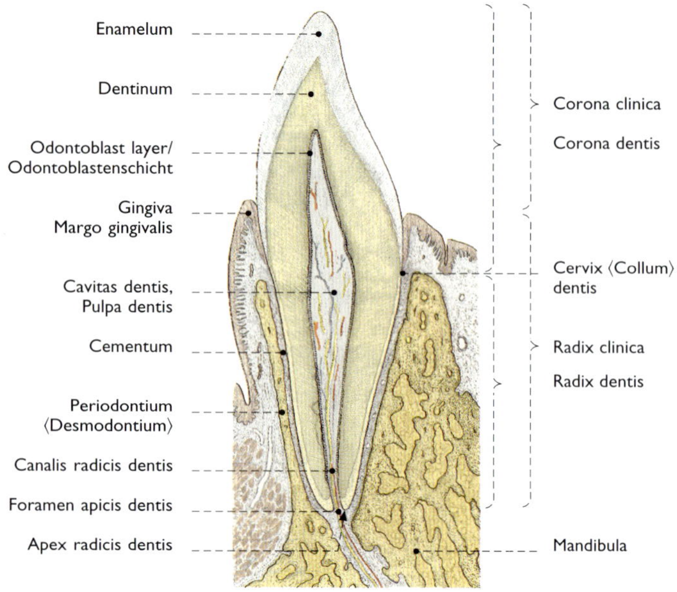

e

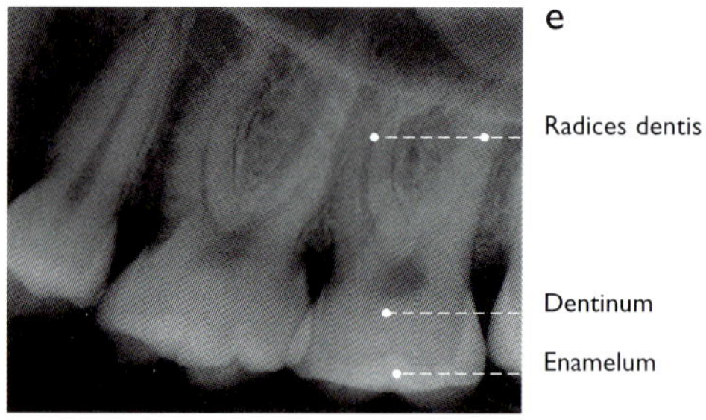

f

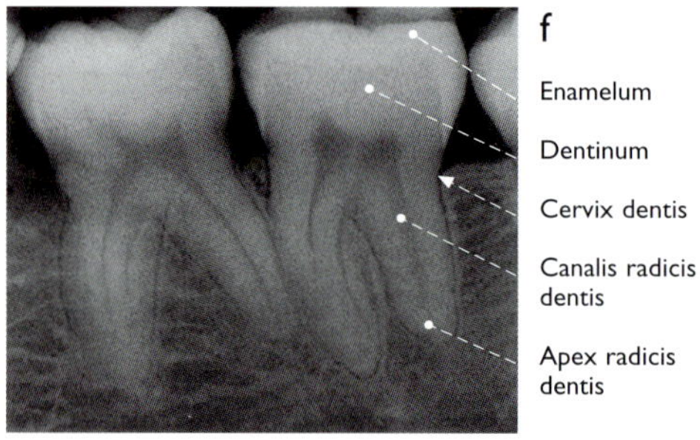

1 Dentes incisivi mediales superior et inferior
2 Dentes incisivi laterales superior et inferior
3 Dentes canini superior et inferior
4 Dentes praemolares I superior et inferior
5 Dentes praemolares II superior et inferior
6 Dentes molares I superior et inferior
7 Dentes molares II superior et inferior
8 Dentes serotini (molares III) superior et inferior

40 Dentes permanentes

40 Permanent teeth

a–c Teeth of the upper and lower jaws (90%)
- a Vestibular surface (labial and buccal surfaces)
- b Lingual surface (palatinal surface)
- c Distal surface

d Lower incisor tooth in situ (400%), lateral cut surface, schematized sagittal section

e, f Lateral radiographs of the first two molars (140%)
- e in the upper jaw
- f in the lower jaw

40 Bleibende Zähne

a–c Zähne von Ober- und Unterkiefer (90%)
- a Facies vestibularis (Facies buccalis, Facies labialis)
- b Facies lingualis (Facies palatinalis)
- c Facies distalis

d Unterer Schneidezahn in situ (400%), Seitenansicht, schematisierter Sagittalschnitt

e, f Seitliche Röntgenbilder der beiden ersten Backenzähne (140%)
- e im Oberkiefer
- f im Unterkiefer

a

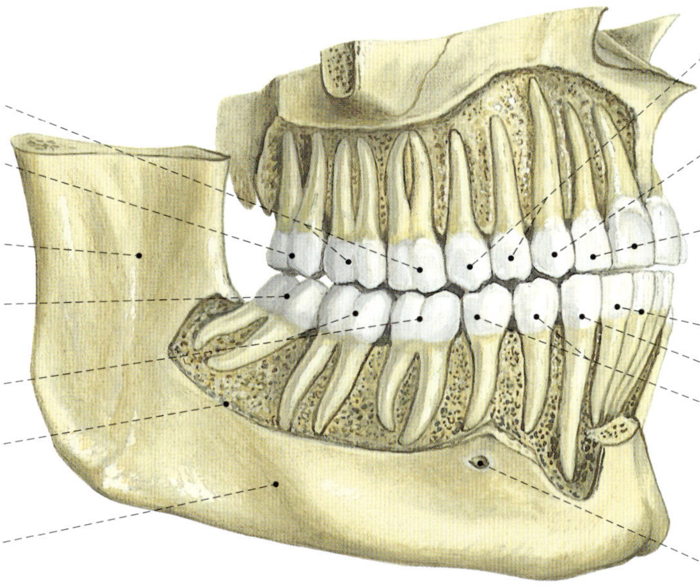

b

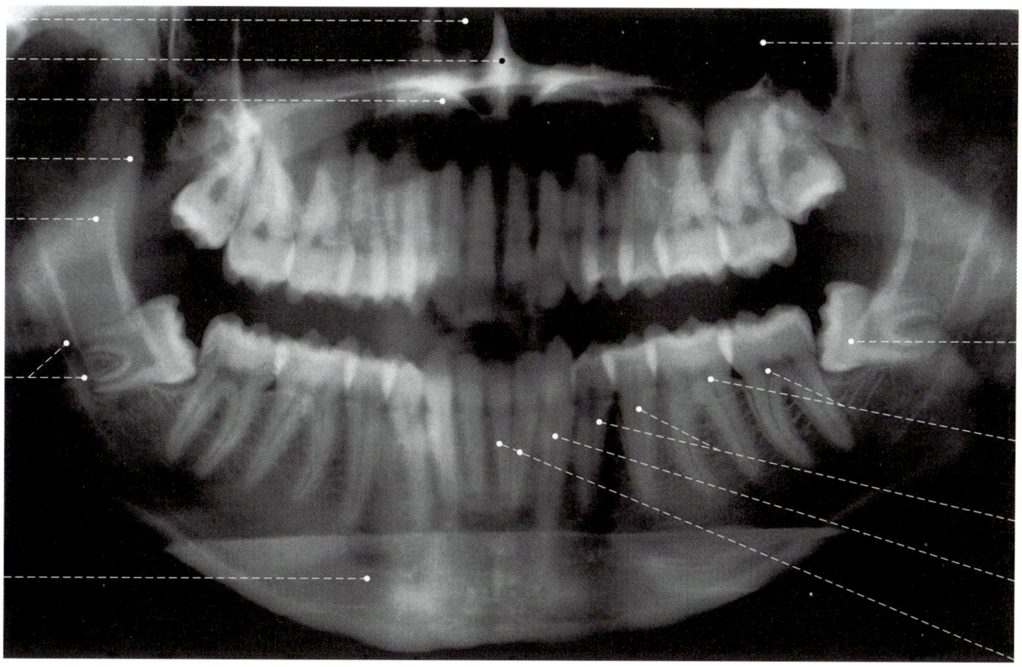

41 Dentes permanentes

41 Permanent teeth (90%)

a Teeth of the upper and lower jaws, right lateral aspect. The vestibular roots were exposed by removal of the external alveolar walls.

b Panoral radiograph of the whole dentition of the upper and lower jaws of an adult

41 Bleibende Zähne (90%)

a Zähne des Ober- und Unterkiefers, rechte Seitenansicht. Die vestibulären Zahnwurzeln sind durch Abmeisseln der äusseren Alveolenwand freigelegt.

b Panorama-Röntgenaufnahme aller Ober- und Unterkieferzähne eines Erwachsenen

a

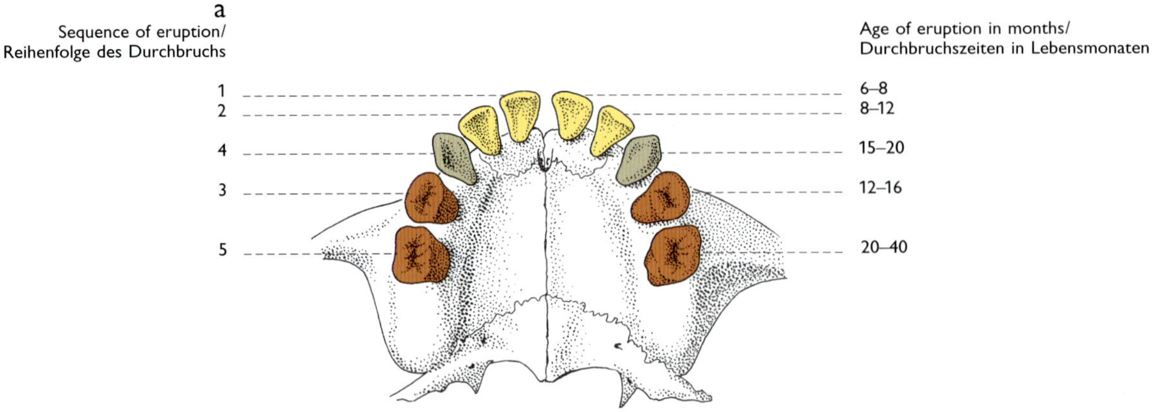

b

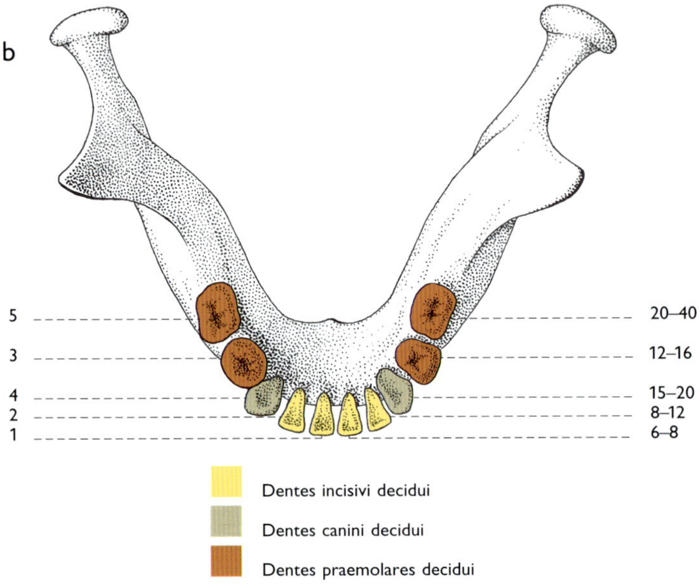

c

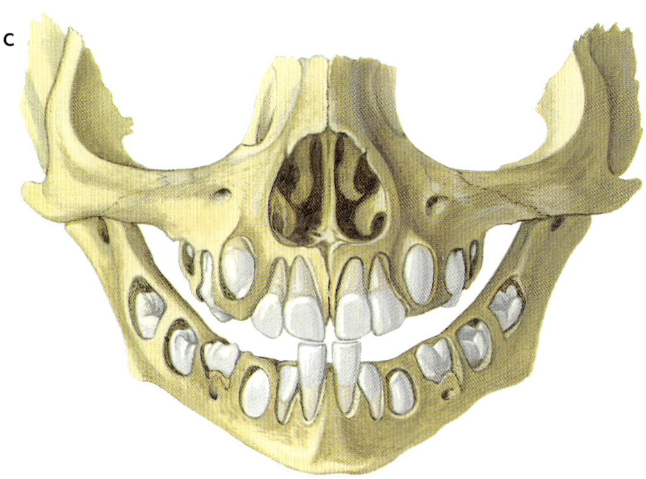

42 Odontogenesis

42 Deciduous dentition

a Upper jaw, inferior aspect
b Lower jaw, superior aspect
c Partially erupted deciduous dentition of a 1-year-old child (100%), frontal aspect

42 Milchgebiss

a Oberkiefer, Ansicht von unten
b Unterkiefer, Ansicht von oben
c Teilweise durchgebrochenes Milchgebiss eines 1jährigen Kindes (100%), Frontalansicht

a

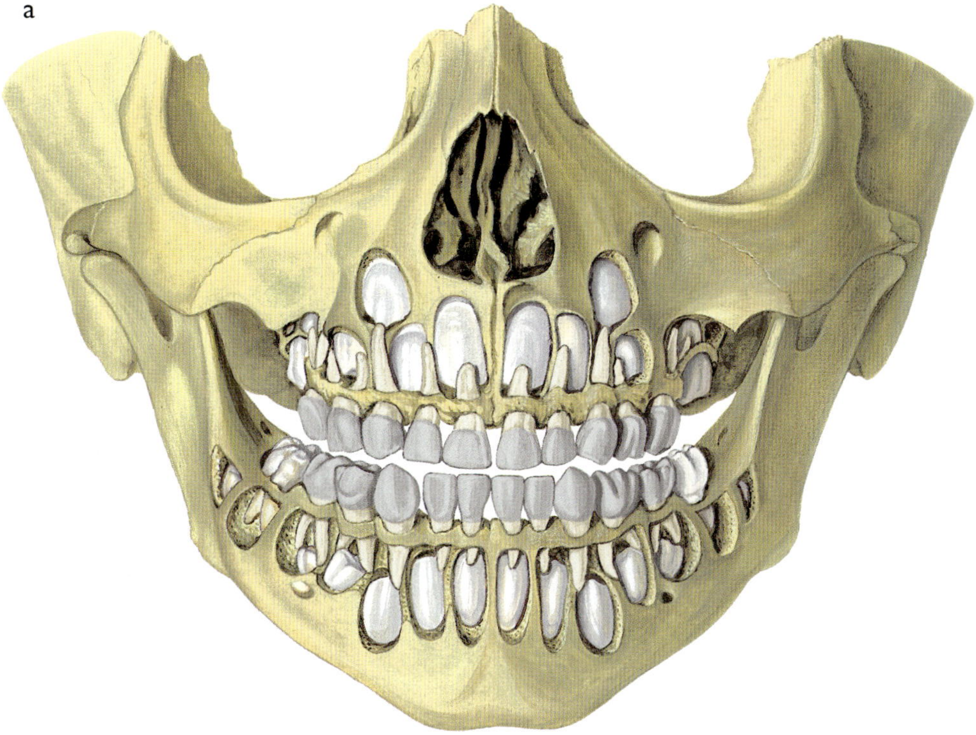

b

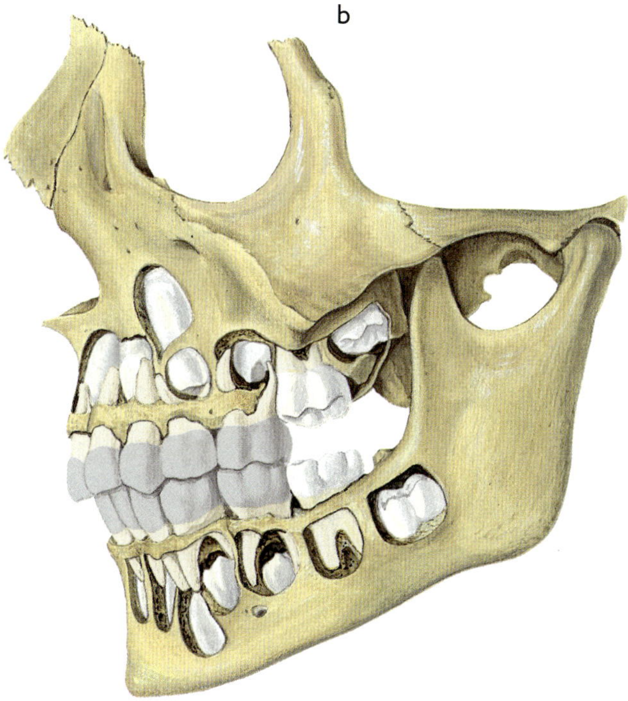

c

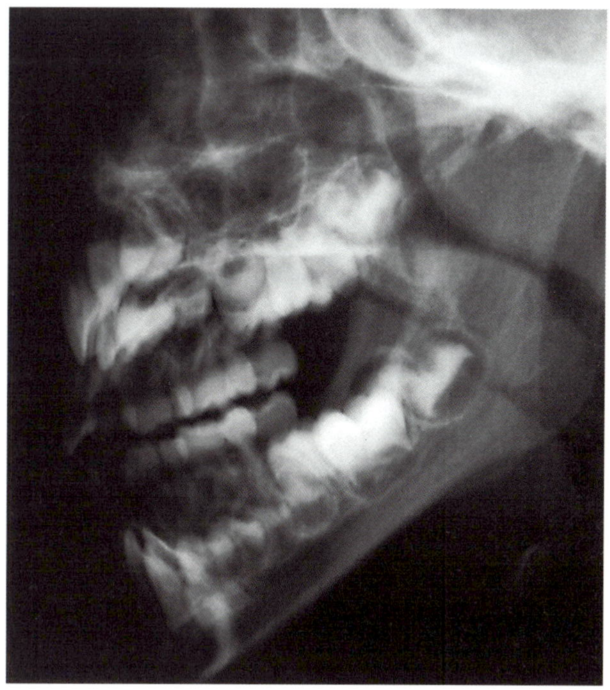

43 Odontogenesis

43 Deciduous dentition and precursors of the permanent teeth of a 5½-year-old boy (100%)

The first molar (the 6th-year molar) has erupted in the lower jaw, and is erupting in the upper jaw.

a Frontal aspect
b Left lateral aspect
c Lateral radiograph

43 Milchgebiss und Zahnanlagen des definitiven Gebisses eines 5½jährigen Knaben (100%)

Der erste Molar (6-Jahr-Molar) ist im Unterkiefer bereits durchgebrochen, im Oberkiefer befindet er sich im Durchbruch.

a Frontalansicht
b Linke Seitenansicht
c Seitliche Röntgenaufnahme

a

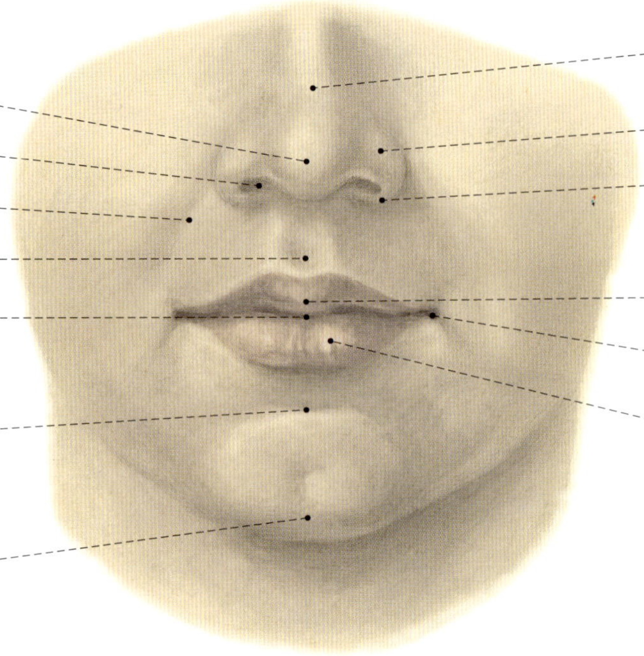

b

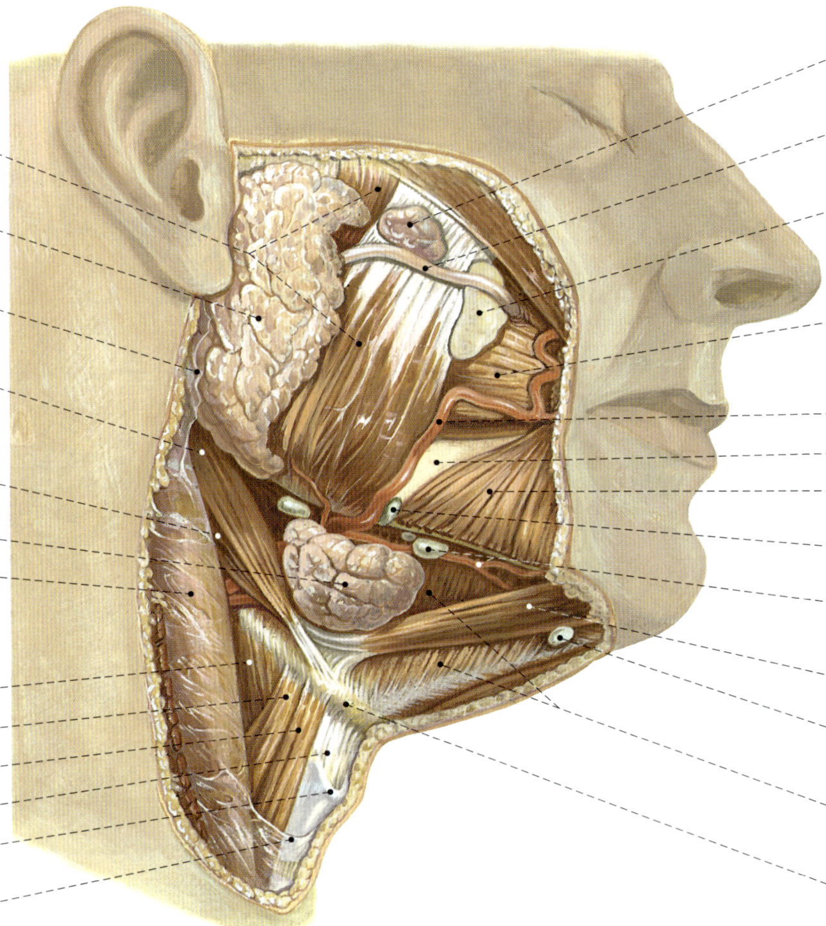

44 Regiones nasalis et facialis, glandulae parotidea et submandibularis

44 Lower part of the face, parotid and submandibular glands (70%)

a Lower part of the face, frontal aspect
b Parotid and submandibular glands, right lateral aspect

44 Untere Gesichtspartie, Ohr- und Unterkieferspeicheldrüse (70%)

a Untere Gesichtsregion, Frontalansicht
b Ohr- und Unterkieferspeicheldrüse, rechte Seitenansicht

a

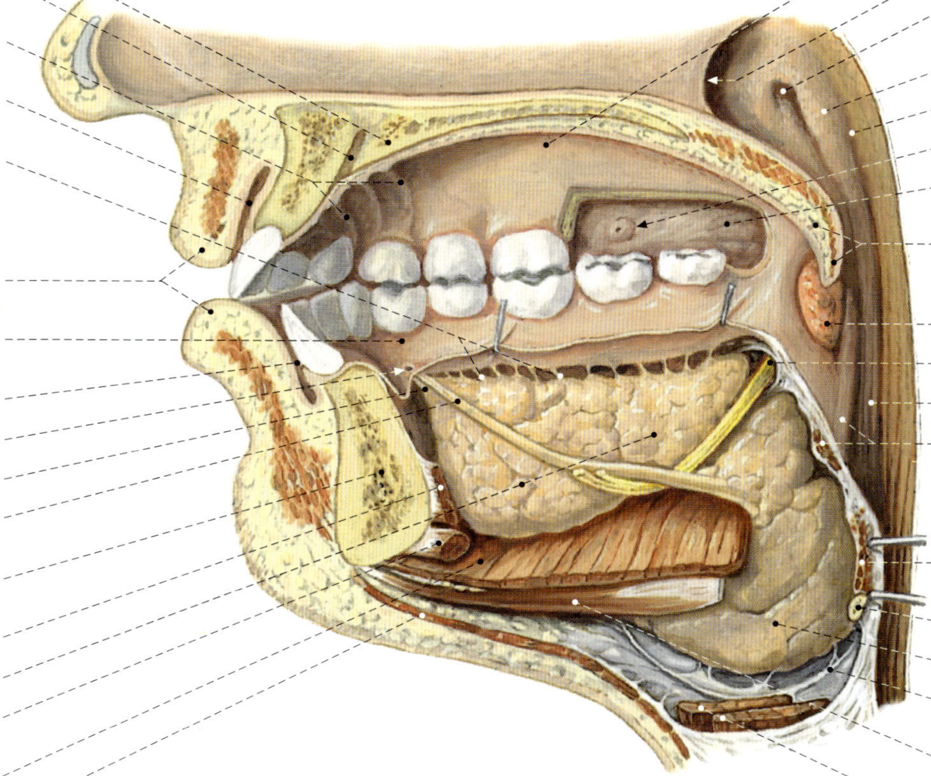

b

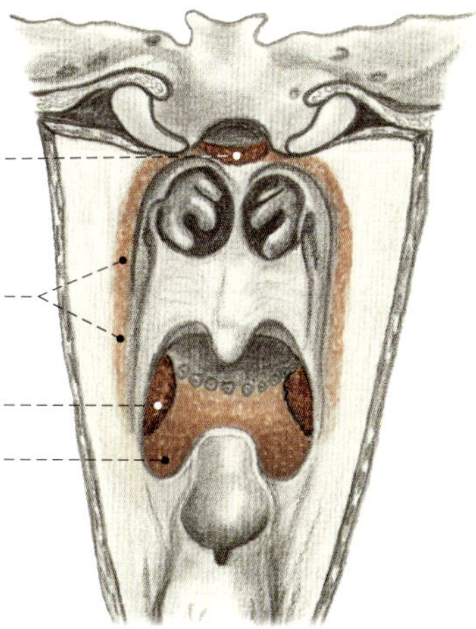

45 Glandulae parotidea, submandibularis et sublingualis

45 Parotid, submandibular and sublingual glands

a Medial aspect of the right half of a head, bisected in the median plane (100%). A window was cut into the posterior part of the lateral wall of the palate in order to demonstrate the parotid papilla. The ventrolateral wall of the pharynx together with the posterior part of the hyoid bone are retracted dorsomedially with hooks, and the submandibular and sublingual glands are shown.

b Waldeyer's ring of lymphoid tissue, consisting of the lingual, palatine and pharyngeal tonsils and minor lymphoid masses, dorsal aspect

45 Ohr-, Unterkiefer- und Unterzungenspeicheldrüse

a Medialansicht der rechten Kopfhälfte nach Medianschnitt (100%). Fensterschnitt im hinteren Teil der seitlichen Gaumenwand zur Darstellung der Papilla parotidea. Die ventrolaterale Pharynxwand ist in Höhe des Zungenbeins durch Haken dorsomedialwärts gezogen, die Glandulae submandibularis et sublingualis sind freipräpariert.

b Waldeyer-Rachenring aus lymphatischem Gewebe, bestehend aus Tonsilla lingualis, Tonsillae palatinae, Tonsilla pharyngea und kleineren Lympheinheiten, Dorsalansicht

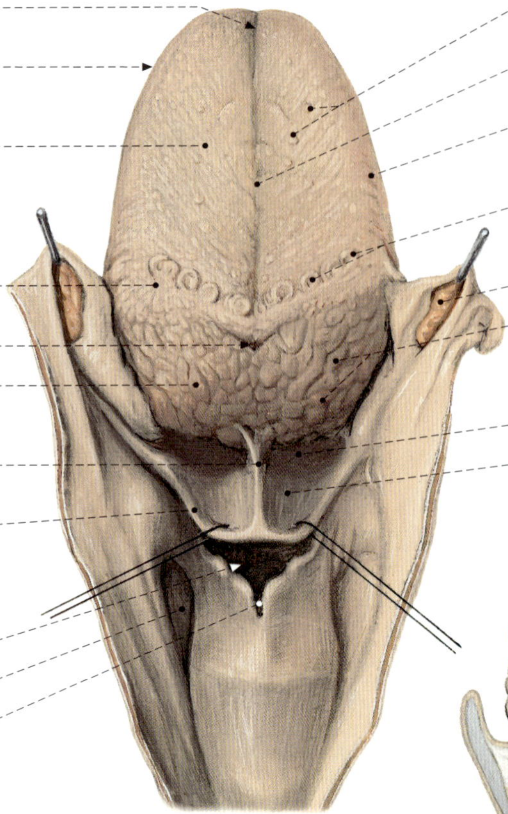

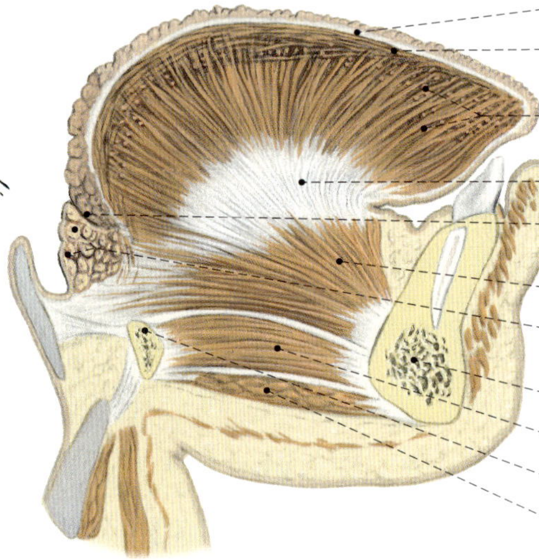

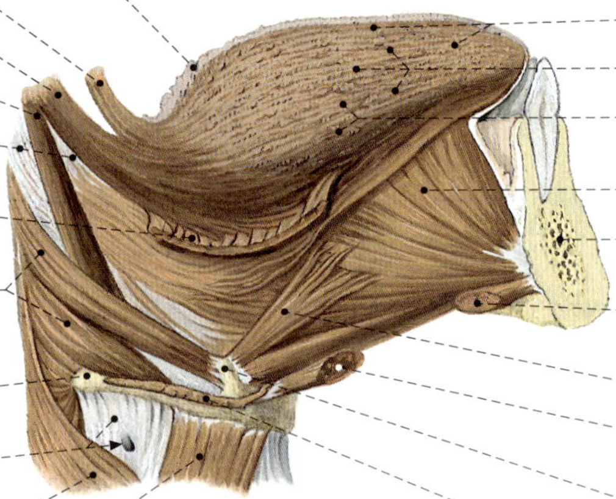

46 Dorsum et Musculi linguae

46 Dorsum and muscles of the tongue (70%)

a Dorsal aspect of the tongue and the ventral wall of the pharynx, opened along the mid-dorsal line
b Medial aspect of the tongue sectioned in the sagittal plane
c Lateral aspect of the tongue after removal of the mucous membrane and the lingual aponeurosis

46 Zungenrücken und Zungenmuskeln (70%)

a Dorsalansicht des Zungenrückens und der ventralen Schlundwand. Der Schlund ist dorsomedian eröffnet.
b Medialansicht der sagittal geschnittenen Zunge
c Seitenansicht der Zunge nach Entfernung der Schleimhaut und der Zungenaponeurose

a

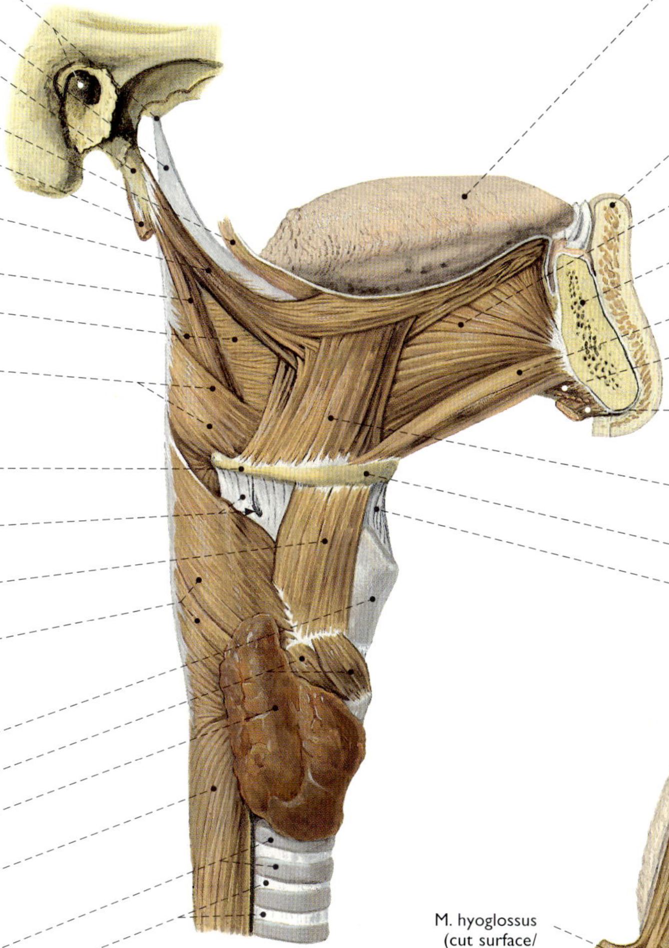

b

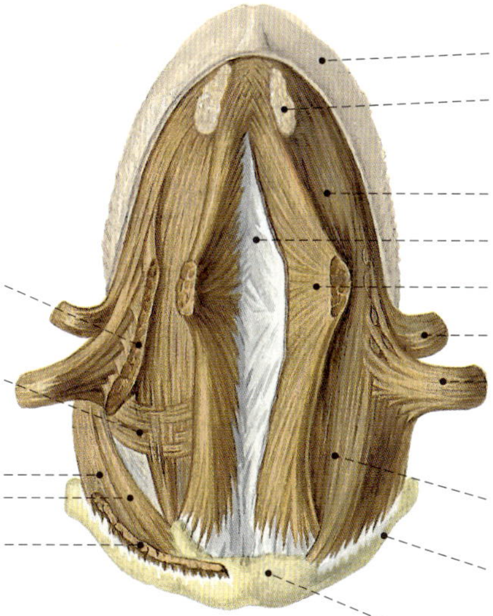

c

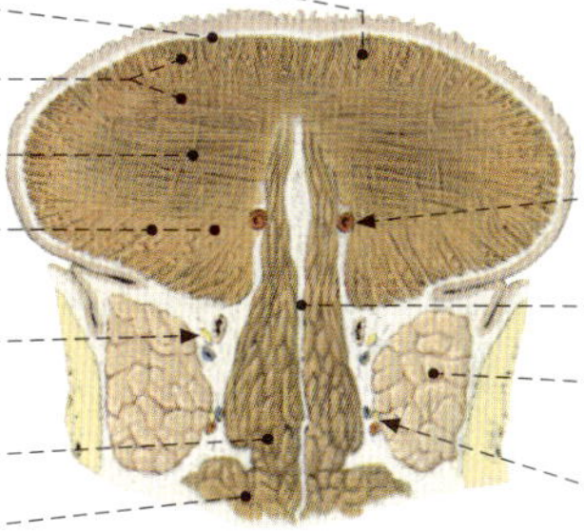

47 Musculi linguae, pharyngis et laryngis

47 Muscles of the tongue, the pharynx and the larynx (70%)

a Tongue, mouth floor and neck, lateral aspect
b The tongue attached to the hyoid bone, basal aspect
c The tongue and the sublingual region, coronal section

47 Zungen-, Schlund- und Kehlkopfmuskeln (70%)

a Zunge, Mundboden und Hals, Seitenansicht
b Am Zungenbein befestigte Zunge, Basalansicht
c Zunge und Unterzungenregion, Frontalschnitt

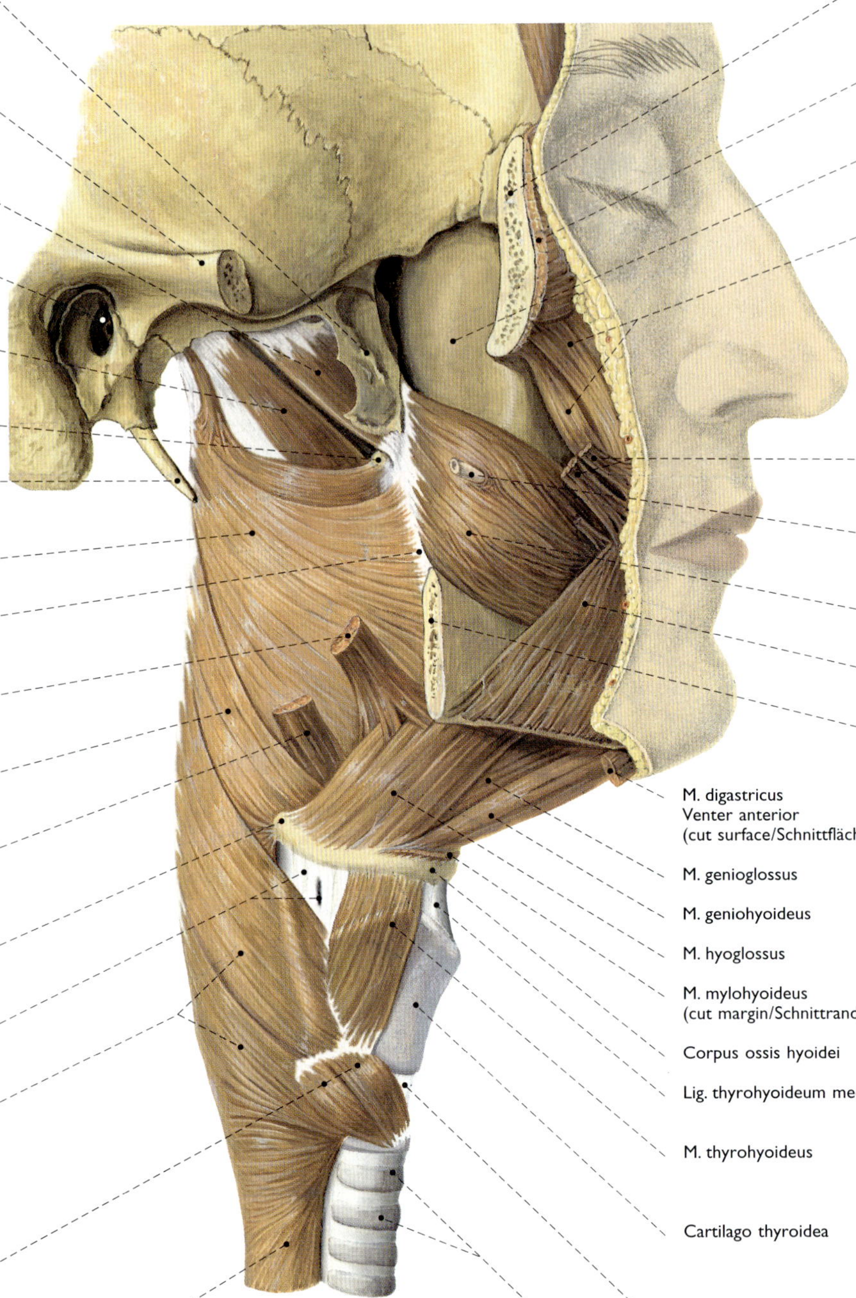

48 Tunica muscularis pharyngis

48 Muscular wall of the pharynx (75%)
Right lateral aspect

48 Muskelwand des Schlundes (75%)
Rechte Seitenansicht

a

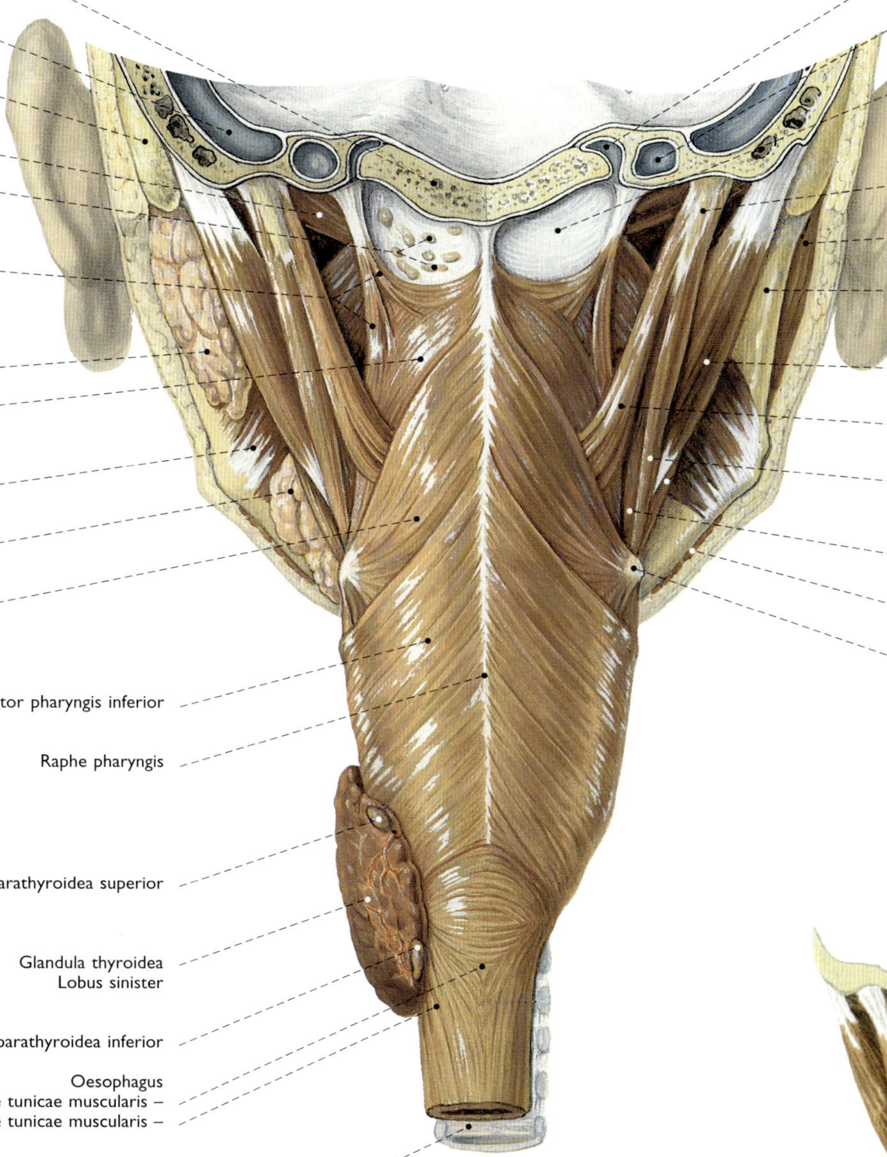

b

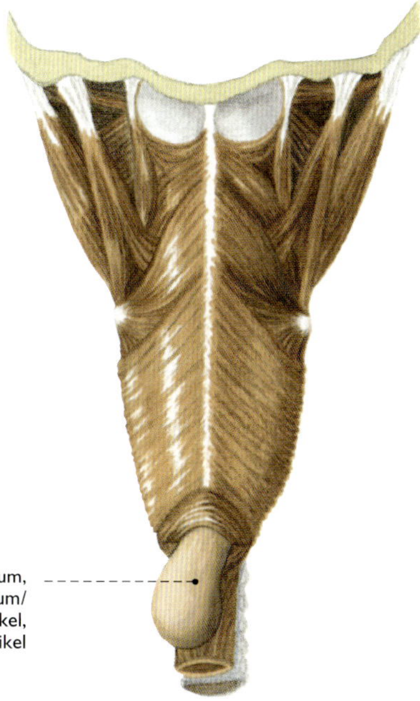

49 Tunica muscularis pharyngis

49 Muscular wall of the pharynx

a Dorsal aspect (70%)
b Representation of a pharynx diverticulum in the Laimer's triangle, dorsal aspect

49 Muskelwand des Schlundes

a Dorsalansicht (70%)
b Darstellung eines Pulsionsdivertikels im Laimer-Dreieck, Dorsalansicht

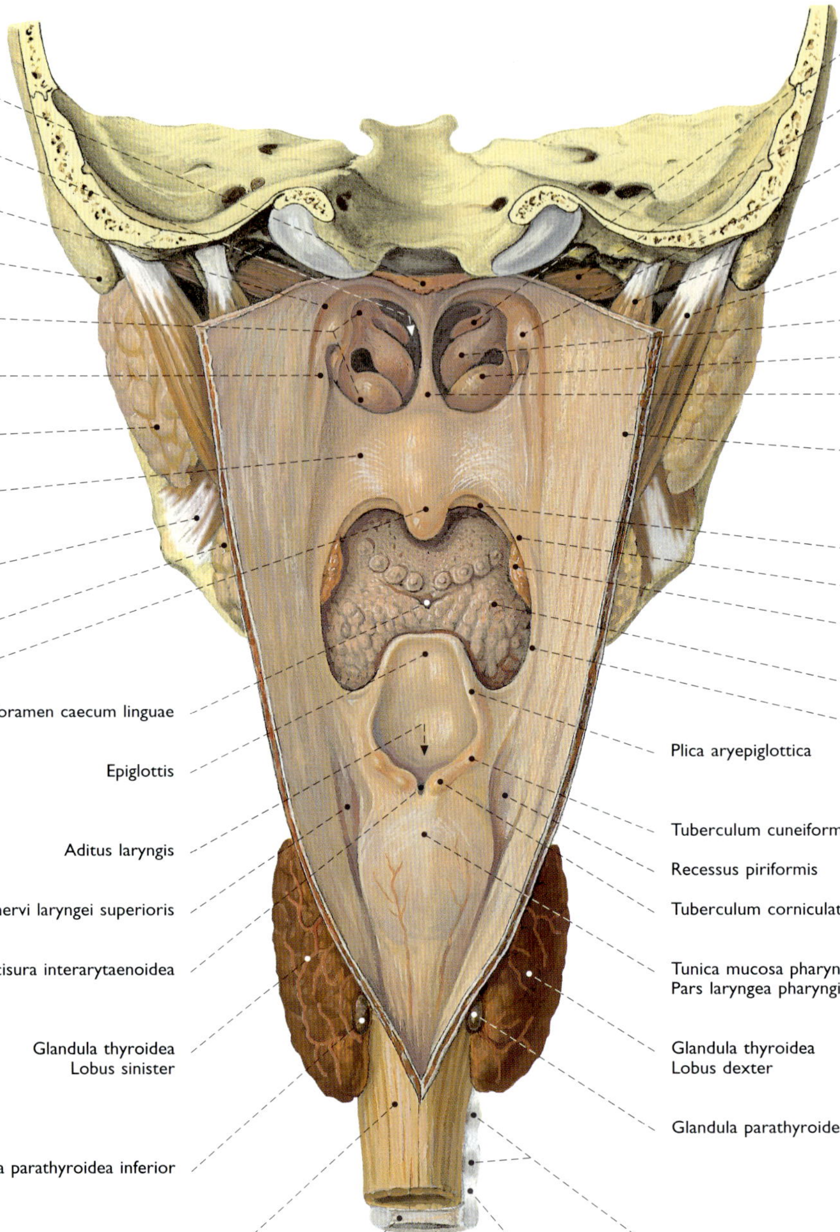

50 Cavitas pharyngis

50 Pharyngeal cavity (70%)

The dorsal wall of the pharynx was cut along the mid-dorsal line and opened. Dorsal aspect

50 Schlundhöhle (70%)

Die dorsale Schlundwand ist median durchtrennt und aufgeklappt. Dorsalansicht

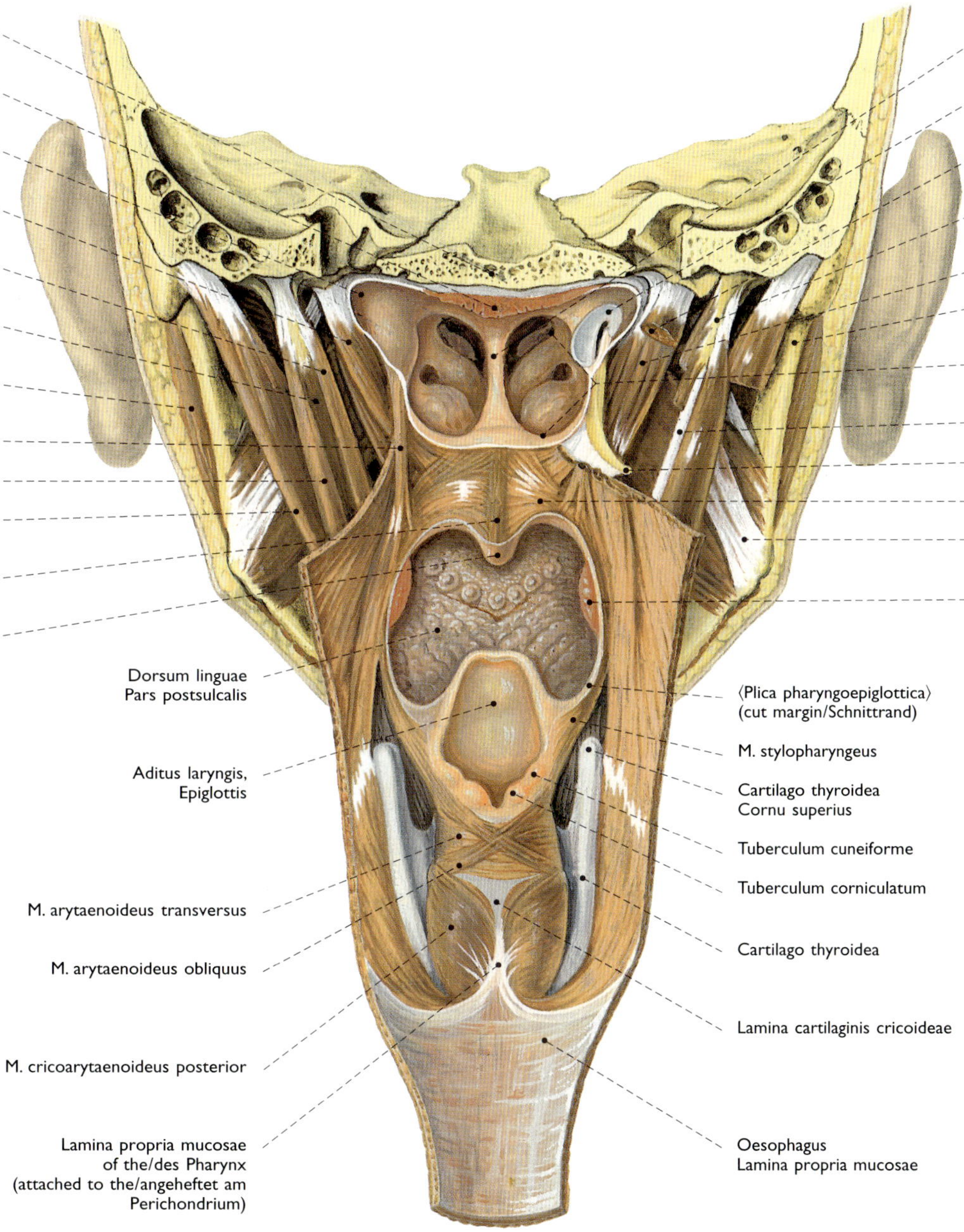

51 Cavitas pharyngis

51 Pharyngeal cavity (70%)

The dorsal wall of the pharynx was cut along the mid-dorsal line and opened. The mucous membrane and the superior pharyngeal constrictor muscle were partially removed in order to expose the different muscles. Dorsal aspect

51 Schlundhöhle (70%)

Die dorsale Schlundwand ist median durchtrennt und aufgeklappt. Schleimhaut und Musculus constrictor pharyngis superior sind grossenteils entfernt, um die verschiedenen Muskeln freizulegen. Dorsalansicht

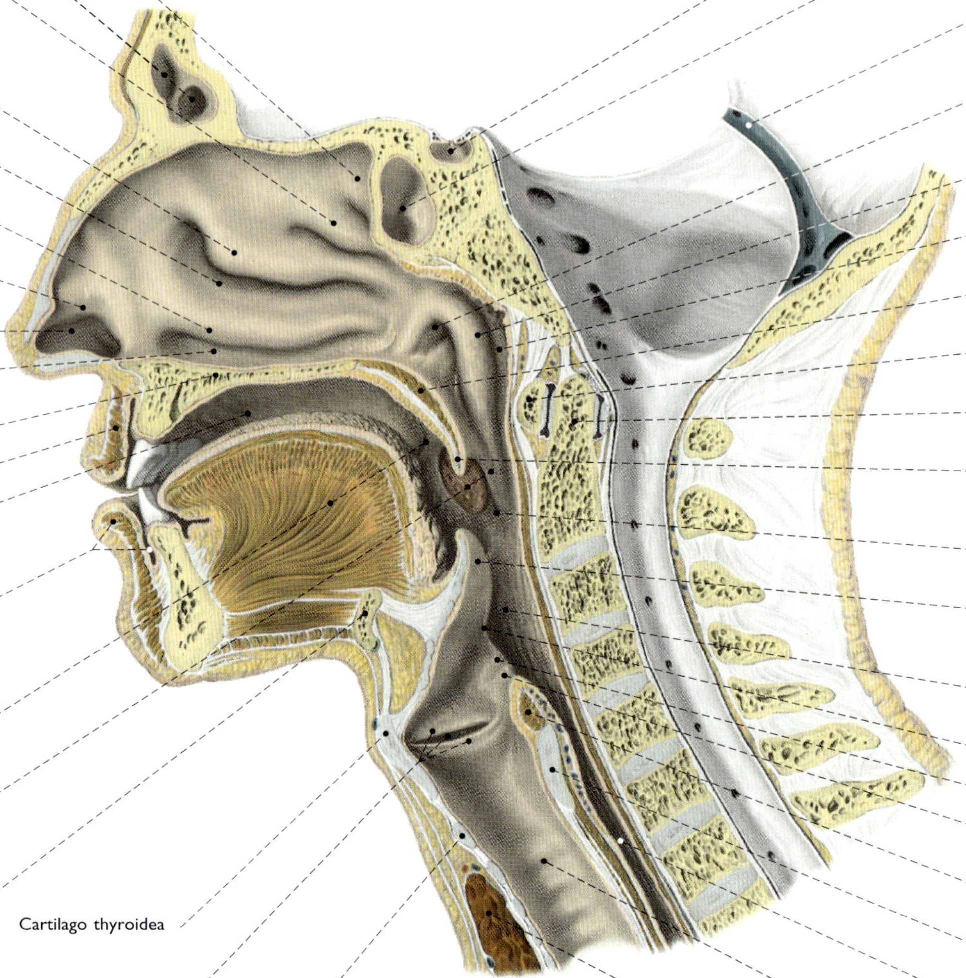

52 Caput et ⟨Collum⟩

52 Alimentary and respiratory systems in the head and neck (70%)

Median sagittal section through the head and neck, medial aspect of the right half

52 Verdauungs- und Atmungssystem in Kopf und Hals (70%)

Medianer Sagittalschnitt durch Kopf und Hals, Medialansicht der rechten Hälfte

a

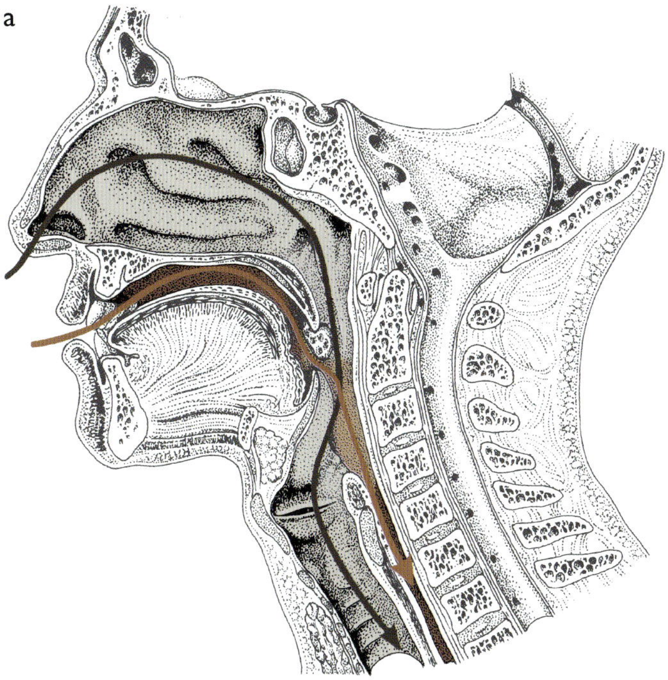

b

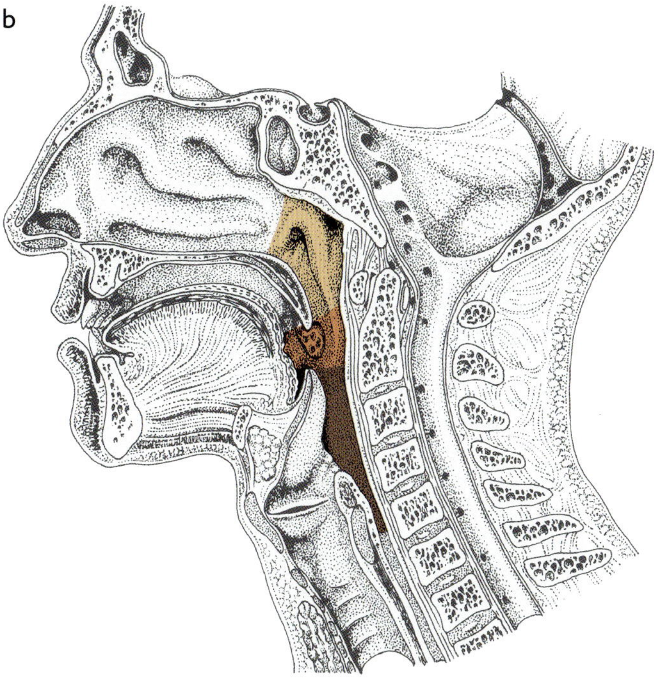

c d

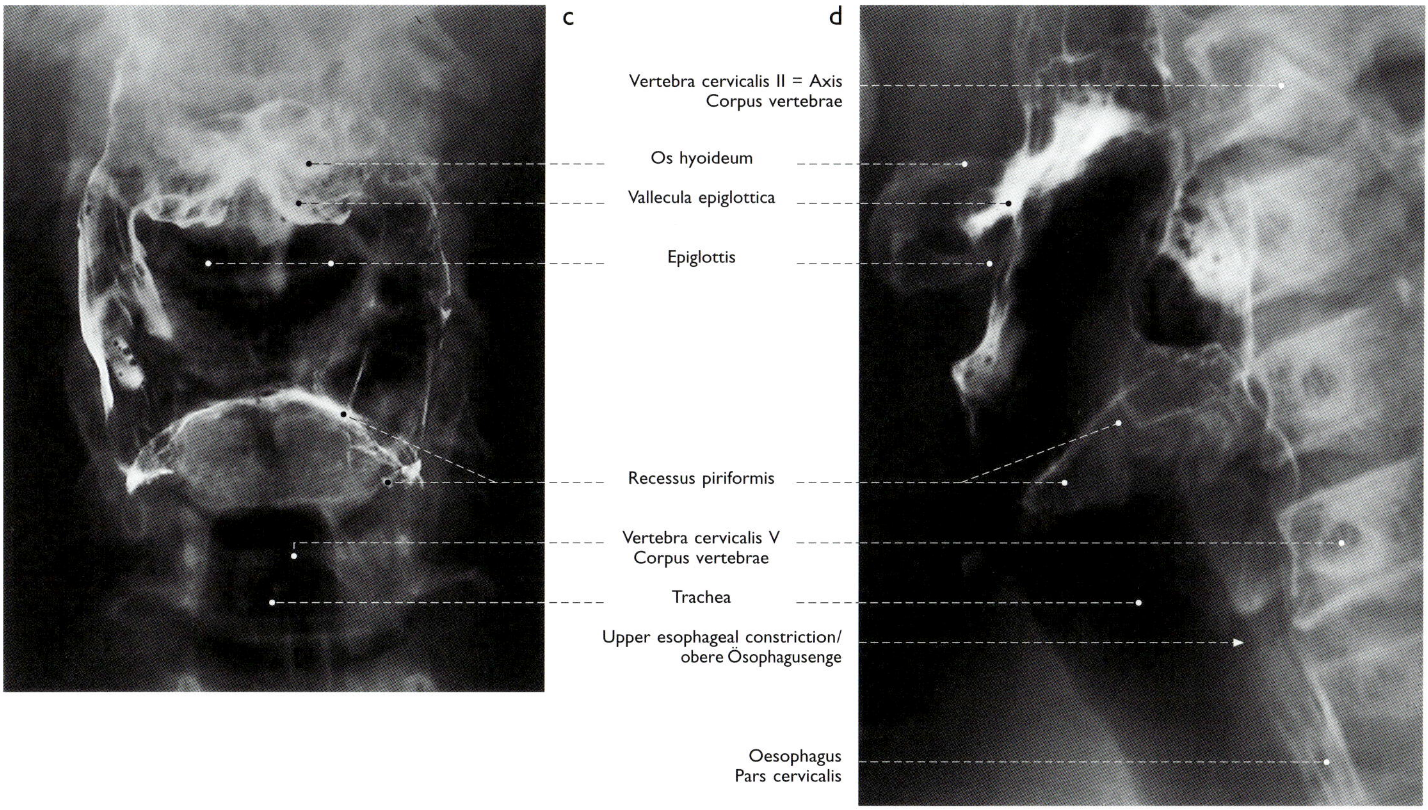

53 Caput et ⟨Collum⟩

53 Alimentary and respiratory systems in the head and neck

a, b Midsagittal sections through the head and neck
a Crossing of the digestive and respiratory systems in the pharynx
b Nasal, oral and laryngeal parts of the pharynx, shown by different colors
c, d Radiographs of the pharynx after oral application of contrast medium
c Posteroanterior radiograph
d Lateral radiograph

53 Verdauungs- und Atmungssystem in Kopf und Hals

a, b Mediane Sagittalschnitte durch Kopf und Hals
a Kreuzung des Verdauungs- und Atmungstrakts im Pharynx
b Naso-, Oro- und Laryngopharynx (Epi-, Meso- und Hypopharynx), durch verschiedene Farben dargestellt
c, d Röntgendarstellungen des Rachens nach oraler Kontrastmittelgabe
c Posteroanteriore Röntgenaufnahme
d Seitliche Röntgenaufnahme

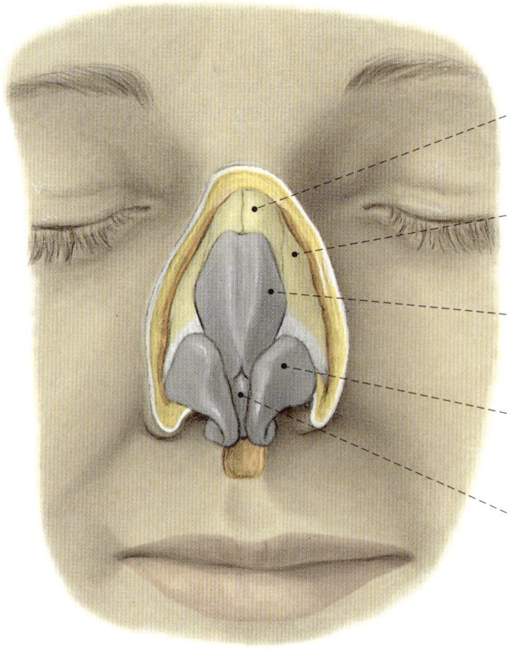

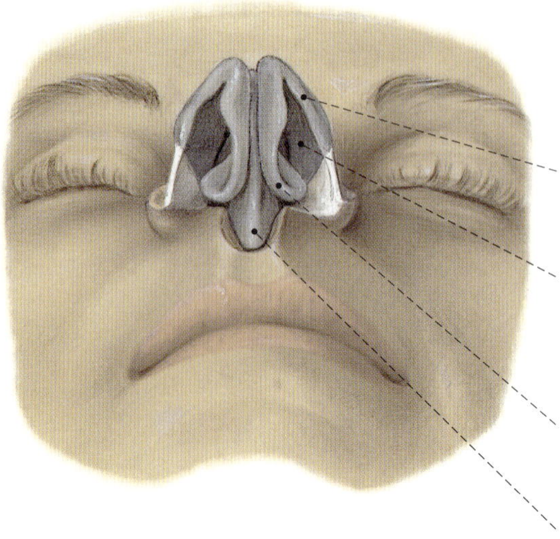

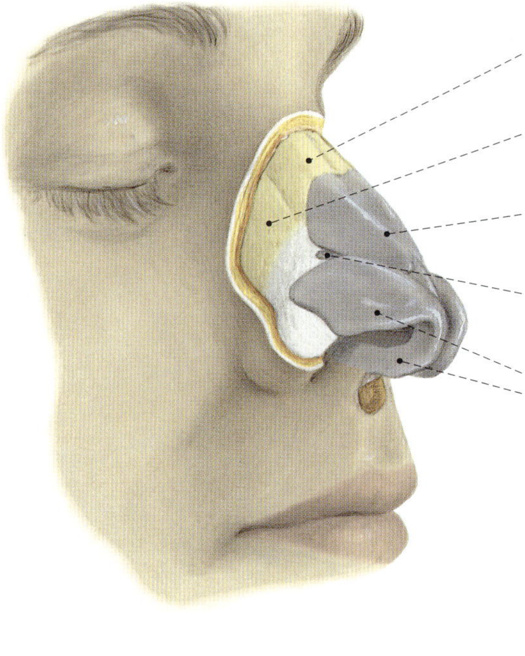

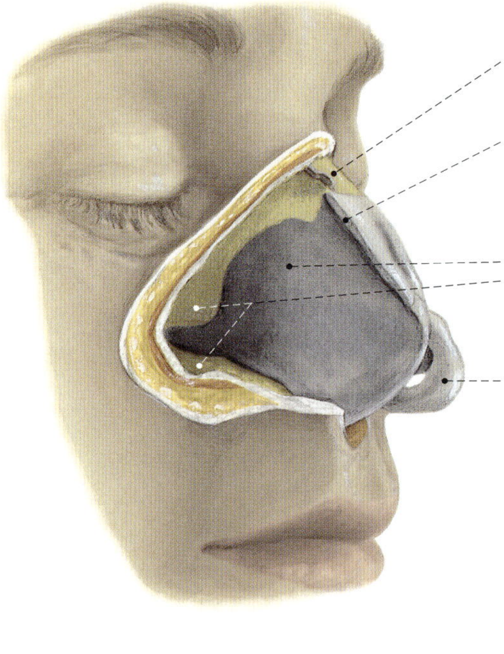

54 Nasus externus

54 Skeleton of the external nose (80%)

- a Frontal aspect
- b Inferior aspect
- c Lateral aspect
- d The nasal septum. The right lateral wall of the external nose was removed. Lateral aspect

54 Skelett der äusseren Nase (80%)

- a Frontalansicht
- b Ansicht von unten
- c Seitenansicht
- d Nasenscheidewand. Die rechte Seitenwand der äusseren Nase ist abgetragen. Seitenansicht

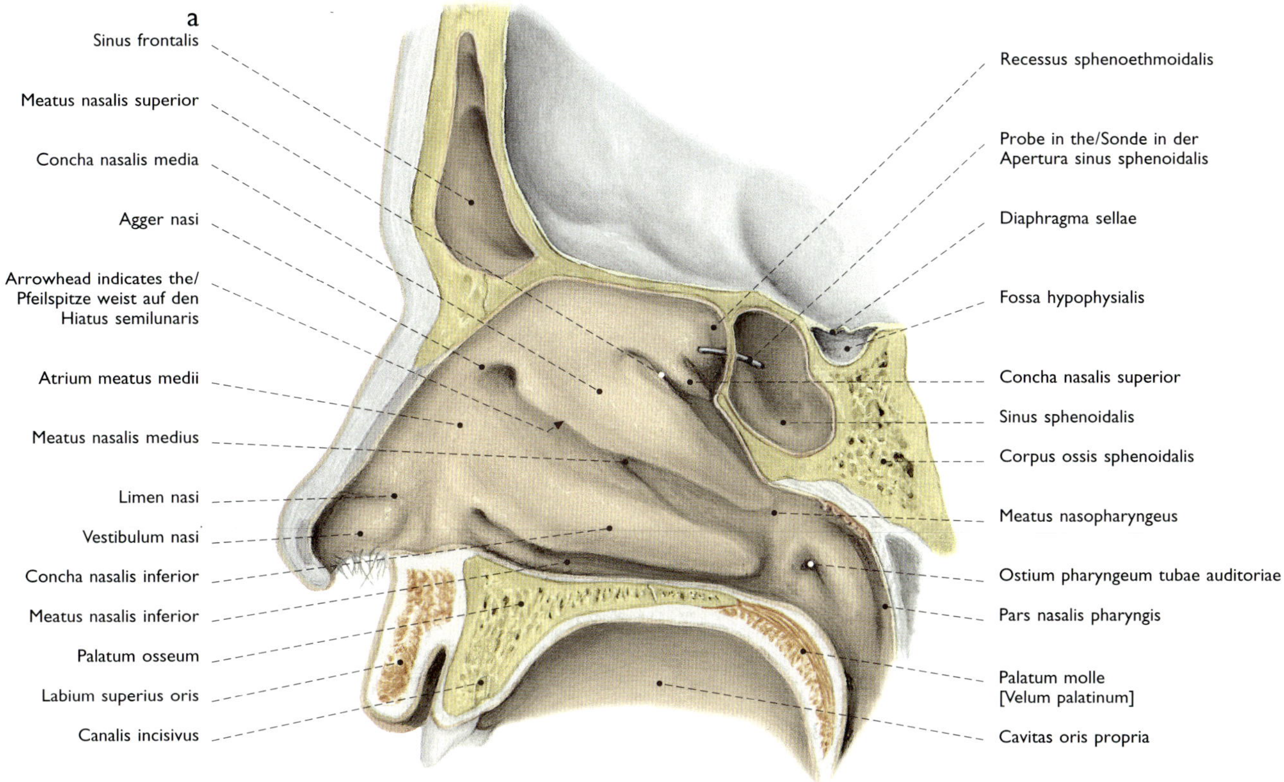

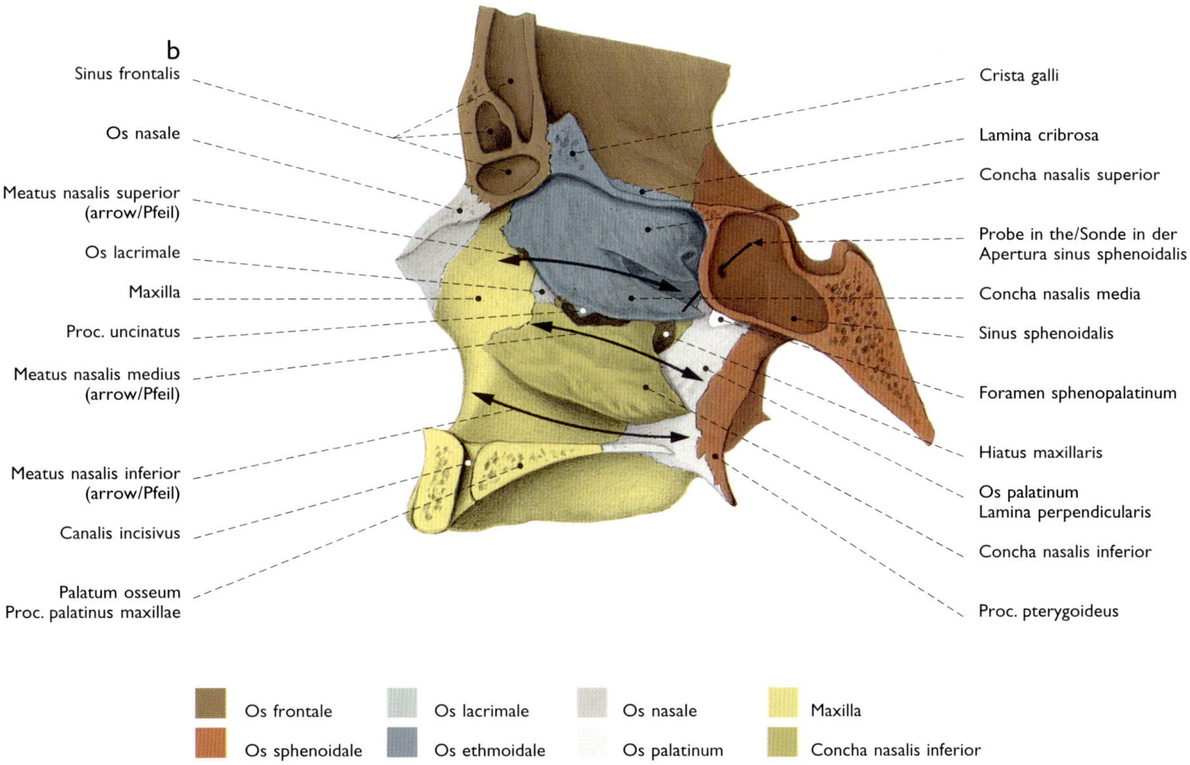

55 Cavitas nasi

55 Lateral wall of the nasal cavity (75%)

Paramedian section to the right of the nasal septum, medial aspect

a Nasal vestibule and nasal mucous membrane

b Bony lateral wall of the nasal cavity. The individual bones are indicated by different colors.

55 Seitenwand der Nasenhöhle (75%)

Paramedianschnitt rechts von der Nasenscheidewand, Medialansicht

a Nasenvorhof und Schleimhautrelief

b Knöcherne seitliche Nasenwand. Die einzelnen Knochen sind in verschiedenen Farben dargestellt.

a

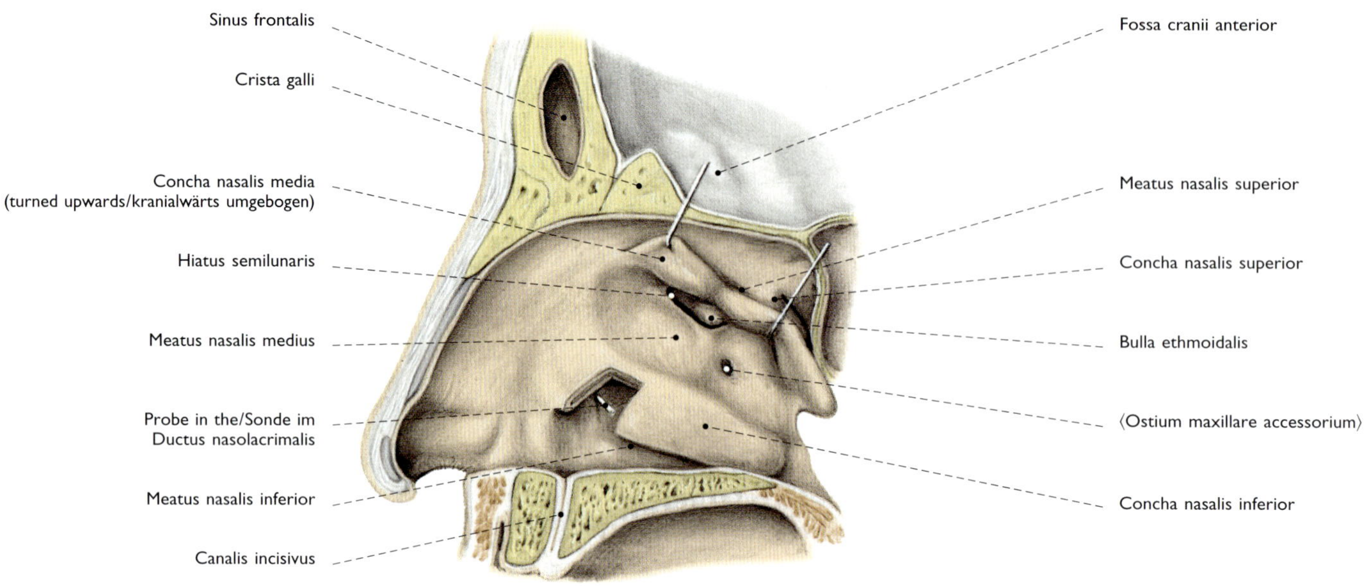

b

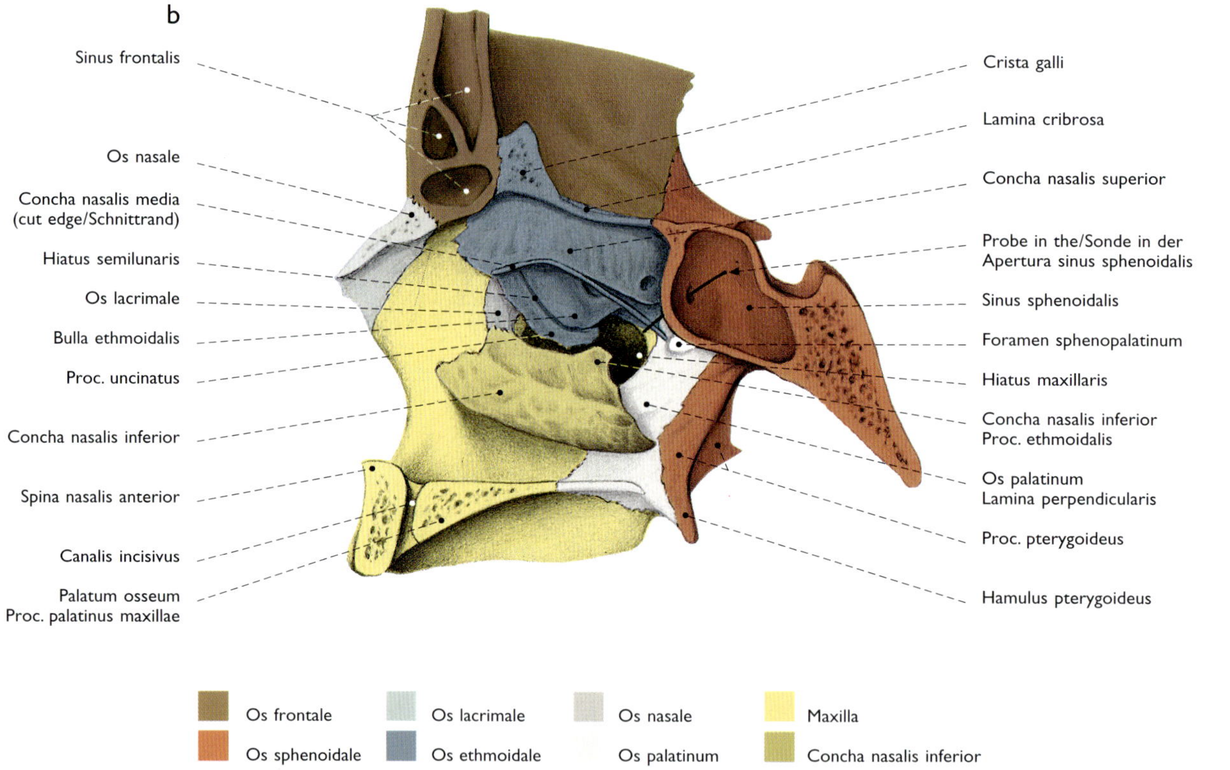

56 Cavitas nasi

56 Lateral wall of the nasal cavity (75%)

Paramedian section to the right of the nasal septum, medial aspect

a Nasal vestibule and nasal mucous membrane. A quadrangular segment was excised from the anterior part of the inferior concha, the middle concha is turned upwards by two hooks.

b Bony lateral wall of the nasal cavity. The middle concha was removed, the incisive canal opened. The individual bones are indicated by different colors.

56 Seitenwand der Nasenhöhle (75%)

Paramedianschnitt rechts von der Nasenscheidewand, Medialansicht

a Nasenvorhof und Nasenschleimhaut. Ein viereckiger Bezirk im vorderen Bereich der unteren Nasenmuschel ist ausgeschnitten, die mittlere Nasenmuschel mittels zweier Haken nach oben geklappt.

b Knöcherne seitliche Nasenwand. Die mittlere Nasenmuschel ist abgetragen, der Canalis incisivus aufgemeisselt. Die verschiedenen Knochen sind in unterschiedlichen Farben dargestellt.

a

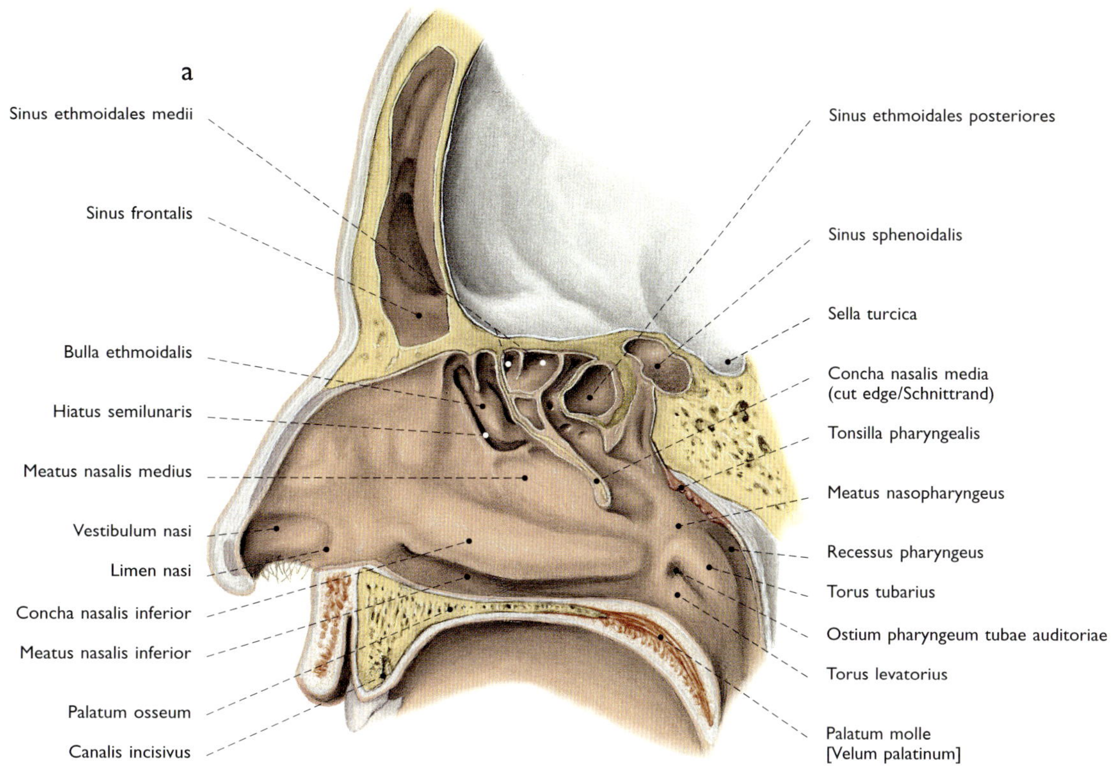

b

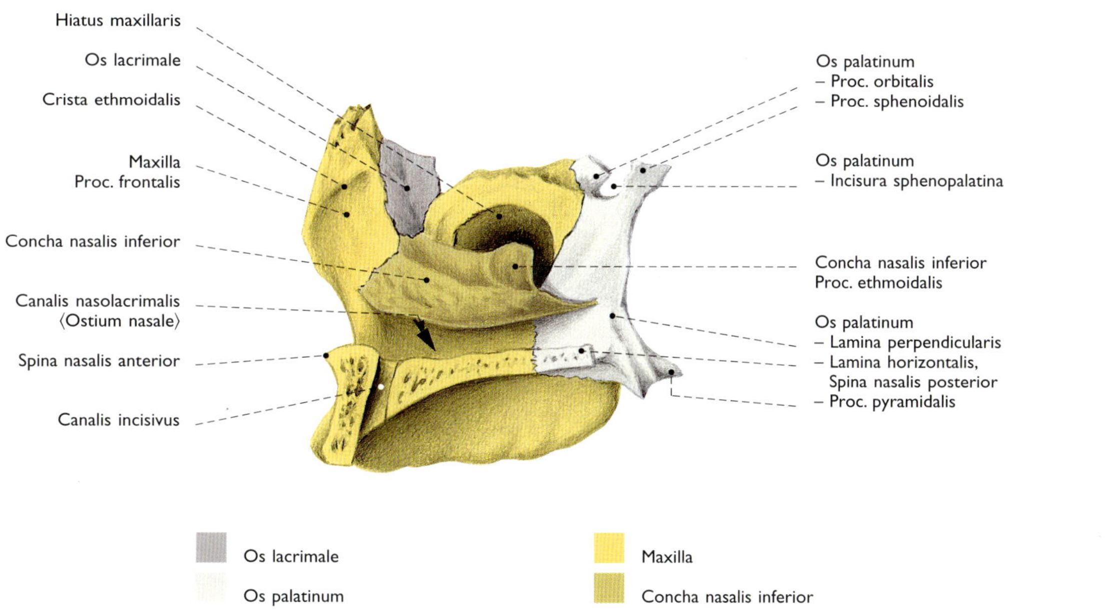

57 Cavitas nasi

57 Lateral wall of the nasal cavity (75%)

Paramedian section to the right of the nasal septum, medial aspect

a Nasal vestibule and nasal mucous membrane. The middle concha was partially removed, the posterior ethmoidal sinuses were opened.

b Bony lateral wall of the nasal cavity. The nasal, frontal, ethmoid and sphenoid bones were removed. The individual bones are indicated by different colors.

57 Seitenwand der Nasenhöhle (75%)

Paramedianschnitt rechts von der Nasenscheidewand, Medialansicht

a Nasenvorhof und Nasenschleimhaut. Die mittlere Nasenmuschel ist teilweise abgetragen, die hinteren Siebbeinzellen sind eröffnet.

b Knöcherne seitliche Nasenwand. Die Ossa nasale, frontale, ethmoidale et sphenoidale sind entfernt. Die verschiedenen Knochen sind in unterschiedlichen Farben dargestellt.

a

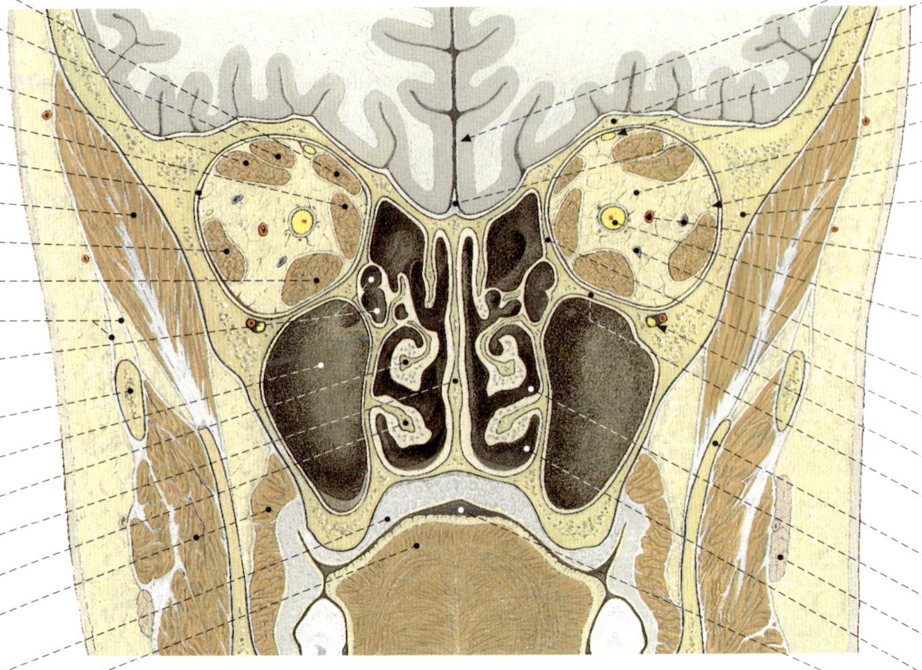

b

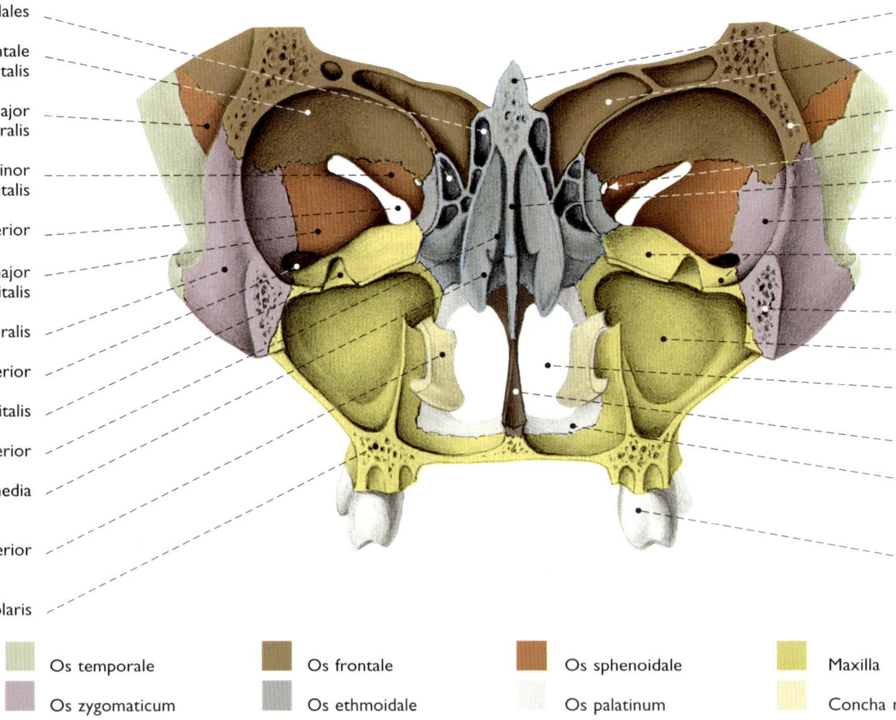

58 Cavitas nasi et Orbita

58 Nasal and orbital cavities (75%)

Coronal section, posterior portion, ventral aspect

a Section of the head of an adult male behind the crista galli

b Section of the skull passing the crista galli. The individual bones are indicated by different colors.

58 Nasen- und Augenhöhle (75%)

Frontalschnitt, hinterer Schnittblock, Ventralansicht

a Schnitt durch den Kopf eines erwachsenen Mannes hinter der Crista galli

b Schnitt durch den Gesichtsschädel im Bereich der Crista galli. Die verschiedenen Knochen sind in unterschiedlichen Farben dargestellt.

a

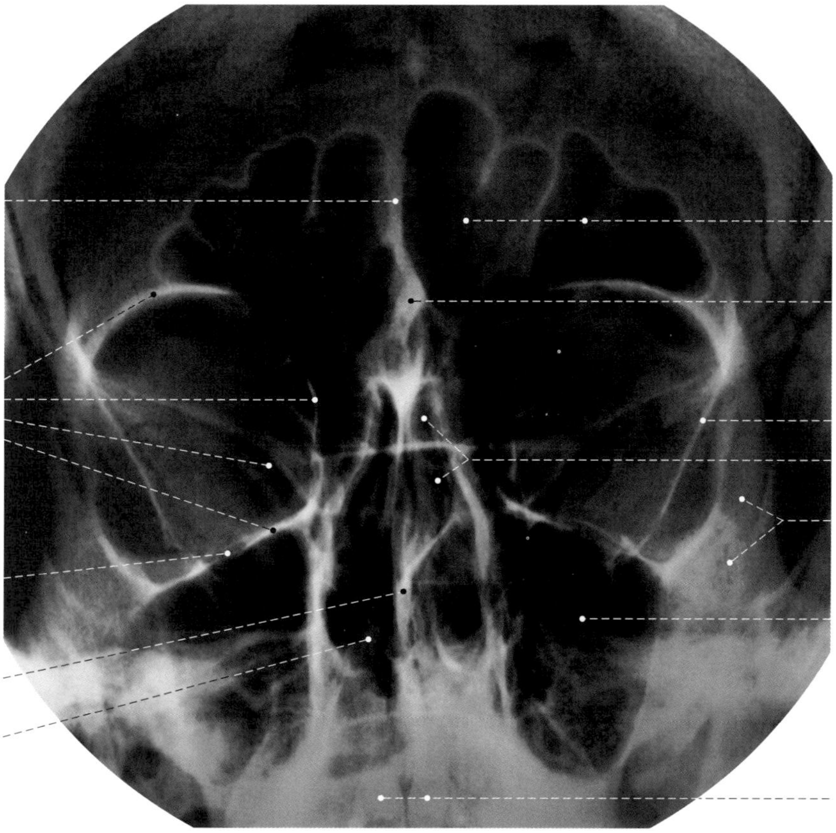

b

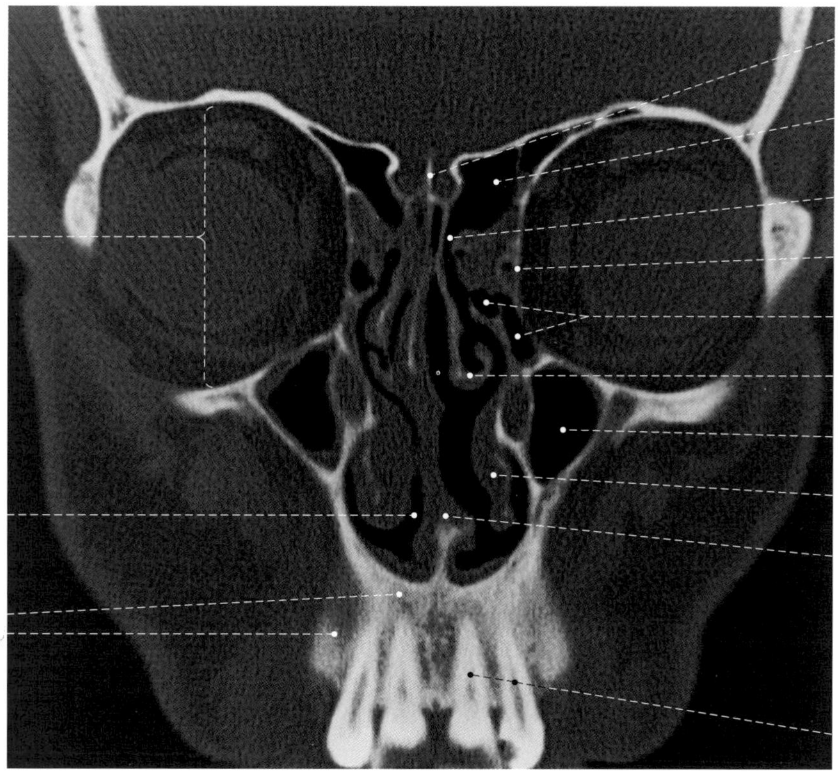

59 Cavitas nasi et Sinus paranasales

59 Nasal cavity and paranasal sinuses (90%)

a Posteroanterior radiograph of the skull
b Coronal computed tomogram (CT) through the viscerocranium

59 Nasenhöhle und Nasennebenhöhlen (90%)

a Posteroanteriore Röntgenaufnahme des Schädels
b Frontales Computertomogramm (CT) durch den Gesichtsschädel

a

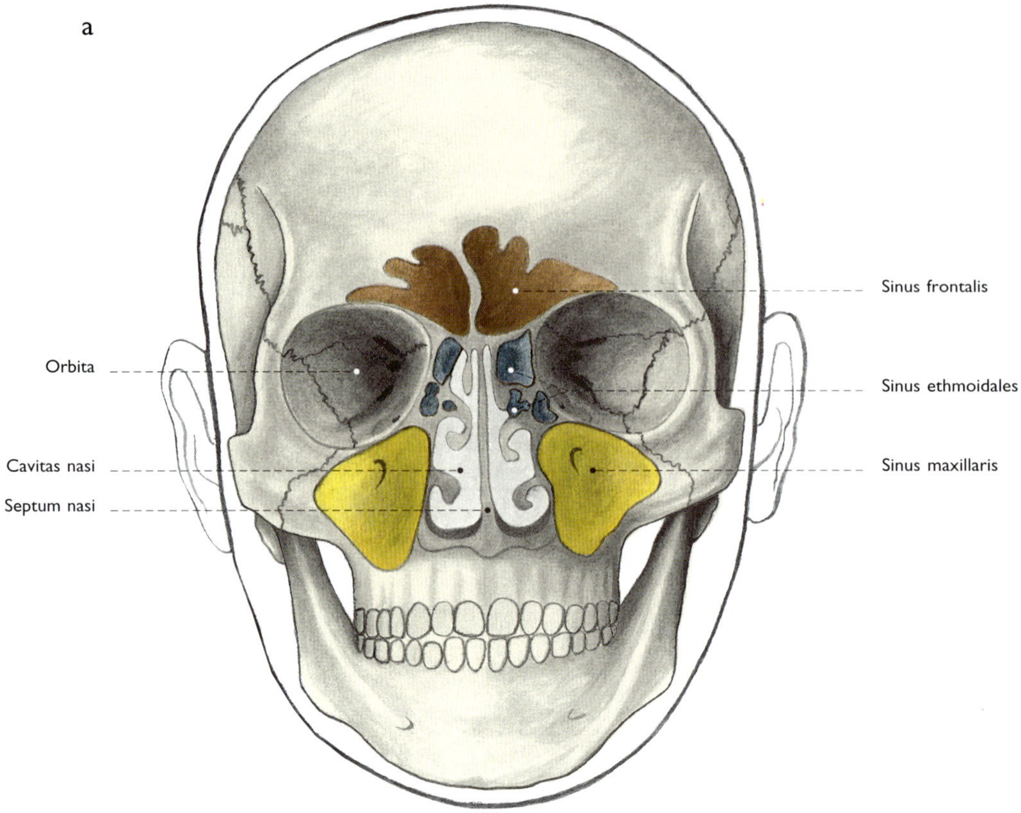

b

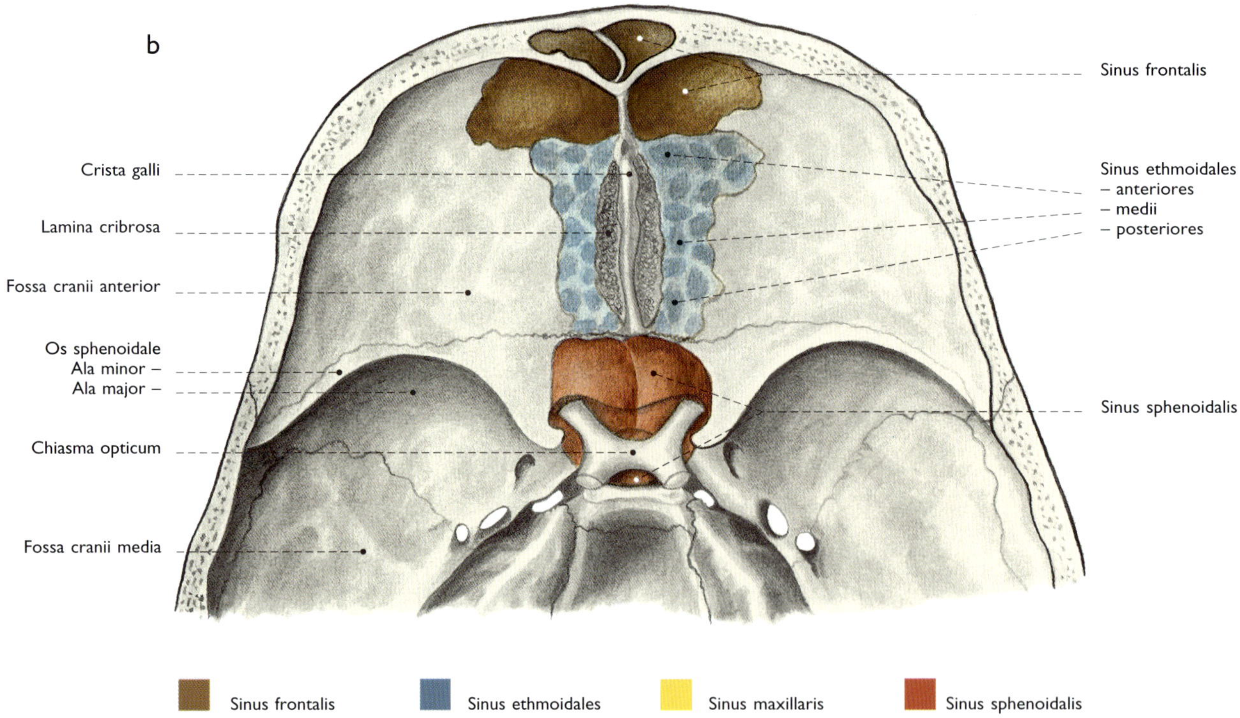

60 Cavitas nasi et Sinus paranasales

60 Nasal cavity and paranasal sinuses

Projections of the paranasal sinuses
a onto the face
b onto the anterior cranial fossa

60 Nasenhöhle und Nasennebenhöhlen

Projektionen der Nasennebenhöhlen
a auf das äussere Gesicht
b auf die vordere Schädelgrube

a

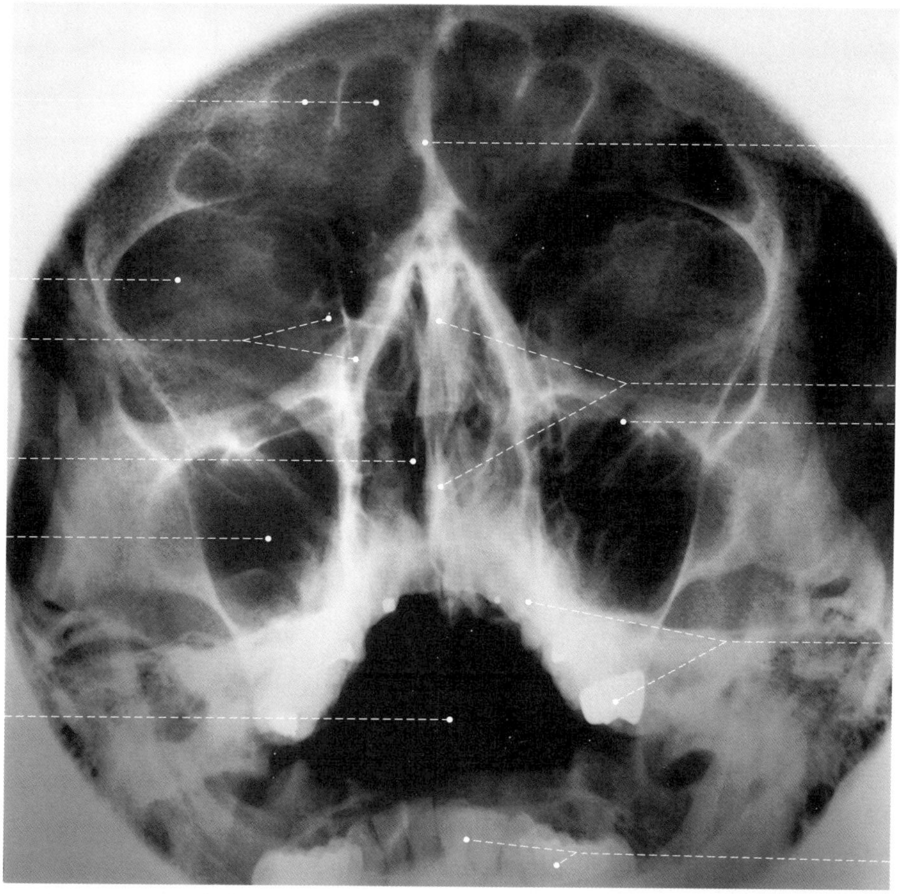

b

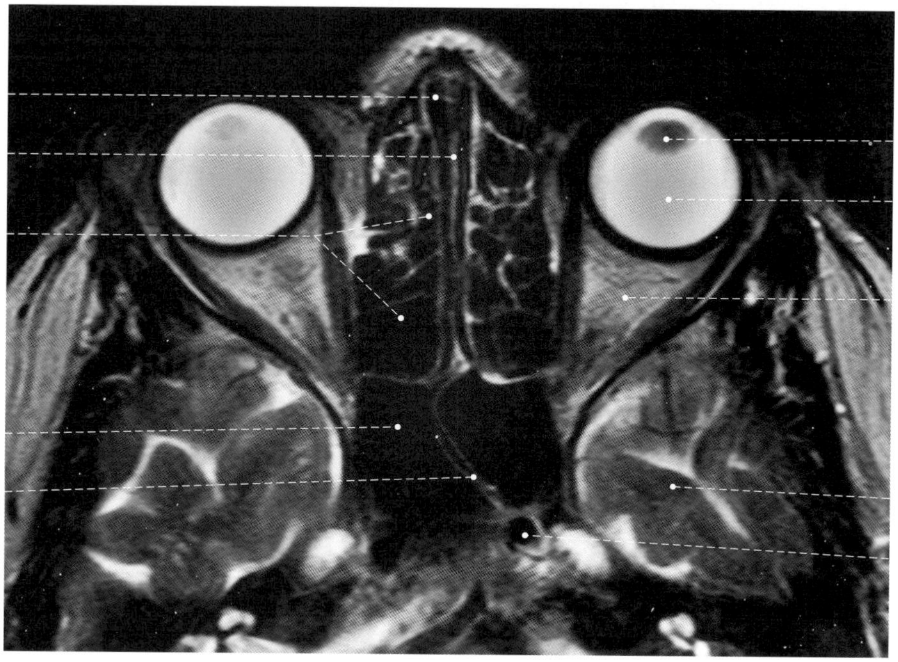

61 Cavitas nasi et Sinus paranasales

61 Nasal cavity and paranasal sinuses (90%)

a Occipitomental radiograph of the skull with the mouth wide open

b Transverse magnetic resonance image (MRI, T_2-weighted) through the orbits and the ethmoidal and sphenoidal sinuses, inferior aspect

61 Nasenhöhle und Nasennebenhöhlen (90%)

a Okzipitomentale Röntgenaufnahme des Schädels bei weit geöffnetem Mund

b Horizontales Kernspintomogramm (MRI, T_2-gewichtet) durch die Orbitae sowie die Siebbein- und Keilbeinhöhlen, Kaudalansicht

a

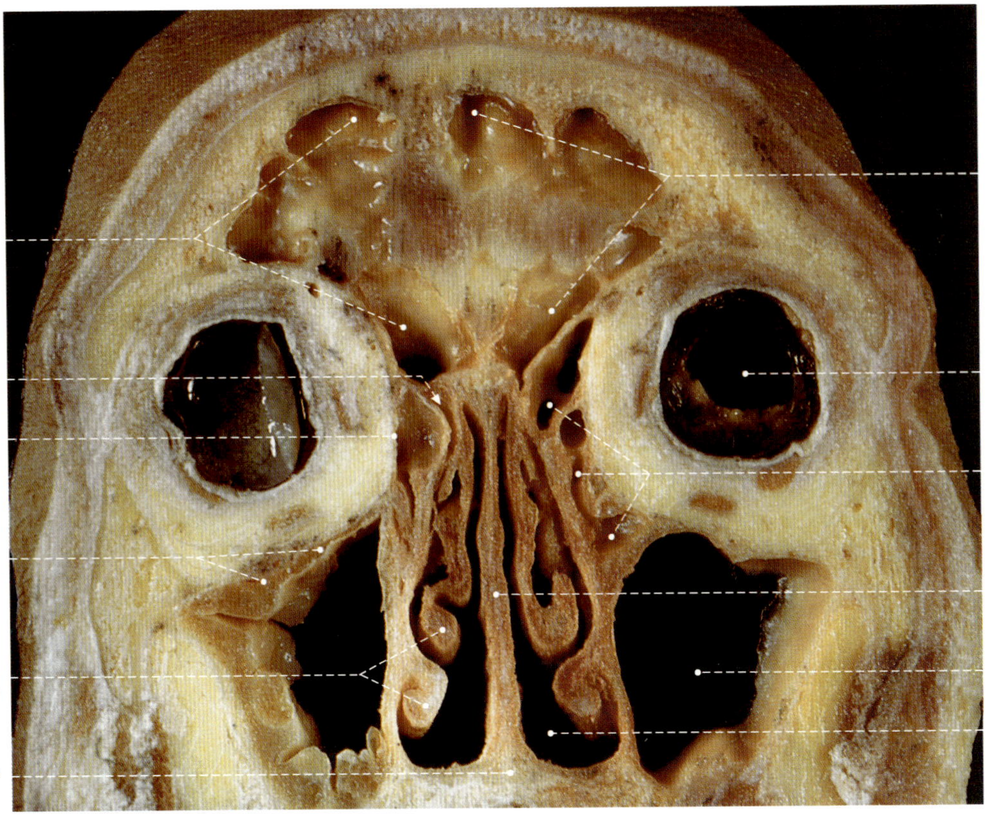

b

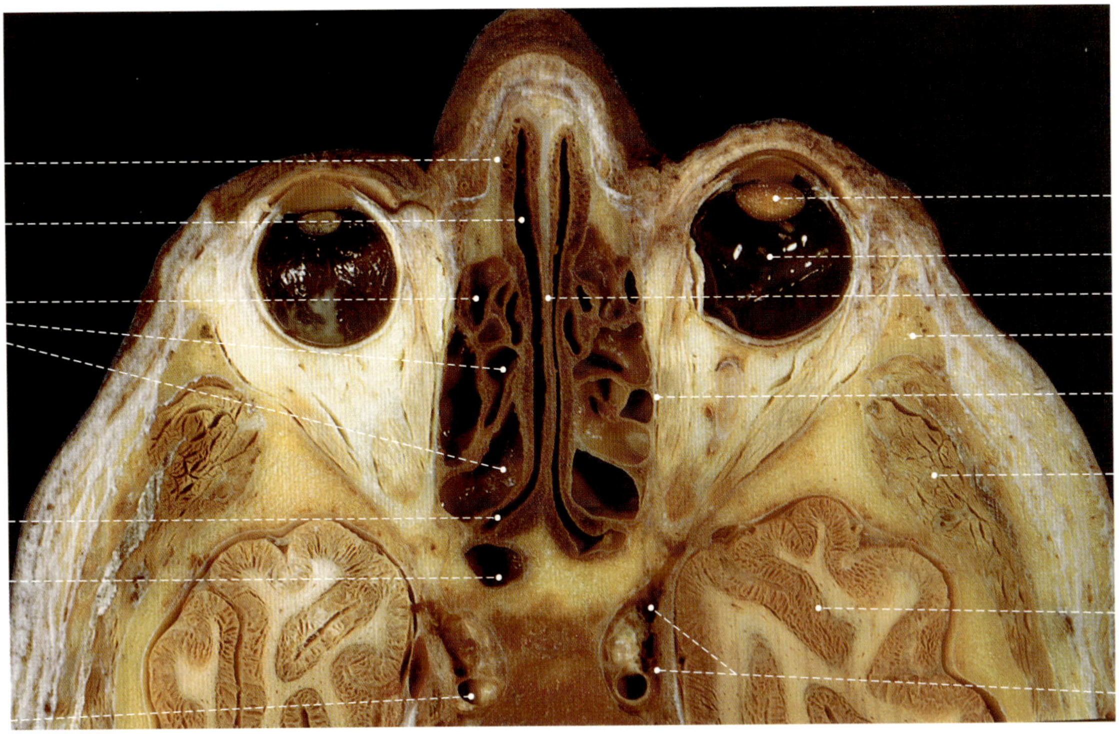

62 Cavitas nasi et Sinus paranasales

62 Nasal cavity and paranasal sinuses (100%)

Anatomical sections

a Coronal section through both eyeballs and the frontal, ethmoidal and maxillary sinuses, ventral aspect

b Transverse section through both eyeballs and the ethmoidal and sphenoidal sinuses, superior aspect

62 Nasenhöhle und Nasennebenhöhlen (100%)

Anatomische Schnitte

a Frontalschnitt durch beide Augäpfel sowie die Stirn-, Siebbein- und Kieferhöhlen, Ventralansicht

b Horizontalschnitt durch beide Augäpfel sowie die Siebbein- und Keilbeinhöhlen, Kranialansicht

a

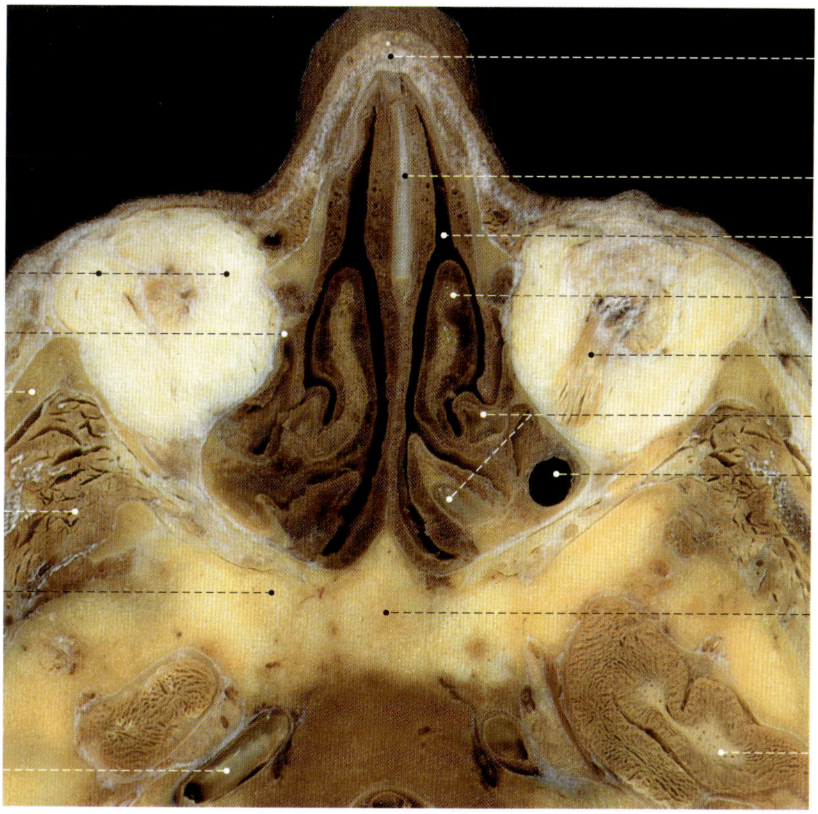

b

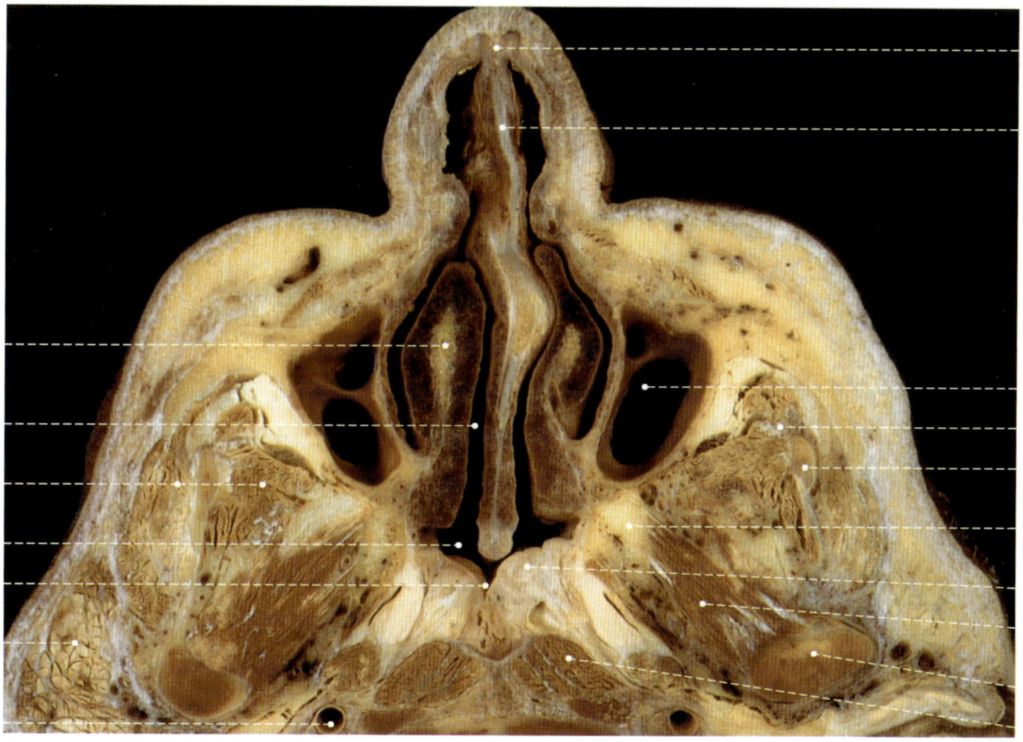

63 Cavitas nasi et Sinus paranasales

63 Nasal cavity and paranasal sinuses (100%)

Transverse anatomical sections
- a through the caudal part of the orbits and the apex of the maxillary sinus
- b through the middle part of the nasal cavity and the maxillary sinuses
- a, b Superior aspect

63 Nasenhöhle und Nasennebenhöhlen (100%)

Horizontale anatomische Schnitte
- a durch den kaudalen Teil der Augenhöhlen und die kraniale Spitze der Kieferhöhle
- b durch den mittleren Bereich der Nasen- und Kieferhöhlen
- a, b Kranialansicht

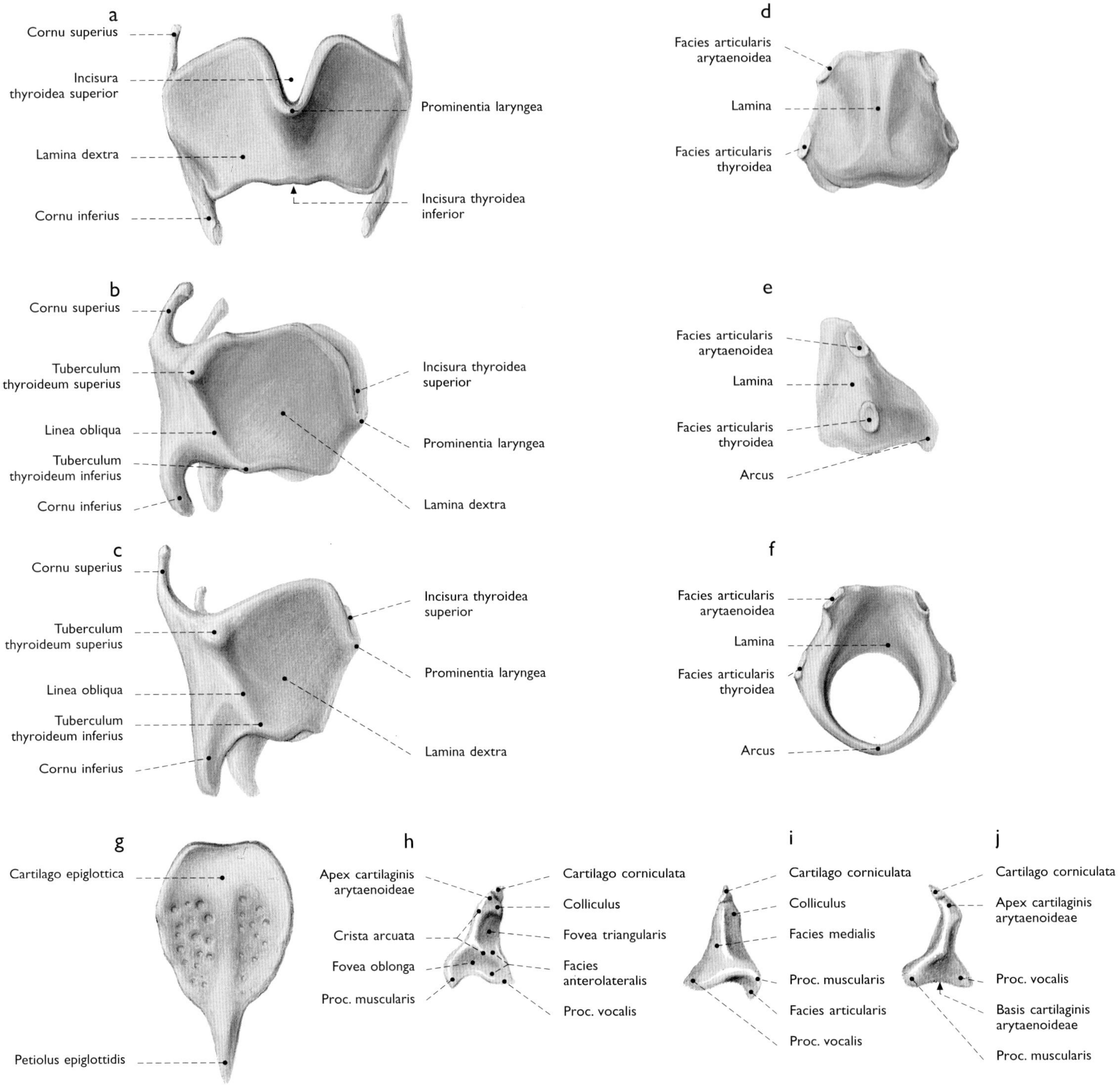

64 Cartilagines thyroidea, cricoidea, epiglottica et arytaenoidea

64 Thyroid, cricoid, epiglottic and arytenoid cartilages (90%)

- a–c Thyroid cartilage
  - a Male larynx, ventral aspect
  - b Male larynx, right lateral aspect
  - c Female larynx, right lateral aspect
- d–f Cricoid cartilage
  - d Dorsal aspect
  - e Right lateral aspect
  - f Superior aspect
- g Epiglottic cartilage, dorsal aspect
- h–j Right arytenoid cartilage
  - h Ventrolateral aspect
  - i Dorsomedial aspect
  - j Right lateral aspect

64 Schild-, Ring-, Kehldeckel- und Stellknorpel (90%)

- a–c Schildknorpel
  - a Kehlkopf eines Mannes, Ventralansicht
  - b Kehlkopf eines Mannes, rechte Seitenansicht
  - c Kehlkopf einer Frau, rechte Seitenansicht
- d–f Ringknorpel
  - d Dorsalansicht
  - e Rechte Seitenansicht
  - f Kranialansicht
- g Kehldeckelknorpel, Dorsalansicht
- h–j Rechter Stellknorpel
  - h Ventrolateralansicht
  - i Dorsomedialansicht
  - j Rechte Seitenansicht

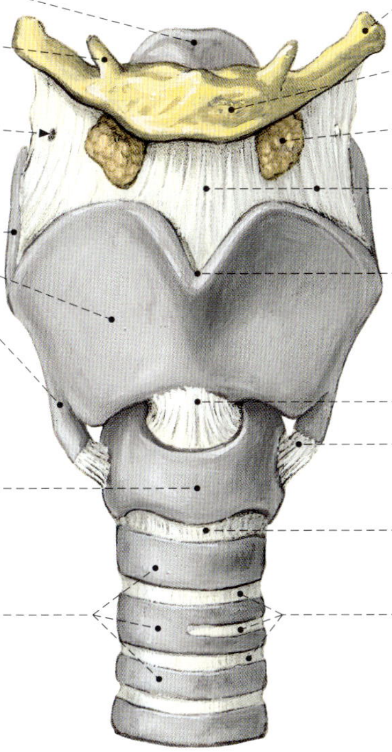

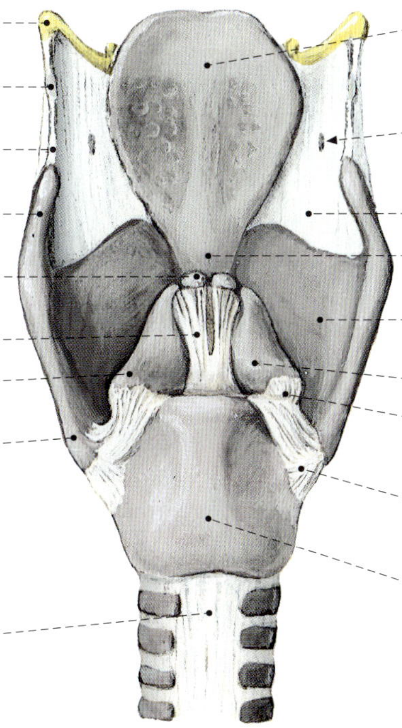

65 Cartilagines laryngeales

65 Laryngeal cartilages and their connections (100%)
a Ventral aspect
b Dorsal aspect

65 Kehlkopfknorpel und Verbindungen (100%)
a Ventralansicht
b Dorsalansicht

a

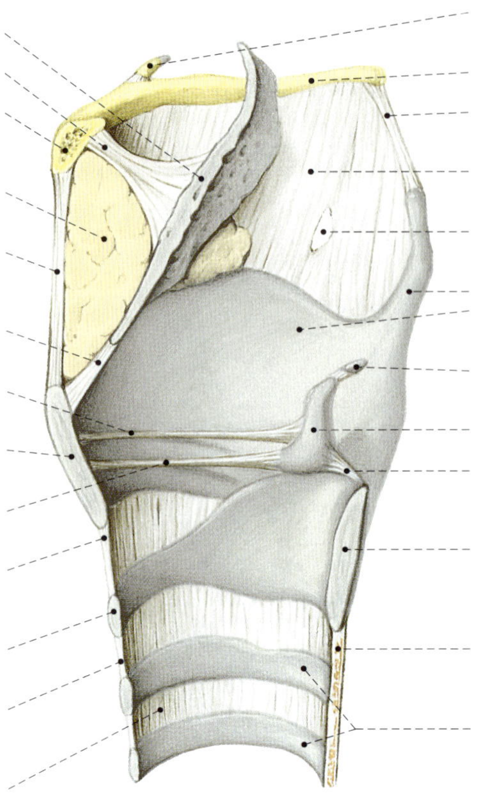

b

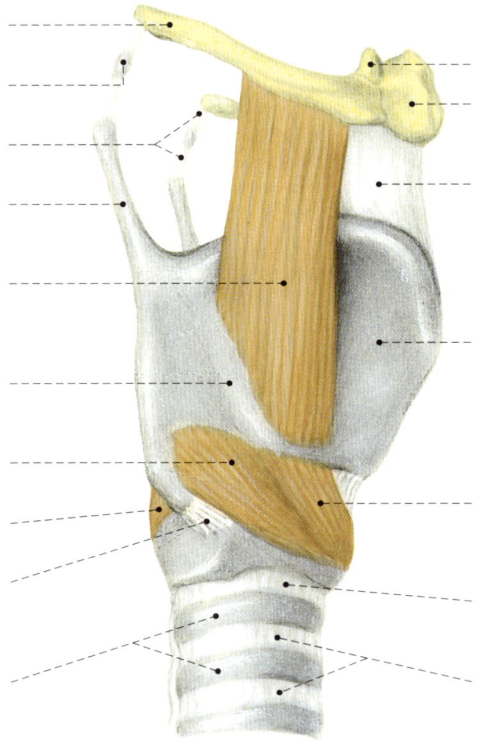

66 Larynx

66 Ligaments and muscles of the larynx (100%)

a Median section through the hyoid bone and the skeleton of the larynx, medial aspect
b Thyrohyoid and cricothyroid muscles, right lateral aspect

66 Bandverbindungen und Muskeln des Kehlkopfs (100%)

a Medianschnitt durch Zungenbein und Kehlkopfskelett, Medialansicht
b Musculi thyrohyoideus et cricothyroideus, rechte Seitenansicht

a

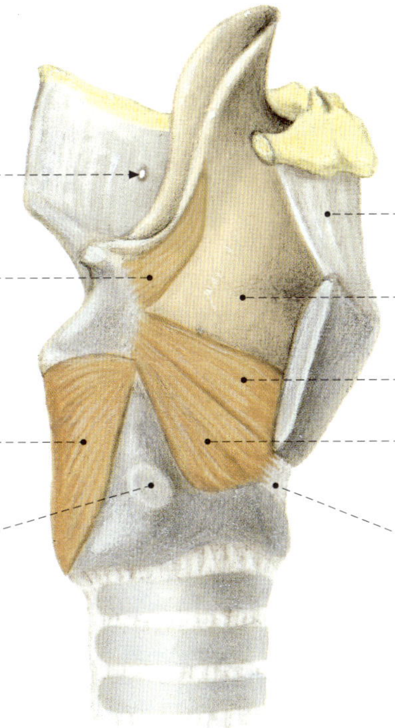

b

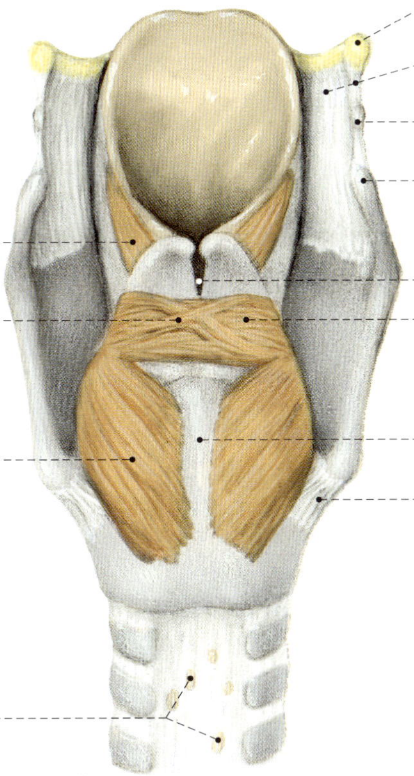

67 Larynx

67 Inner muscles of the larynx (100%)

a The right lamina of the thyroid cartilage was partially removed, the arytenoid muscle is omitted. Right lateral aspect

b The pharyngeal mucosa was completely removed. Dorsal aspect

67 Innere Kehlkopfmuskeln (100%)

a Die rechte Schildknorpelplatte ist teilweise abgetragen, der Musculus arytaenoideus weggelassen. Rechte Seitenansicht

b Die Rachenschleimhaut ist vollständig entfernt. Dorsalansicht

a

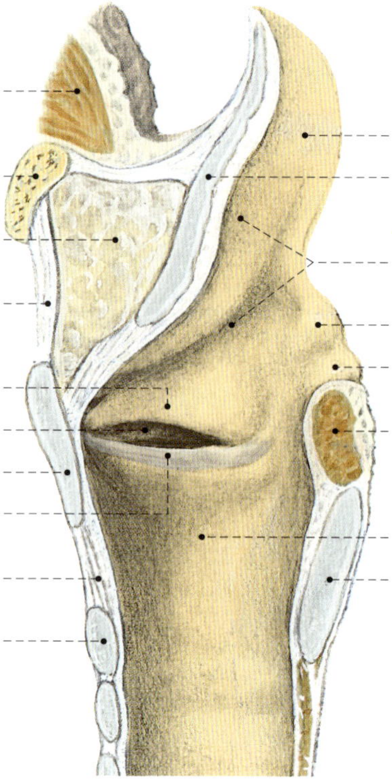

b

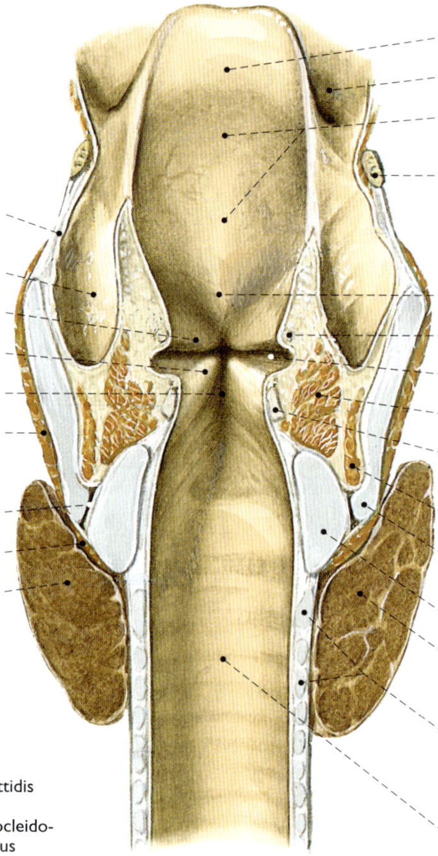

c

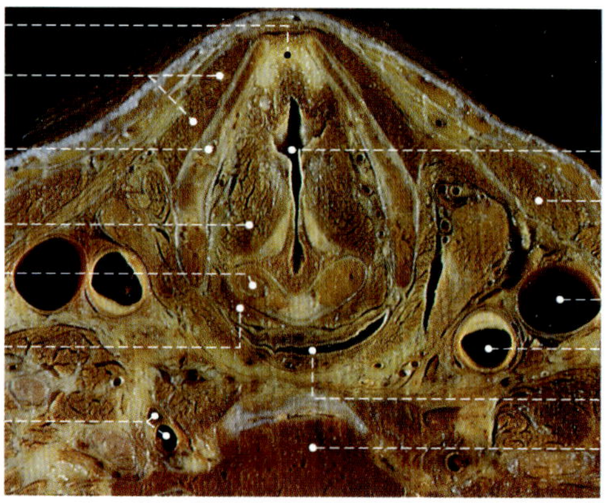

68 Larynx

68 Larynx

a Median section through the larynx (100%), medial aspect
b Anterior segment of a frontal (coronal) section through the larynx (100%), dorsal aspect
c Transverse anatomical section through the glottis (75%), inferior aspect

68 Kehlkopf

a Medianschnitt durch den Kehlkopf (100%), Medialansicht
b Ventraler Schnittblock eines Frontalschnitts durch den Kehlkopf (100%), Dorsalansicht
c Anatomischer Horizontalschnitt durch die Glottis (75%), Kaudalansicht

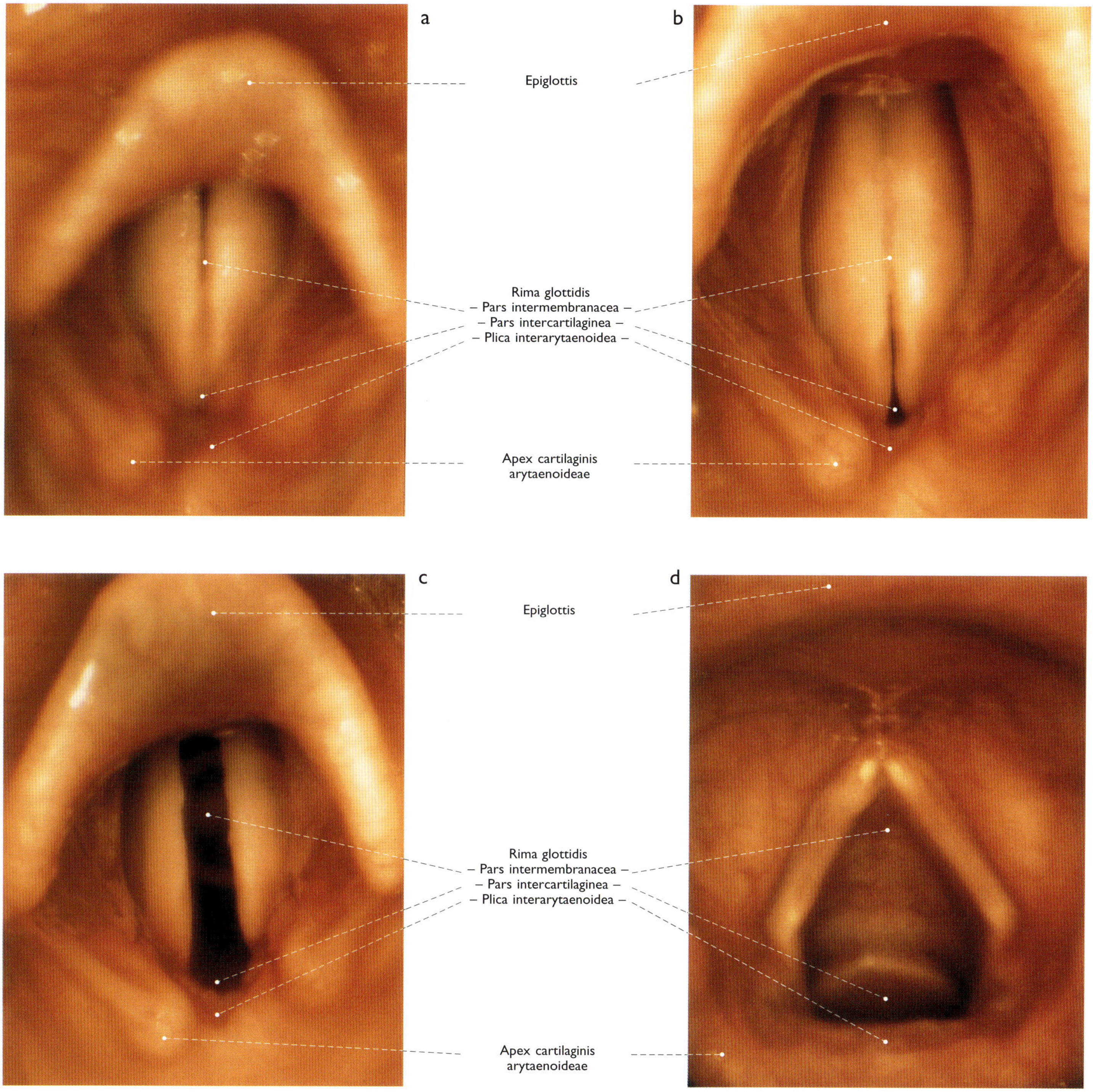

69 Larynx

69 Laryngoscopic view of the interior of the larynx in a living adult (100%)

- a During loud phonation
- b During whispering
- c During quiet respiration
- d During forced inspiration

69 Spiegelbild des Kehlkopfs beim lebenden Erwachsenen (100%)

- a Stellung bei lauter Phonation
- b Stellung bei leiser Phonation (Flüstersprache)
- c Stellung bei ruhiger Respiration
- d Stellung bei forcierter Einatmung

a

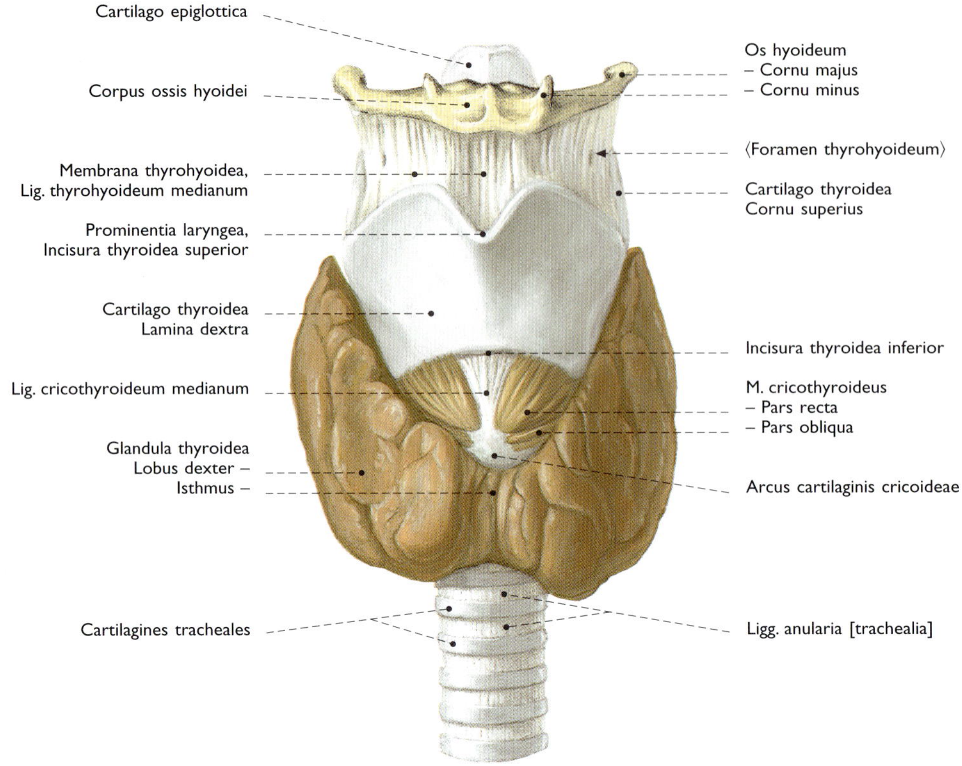

b

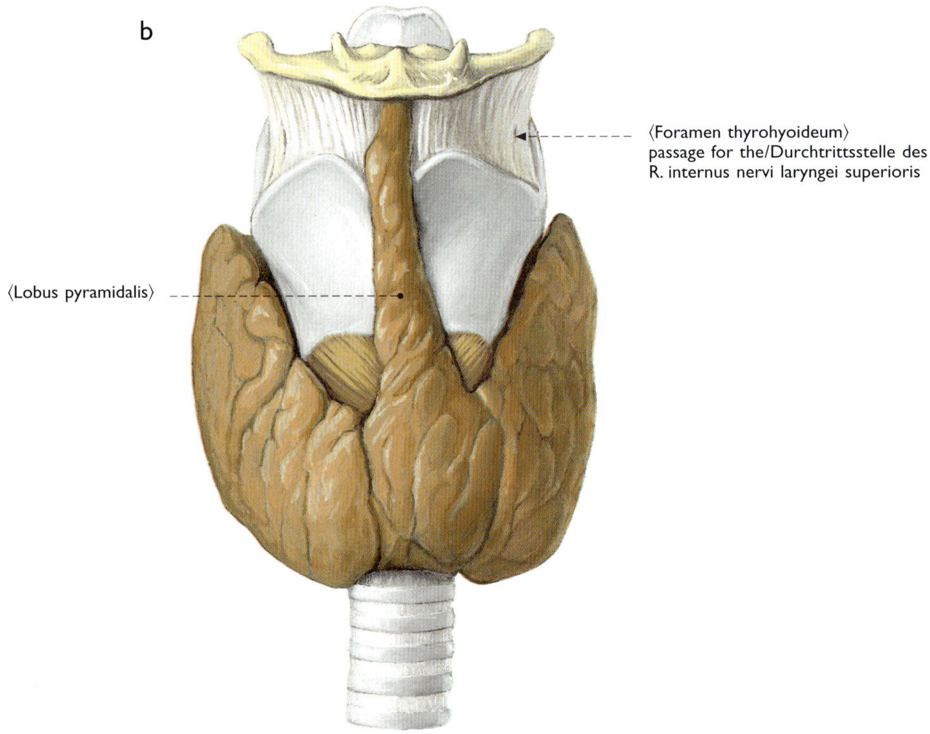

70 Glandula thyroidea et Larynx

70 Thyroid gland and larynx (80%)
Ventral aspect
a Normal situs
b Pyramidal lobe as a variation

70 Schilddrüse und Kehlkopf (80%)
Ventralansicht
a Typischer Situs
b Lobus pyramidalis (Variation)

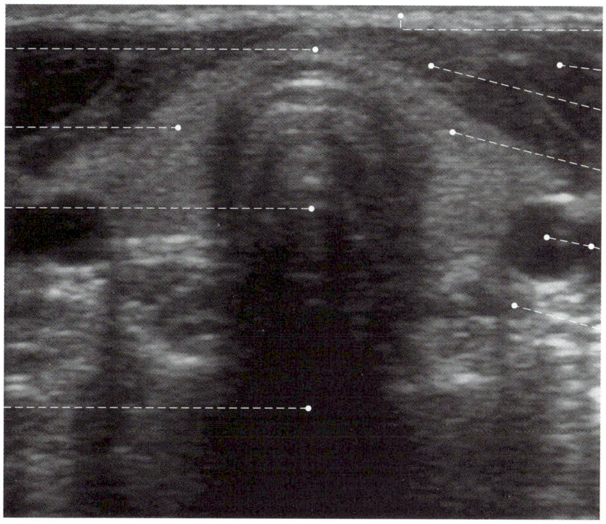

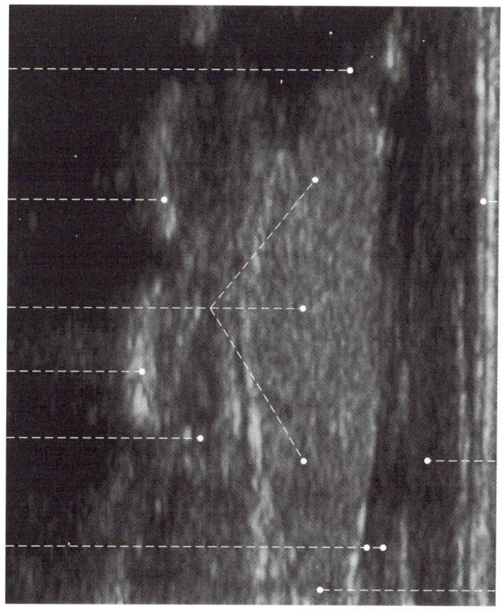

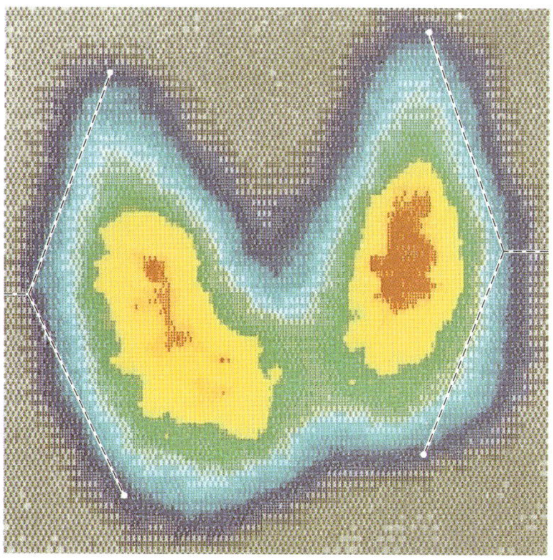

71 Glandula thyroidea

71 Thyroid gland (100%)

- a Transverse ultrasound image
- b Longitudinal ultrasound image through one of the two thyroid lobes
- c Scan of a thyroid gland with normal function after application of ^{99m}Tc

71 Schilddrüse (100%)

- a Horizontales Ultraschallbild
- b Longitudinales Ultraschallbild durch einen der beiden Schilddrüsenlappen
- c Szintigramm einer Schilddrüse mit normaler Funktion nach Gabe von ^{99m}Tc

a

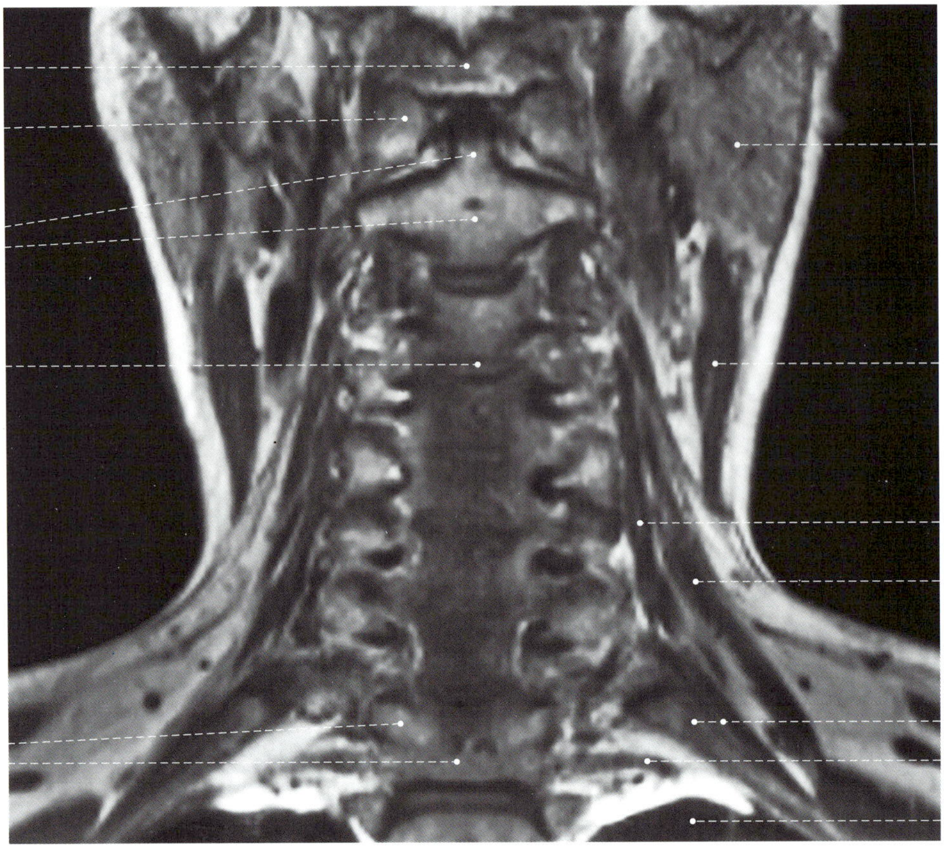

b

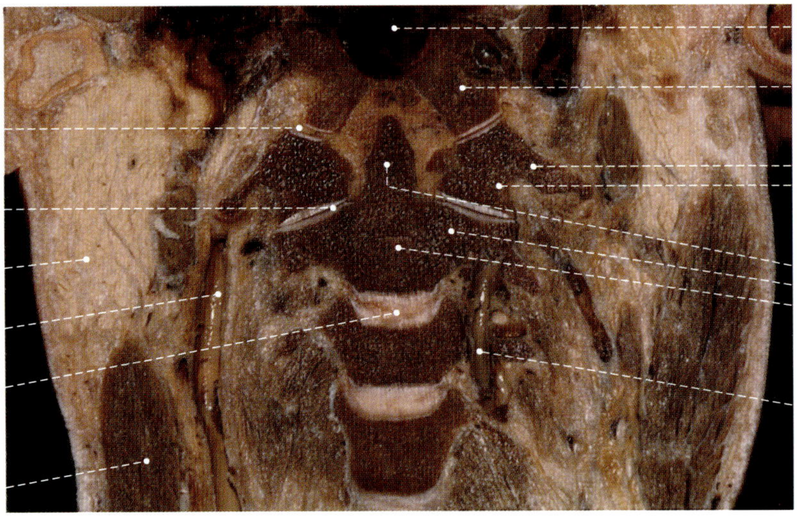

72 ⟨Collum⟩

72 Neck (80%)

Coronal sections through the neck in the plane of the vertebral bodies

a Magnetic resonance image (MRI, T_1-weighted)
b Anatomical section, ventral aspect

72 Hals (80%)

Frontalschnitte durch den Hals in der Ebene der Wirbelkörper

a Kernspintomogramm (MRI, T_1-gewichtet)
b Anatomischer Schnitt, Ventralansicht

a

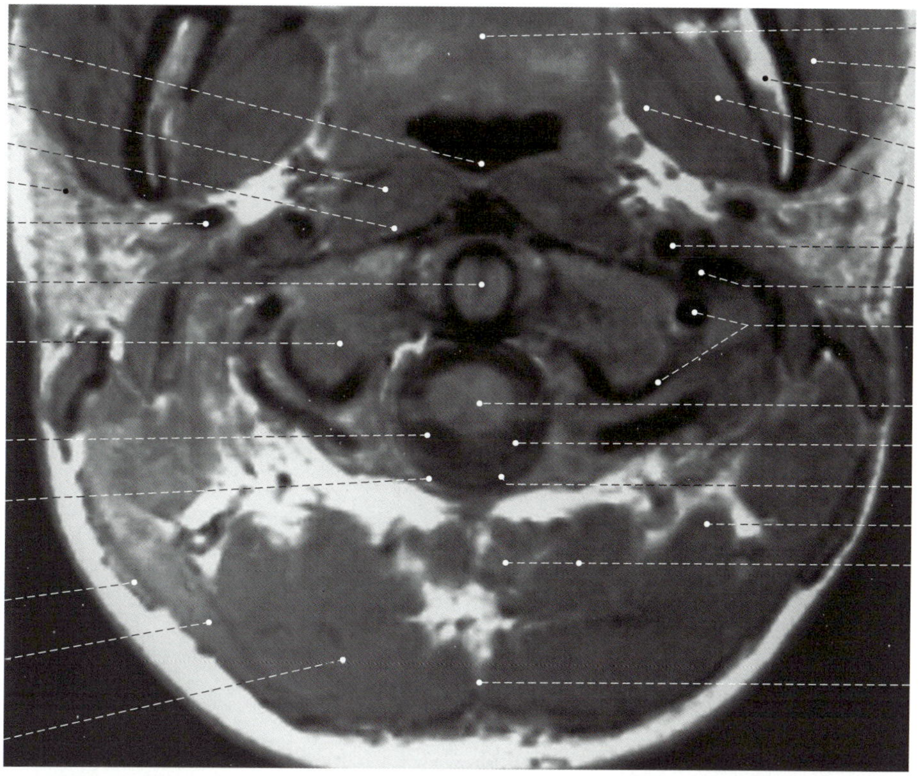

b

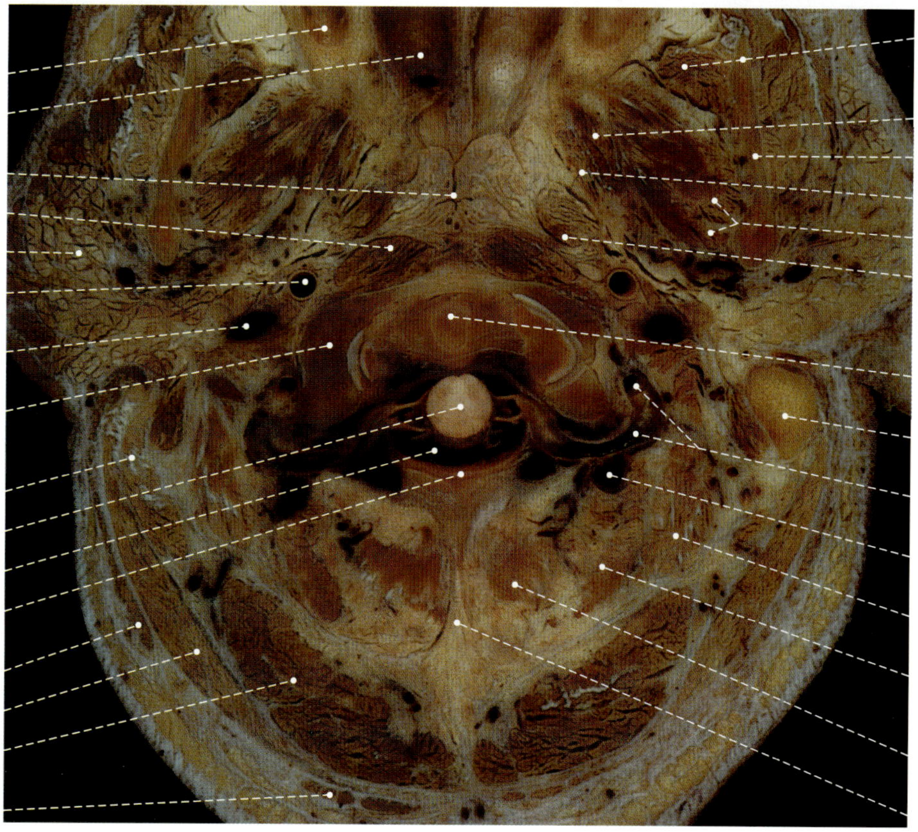

73 ⟨Collum⟩

73 Neck (90%)

Transverse sections through the neck at the level of the first cervical vertebra (C I, atlas)

a Magnetic resonance image (MRI, T_1-weighted), inferior aspect
b Anatomical section, superior aspect

73 Hals (90%)

Querschnitte durch den Hals in Höhe des ersten Halswirbels (C I, Atlas)

a Kernspintomogramm (MRI, T_1-gewichtet), Kaudalansicht
b Anatomischer Schnitt, Kranialansicht

a

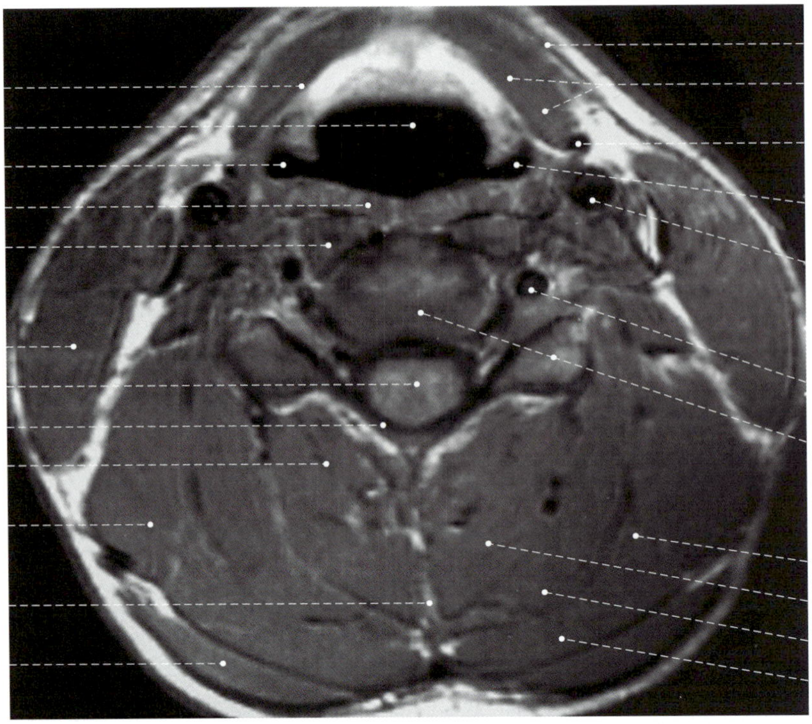

b

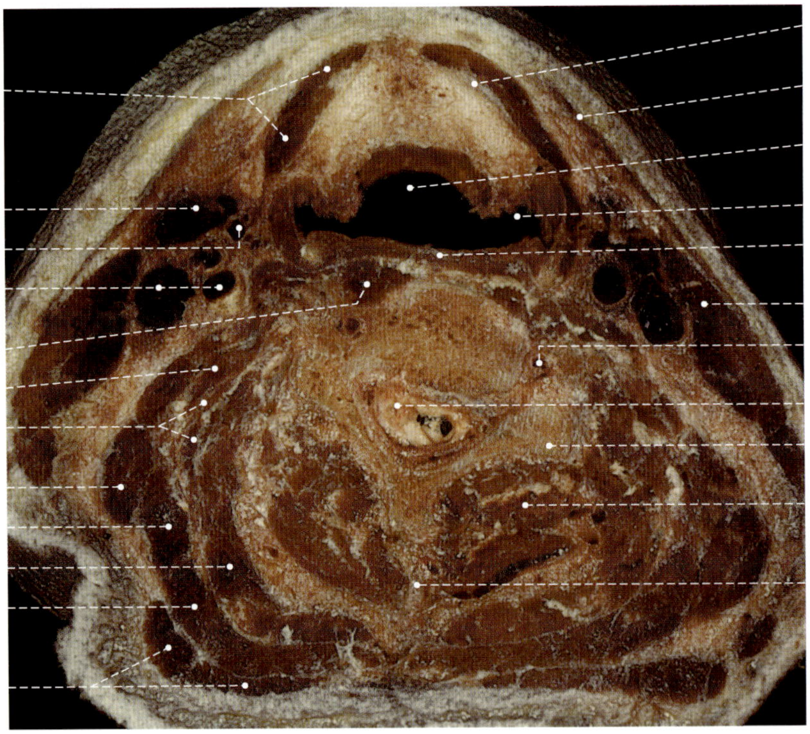

74 ⟨Collum⟩

74 Neck (90%)

Transverse sections through the neck at the level of the fifth cervical vertebra (C V), inferior aspect

a Magnetic resonance image (MRI, T_1-weighted)

b Anatomical section

74 Hals (90%)

Querschnitte durch den Hals in Höhe des fünften Halswirbels (C V), Kaudalansicht

a Kernspintomogramm (MRI, T_1-gewichtet)

b Anatomischer Schnitt

a

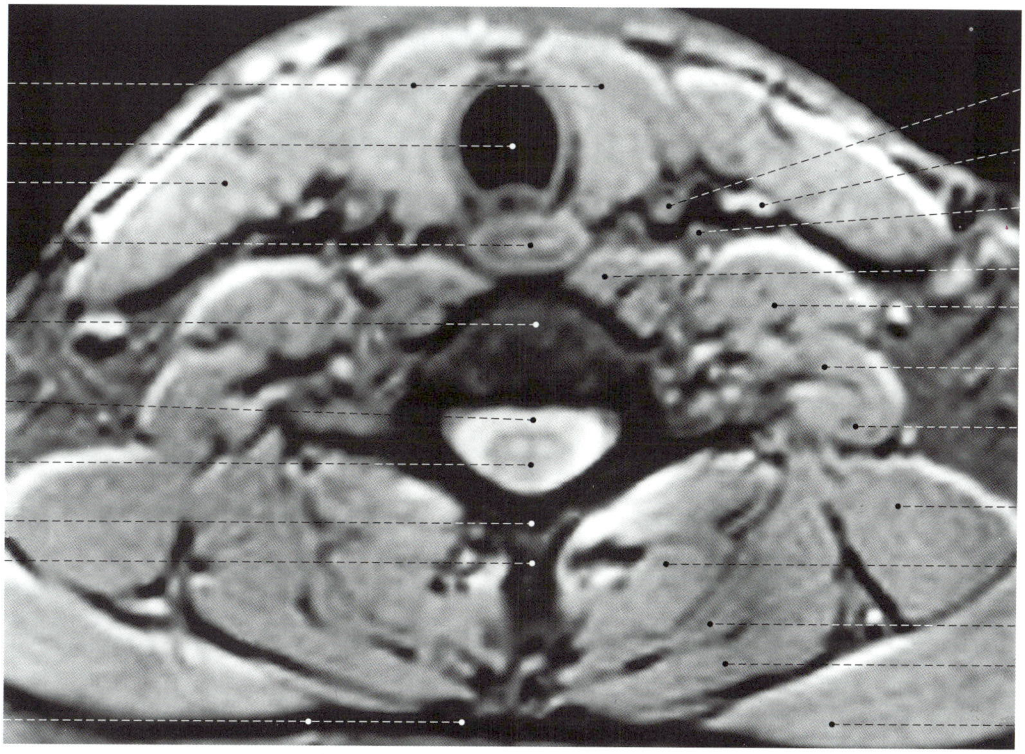

b

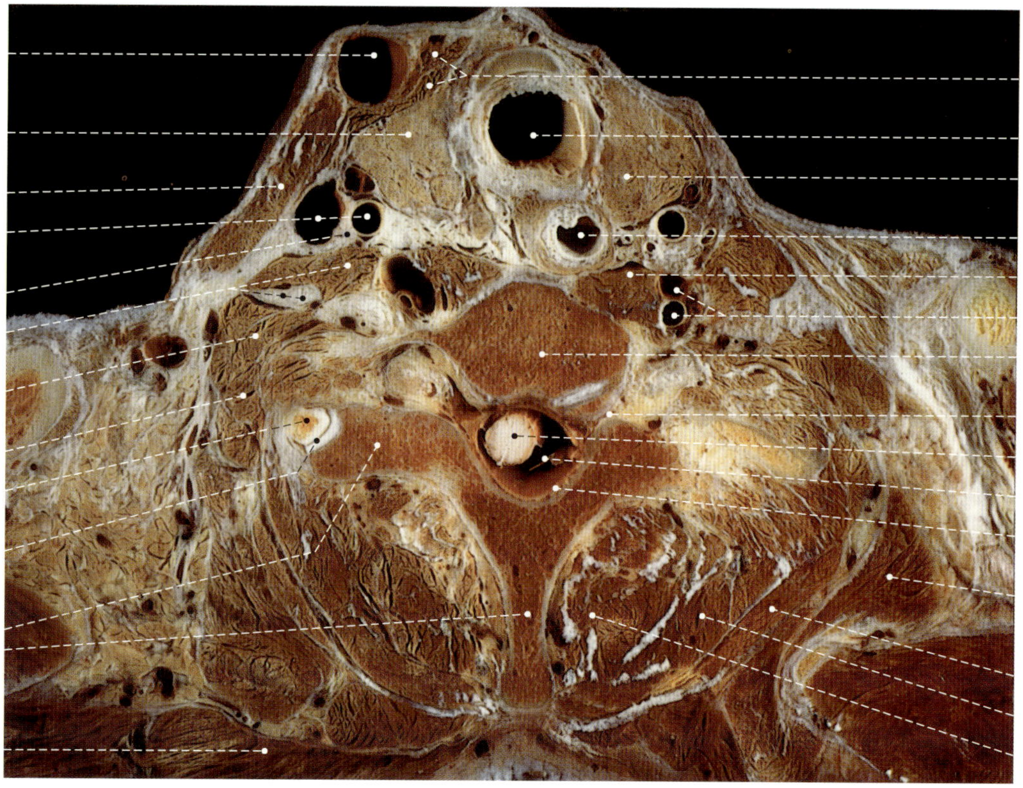

75 ⟨Collum⟩

75 Neck (90%)

Transverse sections through the neck at the level of the seventh cervical vertebra (C VII, a) or the first thoracic vertebra (Th I, b), respectively, inferior aspect

a Magnetic resonance image (MRI, T_2-weighted)
b Anatomical section

75 Hals (90%)

Querschnitte durch den Hals in Höhe des siebten Halswirbels (C VII, a) bzw. des ersten Brustwirbels (Th I, b), Kaudalansicht

a Kernspintomogramm (MRI, T_2-gewichtet)
b Anatomischer Schnitt

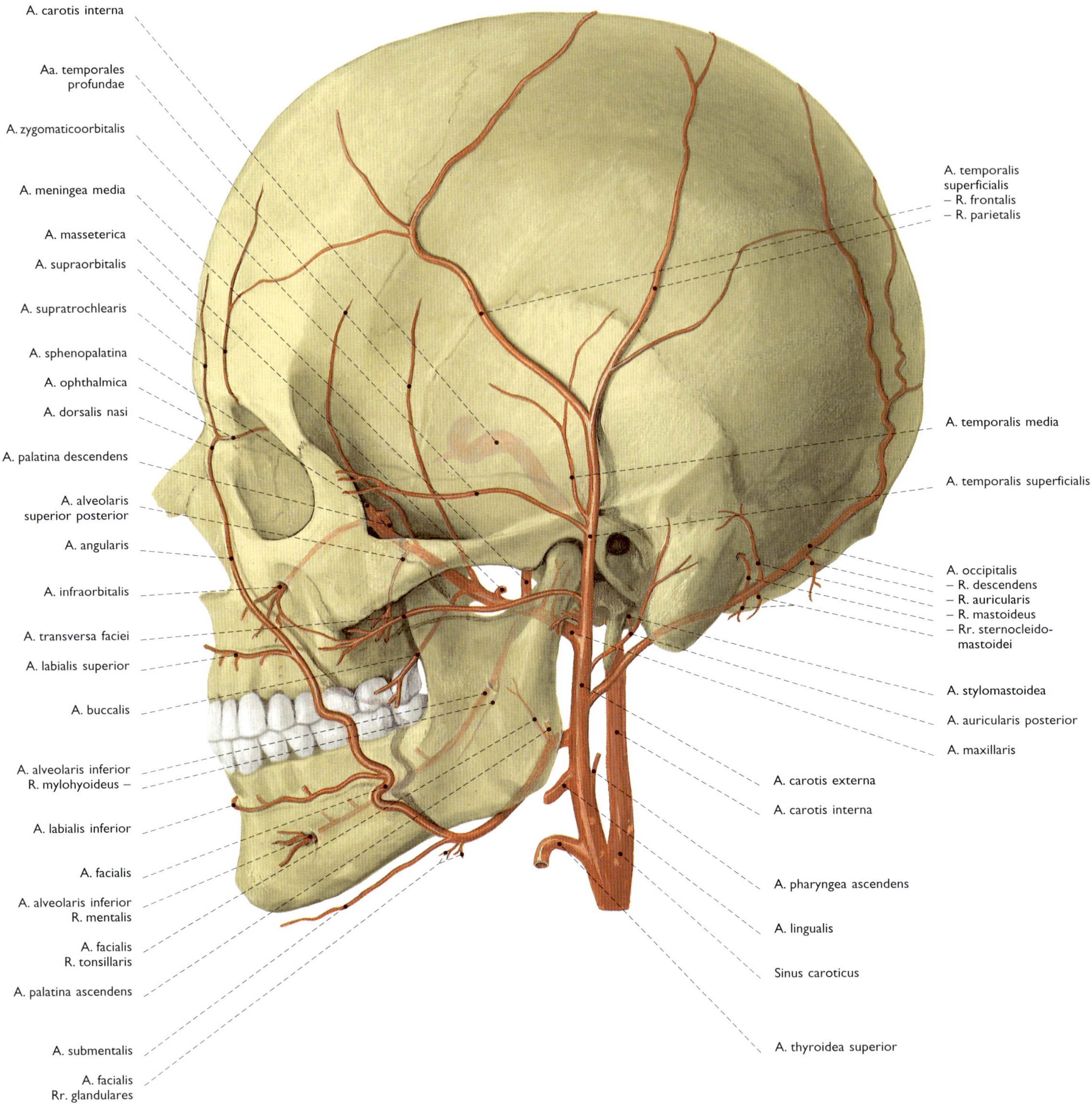

76 ⟨Arteriae capitis⟩

76 Arteries of the skull (75%)
Schematic representation, left lateral aspect

76 Kopfarterien (75%)
Schematische Darstellung, linke Seitenansicht

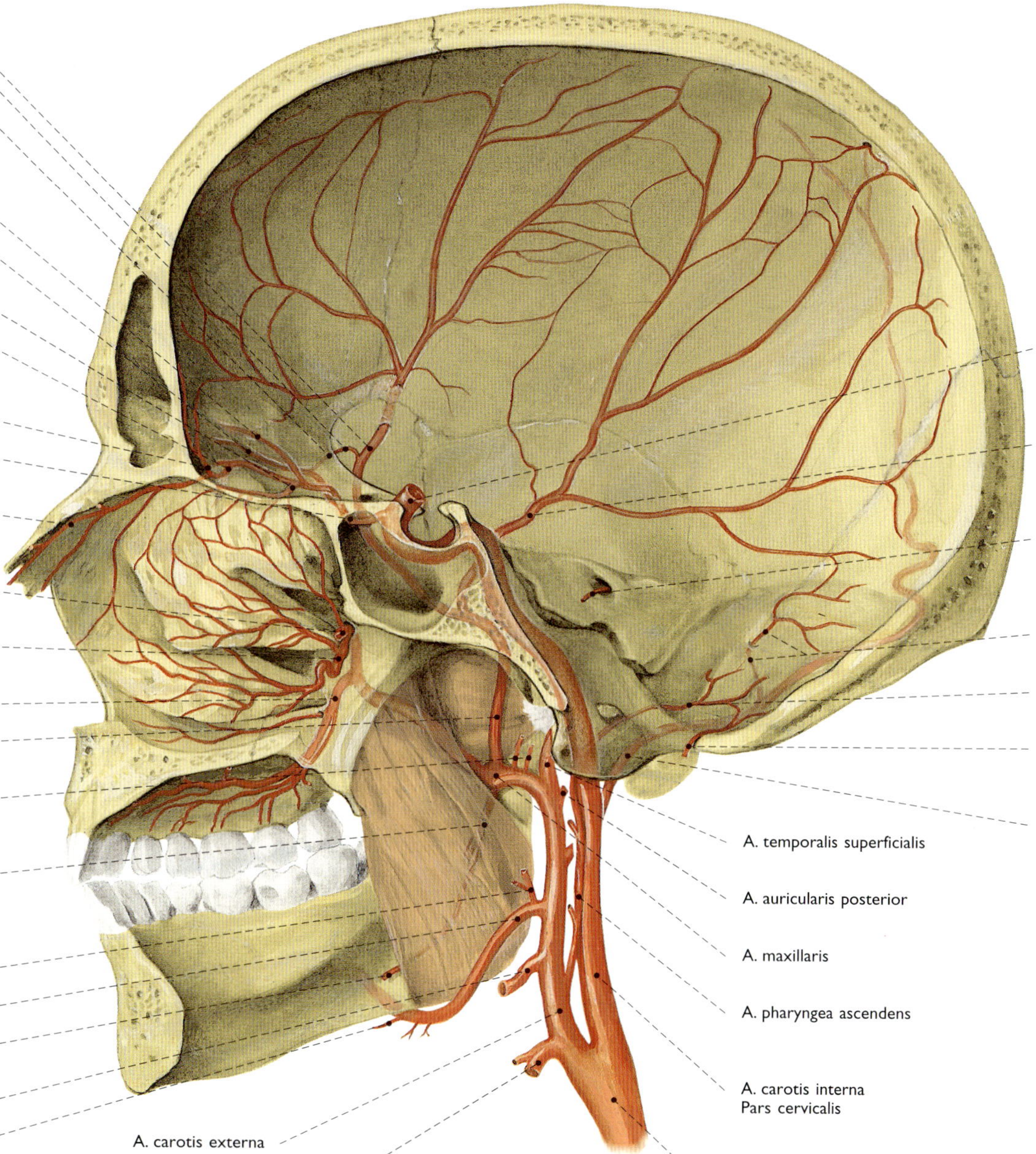

77 ⟨Arteriae capitis⟩

77 **Arteries of the skull** (75%)
Schematic representation, median section, medial aspect of the right half of the skull

77 **Kopfarterien** (75%)
Schematische Darstellung, Medianschnitt, Medialansicht der rechten Kopfhälfte

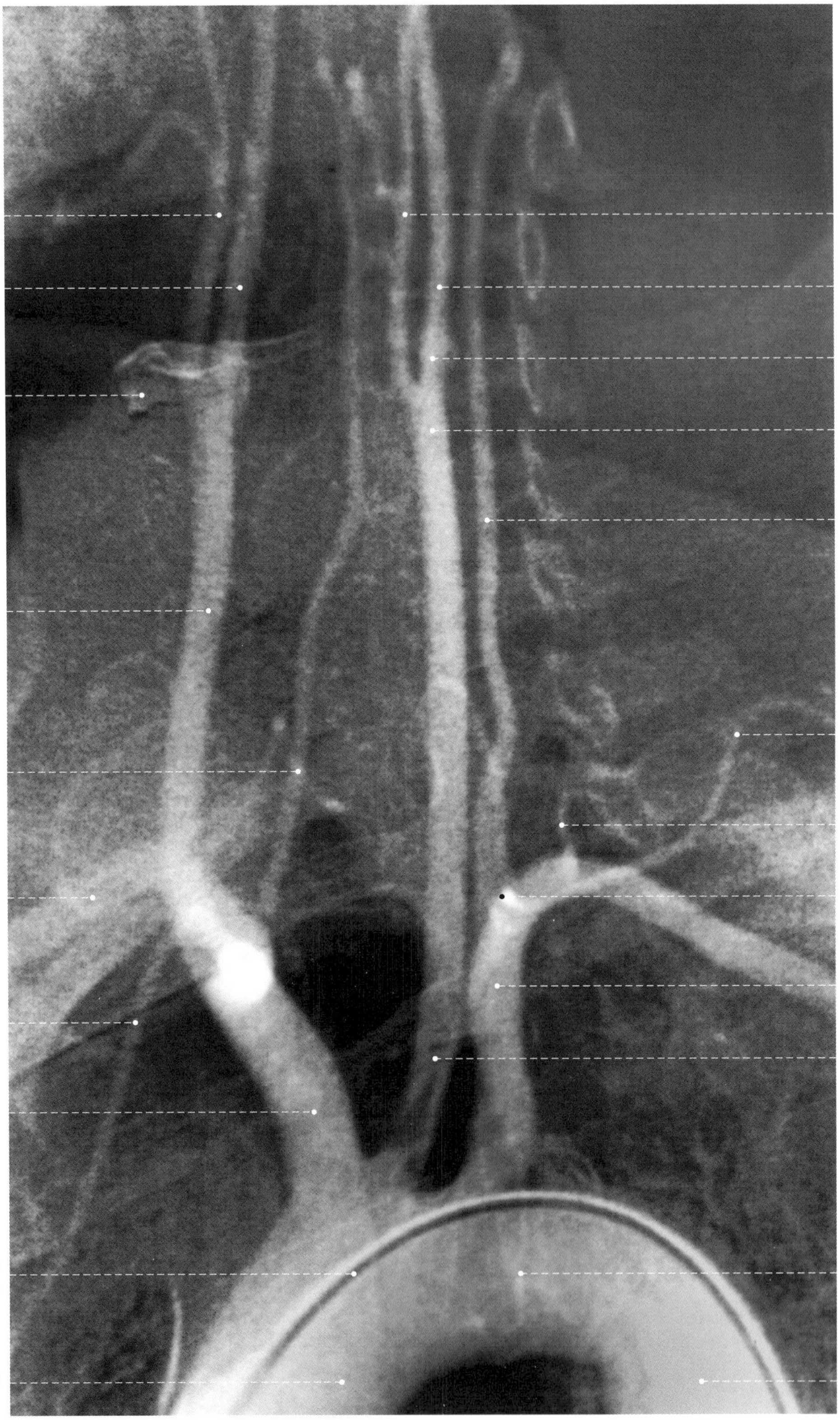

78 ⟨Arteriae colli et capitis⟩

78 Arteries of the neck and head (90%)
Arteriogram, left anterior oblique view (LAO projection)

78 Hals- und Kopfarterien (90%)
Arteriogramm, Schrägansicht von links vorne (2. Schrägdurchmesser, "Boxerstellung")

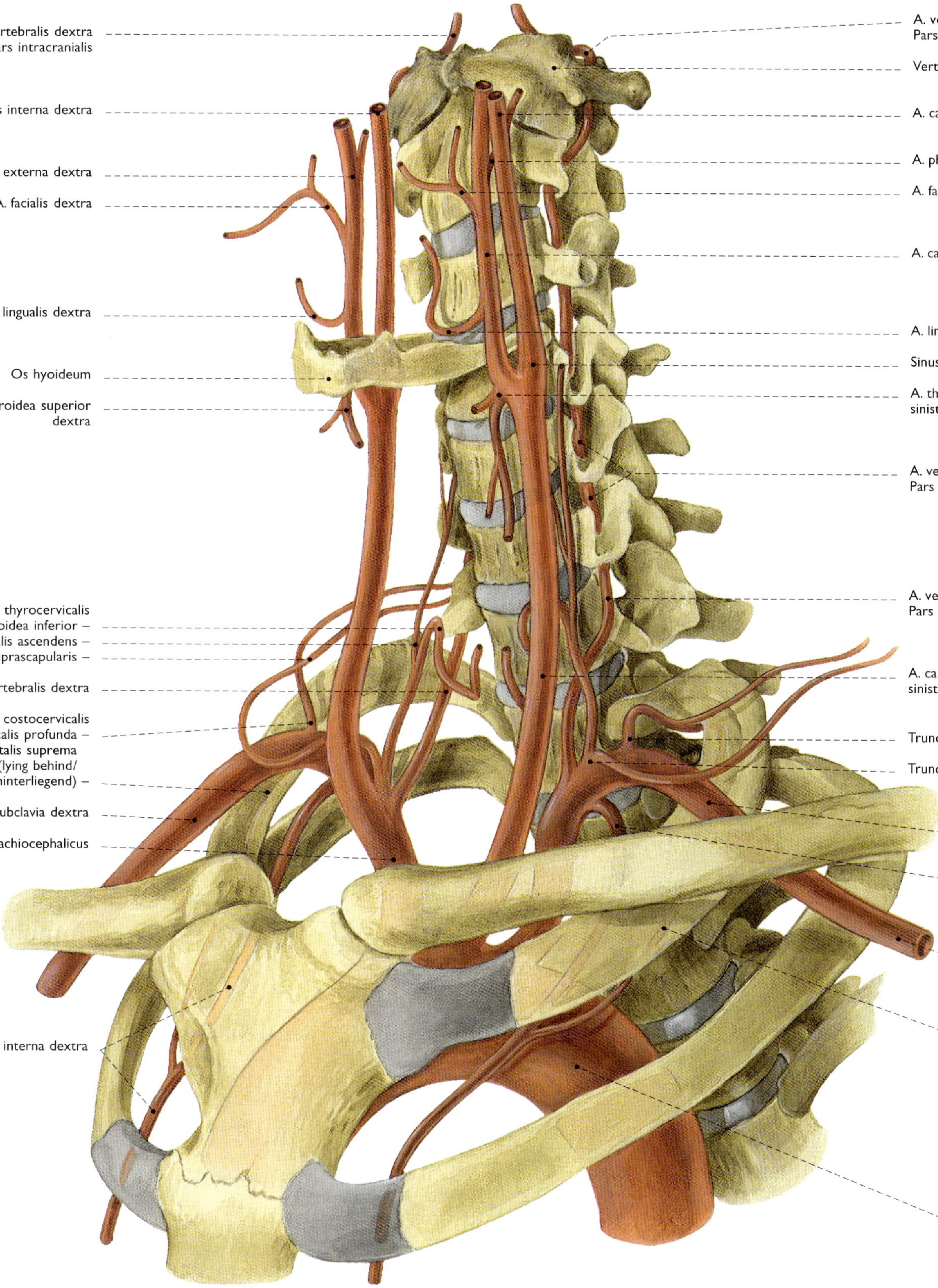

79 ⟨Arteriae colli et capitis⟩

79 Arteries of the neck and head (90%)
Left anterior oblique view (LAO projection)

79 Hals- und Kopfarterien (90%)
Schrägansicht von links vorne
(2. Schrägdurchmesser, "Boxerstellung")

a

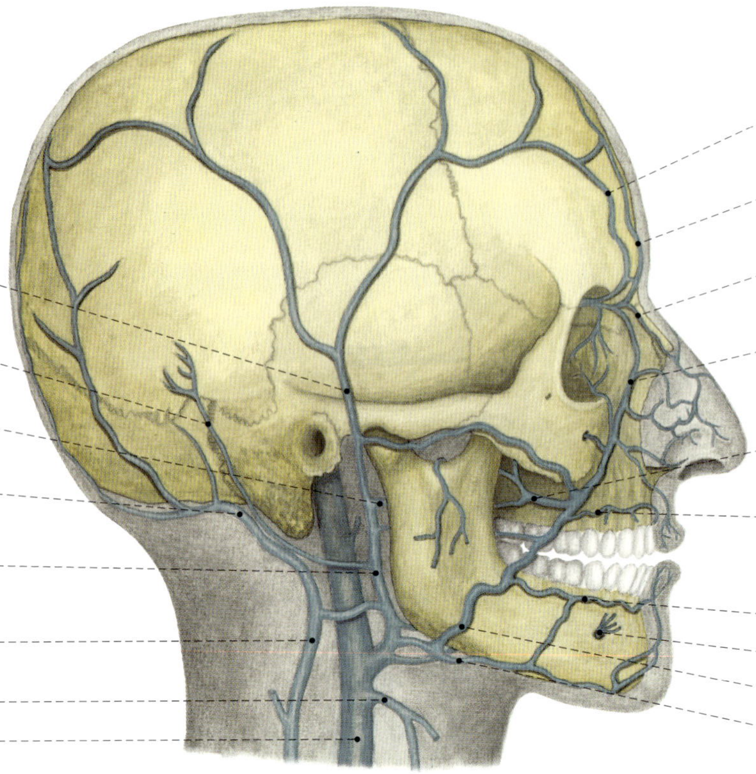

b

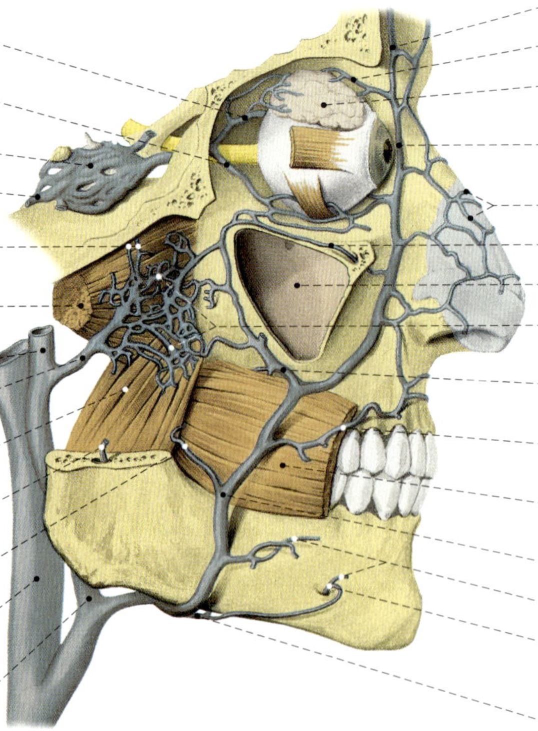

80 ⟨Venae capitis⟩

80 Facial veins

Right lateral aspect

a Superficial veins (50%)

b Deep veins (65%)

80 Venen des Gesichts

Rechte Seitenansicht

a Oberflächliche Venen (50%)

b Tiefe Venen (65%)

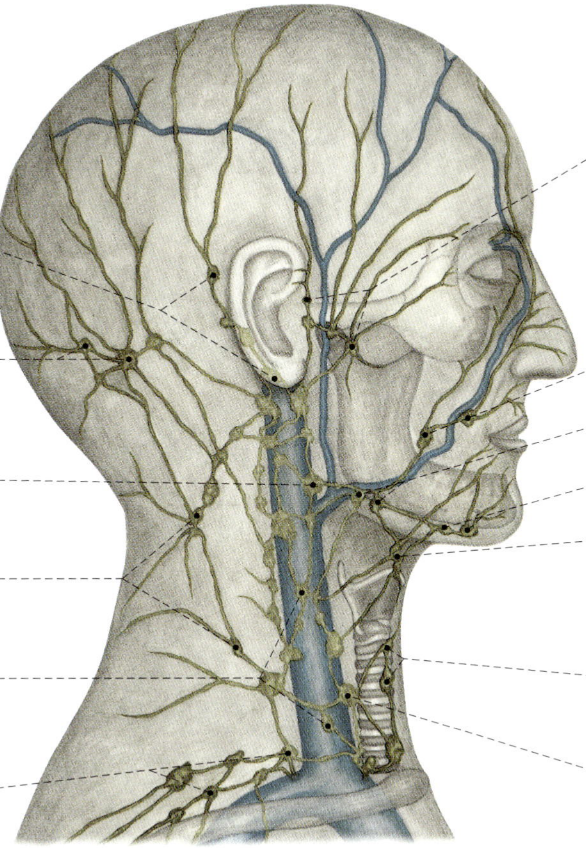

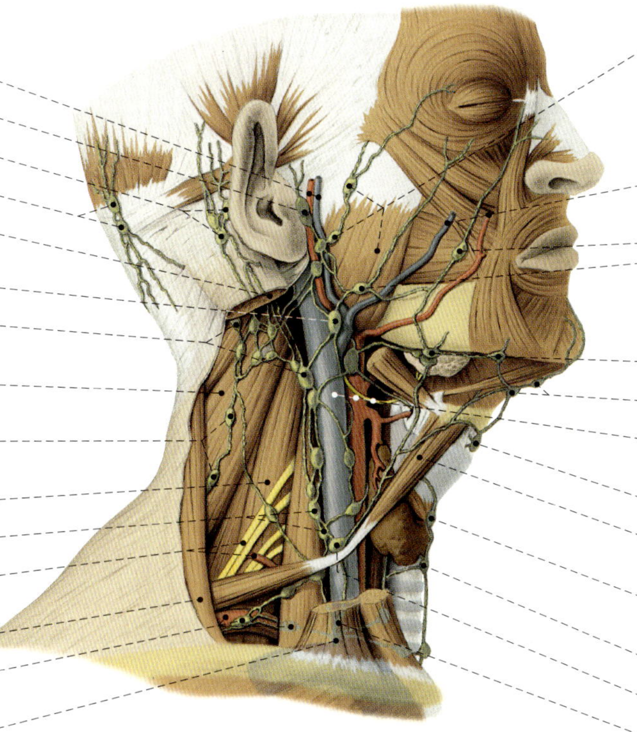

81 ⟨Nodi lymphatici capitis et colli⟩

81 Lymph vessels and lymph nodes of the head and neck (40%)
Schematic representations, right lateral aspect

81 Lymphgefässe und Lymphknoten von Kopf und Hals (40%)
Schematische Darstellungen, rechte Seitenansicht

a

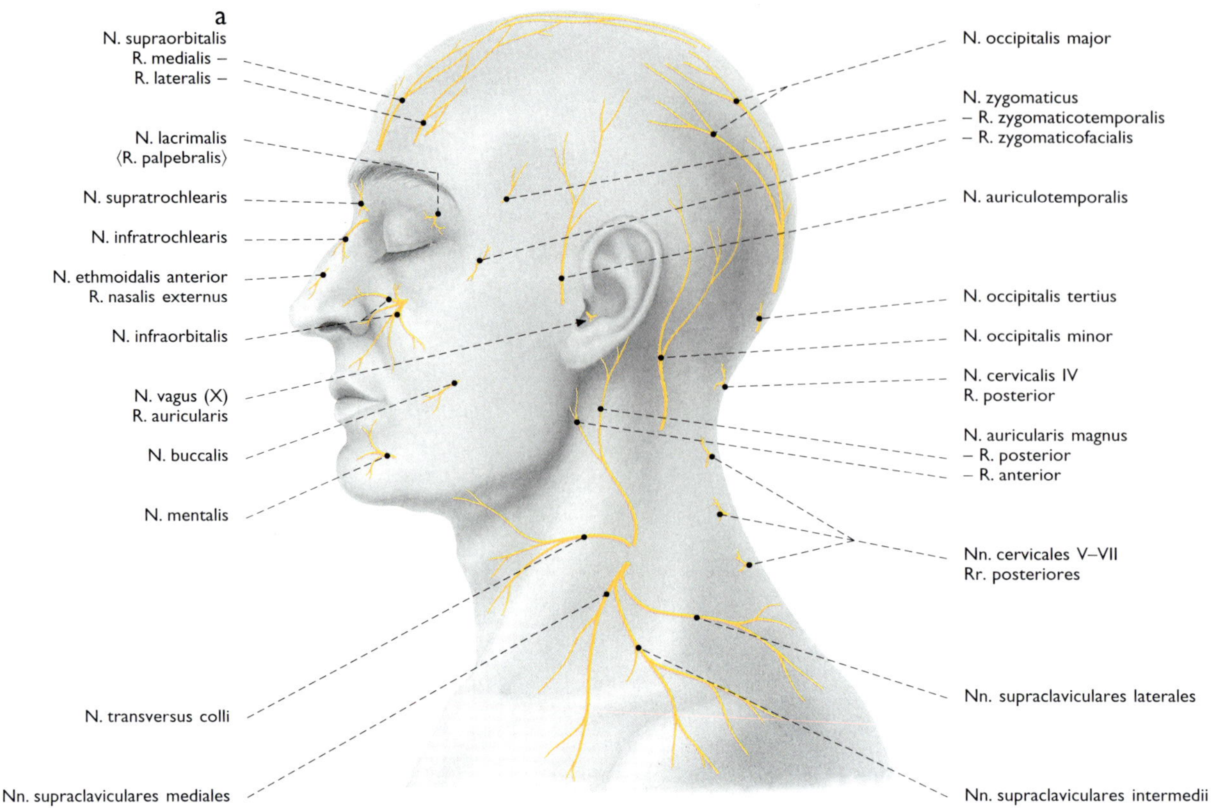

b

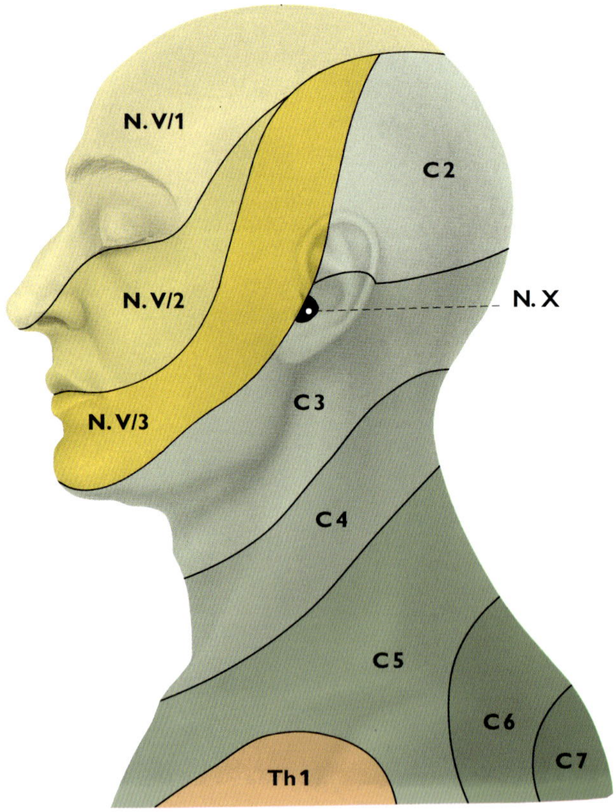

N. V = N. trigeminus
N. V/1 = N. ophthalmicus
N. V/2 = N. maxillaris
N. V/3 = N. mandibularis
N. X = N. vagus

82 ⟨Nervi cutanei capitis et colli⟩

82 Cutaneous and segmental innervation of the head and neck (30%)

Schematic representations, left lateral aspect

a Cutaneous nerves
b Segmental innervation

82 Hautnerven und segmentale Innervation an Kopf und Hals (30%)

Schematische Darstellungen, linke Seitenansicht

a Hautnerven
b Segmentale Innervation

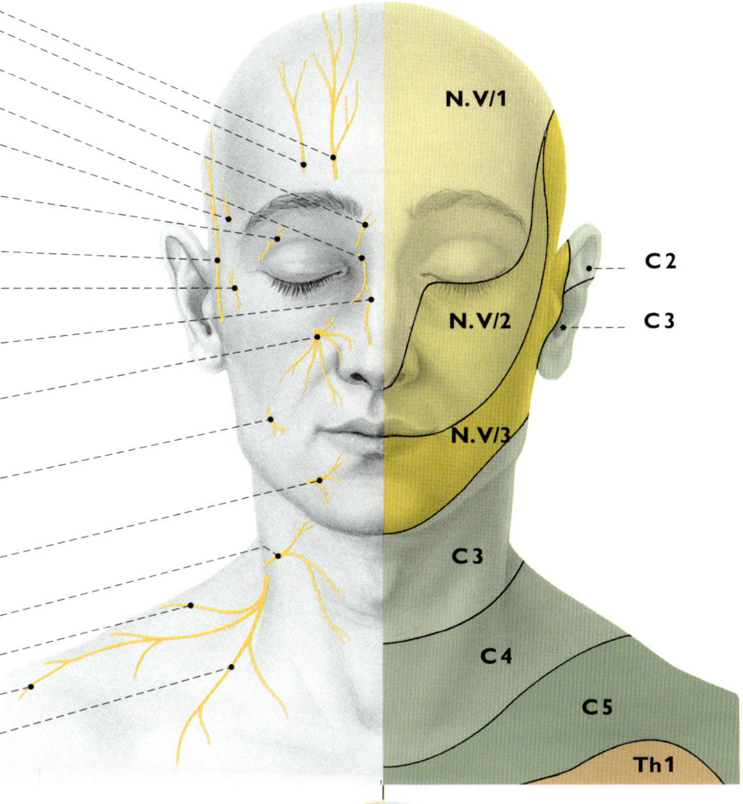

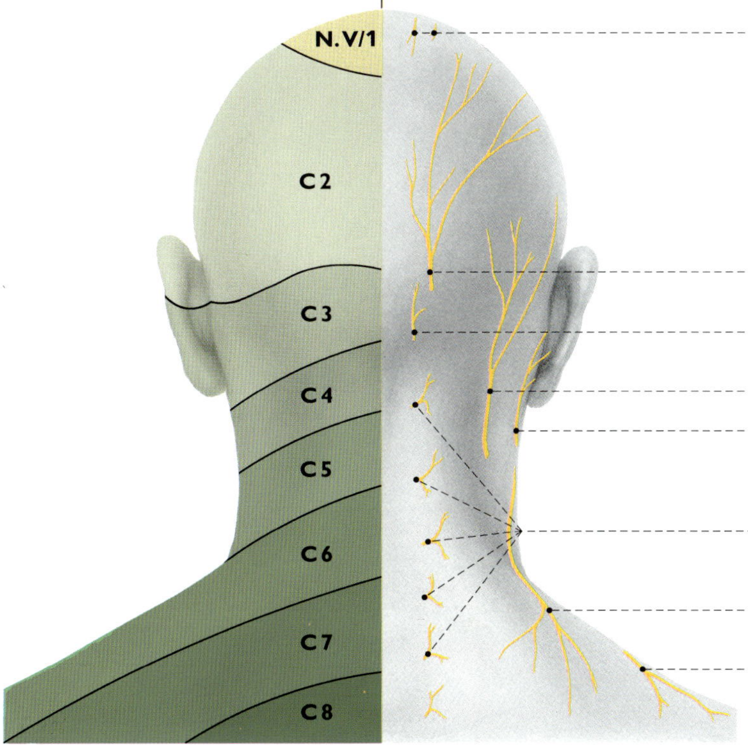

83 ⟨Nervi cutanei capitis et colli⟩

83 Cutaneous and segmental innervation of the head and neck (30%)

Schematic representations

a Frontal aspect

b Dorsal aspect

83 Hautnerven und segmentale Innervation an Kopf und Hals (30%)

Schematische Darstellungen

a Frontalansicht

b Dorsalansicht

a

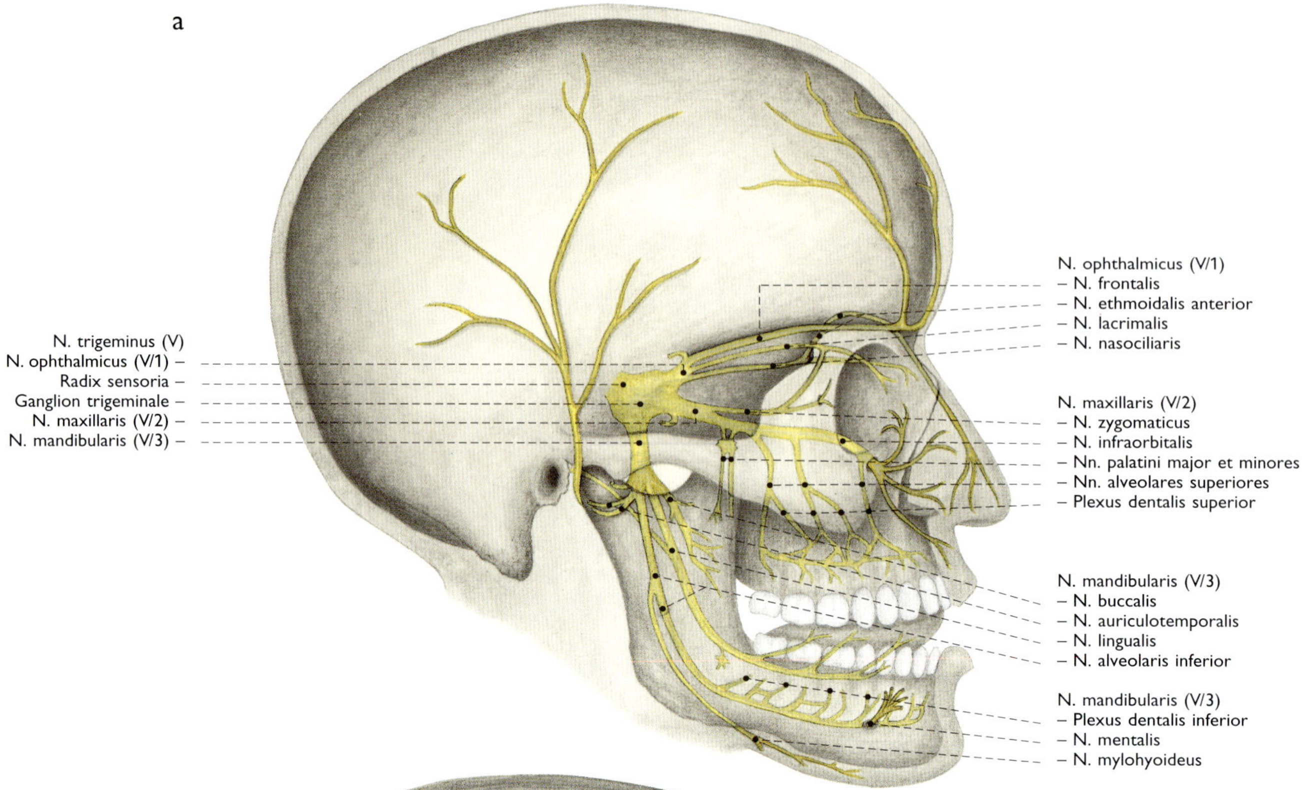

b

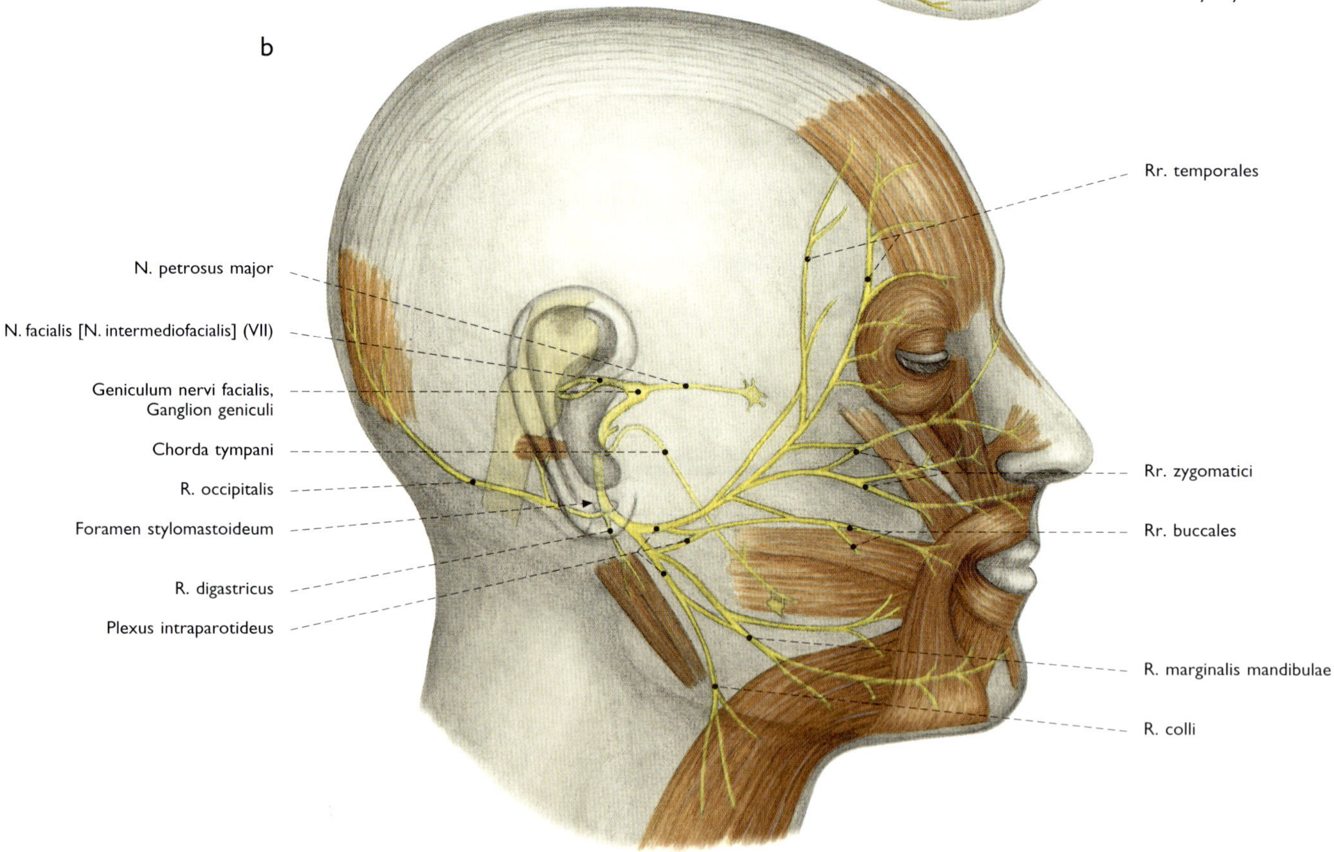

84 ⟨Nervi capitis⟩

84 Nerves of the head (50%)

Schematic representations, right lateral aspect

- a Ramification of the trigeminal nerve (N. V) in the deep facial region
- b Ramification of the facial nerve (N. VII) in the superficial facial region

84 Kopfnerven (50%)

Schematische Darstellungen, rechte Seitenansicht

- a Verzweigung des Nervus trigeminus (N. V) in der tiefen Gesichtsregion
- b Verzweigung des Nervus facialis (N. VII) in der oberflächlichen Gesichtsregion

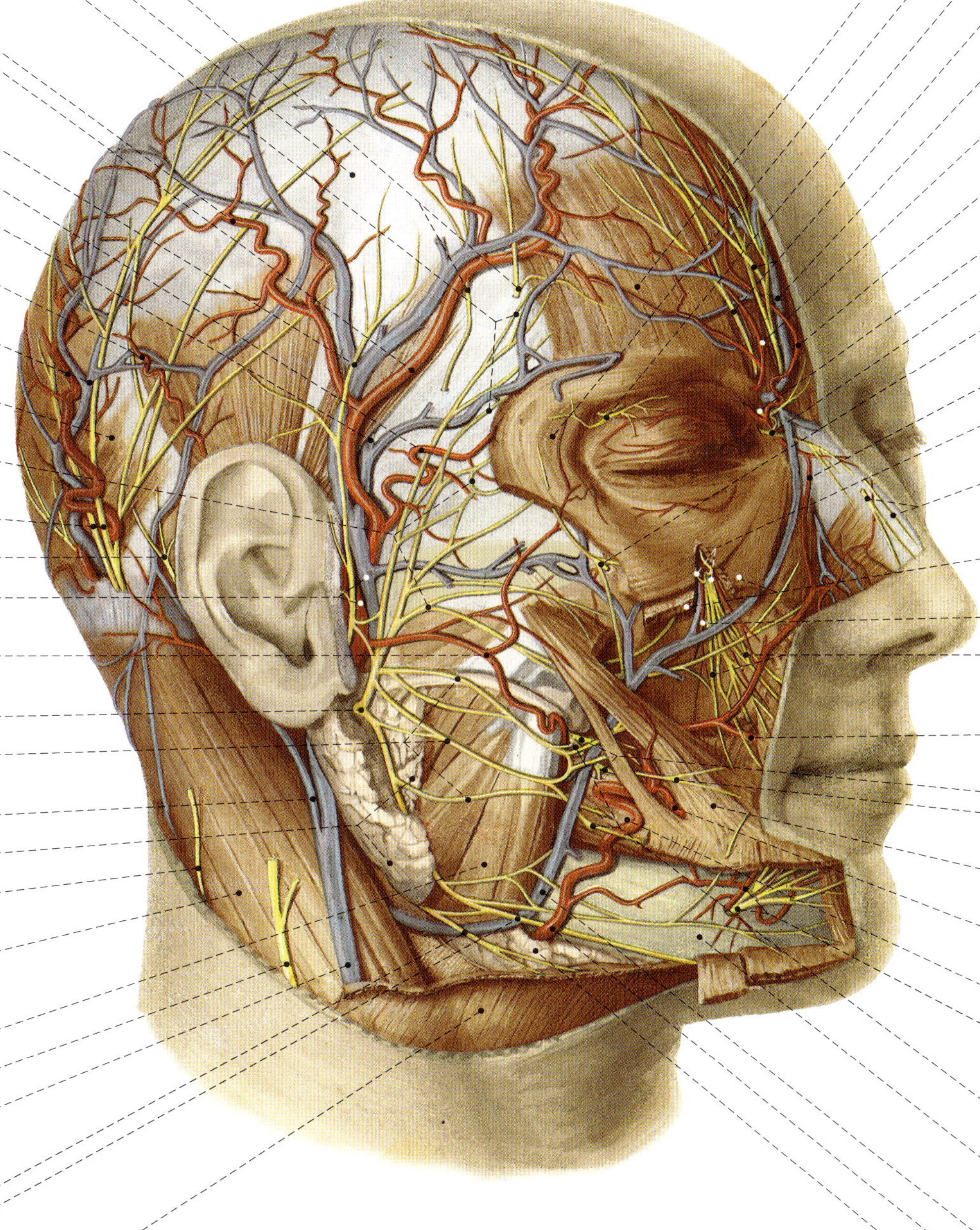

85 Regio facialis superficialis

85 Superficial blood vessels and nerves of the head (60%)
Right lateral aspect

85 Oberflächliche Blutgefässe und Nerven des Kopfes (60%)
Rechte Seitenansicht

a

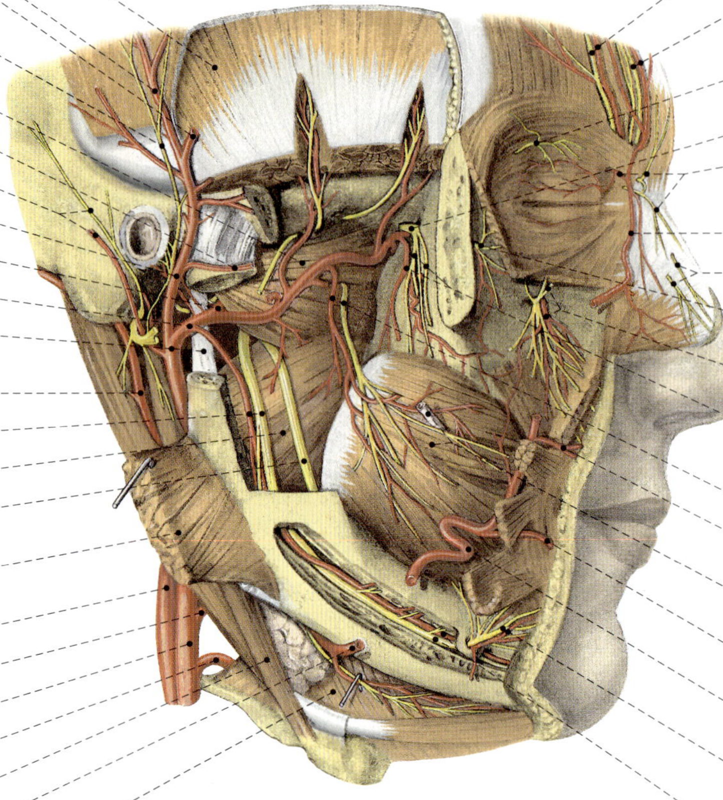

b

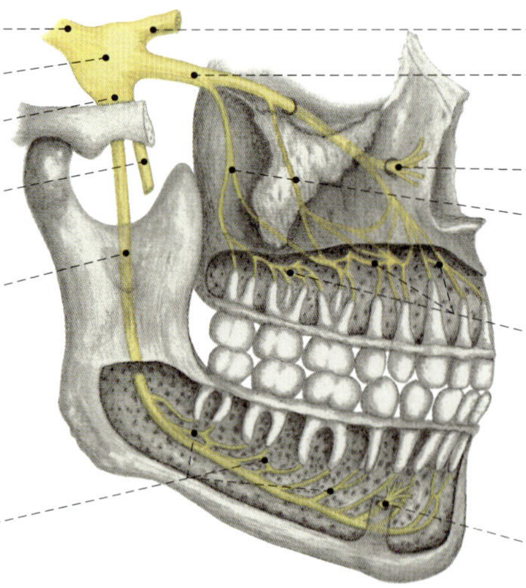

86 Regio facialis profunda

86 Arteries and nerves of the deep lateral facial region (65%)

Right lateral aspect

a The zygomatic arch and the ramus of the mandible were removed, the mandibular canal was opened.

b Innervation of the teeth in the upper and lower jaws by branches of the trigeminal nerve (N. V), schematic representation

86 Arterien und Nerven der tiefen seitlichen Gesichtsregion (65%)

Rechte Seitenansicht

a Jochbogen und Unterkieferast sind entfernt, der Mandibularkanal ist eröffnet.

b Innervation der Ober- und Unterkieferzähne durch Äste des Nervus trigeminus (N. V), schematische Darstellung

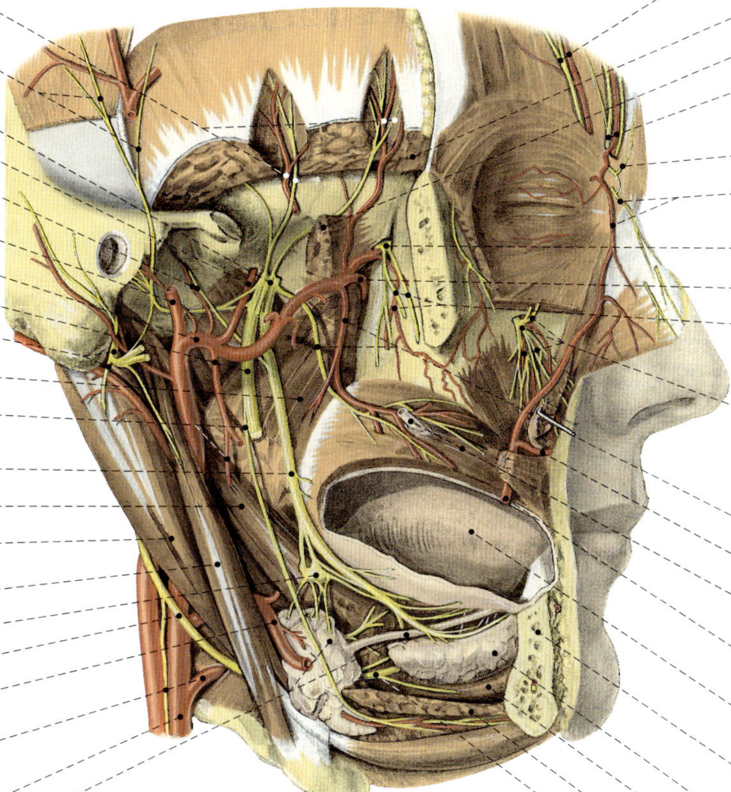

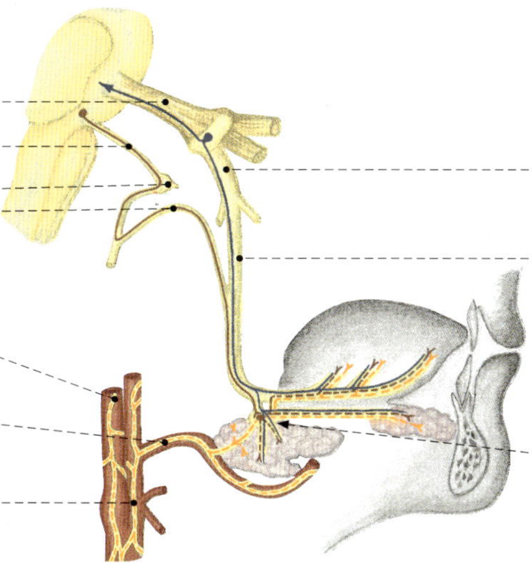

87 Regio facialis profunda

87 Arteries and nerves of the deep lateral facial region

Right lateral aspect

a The zygomatic arch, the right half of the mandible and parts of the lateral pterygoid muscle were removed (65%).

b Submandibular ganglion and innervation of the submandibular and sublingual glands (50%), schematic representation

87 Arterien und Nerven der tiefen seitlichen Gesichtsregion

Rechte Seitenansicht

a Jochbogen, rechte Unterkieferhälfte und der ansatznahe Teil des Musculus pterygoideus lateralis sind abgetragen (65%).

b Ganglion submandibulare und Innervation der Unterkiefer- und Unterzungenspeicheldrüse (50%), schematische Darstellung

a

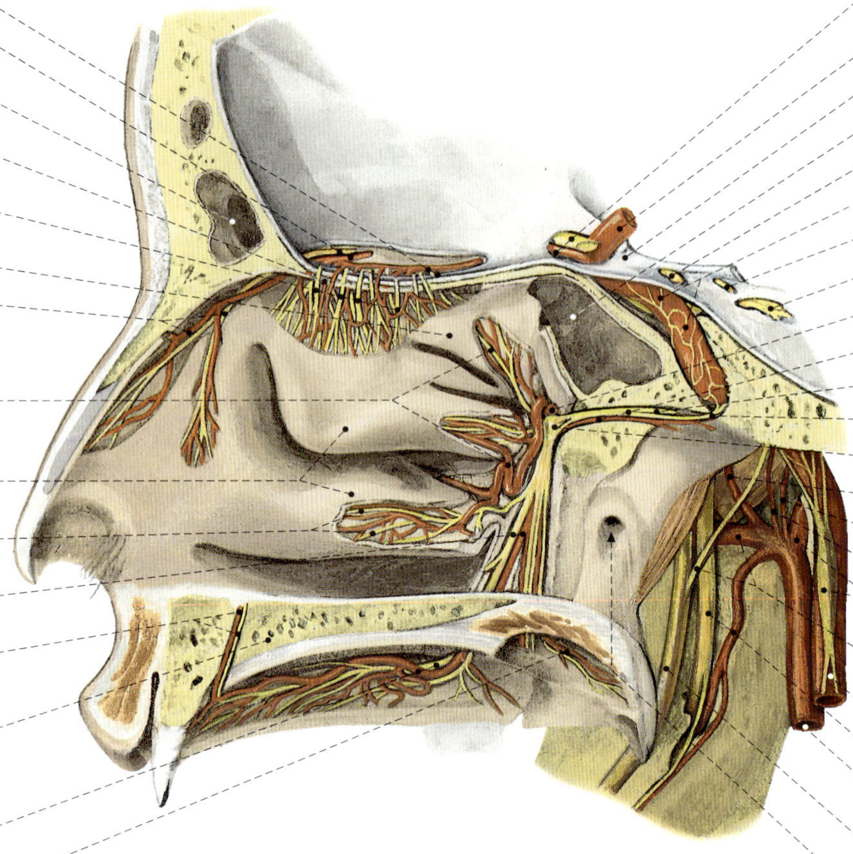

b

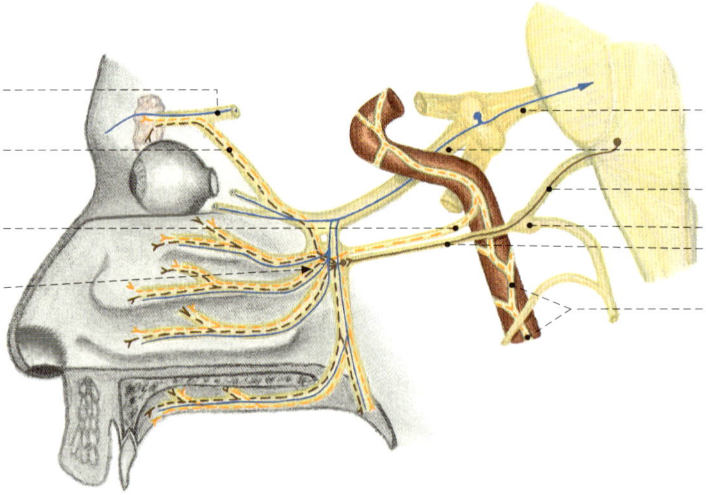

88 Regio facialis profunda

88 Arteries and nerves of the deep medial facial region

Medial aspect

a Lateral wall of the nasal cavity, palate, infratemporal and pterygopalatine fossae (90%)

b Pterygopalatine ganglion and innervation of the lacrimal gland and the mucous membranes of the nose and the palate (50%), schematic representation

88 Arterien und Nerven der tiefen medialen Gesichtsregion

Medialansicht

a Seitliche Wand der Nasenhöhle, Gaumen, Unterschläfen- und Flügelgaumengrube (90%)

b Ganglion pterygopalatinum und Innervation der Tränendrüse sowie der Nasen- und Gaumenschleimhaut (50%), schematische Darstellung

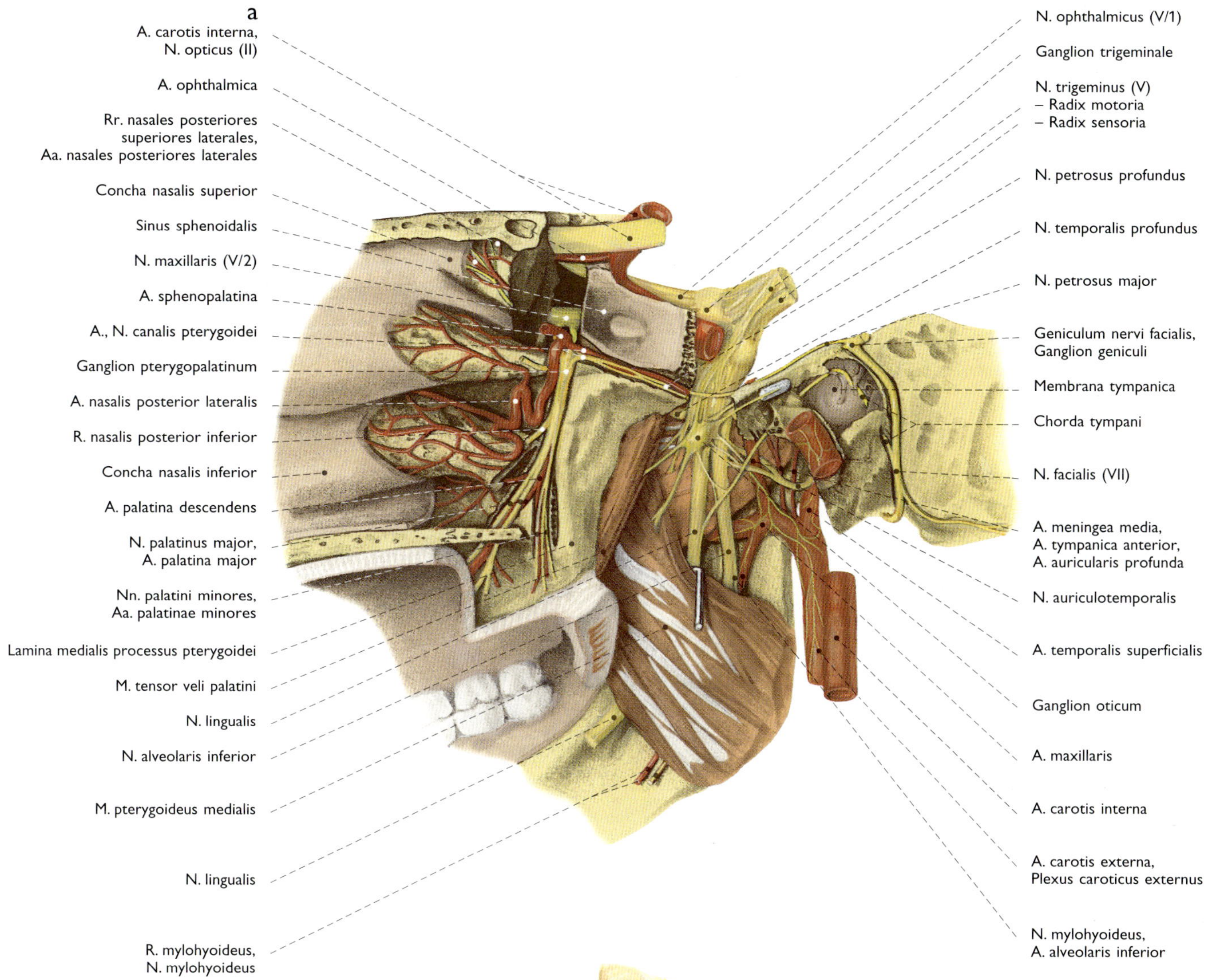

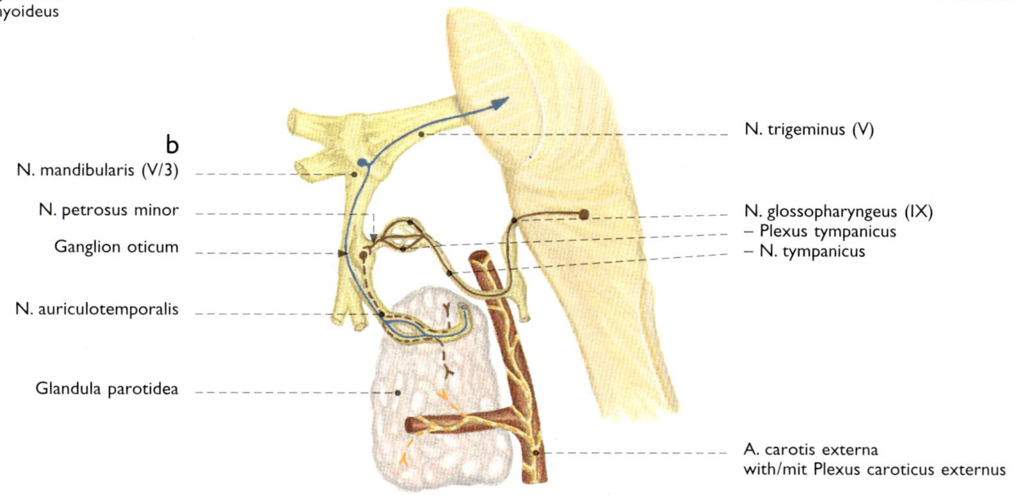

89 Regio facialis profunda

89 Arteries and nerves of the deep medial facial region

Medial aspect

a Lateral wall of the nasal cavity, infratemporal and pterygopalatine fossae.
The facial canal was opened (90%).

b Otic ganglion and innervation of the parotid gland (50%), schematic representation

89 Arterien und Nerven der tiefen medialen Gesichtsregion

Medialansicht

a Seitliche Wand der Nasenhöhle, Unterschläfen- und Flügelgaumengrube. Der Fazialiskanal ist eröffnet (90%).

b Ganglion oticum und Innervation der Ohrspeicheldrüse (50%), schematische Darstellung

a

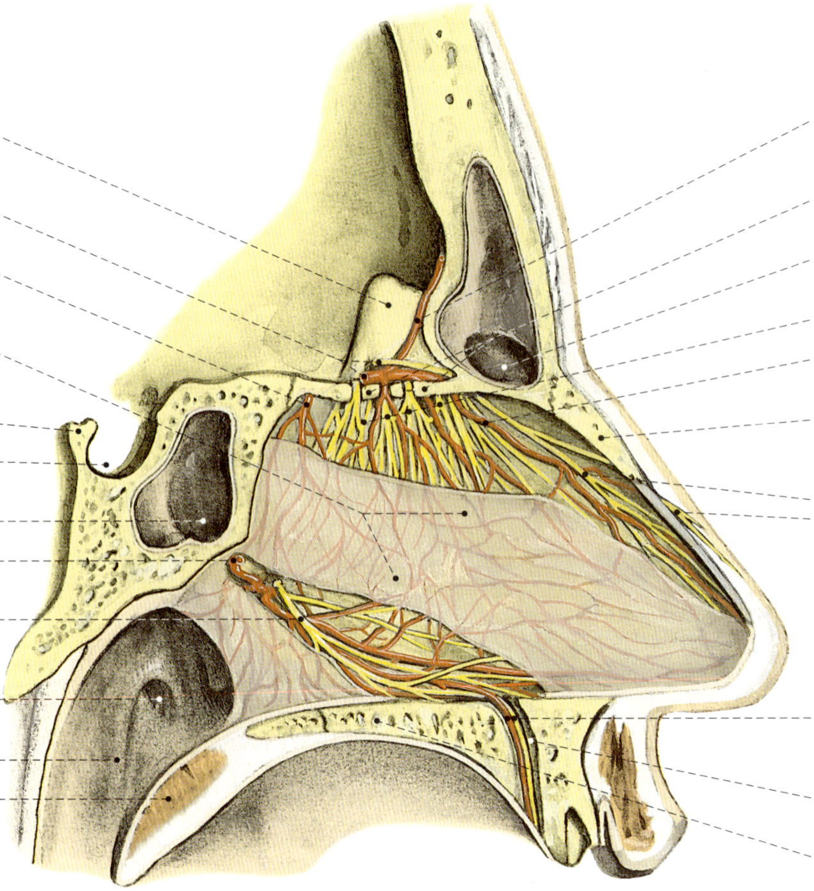

b c

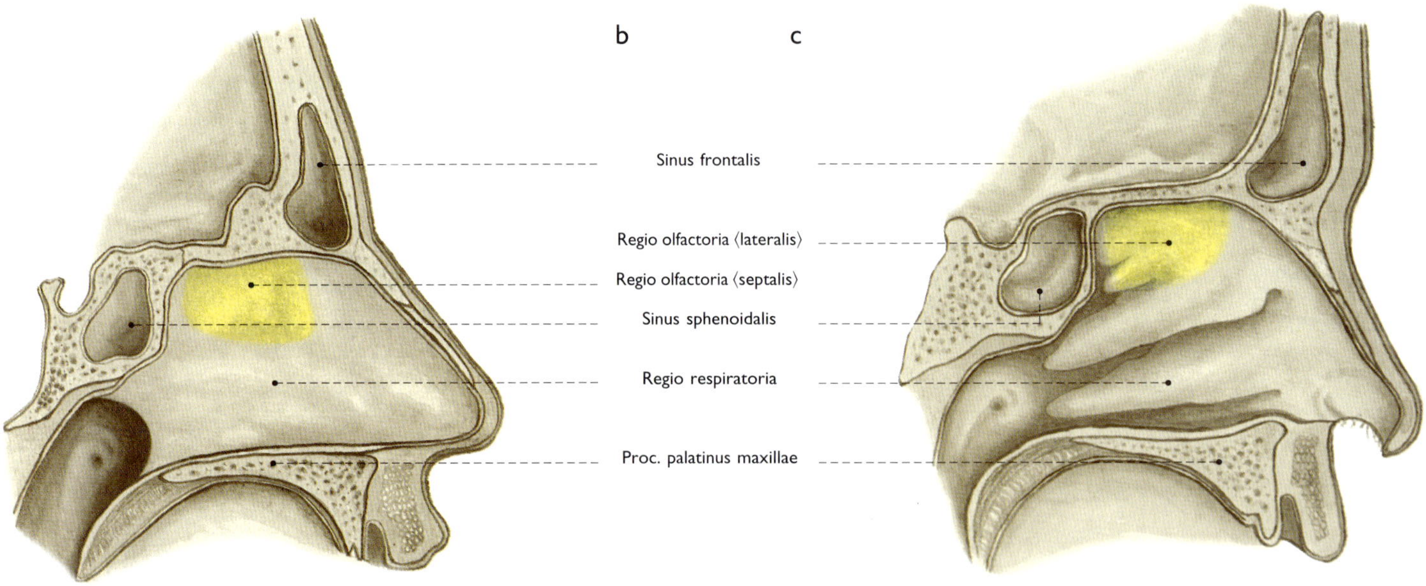

90 Regio facialis profunda

90 Arteries and nerves of the deep medial facial region

Medial aspect

a Nasal septum, sagittal section slightly to the right of the median plane. The nasal mucous membrane was partially removed (90%).

b, c Olfactory regions (50%), marked by yellow color,

b at the nasal septum

c at the lateral wall of the nasal cavity

90 Arterien und Nerven der tiefen medialen Gesichtsregion

Medialansicht

a Nasenscheidewand, Sagittalschnitt etwas rechts der Medianebene. Die Nasenschleimhaut ist teilweise abgetragen (90%).

b, c Riechregionen (50%), gelb hervorgehoben,

b am Nasenseptum

c an der Seitenwand der Nasenhöhle

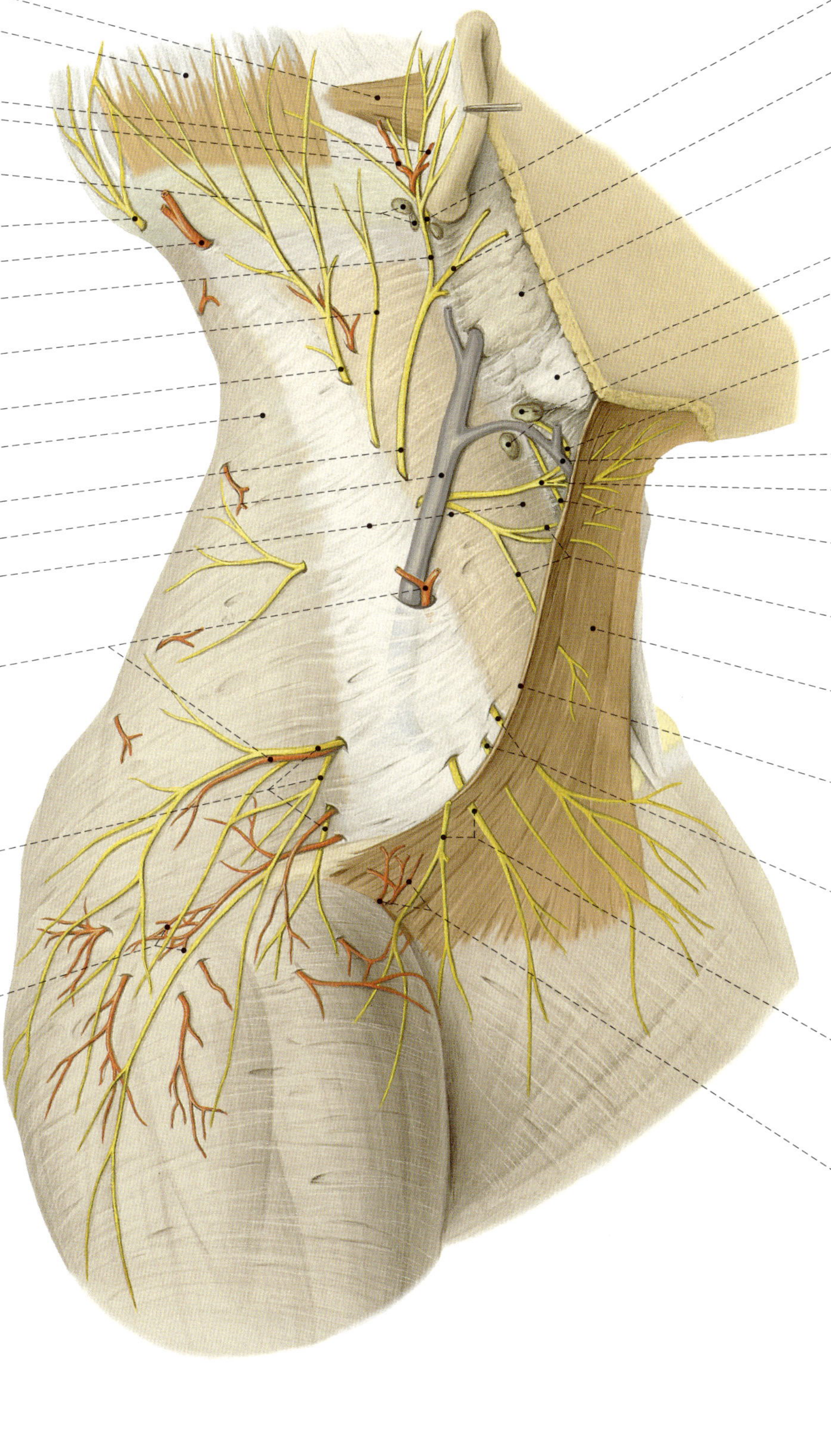

91 ⟨Collum⟩

91 Subcutaneous blood vessels and nerves of the neck region (50%)
Right lateral aspect

91 Subkutane Blutgefässe und Nerven im Halsbereich (50%)
Rechte Seitenansicht

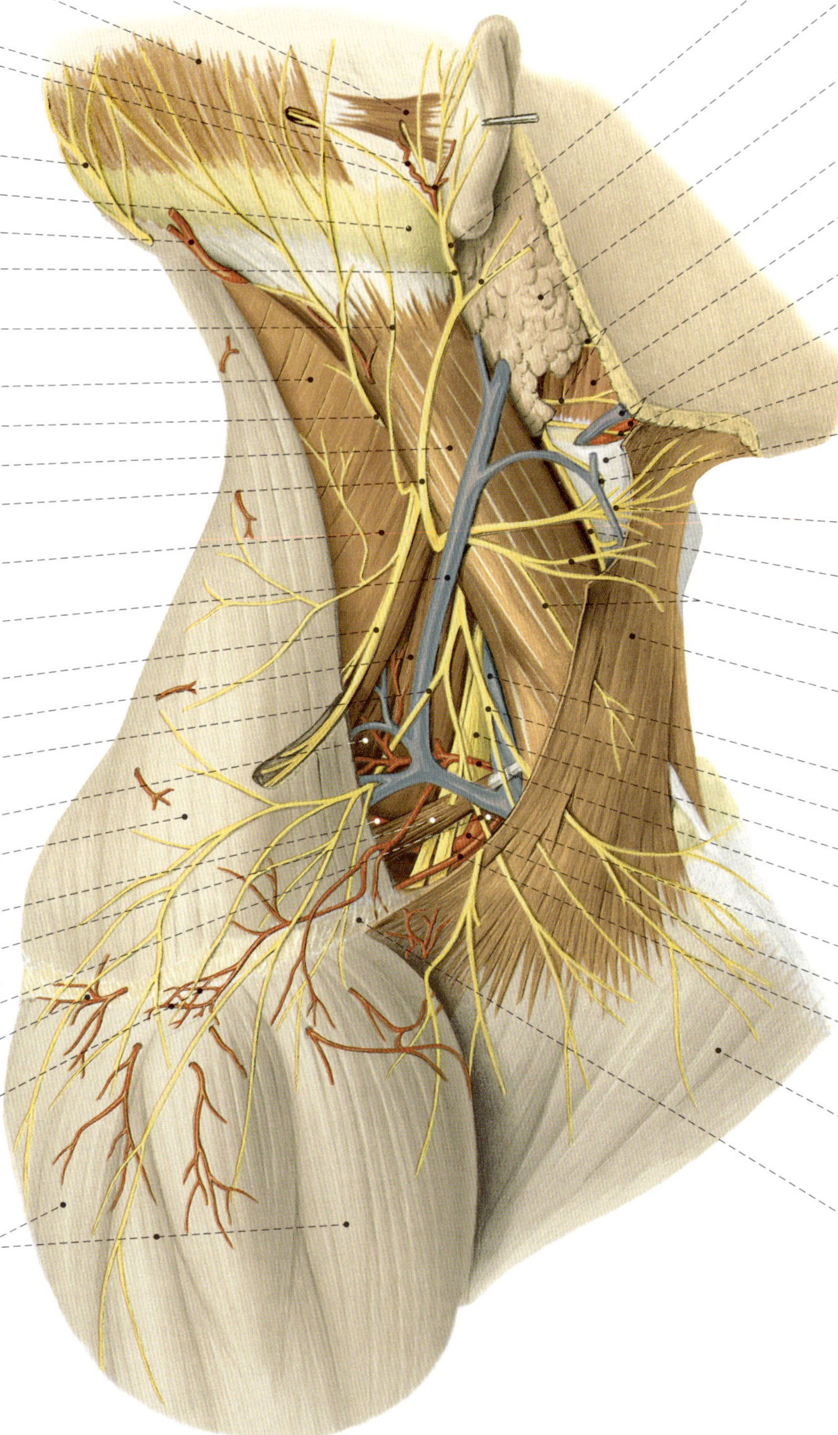

92 ⟨Collum⟩

92 Superficial blood vessels and nerves of the neck region (50%)
Right lateral aspect

92 Oberflächliche Blutgefässe und Nerven im Halsbereich (50%)
Rechte Seitenansicht

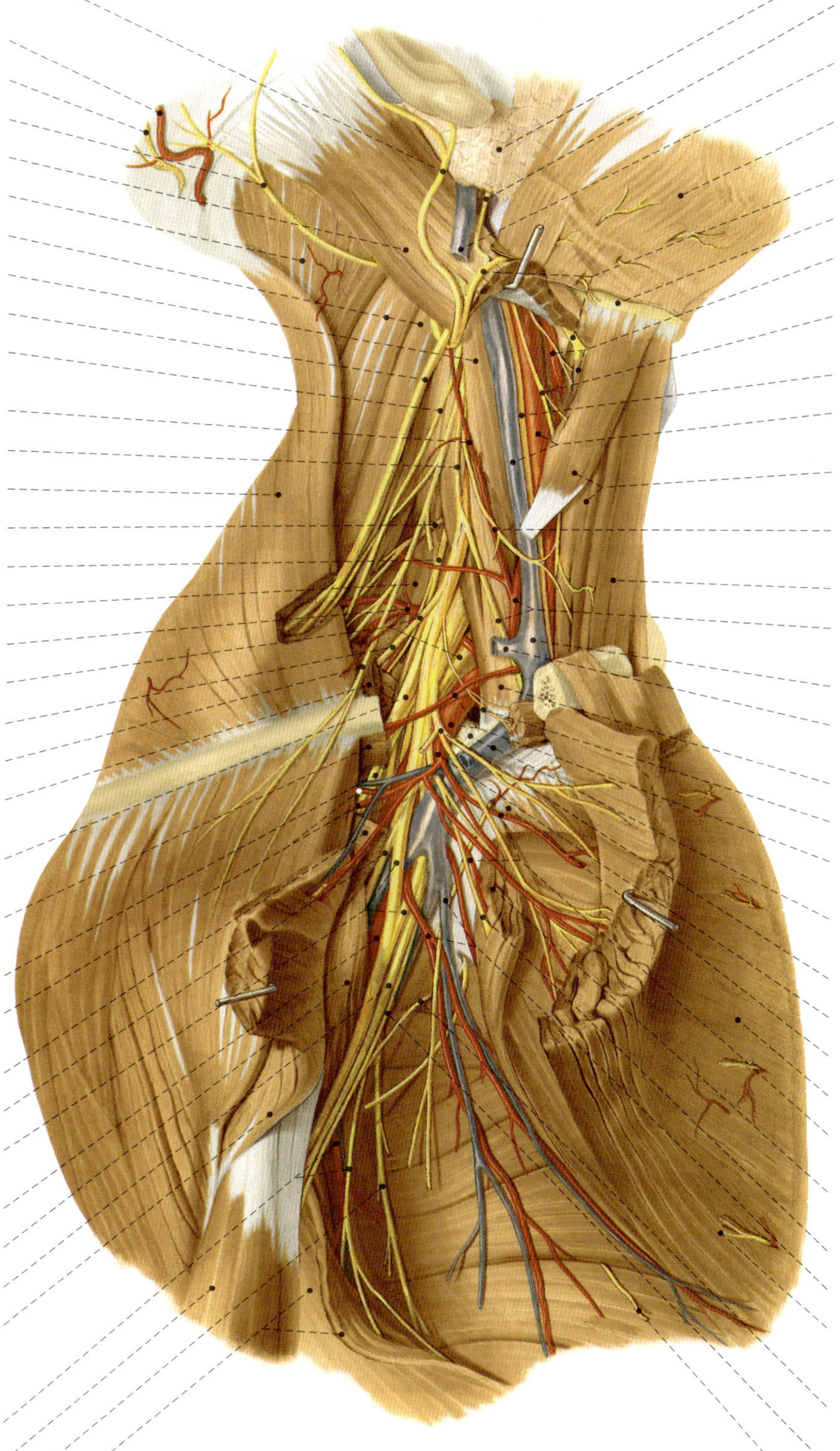

93 ⟨Collum⟩ et Thorax

93 Blood vessels and nerves of the neck, axilla and thorax (50%)

The platysma and the sternocleidomastoid muscle were removed. Right lateral aspect

93 Blutgefässe und Nerven des Halses, der Achselhöhle und der Brustwand (50%)

Das Platysma und der Musculus sternocleidomastoideus sind entfernt. Rechte Seitenansicht

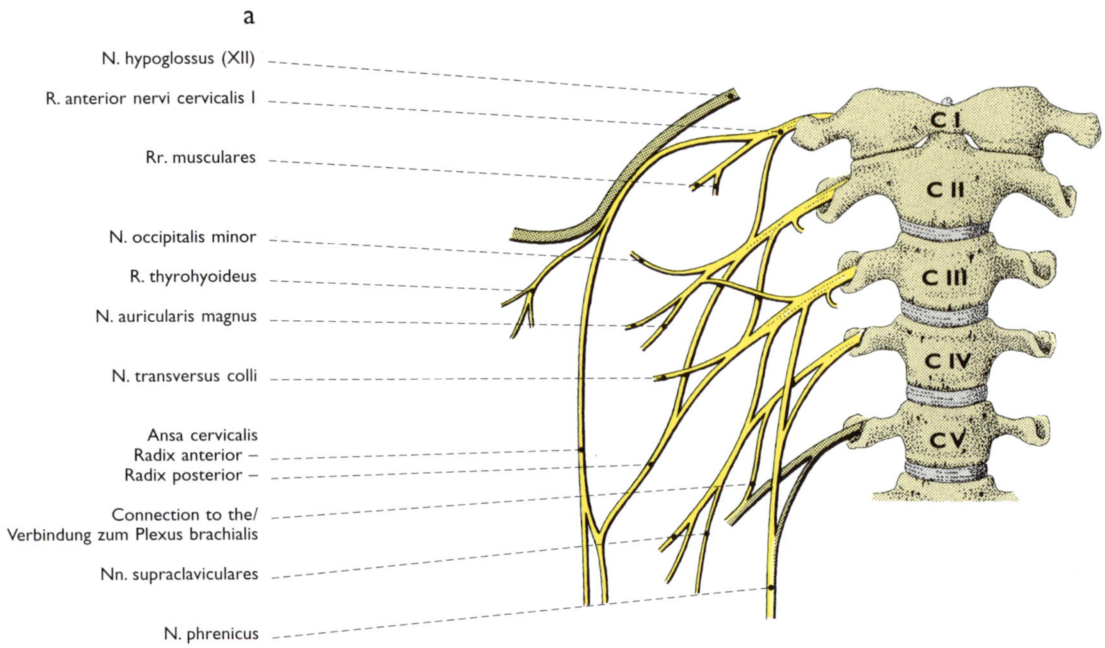

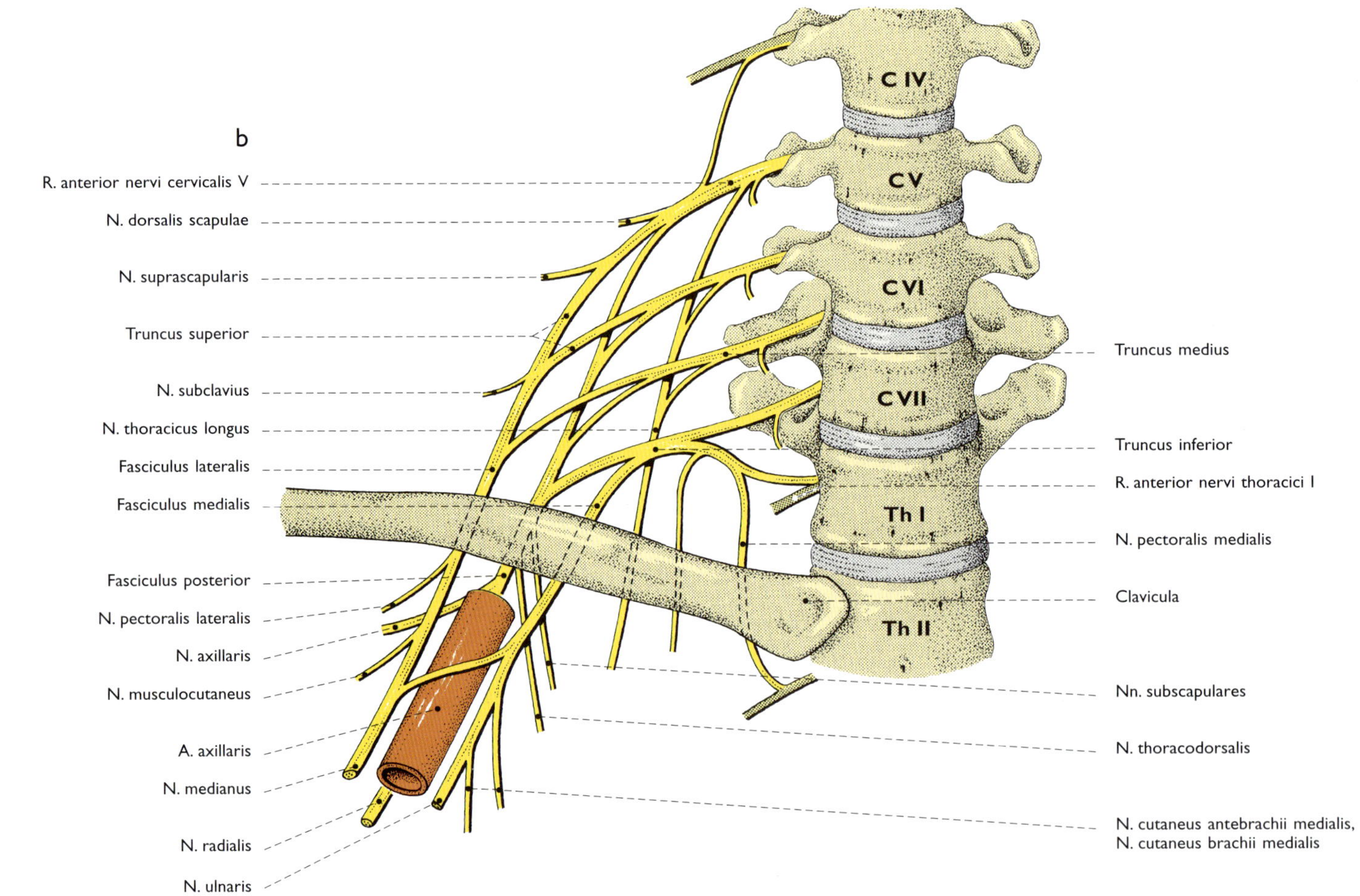

94 Plexus cervicalis et brachialis

94 Cervical and brachial plexuses

Schematic representations, ventral aspect
a Cervical plexus
b Brachial plexus

94 Hals- und Armgeflechte

Schematische Darstellungen, Ventralansicht
a Halsgeflecht
b Armgeflecht

a

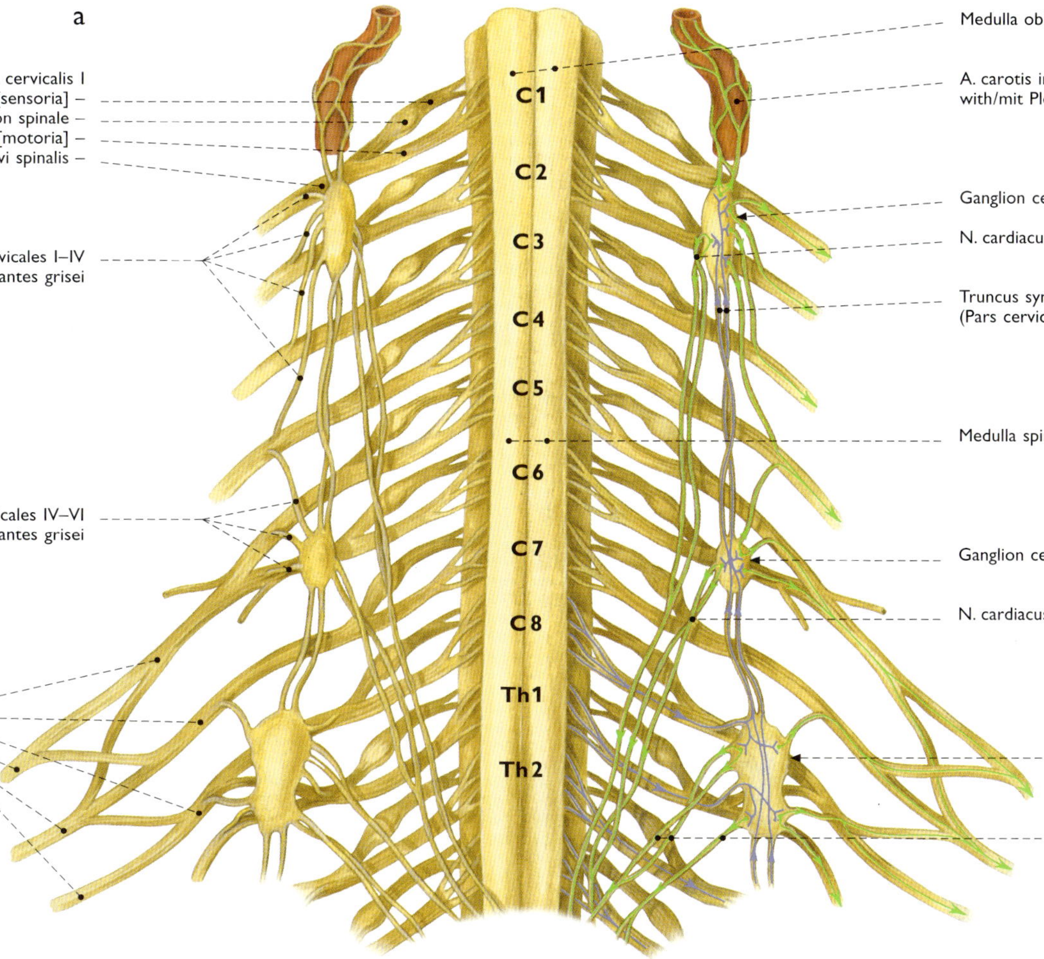

b

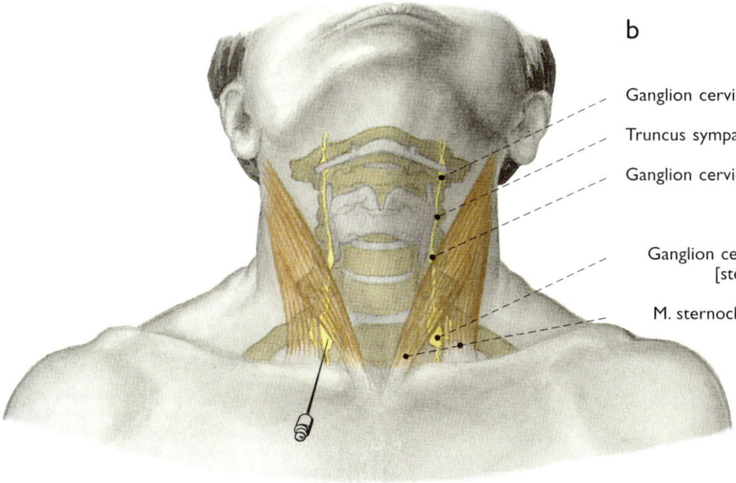

c

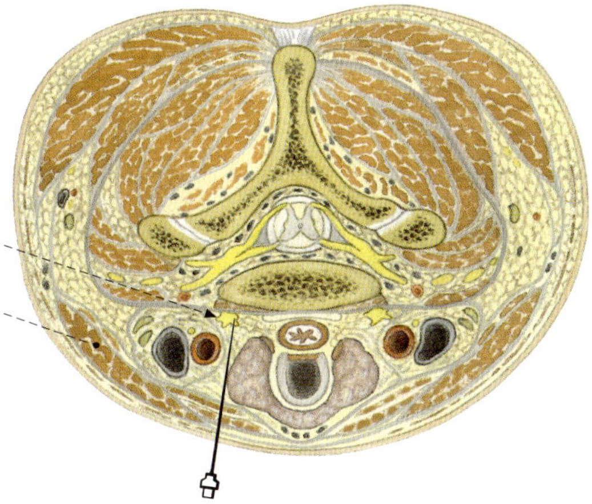

95 ⟨Collum⟩

95 Autonomic nervous system in the neck

a Essential circuitry and origins of the sympathetic nervous system in the neck and upper thorax regions (60%), schematic representation, ventral aspect

b, c Puncture of the cervicothoracic (stellate) ganglion

b Ventral aspect (25%)

c Transverse section at the level of the first thoracic vertebra (Th I) (50%), superior aspect

95 Autonomes Nervensystem im Hals

a Verschaltung und Ursprünge des Sympathikus im Hals und oberen Brustbereich (60%), schematische Darstellung, Ventralansicht

b, c Punktion des Ganglion cervicothoracicum [stellatum]

b Ventralansicht (25%)

c Horizontalschnitt in Höhe des ersten Brustwirbels (Th I) (50%), Kranialansicht

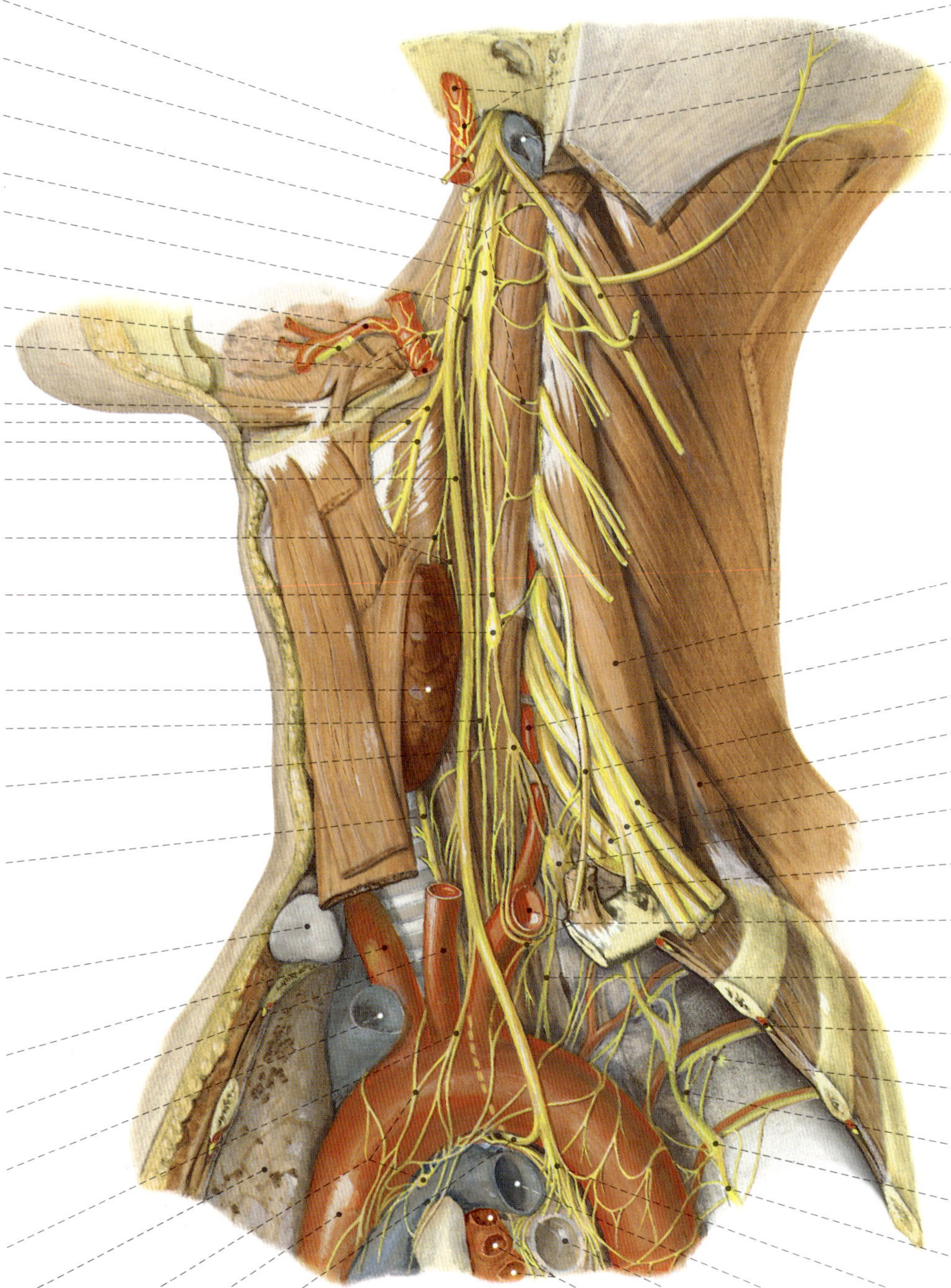

96 ⟨Collum⟩ et Thorax

96 Autonomic nervous system in the neck and the upper part of the thorax (60%)
Left lateral aspect

96 Autonomes Nervensystem im Hals und oberen Brustbereich (60%)
Linke Seitenansicht

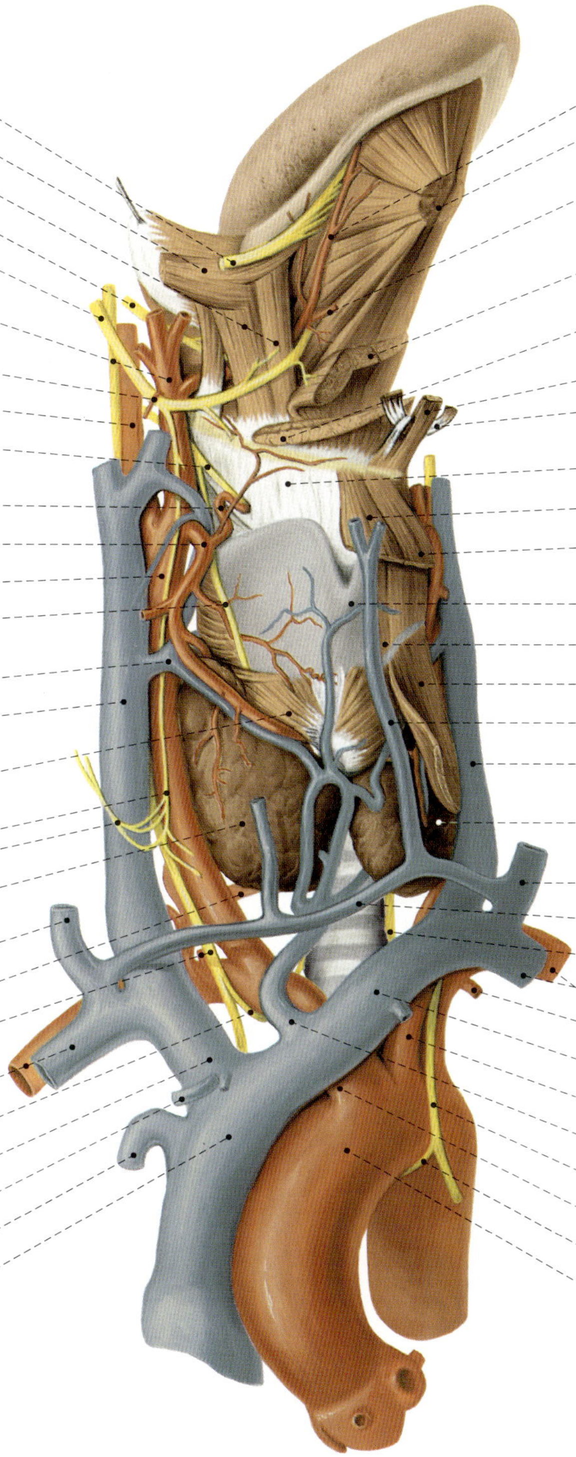

97 ⟨Viscera colli⟩

97 Blood vessels and nerves of the cervical viscera and the tongue (70%)
Right ventrolateral aspect

97 Blutgefässe und Nerven der Halseingeweide und der Zunge (70%)
Rechte Ventrolateralansicht

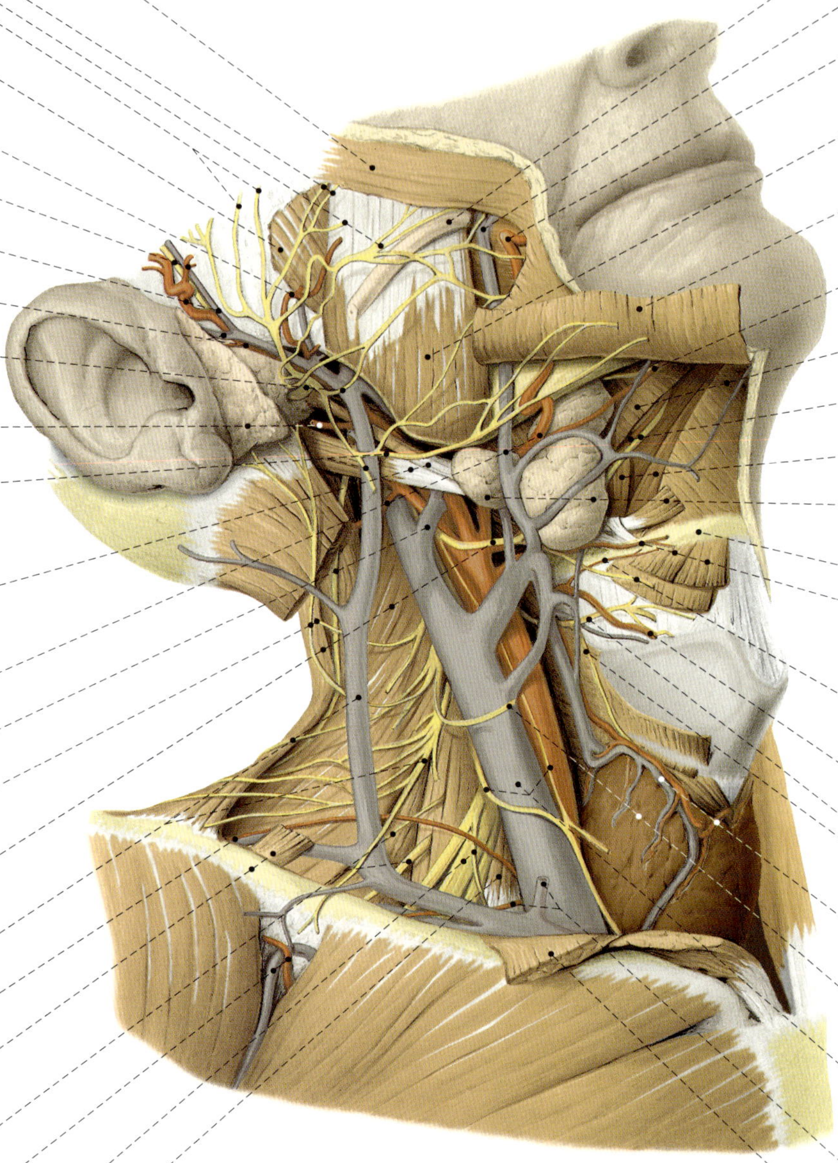

98 ⟨Collum⟩

98 Blood vessels and nerves of the neck (60%)

The parotid gland, the sternocleidomastoid muscle, the supra- and infrahyoid muscles were partially removed. Right ventrolateral aspect

98 Blutgefässe und Nerven des Halses (60%)

Ohrspeicheldrüse, Musculus sternocleidomastoideus sowie supra- und infrahyale Muskeln sind teilweise entfernt. Rechte Ventrolateralansicht

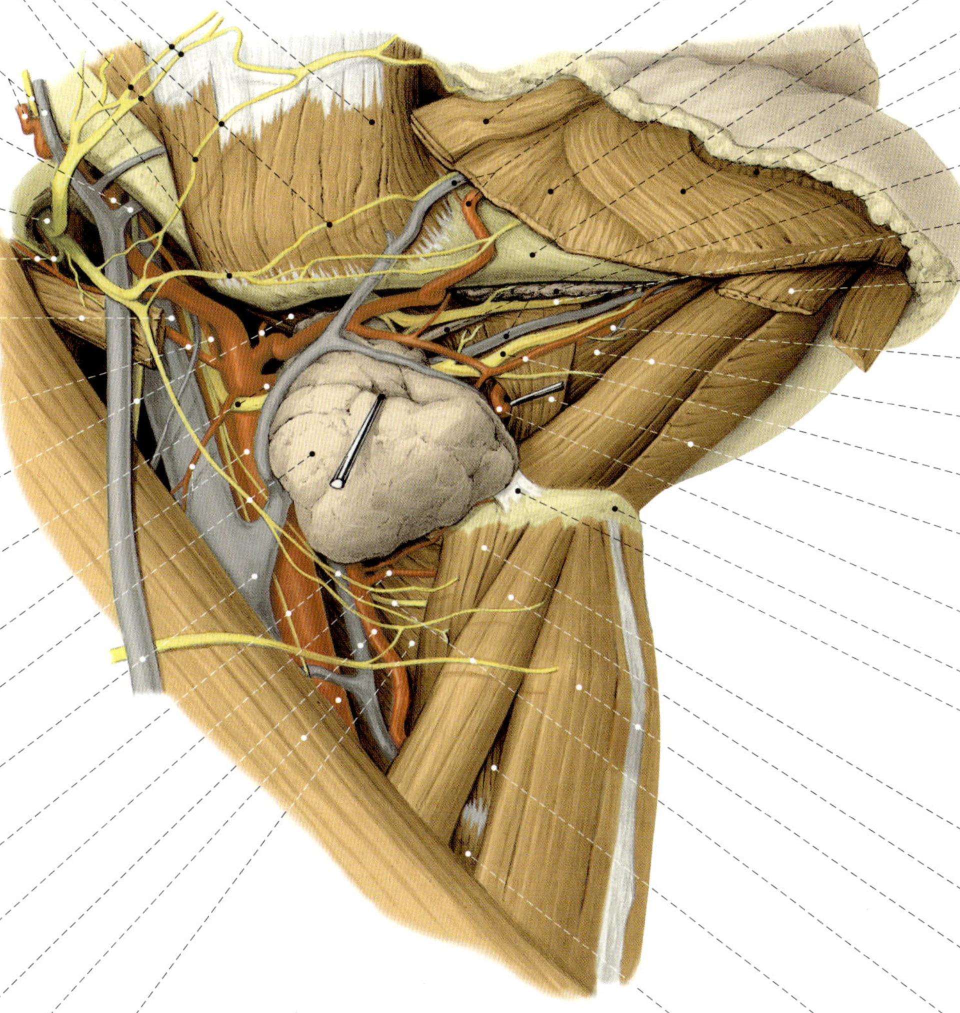

99 ⟨Collum⟩

99 Blood vessels and nerves in the anterior triangle of the neck (100%)

The parotid gland and parts of the mylohyoid muscle and the platysma were removed.
Right lateral aspect

99 Blutgefässe und Nerven im vorderen Halsdreieck (100%)

Die Ohrspeicheldrüse sowie Teile des Musculus mylohyoideus und des Platysma sind entfernt.
Rechte Seitenansicht

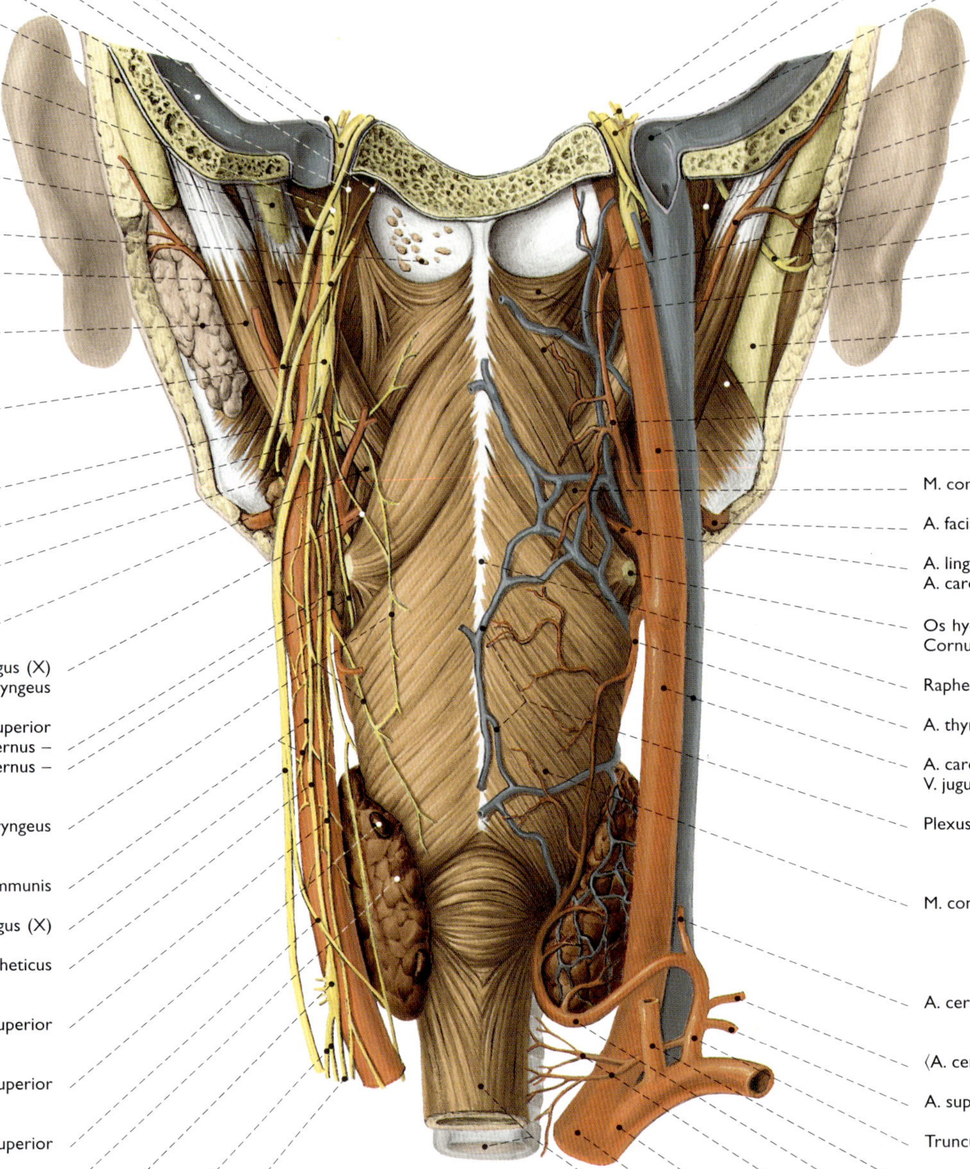

100 Pharynx

100 Blood vessels and nerves on the dorsolateral wall of the pharynx (70%)

The right side especially emphasizes the blood vessels, the left side the nerves. Dorsal aspect

100 Blutgefässe und Nerven an der dorsolateralen Rachenwand (70%)

Die Blutgefässe sind vornehmlich auf der rechten, die Nerven auf der linken Seite dargestellt. Dorsalansicht

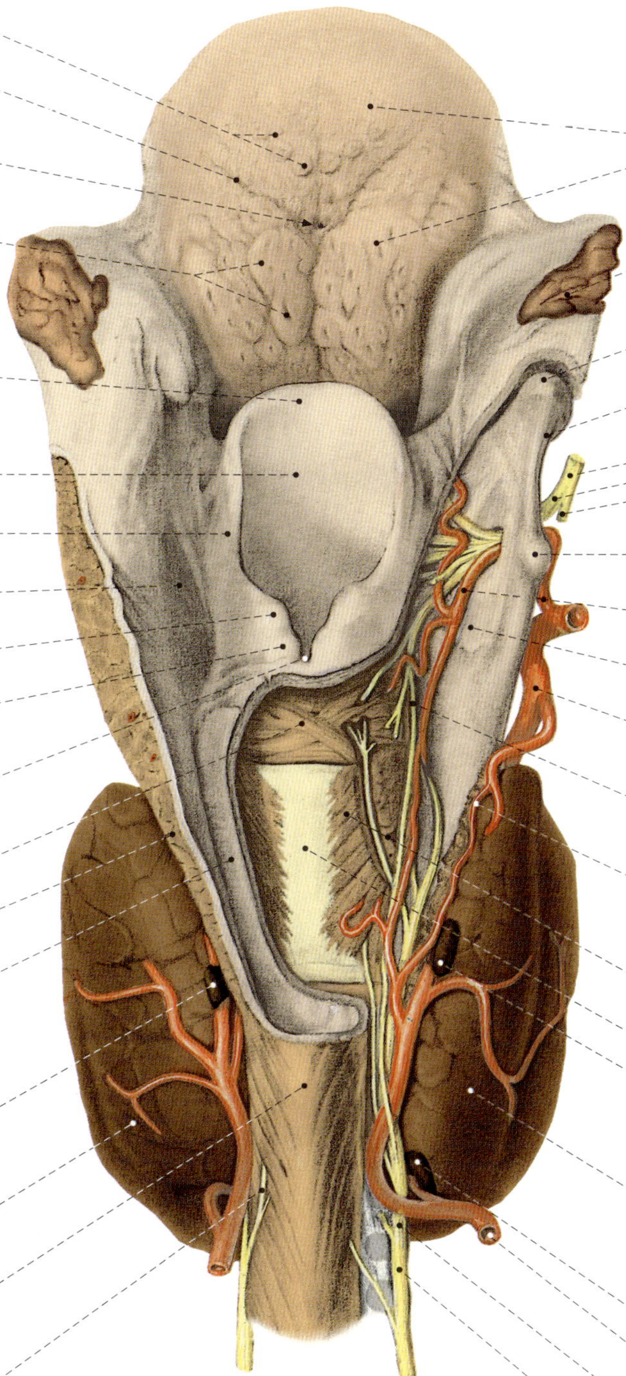

101 Larynx et Glandula thyroidea

101 Arteries and nerves of the larynx and thyroid gland (100%)

The dorsal wall of the pharynx was removed completely, the ventral one partially.
Dorsal aspect

101 Arterien und Nerven des Kehlkopfs und der Schilddrüse (100%)

Die Hinterwand des Rachens ist vollständig, die Vorderwand teilweise entfernt.
Dorsalansicht

Organa thoracis

Thoracic Viscera

Brusteingeweide

a

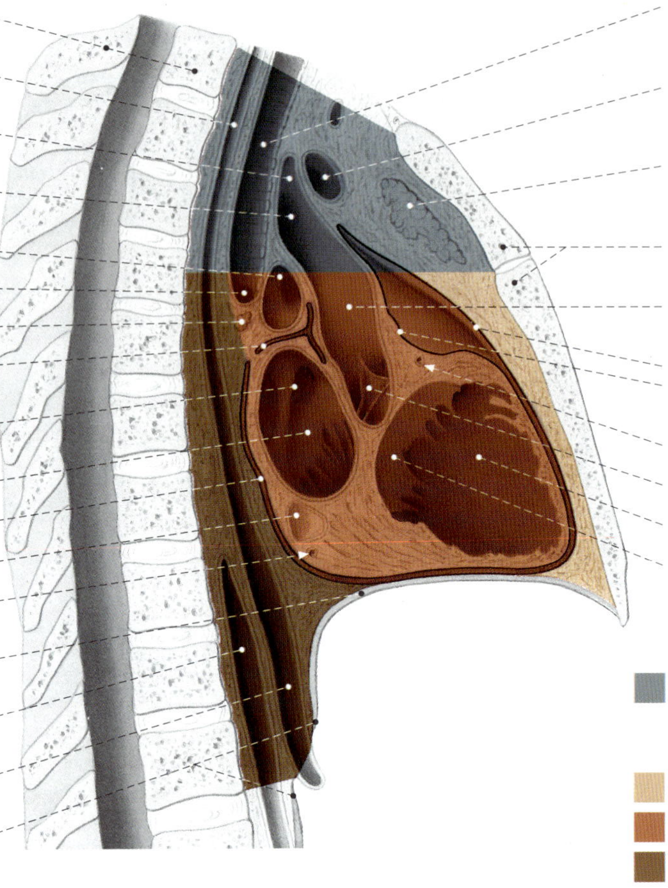

b

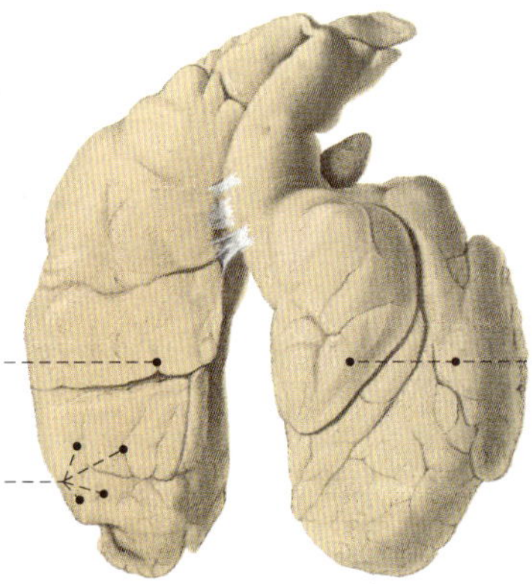

c

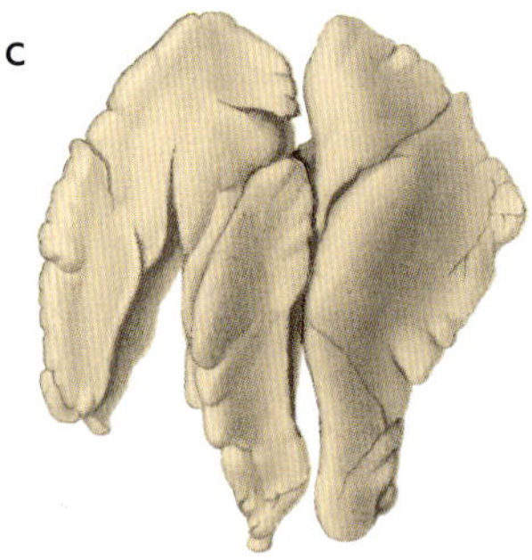

104 Mediastinum et Thymus

104 Mediastinum and thymus

a Subdivision of the mediastinum, median section (40%), medial aspect of the left half
b Thymus of a 3-year-old child (75%), ventral aspect
c Thymus of a newborn child (75%), ventral aspect

104 Mediastinum und Thymus

a Gliederung des Mediastinum, Medianschnitt (40%), Medialansicht der linken Hälfte
b Thymus eines 3jährigen Kindes (75%), Ventralansicht
c Thymus eines Neugeborenen (75%), Ventralansicht

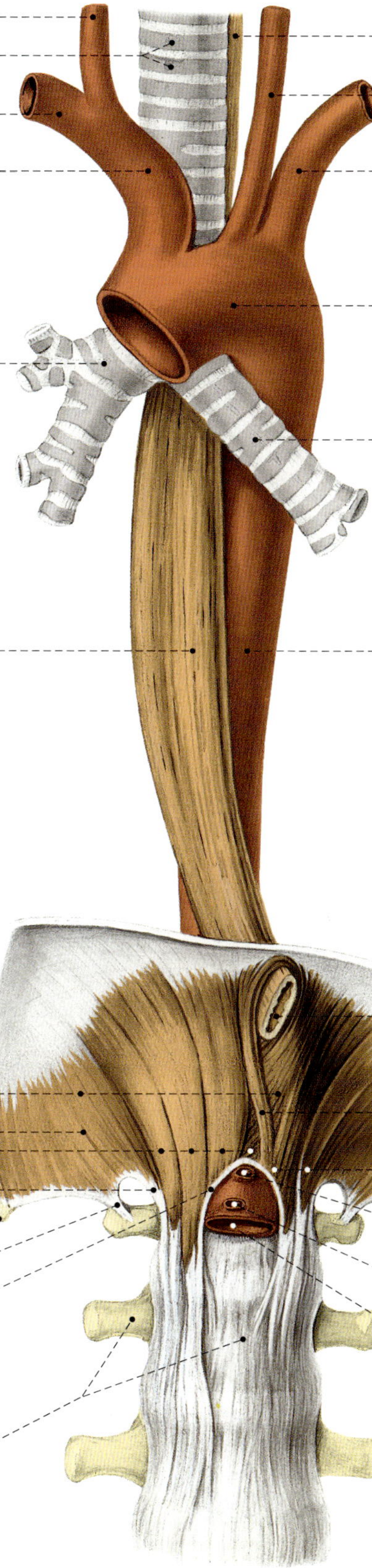

105 Oesophagus, Trachea et Aorta

105 Esophagus and adjacent organs (50%)
Ventral aspect

105 Speiseröhre und benachbarte Organe (50%)
Ventralansicht

a

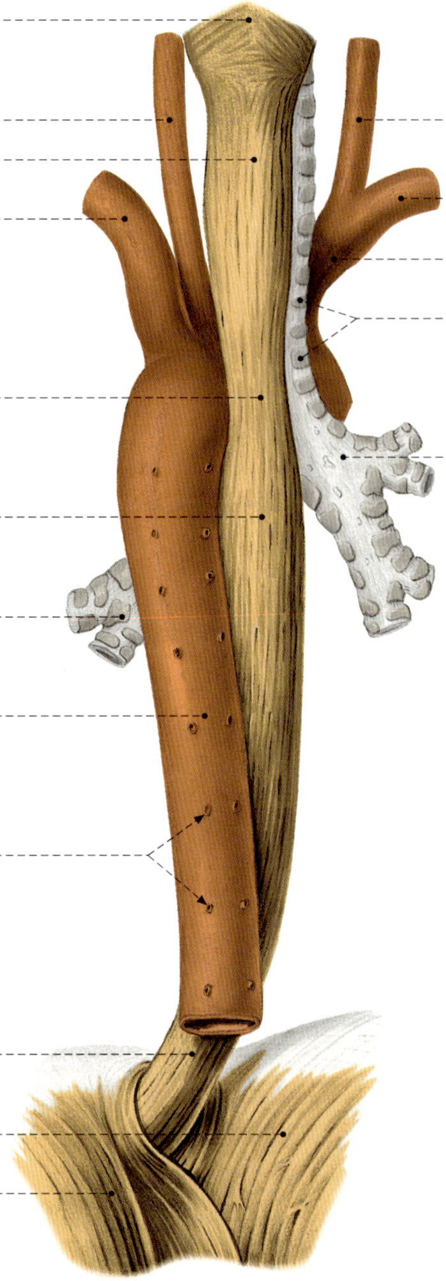

b

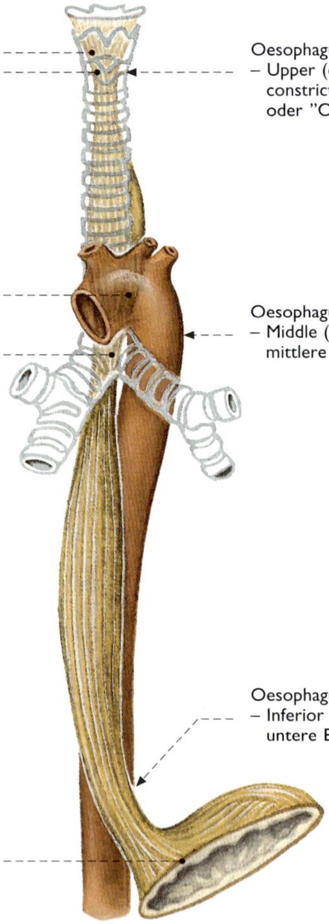

106 Oesophagus, Trachea et Aorta

106 Esophagus and adjacent organs

a Dorsal aspect (50%)
b Typical narrowings of the esophagus (35%), schematic repesentation, ventral aspect

106 Speiseröhre und benachbarte Organe

a Dorsalansicht (50%)
b Typische Engstellen der Speiseröhre (35%), schematische Darstellung, Ventralansicht

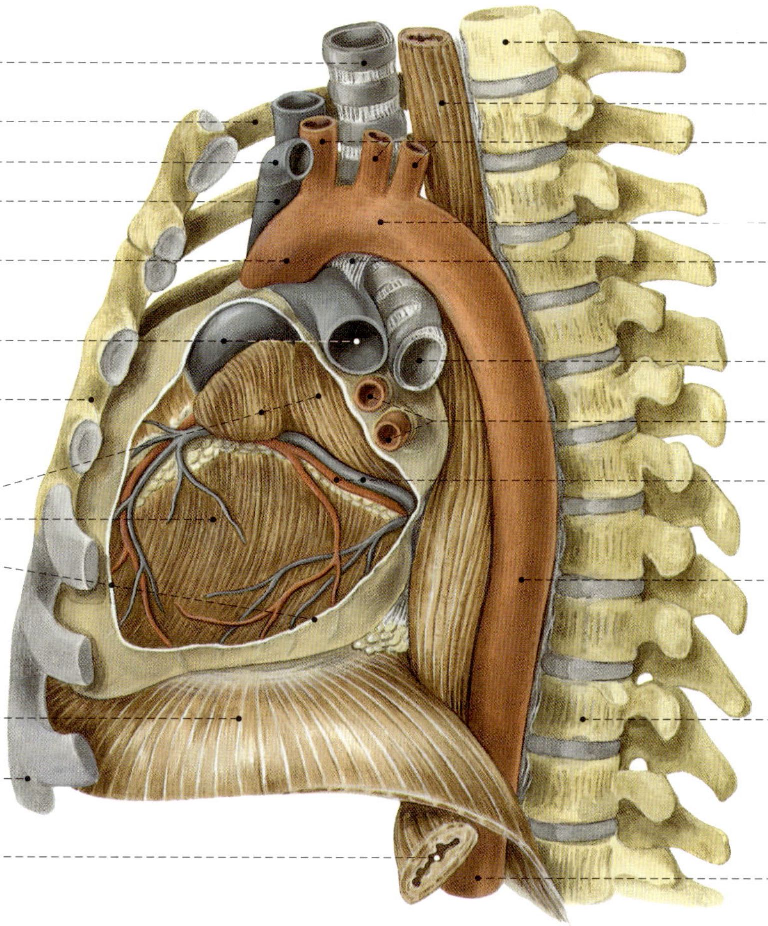

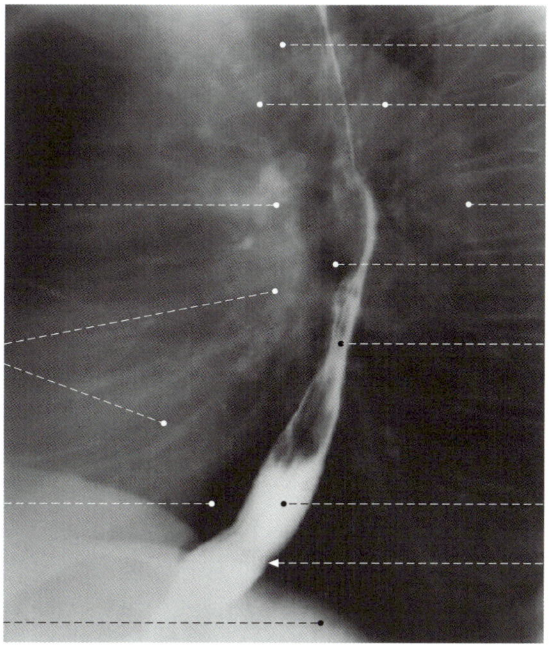

107 Oesophagus, Trachea, Cor et Aorta

107 Esophagus and adjacent organs (35%)

a Left lateral aspect of the mediastinum. The pericardium was partially removed.
b Lateral radiograph. The esophagus contains contrast medium.

107 Speiseröhre und benachbarte Organe (35%)

a Linke Seitenansicht des Mediastinum. Der Herzbeutel ist teilweise entfernt.
b Seitliche Röntgenaufnahme. Die Speiseröhre enthält Kontrastbrei.

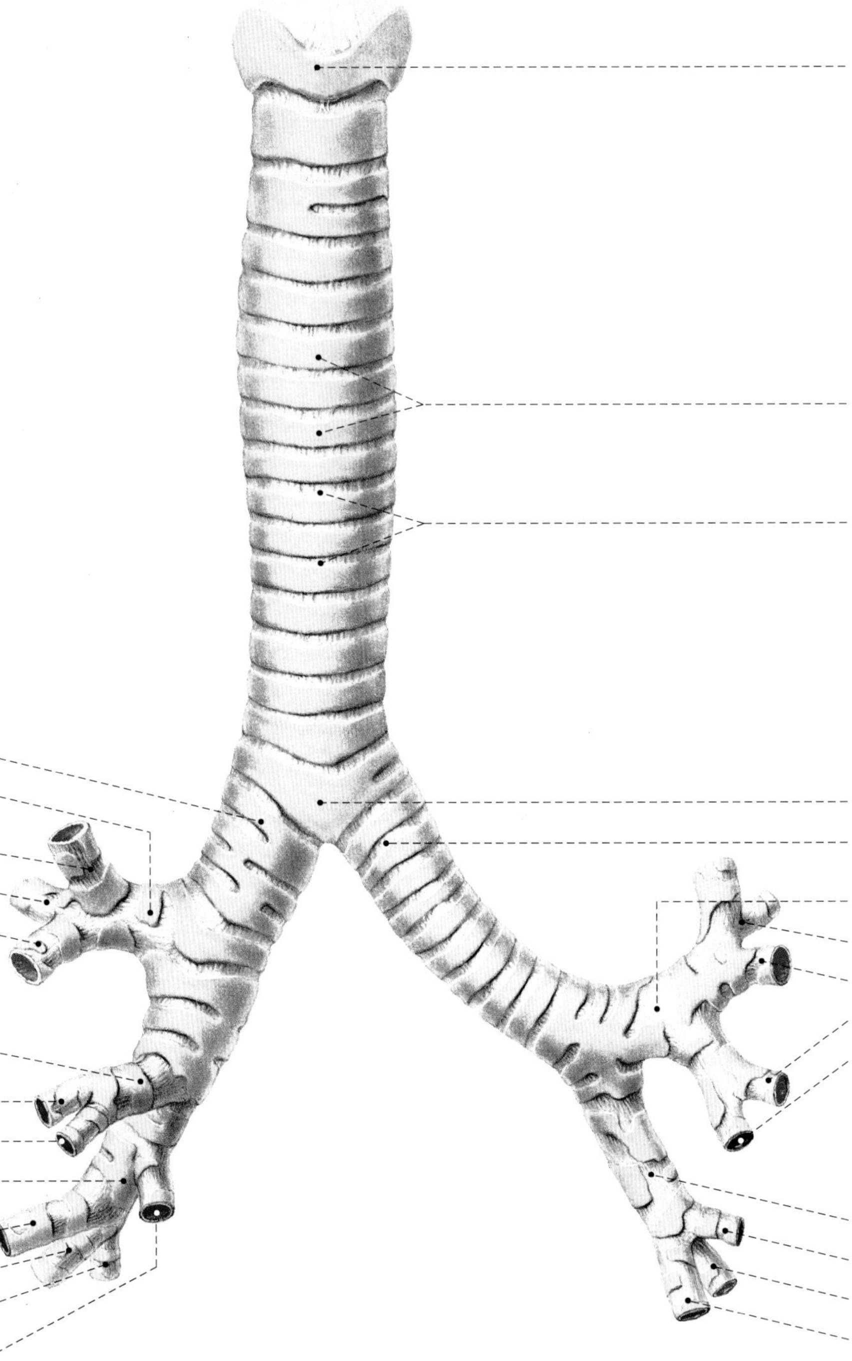

108 Trachea et Bronchi

108 Trachea and bronchi (90%)
Ventral aspect

108 Luftröhre und Bronchien (90%)
Ventralansicht

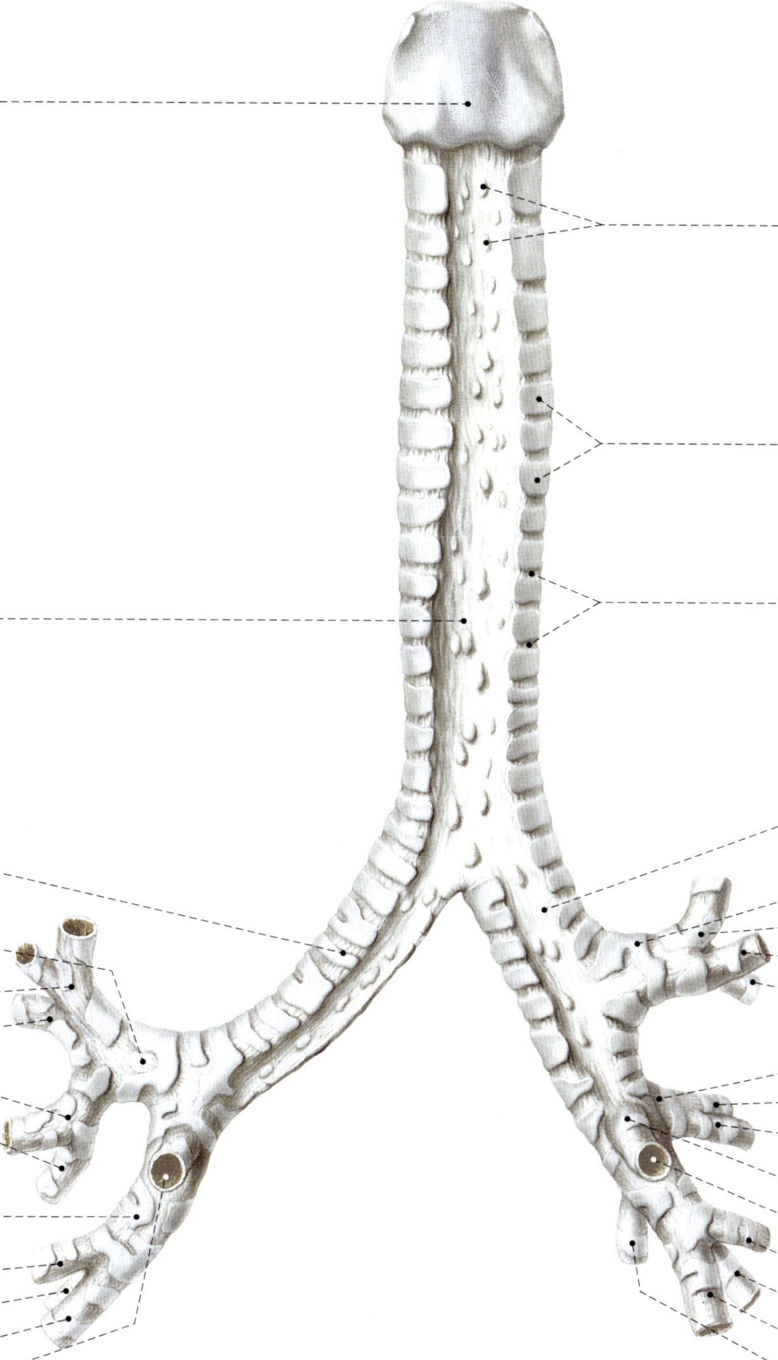

109 Trachea et Bronchi

109 Trachea and bronchi (90%)
Dorsal aspect

109 Luftröhre und Bronchien (90%)
Dorsalansicht

a

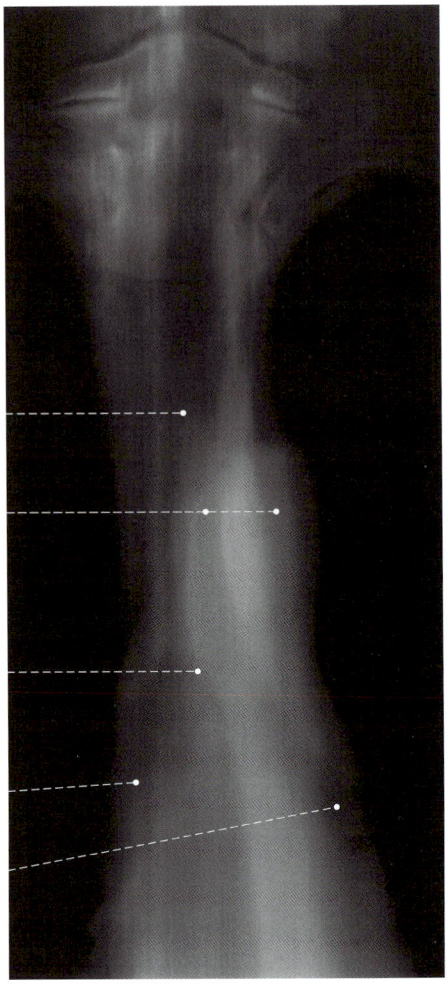

b

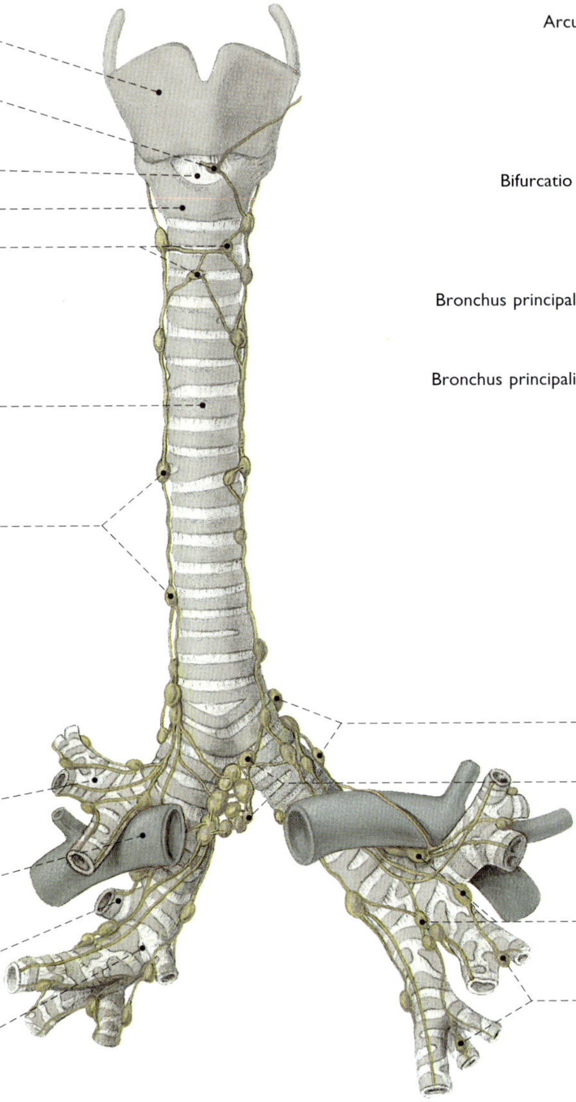

110 Trachea et Bronchi

110 Trachea and bronchi

a Posteroanterior tomogram of the trachea (50%)
b Lymph vessels and lymph nodes of the trachea and the bronchi (70%), ventral aspect

110 Luftröhre und Bronchien

a Posteroanteriores Schichtbild (Tomogramm) der Trachea (50%)
b Lymphgefässe und Lymphknoten im Bereich der unteren Atemwege (70%), Ventralansicht

a

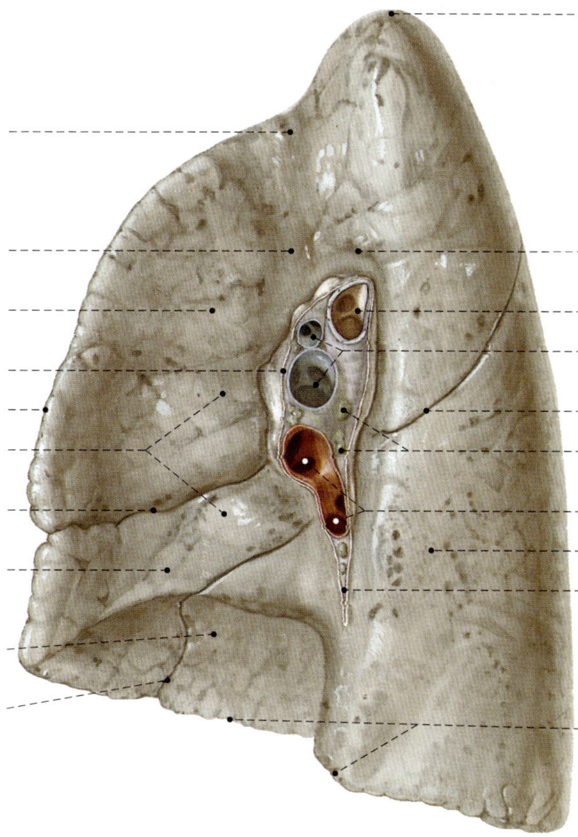

b

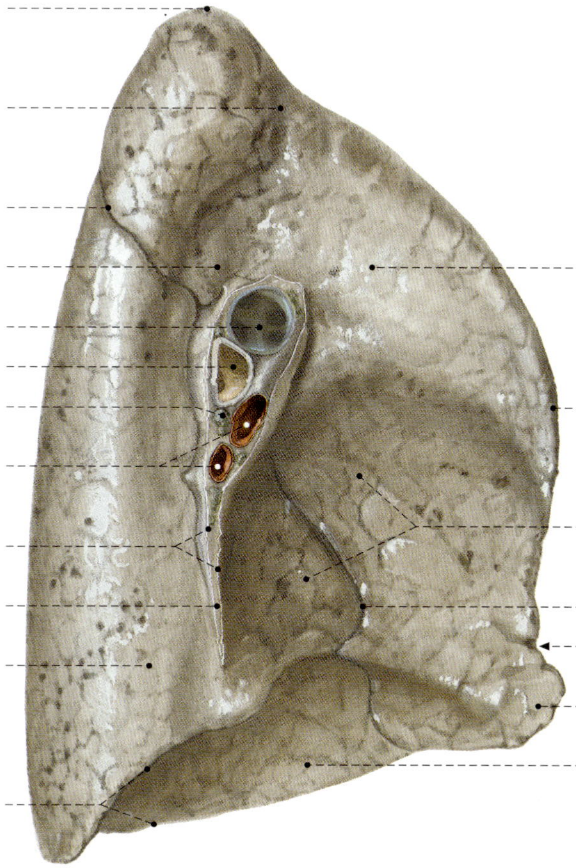

111 Pulmones dexter et sinister

111 Right und left lungs (40%)

a Right lung, mediastinal aspect
b Left lung, mediastinal aspect

111 Rechte und linke Lunge (40%)

a Rechte Lunge, Mediastinalansicht
b Linke Lunge, Mediastinalansicht

a

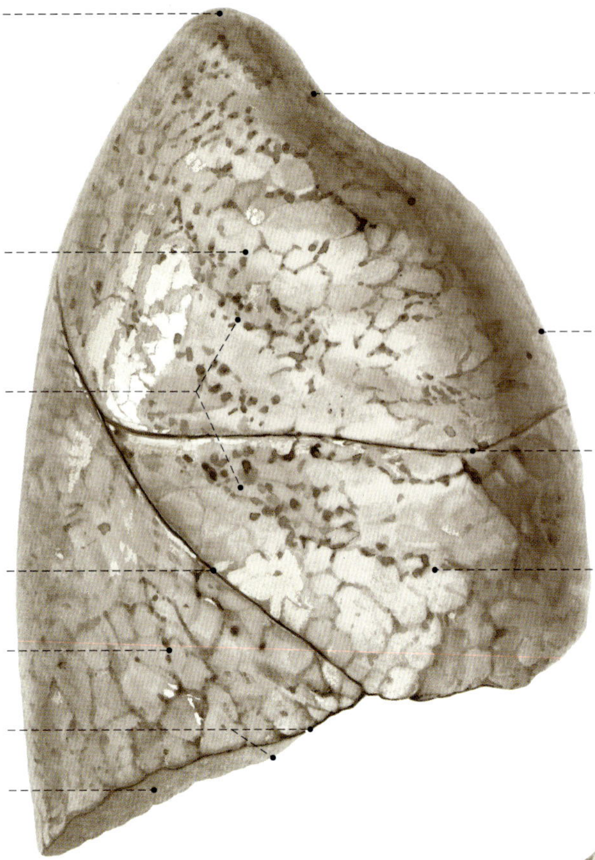

b

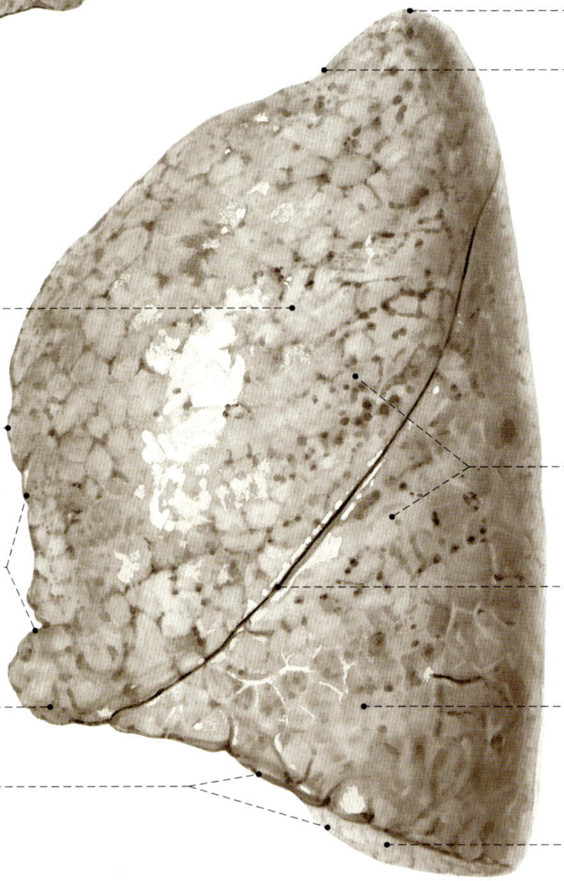

112 Pulmones dexter et sinister

112 Right und left lungs (40%)

a Right lung, lateral aspect
b Left lung, lateral aspect

112 Rechte und linke Lunge (40%)

a Rechte Lunge, Seitenansicht
b Linke Lunge, Seitenansicht

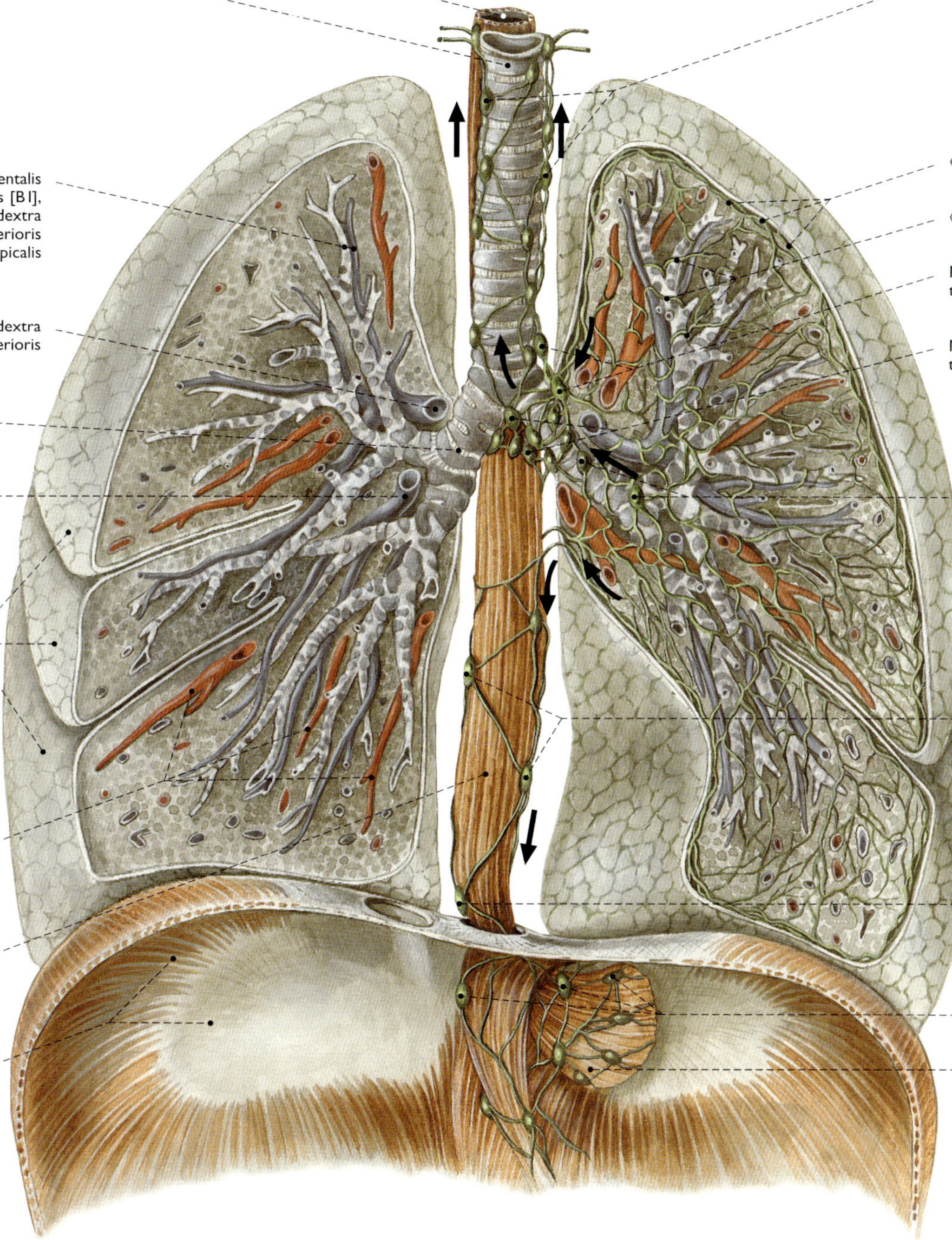

113 Pulmones dexter et sinister

113 Right and left lungs (80%)
Lymph vessels, lymph nodes, and lymphatic drainage from the left lung, ventral aspect

113 Rechte und linke Lunge (80%)
Lymphgefässe, Lymphknoten und Lymphabfluss aus der linken Lunge, Ventralansicht

a

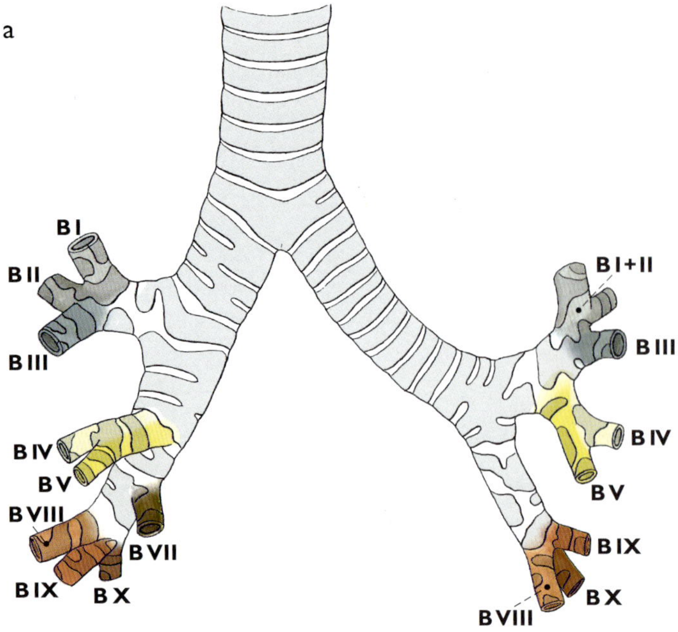

b

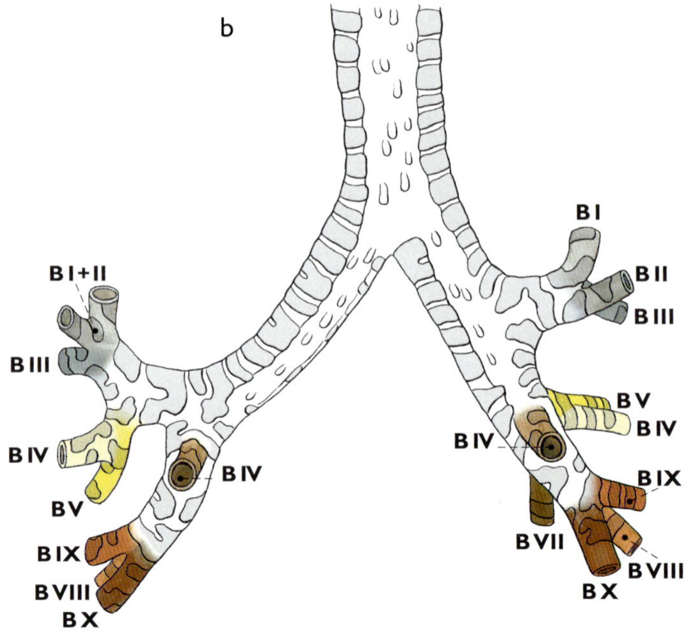

Branches of the two principal bronchi/Verzweigungen der beiden Hauptbronchien

Bronchus principalis dexter

Bronchus lobaris superior dexter
- B I Bronchus segmentalis apicalis
- B II Bronchus segmentalis posterior
- B III Bronchus segmentalis anterior

Bronchus lobaris medius dexter
- B IV Bronchus segmentalis lateralis
- B V Bronchus segmentalis medialis

Bronchus lobaris inferior dexter
- B VI Bronchus segmentalis superior
- B VII Bronchus segmentalis basalis medialis [cardiacus]
- B VIII Bronchus segmentalis basalis anterior
- B IX Bronchus segmentalis basalis lateralis
- B X Bronchus segmentalis basalis posterior

Bronchus principalis sinister

Bronchus lobaris superior sinister
- B I+II Bronchus segmentalis apicoposterior
- B III Bronchus segmentalis anterior
- B IV Bronchus lingularis superior
- B V Bronchus lingularis inferior

Bronchus lobaris inferior sinister
- B VI Bronchus segmentalis superior
- B VIII Bronchus segmentalis basalis anterior
- B IX Bronchus segmentalis basalis lateralis
- B X Bronchus segmentalis basalis posterior

c

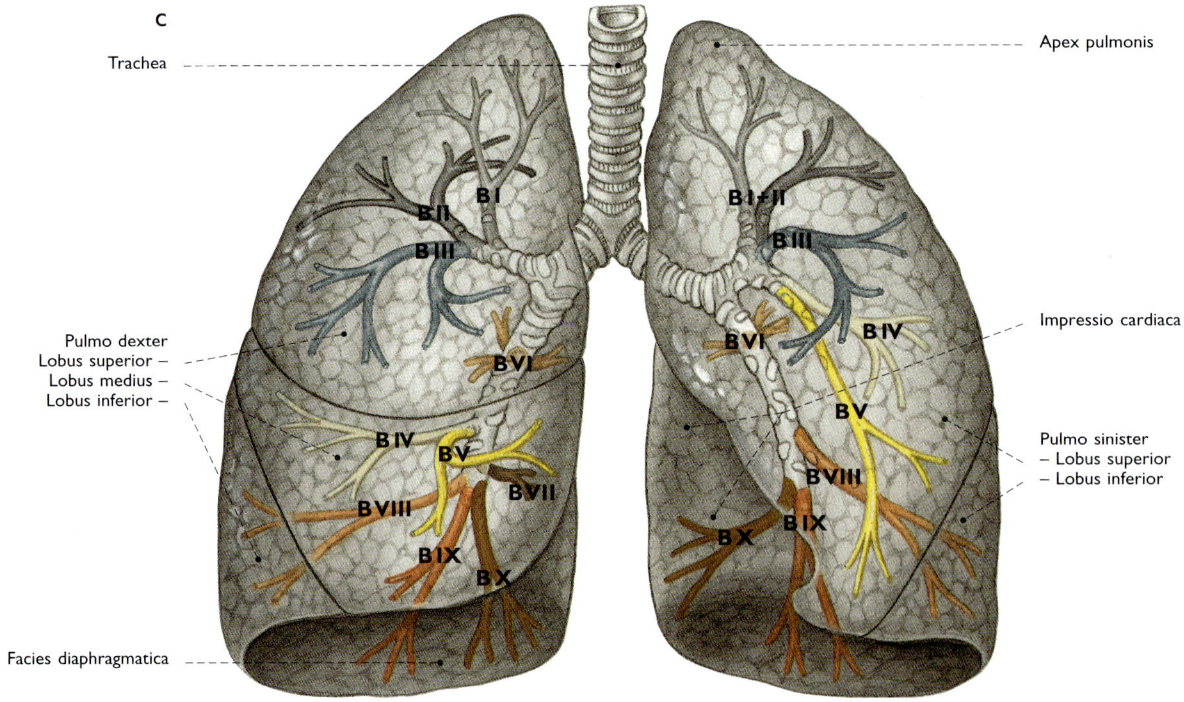

114 Arbor bronchialis

114 Bronchial tree

Schematic representations

a Bronchial tree (70%), ventral aspect
b Bronchial tree (70%), dorsal aspect
c Subdivision of the bronchial tree in the right and left lungs (50%), ventral aspect

114 Bronchialbaum

Schematische Darstellungen

a Bronchialbaum (70%), Ventralansicht
b Bronchialbaum (70%), Dorsalansicht
c Verzweigung des Bronchialbaums in der rechten und linken Lunge (50%), Ventralansicht

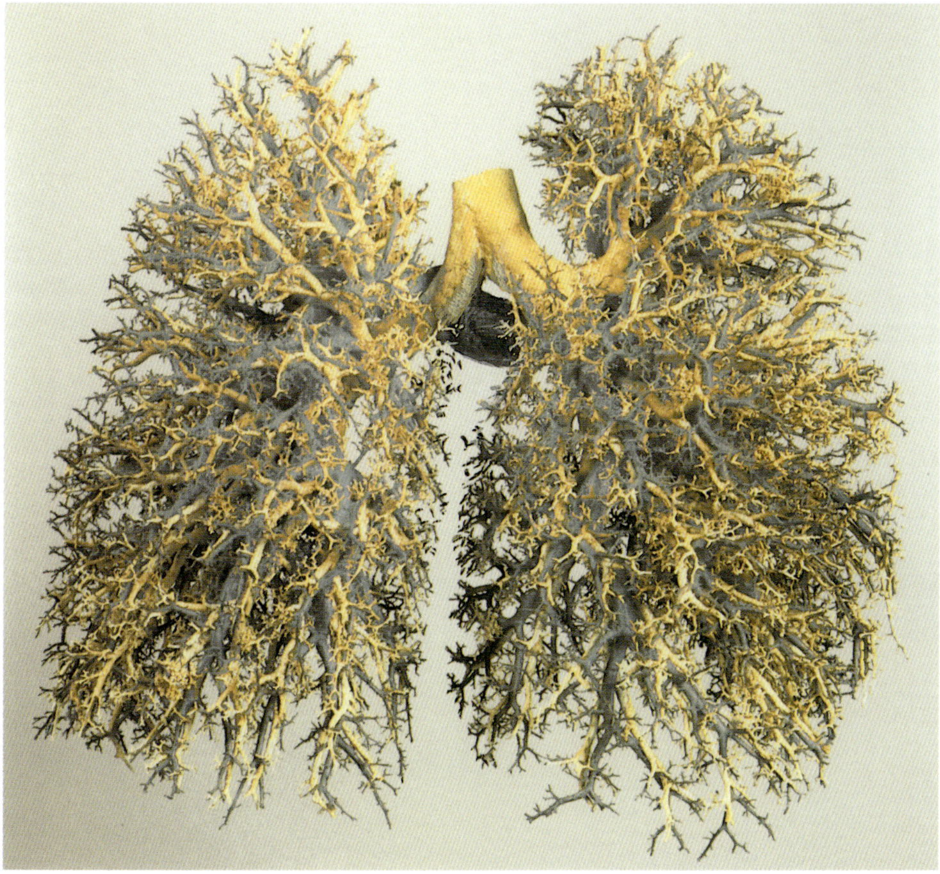

a

Yellow: Trachea and bronchial tree
Blue: Truncus pulmonalis and its ramifications

Gelb: Trachea und Bronchialbaum
Blau: Truncus pulmonalis und seine Verzweigungen

b

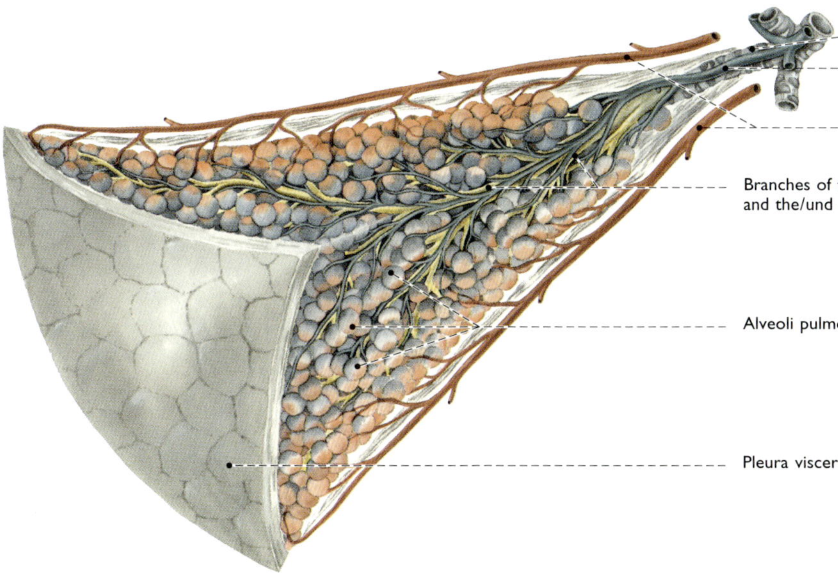

115 Arbor bronchialis

115 Bronchial tree and bronchopulmonary segment

a Bronchial tree and branches of the pulmonary arteries, corrosion preparation (70%), dorsal aspect (Anatomical Collection, Basel)
b Bronchopulmonary segment, schematic representation

115 Bronchialbaum und Lungensegment

a Bronchialbaum und Aufzweigungen des Truncus pulmonalis, Korrosionspräparat (70%), Dorsalansicht (Anatomische Sammlung, Basel)
b Lungensegment, schematische Darstellung

a

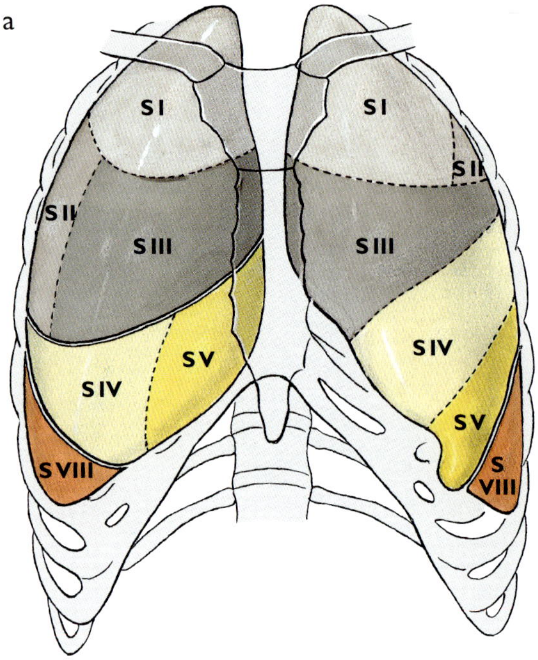

b

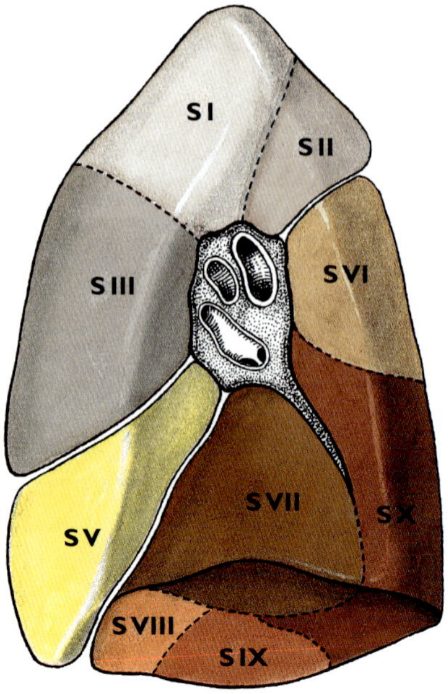

c

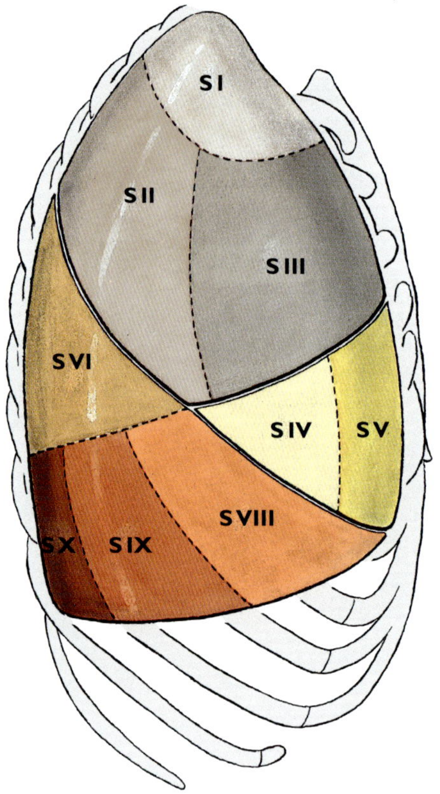

Pulmo dexter

Lobus superior

S I Segmentum apicale
S II Segmentum posterius
S III Segmentum anterius

Lobus medius

S IV Segmentum laterale
S V Segmentum mediale

Lobus inferior

S VI Segmentum superius
S VII Segmentum basale mediale [cardiacum]
S VIII Segmentum basale anterius
S IX Segmentum basale laterale
S X Segmentum basale posterius

Segment borders represented as dashed lines, lobe borders as solid lines/
Segmentgrenzen gestrichelt, Lappengrenzen ausgezogen

116 Segmenta bronchopulmonalia

116 Bronchopulmonary segments (25%)

Schematic representations. Diverse segments are indicated by different colors.

a Right and left lungs, ventral aspect
b Right lung, mediastinal aspect
c Right lung, lateral aspect

116 Lungensegmente (25%)

Schematische Darstellungen. Die verschiedenen Segmente sind durch unterschiedliche Farben dargestellt.

a Rechte und linke Lunge, Ventralansicht
b Rechte Lunge, Mediastinalansicht
c Rechte Lunge, Seitenansicht

a

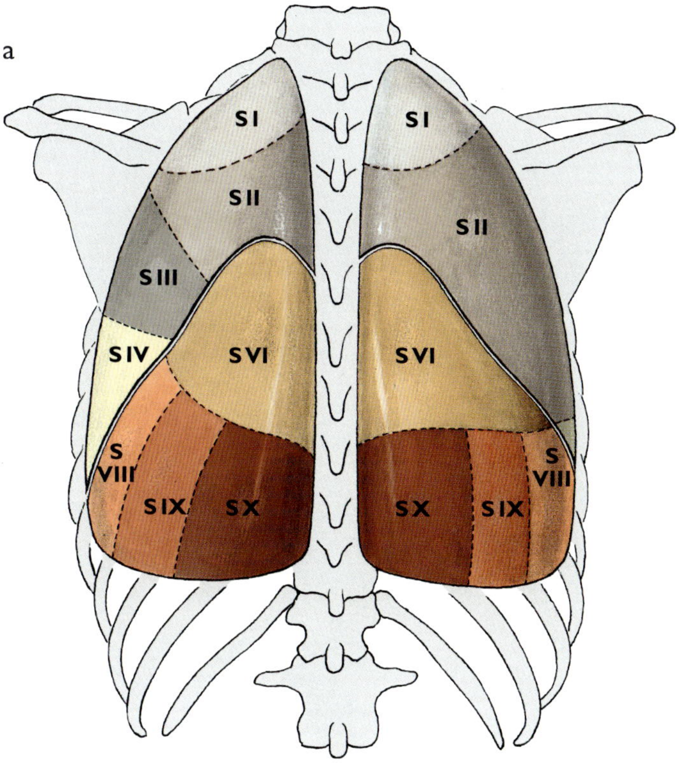

b

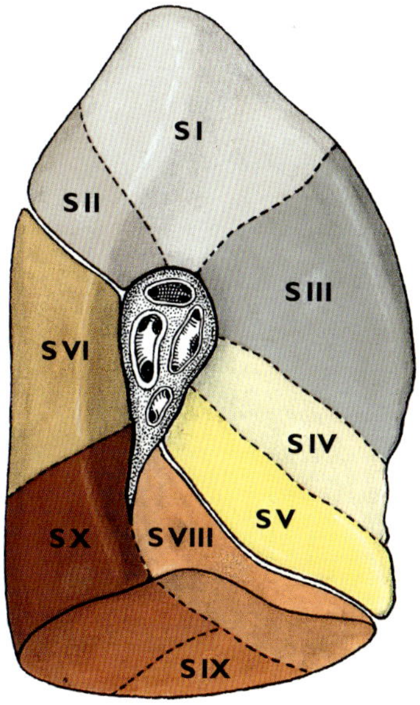

c

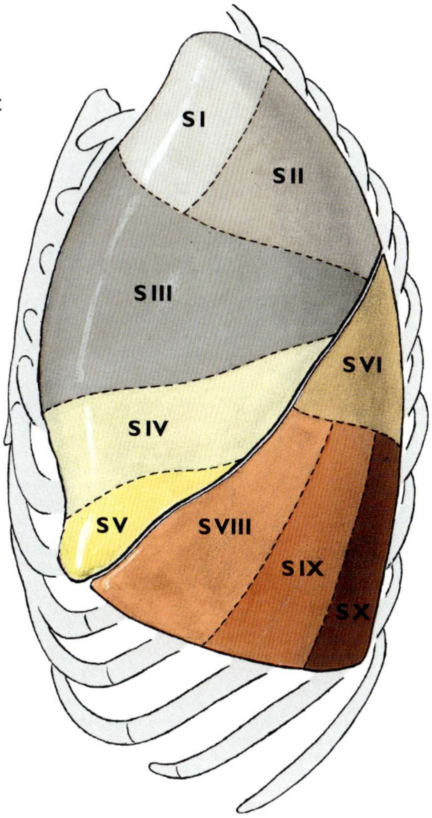

Pulmo sinister

Lobus superior
SI+II Segmentum apicoposterius
SIII Segmentum anterius
SIV Segmentum lingulare superius
SV Segmentum lingulare inferius

Lobus inferior
SVI Segmentum superius
SVIII Segmentum basale anterius
SIX Segmentum basale laterale
SX Segmentum basale posterius

Segment borders represented as dashed lines, lobe borders as solid lines/
Segmentgrenzen gestrichelt, Lappengrenzen ausgezogen

117 Segmenta bronchopulmonalia

117 Bronchopulmonary segments (25%)

Schematic representations. Diverse segments are indicated by different colors.

a Right and left lungs, dorsal aspect
b Left lung, mediastinal aspect
c Left lung, lateral aspect

117 Lungensegmente (25%)

Schematische Darstellungen. Die verschiedenen Segmente sind durch unterschiedliche Farben dargestellt.

a Rechte und linke Lunge, Dorsalansicht
b Linke Lunge, Mediastinalansicht
c Linke Lunge, Seitenansicht

a

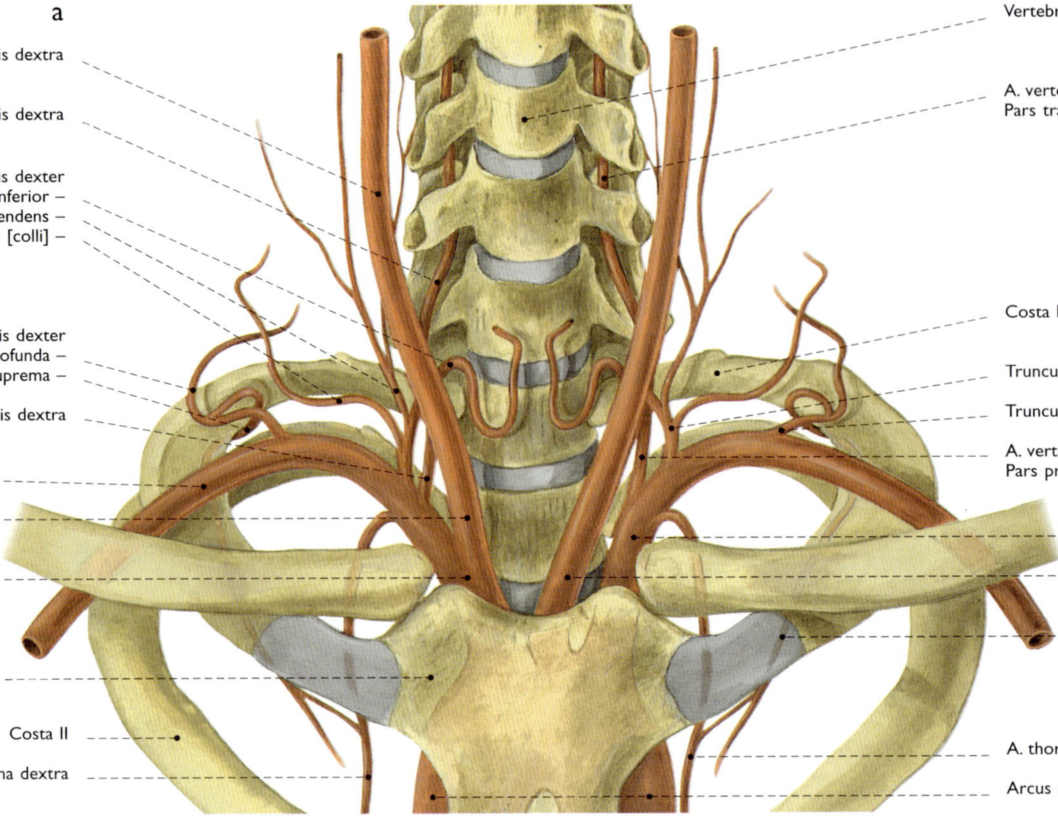

b

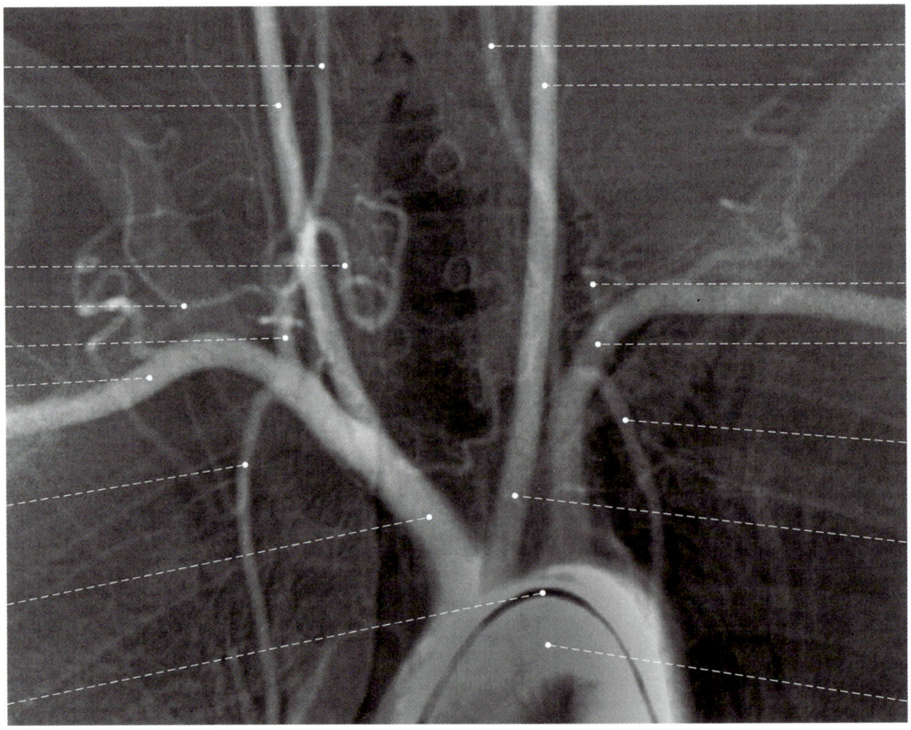

118 Thorax et ⟨Collum⟩

118 Arteries of the upper thorax and the lower neck (65%)

a Schematic representation, ventral aspect
b Anteroposterior aortogram

118 Arterien der oberen Thorax- und unteren Halsregion (65%)

a Schematische Darstellung, Ventralansicht
b Anteroposteriores Aortogramm

a

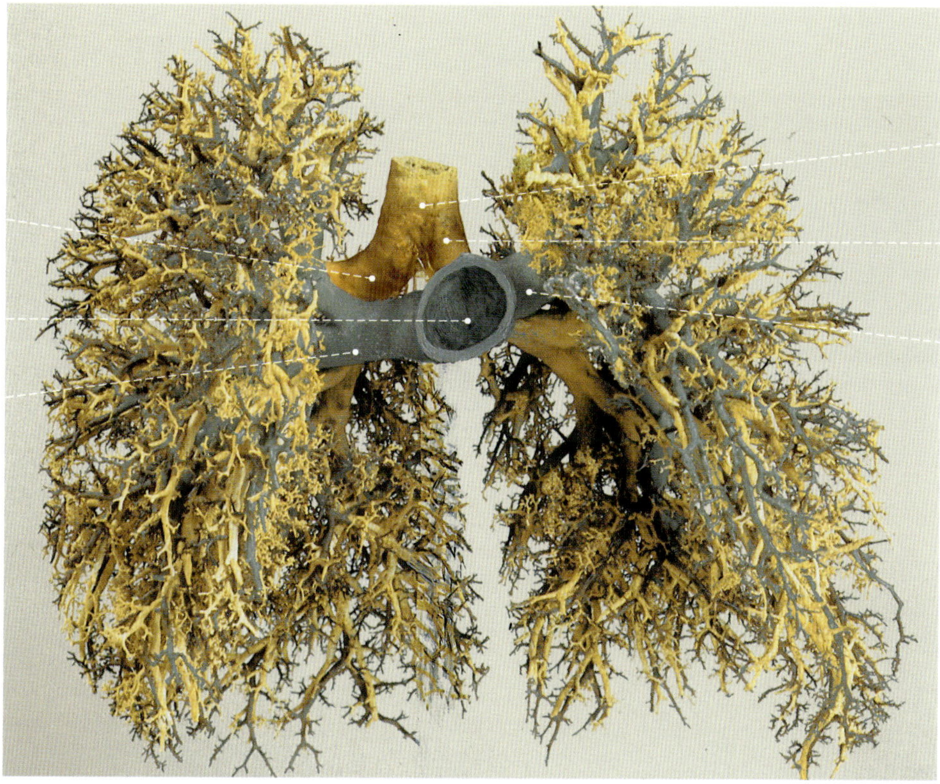

Yellow: Trachea and bronchial tree
Blue: Truncus pulmonalis and its ramifications

Gelb: Trachea und Bronchialbaum
Blau: Truncus pulmonalis und seine Verzweigungen

b

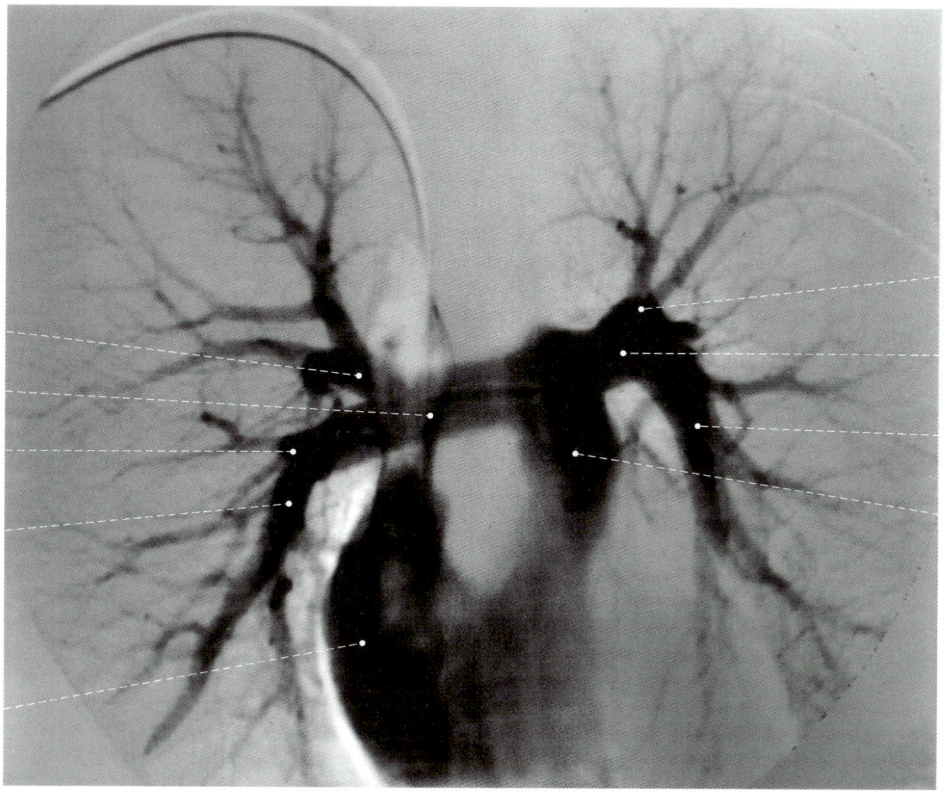

119 Arbor bronchialis

119 Bronchial tree and pulmonary arteries (70%)

a Bronchial tree, pulmonary trunk and branches of the pulmonary arteries, corrosion preparation, ventral aspect
b Dextrocardiogram and pulmonary arteriogram after injection of contrast medium into the right ventricle of the heart, ventral aspect

119 Bronchialbaum und Lungenarterien (70%)

a Bronchialbaum, Truncus pulmonalis und Aufzweigungen der Lungenarterien, Korrosionspräparat, Ventralansicht
b Röntgenologische Kontrastmitteldarstellung des rechten Herzventrikels (Dextrogramm) sowie der Lungenarterien, Ventralansicht

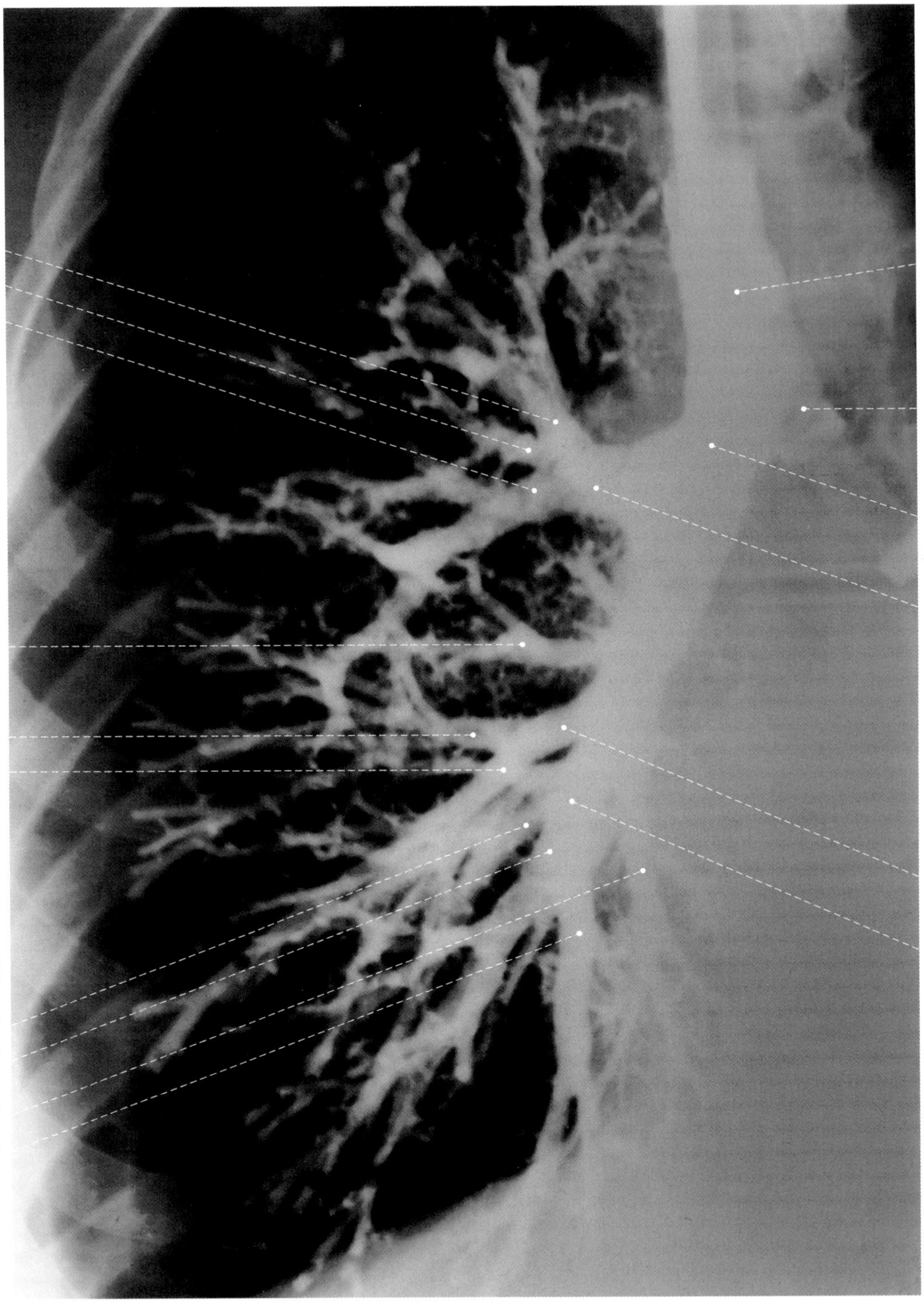

120 Arbor bronchialis

120 Right bronchial tree (75%)
Bronchography, anteroposterior radiograph

120 Rechter Bronchialbaum (75%)
Bronchographie, anteroposteriore Röntgenaufnahme

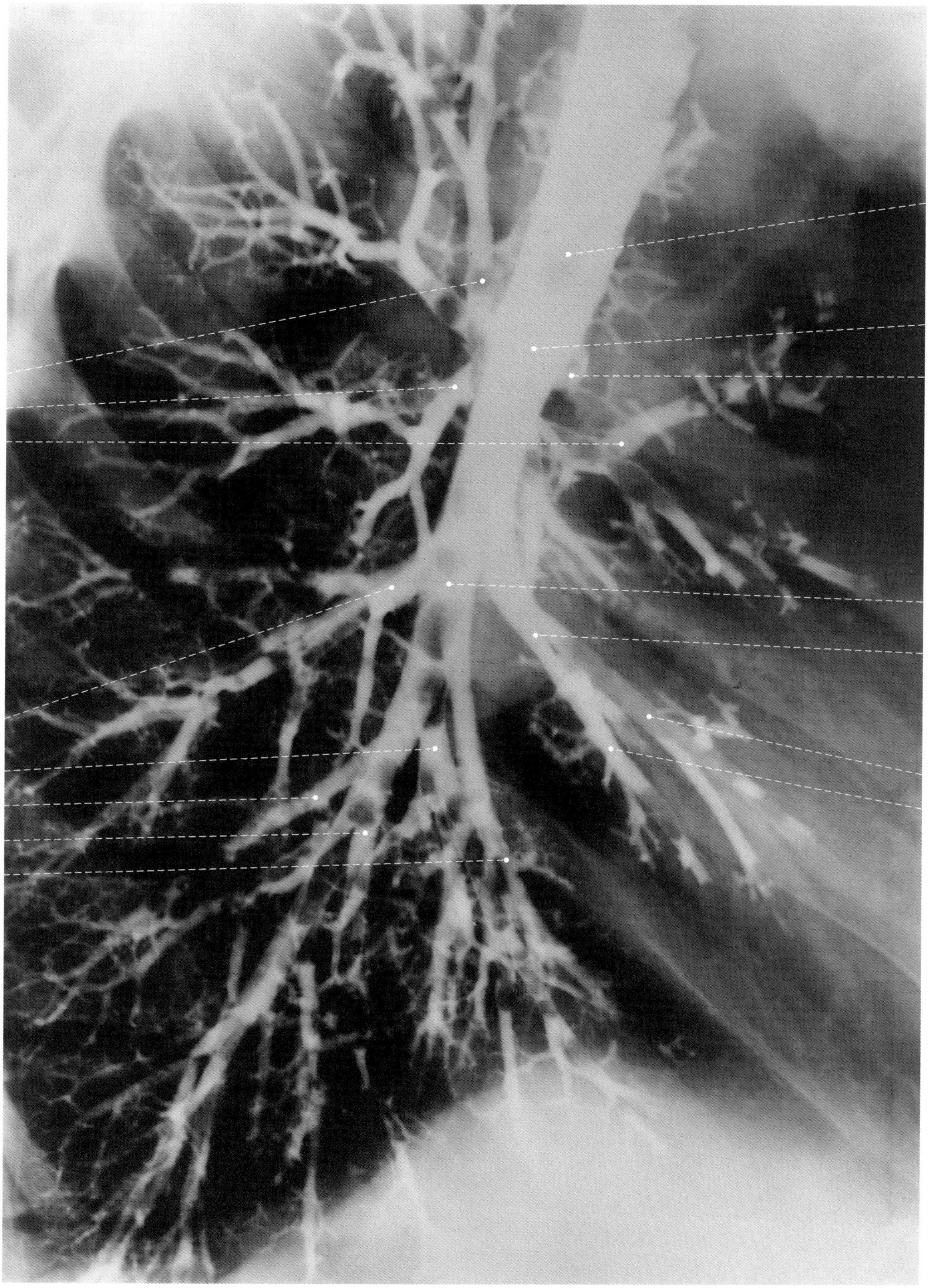

121 Arbor bronchialis

121 Right bronchial tree (70%)
Bronchography, lateral radiograph

121 Rechter Bronchialbaum (70%)
Bronchographie, seitliche Röntgenaufnahme

a

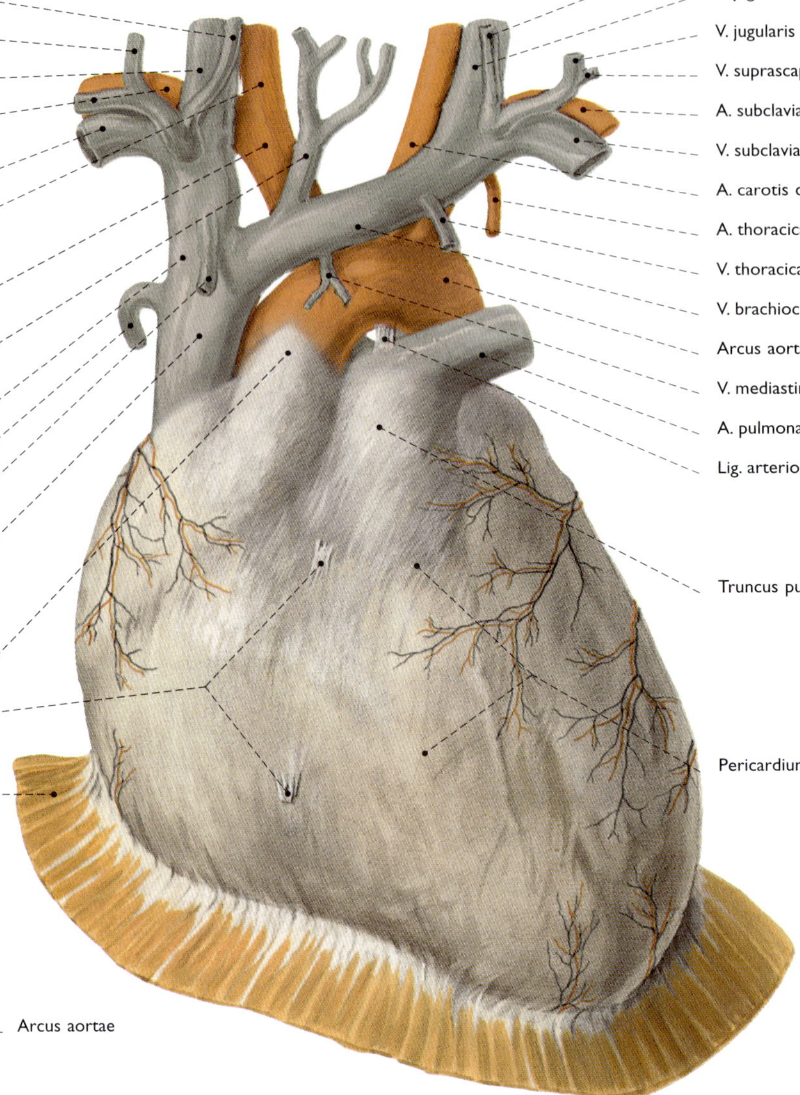

b

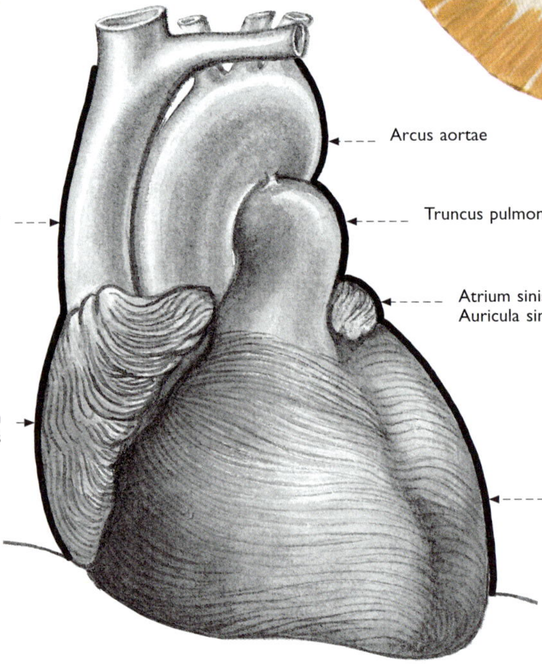

122 Pericardium fibrosum

122 Pericardial sac and silhouette of the heart

Ventral aspect

- a Pericardial sac and large blood vessels close to the heart (75%)
- b Silhouette of the heart marked by the black marginal line (50%), schematic representation

122 Herzbeutel und Herzsilhouette

Ventralansicht

- a Herzbeutel und grosse herznahe Gefässe (75%)
- b Herzsilhouette, hervorgehoben durch die schwarze Randlinie (50%), schematische Darstellung

a

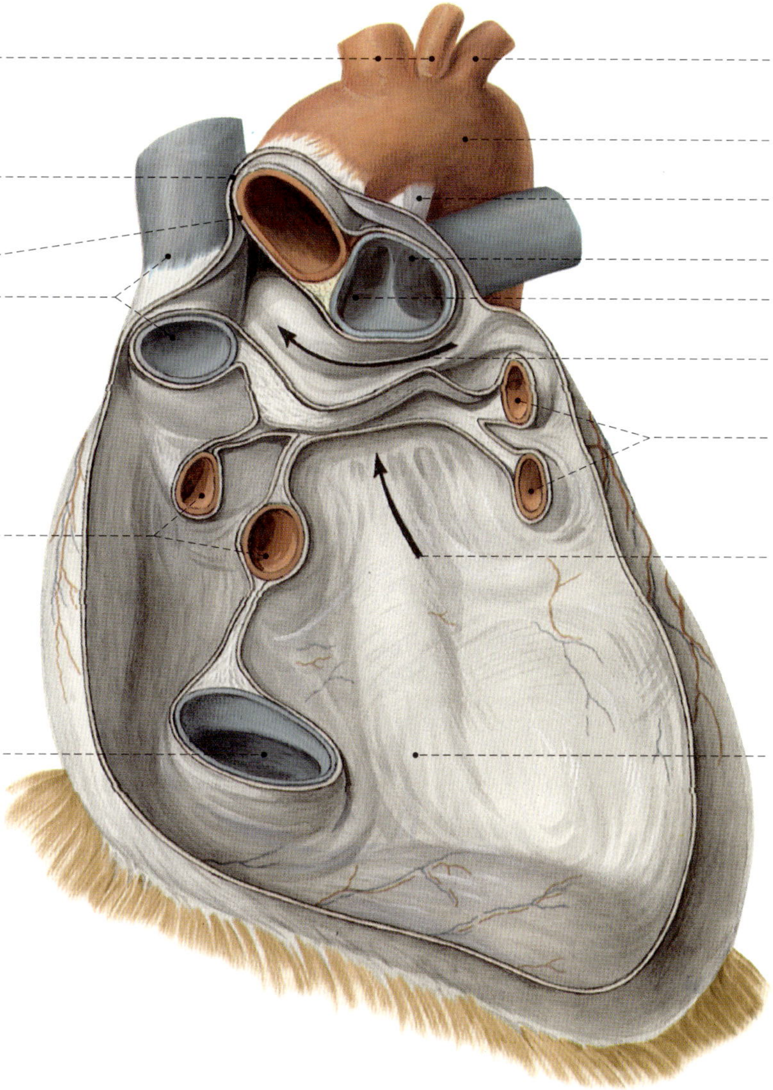

b

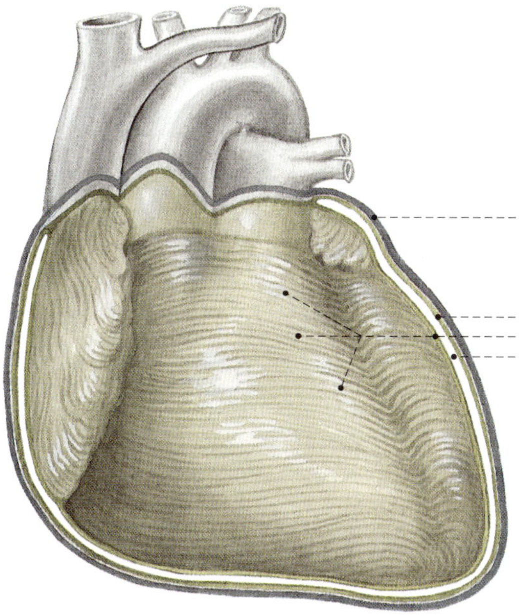

123 Pericardium

123 Pericardial sac

Ventral aspect

a Posterior wall of the pericardial cavity (75%)

b Construction of the serous and fibrous pericardium (50%), schematic representation

123 Herzbeutel

Ventralansicht

a Hinterwand des Herzbeutels (75%)

b Konstruktion des serösen und fibrösen Perikards (50%), schematische Darstellung

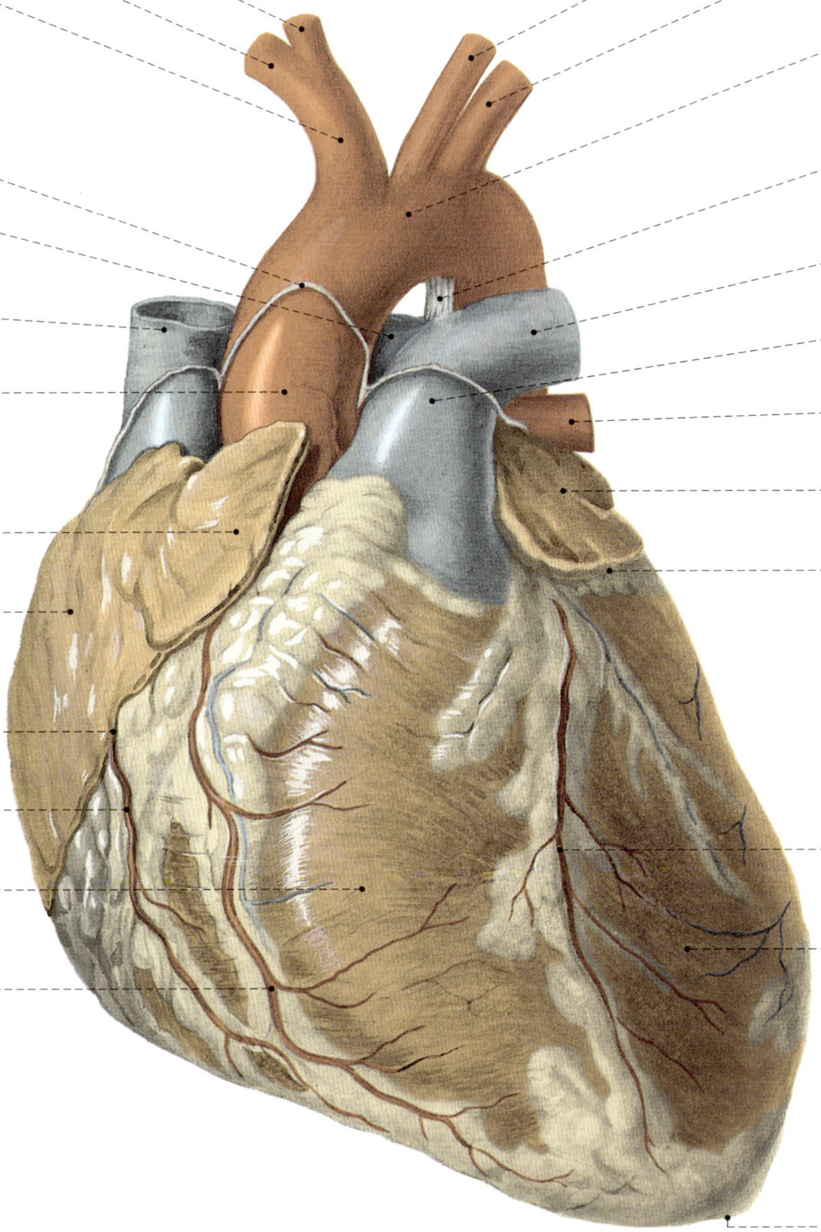

124 Cor

124 Heart and large blood vessels

(100%)

The fibrous pericardium and the parietal layer of the serous pericardium were removed. Ventral aspect

124 Herz und herznahe Gefässe

(100%)

Das fibröse Perikard und die wandständige Schicht des serösen Perikards sind entfernt. Ventralansicht

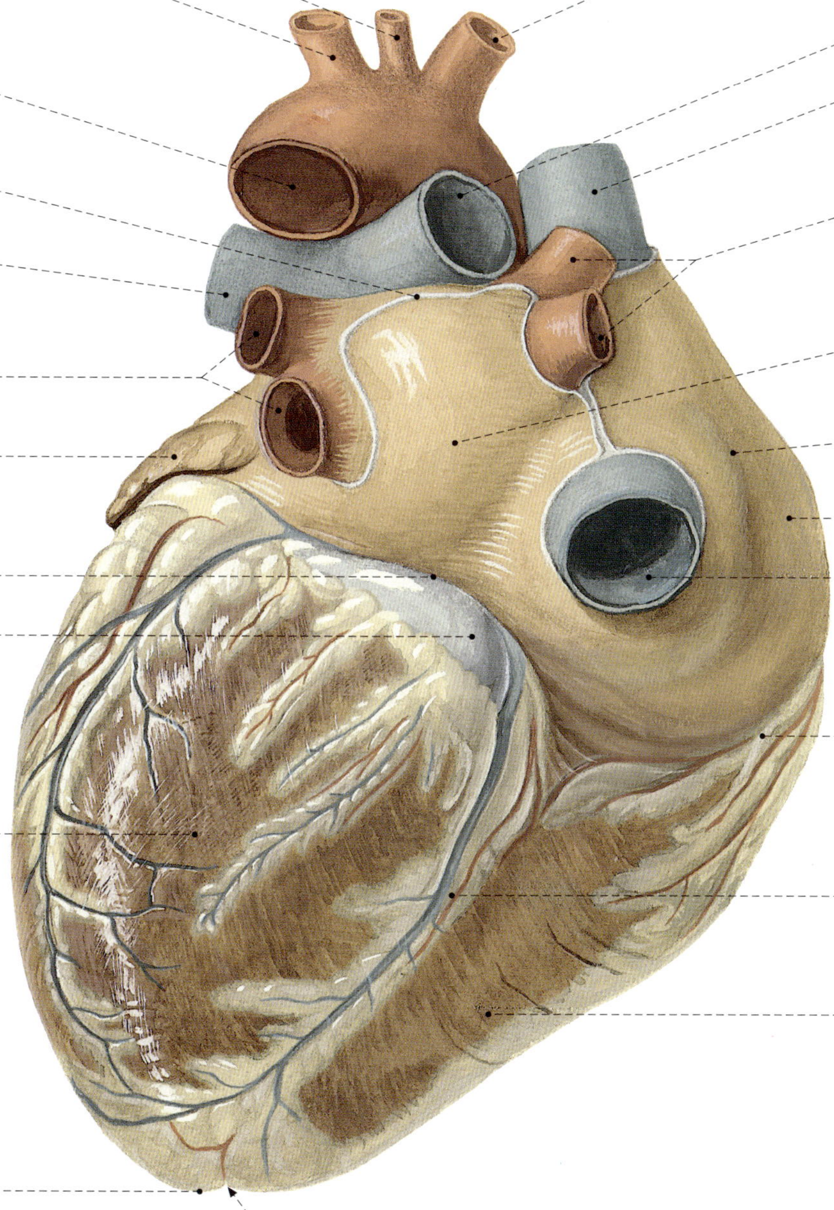

125 Cor

125 Heart and large blood vessels
(100%)
The fibrous pericardium and the parietal layer of the serous pericardium were removed. Dorsal aspect

125 Herz und herznahe Gefässe
(100%)
Das fibröse Perikard und die wandständige Schicht des serösen Perikards sind entfernt. Dorsalansicht

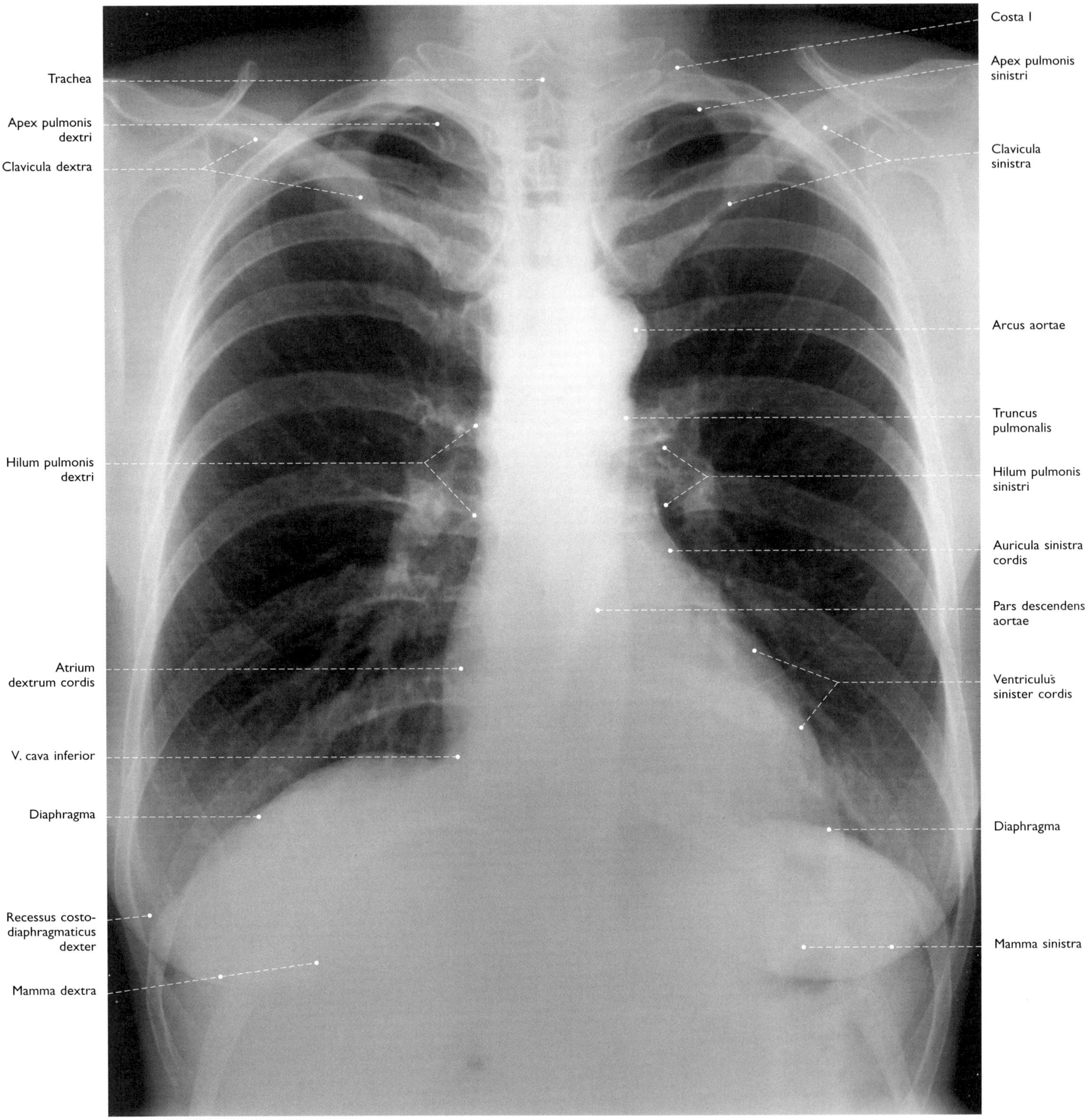

126 Thorax

126 Thoracic organs (55%)
Posteroanterior radiograph

126 Brustorgane (55%)
Posteroanteriore Röntgenaufnahme

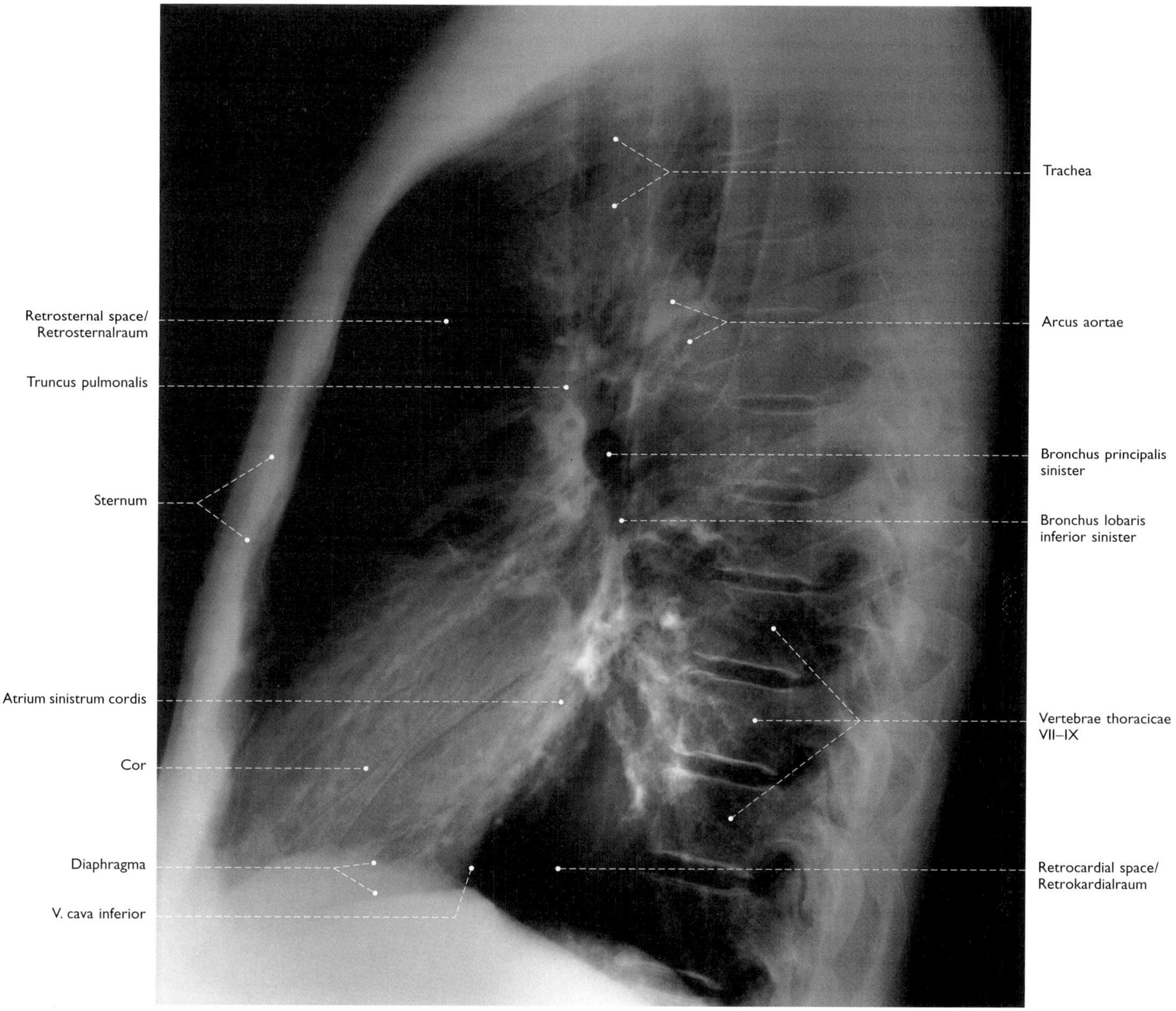

127 Thorax

127 Thoracic organs (55%)
Lateral radiograph

127 Brustorgane (55%)
Seitliche Röntgenaufnahme

a

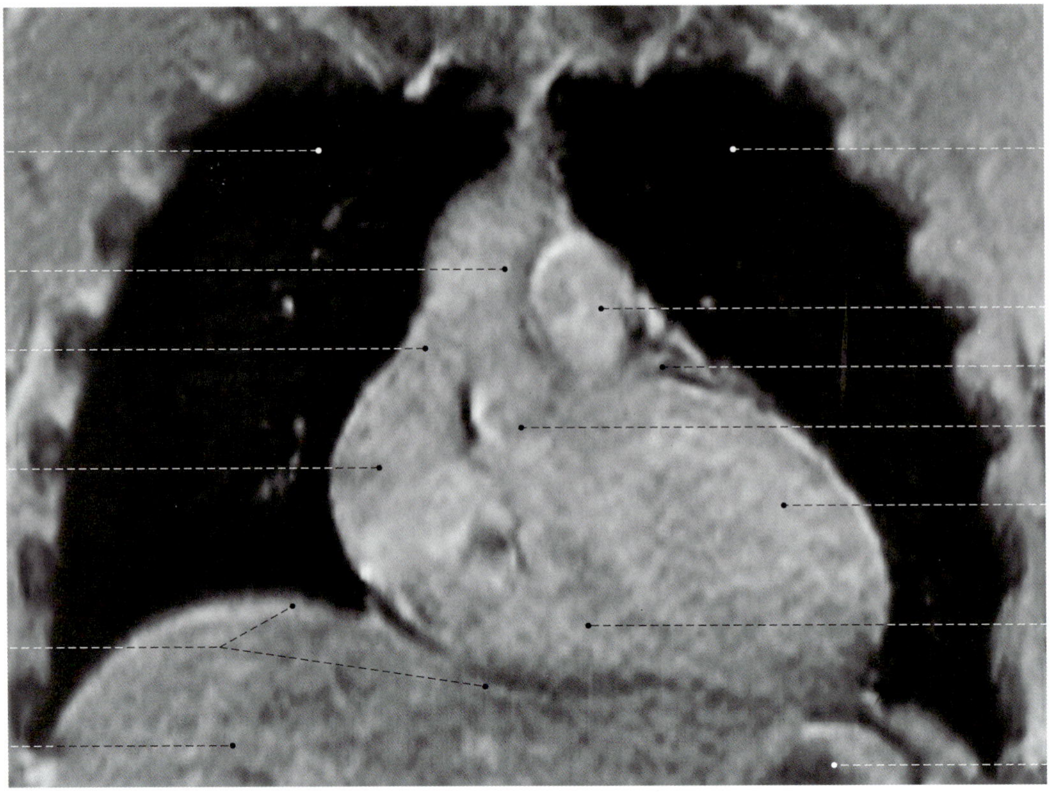

b

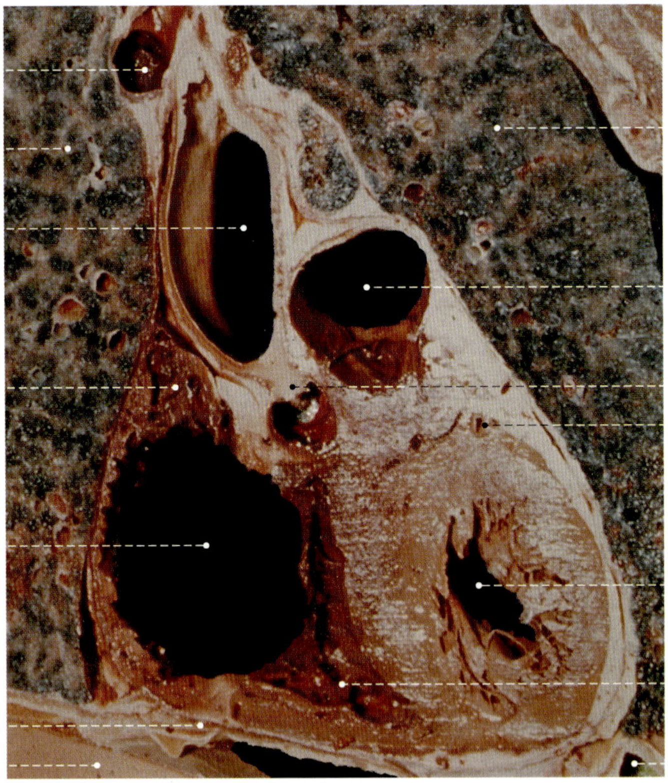

128 Cor

128 Heart and large blood vessels (55%)

Coronal sections through the thorax in the plane of ventral parts of the heart, ventral aspect

a Magnetic resonance image (MRI, T_2-weighted)
b Anatomical section

128 Herz und herznahe Gefässe (55%)

Frontalschnitte durch den Thorax im Bereich ventraler Herzabschnitte, Ventralansicht

a Kernspintomogramm (MRI, T_2-gewichtet)
b Anatomischer Schnitt

a

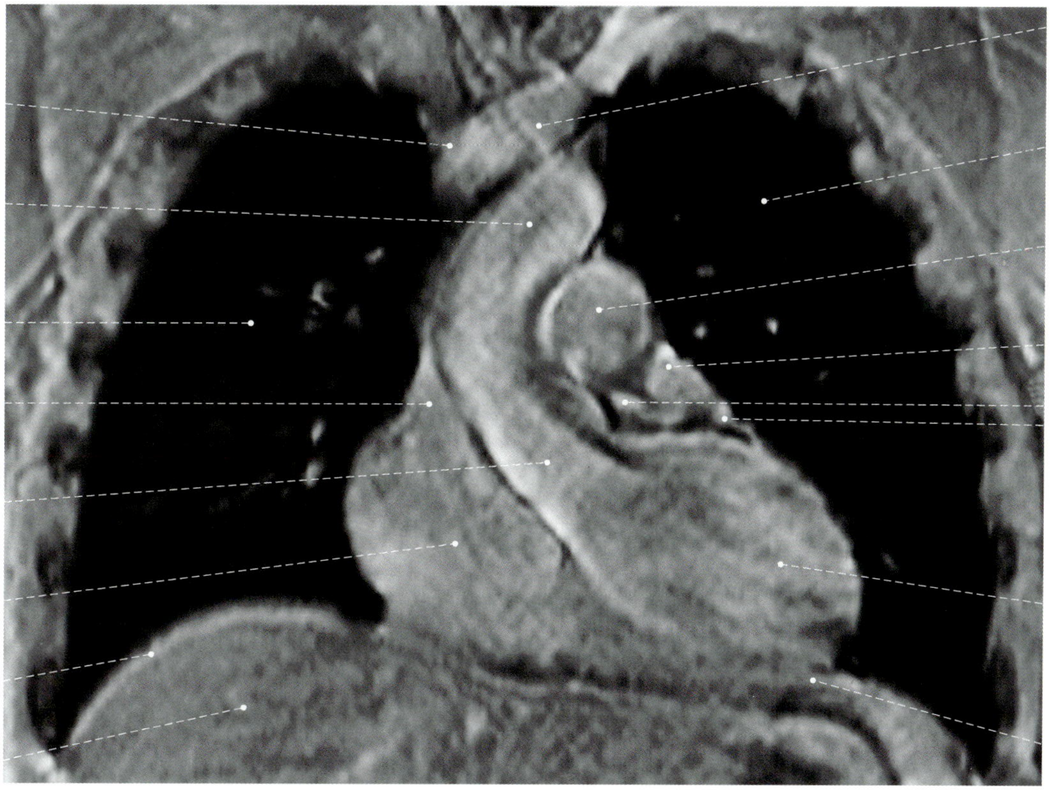

b

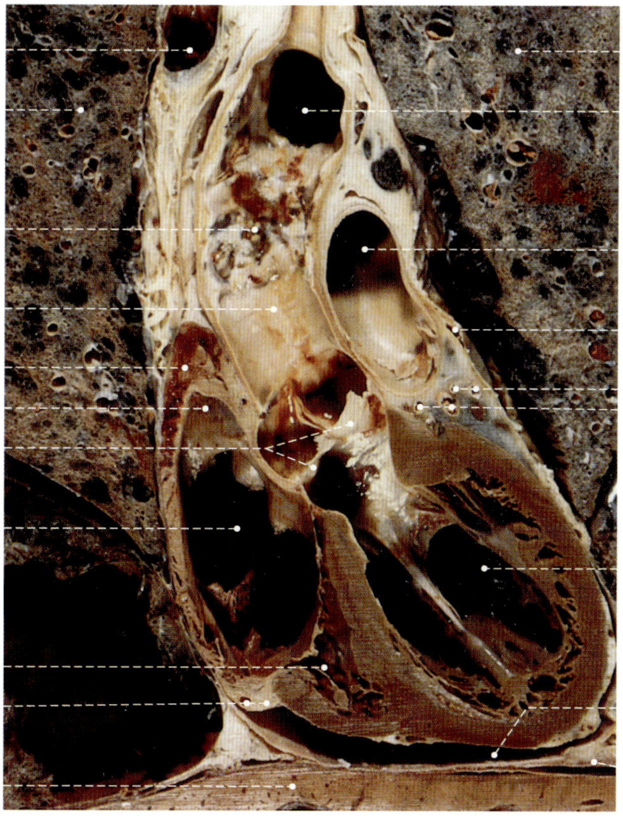

129 Cor

129 Heart and large blood vessels (55%)

Coronal sections through the thorax and the heart cut more dorsally than in Fig. 128, ventral aspect

a Magnetic resonance image (MRI, T_2-weighted)
b Anatomical section

129 Herz und herznahe Gefässe (55%)

Frontalschnitte durch den Thorax und das Herz etwas dorsaler als in Abb. 128, Ventralansicht

a Kernspintomogramm (MRI, T_2-gewichtet)
b Anatomischer Schnitt

a

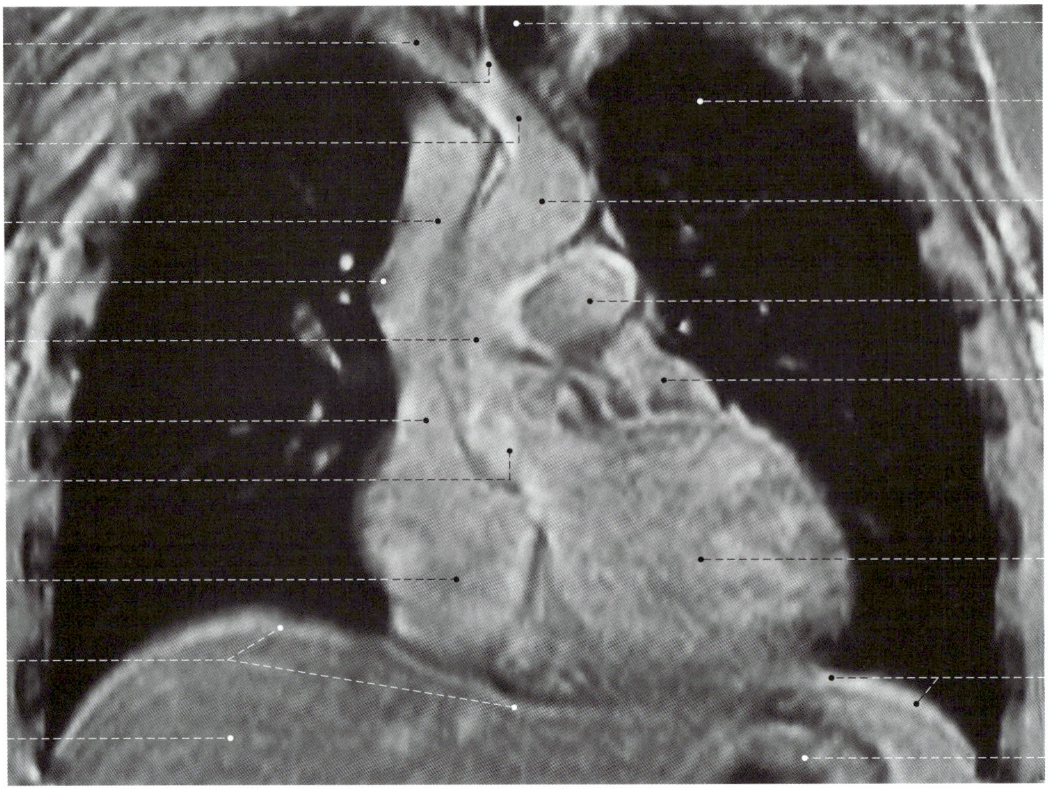

b

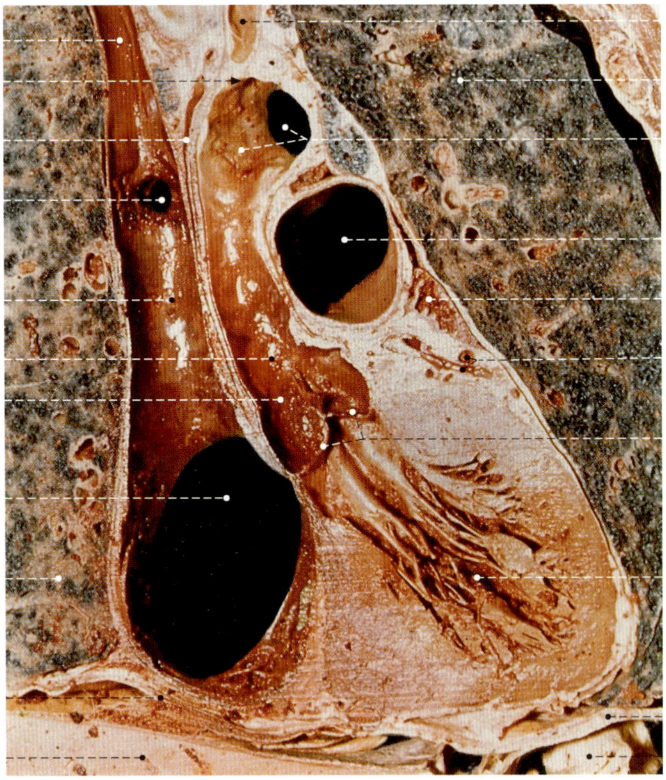

130 Cor

130 Heart and large blood vessels (55%)

Coronal sections through the thorax and the heart, cut a little more dorsally than in Fig. 129, ventral aspect

a Magnetic resonance image (MRI, T_2-weighted)

b Anatomical section

130 Herz und herznahe Gefässe (55%)

Frontalschnitte durch den Thorax und das Herz etwas dorsaler als in Abb. 129, Ventralansicht

a Kernspintomogramm (MRI, T_2-gewichtet)

b Anatomischer Schnitt

a

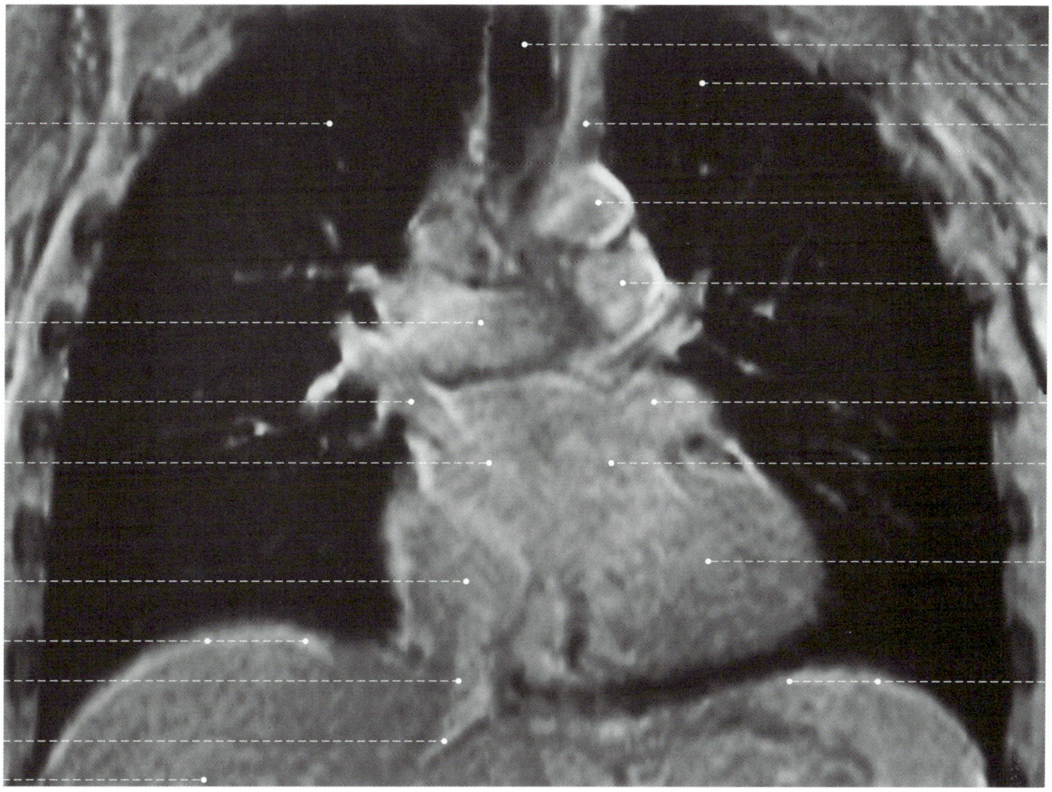

b

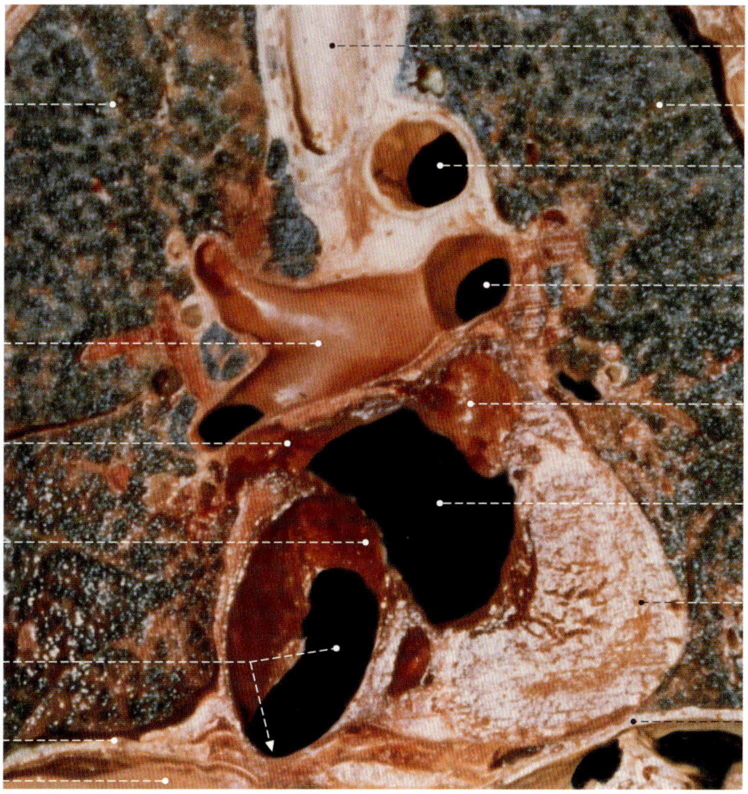

131 Cor

131 Heart and large blood vessels (55%)

Coronal sections through the thorax in the plane of dorsal parts of the heart, ventral aspect

a Magnetic resonance image (MRI, T_2-weighted)

b Anatomical section

131 Herz und herznahe Gefässe (55%)

Frontalschnitte durch den Thorax im Bereich dorsaler Herzabschnitte, Ventralansicht

a Kernspintomogramm (MRI, T_2-gewichtet)

b Anatomischer Schnitt

a

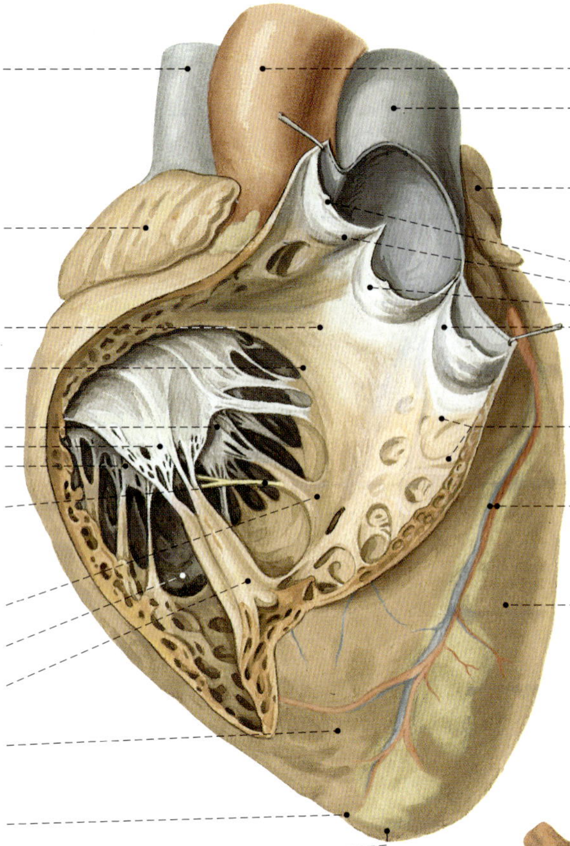

b

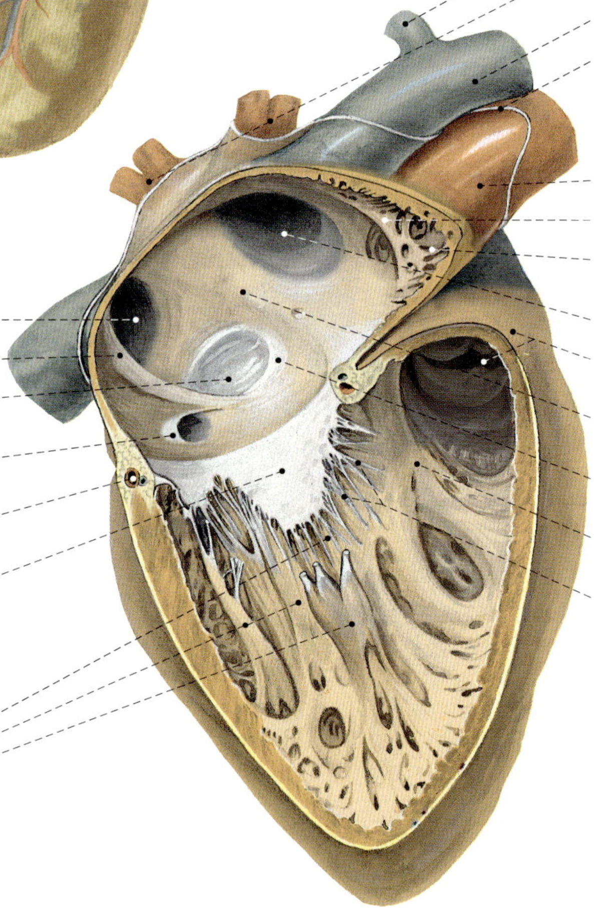

132 Cor

132 Heart (70%)

a Right ventricle and ostium trunci pulmonalis after having cut a window in the anterior walls of the right ventricle and the ostium trunci pulmonalis, ventral aspect

b Right atrium and right ventricle of the heart after removal of the lateral walls of the right atrium and the right ventricle, lateral aspect

132 Herz (70%)

a Rechter Ventrikel und Ostium trunci pulmonalis nach Fensterung der ventralen Herz- und Gefässwand, Ventralansicht

b Rechter Vorhof und rechte Kammer des Herzens nach Entfernung der seitlichen Wände des rechten Vorhofs und der rechten Kammer, Seitenansicht

a

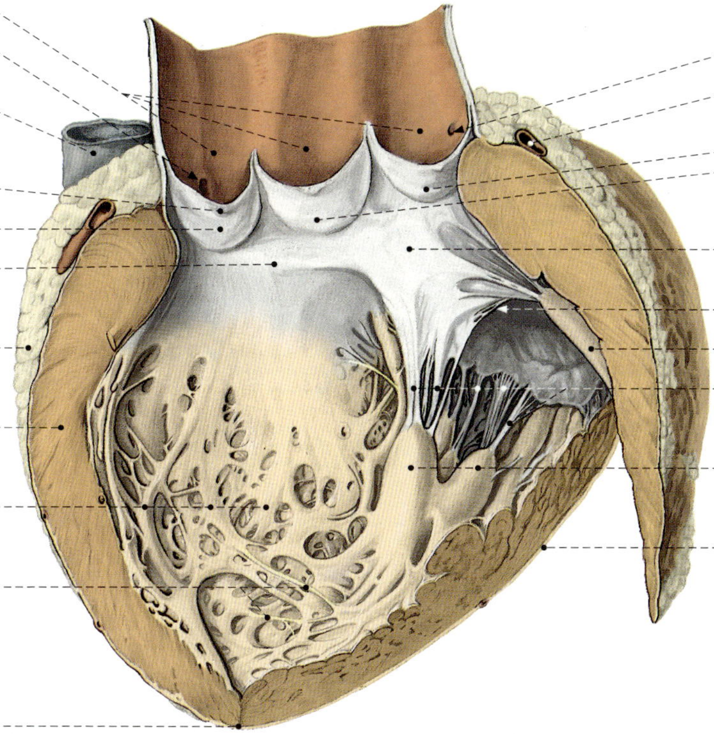

b

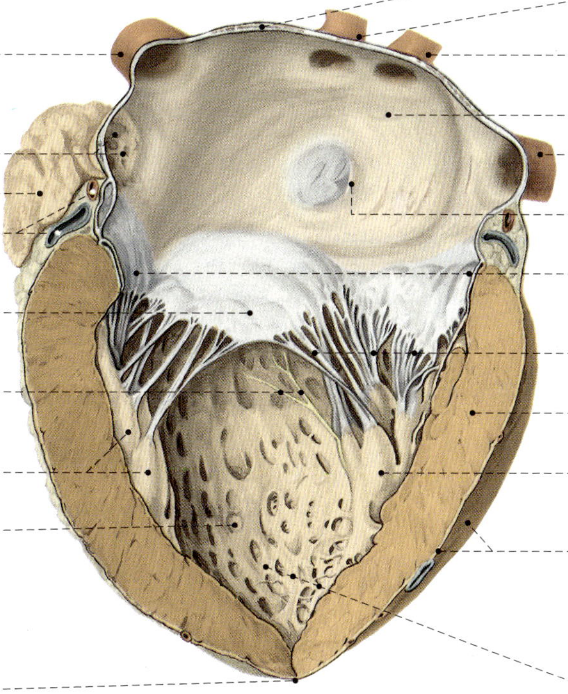

133 Cor

133 Heart (70%)

a Internal aspect of the left ventricle and the ascending aorta (effluent blood pathway). Incision from the apex of the heart to the point between the right and left semilunar cusps of the aortic valve. A second longitudinal section exposes the affluent blood pathway of the left ventricle.

b Internal aspect of the left atrium and the affluent blood pathway of the left ventricle. Longitudinal incision along the rounded (left) margin of the heart

133 Herz (70%)

a Innenansicht des linken Ventrikels und der aufsteigenden Aorta (Ausströmungsbahn). Längsschnitt von der Herzspitze zur Grenze zwischen der rechten und linken Tasche der Aortenklappe. Ein zweiter Längsschnitt eröffnet die Einströmungsbahn der linken Kammer.

b Innenansicht des linken Vorhofs und der Einströmungsbahn der linken Kammer. Längsschnitt entlang des stumpfen (linken) Herzrands

a

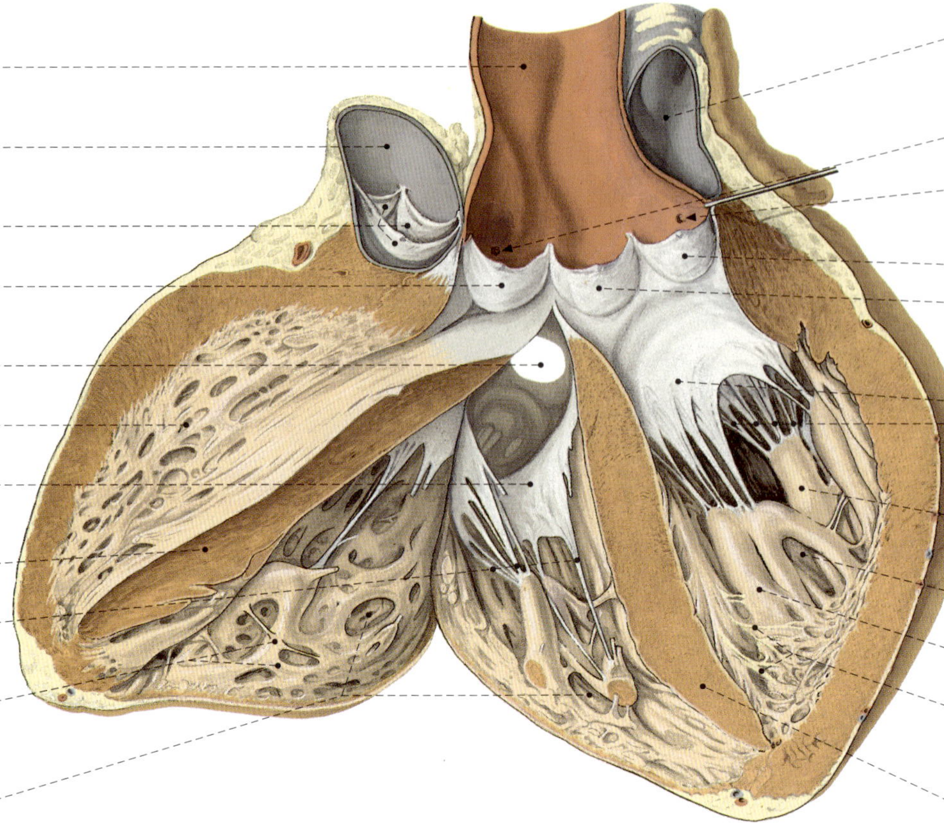

b

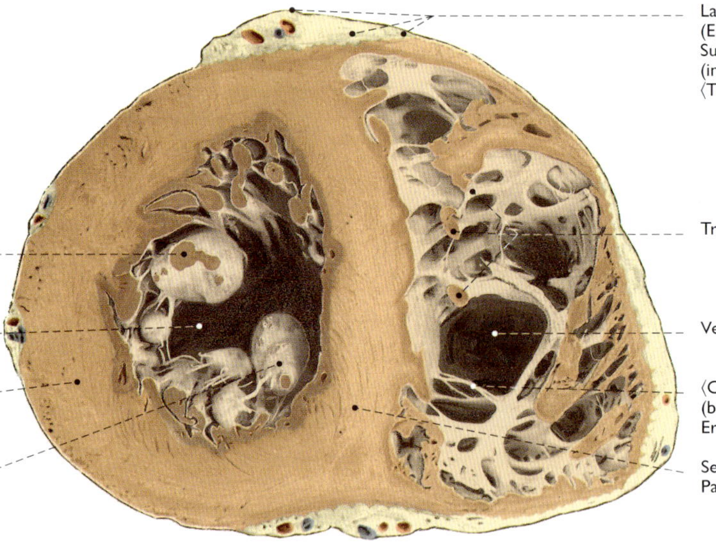

134 Cor

134 Heart

a Internal aspect of both ventricles of the heart and the left effluent blood pathway. Longitudinal section perpendicular to the interventricular septum (70%)

b Superior view into both ventricles of the heart, opened by a transverse section (80%)

134 Herz

a Innenansicht der beiden Herzkammern und der Ausströmungsbahn der linken Kammer. Längsschnitt senkrecht zur Kammerseptumebene (70%)

b Kranialeinblick in beide, durch Transversalschnitt eröffnete Herzkammern (80%)

a

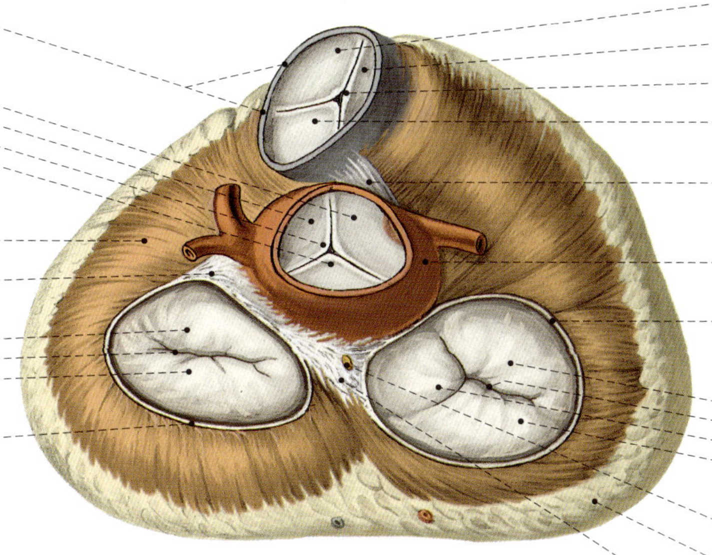

b

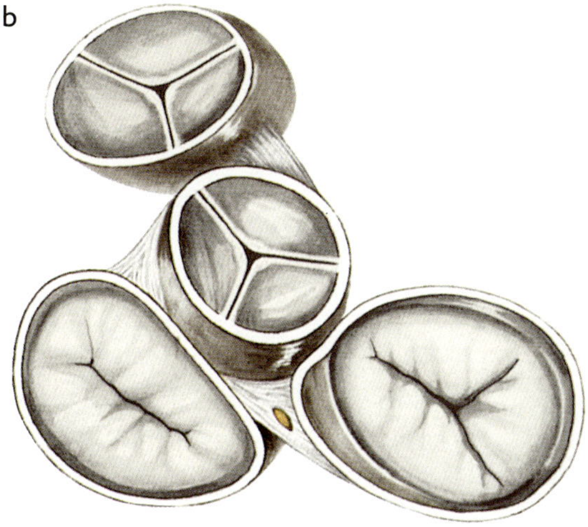

c

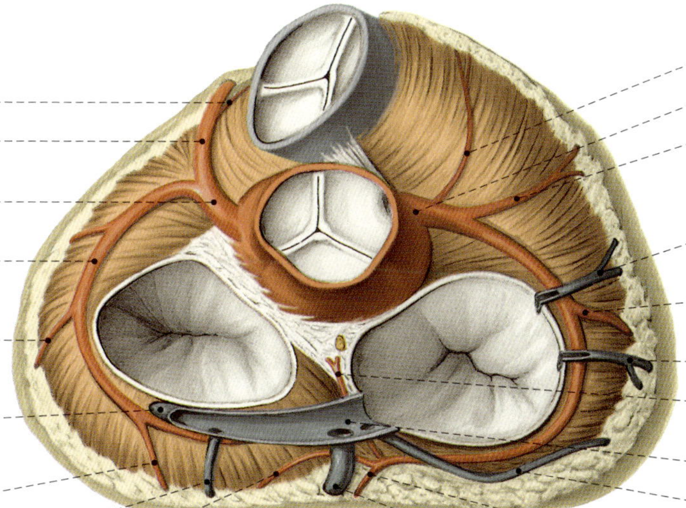

135 Cor

135 Heart (80%)

Superior aspect

- a Base of the ventricles with the atrioventricular and arterial orifices
- b Fibrous base of the heart (anuli fibrosi et trigona fibrosa)
- c Origin of the coronary vessels

135 Herz (80%)

Kranialansicht

- a Ventrikelbasis mit atrioventrikulären und arteriellen Ostien
- b Herzskelett (Anuli fibrosi et Trigona fibrosa)
- c Ursprung der Koronargefässe

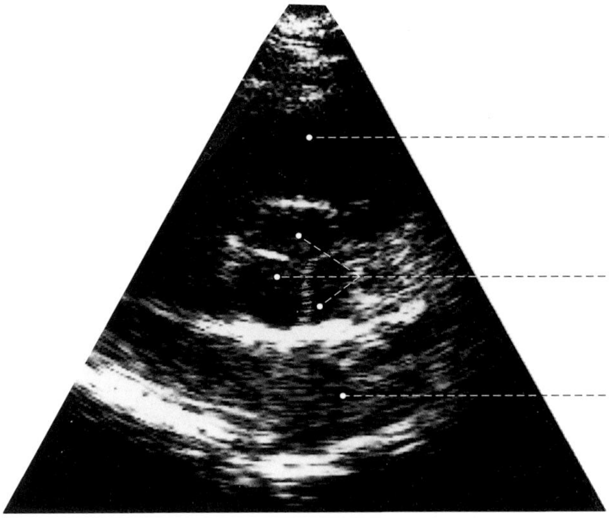

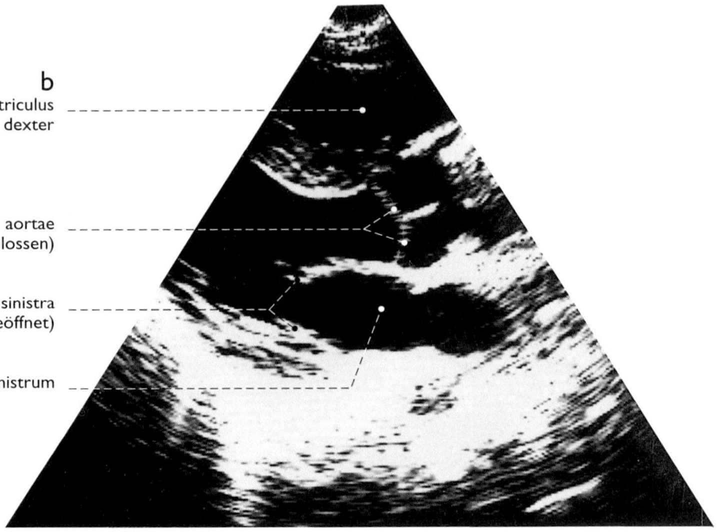

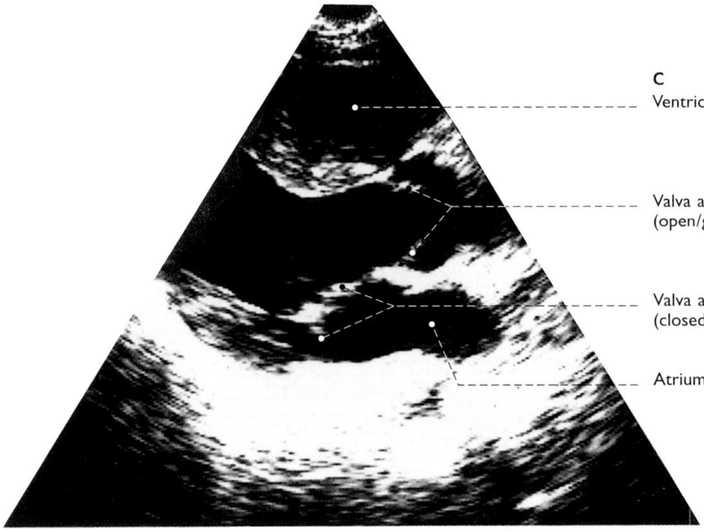

136 Cor

136 Heart

Ultrasound images (echocardiograms)

- a Transverse section through the aortic valve along the short axis. The aortic valve is closed. Inferior aspect
- b, c Longitudinal section through the heart along the parasternal long axis of the left ventricle
  - b when, during the filling phase of the ventricular diastole, the aortic valve is closed and the mitral valve open
  - c when, during the ejection phase of the ventricular systole, the aortic valve is open and the mitral valve closed

136 Herz

Ultraschallbilder (Echokardiogramme)

- a Querschnitt durch die Aortenwurzel entlang der kurzen Achse. Die Aortenklappe ist geschlossen. Kaudalansicht
- b, c Längsschnitt durch das Herz entlang der parasternalen Längsachse des linken Ventrikels
  - b bei geschlossener Aortenklappe und geöffneter Mitralklappe während der Füllungsphase der Kammerdiastole
  - c bei geöffneter Aortenklappe und geschlossener Mitralklappe während der Austreibungsphase der Kammersystole

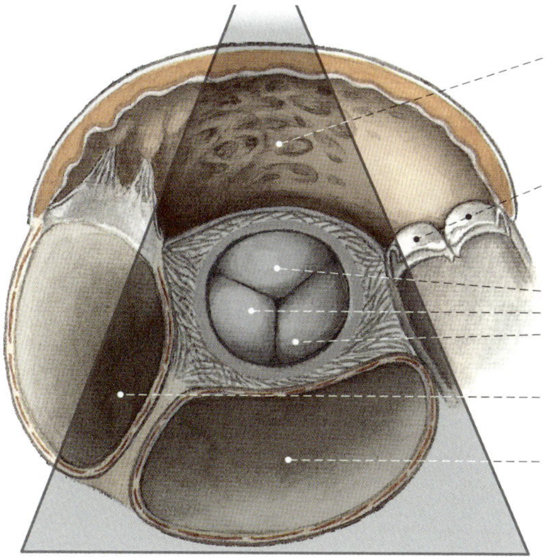

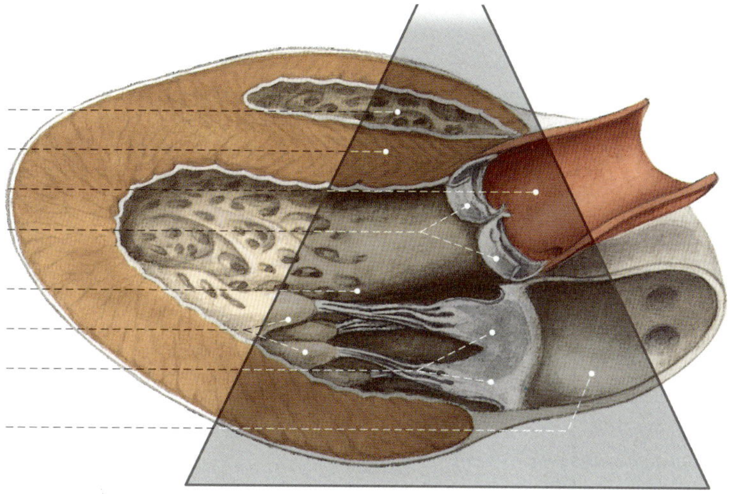

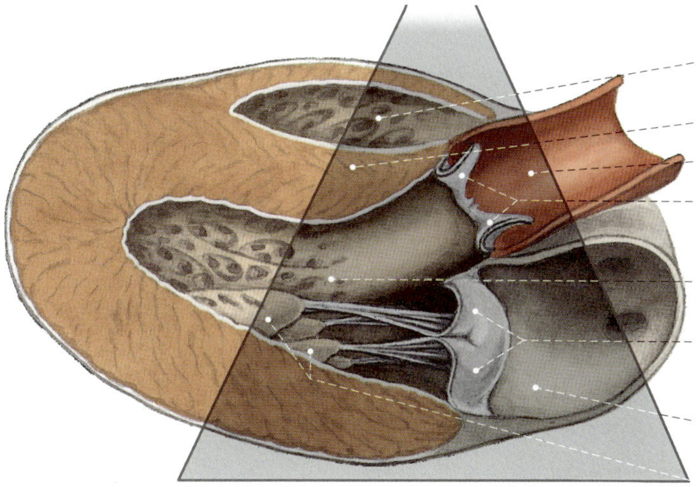

137 Cor

137 Heart

Sections through the heart in analogy to the echocardiograms of Fig. 136. The sectors shown in these echocardiograms are indicated by lines.

a Transverse section just below the aortic valve. The aortic valve is closed. Inferior aspect

b, c Longitudinal section through the left ventricle

b when, during the filling phase of the ventricular diastole, the aortic valve is closed and the mitral valve open

c when, during the ejection phase of the ventricular systole, the aortic valve is open and the mitral valve closed

137 Herz

Schnitte durch das Herz analog zu den Echokardiogrammen in Abb. 136. Die in diesen Echokardiogrammen dargestellten Ausschnitte sind durch Linien markiert.

a Querschnitt durch das Herz in Höhe der Aortenwurzel bei geschlossener Aortenklappe, Kaudalansicht

b, c Längsschnitt durch den linken Ventrikel

b bei geschlossener Aortenklappe und geöffneter Mitralklappe während der Füllungsphase der Kammerdiastole

c bei geöffneter Aortenklappe und geschlossener Mitralklappe während der Austreibungsphase der Kammersystole

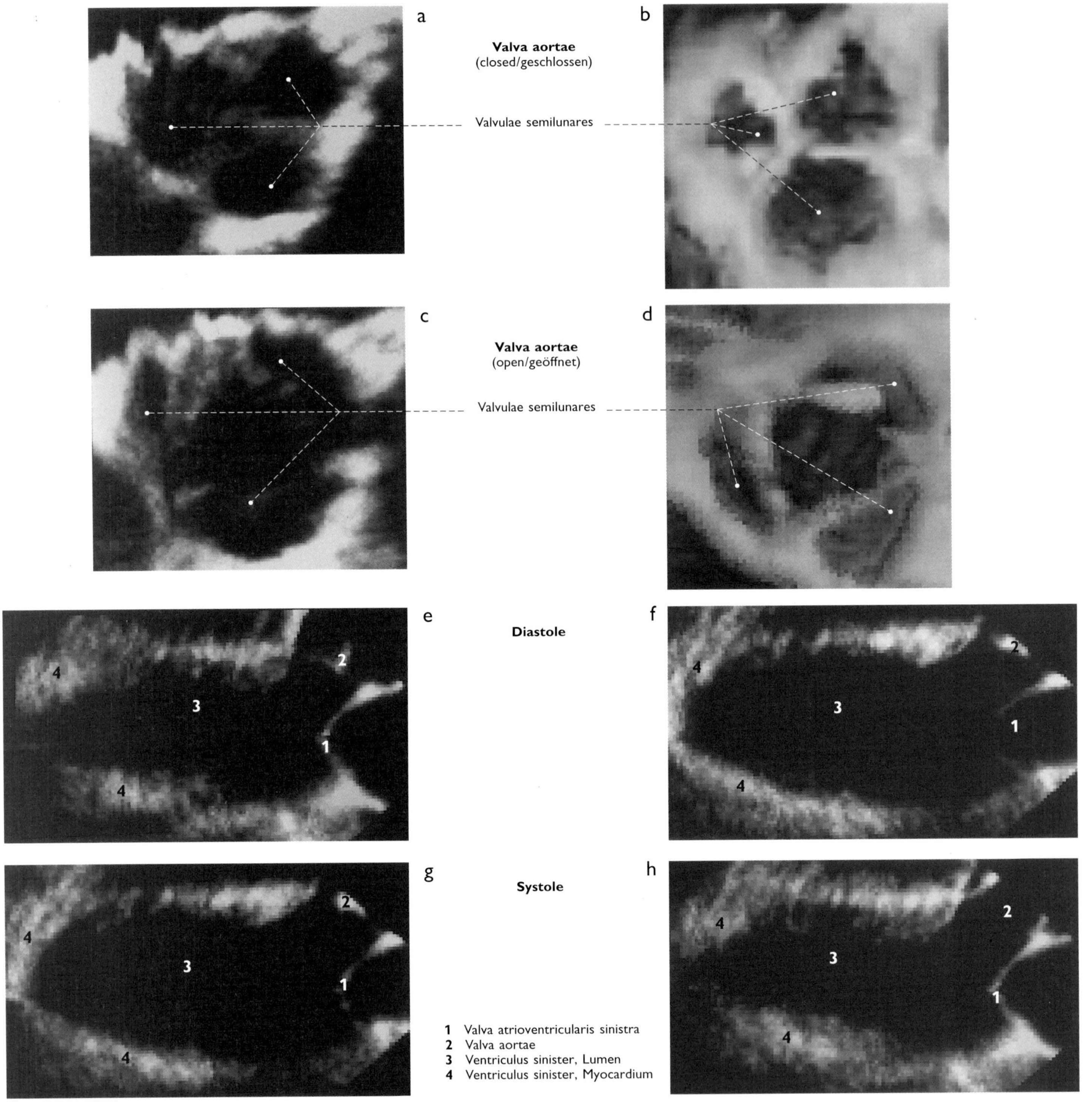

138 Cor

138 Heart

Ultrasound images (echocardiograms)

- a–d Transverse sections through the aortic valve along the short axis when the aortic valve is
- a, b closed
- c, d open
- e–h Longitudinal sections through the left ventricle
- e during the relaxation phase of the diastole when the mitral and aortic valves are closed
- f during the filling phase of the diastole when the mitral valve is open
- g during the contraction phase of the systole when the mitral and aortic valves are closed
- h during the ejection phase of the systole when the aortic valve is open

138 Herz

Ultraschallbilder (Echokardiogramme)

- a–d Querschnitte durch die Aortenwurzel entlang der kurzen Achse
- a, b bei geschlossener Aortenklappe
- c, d bei geöffneter Aortenklappe
- e–h Längsschnitte durch den linken Ventrikel
- e während der Entspannungsphase der Diastole bei geschlossener Mitral- und Aortenklappe
- f während der Füllungsphase der Diastole bei geöffneter Mitralklappe
- g während der Anspannungsphase der Systole bei geschlossener Mitral- und Aortenklappe
- h während der Austreibungsphase der Systole bei geöffneter Aortenklappe

a

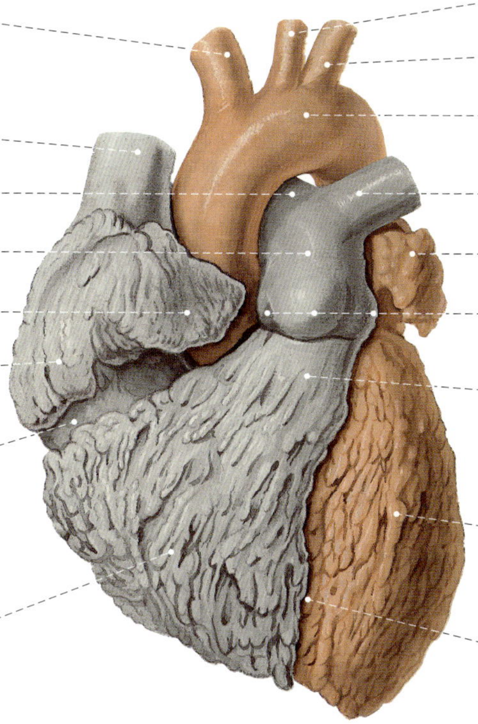

b

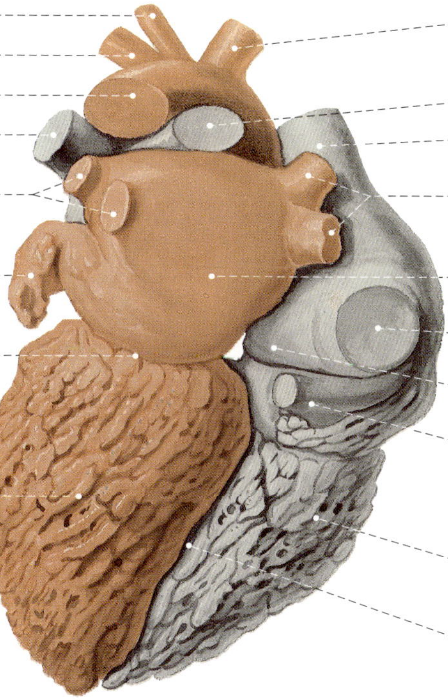

Right heart **blue**, left heart **red**

Rechte Herzräume **blau**, linke Herzräume **rot**

139 Cor

139 Cast of the cavities of the heart (80%)

a Ventral aspect
b Dorsal aspect

139 Ausguss der Binnenräume des Herzens (80%)

a Ventralansicht
b Dorsalansicht

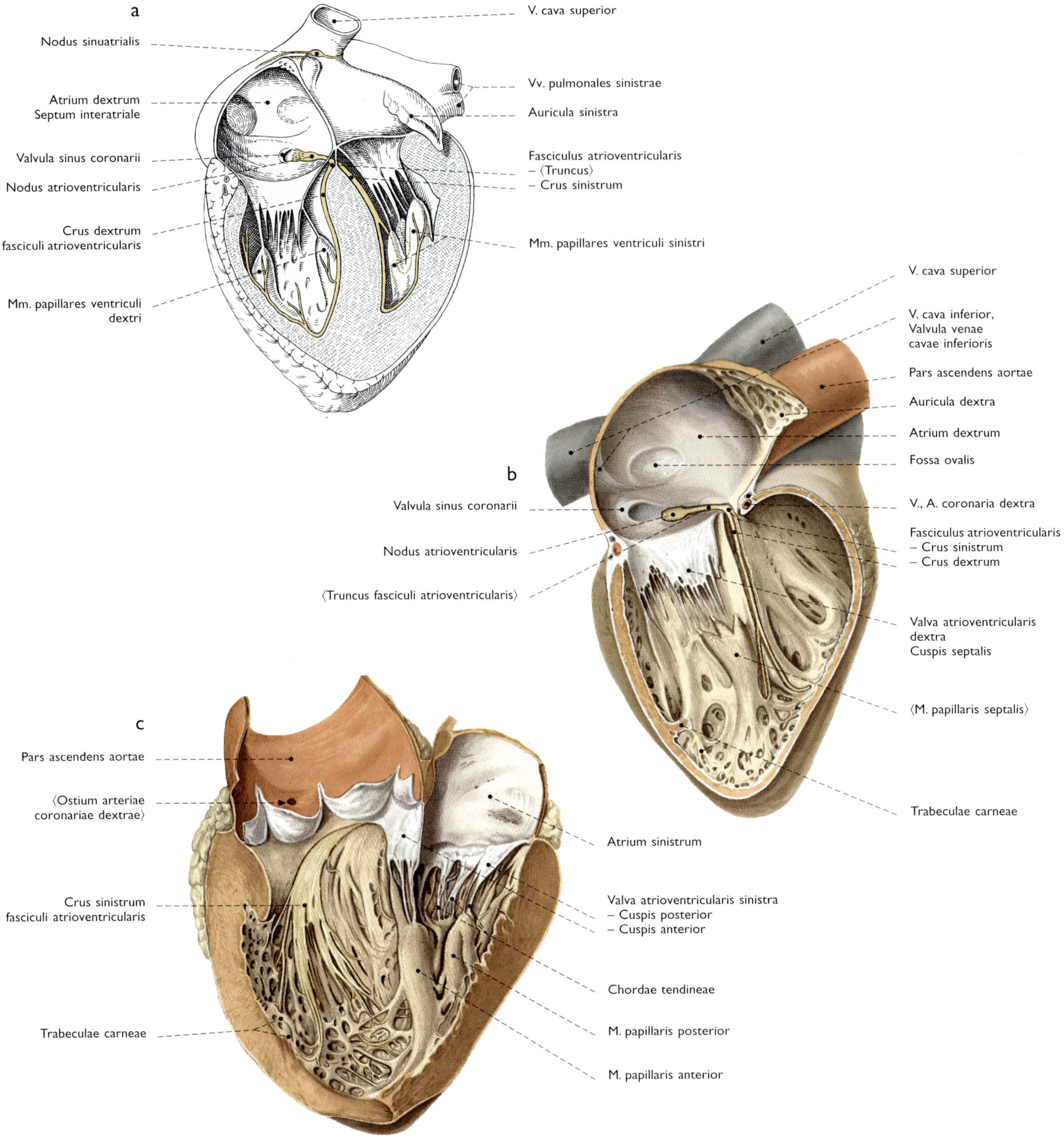

140 Systema conducens cordis

140 Conducting system of the heart (40%)

a The right atrium and both ventricles are opened by a longitudinal section perpendicular to the plane of the cardiac septum. Ventral aspect, schematic representation
b View from the right to the cardiac septum
c View from the left to the cardiac septum

140 Erregungsleitungssystem des Herzens (40%)

a Rechter Vorhof und beide Kammern sind durch einen Längsschnitt senkrecht zur Herzscheidewand eröffnet. Ventralansicht, schematische Darstellung
b Blick von rechts auf die Herzscheidewand
c Blick von links auf die Herzscheidewand

a

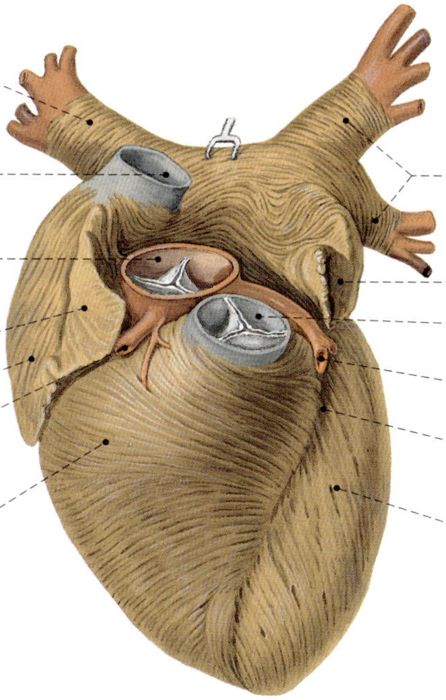

b

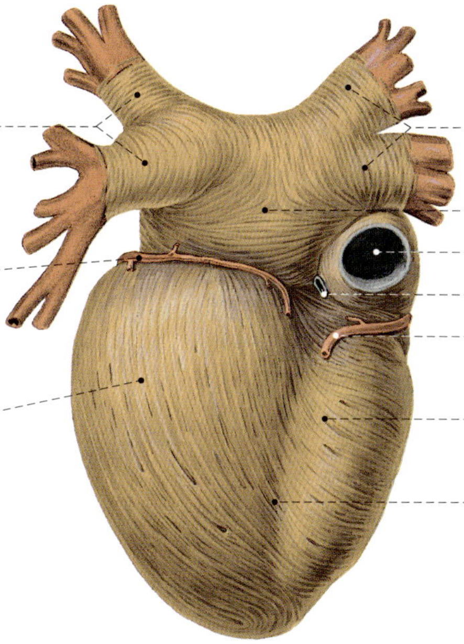

c

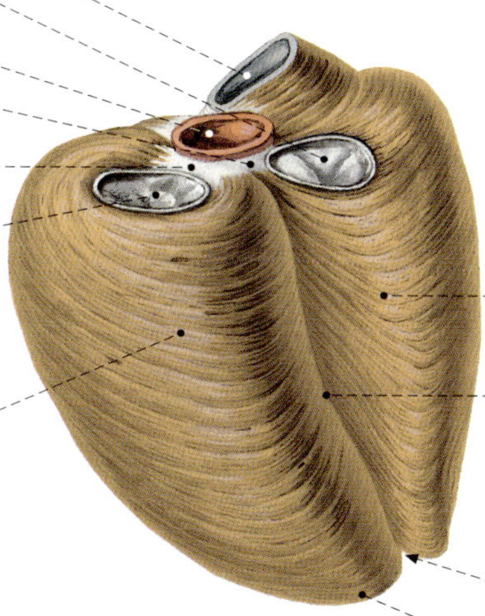

141 Myocardium

141 Myocardium (45%)

a, b Superficial layer of the muscular fibres of the heart after removal of the visceral layer of the serous pericardium (epicardium), the cardial fat and connective tissue
a Ventral aspect
b Dorsal aspect
c Middle (circular) layer of the myocardium of the ventricles, dorsal aspect

141 Herzmuskulatur (45%)

a, b Oberflächliche Schicht der Herzmuskulatur nach Entfernung der Lamina visceralis des Pericardium serosum (Epikard) und des subepikardialen Fett- und Bindegewebes
a Ventralansicht
b Dorsalansicht
c Mittlere (zirkuläre) Schicht der Herzkammermuskulatur, Dorsalansicht

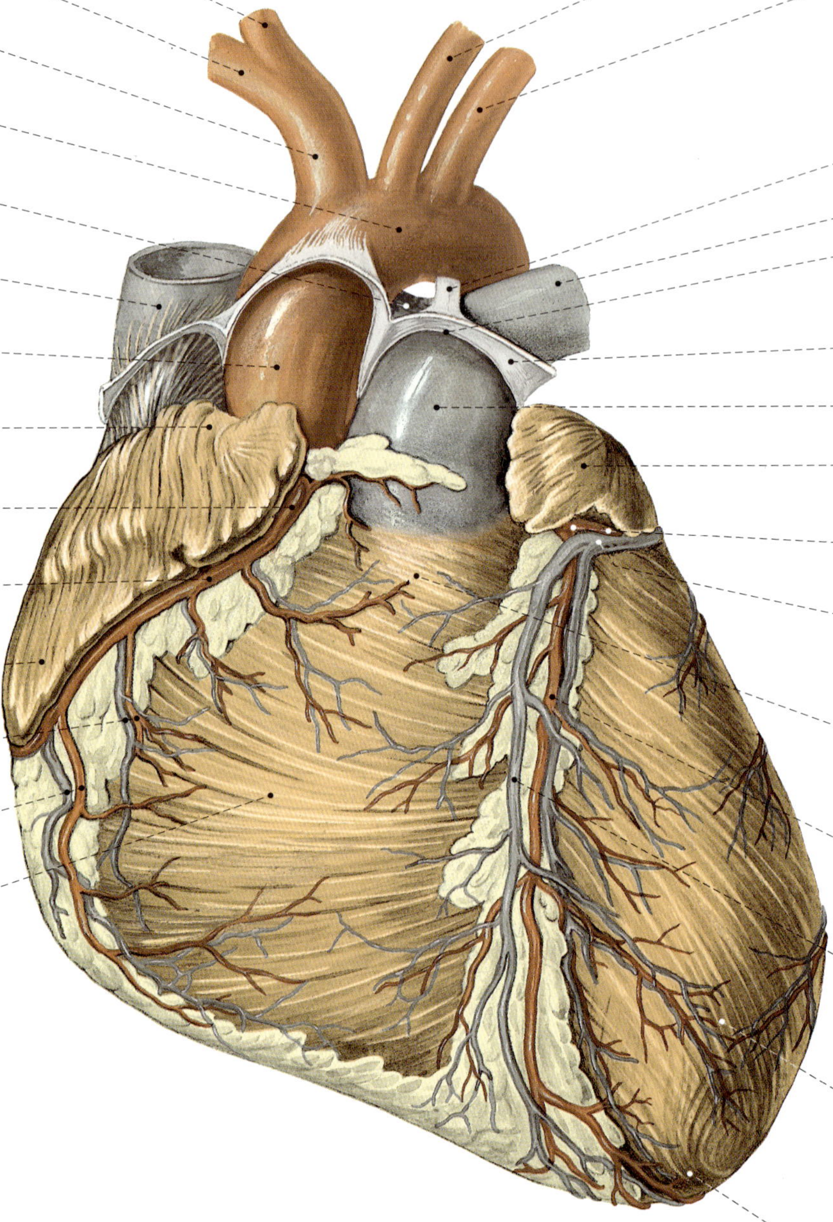

142 Arteriae et Venae coronariae

142 Arteries and veins of the heart
(100%)
The epicardium was removed. Ventral aspect

142 Arterien und Venen des Herzens
(100%)
Das Epikard ist abpräpariert. Ventralansicht

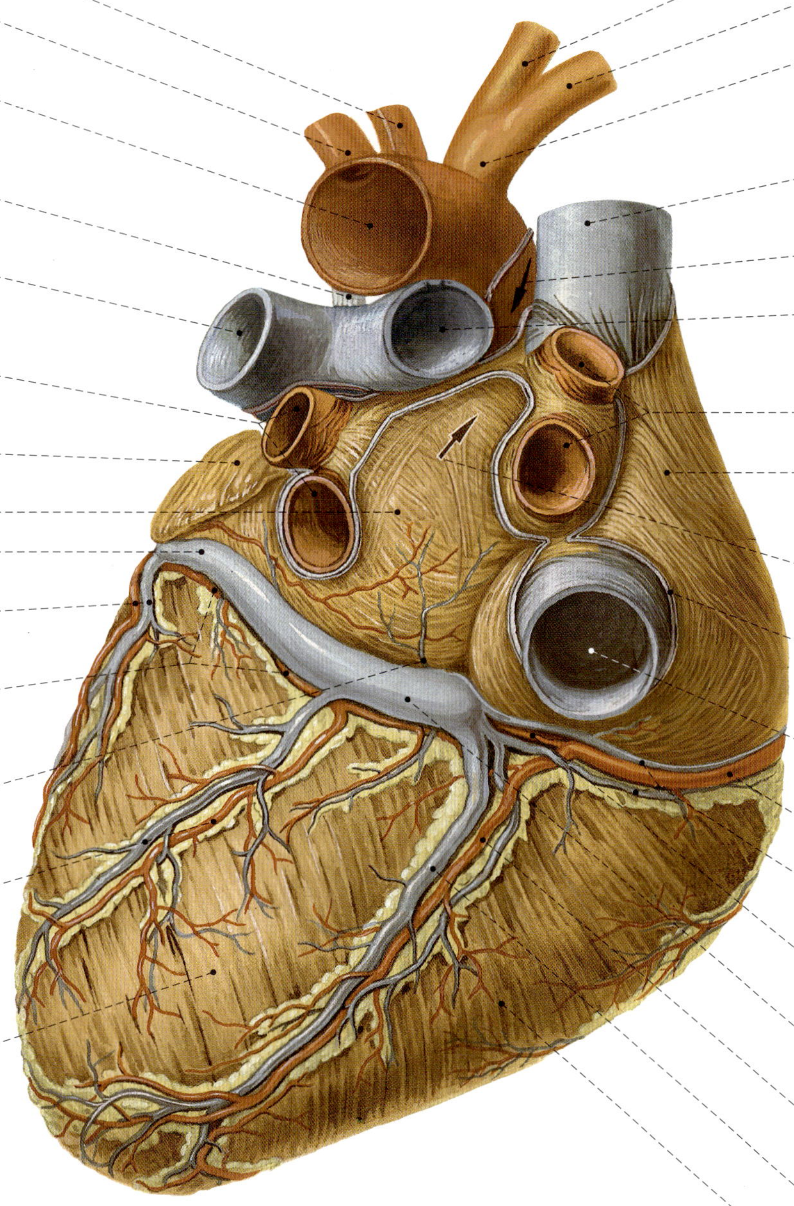

143 Arteriae et Venae coronariae

143 Arteries and veins of the heart
(100%)
The epicardium was removed. Base of the heart and diaphragmatic face, dorsal aspect

143 Arterien und Venen des Herzens
(100%)
Das Epikard ist entfernt. Basis cordis und Facies diaphragmatica, Dorsalansicht

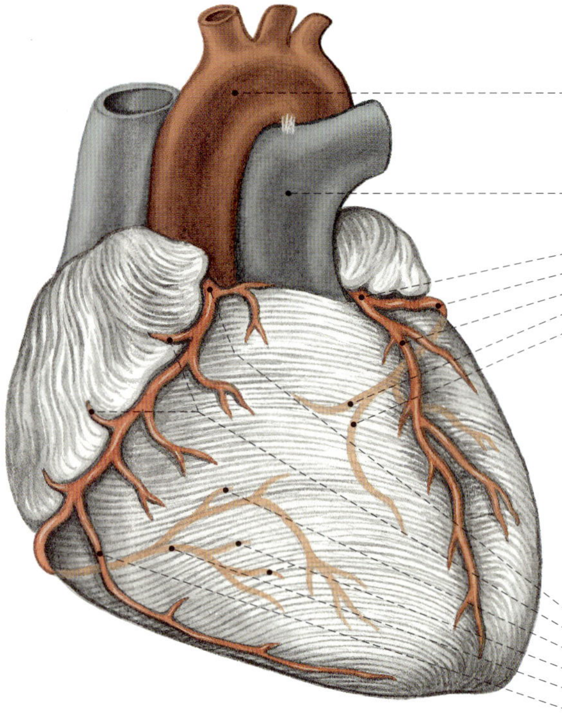

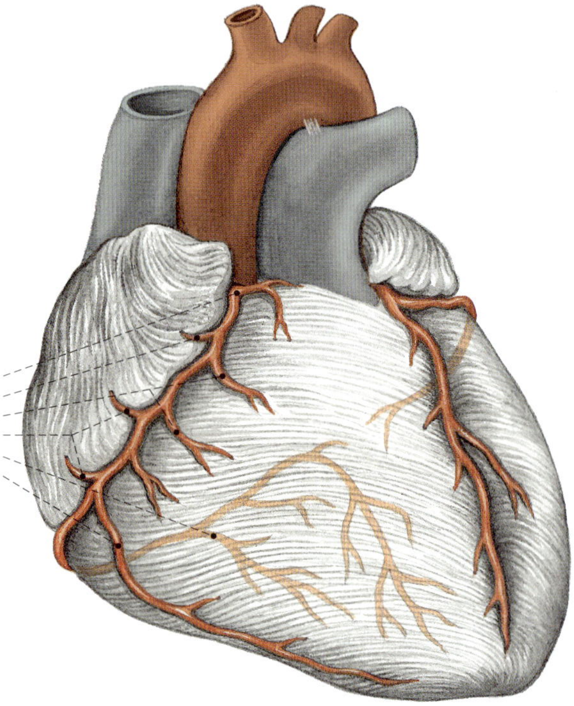

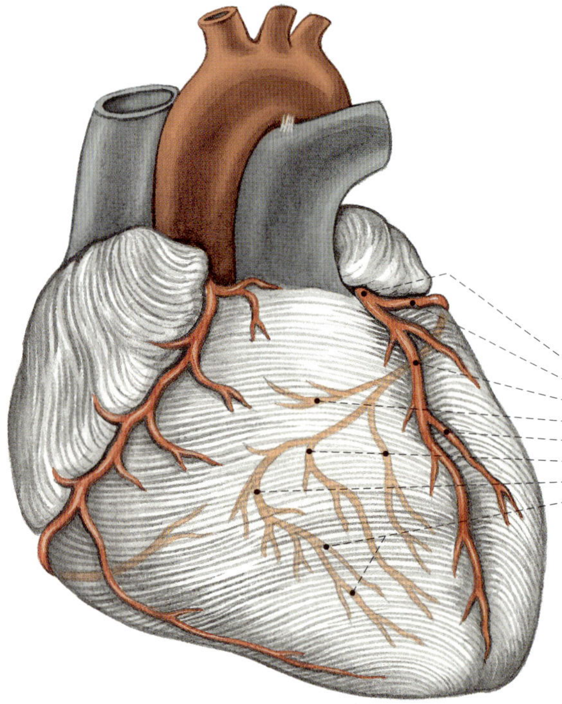

144 Cor

144 Coronary arteries of the heart (70%)

Schematic representations, ventral aspect

a Commonest arrangement
b Right dominance, the right coronary artery supplying the posterior walls of both ventricles
c Left dominance, the left coronary artery supplying the posterior walls of both ventricles

144 Herzkranzarterien (70%)

Schematische Darstellungen, Ventralansicht

a Normalversorgungstyp (ausgeglichener Versorgungstyp)
b Rechtsversorgungstyp. Die rechte Herzkranzarterie versorgt die Rückwand beider Ventrikel.
c Linksversorgungstyp. Die linke Herzkranzarterie versorgt die Rückwand beider Ventrikel.

a

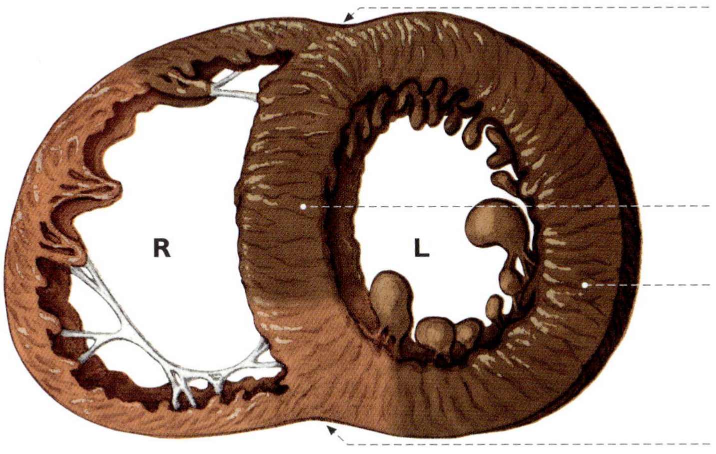

b

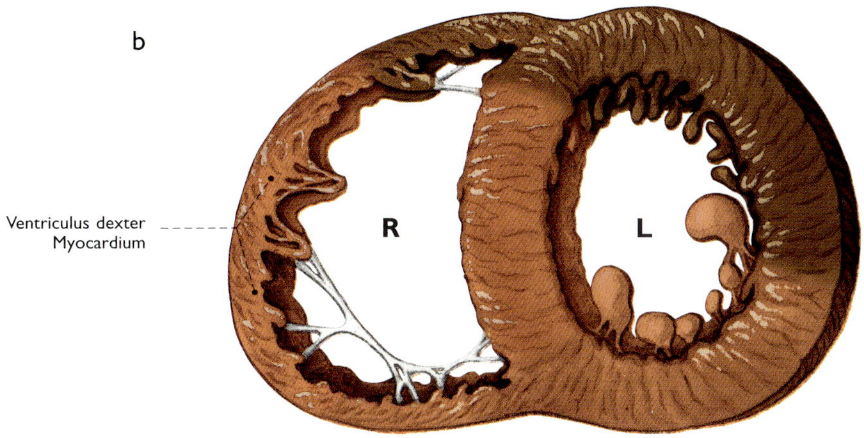

c

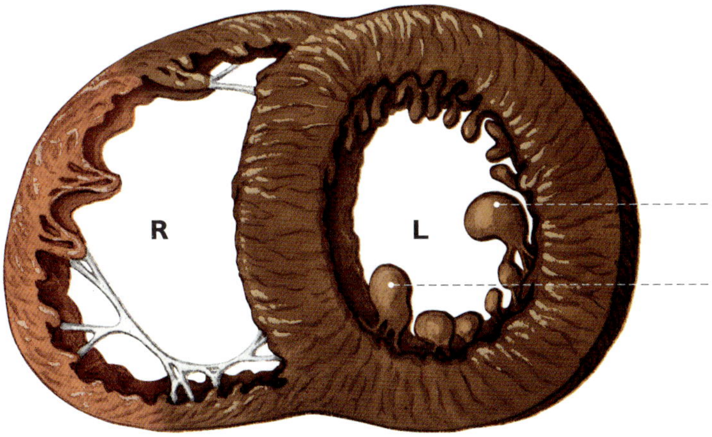

145 Cor

145 Coronary arteries of the heart (80%)

Circulation areas of both coronary arteries in correspondence to Fig. 144, transverse section through both ventricles rectangular to the cardiac axis, schematic representations, inferior aspect from the apex of the heart (according to Bargmann, 1963, and Töndury, 1970)

a Commonest arrangement
b Right dominance
c Left dominance

The circulation areas of the right coronary artery are marked by **red**, those of the left coronary artery by **brown** color.

145 Herzkranzarterien (80%)

Versorgungsgebiete beider Herzkranzarterien entsprechend Abb. 144, Querschnitt durch beide Herzventrikel senkrecht zur Herzachse, schematische Darstellungen, Kaudalansicht von der Herzspitze her (nach Bargmann, 1963, und Töndury, 1970)

a Normalversorgungstyp (ausgeglichener Versorgungstyp)
b Rechtsversorgungstyp
c Linksversorgungstyp

Die Versorgungsgebiete der rechten Herzkranzarterie sind **rot**, die der linken Herzkranzarterie **braun** dargestellt.

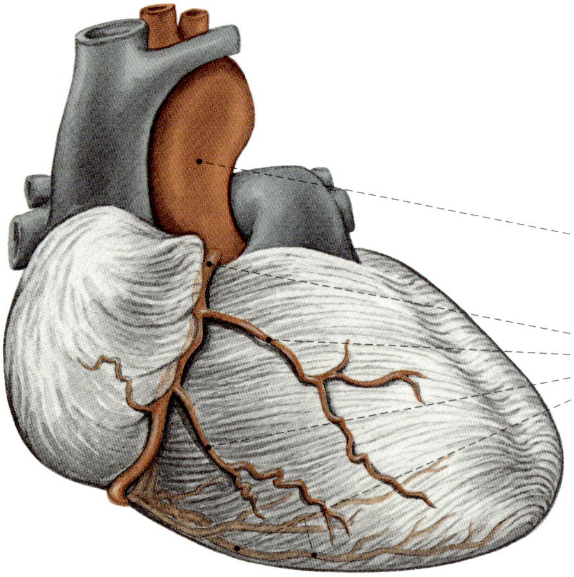

a

Pars ascendens aortae

A. coronaria dextra
– ⟨R. anterior ventriculi dextri⟩ –
– R. marginalis dexter –
– R. interventricularis posterior –

b

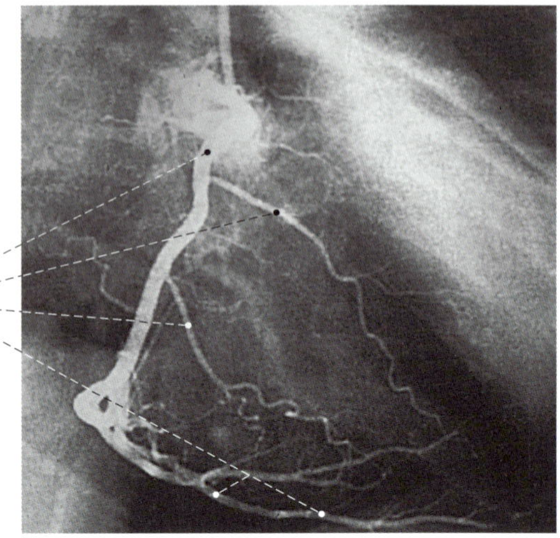

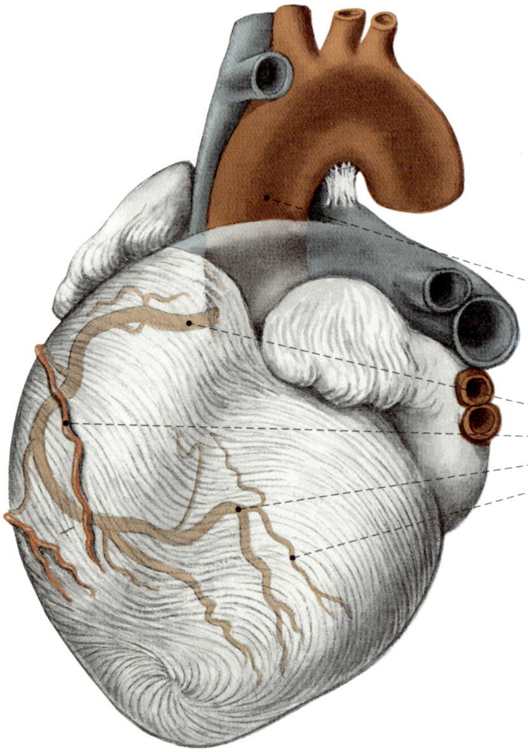

c

Pars ascendens aortae

A. coronaria dextra
– R. marginalis dexter –
– R. interventricularis posterior –
– (R. posterolateralis dexter) –

d

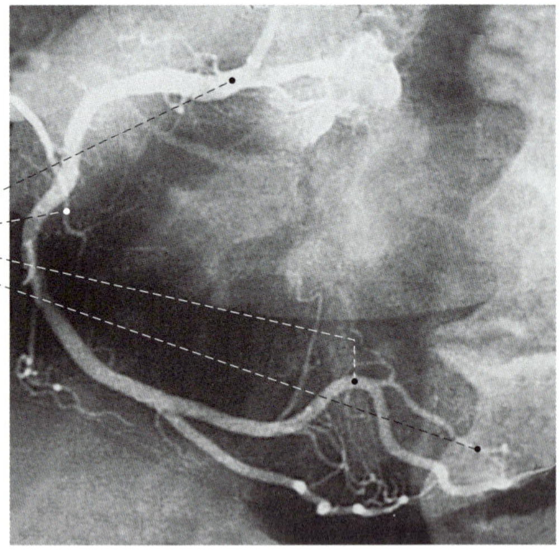

146 Cor

146 Coronary arteries of the heart (50%)

Right coronary artery

a, b Right anterior oblique view (RAO projection)
c, d Left anterior oblique view (LAO projection)
a, c Schematic representations
b, d Selective coronary angiograms

146 Herzkranzarterien (50%)

Rechte Herzkranzarterie

a, b Ansicht von schräg rechts vorne (1. Schrägdurchmesser, "Fechterstellung")
c, d Ansicht von schräg links vorne (2. Schrägdurchmesser, "Boxerstellung")
a, c Schematische Darstellungen
b, d Selektive Koronarangiogramme

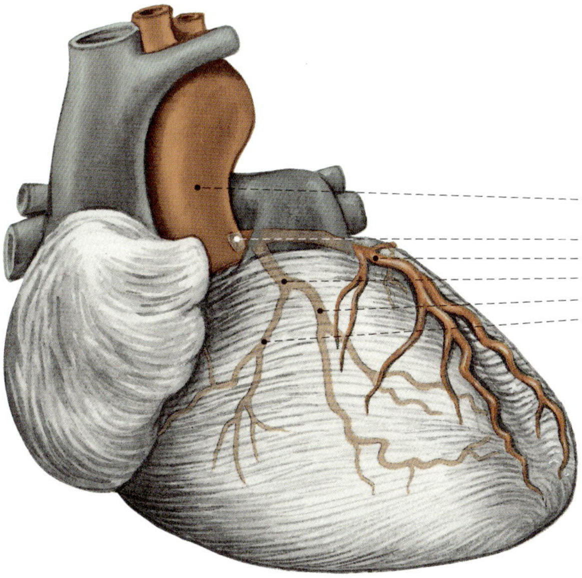

a

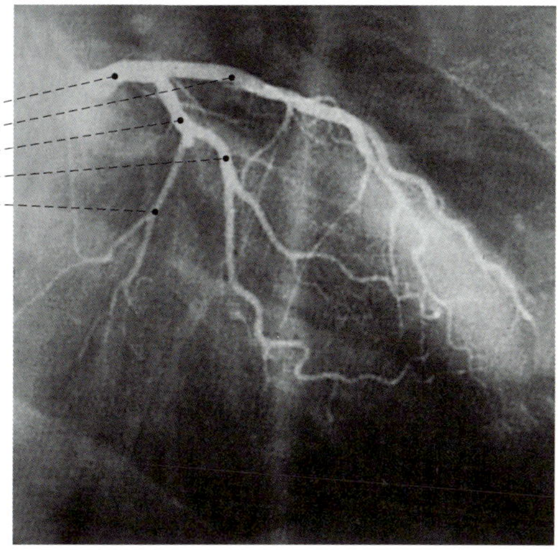

b

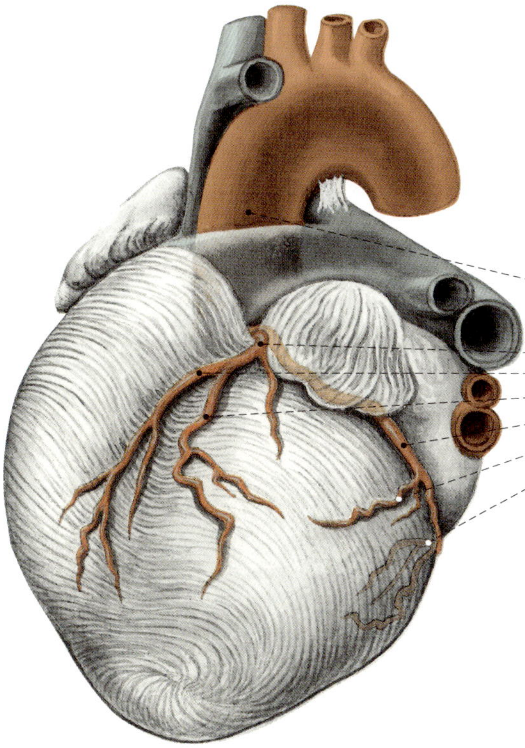

c

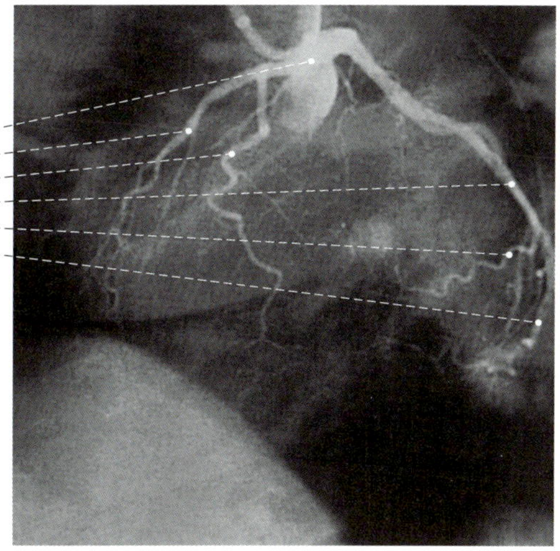

d

147 Cor

147 Coronary arteries of the heart (50%)

Left coronary artery
a, b Right anterior oblique view (RAO projection)
c, d Left anterior oblique view (LAO projection)
a, c Schematic representations
b, d Selective coronary angiograms

147 Herzkranzarterien (50%)

Linke Herzkranzarterie
a, b Ansicht von schräg rechts vorne (1. Schrägdurchmesser, "Fechterstellung")
c, d Ansicht von schräg links vorne (2. Schrägdurchmesser, "Boxerstellung")
a, c Schematische Darstellungen
b, d Selektive Koronarangiogramme

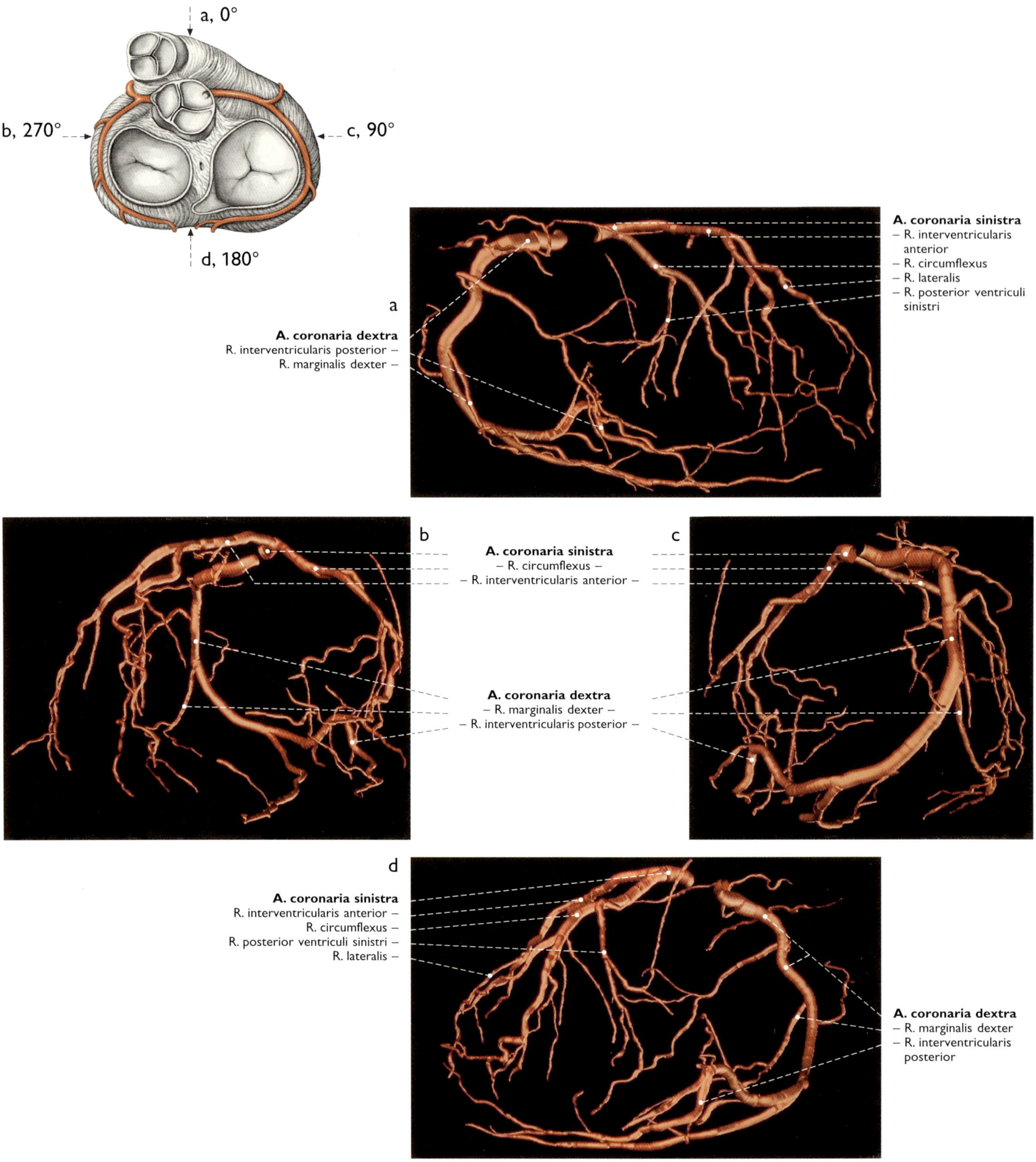

148 Cor

148 Coronary arteries of the heart (50%)

Three-dimensional reconstruction of the course of the right and left coronary arteries

a seen from ventral (0°)
b seen from left lateral (270°)
c seen from right lateral (90°)
d seen from dorsal (180°)
as indicated in the sketch above

148 Herzkranzarterien (50%)

Dreidimensionale Rekonstruktion des Verlaufs der rechten und linken Herzkranzarterie

a von ventral gesehen (0°)
b von links lateral gesehen (270°)
c von rechts lateral gesehen (90°)
d von dorsal gesehen (180°),
wie in der obenstehenden Skizze dargestellt

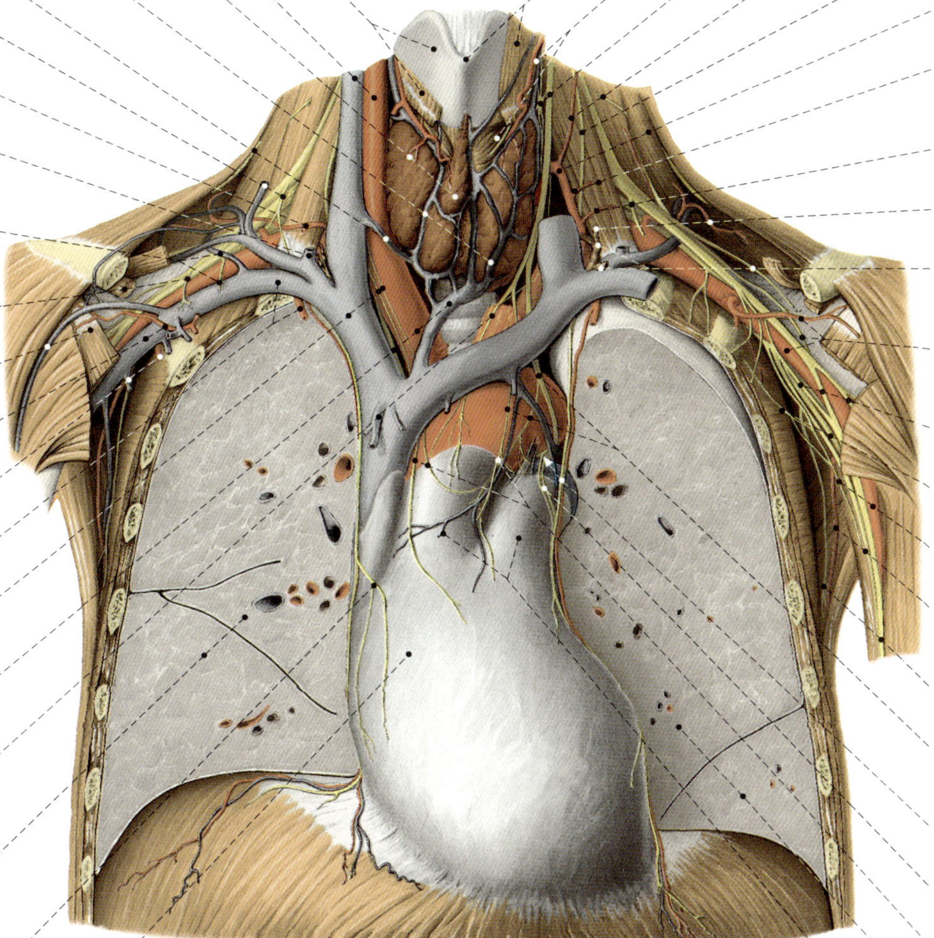

149 Thorax

149 Blood vessels and nerves of the neck, mediastinum and axillary region (40%)

The sternocleidomastoid and the strap muscles, the anterior thoracic wall and the anterior parts of the lungs were removed. Ventral aspect

149 Blutgefässe und Nerven im Bereich des Halses, im Mediastinum und in der Achselhöhle (40%)

Der Musculus sternocleidomastoideus und die infrahyalen Muskeln sind entfernt, die vordere Brustwand und die vorderen Lungenpartien abgetragen. Ventralansicht

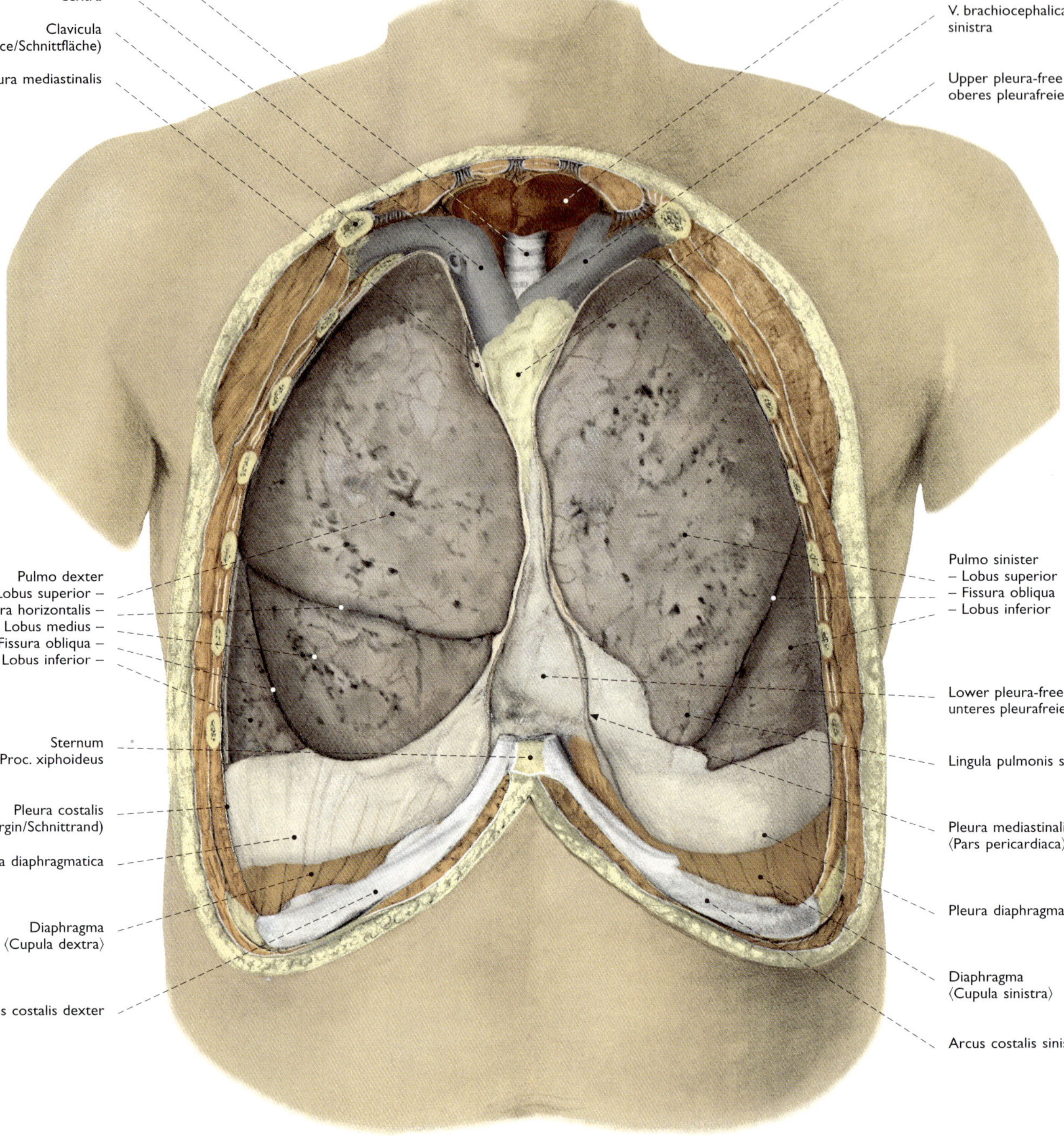

150 Thorax

150 Thoracic organs in situ (40%)
The anterior thoracic wall was removed.
Ventral aspect

150 Brustorgane in situ (40%)
Die vordere Brustwand ist entfernt.
Ventralansicht

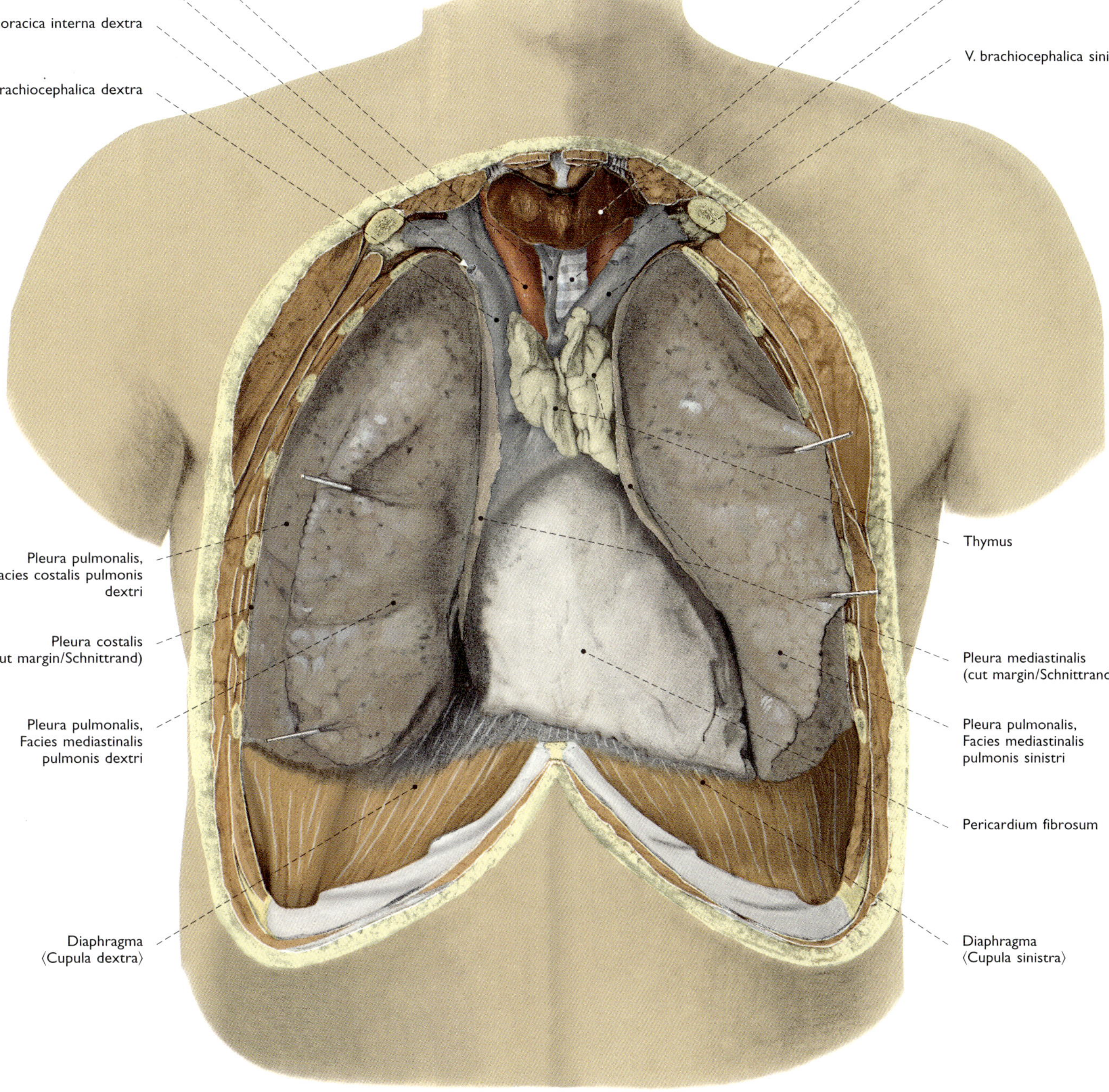

151 Thorax

151 Thoracic organs (40%)

From the same preparation used for Fig. 150 after removal of the diaphragmatic pleura. Both lungs were displaced laterally and rotated by hooks inserted along their anterior margins. The ventral aspect of the fibrous pericardium is revealed, the thymus and the large vessels are exposed in the upper triangle free of pleura. Ventral aspect

151 Brustorgane (40%)

Gleiches Präparat wie Abb. 150 nach Entfernung der Pleura diaphragmatica. Beide Lungen sind seitwärts umgeklappt, so dass ihre mediastinale Fläche und die Ventralfläche des Herzbeutels sichtbar werden. Der Thymus und die grossen Gefässe sind im oberen pleurafreien Dreieck freigelegt. Ventralansicht

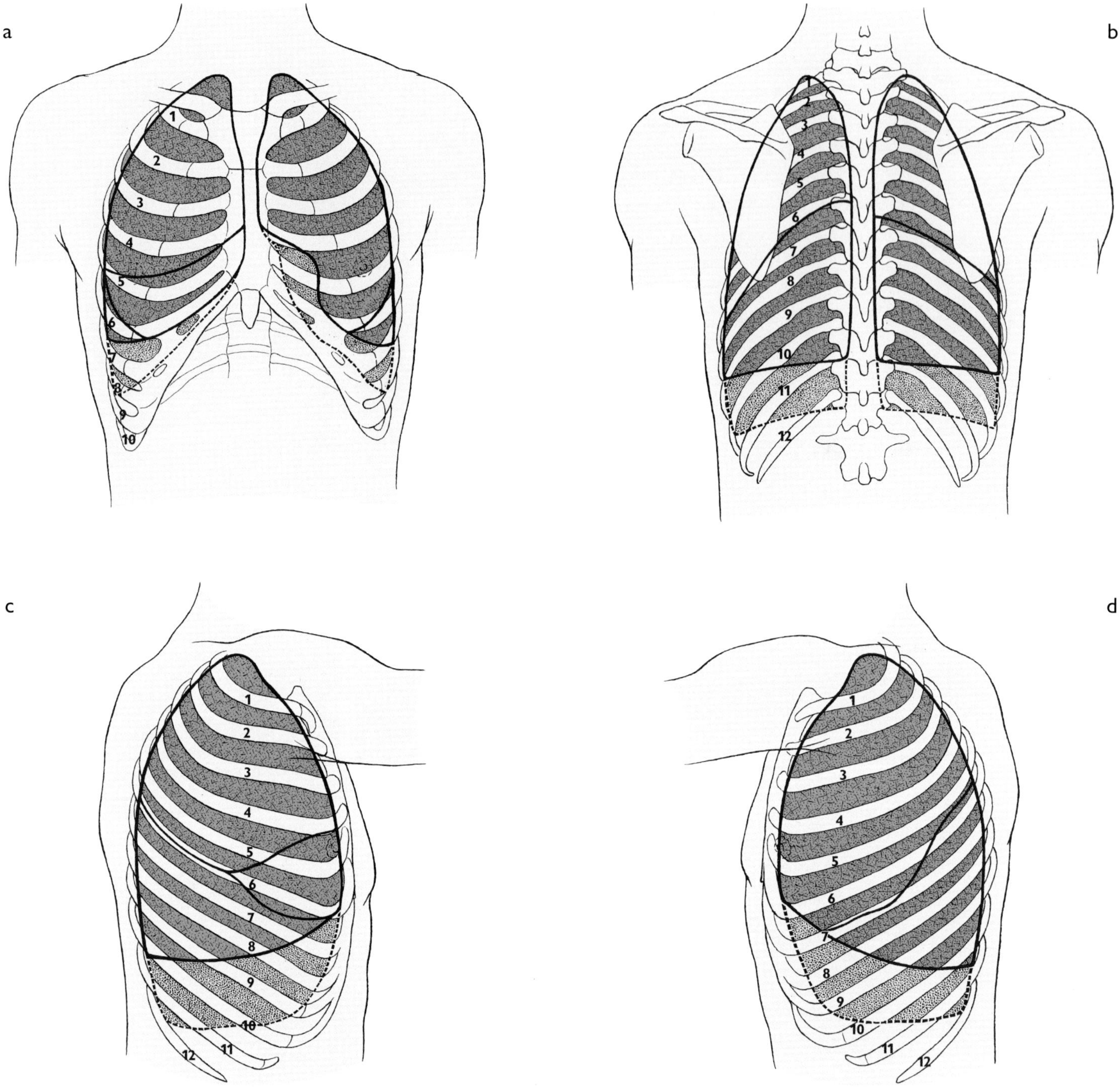

······ Pleural borders
—— Borders of the lungs and their lobes

······ Pleuragrenzen
—— Lungen- und Lungenlappengrenzen

152 Pleura et Pulmones

152 Typical borders of the pleura and the lungs

Schematic representations. The borders of the pleura are indicated by a dashed line, the lung borders by a continuous line.

a Ventral aspect
b Dorsal aspect
c Right lateral aspect
d Left lateral aspect

152 Typische Pleura- und Lungengrenzen

Schematische Darstellungen. Die Pleuragrenzen sind als gebrochene Linie, die Lungengrenzen als durchgezogene Linie dargestellt.

a Ventralansicht
b Dorsalansicht
c Rechte Seitenansicht
d Linke Seitenansicht

a

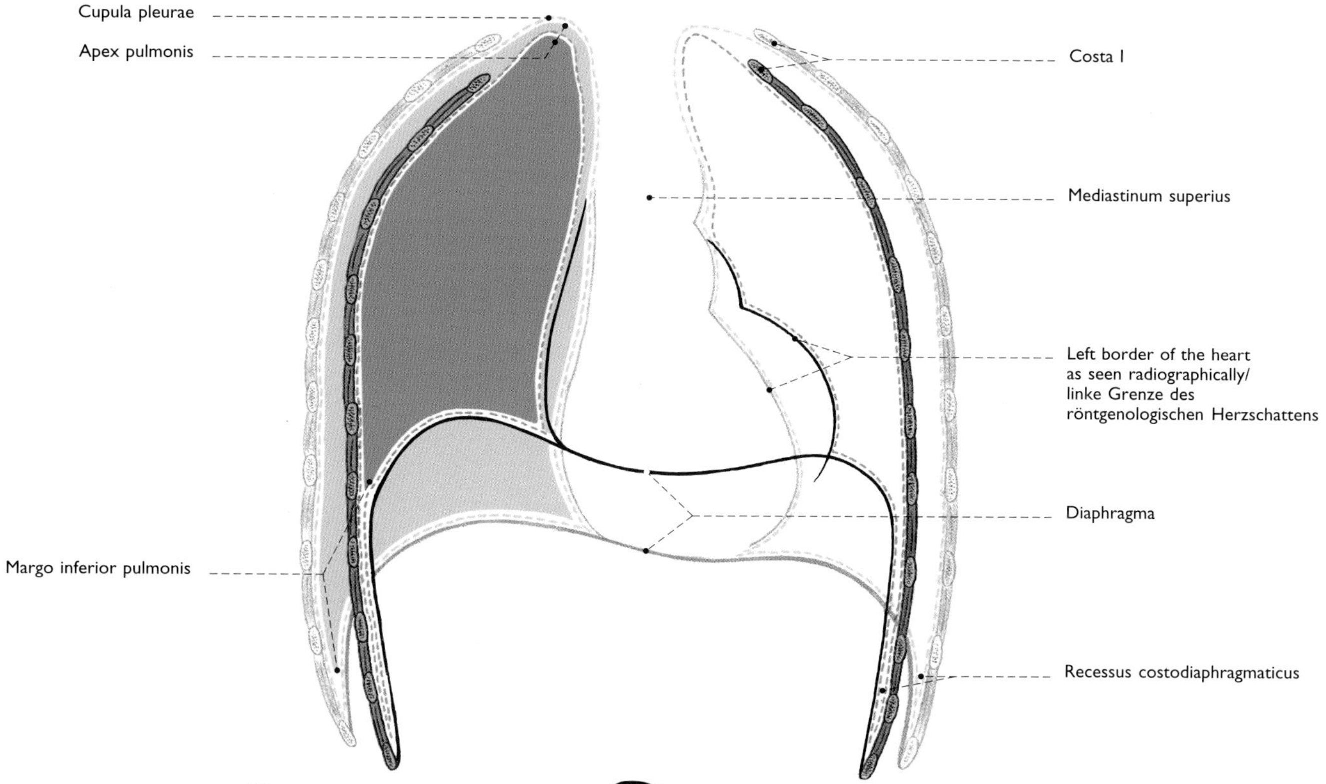

b

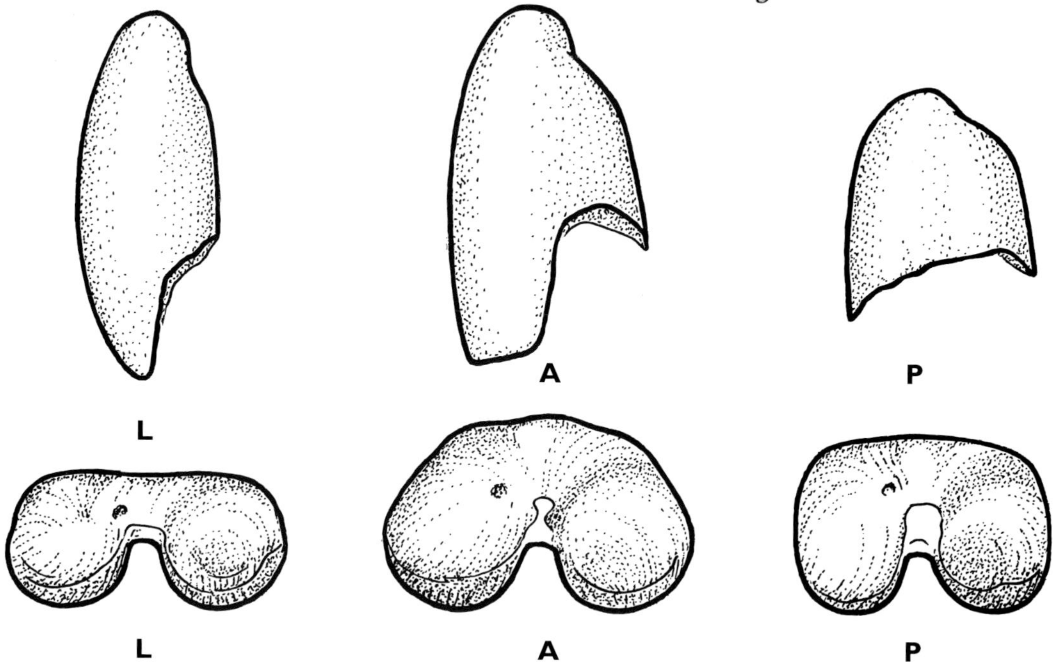

153 Cavitas pleuralis et Cavitas thoracis

153 Pleural and thoracic cavities

Schematic representations

a Outline of the parietal pleura (dashed line) in the expiratory phase given in dark blue, in the inspiratory phase in middle blue. Projection of the right lung during expiration shown in dark blue, inspiratory increase in middle blue, coronal section of the thorax

b Shapes of the thorax of leptosome (L), athletic (A), and pyknic (P) subjects, schematic outlines, lateral and caudal aspects

153 Pleura- und Thoraxraum

Schematische Darstellungen

a Umriss der Pleura parietalis (gestrichelt) in der exspiratorischen Phase dunkelblau, in der inspiratorischen Phase hellblau eingezeichnet. Projektionsfläche der rechten Lunge während der Exspiration dunkelblau, inspiratorischer Zuwachs hellblau dargestellt, Frontalschnitt des Thorax

b Thoraxformen von Individuen mit leptosomem (L), athletischem (A) und pyknischem (P) Körperbau, Umrisszeichnungen, Seiten- und Kaudalansichten

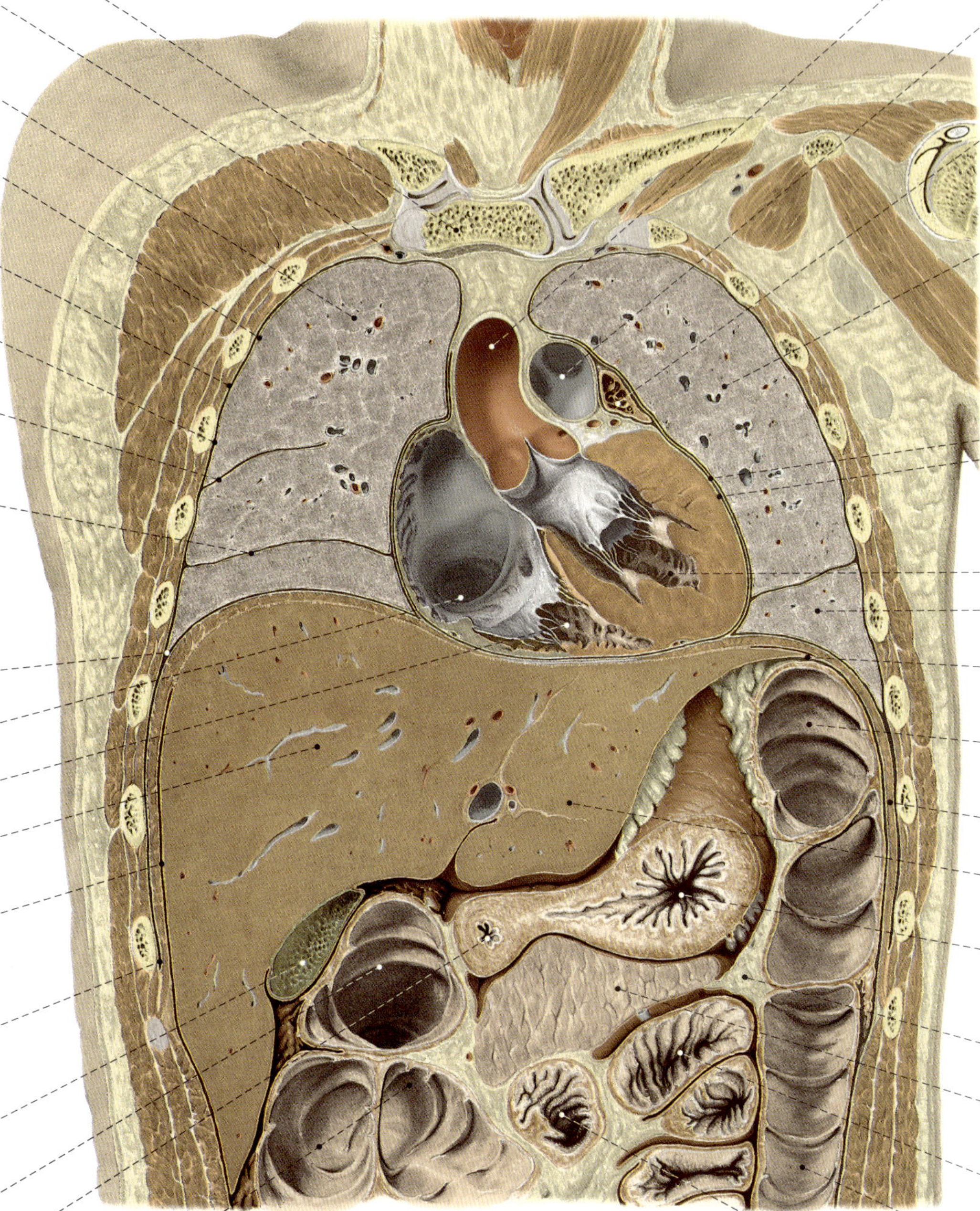

154 Thorax et Abdomen

154 Upper part of the trunk (40%)

Coronal section through the trunk in the plane of the sternoclavicular joints, drawing of an anatomical section, ventral aspect

154 Oberer Teil des Rumpfes (40%)

Frontalschnitt durch den Rumpf im Bereich der medialen Schlüsselbeingelenke, Zeichnung eines anatomischen Schnitts, Ventralansicht

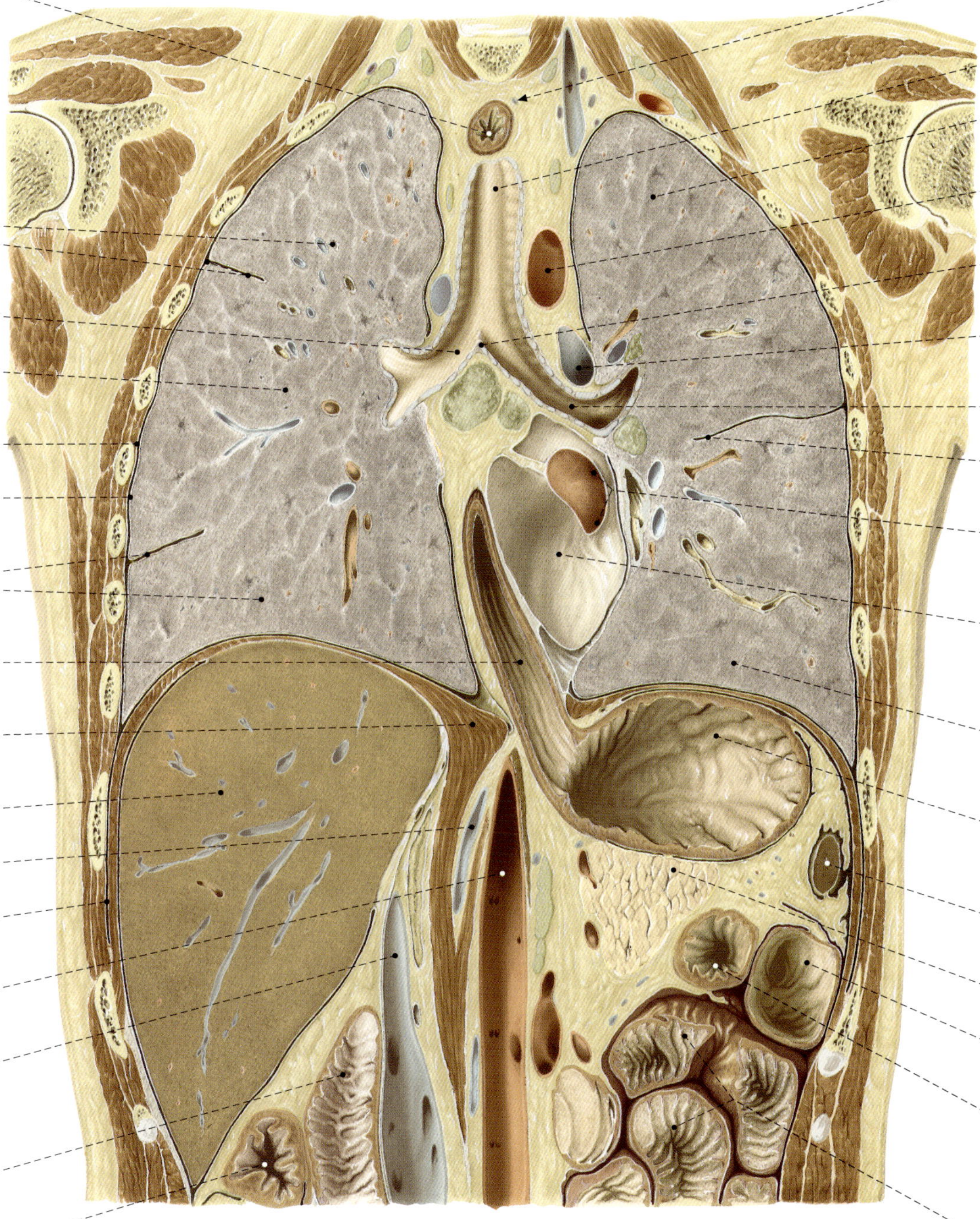

155 Thorax et Abdomen

155 Upper part of the trunk (40%)

Coronal section through the trunk in the plane of the tracheal bifurcation and the esophageal hiatus, drawing of an anatomical section, ventral aspect

155 Oberer Teil des Rumpfes (40%)

Frontalschnitt durch den Rumpf im Bereich der Bifurcatio tracheae und des Hiatus oesophageus, Zeichnung eines anatomischen Schnitts, Ventralansicht

a

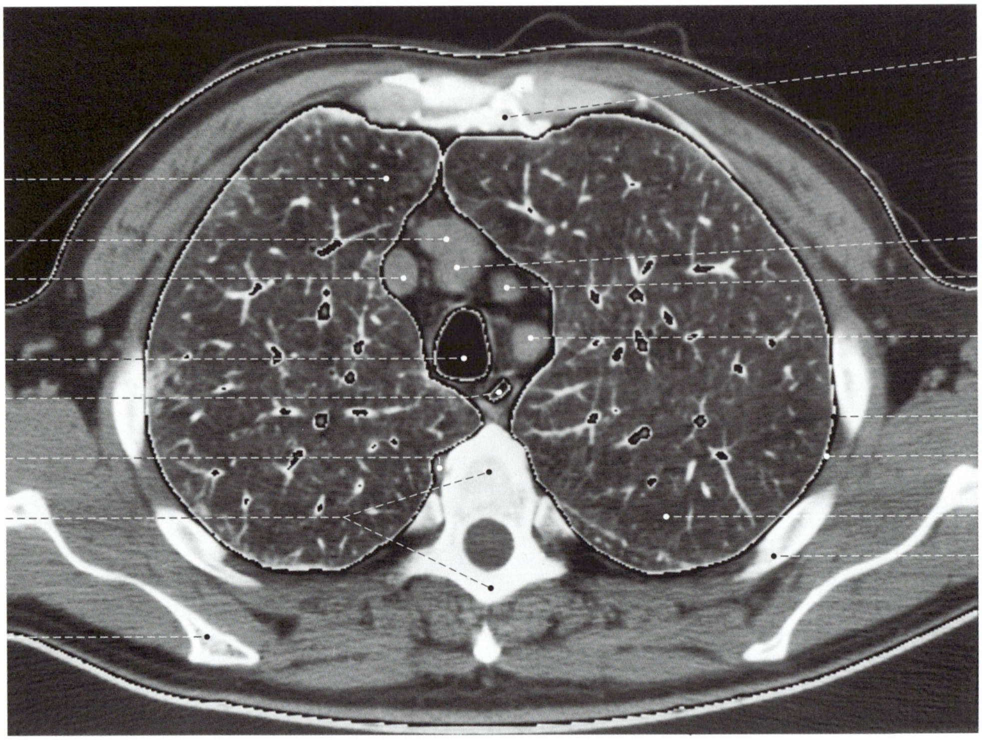

b

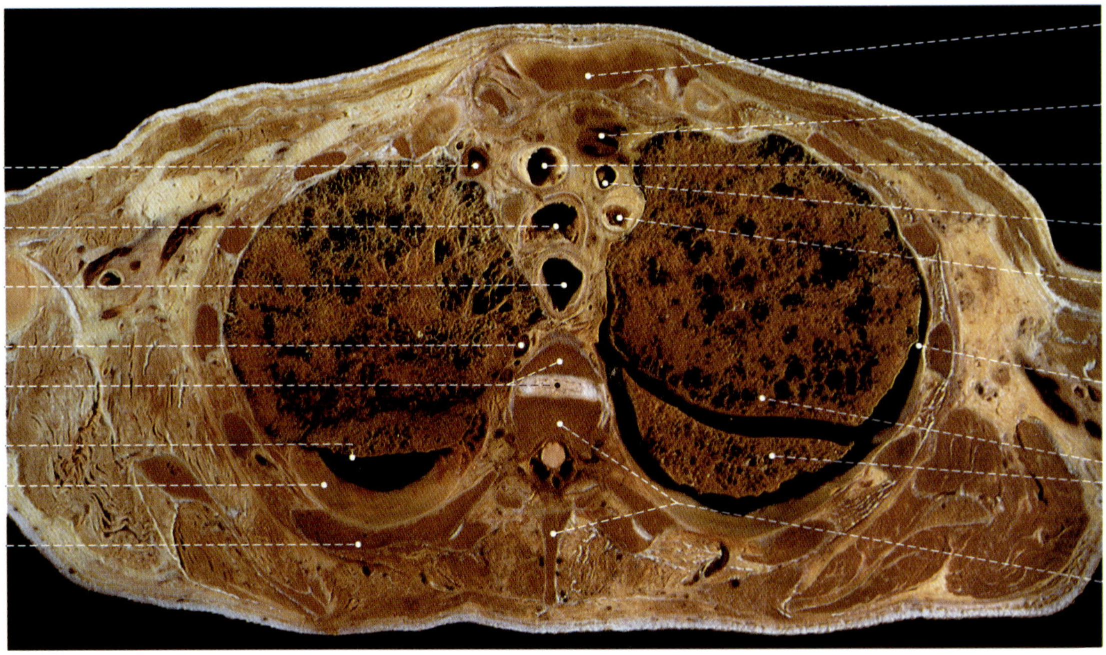

156 Thorax

156 Thorax (50%)

Transverse sections through the lungs and the upper mediastinum at the level of the fourth (Th IV, a) and the third/fourth (Th III/IV, b) thoracic vertebral bodies, respectively, inferior aspect

a Computed tomogram (CT) after injection of contrast medium (combined mediastinal and lung window settings)

b Anatomical section

156 Brustkorb (50%)

Horizontalschnitte durch die Lungen und das obere Mediastinum auf Höhe des vierten (Th IV, a) bzw. der dritten/vierten (Th III/IV, b) Brustwirbelkörpers, Kaudalansicht

a Computertomogramm (CT) nach Kontrastmittelgabe (kombinierte Mediastinum- und Lungendarstellung)

b Anatomischer Querschnitt

a

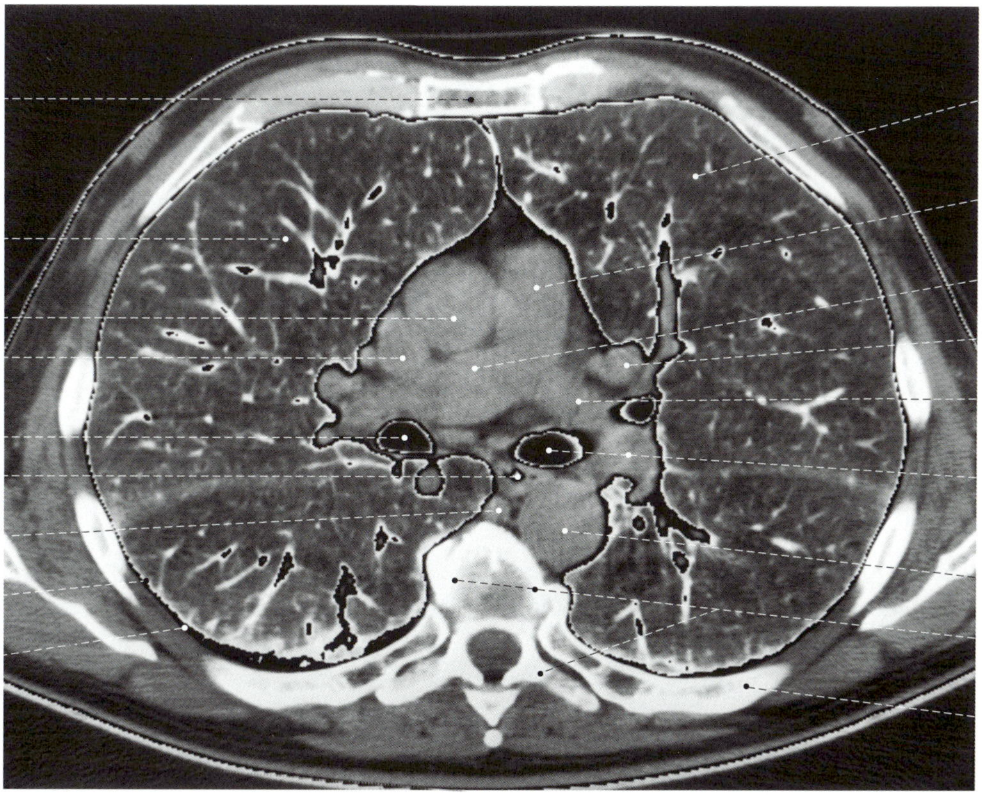

b

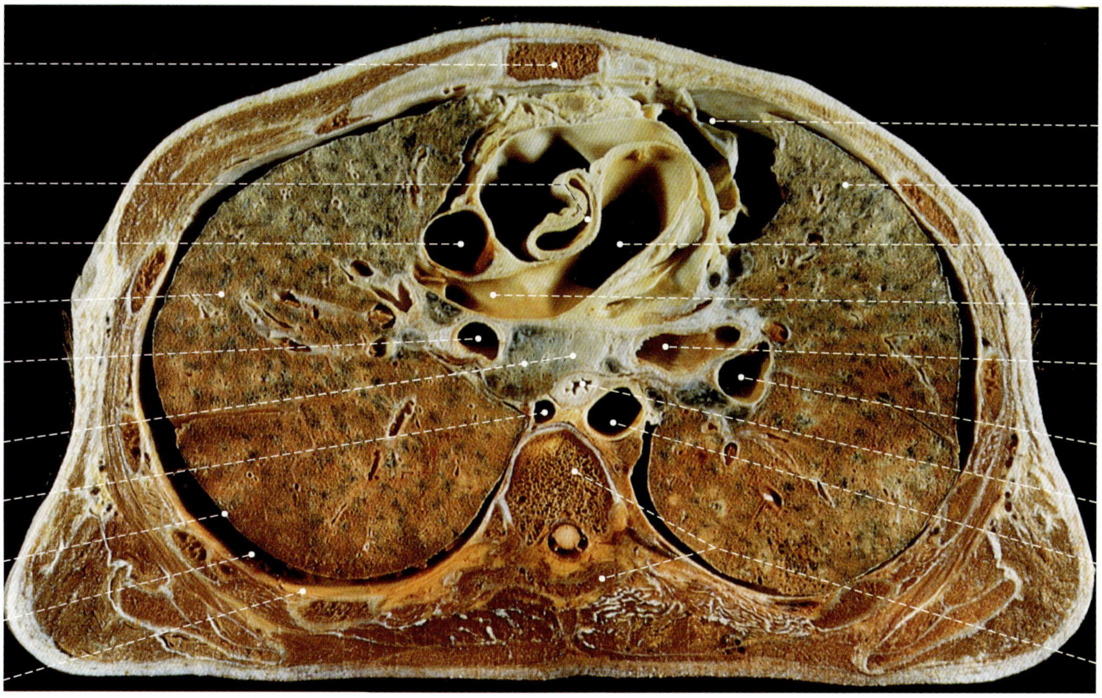

157 Thorax

157 Thorax (50%)

Transverse sections through the lungs and the mediastinum at the level of the sixth thoracic vertebral body (Th VI), inferior aspect

a Computed tomogram (CT) after injection of contrast medium (combined mediastinal and lung window settings)

b Anatomical section

157 Brustkorb (50%)

Horizontalschnitte durch Lunge und Mediastinum auf Höhe des sechsten Brustwirbelkörpers (Th VI), Kaudalansicht

a Computertomogramm (CT) nach Kontrastmittelgabe (kombinierte Mediastinum- und Lungendarstellung)

b Anatomischer Querschnitt

a

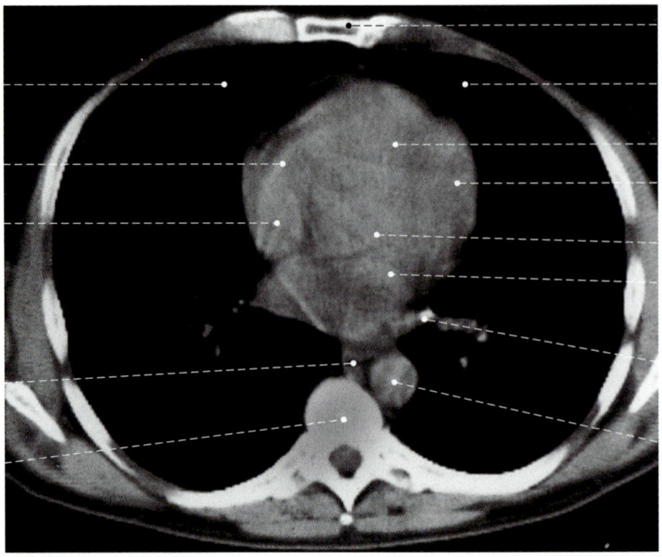

b

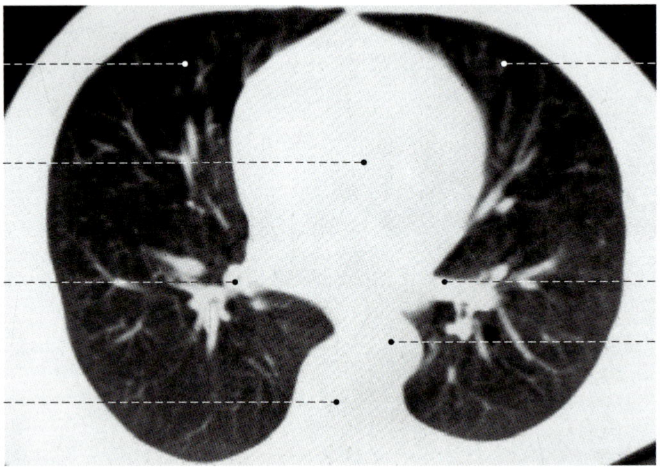

c

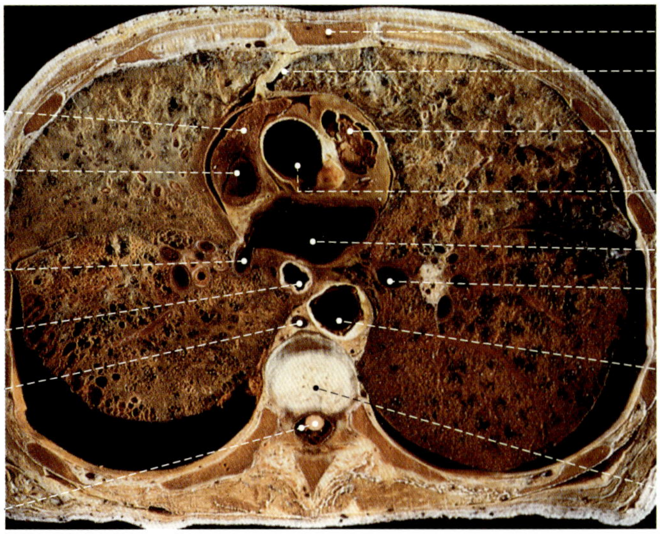

158 Thorax

158 Thorax (35%)

Transverse sections at the level of the seventh thoracic vertebral body (Th VII), inferior aspect

a, b Computed tomograms (CT)
a Mediastinal window setting
b Lung window setting
c Anatomical section

158 Brustkorb (35%)

Querschnitte auf Höhe des siebten Brustwirbelkörpers (Th VII), Kaudalansicht

a, b Computertomogramme (CT)
a Mediastinumdarstellung
b Lungendarstellung
c Anatomischer Schnitt

a

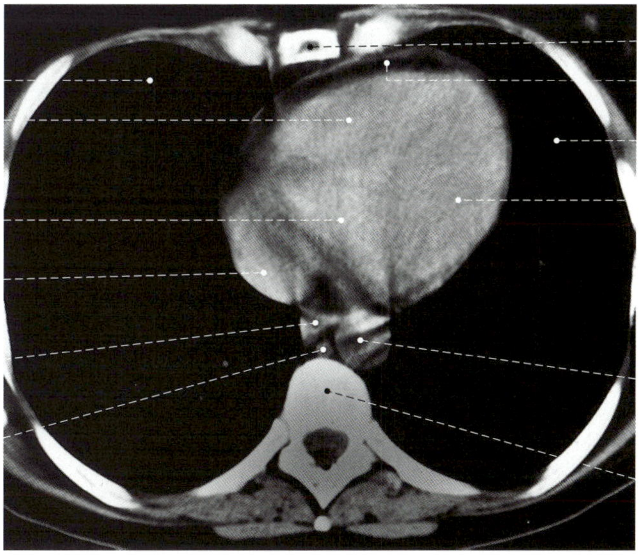

b

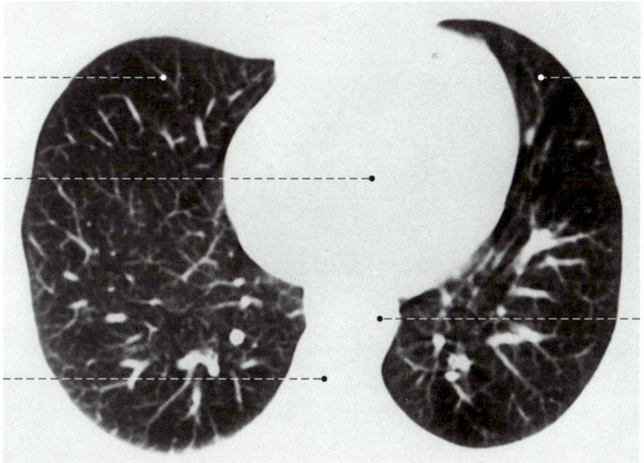

c

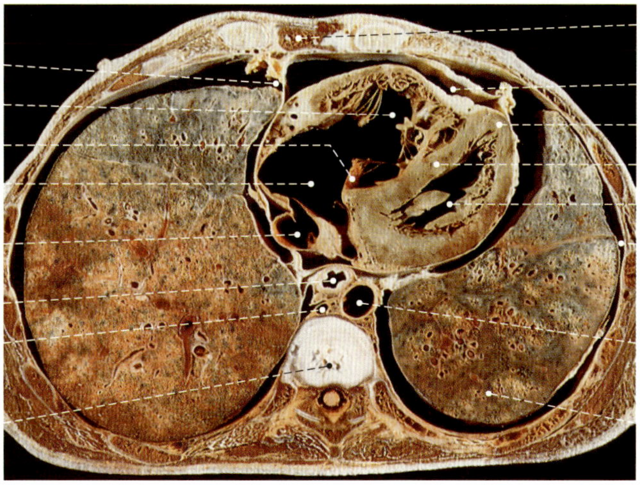

159 Thorax

159 Thorax (35%)

Transverse sections at the level of the eighth thoracic vertebral body (Th VIII), inferior aspect

a, b Computed tomograms (CT)
a Mediastinal window setting
b Lung window setting
c Anatomical section

159 Brustkorb (35%)

Querschnitte auf Höhe des achten Brustwirbelkörpers (Th VIII), Kaudalansicht

a, b Computertomogramme (CT)
a Mediastinumdarstellung
b Lungendarstellung
c Anatomischer Schnitt

a

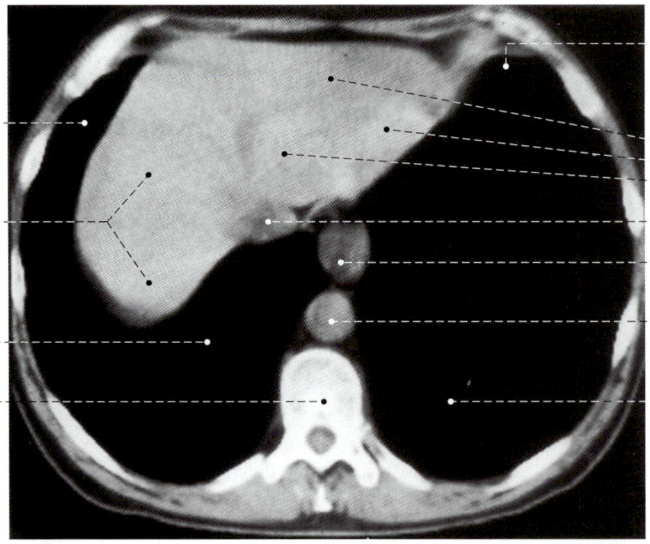

b

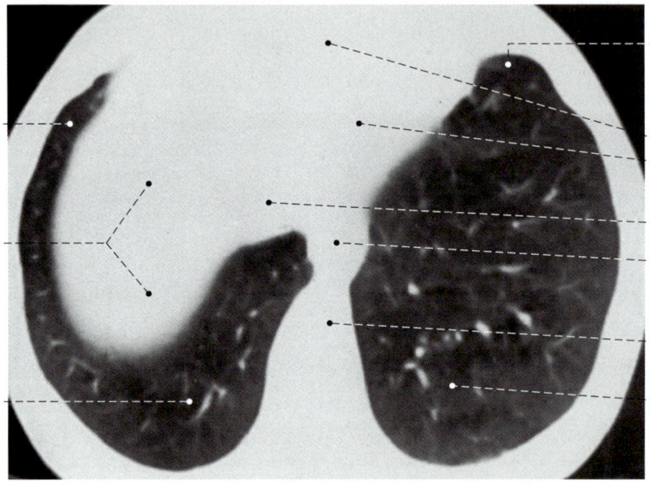

c

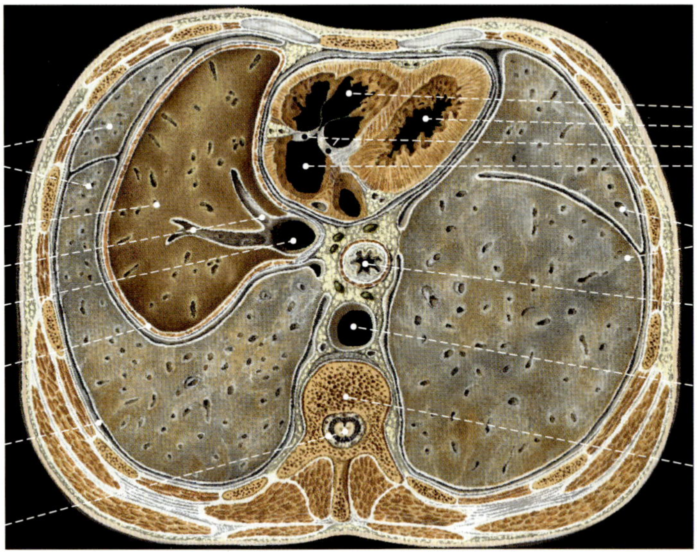

160 Thorax

160 Thorax (35%)

Transverse sections at the level of the ninth thoracic vertebral body (Th IX), inferior aspect

a, b Computed tomograms (CT)
a Mediastinal window setting
b Lung window setting
c Anatomical section (painted)

160 Brustkorb (35%)

Querschnitte auf Höhe des neunten Brustwirbelkörpers (Th IX), Kaudalansicht

a, b Computertomogramme (CT)
a Mediastinumdarstellung
b Lungendarstellung
c Anatomischer Schnitt (gezeichnet)

a

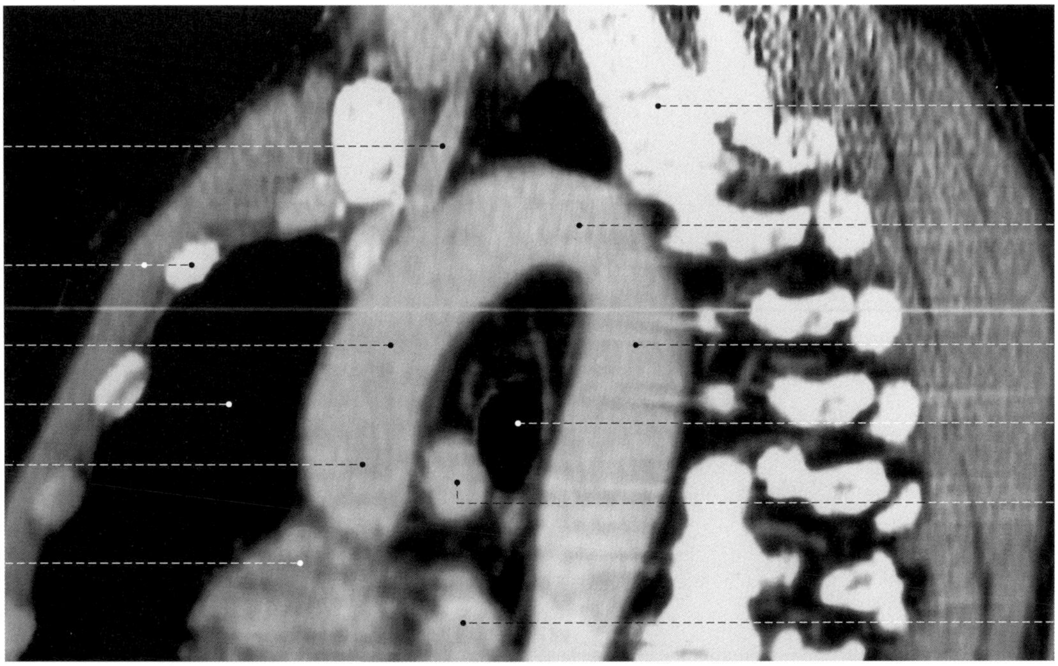

b

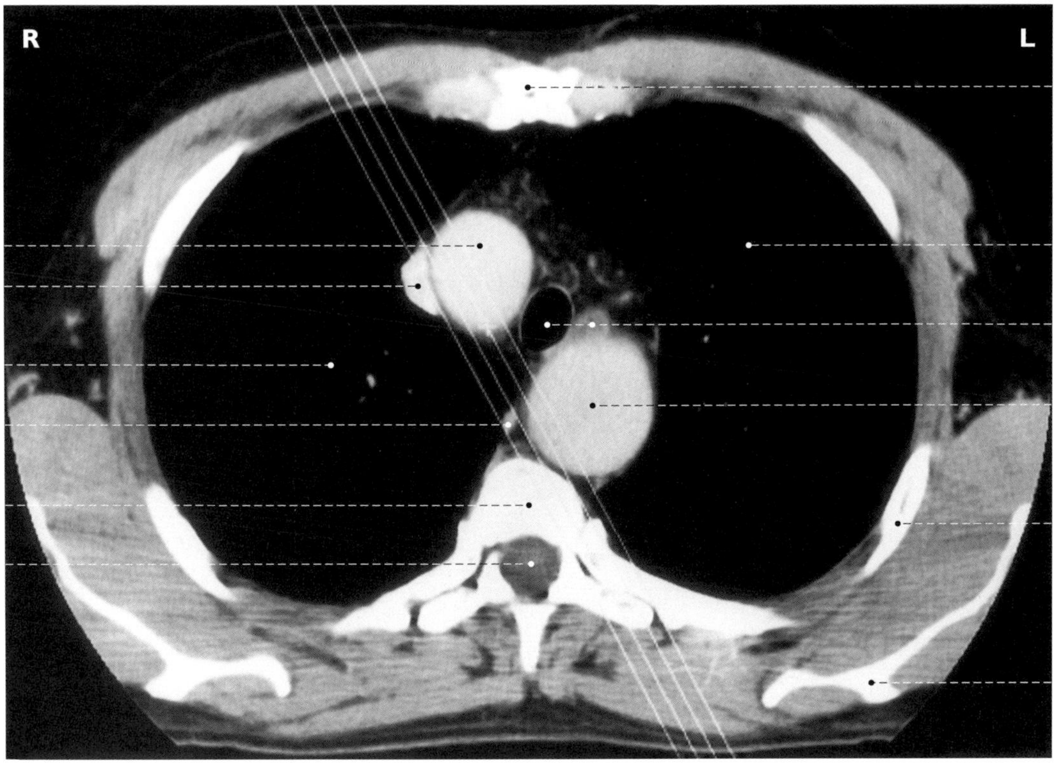

161 Thorax

161 Thorax (50%)

Computed tomograms (CT) showing the aorta

a Oblique section through the aortic arch from right anterior to left posterior as indicated in b (sagittal reconstruction)

b Transverse section at the level of the sternal angle and the body of the fourth thoracic vertebra (Th IV) as indicated in a

161 Brustkorb (50%)

Computertomogramme (CT) zur Darstellung der Aorta

a Schrägschnitt durch den Aortenbogen von rechts vorne nach links hinten, wie in b angegeben (sagittale Rekonstuktion)

b Horizontaler Querschnitt auf Höhe des Brustbeinwinkels und des vierten Brustwirbelkörpers (Th IV), wie in a angegeben

a

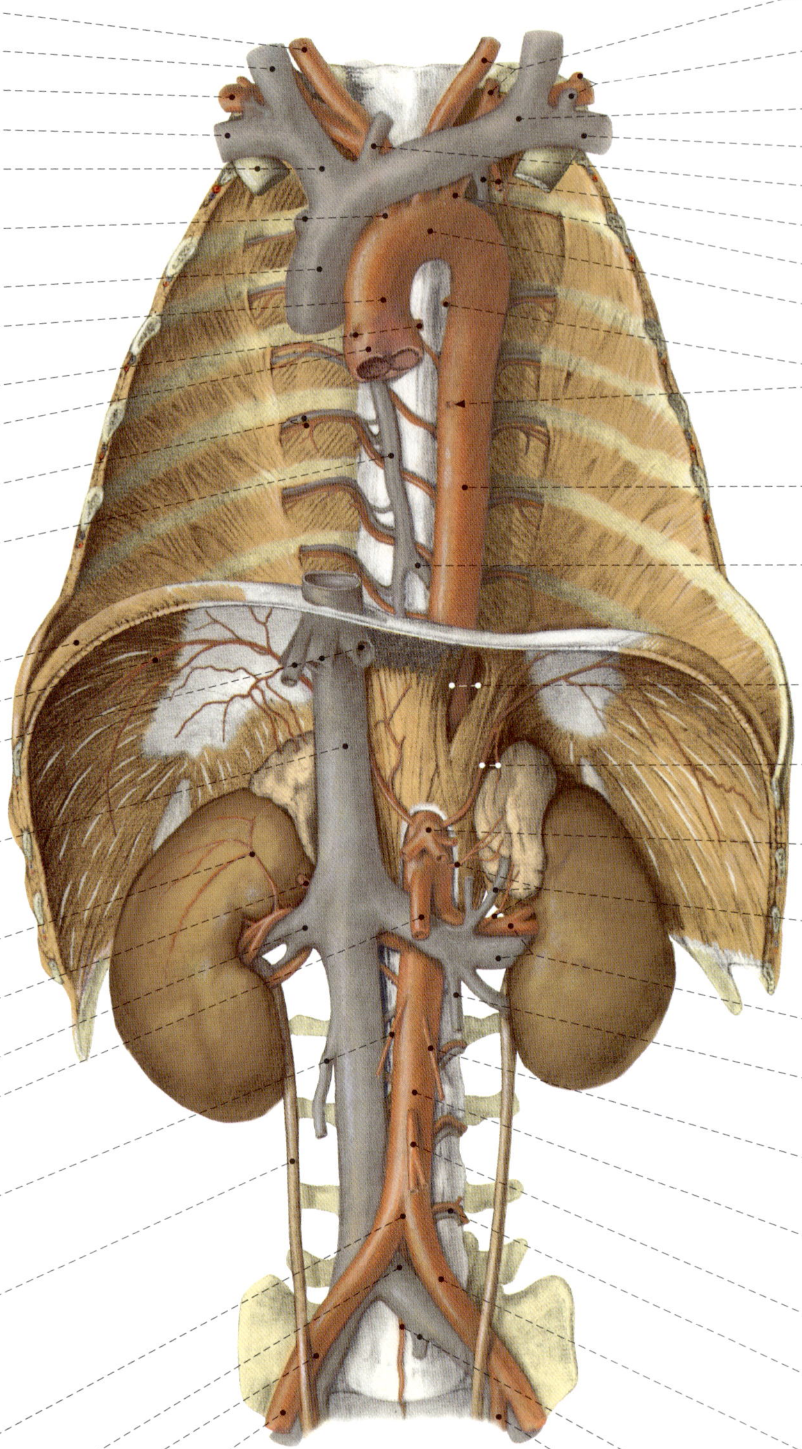

162 Thorax et Abdomen

162 Blood vessels on the anterior side of the dorsal body wall
Ventral aspect (40%)

162 Blutgefässe an der Vorderseite der dorsalen Rumpfwand
Ventralansicht (40%)

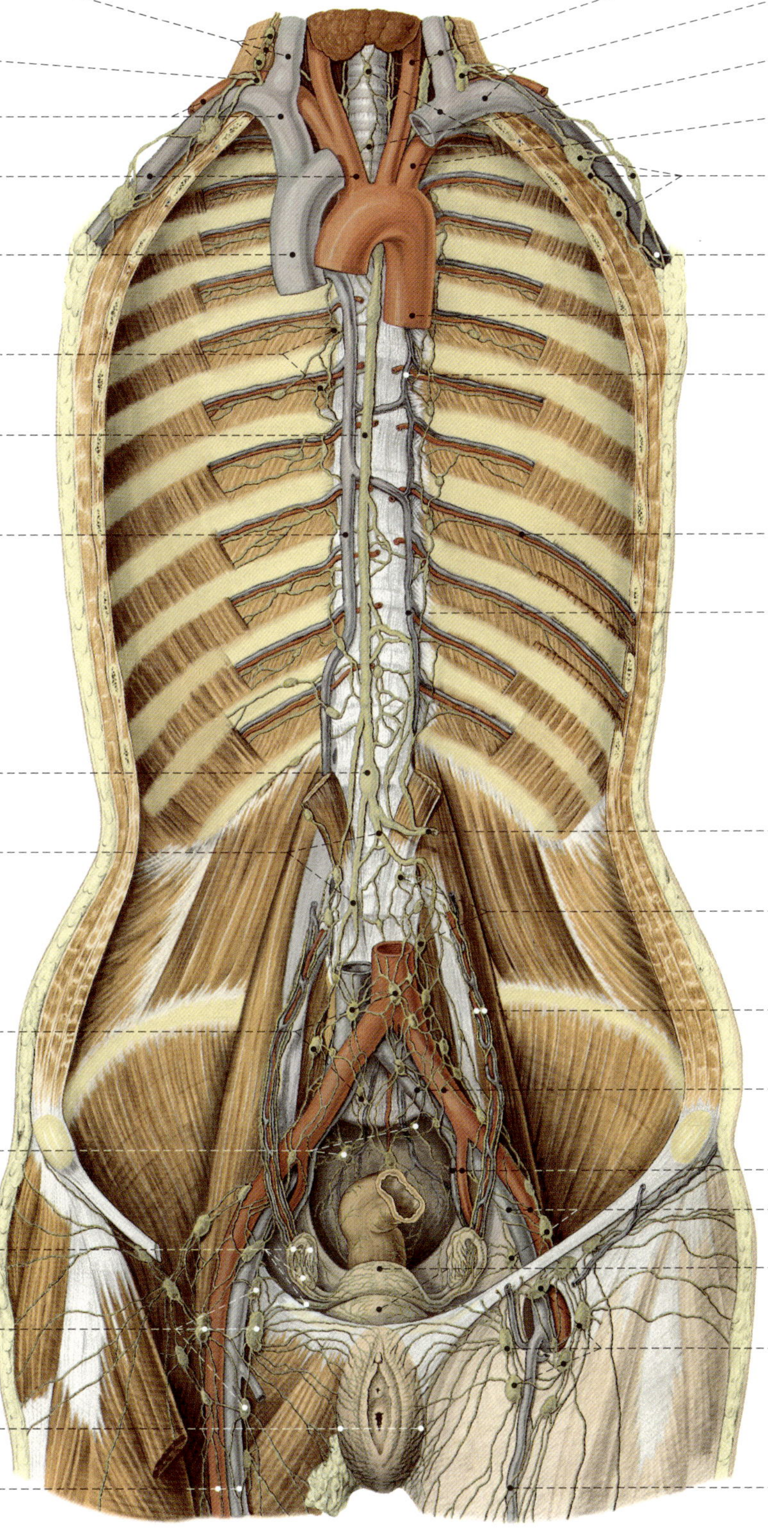

163 ⟨Nodi lymphatici thoracis, abdominis et pelvis⟩

163 Lymph vessels and lymph nodes of the thorax, abdomen and pelvis of a female (30%)

Lymph vessels and lymph nodes on the anterior side of the dorsal body wall, pelvis and inguinal region, ventral aspect

163 Lymphsystem im Brust-, Bauch- und Beckenbereich einer Frau (30%)

Lymphgefässe und Lymphknoten an der Vorderseite der dorsalen Rumpfwand, im Becken und in der Leistenregion, Ventralansicht

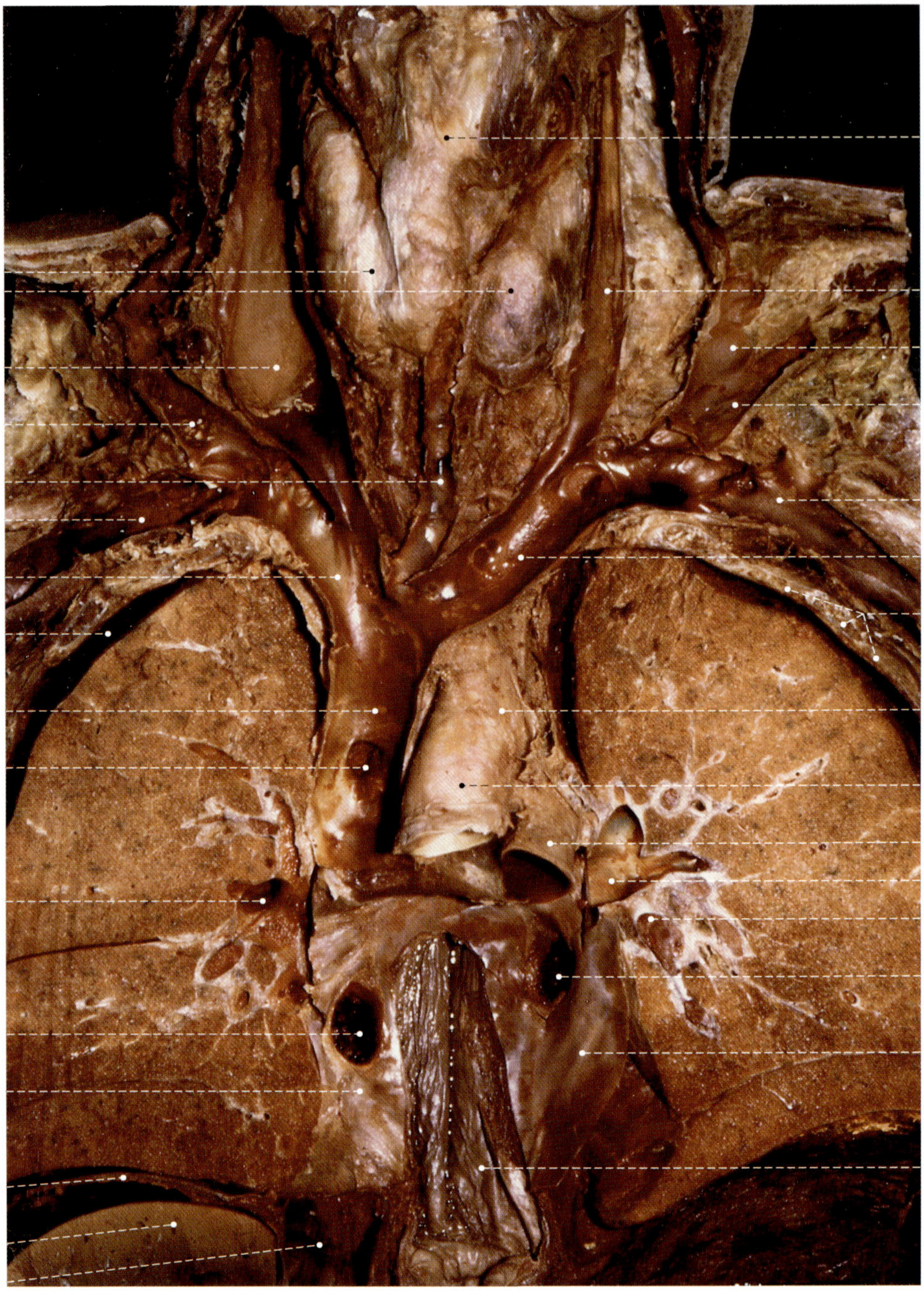

164 Mediastinum and large vein trunks (55%)

Coronal section through the thorax and the lower neck in the plane of the superior caval and the internal jugular veins, anatomical section, ventral aspect

164 Mediastinum und grosse Venenstämme (55%)

Frontalschnitt durch den Thorax und den unteren Hals in der Ebene der oberen Hohlvene und der Venae jugulares internae, anatomischer Schnitt, Ventralansicht

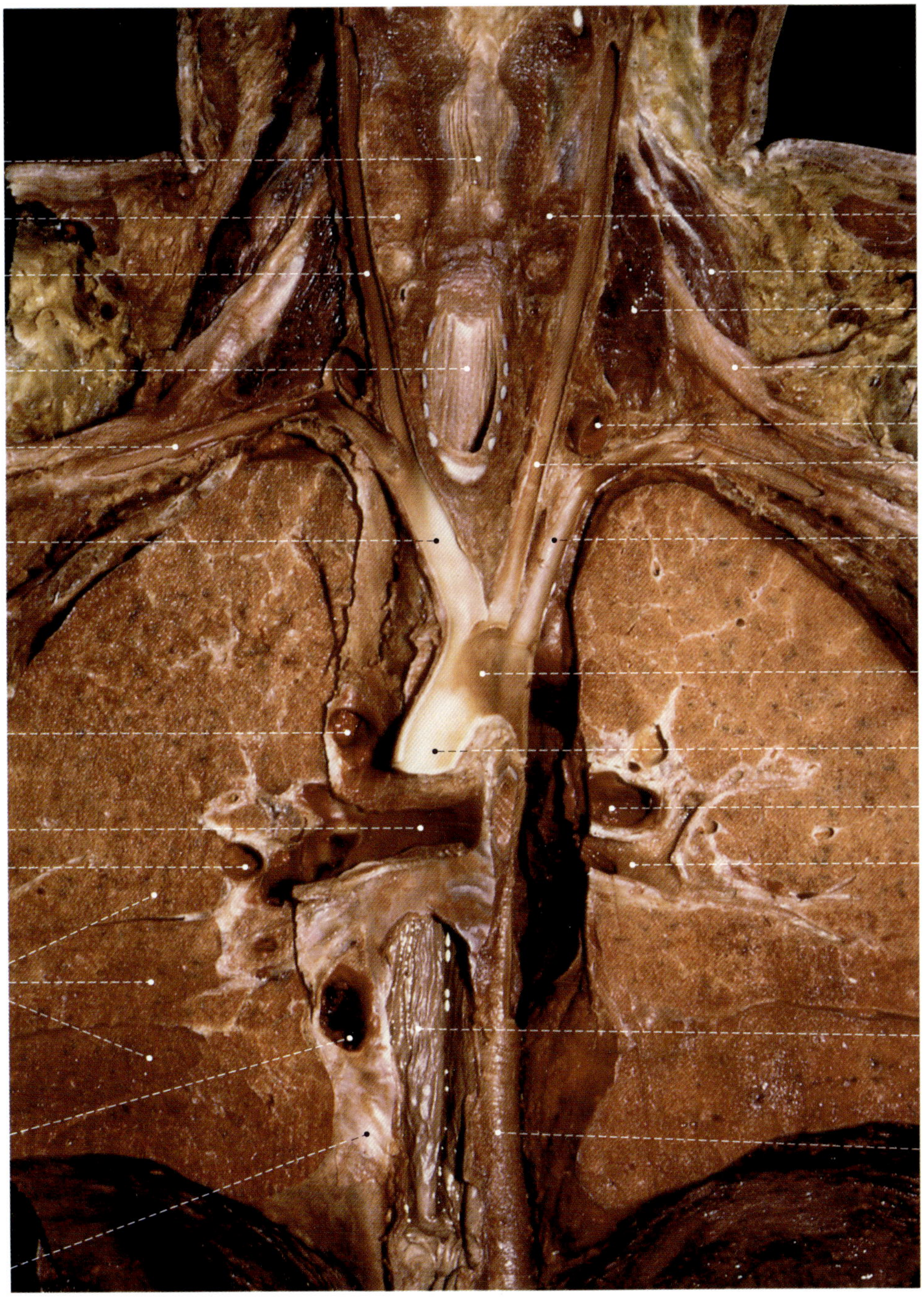

165 Thorax

165 Mediastinum and large artery trunks (55%)

Coronal section through the thorax and the lower neck in the plane of the aortic arch and the common carotid arteries, anatomical section, ventral aspect

165 Mediastinum und grosse Arterienstämme (55%)

Frontalschnitt durch den Thorax und den unteren Hals in der Ebene des Aortenbogens und der gemeinsamen Kopfschlagadern, anatomischer Schnitt, Ventralansicht

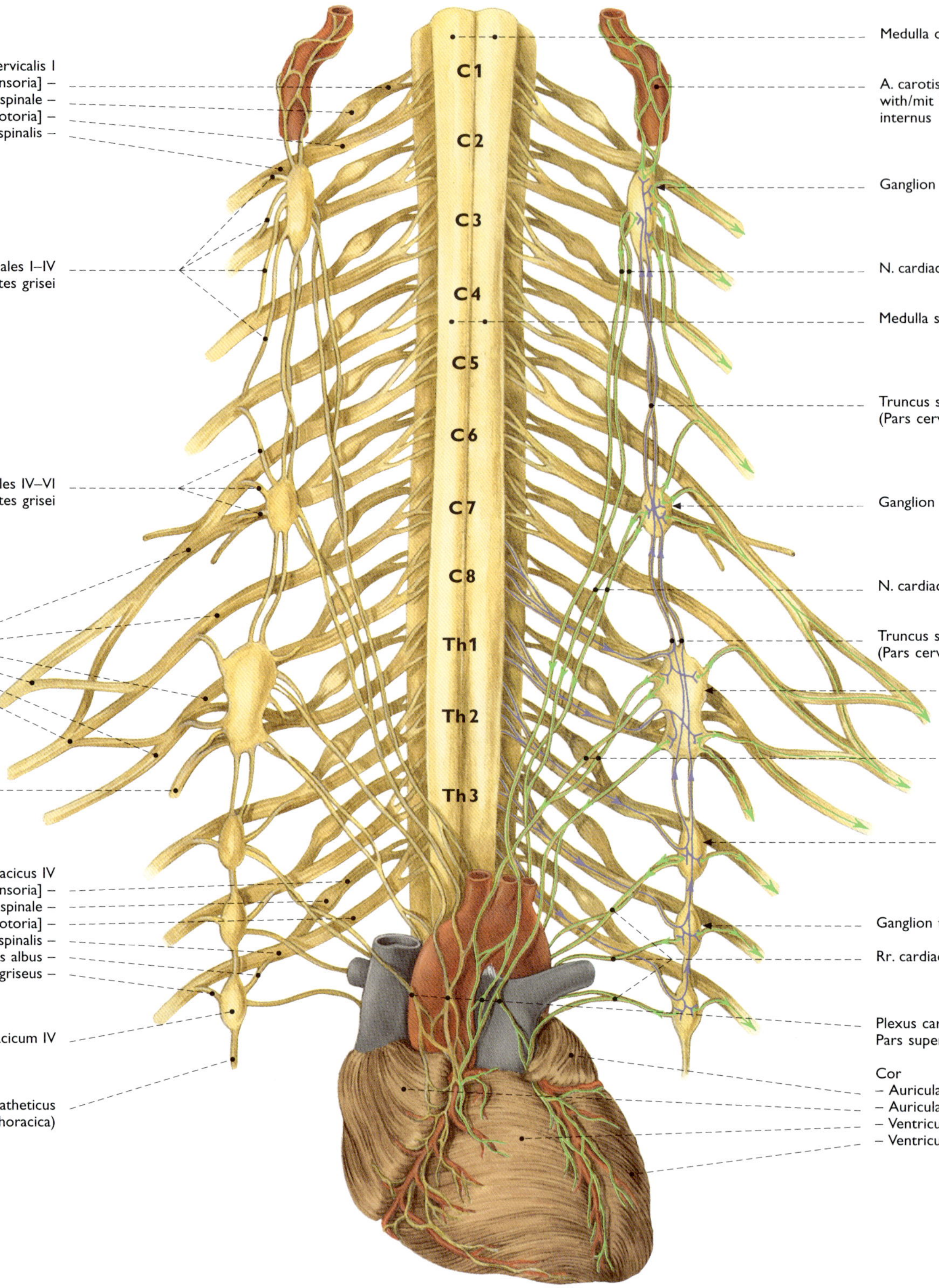

166 ⟨Collum⟩ et Thorax

166 Sympathetic nervous system of the neck and the upper thorax

Schematic representation

On the left side, the preganglionic neurons of the sympathetic efferent nervous system are indicated by blue lines, the postganglionic neurons by green ones.

166 Sympathisches Nervensystem im Hals und oberen Brustraum

Schematische Darstellung

Links sind die präganglionären Neurone des sympathischen efferenten Nervensystems durch blaue Linien, die postganglionären Neurone durch grüne Linien markiert.

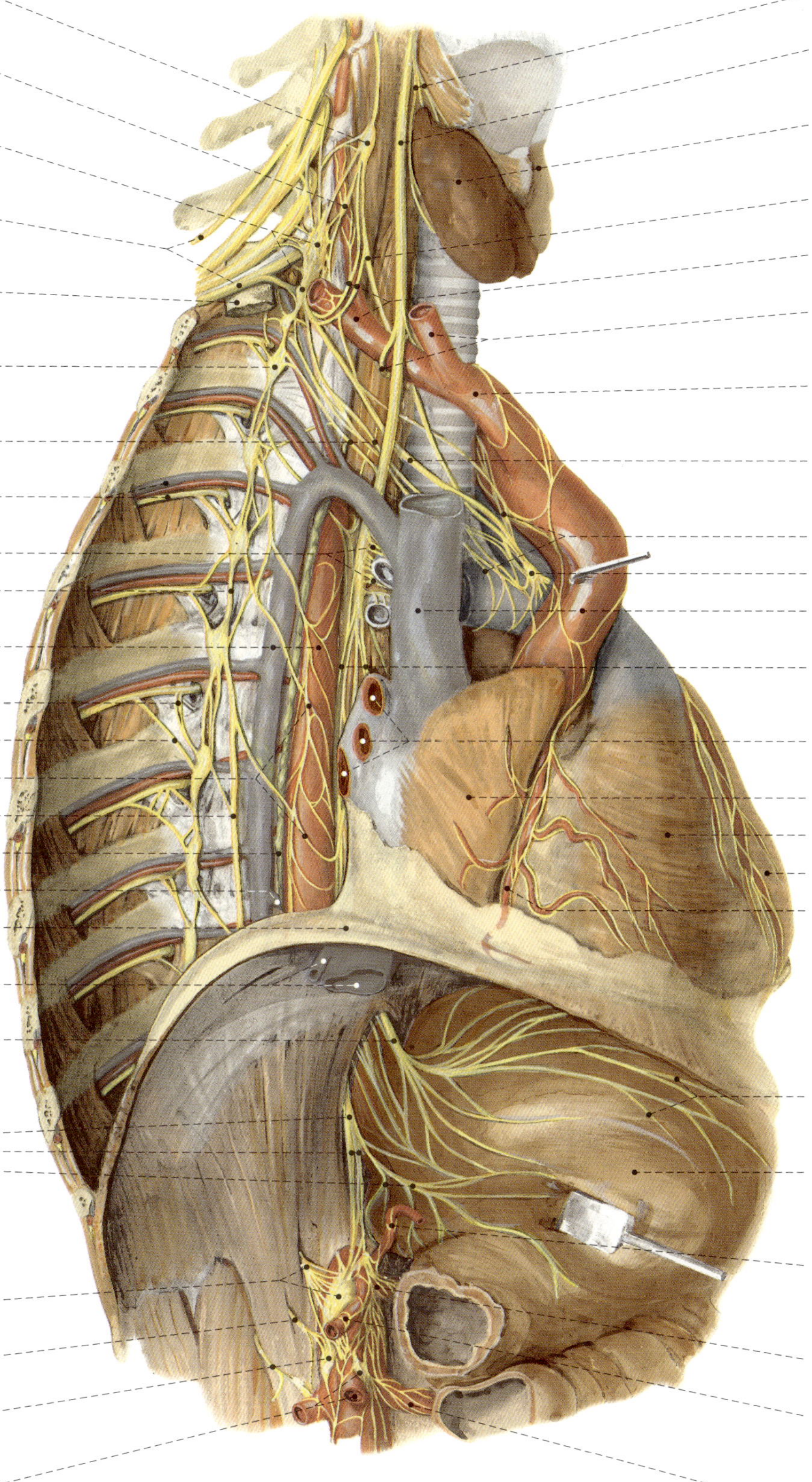

167 Thorax et Abdomen

167 Autonomic nervous system in the thorax and upper abdomen (50%)
Anterolateral aspect

167 Autonomes Nervensystem im Brustraum und Oberbauch (50%)
Ventrolateralansicht

Organa abdominis et pelvis

Abdominal and Pelvic Viscera

Bauch- und Beckeneingeweide

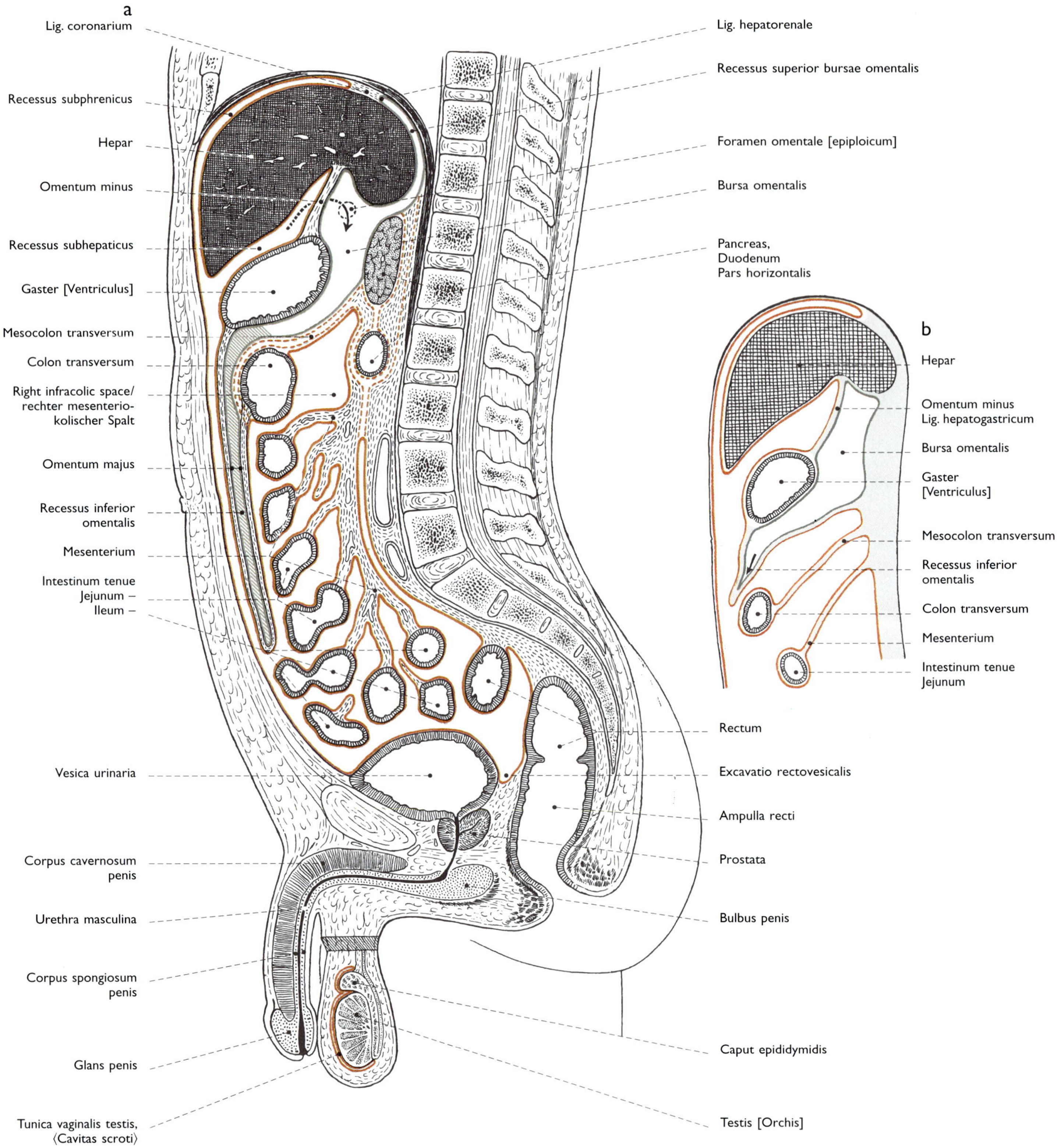

170 Cavitas peritonealis

170 Peritoneal cavity

Schematized median sections

a Adult stage of a male. Step-cut in the scrotum in order to expose the right cavity of the tunica vaginalis

b Fetal stage

170 Peritonealhöhle

Schematisierte Medianschnitte

a Erwachsenenstadium eines Mannes. Stufenschnitt im Skrotalbereich zur Darstellung der rechten Cavitas scroti

b Fetalstadium

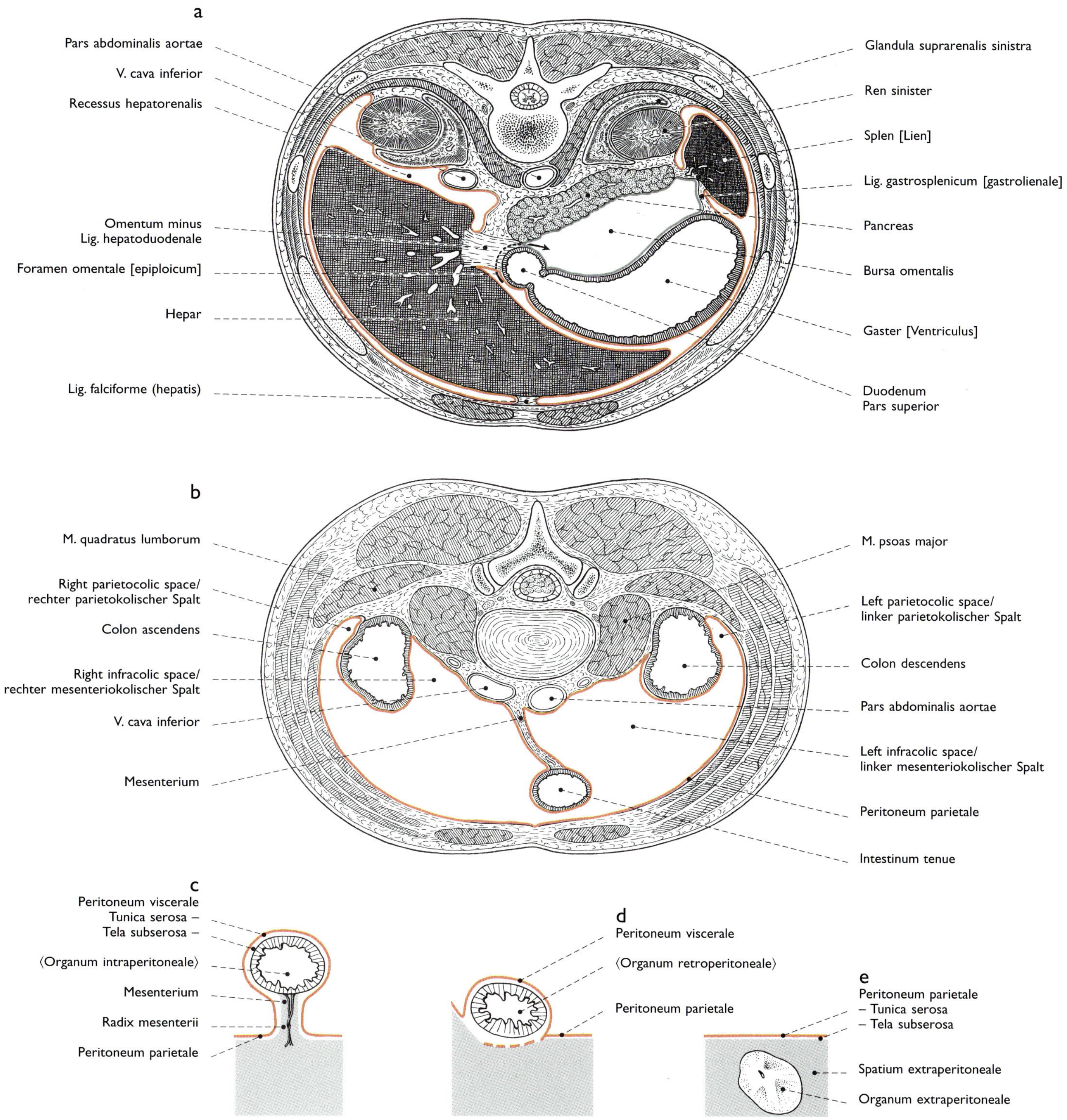

171 Cavitas peritonealis

171 Peritoneal cavity

a, b Superior aspect of schematized transverse sections at the level of the
- a upper (supramesocolic) abdominal region (upper edge of the epiploic foramen)
- b lower (inframesocolic) abdominal region

c–e Schematic representations of the
- c intraperitoneal position
- d (secondary) retroperitoneal position
- e extraperitoneal position

171 Peritonealhöhle

a, b Kranialansicht schematisierter Transversalschnitte im Bereich des
- a oberen (supramesokolischen) Bauchsitus (Höhe der oberen Begrenzung des Foramen omentale)
- b unteren (inframesokolischen) Bauchsitus

c–e Schematische Darstellung der
- c intraperitonealen Lage
- d (sekundär) retroperitonealen Lage
- e extraperitonealen Lage

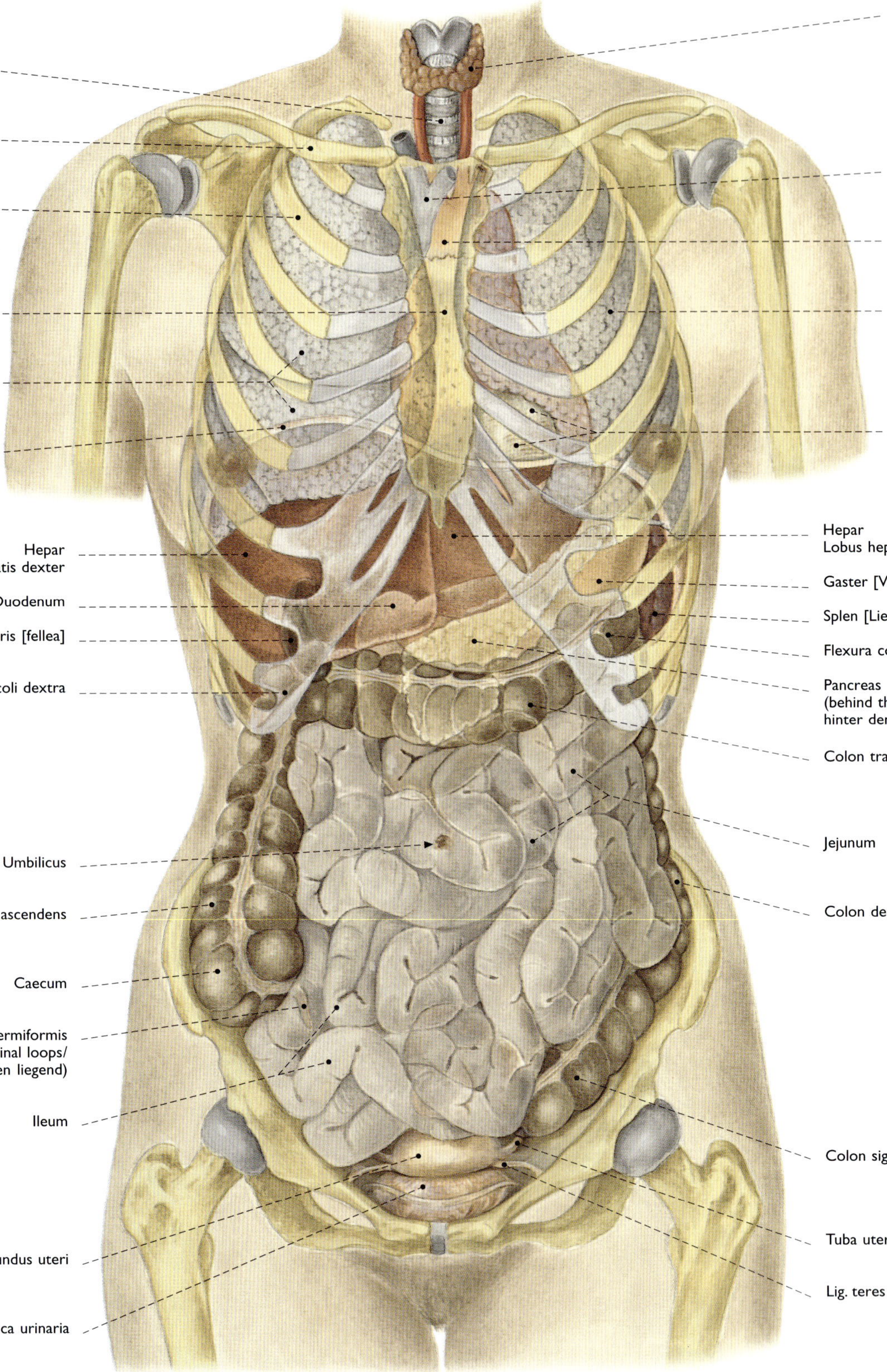

172 Thorax et Abdomen

172 Thoracic and abdominal viscera
(30%)
Surface projections onto the chest and anterior abdominal walls, ventral aspect

172 Brust- und Bauchorgane
(30%)
Projektion auf die vordere Leibeswand, Ventralansicht

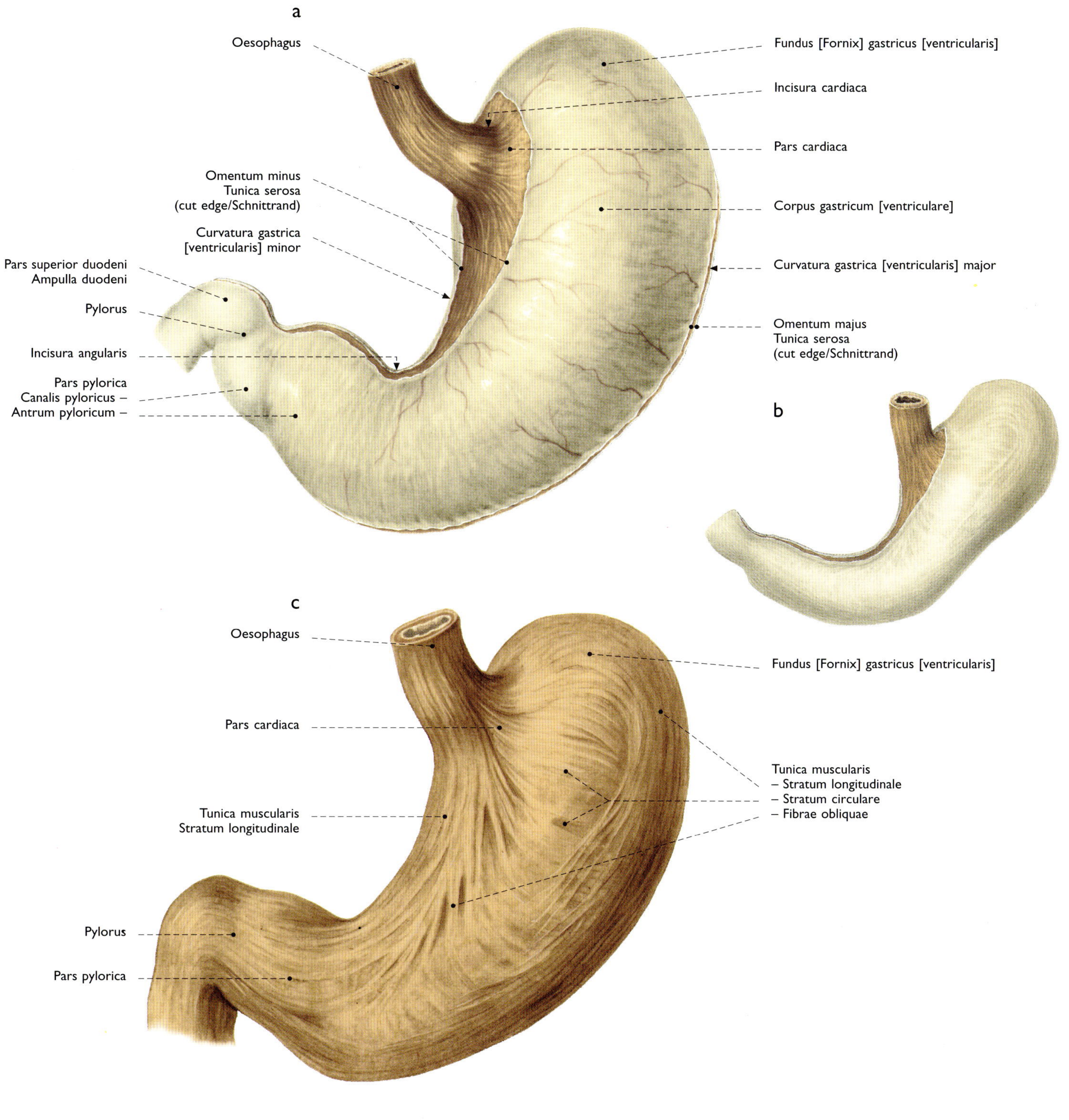

173 Gaster [Ventriculus]

173 Stomach (55%)

Ventral aspect

a External aspect of a completely filled stomach with distended walls
b Empty stomach with extremely contracted walls
c Muscle tracts in the anterior wall of the stomach after removal of the serosa and subserosa

173 Magen (55%)

Ventralansicht

a Aussenansicht eines stark gefüllten Magens mit gedehnten Wänden
b Leerer Magen mit extrem kontrahierten Wänden
c Muskelzüge an der Magenvorderwand nach Entfernung der Tunica serosa und Tela subserosa

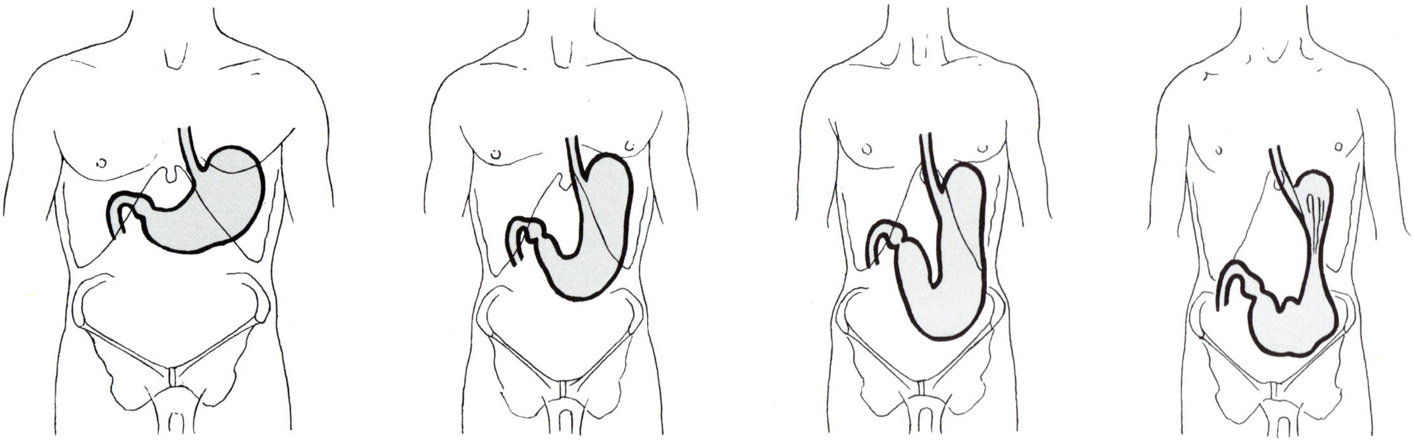

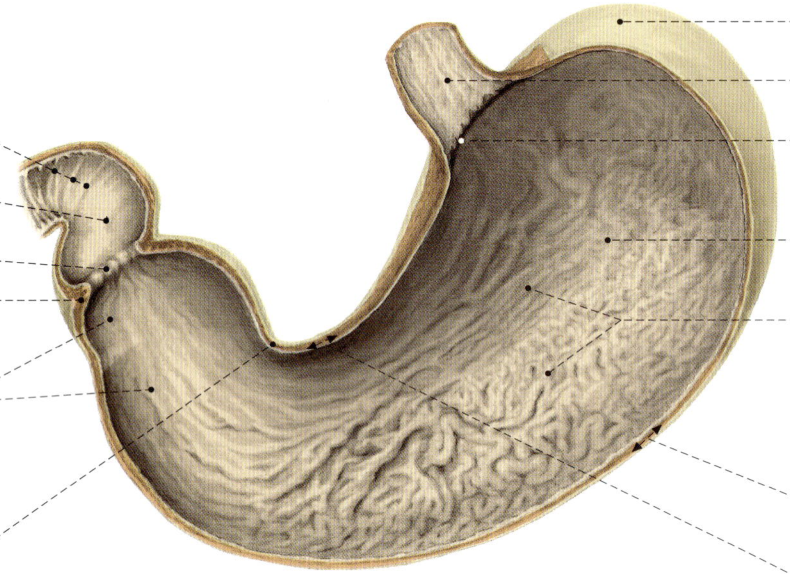

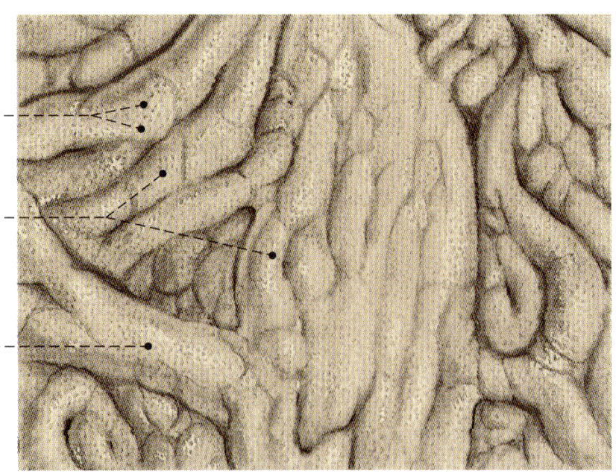

174 Gaster [Ventriculus]

174 Stomach

a–d Schematic representations of some important functional forms of the stomach
a Hypertonic stomach
b Orthotonic stomach
c Hypotonic stomach
d Atonic stomach
e, f Internal aspect of the stomach
e Internal aspect of the posterior wall after removal of the anterior wall (50%)
f Mucosa of the body of the stomach (300%)

174 Magen

a–d Schematische Darstellung einiger wichtiger funktioneller Magenformen
a Hypertonischer Magen
b Orthotonischer Magen
c Hypotonischer Magen
d Atonischer Magen
e, f Innenansicht des Magens
e Innenansicht der dorsalen Wand nach Entfernung der ventralen Wand (50%)
f Schleimhaut der Korpusregion (300%)

a

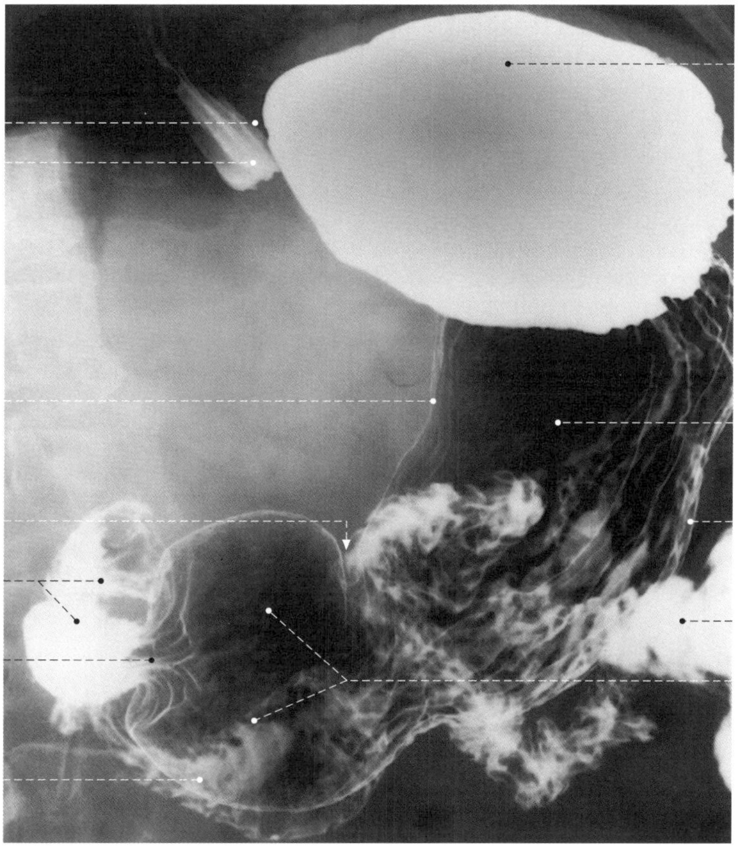

b

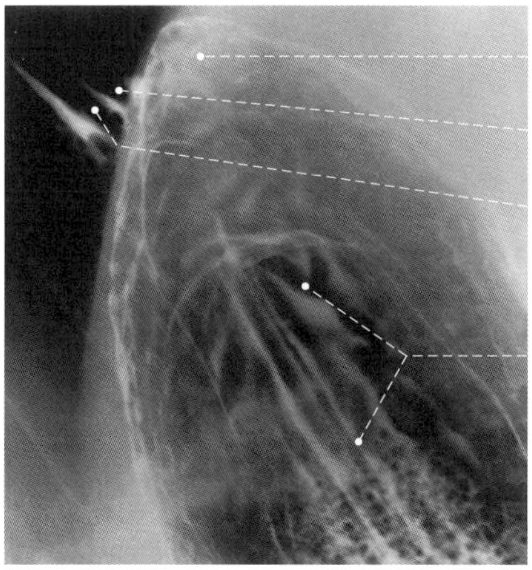

c

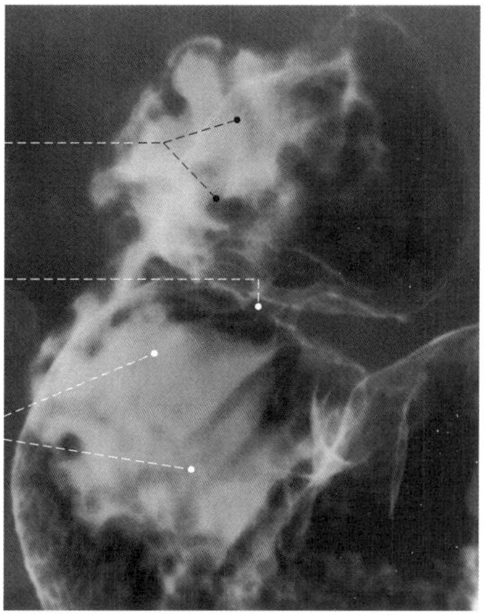

175 Gaster [Ventriculus]

175 Stomach (50%)

Double-contrast radiographs after barium meal and distension by air

a Posteroanterior radiograph of the stomach and the duodenal ampulla
b Spot film radiograph of the gastric fundus
c Oblique radiograph of the pyloric part of the stomach and the duodenal ampulla from right anterior (RAO projection)

175 Magen (50%)

Röntgenaufnahmen in Doppelkontrastdarstellung nach Bariumbreigabe und Luftblähung

a Posteroanteriores Bild des Magens und der Ampulla duodeni
b Zielaufnahme des Magenfundus
c Schrägaufnahme der Pars pylorica des Magens und der Ampulla duodeni von rechts vorne (1. Schrägdurchmesser, "Fechterstellung")

a

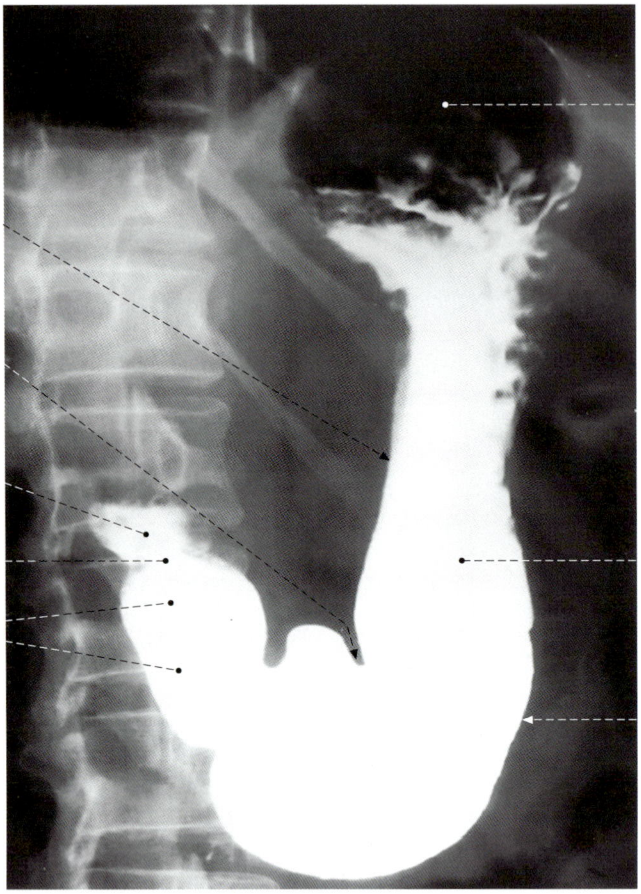

b

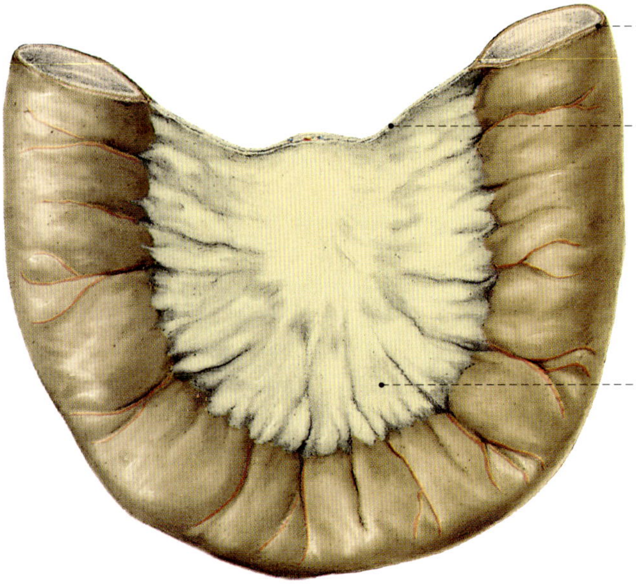

c

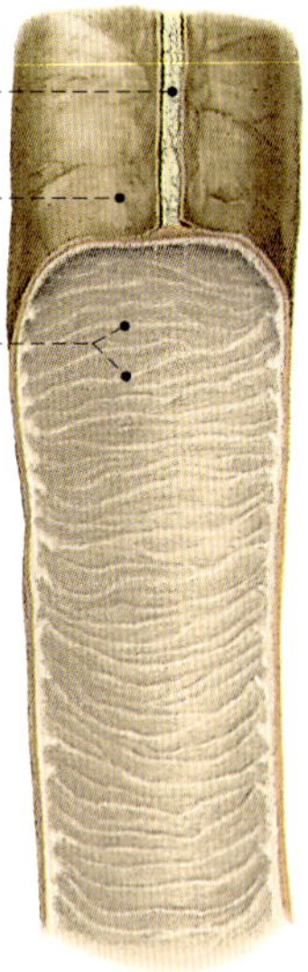

176 Gaster [Ventriculus] et Intestinum tenue

176 Stomach and small intestine (50%)

a Posteroanterior radiograph of the stomach and the duodenal ampulla after barium meal
b Jejunal loop with its mesentery, external aspect
c Jejunum partially opened

176 Magen und Dünndarm (50%)

a Posteroanteriores Röntgenbild des Magens und der Ampulla duodeni nach Bariumbreifüllung
b Jejunalschlinge mit Mesenterium, Aussenansicht
c Jejunum teilweise eröffnet

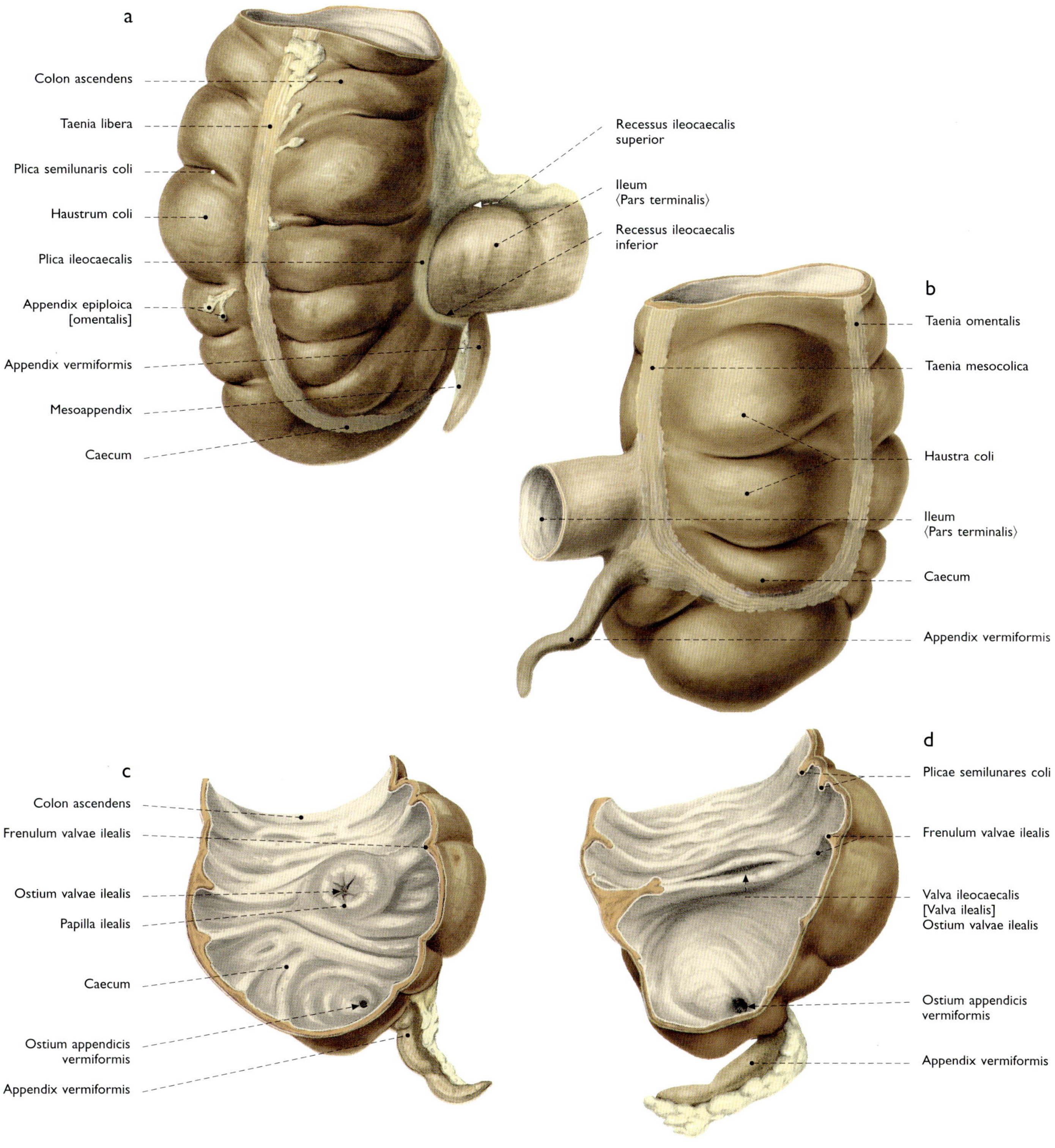

177 Caecum et Appendix vermiformis

177 Cecum and vermiform appendix

(70%)

a Ventral aspect
b Dorsal aspect
c, d Ileocecal valve and ostium of the appendix after removal of the ventrolateral wall of the large intestine
c View during life
d View of the relaxed intestine after death

177 Blinddarm und Wurmfortsatz

(70%)

a Ventralansicht
b Dorsalansicht
c, d Valva ileocaecalis und Öffnung des Wurmfortsatzes nach Entfernung der ventrolateralen Dickdarmwand
c Vitalaspekt
d Aspekt des erschlafften Darms nach dem Tod

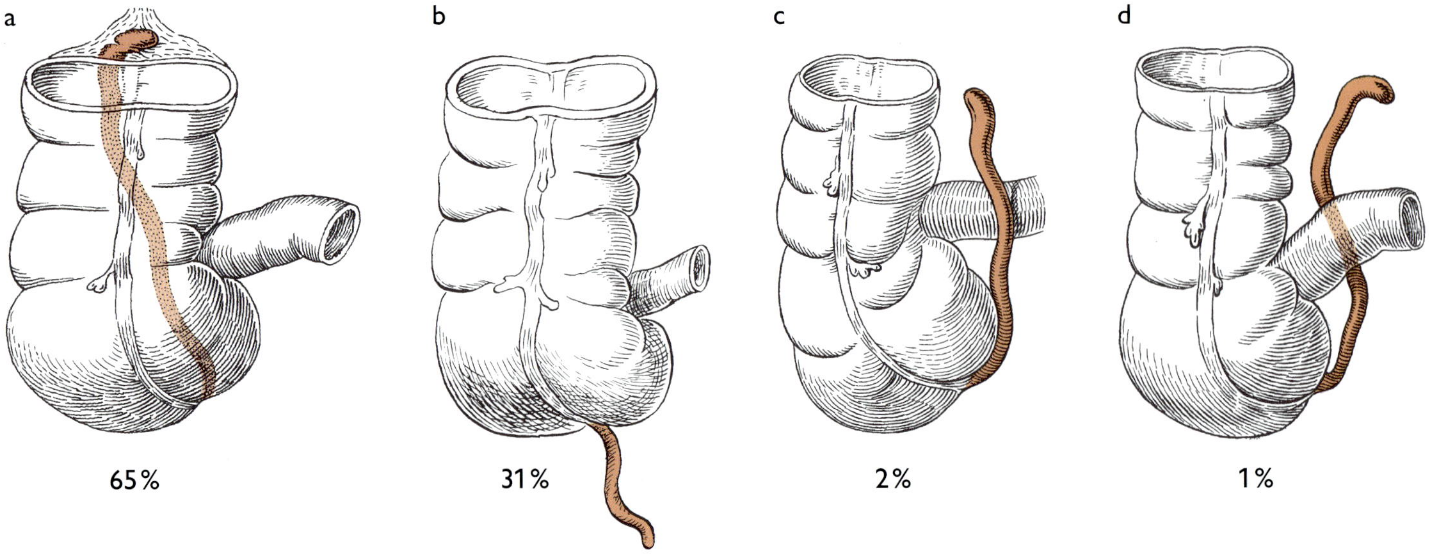

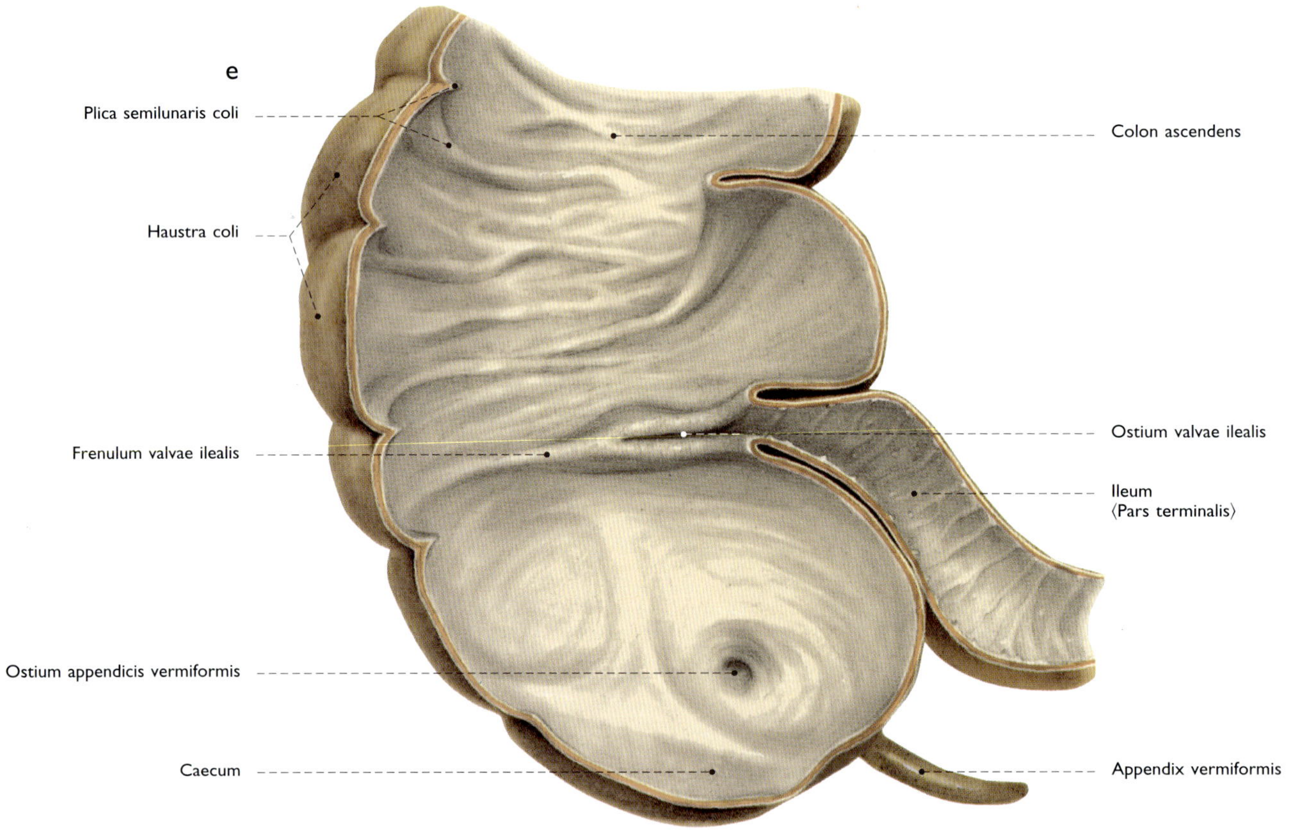

178 Caecum et Appendix vermiformis

178 Cecum and vermiform appendix

a–d Four possible positions of the vermiform appendix. The percentile numbers beneath the pictures indicate the approximate frequency.

- a Retrocecal and retrocolic position
- b Pendulous position into the lesser pelvis
- c Position in front of the terminal part of the ileum
- d Retroileal position
- e Internal aspect of the dorsal wall of the cecum after removal of the ventral intestinal wall (80%)

178 Blinddarm und Wurmfortsatz

a–d Vier mögliche Lagen des Wurmfortsatzes. Die Prozentzahlen unter den Abbildungen geben die ungefähre Häufigkeit an.

- a Retrozäkale und retrokolische Lage
- b Ins kleine Becken herabhängende Lage
- c Präileale Lage
- d Retroileale Lage
- e Innenansicht der Hinterwand des Caecum (des Blinddarms) nach Entfernung der ventralen Darmwand (80%)

a

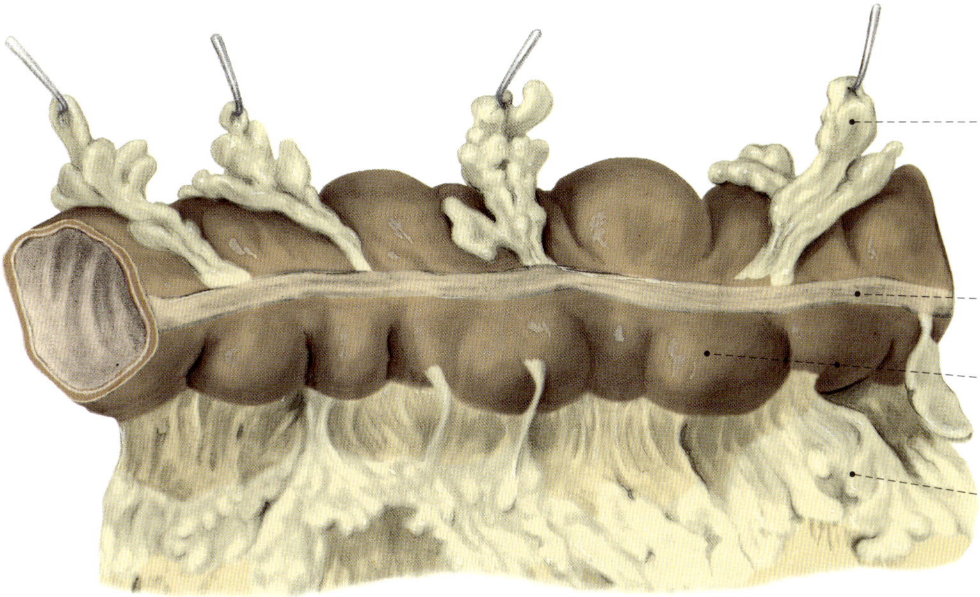

b

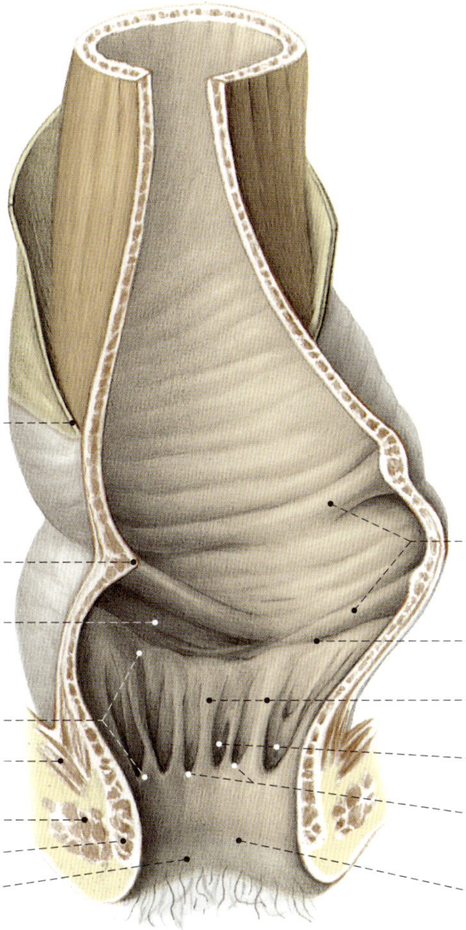

179 Colon transversum et Colon rectum

179 Transverse colon and rectum (70%)

a Dorsal aspect of the middle portion of the transverse colon
b Interior of the rectum revealed by incising along the midventral line

179 Querkolon und Mastdarm (70%)

a Dorsalansicht des mittleren Drittels des Querkolons
b Innenansicht des durch Längsschnitt medioventral eröffneten Mastdarms

a

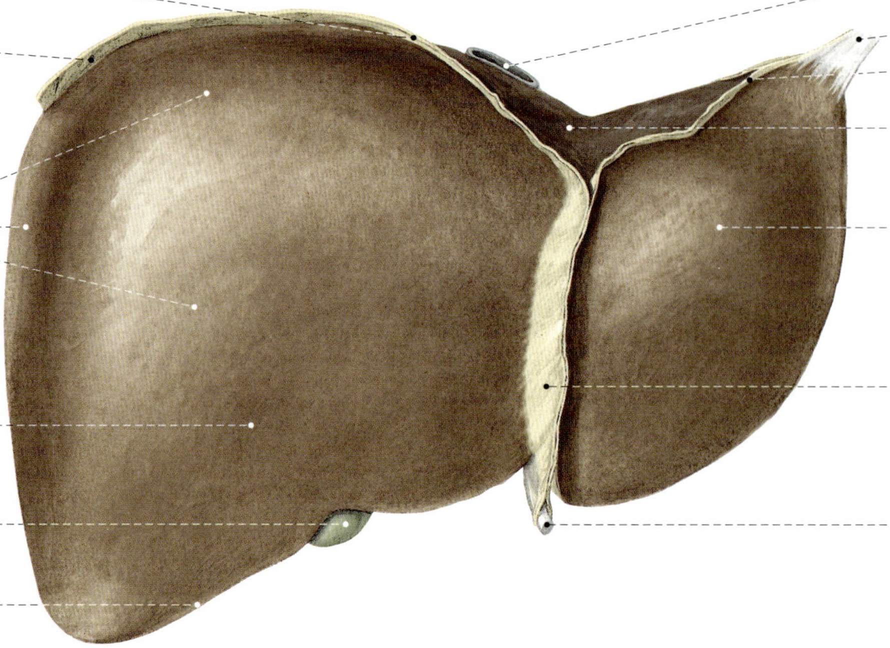

b

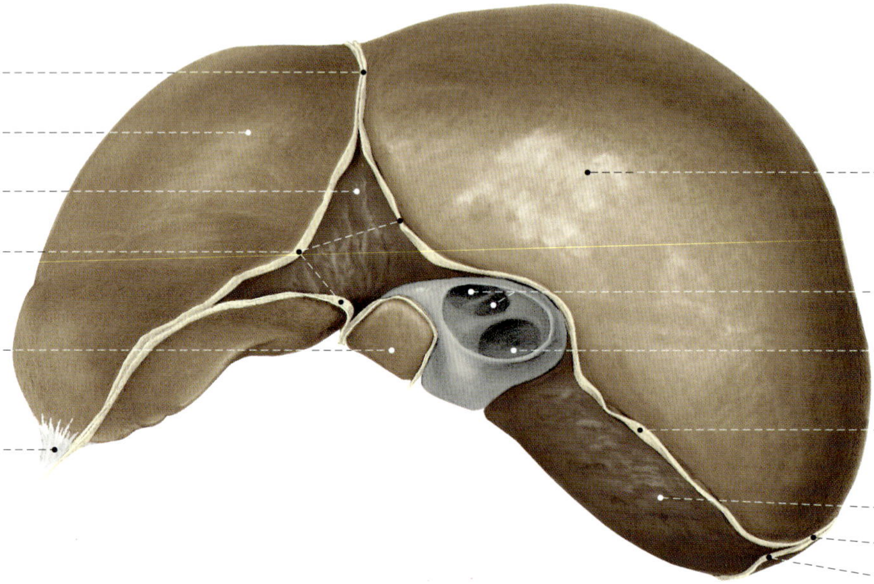

180 Hepar

180 Liver (45%)
Diaphragmatic surface
a Ventral aspect
b Superior aspect

180 Leber (45%)
Facies diaphragmatica
a Ventralansicht
b Kranialansicht

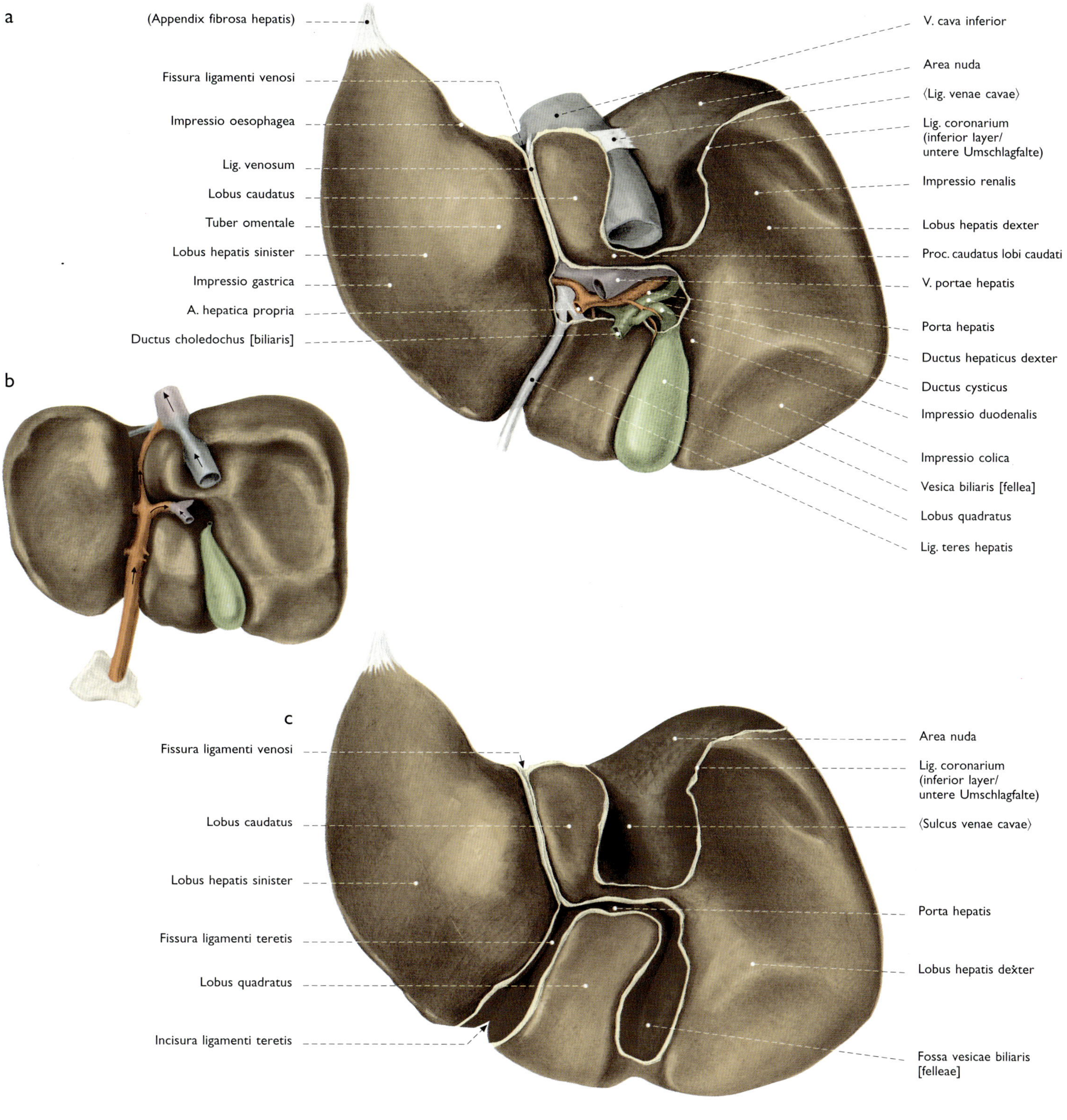

181 Hepar

181 Liver (45%)

Visceral surface, dorsal aspect

a Liver of an adult
b Liver of a newborn child
c Liver of an adult after removal of the gallbladder, the inferior caval and portal veins, and the remnants of embryonal vessels

181 Leber (45%)

Facies visceralis, Dorsalansicht

a Leber eines Erwachsenen
b Leber eines Neugeborenen
c Leber eines Erwachsenen nach Entfernung der Gallenblase, der unteren Hohlvene, der Portalgebilde und der ligamentösen Reste embryonaler Gefässe

a

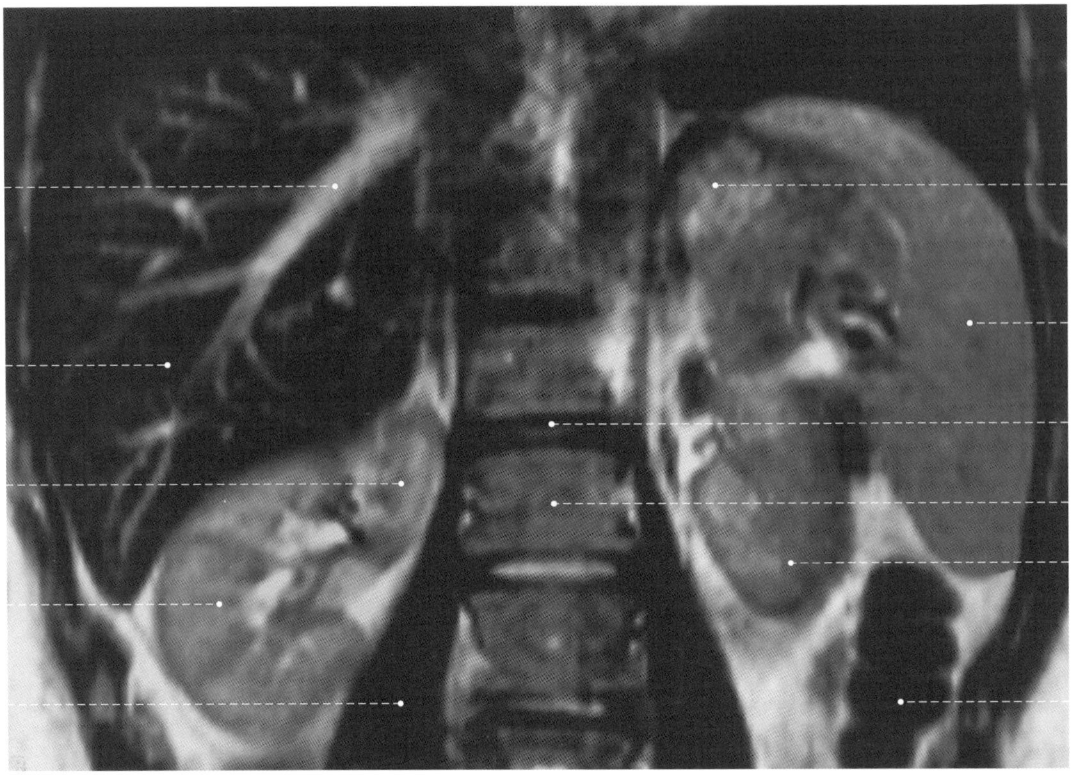

b

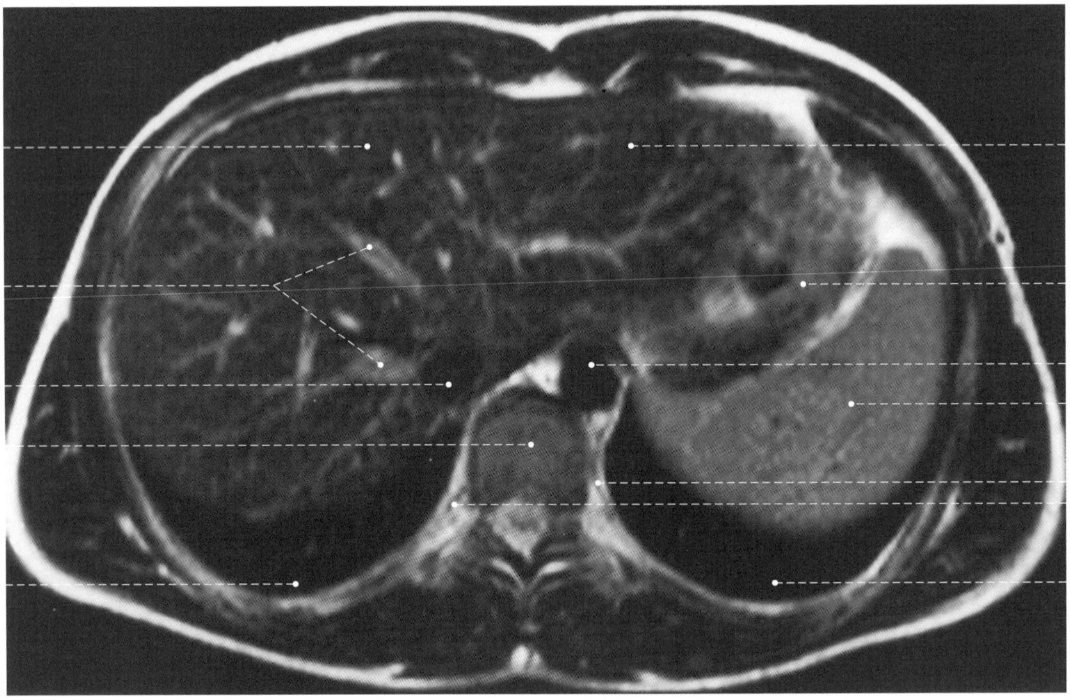

182 Hepar

182 Liver and adjacent organs (50%)

Magnetic resonance images (MRI, T_2-weighted)

a Coronal section through the abdomen in the plane of the lumbar vertebral bodies, ventral aspect

b Transverse section through the epigastrium at the level of the eleventh thoracic vertebral body, inferior aspect

182 Leber und benachbarte Organe (50%)

Kernspintomogramme (MRI, T_2-gewichtet)

a Frontalschnitt durch den Bauchraum in der Ebene der Lendenwirbelkörper, Ventralansicht

b Horizontalschnitt durch den Oberbauch in Höhe des elften Brustwirbelkörpers, Kaudalansicht

a

Blue: Portal vein with ramifications
Red: Hepatic artery with branches
Yellow: Bile ducts and hepatic ducts

Blau: V. portae hepatis und Aufzweigungen
Rot: A. hepatica propria und Äste
Gelb: Gallengangsystem und Ductus hepatici

b

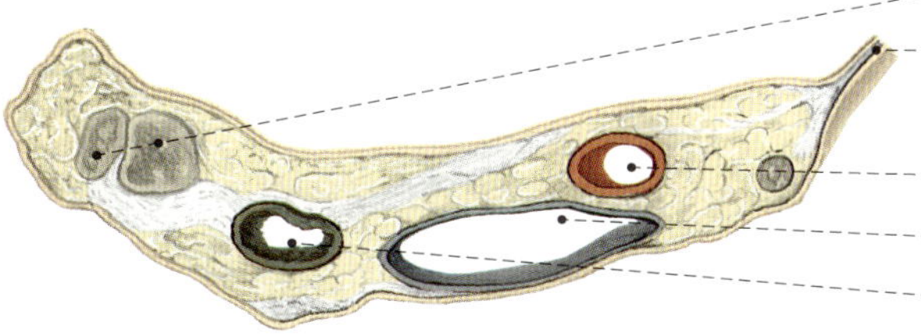

c

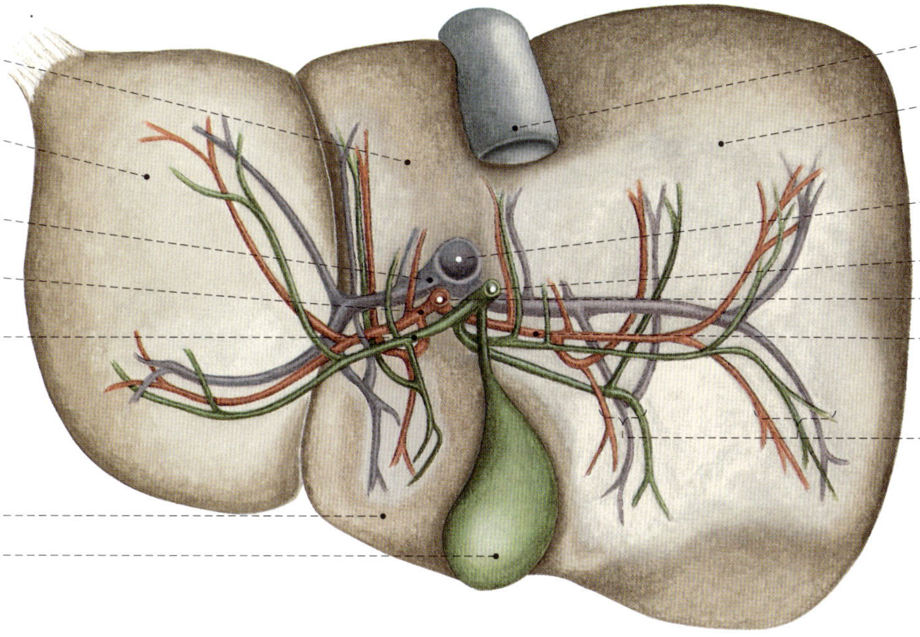

183 ⟨Vasa hepatica⟩, Ductuli biliferi et Ductus hepatici

183 Hepatic blood vessels and intrahepatic bile duct system

a Corrosion cast (50%), ventral aspect
b Hepatoduodenal ligament (150%), transverse section, inferior aspect
c Distribution pattern of the blood vessels and bile ducts in the liver (50%), schematic representation, dorsal aspect

183 Blutgefässe der Leber und intrahepatische Gallenwege

a Korrosionspräparat (50%), Ventralansicht
b Ligamentum hepatoduodenale (150%), Horizontalschnitt, Kaudalansicht
c Verzweigungsmuster der Blutgefässe und Gallengänge in der Leber (50%), schematische Darstellung, Dorsalansicht

a

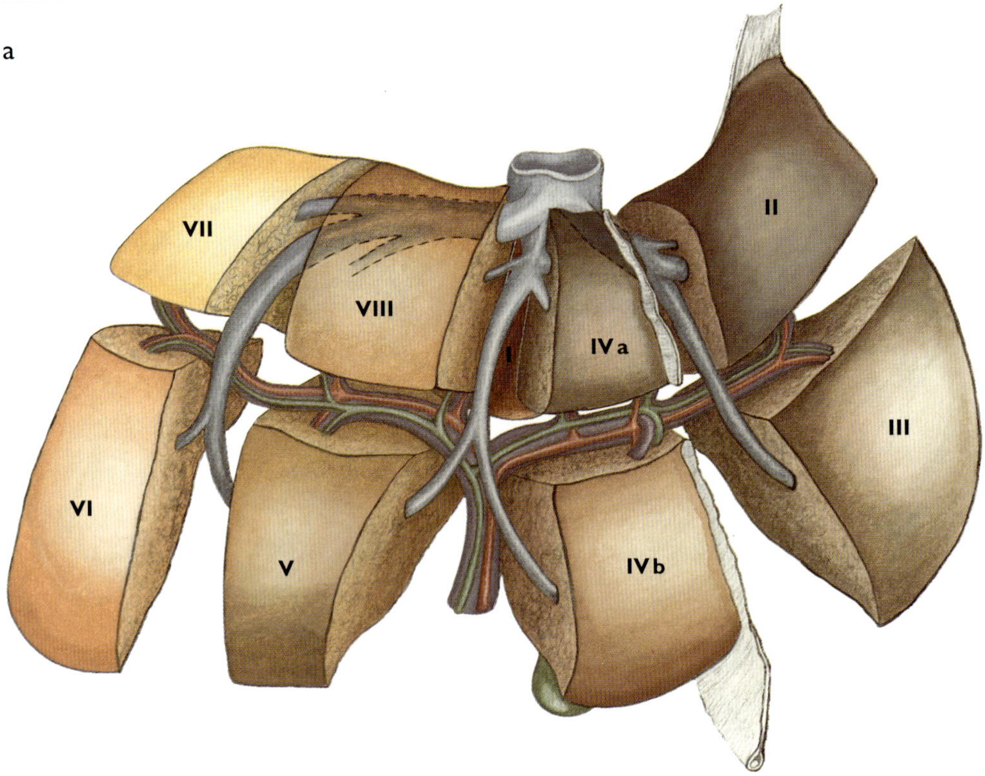

b

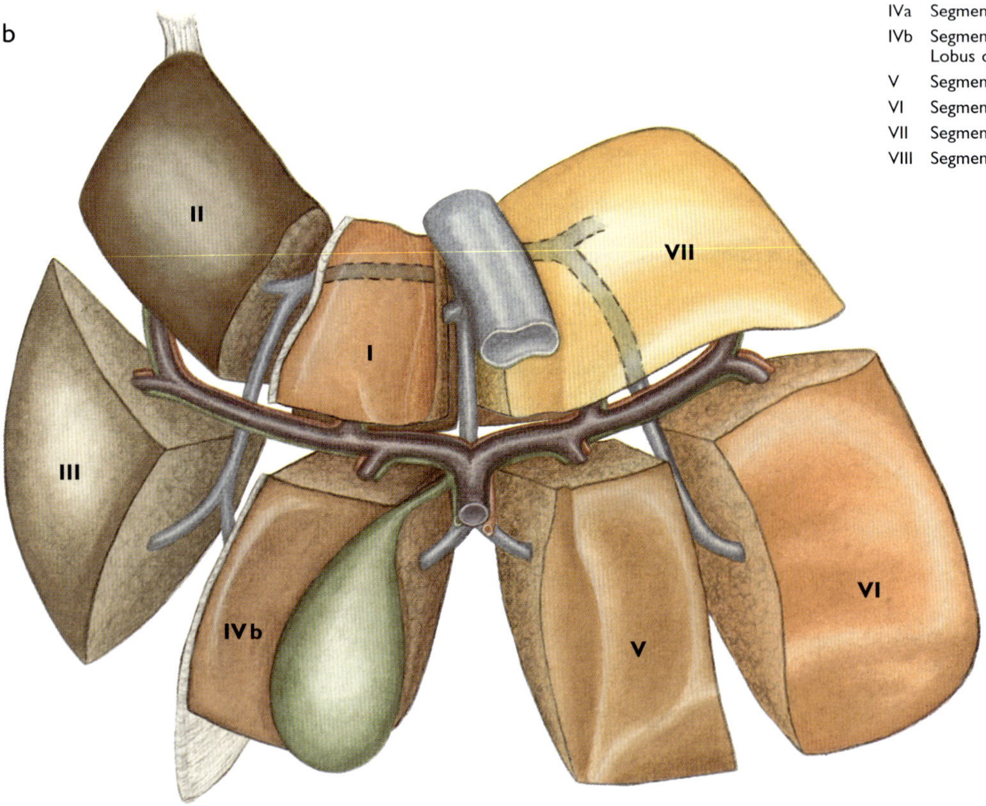

I	Lobus caudatus	A. hepatica propria R. sinister
II	Segmentum laterale superius	
III	Segmentum laterale inferius	
IVa	Segmentum mediale superius	
IVb	Segmentum mediale inferius, Lobus quadratus	
V	Segmentum anterius inferius	A. hepatica propria R. dexter
VI	Segmentum posterius inferius	
VII	Segmentum posterius superius	
VIII	Segmentum anterius superius	

184 ⟨Segmenta hepatis⟩

184 Segmentation of the liver (50%)

Schematic representations (according to Heberer, Köle and Tscherne, 1986)

a Diaphragmatic surface, ventral aspect
b Visceral surface, dorsal aspect

184 Segmentgliederung der Leber (50%)

Schematische Darstellungen (nach Heberer, Köle und Tscherne, 1986)

a Facies diaphragmatica, Ventralansicht
b Facies visceralis, Dorsalansicht

a

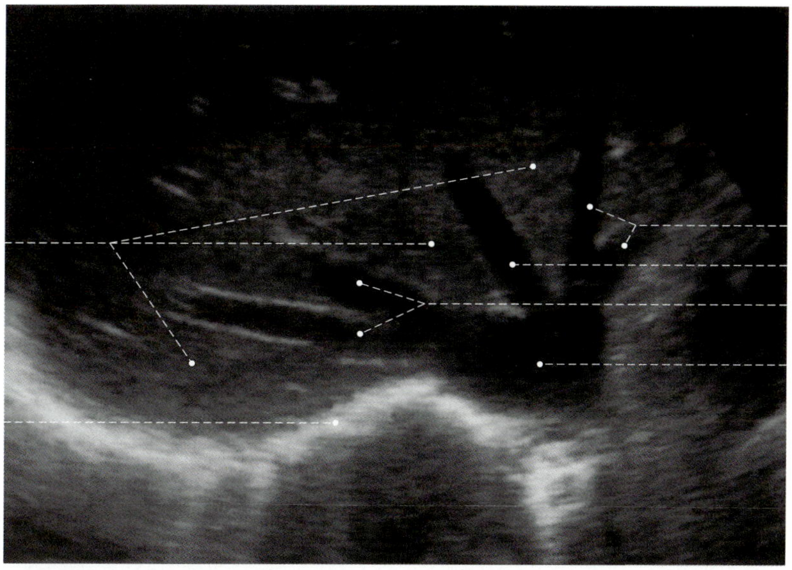

b

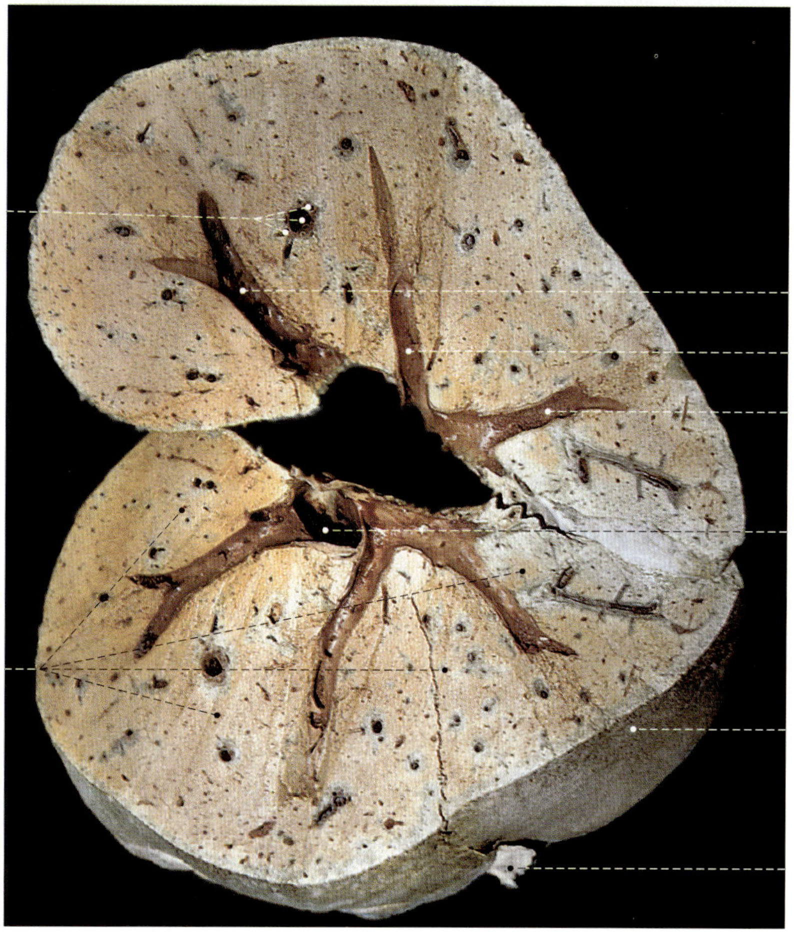

185 Hepar

185 Liver

a Ultrasound image showing the liver veins (40%), subcostal oblique section

b Section through the upper part of a human liver, opening the four cranial liver segments. The ventral part of the liver is turned upwards (30%). Ventral aspect

185 Leber

a Ultraschallbild der Lebervenen ("Lebervenenstern") (40%), subkostaler Schrägschnitt

b Schnitt durch den oberen Teil einer menschlichen Leber. Die vier kranialen Lebersegmente sind eröffnet, der ventrale Teil der Leber ist nach oben geklappt (30%). Ventralansicht

a

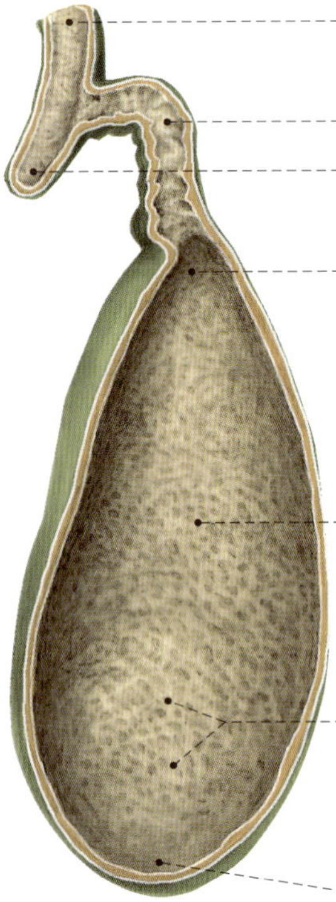

b

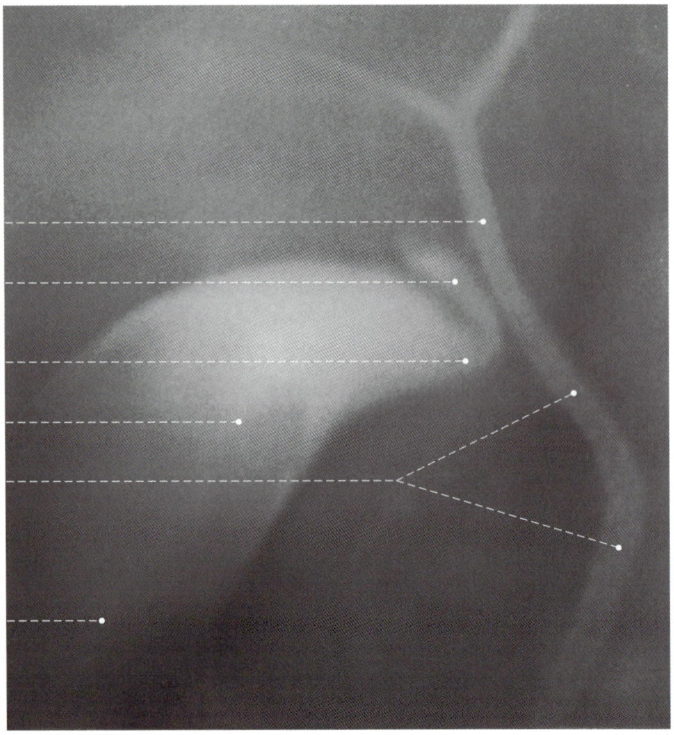

c

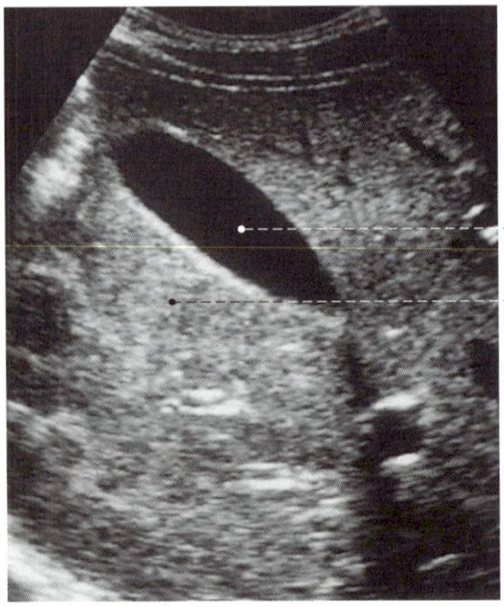

d

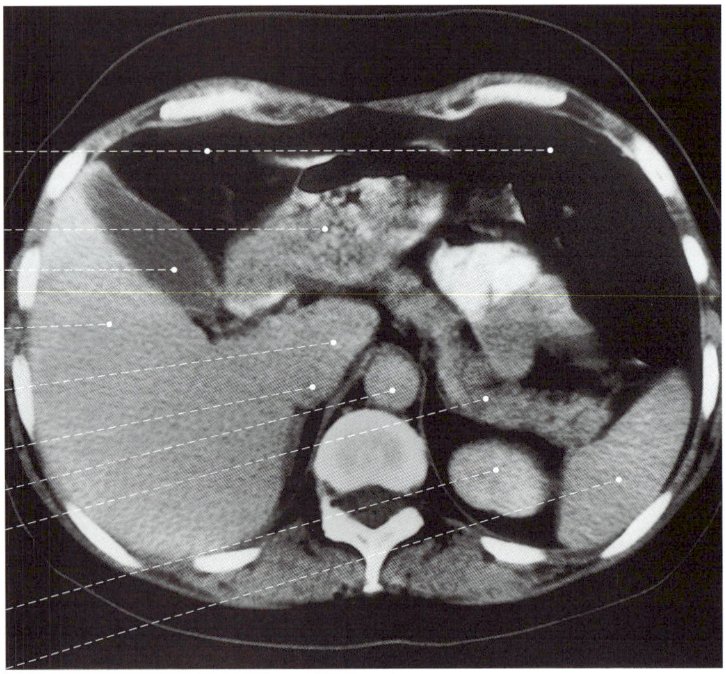

186 Vesica biliaris [fellea]

186 Gallbladder and extrahepatic bile duct system

a Gallbladder and bile ducts after removal of their dorsal walls (90%), dorsal aspect
b Cholecystocholangiography after intravenous injection of contrast medium (100%), posteroanterior radiograph
c Ultrasound image of the gallbladder (50%), subcostal oblique section
d Transverse computed tomogram (CT) at the level of the twelfth thoracic vertebral body (35%), inferior aspect

186 Gallenblase und extrahepatische Gallenwege

a Gallenblase und Gallenwege nach Entfernung ihrer dorsalen Wandungen (90%), Dorsalansicht
b Cholezystcholangiographie nach intravenöser Kontrastmittelgabe (100%), posteroanteriore Röntgenaufnahme
c Ultraschallbild der Gallenblase (50%), subkostaler Schrägschnitt
d Horizontales Computertomogramm (CT) in Höhe des zwölften Brustwirbelkörpers (35%), Kaudalansicht

a

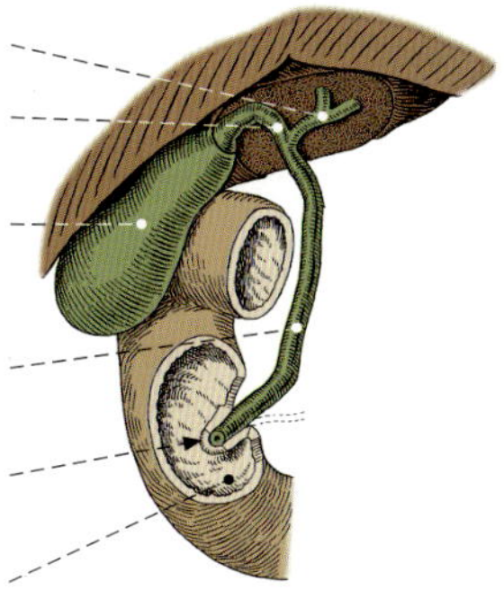

b

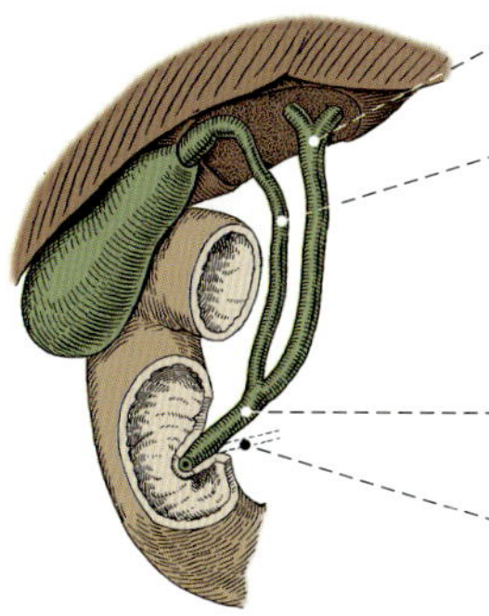

c

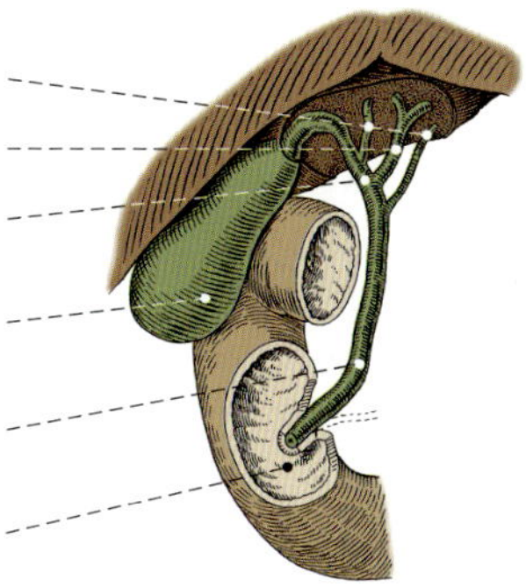

d

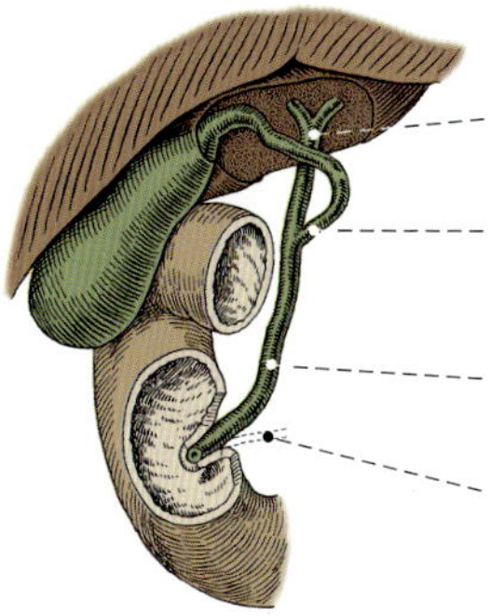

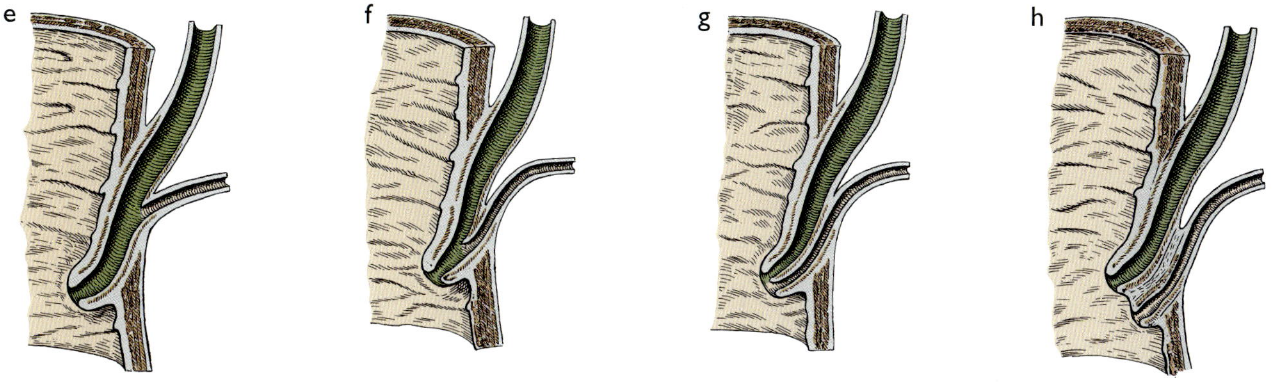

187 Ductus hepatici, Ductus cysticus, Ductus choledochus [biliaris] et Ductus pancreaticus

187 Extrahepatic bile duct system

a–d Variations of the hepatic and cystic ducts, schematic representations
- a The hepatic and cystic ducts joining at a high level close to the liver
- b The hepatic and cystic ducts joining at a low level distant from the liver
- c Accessory hepatic ducts joining the main ducts at a high level
- d The cystic duct winding ventrodorsally around the common hepatic duct prior to their junction

e–h Variations of the orifices of the bile and pancreatic ducts, schematic representations
- e The main pancreatic duct merging with the bile duct before they penetrate the duodenal wall
- f Common opening of the bile duct and the pancreatic duct joining within the intestinal wall
- g The two independent ducts ending in one common opening at the major duodenal papilla
- h The ducts open separately into the duodenum creating a bipartite papilla

187 Extrahepatische Gallenwege

a–d Variationen von Leber- und Gallengang, schematische Darstellungen
- a Hohe, lebernahe Vereinigung von Ductus hepaticus und Ductus cysticus
- b Tiefe, leberferne Vereinigung von Ductus hepaticus und Ductus cysticus
- c Ductus hepatici accessorii bei hoher Vereinigung des Leber- und Gallenblasengangs
- d Der Ductus cysticus windet sich vor der Vereinigung ventrodorsalwärts um den Ductus hepaticus communis

e–h Variationen der Mündung des Hauptgallen- und Pankreasgangs, schematische Darstellungen
- e Einmündung des Ductus pancreaticus in den Ductus choledochus ausserhalb der Darmwand
- f Gemeinsame Mündung von Ductus choledochus und Ductus pancreaticus, die sich in der Darmwand vereinigen
- g Gemeinsame Mündung beider Gänge, deren Lumina bis an die Spitze der Papilla duodeni major getrennt bleiben
- h Getrennte Mündung von Ductus choledochus und Ductus pancreaticus, Ausbildung einer Papilla bipartita

a

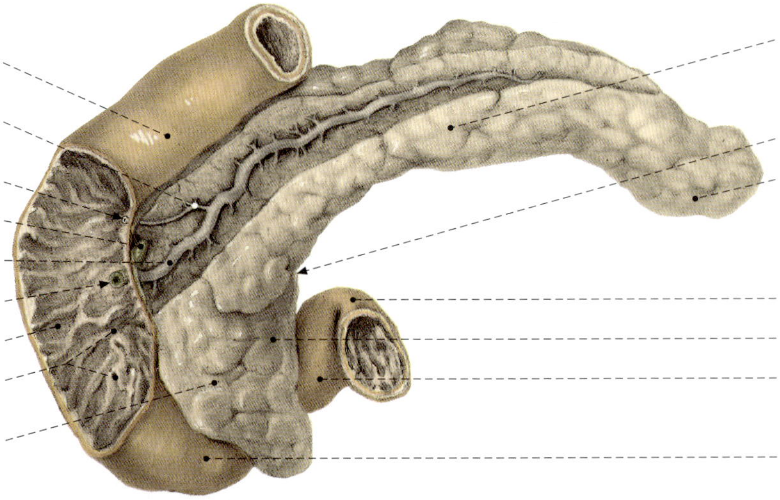

b

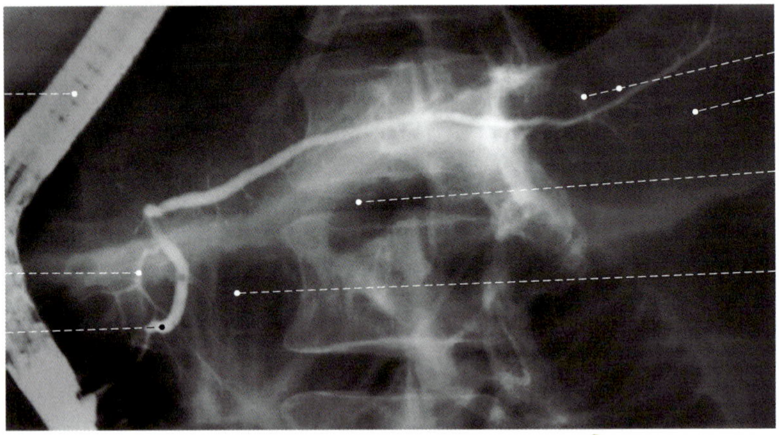

c

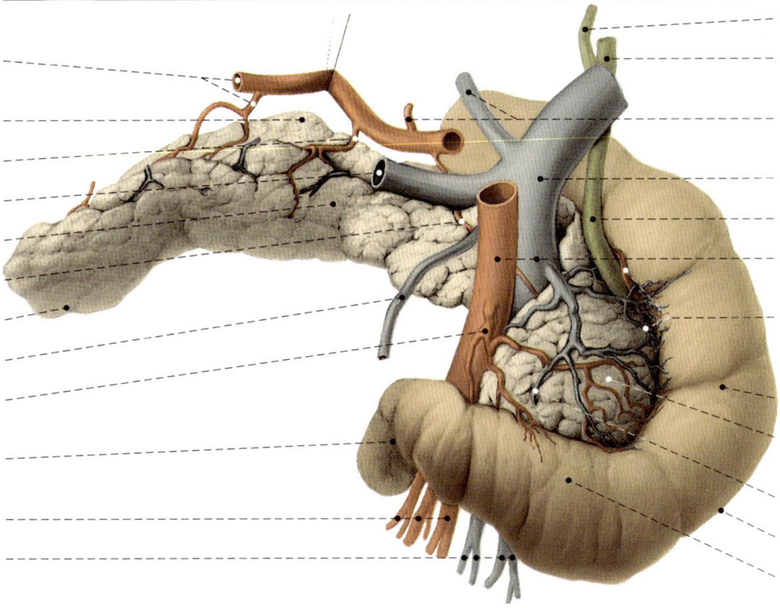

188 Pancreas et Duodenum

188 Pancreas and duodenum (60%)

a The anterior wall of the descending part of the duodenum was removed. The main pancreatic duct and the accessory pancreatic ducts are exposed. Ventral aspect

b Endoscopic retrograde cholangiopancreatography (ERCP) imaging the pancreatic duct, postero-anterior radiograph

c Pancreas, duodenum, bile duct and adjacent blood vessels, dorsal aspect

188 Pankreas und Duodenum (60%)

a Die Vorderwand der Pars descendens duodeni ist abgetragen, Ductus pancreaticus und Ductus pancreaticus accessorius sind freipräpariert. Ventralansicht

b Endoskopisch-retrograde Cholangiopankreatikographie (ERCP) mit Darstellung des Ductus pancreaticus, posteroanteriore Röntgenaufnahme

c Pankreas, Duodenum, Ductus choledochus und benachbarte Blutgefässe, Dorsalansicht

a

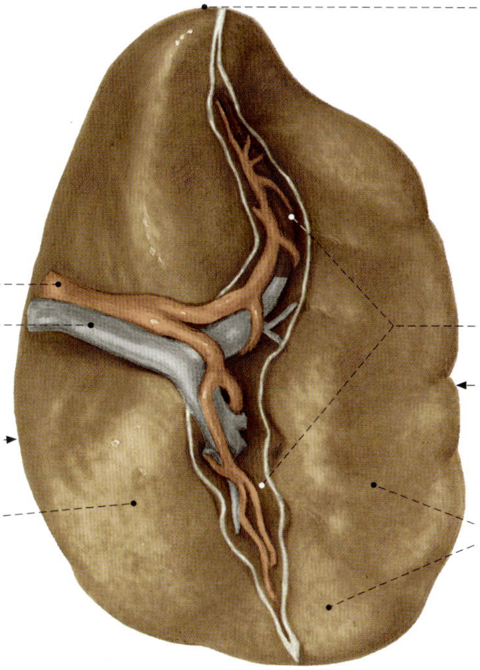

b

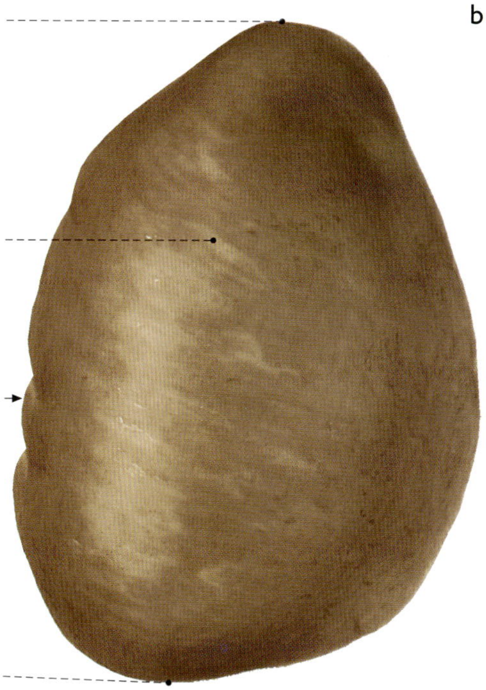

c

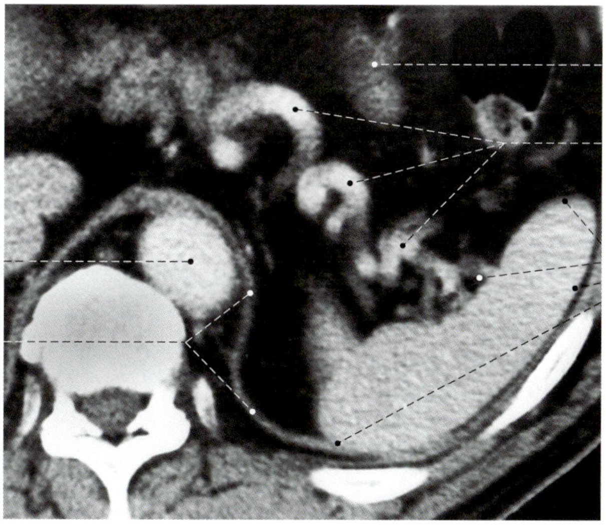

d

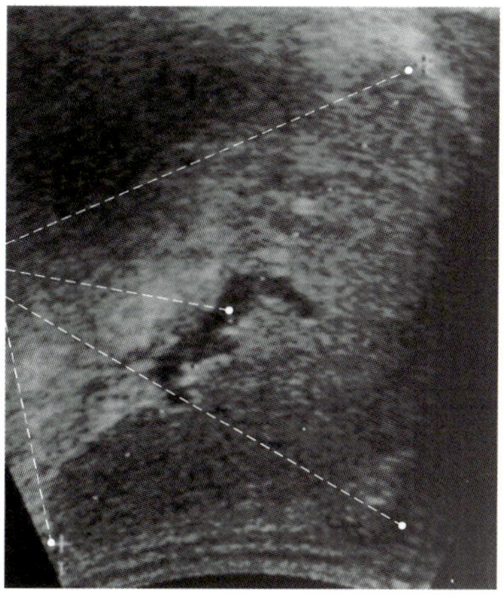

189 Splen [Lien]

189 Spleen

a Visceral surface
b Diaphragmatic surface
c Transverse computed tomogram (CT) after injection of contrast medium (50%), inferior aspect
d Ultrasound image of the spleen (60%), subcostal oblique section

189 Milz

a Facies visceralis
b Facies diaphragmatica
c Horizontales Computertomogramm (CT) nach Kontrastmittelgabe (50%), Kaudalansicht
d Ultraschallbild der Milz (60%), subkostaler Schrägschnitt

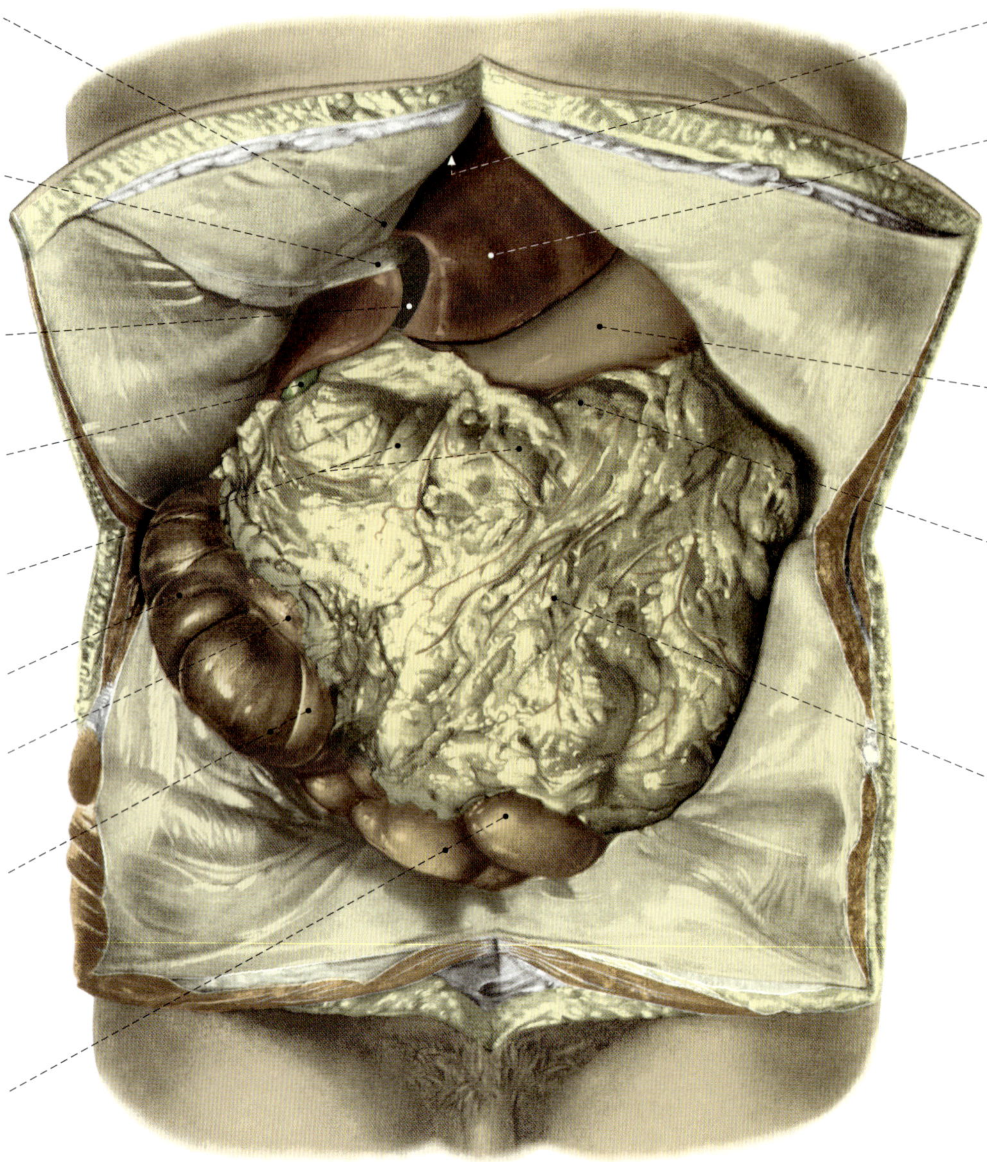

190 Abdomen

190 Superficial abdominal viscera (30%)

The abdominal wall was opened by a cruciate incision and retracted. Ventral aspect

190 Oberflächlicher Bauchsitus (30%)

Die Bauchdecke ist durch einen Kreuzschnitt eröffnet und zurückgeschlagen. Ventralansicht

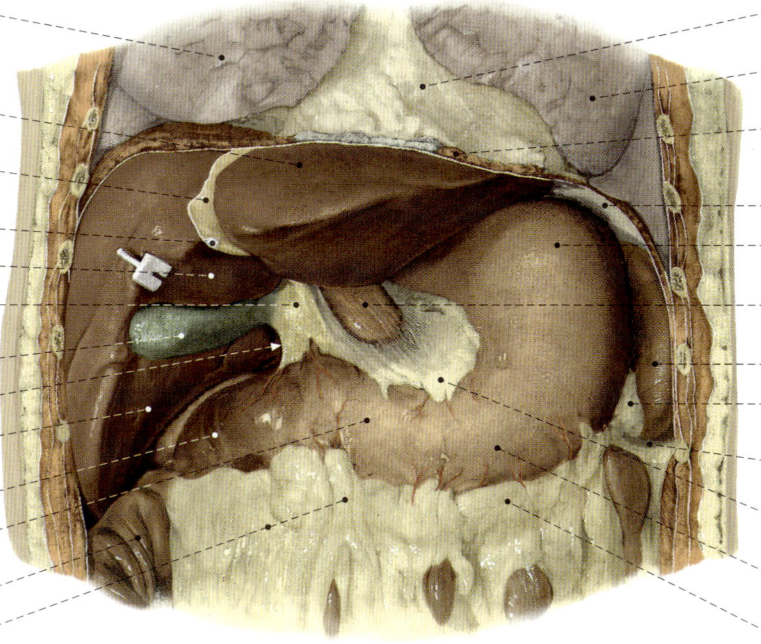

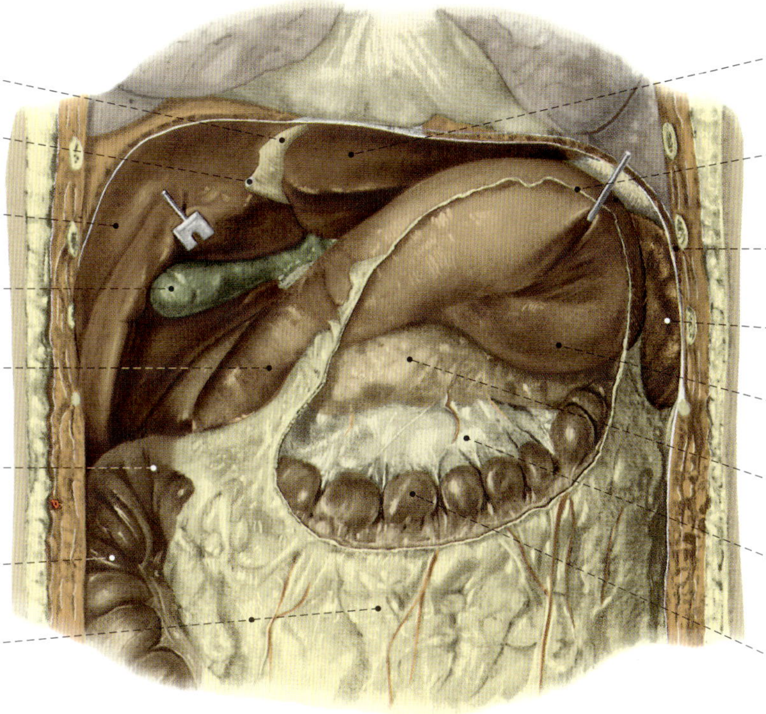

191 Abdomen

191 Viscera of the upper abdomen (35%)

The ventral body wall was removed, the liver is raised by hooks.

a Ventral aspect

b The gastrocolic ligament was additionally cut transversely and the stomach retracted cranially. By this, the dorsal wall of the omental bursa is exposed. Ventral aspect

191 Organe der Oberbauchregion (35%)

Die ventrale Körperwand ist entfernt, die Leber durch Haken angehoben.

a Ventralansicht

b Das Ligamentum gastrocolicum ist zusätzlich quer durchtrennt und der Magen nach kranial gezogen. Hierdurch wird die dorsale Wand der Bursa omentalis sichtbar. Ventralansicht

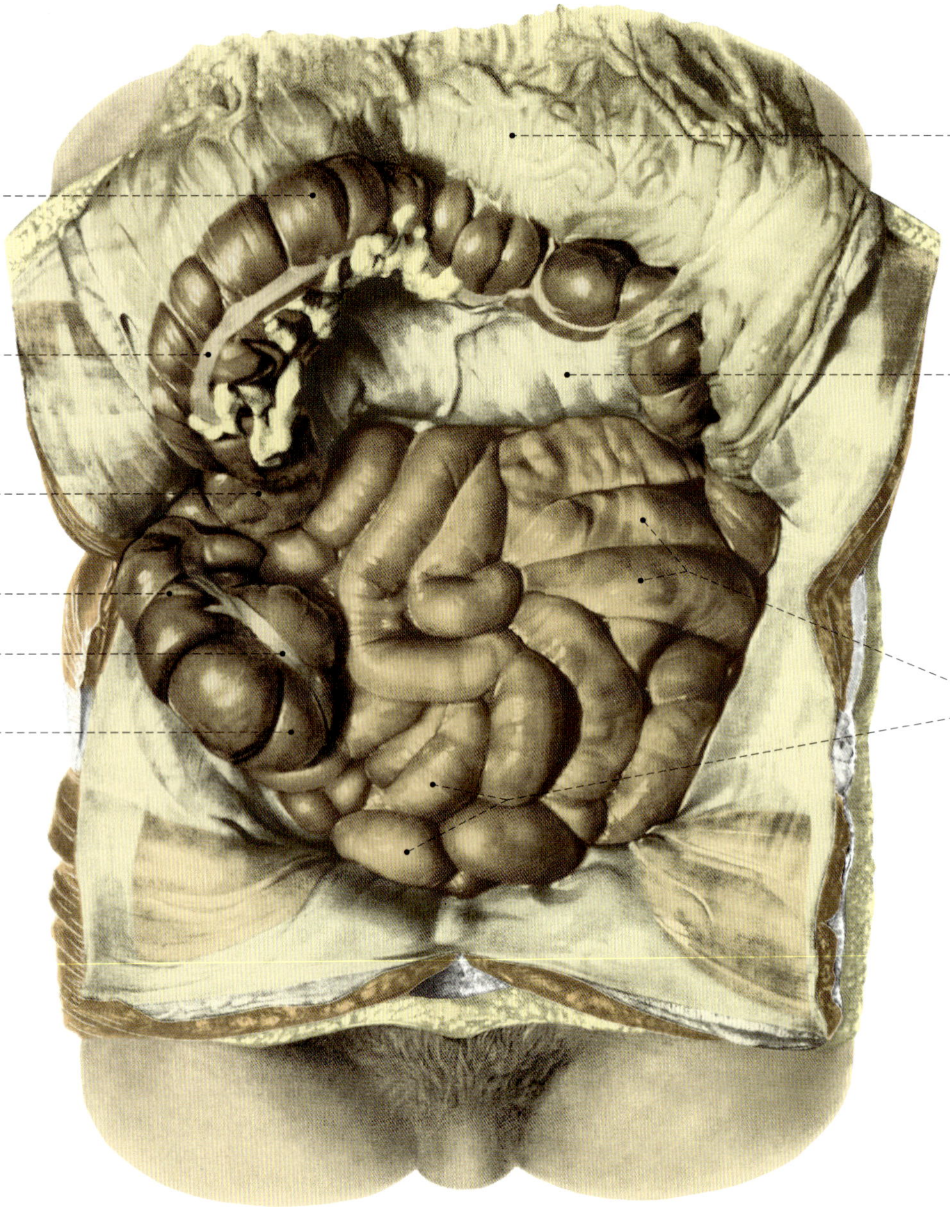

192 Abdomen

192 Intraperitoneal organs of the lower abdomen (30%)

The greater omentum and the transverse colon are retracted upwards. Ventral aspect

192 Intraperitoneale Organe der Unterbauchregion (30%)

Grosses Netz und Querkolon sind hochgeklappt. Ventralansicht

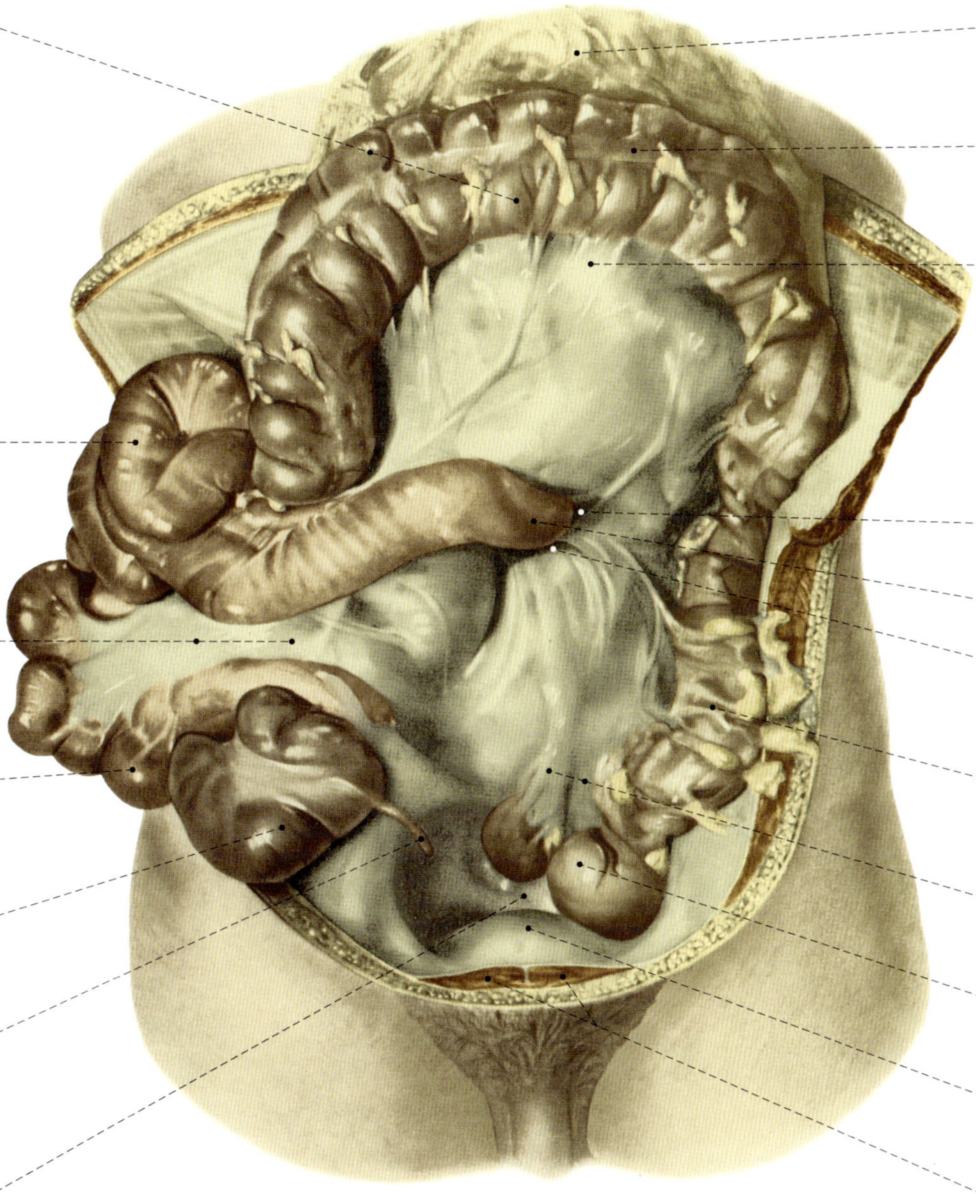

193 Abdomen

193 Intraperitoneal organs of the lower abdomen (30%)

The transverse colon and the greater omentum are turned upwards, the ileum and jejunum drawn to the right. Ventral aspect

193 Intraperitoneale Organe der Unterbauchregion (30%)

Querkolon und grosses Netz sind hochgeklappt, Ileum und Jejunum nach rechts verlagert. Ventralansicht

a

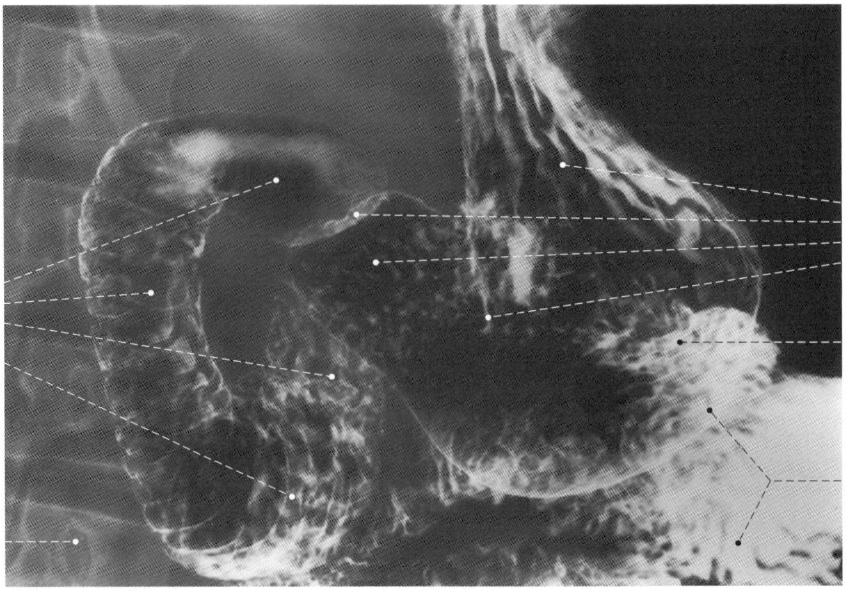

b

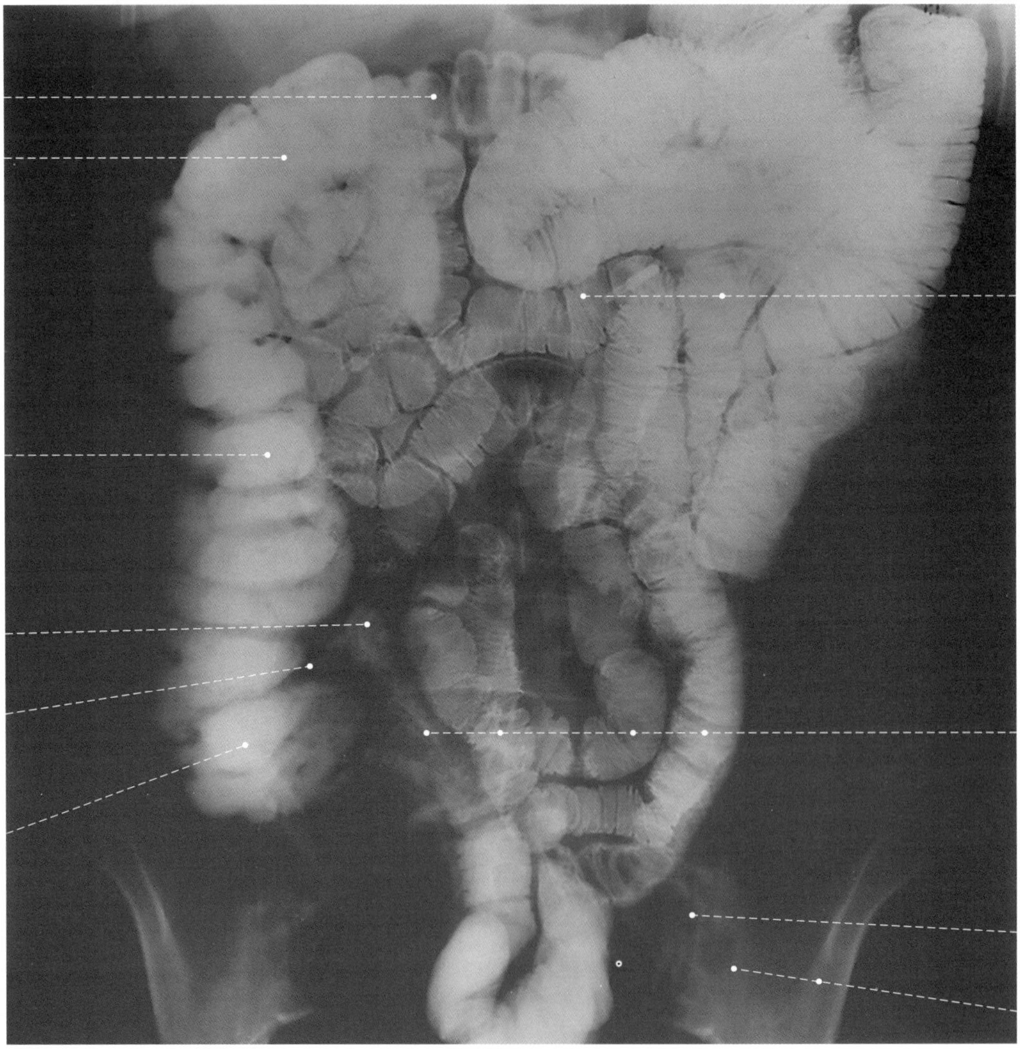

194 Gaster [Ventriculus] et Intestinum tenue

194 Stomach and small intestine (50%)

a Double-contrast radiograph of the stomach and the duodenum after barium meal and distension by air, right anterior oblique (RAO) projection
b Posteroanterior radiograph of the small intestine and the ascending colon after barium meal

194 Magen und Dünndarm (50%)

a Röntgenbild des Magens und Zwölffingerdarms in Doppelkontrastdarstellung, Schrägaufnahme von rechts vorne ("Fechterstellung")
b Posteroanteriores Röntgenbild des Dünn- und des aufsteigenden Dickdarms nach Bariumbreigabe

a

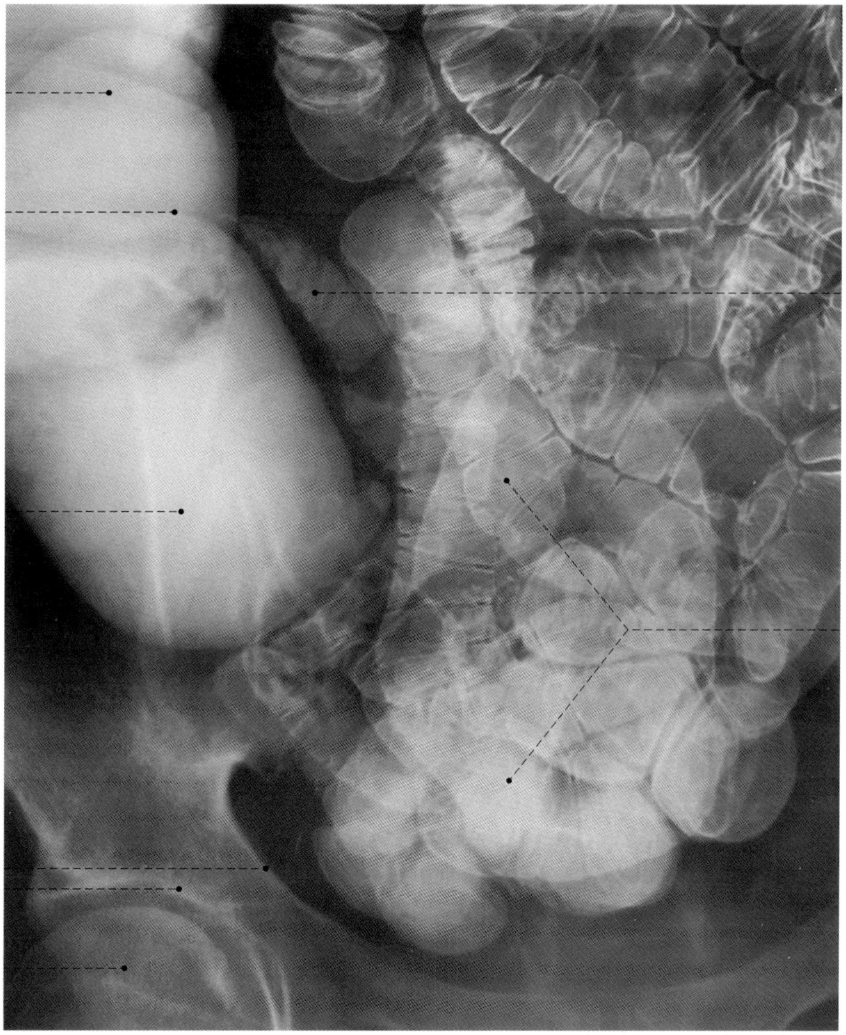

b

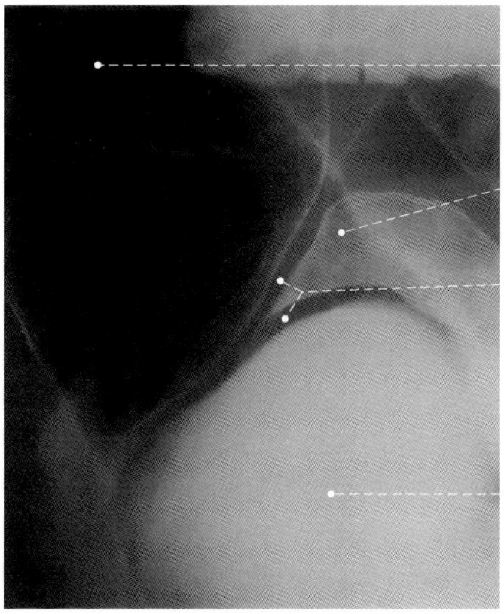

c

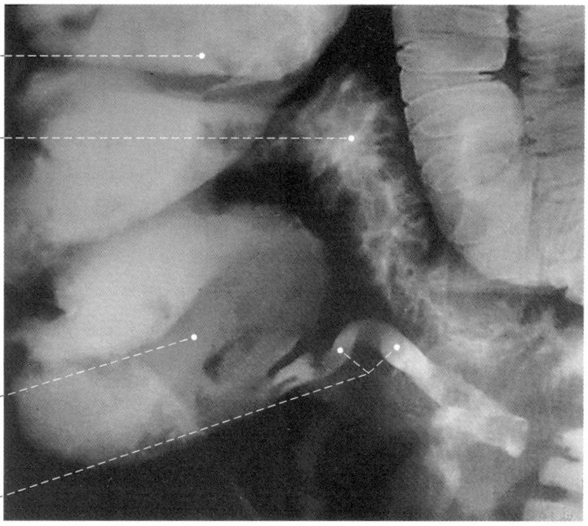

195 Ileum et Caecum

195 Ileum and cecum (50%)

Posteroanterior radiographs of the small intestine and the ascending colon after barium meal and distension by air (double-contrast radiography)

a Ileum, cecum and ascending colon
b Ileocecal valve
c Terminal ileum, cecum and vermiform appendix

195 Dünndarm und Blinddarm (50%)

Posteroanteriore Röntgenbilder der Endabschnitte des Dünndarms und der Anfangsteile des Dickdarms nach Bariumbreigabe und Luftblähung (Doppelkontrastdarstellung)

a Ileum, Blinddarm und aufsteigender Dickdarm
b Valva ileocaecalis
b Endständiges Ileum, Blinddarm und Wurmfortsatz

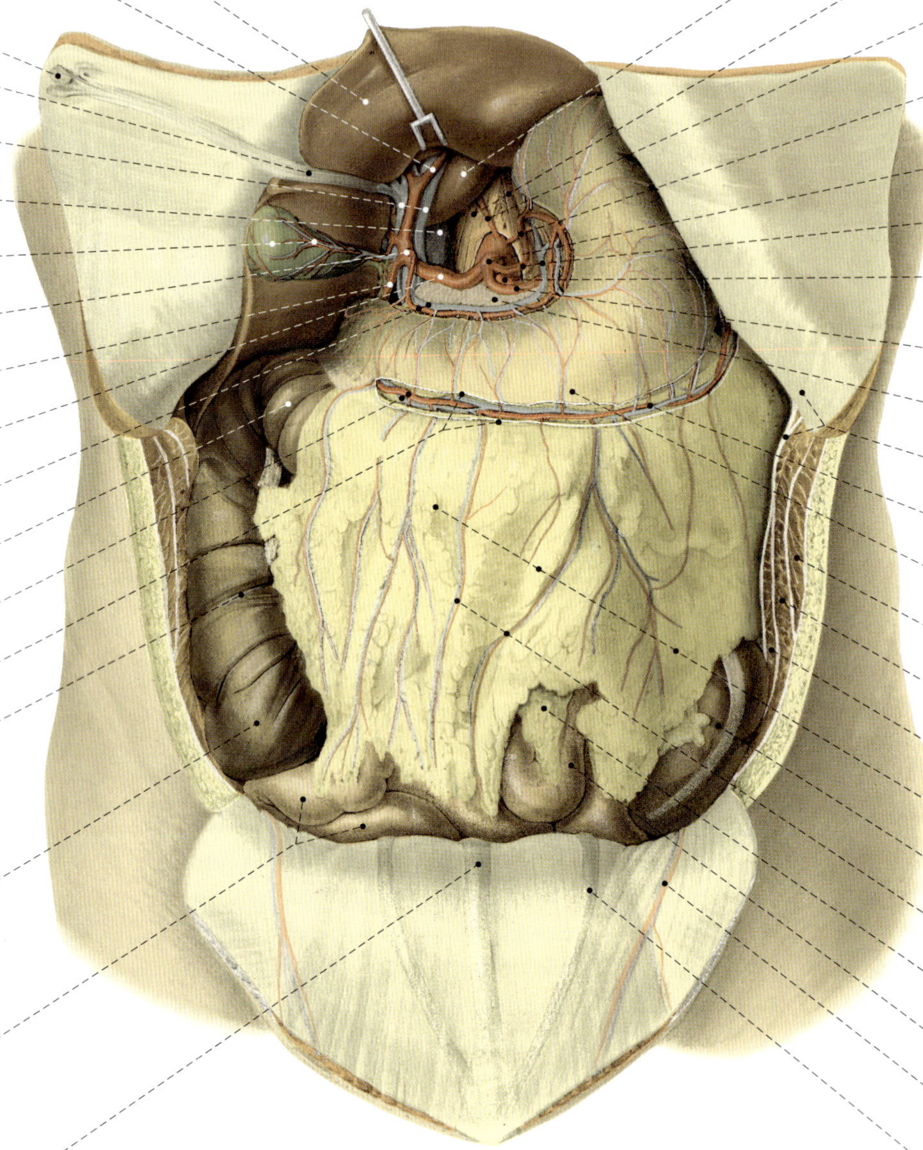

196 Abdomen

196 Celiac trunk and its branches

(30%)

The lesser omentum was removed, the liver is retracted upwards. Ventral aspect

196 Truncus coeliacus und seine Äste

(30%)

Das kleine Netz ist entfernt, die Leber nach oben geklappt. Ventralansicht

a

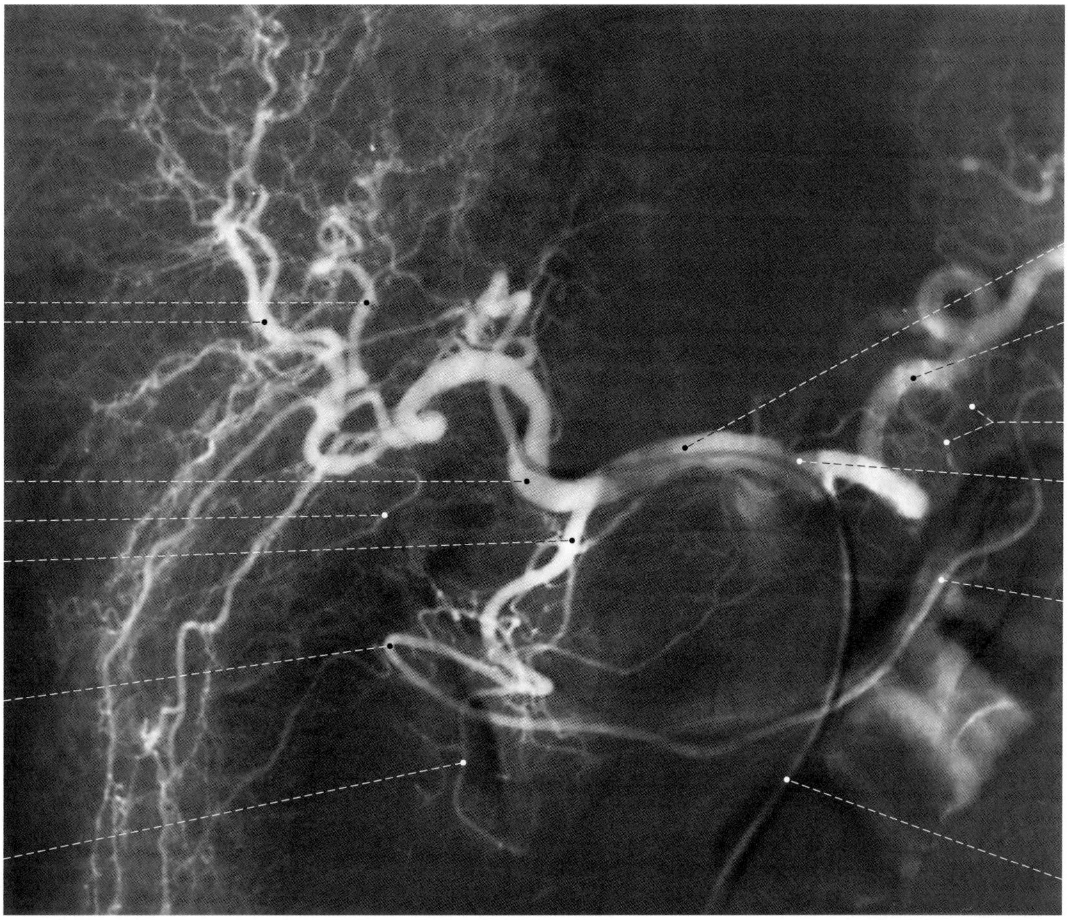

b

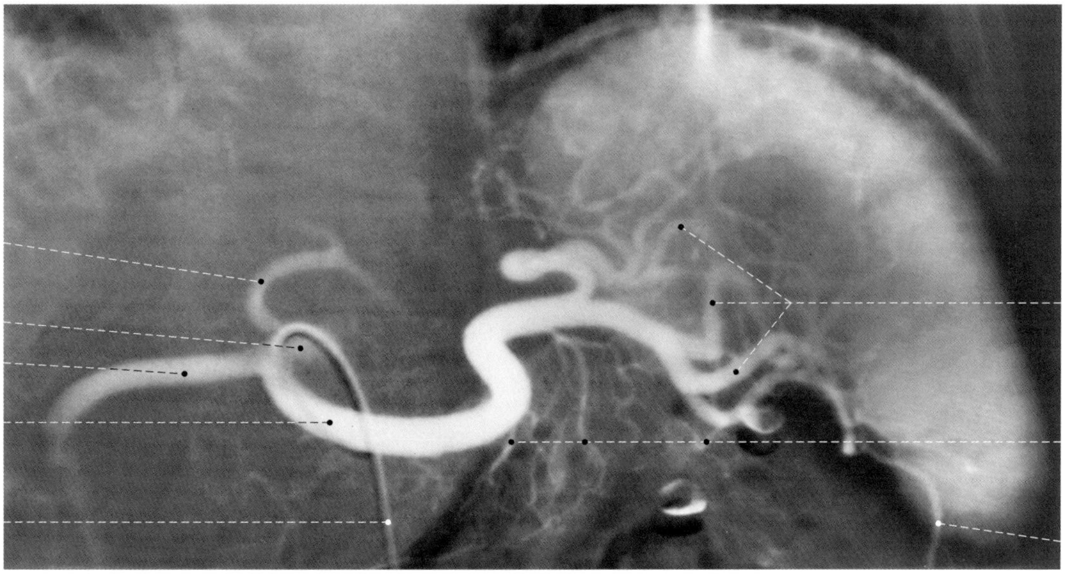

197 Abdomen

197 Celiac trunk and its branches

Anteroposterior radiographs

a Celiacography, selective arteriogram of the celiac trunk (70%)

b Splenography, selective arteriogram of the splenic artery (90%)

197 Truncus coeliacus und seine Äste

Anteroposteriore Röntgenaufnahmen

a Zöliakographie, selektives Arteriogramm des Truncus coeliacus (70%)

b Splenographie, selektives Arteriogramm der Arteria splenica (90%)

a

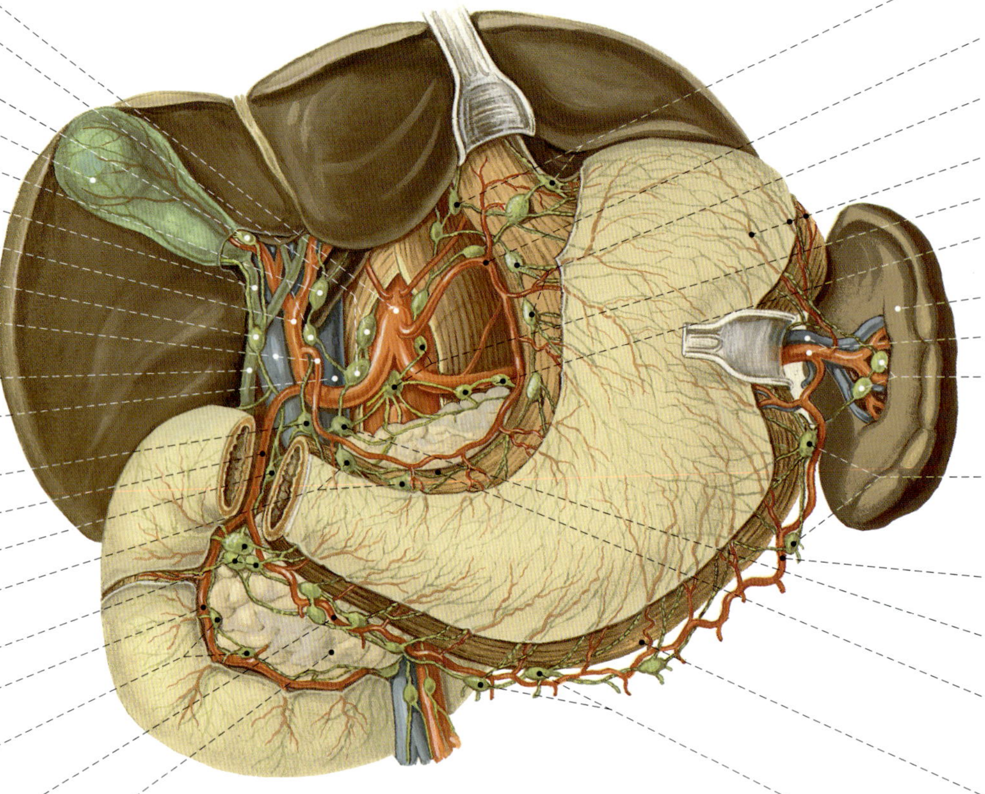

b

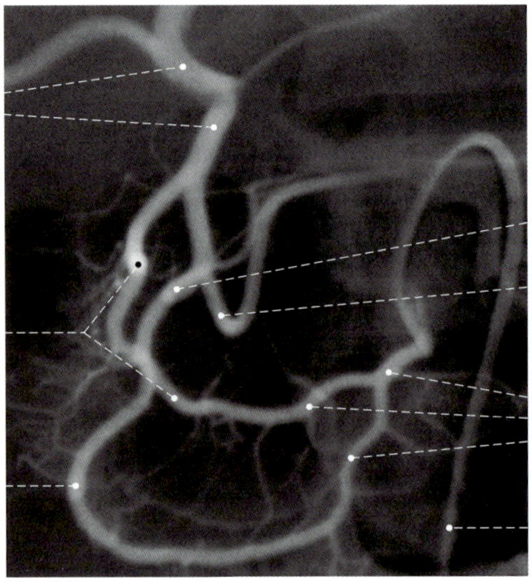

198 ⟨Arteriae et Nodi lymphatici abdominis⟩

198 Arteries, lymph vessels and lymph nodes in the upper abdominal region (40%)

a Ventral aspect
b Arteriogram of the common hepatic artery and the upper mesenteric artery in order to show the arterial supply of the head of the pancreas

198 Arterien, Lymphgefässe und Lymphknoten in der Oberbauchregion (40%)

a Ventralansicht
b Arteriogramm der Arteria hepatica communis und der Arteria mesenterica superior zur Darstellung der arteriellen Gefässversorgung des Pankreaskopfes

a

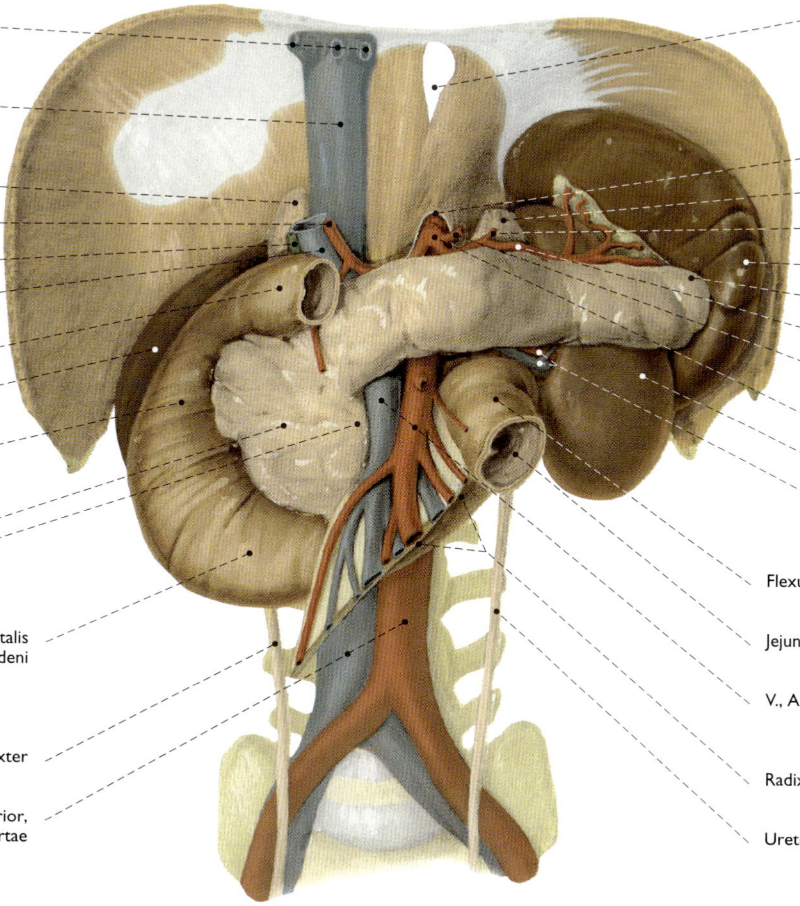

b

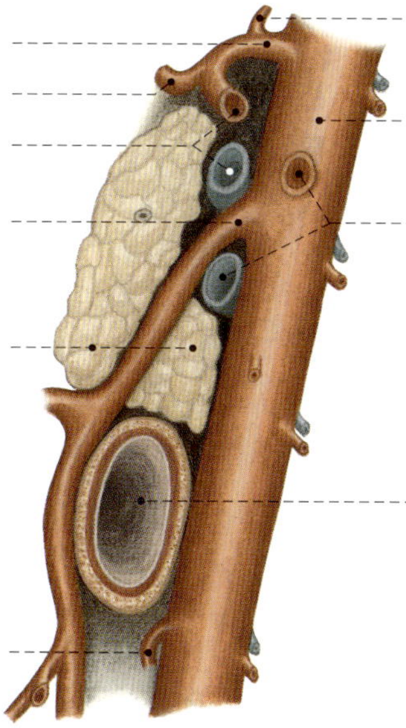

c

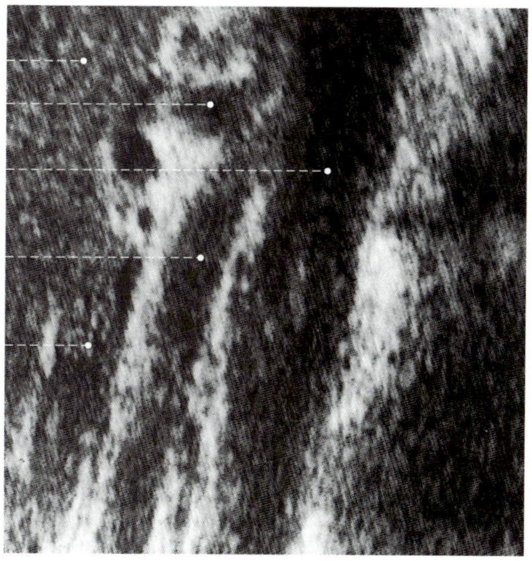

199 Abdomen

199 Retroperitoneal organs of the upper abdomen and spleen

a Ventral aspect (60%)
b Scissor-like arrangement of the abdominal aorta and the superior mesenteric artery around the duodenum (70%), medial aspect of the right half
c Ultrasound image of the abdominal aorta and the superior mesenteric artery (70%), midsagittal longitudinal section

199 Retroperitoneale Oberbauchorgane und Milz

a Ventralansicht (60%)
b Scherenartige Anordnung der Bauchaorta und der oberen Mesenterialarterie um das Duodenum (70%), Medialansicht der rechten Hälfte
c Ultraschallbild der Bauchaorta und der oberen Mesenterialarterie (70%), medianer Oberbauchlängsschnitt

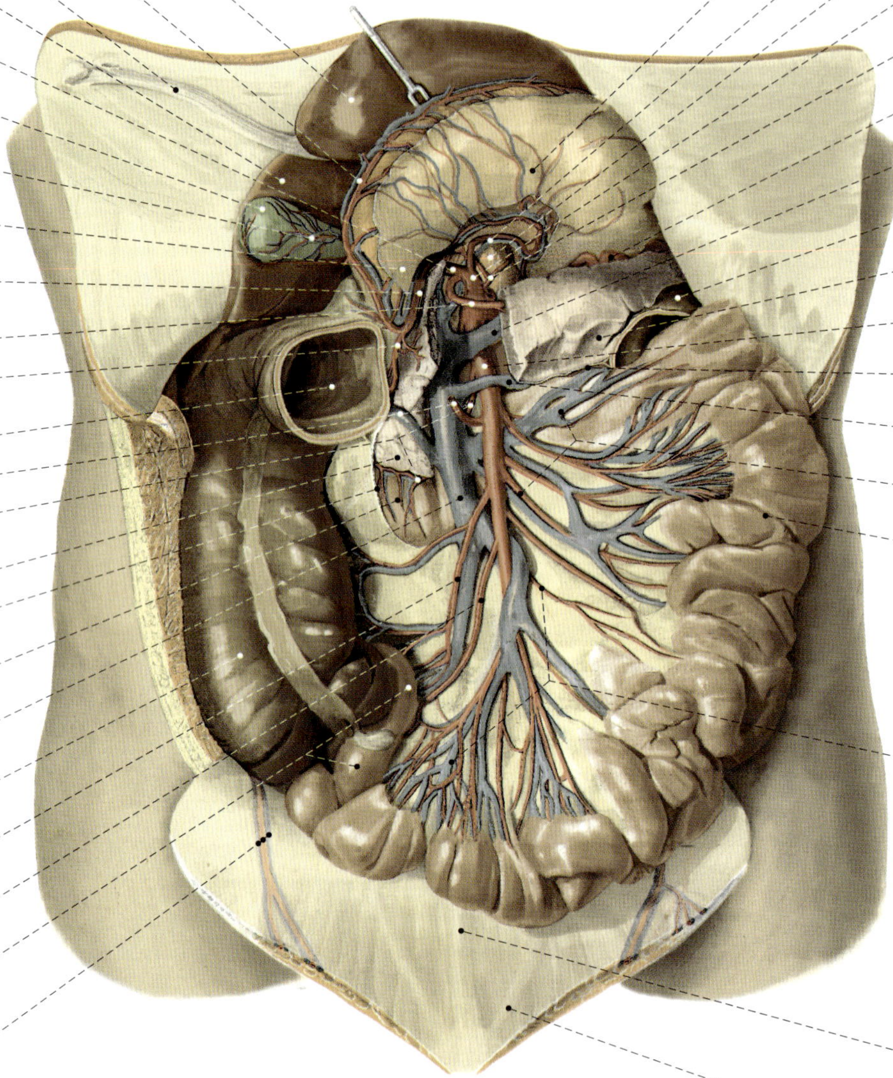

200 Abdomen

200 Celiac trunk and superior mesenterial vessels with their branches (30%)

The greater omentum and the transverse colon were removed, the stomach is retracted upwards and the small intestine drawn to the left. Ventral aspect

200 Truncus coeliacus und Vasa mesenterica superiora mit ihren Ästen (30%)

Grosses Netz und Querkolon sind entfernt, der Magen nach oben geklappt und der Dünndarm nach links verlagert. Ventralansicht

a

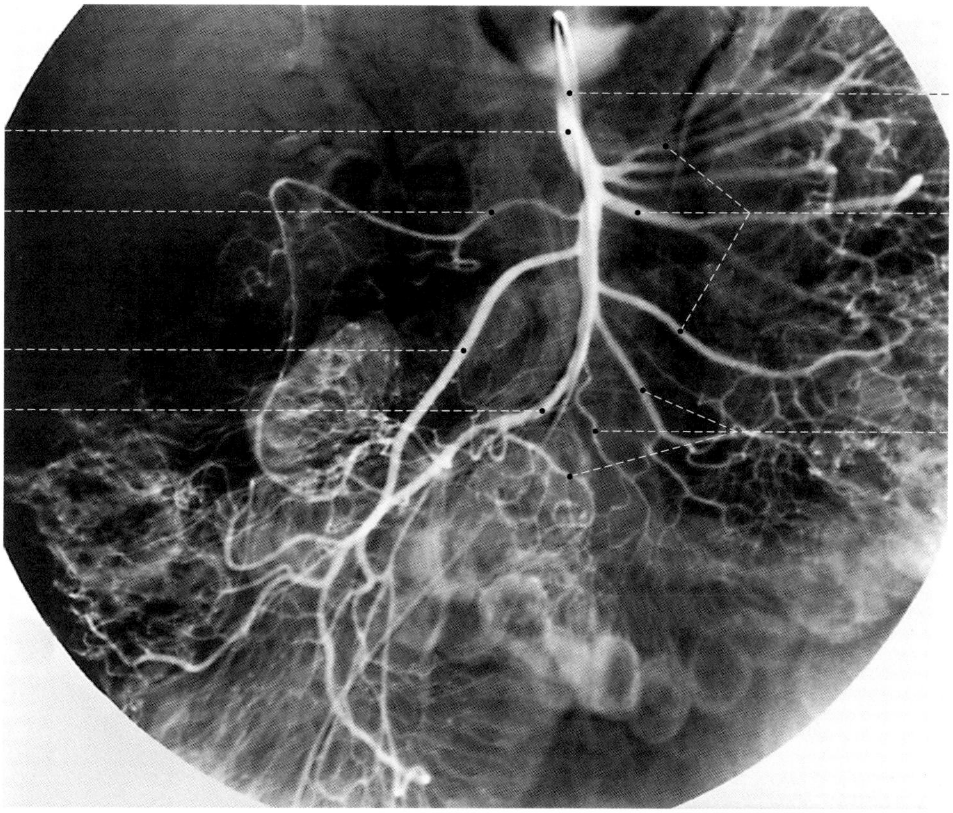

b

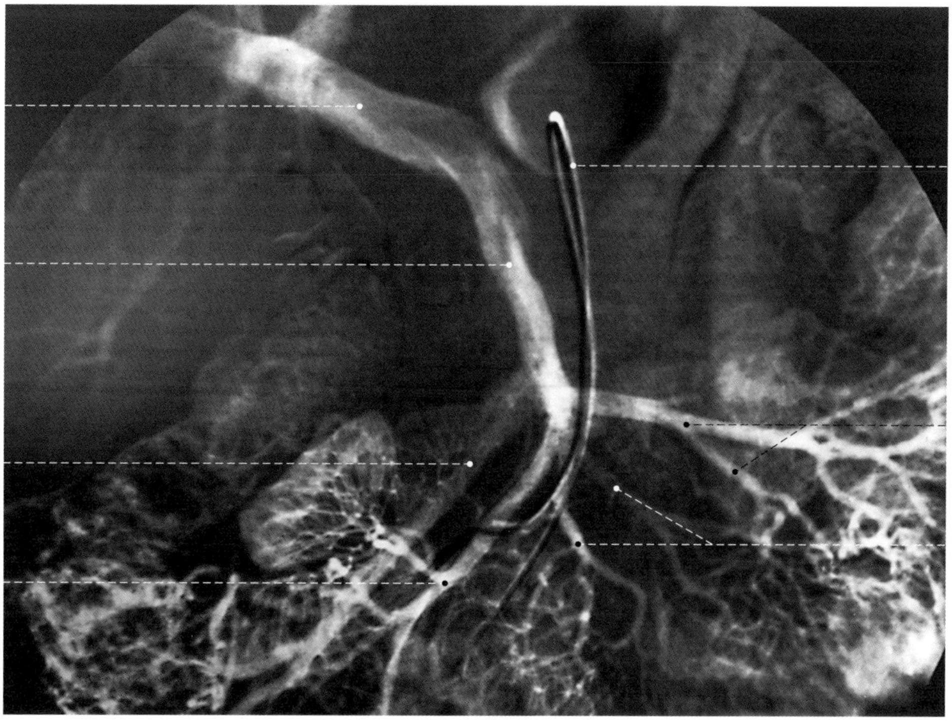

201 Abdomen

201 Superior mesenteric vessels and portal vein (60%)

Posteroanterior radiographs

a Selective arteriogram of the superior mesenteric artery

b Venous phase of the superior mesenteric arteriogram showing the superior mesenteric and portal veins

201 Vasa mesenterica superiora und Vena portae hepatis (60%)

Posteroanteriore Röntgenaufnahmen

a Selektives Arteriogramm der oberen Mesenterialarterie

b Venöse Phase des Arteriogramms der oberen Mesenterialarterie zur Darstellung der Vena mesenterica superior und der Vena portae

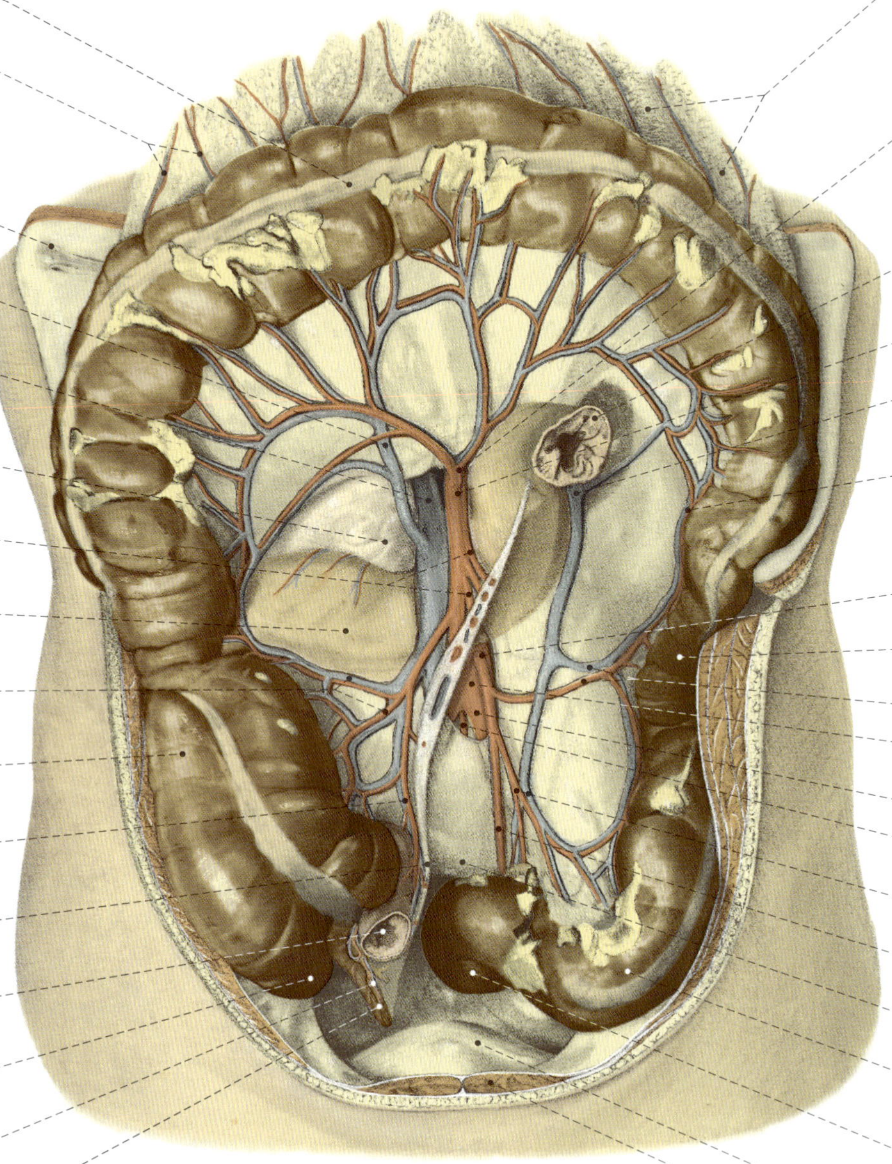

202 Abdomen

202 Vascular supply of the large intestine (30%)

The greater omentum and the transverse colon are retracted upwards. The small intestine was transected both at the duodenojejunal flexure and the ileocaecal valve, and removed along with its mesentery which was severed at the root of the mesentery. Ventral aspect

202 Gefässversorgung des Dickdarms (30%)

Grosses Netz und Querkolon sind hochgeklappt. Der Dünndarm ist an der Flexura duodenojejunalis und der Valva ileocaecalis durchtrennt und samt Mesenterium an der Radix mesenterii abgetragen. Ventralansicht

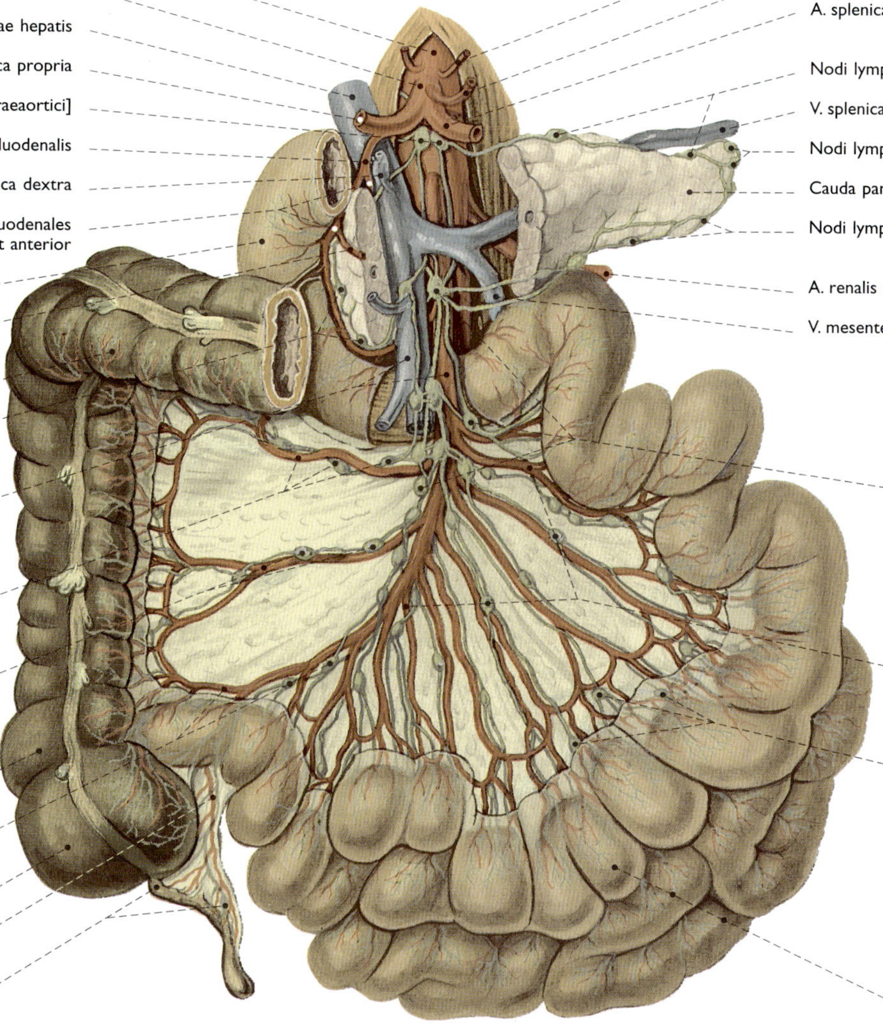

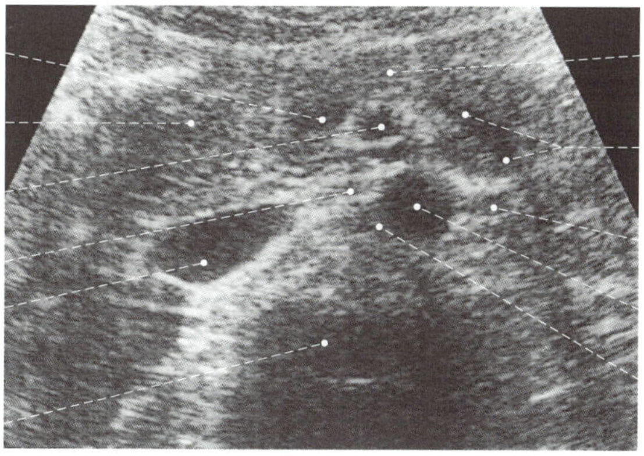

203 ⟨Nodi lymphatici, Arteriae et Venae abdominis⟩

203 Lymph vessels, lymph nodes and blood vessels of the mesentery and retroperitoneal organs of the upper abdominal region

a Ventral aspect (40%)
b Ultrasound image (60%), transverse section through the upper abdomen

203 Lymphgefässe, Lymphknoten und Blutgefässe im Bereich der Mesenterien und retroperitonealen Organe der Oberbauchregion

a Ventralansicht (40%)
b Ultraschallbild (60%), Oberbauchquerschnitt

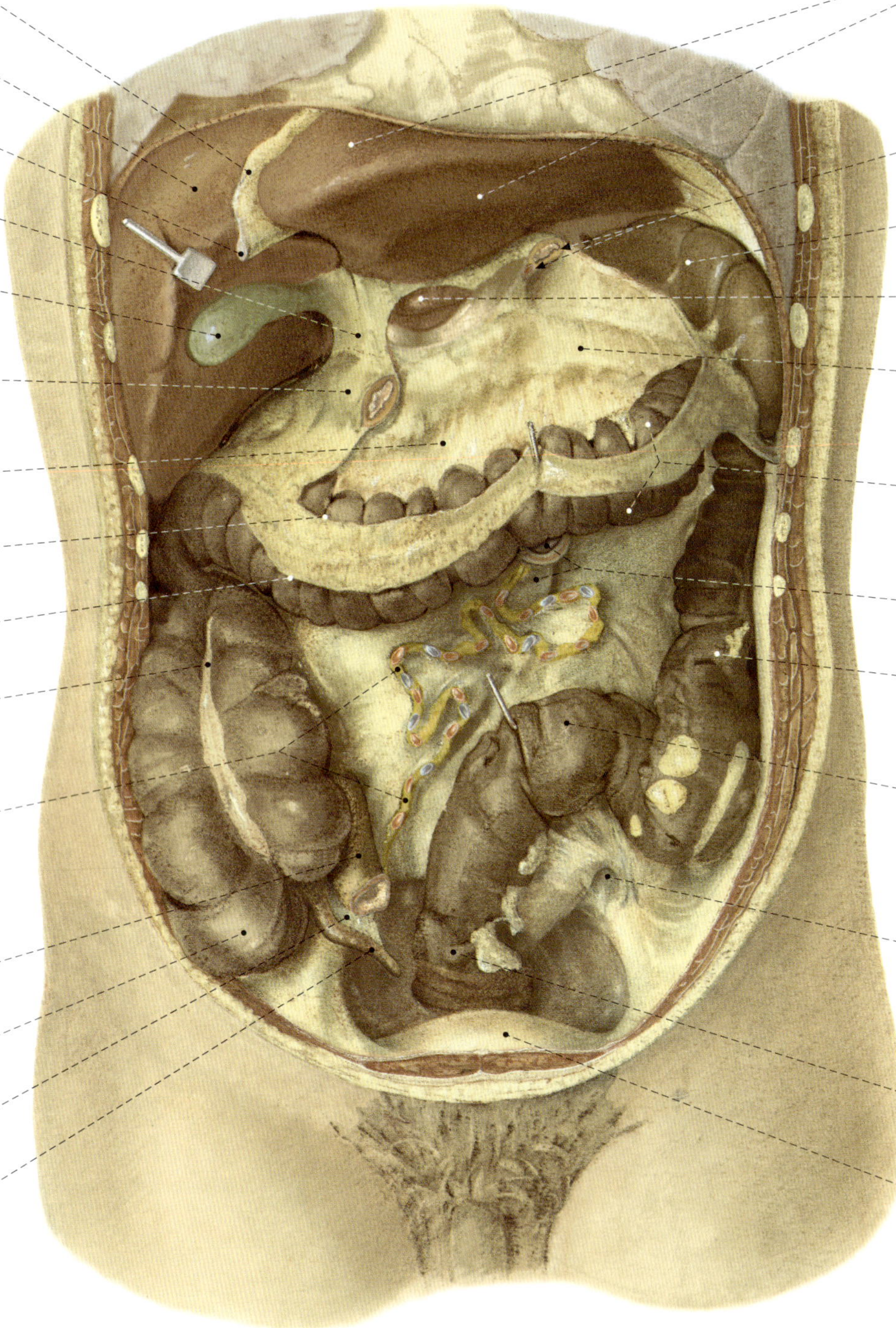

204 Abdomen

204 Large intestine and mesenteries

(30%)

The anterior body wall was removed. The stomach was taken away by severing the hepatogastric, gastrolienal and gastrocolic ligaments. The greater omentum, the jejunum and the ileum with their mesenteries, excepting the terminal ileum, were also removed. Ventral aspect

204 Dickdarm und Mesenterien

(30%)

Die vordere Rumpfwand ist entfernt. Der Magen ist nach Durchtrennung der Ligamenta hepatogastricum, gastrolienale et gastrocolicum entnommen. Grosses Netz, Jejunum und Ileum, ausgenommen das terminale Ileum, sind ebenfalls samt Mesenterium entfernt. Ventralansicht

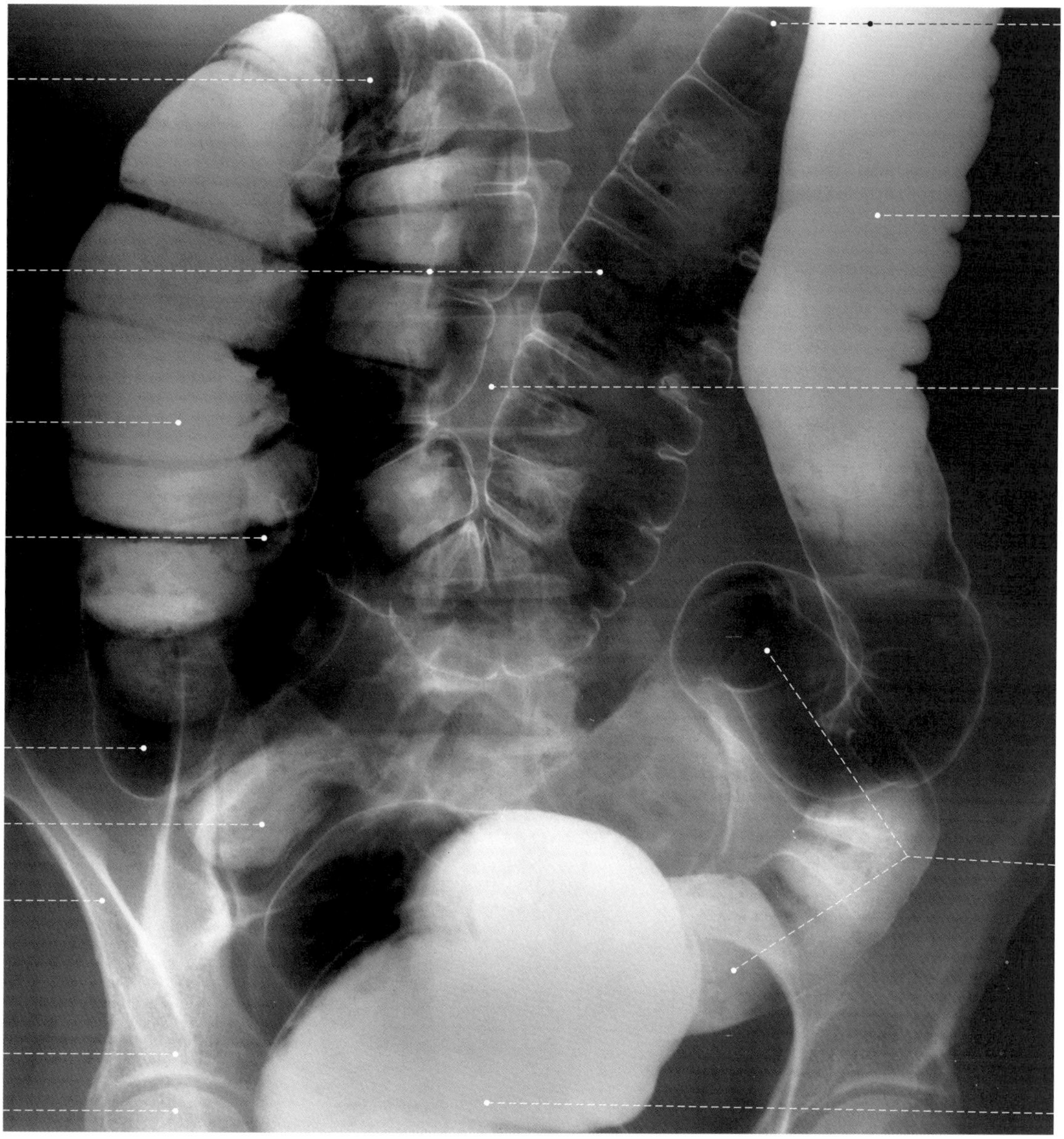

205 Abdomen

205 Large intestine (50%)
Double-contrast barium enema of the large bowel (colon) with a V-shaped, pendulous transverse colon, posteroanterior radiograph

205 Dickdarm (50%)
Röntgenbild des Dickdarms mit einem V-förmigen, herabhängenden Querkolon in Doppelkontrast-darstellung nach Barium-Kontrasteinlauf, posteroanteriore Aufnahme

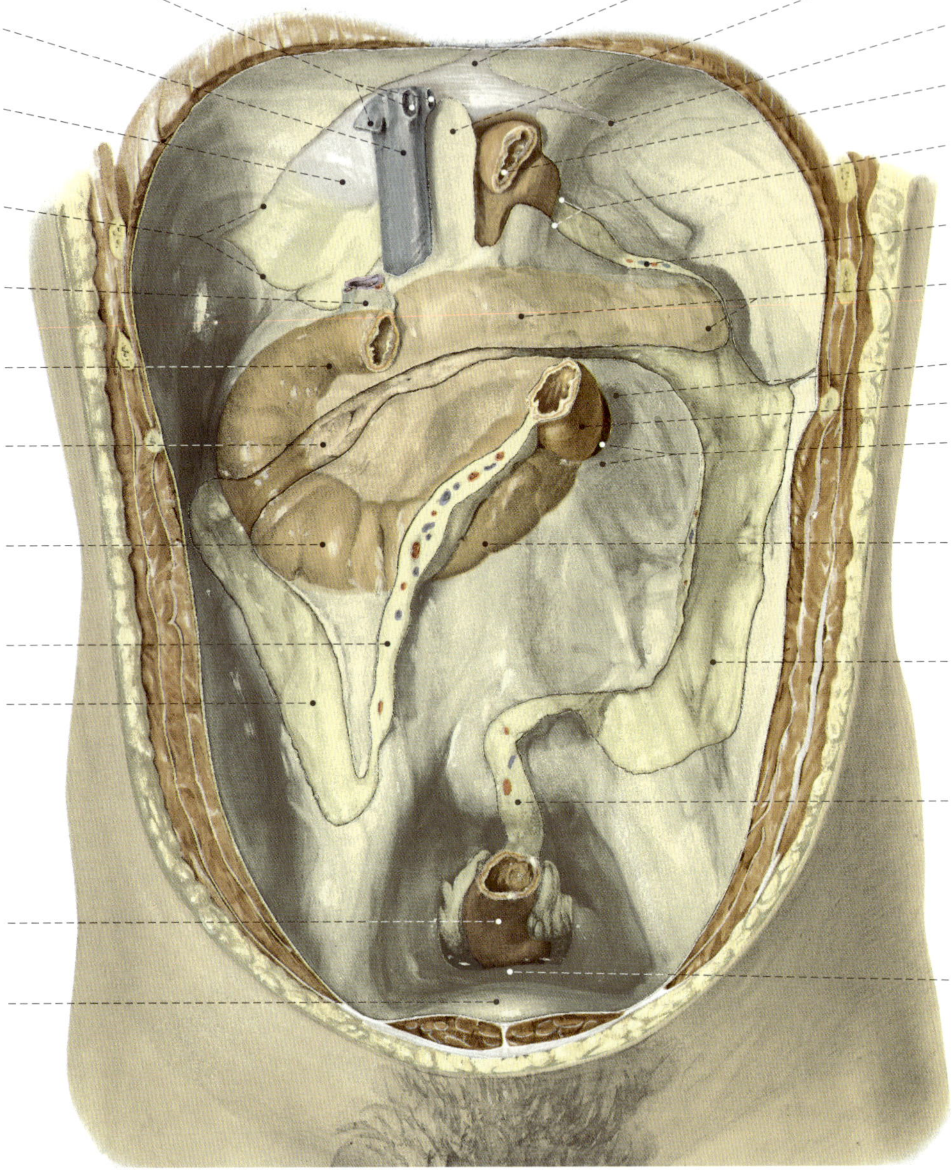

206 Abdomen

206 Posterior abdominal wall (30%)

Duodenum and pancreas remaining in situ. The contact areas of the retroperitoneal parts of the colon and the liver as well as the mesenterial attachment lines are shown. Ventral aspect

206 Hintere Bauchwand (30%)

Duodenum und Pankreas sind in situ belassen. Die Anlagerungsflächen der retroperitonealen Kolonteile, die Verwachsungsfläche der Leber und die Mesenterialwurzeln sind gezeigt. Ventralansicht

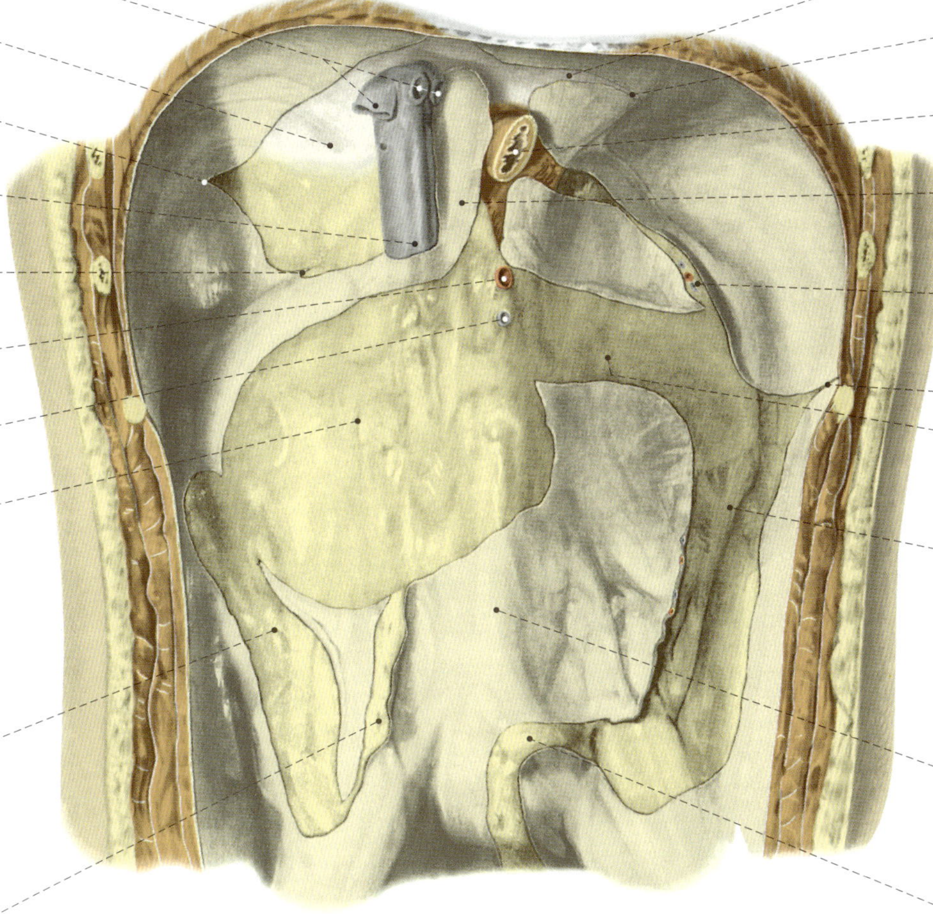

207 Abdomen

207 Posterior abdominal wall (30%)

Lines of peritoneal reflexion after removal of the intraperitoneal organs and contact areas of the retroperitoneal organs. The duodenum and the pancreas were removed. Ventral aspect

207 Hintere Bauchwand (30%)

Umschlaglinien des Peritoneum nach Entfernung der intraperitonealen Organe und Anlagerungsflächen retroperitonealer Organe. Duodenum und Pankreas sind entfernt. Ventralansicht

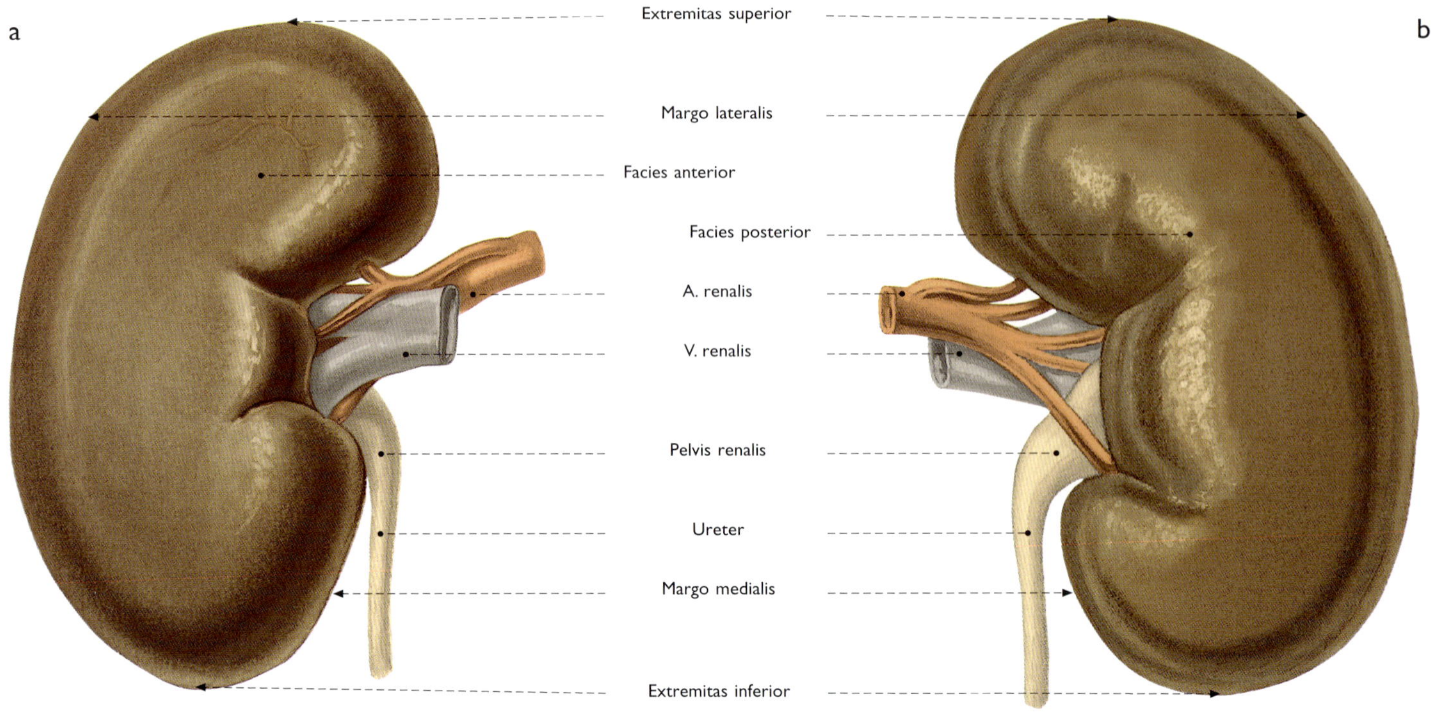

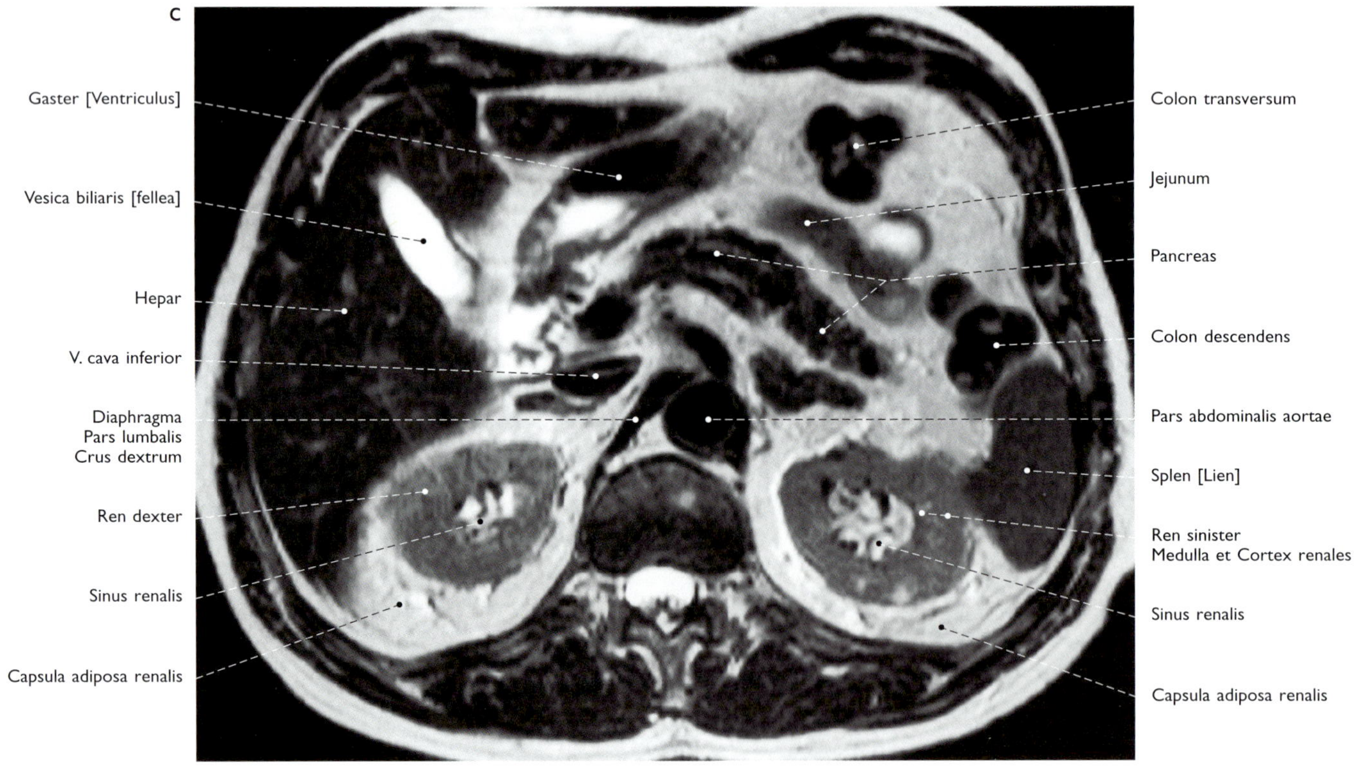

208 Ren [Nephros]

208 Kidney of an adult

a, b The main hilar structures of the right kidney were separated from each other (90%).
a Ventral aspect
b Dorsal aspect
c Transverse magnetic resonance image (MRI, T_2-weighted) through the upper abdomen at the level of the first lumbar vertebral body to show both kidneys and the perirenal fat (40%), inferior aspect

208 Niere eines Erwachsenen

a, b Die Inhaltsgebilde des Hilum der rechten Niere sind freipräpariert (90%).
a Ventralansicht
b Dorsalansicht
c Horizontales Kernspintomogramm (MRI, T_2-gewichtet) durch den Oberbauch in Höhe des ersten Lendenwirbelkörpers zur Darstellung beider Nieren und ihrer Fettkapsel (40%), Kaudalansicht

a

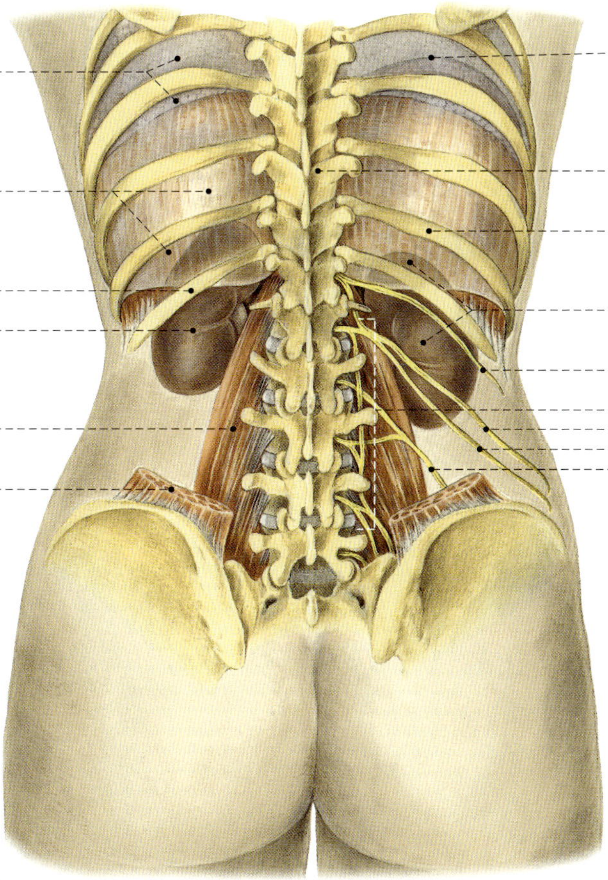

b

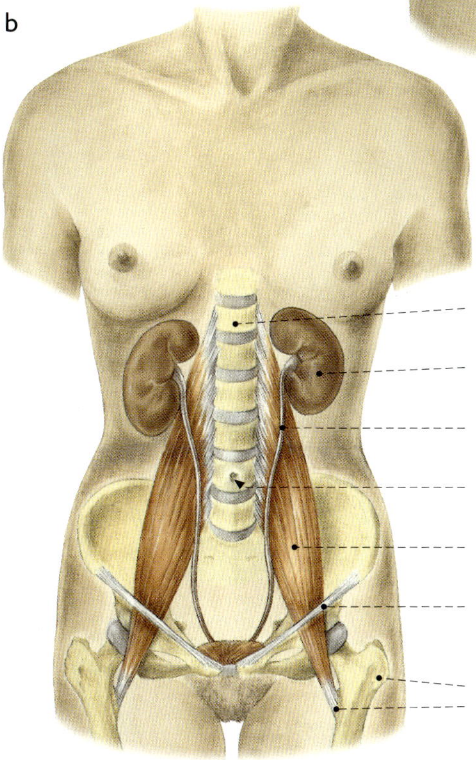

c

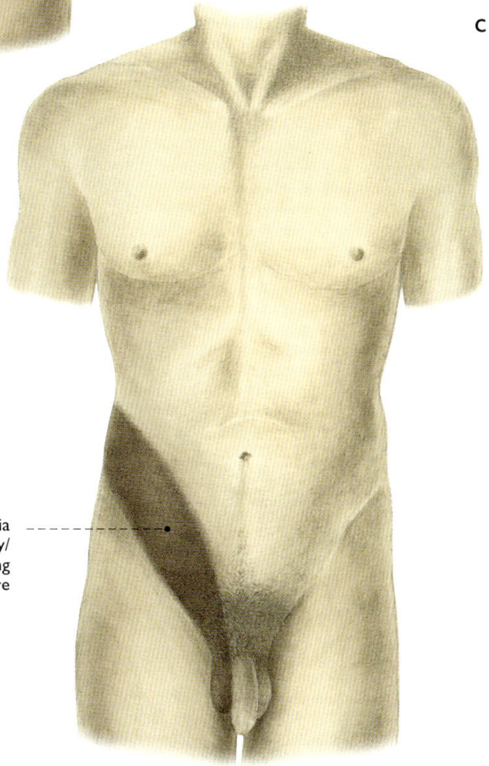

209 Ren [Nephros]

209 Kidneys

a Kidneys and adjacent structures projected to the posterior abdominal wall (25%), dorsal aspect
b Projection of the kidneys and the major psoas muscles onto the anterior abdominal wall (15%), ventral aspect
c Area of hyperalgesia of the right kidney (15%), ventral aspect

209 Nieren

a Projektion der Nieren auf die hintere Bauchwand (25%), Dorsalansicht
b Projektion der Nieren und der grossen Lendenmuskeln auf die vordere Bauchwand (15%), Ventralansicht
c Feld der Schmerzprojektion der rechten Niere (15%), Ventralansicht

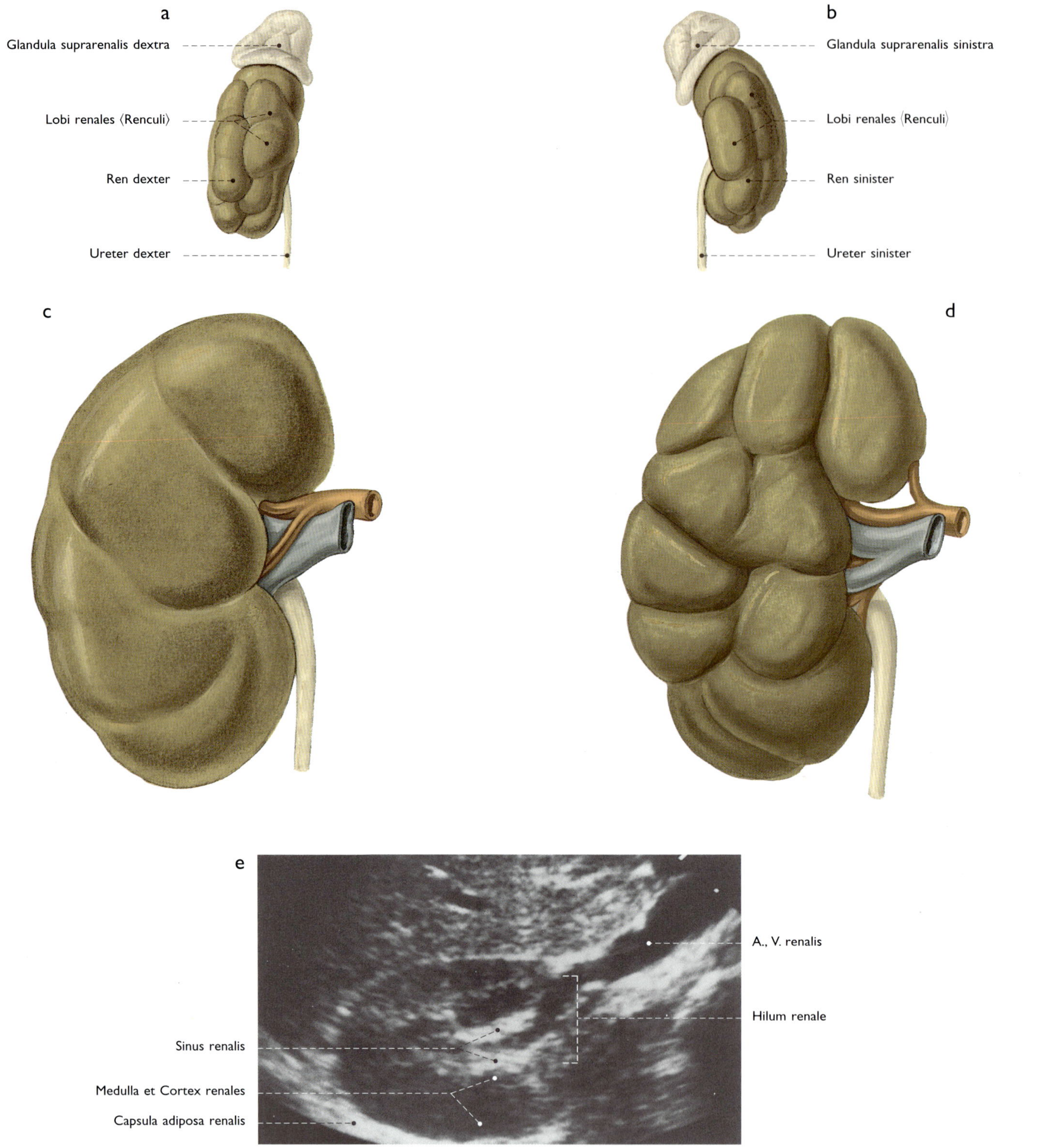

210 Ren fetalis et ⟨Ren lobatus⟩

210 Fetal and lobulated kidneys

a, b Kidneys and suprarenal glands of a human fetus, seventh month of pregnancy (100%), ventral aspect

c, d Right kidney, ventral aspect (80%)

c Faintly persisting fetal lobulation of the renal surface of an adult

d Markedly persisting fetal lobulation of the renal surface of an adult

e Ultrasound image of the right kidney (80%), transverse section

210 Fetale und gefurchte Nieren

a, b Nieren und Nebennieren eines menschlichen Feten, siebter Schwangerschaftsmonat (100%), Ventralansicht

c, d Rechte Niere, Ventralansicht (80%)

c Schwach ausgeprägte fetale renkuläre Gliederung der Nierenoberfläche bei einem Erwachsenen

d Deutlich erhaltene fetale renkuläre Gliederung der Nierenoberfläche bei einem Erwachsenen

e Ultraschallbild der rechten Niere (80%), Horizontalschnitt

a

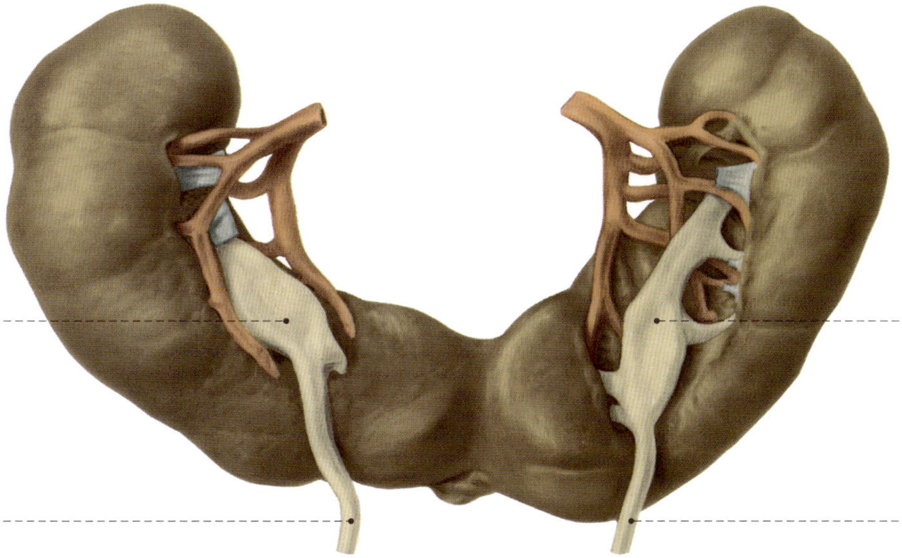

b

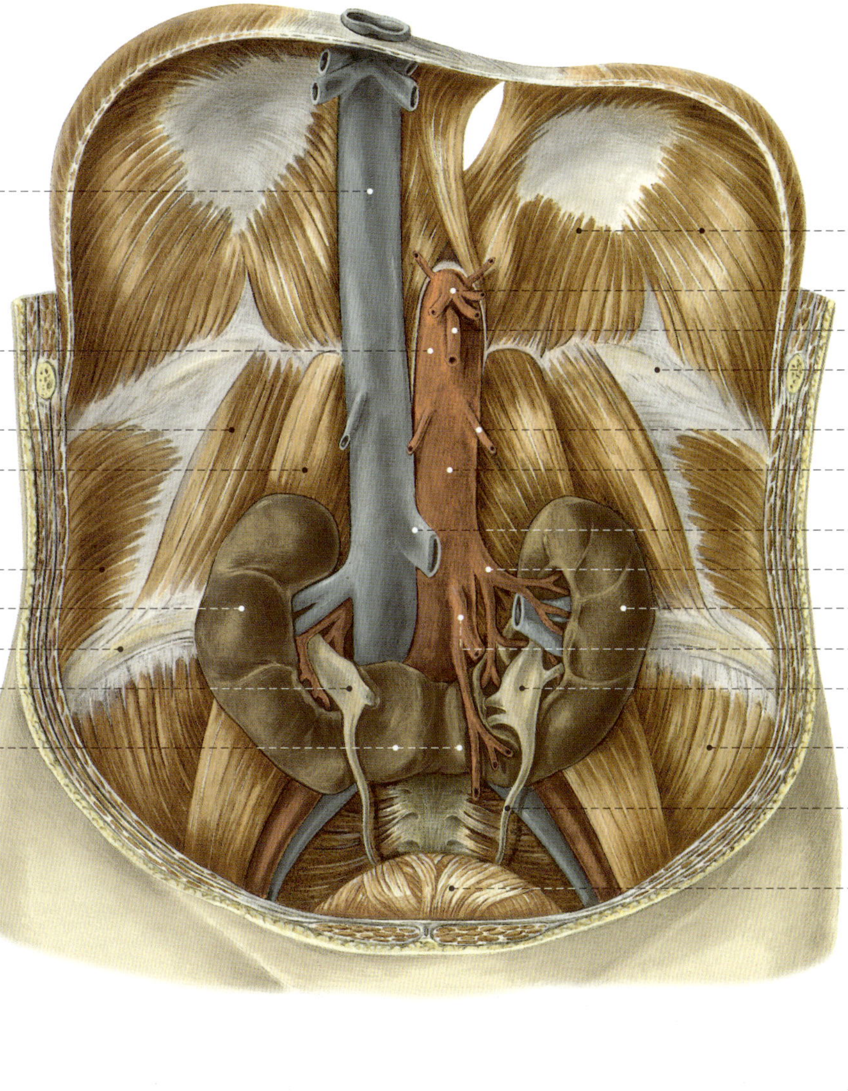

211 ⟨Ren arcuatus⟩

211 Horseshoe kidney

a Isolated horseshoe kidney (60%), ventral aspect
b Horseshoe kidney in situ (30%), ventral aspect

211 Hufeisenniere

a Isolierte Hufeisenniere (60%), Ventralansicht
b Hufeisenniere in situ (30%), Ventralansicht

a

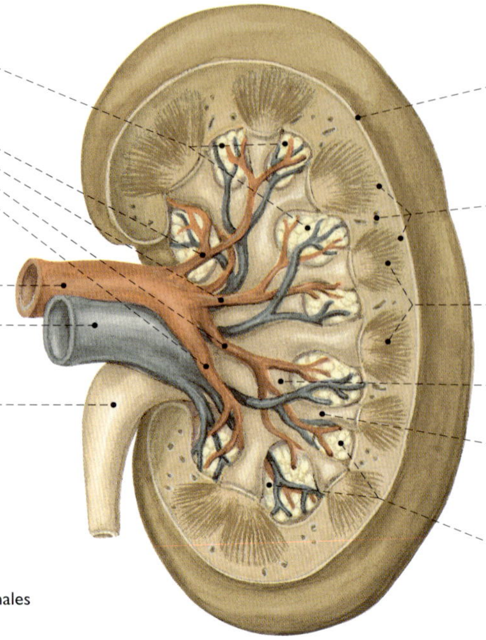

b

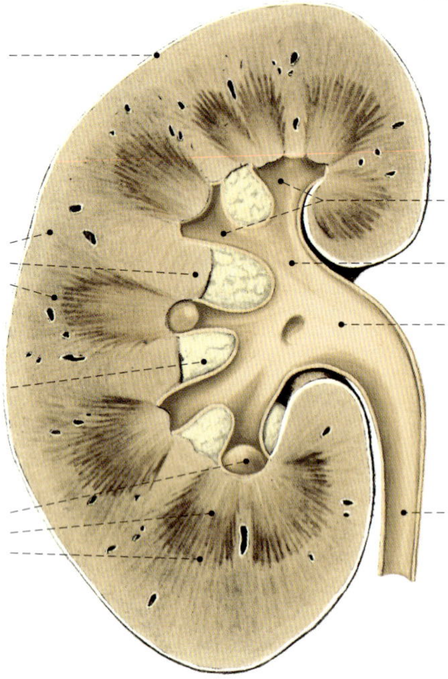

c

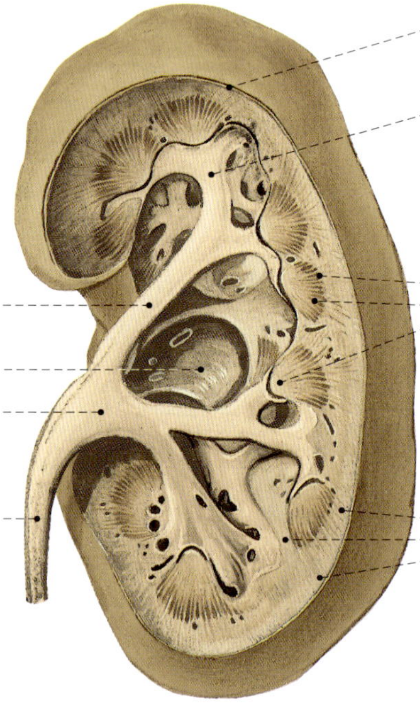

212 Ren [Nephros]

212 Kidney (80%)

Ventral aspect

a Left renal sinus with renal pelvis and renal blood vessels. The renal tissue of the anterior part of the kidney was removed.
b Longitudinal section through the right kidney, cut surface of the posterior half
c Left renal sinus with the renal pelvis and its division into calices. The fatty tissue and the vascular branches in the renal sinus were removed.

212 Niere (80%)

Ventralansicht

a Linker Sinus renalis mit Nierenbecken und Blutgefässen. Das Nierengewebe des vorderen Organteils ist entfernt.
b Längsschnitt durch die rechte Niere, Schnittfläche der hinteren Hälfte
c Linker Sinus renalis mit Nierenbecken und Nierenkelchen. Das Fettgewebe und die Gefässverzweigungen im Sinus renalis sind entfernt.

a

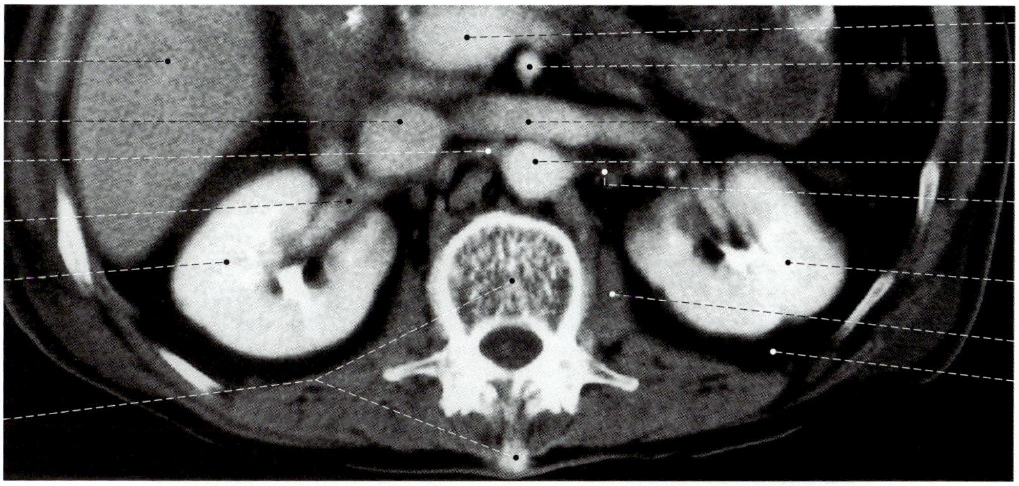

b

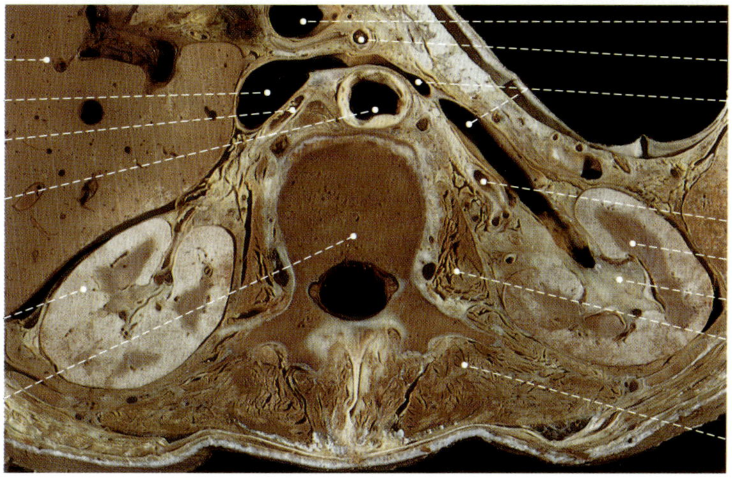

c

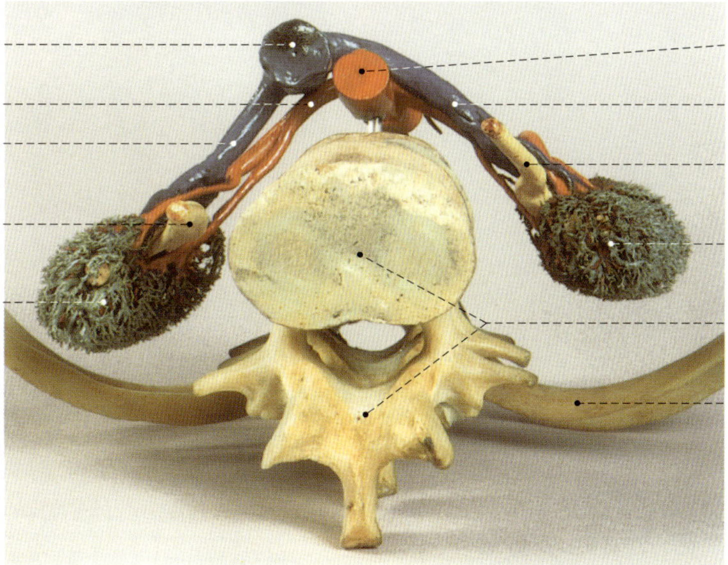

213 Ren [Nephros]

213 Kidneys and renal blood vessels

(40%)

Inferior aspect

- a Transverse computed tomogram (CT) at the level of the second lumbar vertebral body
- b Transverse anatomical section at the same level as in Fig. a
- c Corrosion cast of the renal blood vessels and the ureters of both kidneys of a 15-year-old girl (Anatomical Collection, Basel)

213 Nieren und Nierengefässe

(40%)

Kaudalansicht

- a Horizontales Computertomogramm (CT) auf Höhe des zweiten Lendenwirbelkörpers
- b Anatomischer Querschnitt auf gleicher Höhe wie in Abb. a
- c Korrosionspräparat der Nierengefässe und der Harnleiter beider Nieren eines 15jährigen Mädchens (Anatomische Sammlung, Basel)

a

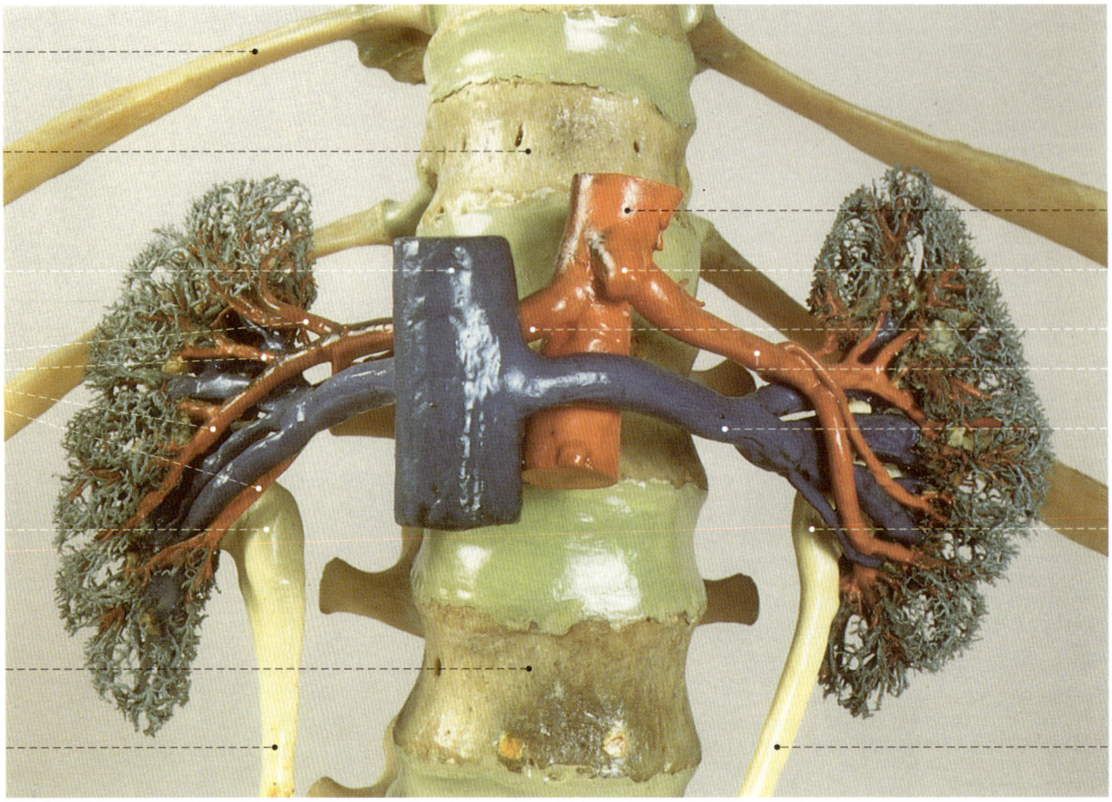

b

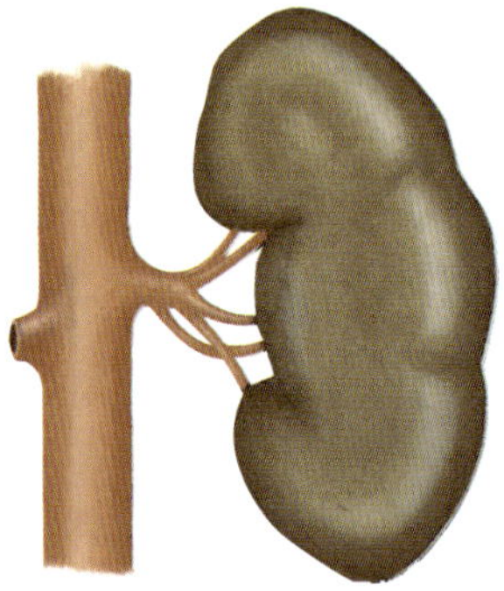

60%

c

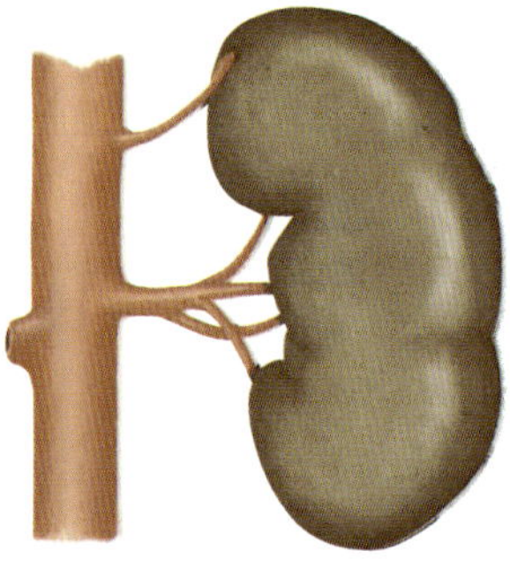

8%

d

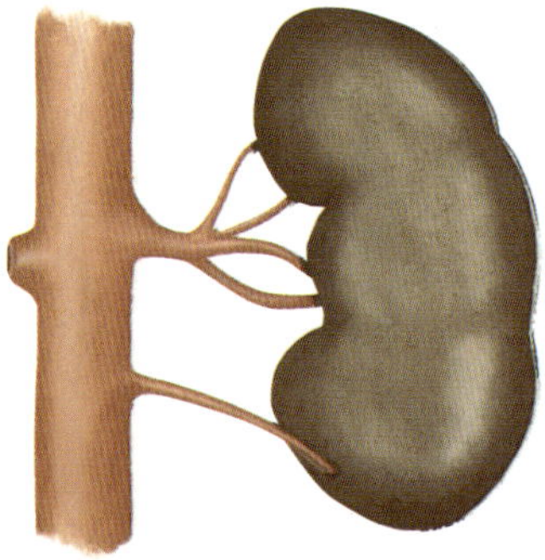

6%

e

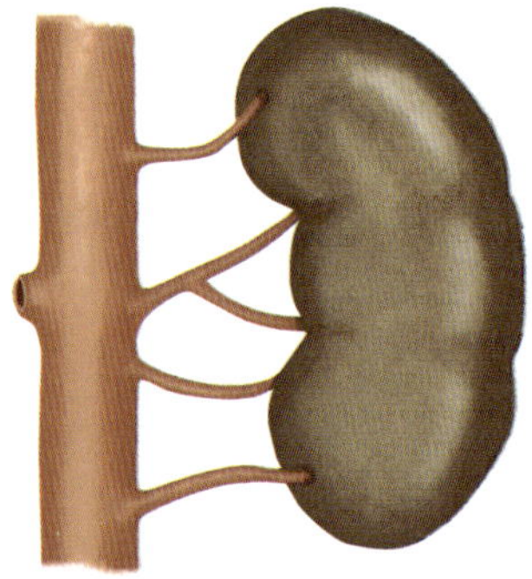

5%

214 Ren [Nephros]

214 Kidneys and renal blood vessels

Ventral aspect

- a Corrosion cast of the renal blood vessels and the ureters of both kidneys of a 15-year-old girl (60%) (Anatomical Collection, Basel)
- b–e Variations of the renal arteries. The percentile numbers beneath the pictures indicate the approximate frequency.
- b 'Normal pattern' with only one renal artery from the aorta
- c Additional upper polar artery from the aorta
- d Additional lower polar artery from the aorta
- c Multiple (more than two) renal arteries from the aorta

214 Nieren und Nierengefässe

Ventralansicht

- a Korrosionspräparat der Nierengefässe und der Harnleiter beider Nieren eines 15jährigen Mädchens (60%) (Anatomische Sammlung, Basel)
- b–e Variationen der Nierenarterien. Die Prozentzahlen unter den Abbildungen geben die ungefähre Häufigkeit an.
- b "Normalsituation" mit nur einer Nierenarterie aus der Aorta
- c Zusätzliche obere Polarterie aus der Aorta
- d Zusätzliche untere Polarterie aus der Aorta
- e Mehrere (mehr als zwei) Nierenarterien aus der Aorta

a

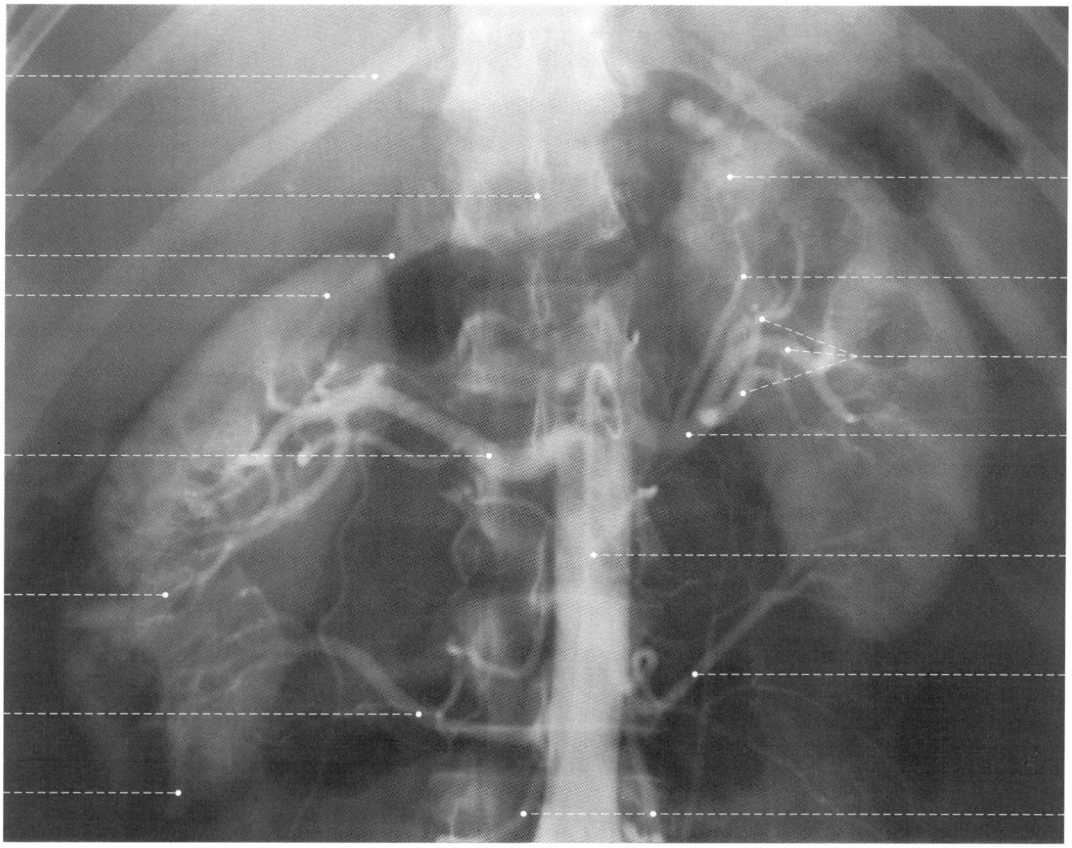

b c d e

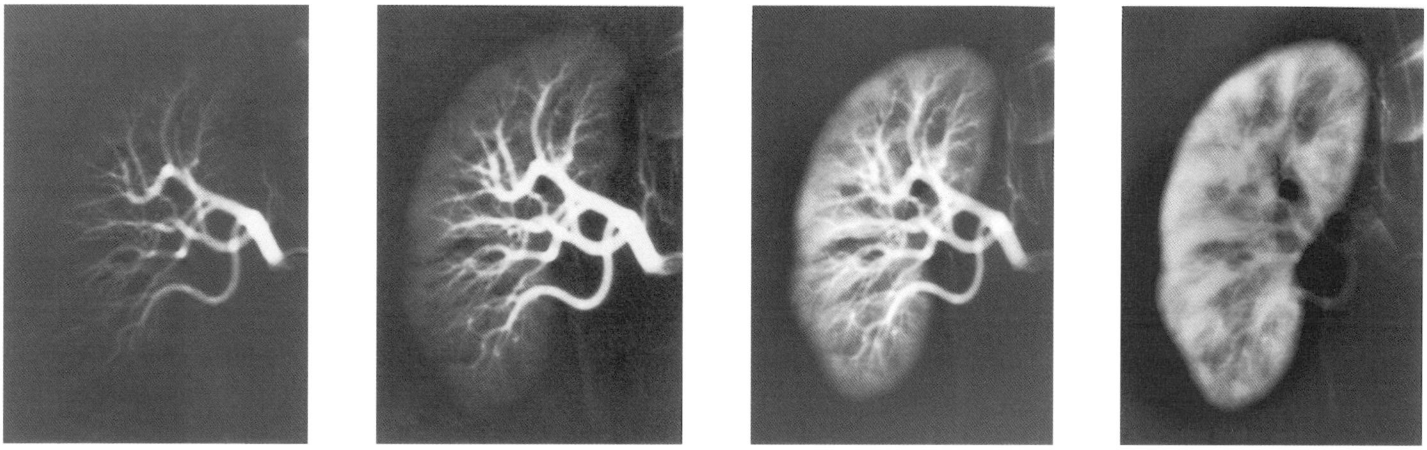

215 Ren [Nephros]

215 Kidneys and renal blood vessels

Ventral aspect

a Renal arteriogram of both renal arteries, at both sides lower polar arteries from the aorta (60%), posteroanterior radiograph

b–e Selective renal arteriogram of the right renal artery, stages of increasing enrichment of contrast medium in the renal parenchyma (30%)

215 Nieren und Nierengefässe

Ventralansicht

a Nierenarteriogramm beider Nierenarterien, untere Polarterien auf beiden Seiten (60%), posteroanteriores Röntgenbild

b–e Selektives Nierenarteriogramm der rechten Nierenarterie, Stadien zunehmender Anreicherung des Kontrastmittels im Nierenparenchym (30%)

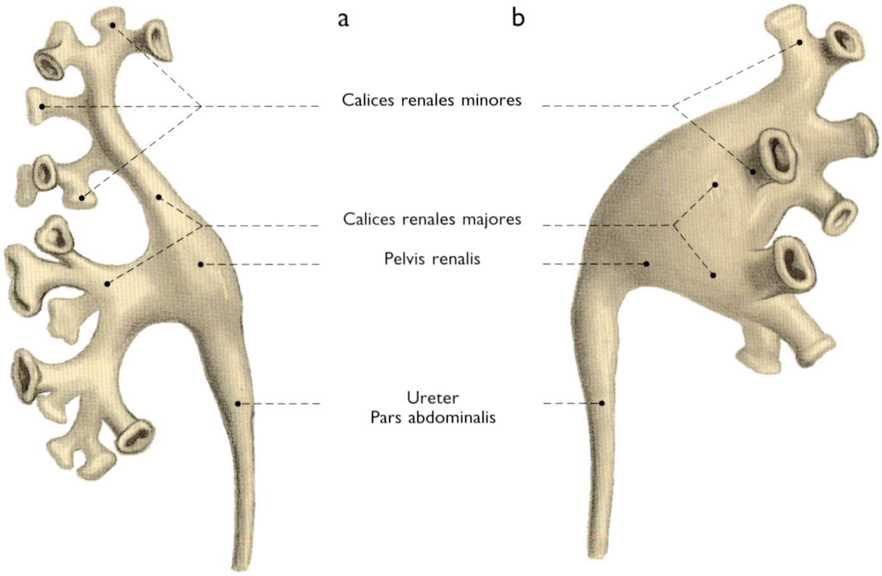

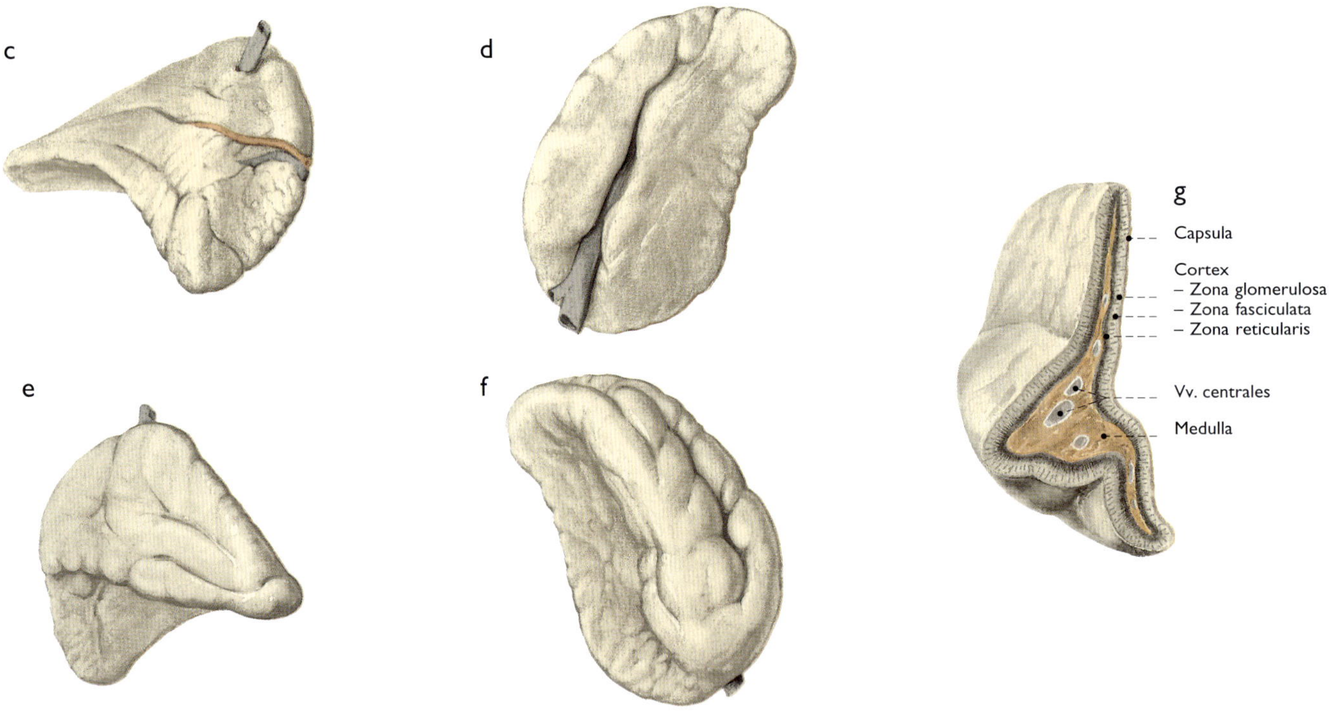

216 Pelvis renalis et Glandula suprarenalis

216 Renal pelvis and suprarenal gland

a, b Renal pelves (100%), ventral aspect
a Right renal pelvis, dendritic type
b Left renal pelvis, ampullary type
c–g Suprarenal glands
c, e Right suprarenal gland (80%)
d, f Left suprarenal gland (80%)
c, d Anterior surface
e, f Posterior surface
g Longitudinal section through a suprarenal gland (200%)

216 Nierenbecken und Nebenniere

a, b Nierenbecken (100%), Ventralansicht
a Rechtes Nierenbecken, dendritischer Typ
b Linkes Nierenbecken, ampullärer Typ
c–g Nebennieren
c, e Rechte Nebenniere (80%)
d, f Linke Nebenniere (80%)
c, d Facies anterior
e, f Facies posterior
g Längsschnitt durch eine Nebenniere (200%)

a

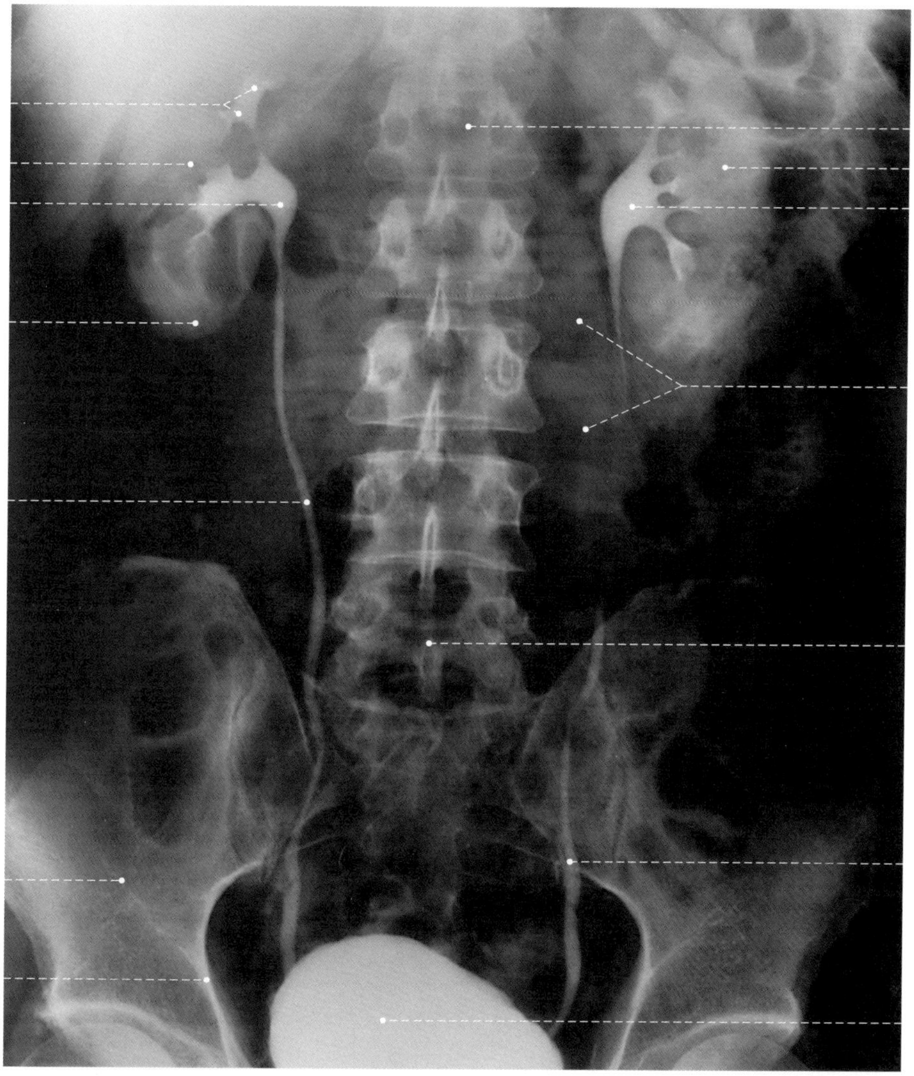

b

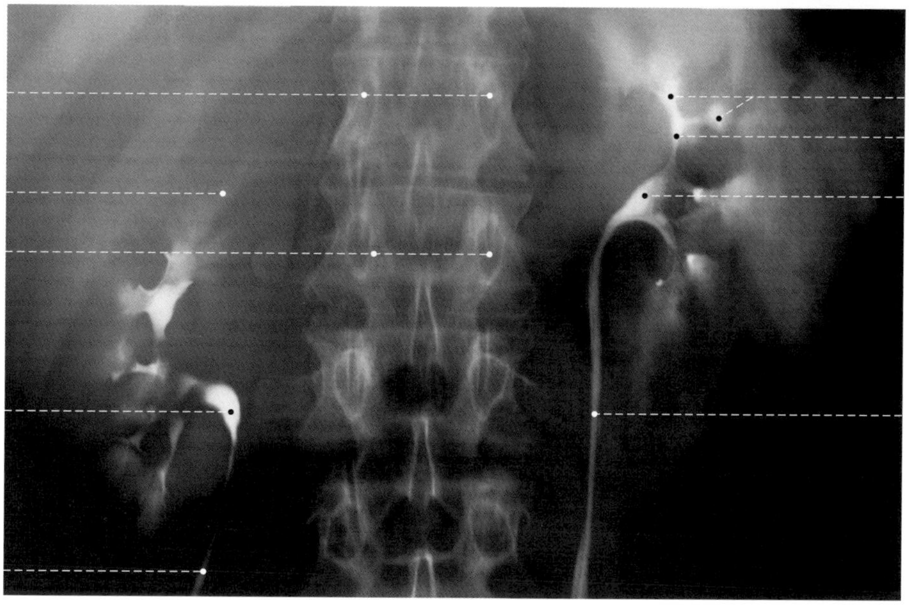

217 Pelvis renalis et Ureter

217 Renal pelvis and Ureter

Intravenous excretion urograms

a Anteroposterior radiograph (30%)

b Anteroposterior tomogram of the renal pelves and calices (40%)

217 Nierenbecken und Harnleiter

Intravenöse Ausscheidungsurogramme

a Anteroposteriores Röntgenbild (30%)

b Anteroposteriores Schichtbild (Tomogramm) von Nierenbecken und Nierenkelchen (40%)

a

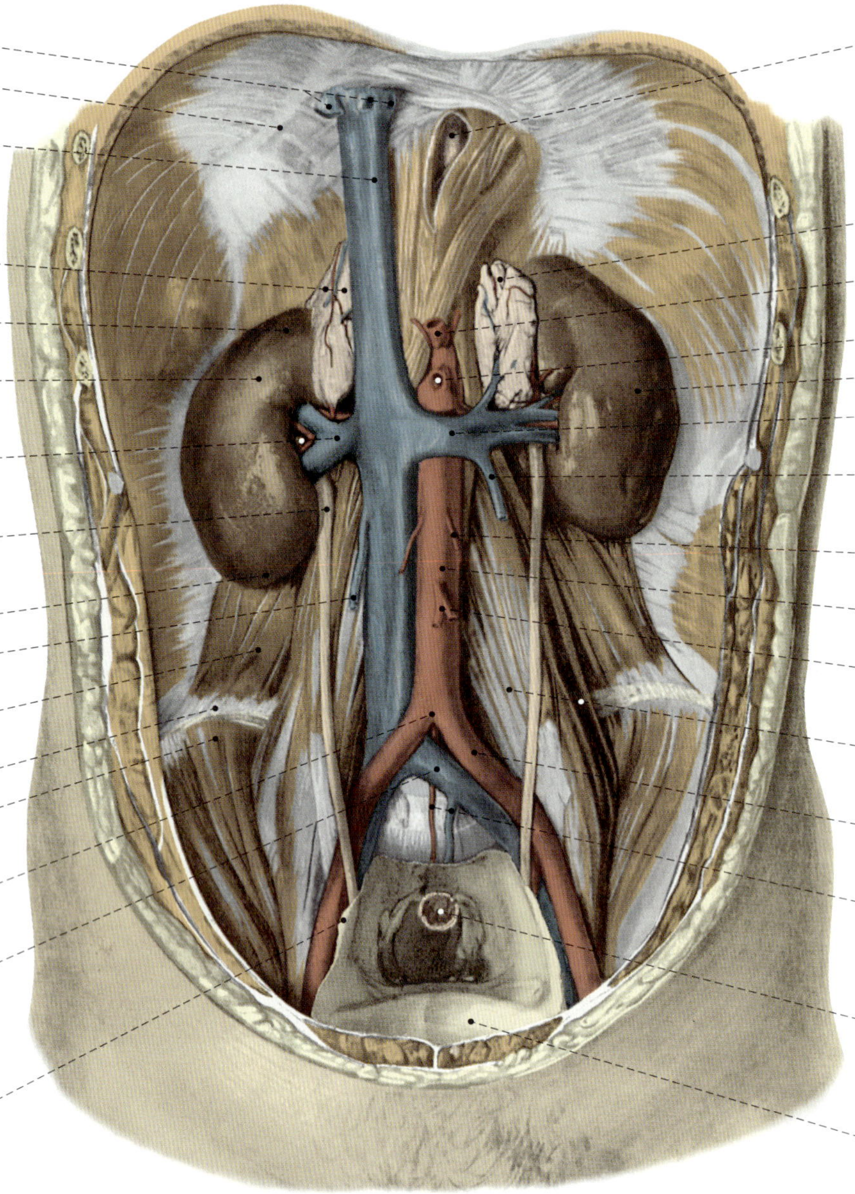

b

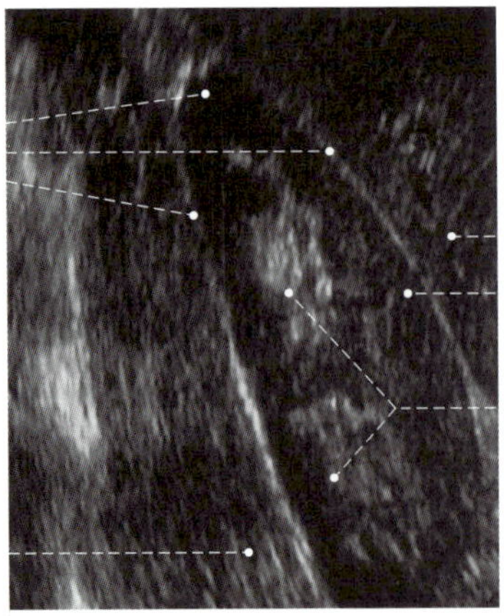

218 Abdomen

218 Urinary organs and great abdominal vessels

a Ventral aspect (30%)
b Ultrasound image (50%), subcostal longitudinal section

218 Harnorgane und grosse Bauchgefässe

a Ventralansicht (30%)
b Ultraschallbild der Niere (50%), subkostaler Längsschnitt

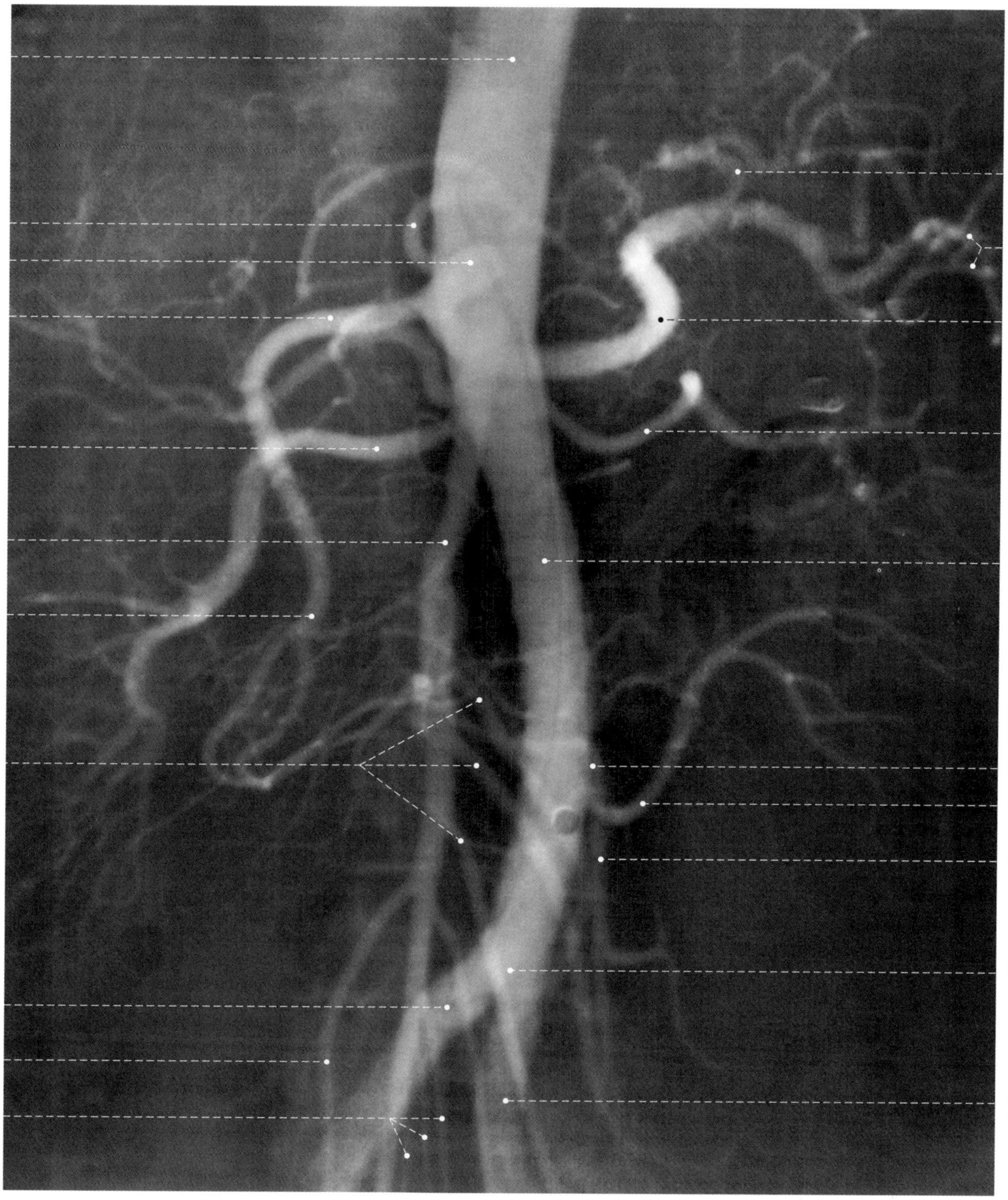

219 Abdomen

219 Large abdominal arteries (80%)

Abdominal aortogram, posteroanterior radiograph. The abdominal aorta is slightly displaced to the left.

219 Grosse Baucharterien (80%)

Aortogramm der Bauchaorta, posteroanteriore Röntgenaufnahme. Die Bauchaorta ist etwas nach links verlagert.

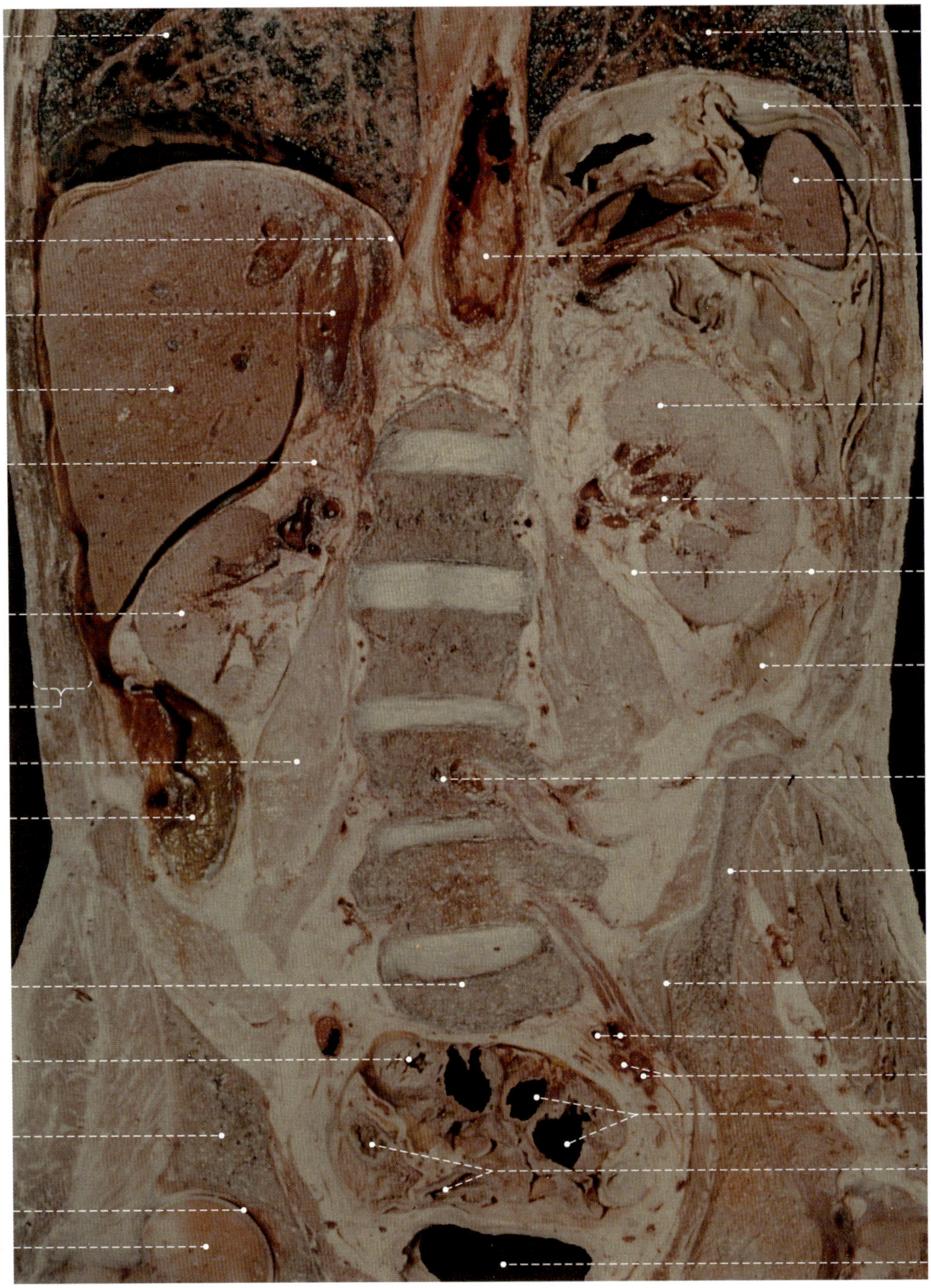

220 Abdomen

220 Abdomen (30%)

Coronal anatomical section through the abdomen in the plane of the lumbar vertebral bodies, ventral aspect

220 Bauch (30%)

Frontaler anatomischer Schnitt durch den Bauchraum in der Ebene der Lendenwirbelkörper, Ventralansicht

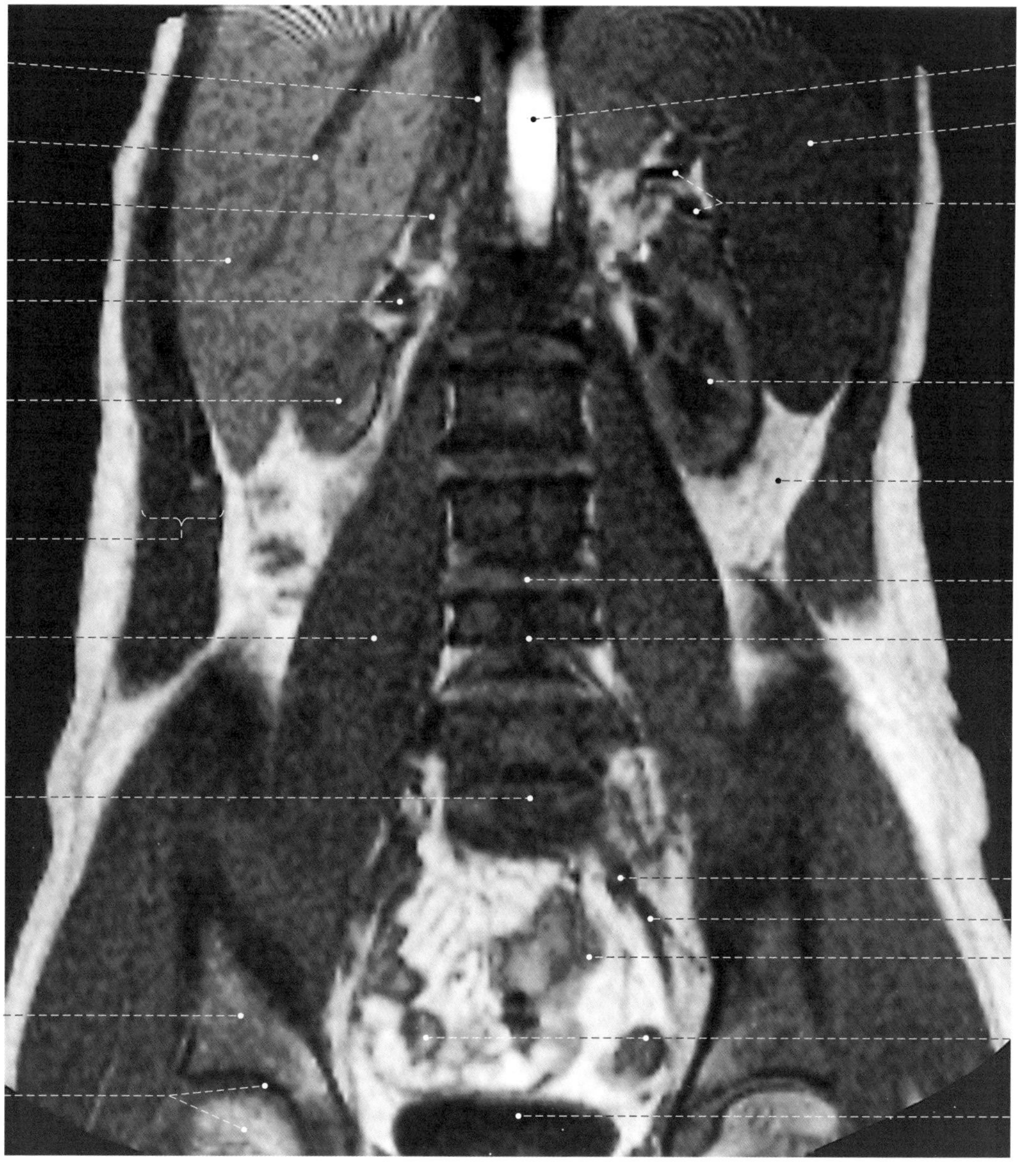

221 Abdomen

221 Abdomen (30%)

Coronal magnetic resonance image (MRI, T_1-weighted) through the abdomen in the plane of the lumbar vertebral bodies and the femoral heads, ventral aspect

221 Bauch (30%)

Frontales Kernspintomogramm (MRI, T_1-gewichtet) durch den Bauchraum in der Ebene der Lendenwirbelkörper und der Oberschenkelköpfe, Ventralansicht

a

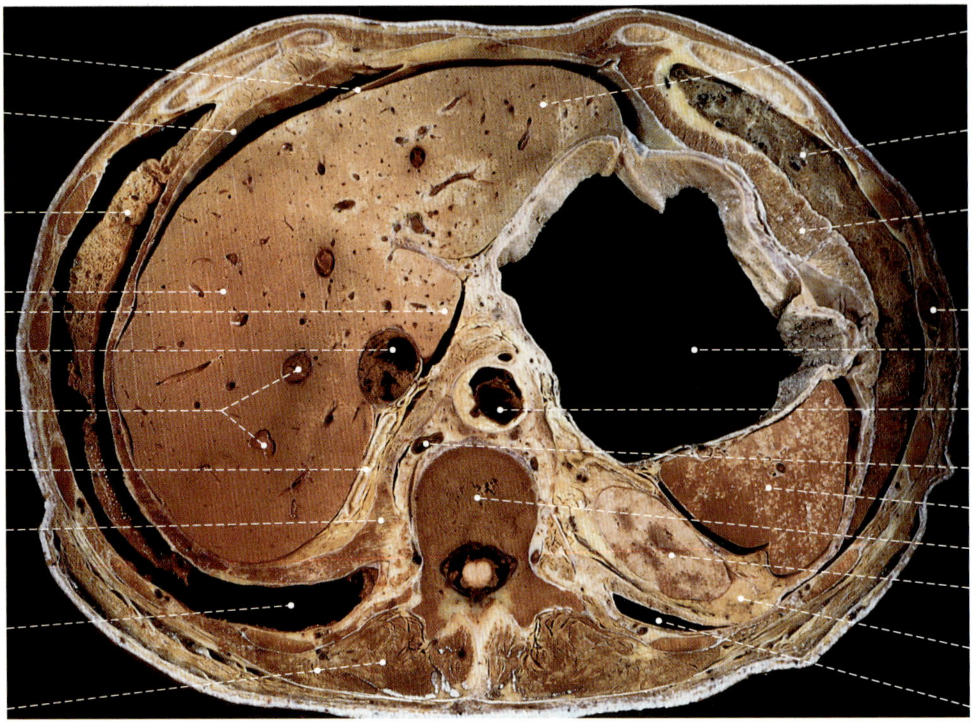

b

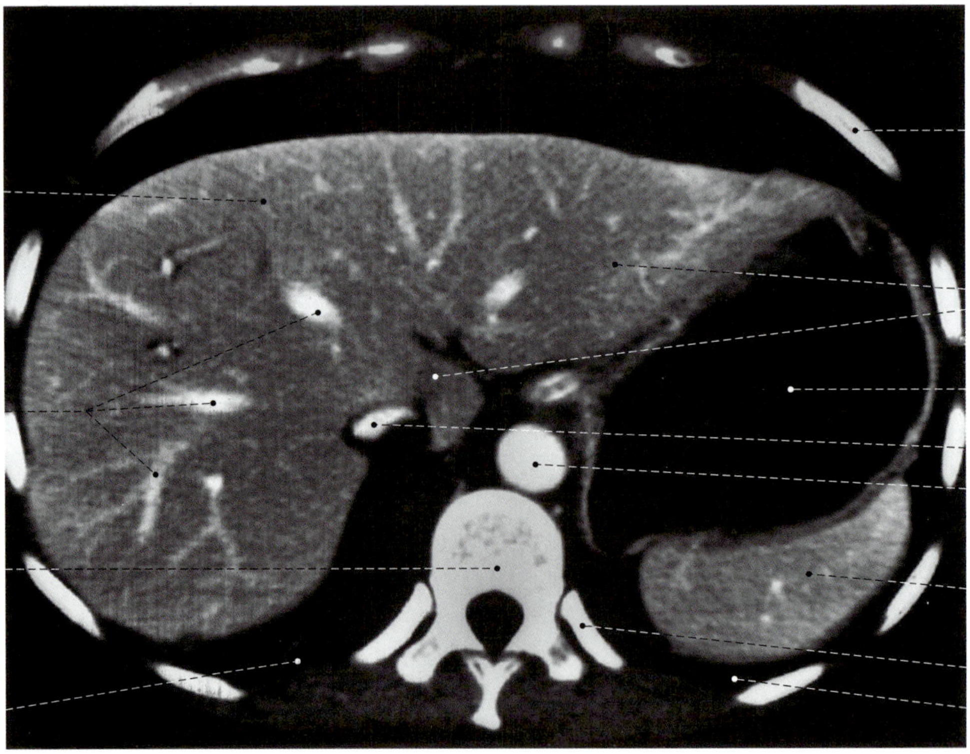

222 Abdomen

222 Abdomen (40%)

Transverse sections through the upper abdomen at the level of the eleventh thoracic vertebral body, inferior aspect

a Anatomical section

b Computed tomogram (CT) after injection of contrast medium

222 Bauch (40%)

Horizontalschnitte durch den Oberbauch in Höhe des elften Brustwirbelkörpers, Kaudalansicht

a Anatomischer Querschnitt

b Computertomogramm (CT) nach Kontrastmittelgabe

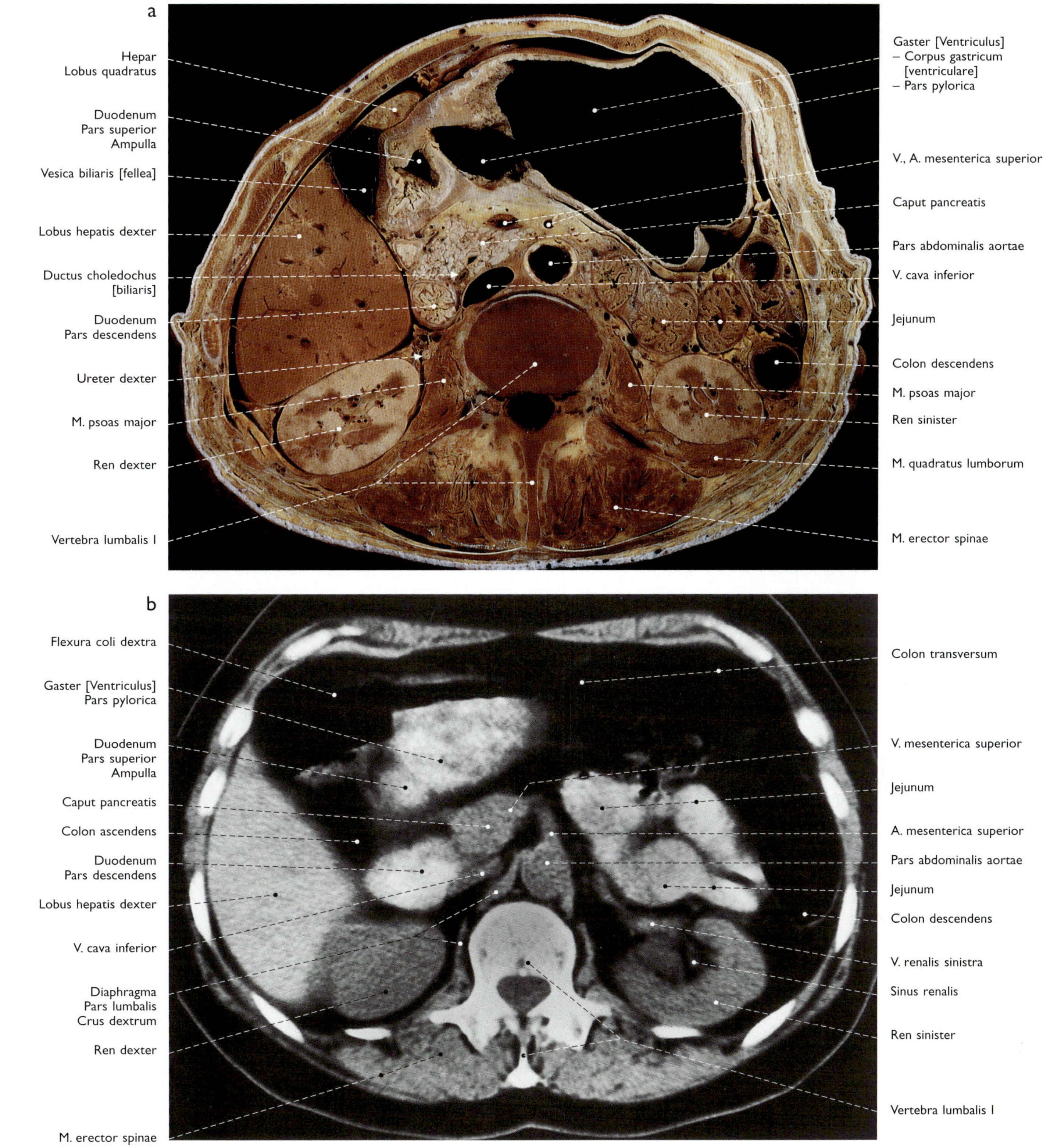

223 Abdomen

223 Abdomen (40%)

Transverse sections through the upper abdomen at the level of the first lumbar vertebral body, inferior aspect

a Anatomical section
b Computed tomogram (CT)

223 Bauch (40%)

Horizontalschnitte durch den Oberbauch in Höhe des ersten Lendenwirbelkörpers, Kaudalansicht

a Anatomischer Querschnitt
b Computertomogramm (CT)

a

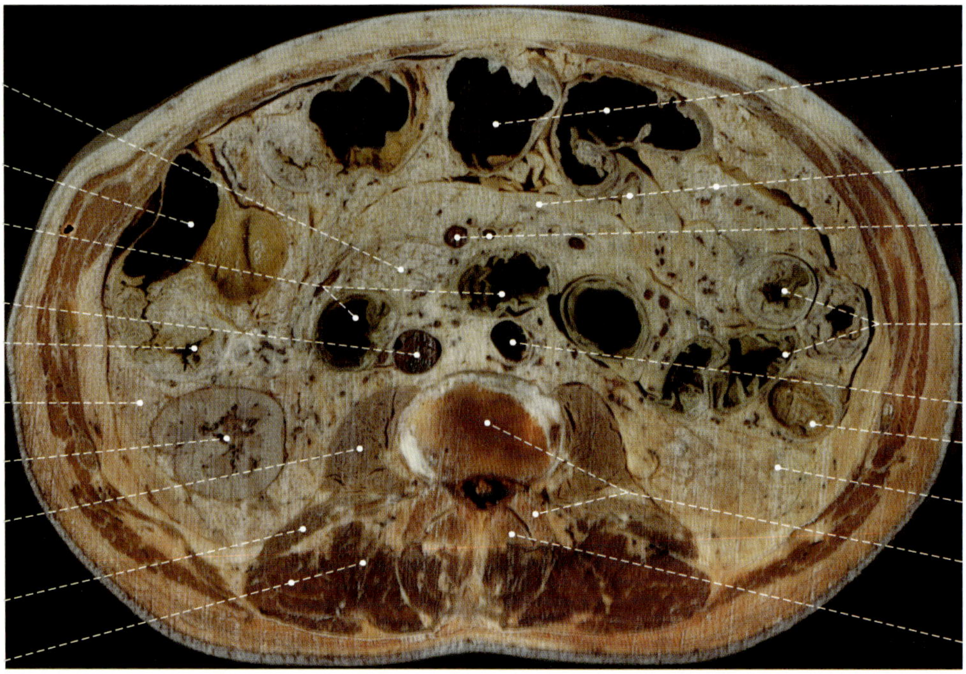

b

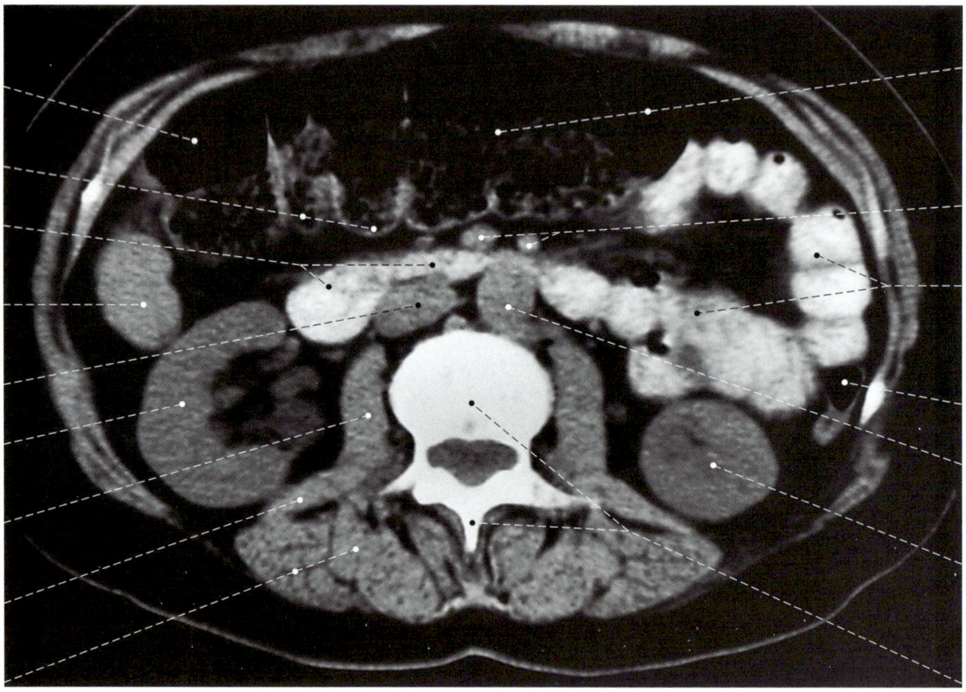

224 Abdomen

224 Abdomen (40%)

Transverse sections through the abdomen at the level of the second (b) and third (a) lumbar vertebral body, inferior aspect

a Anatomical section
b Computed tomogram (CT)

224 Bauch (40%)

Horizontalschnitte durch den Bauch in Höhe des zweiten (b) bzw. dritten (a) Lendenwirbelkörpers, Kaudalansicht

a Anatomischer Querschnitt
b Computertomogramm (CT)

a

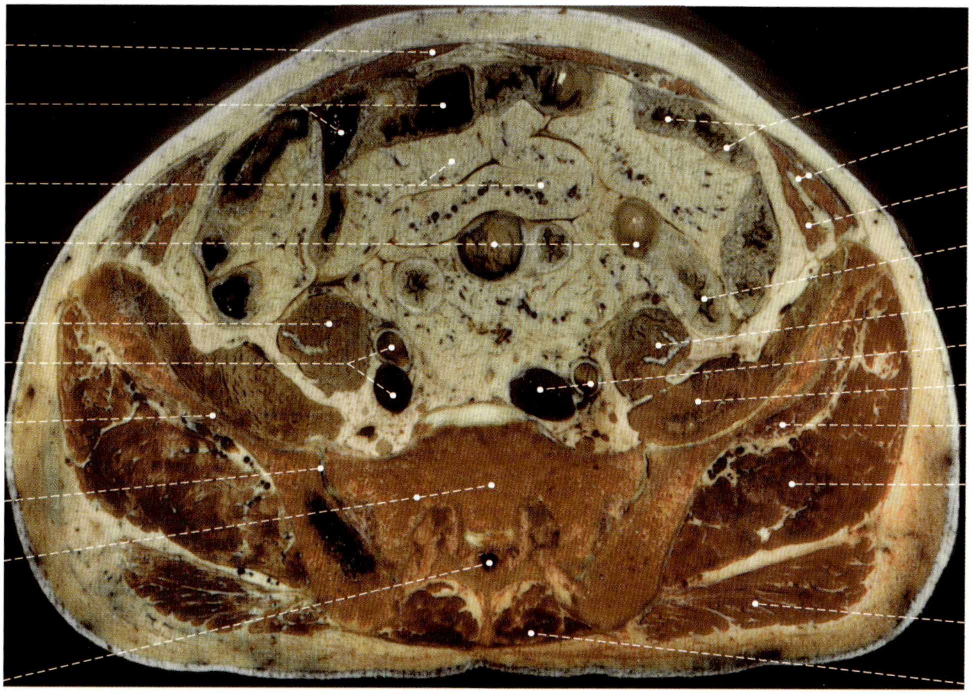

b

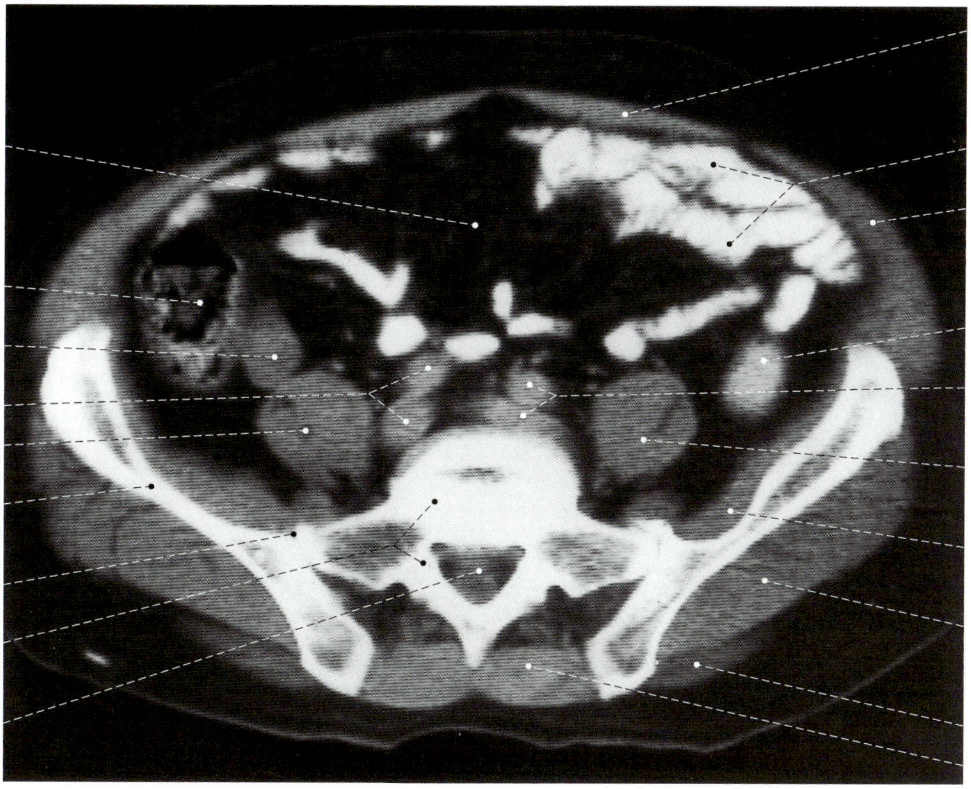

225 Abdomen

225 Abdomen (40%)

Transverse sections through the abdomen at the level of the first sacral vertebra and the sacroiliac joints, inferior aspect

a Anatomical section
b Computed tomogram (CT)

225 Bauch (40%)

Horizontalschnitte durch den Bauch in Höhe des ersten Kreuzbeinwirbels und der Kreuzbein-Darmbein-Gelenke, Kaudalansicht

a Anatomischer Querschnitt
b Computertomogramm (CT)

a

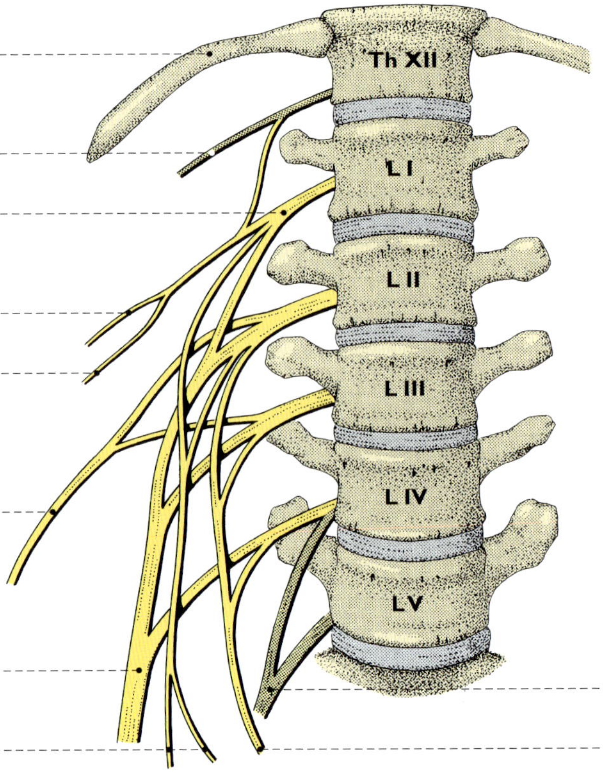

b

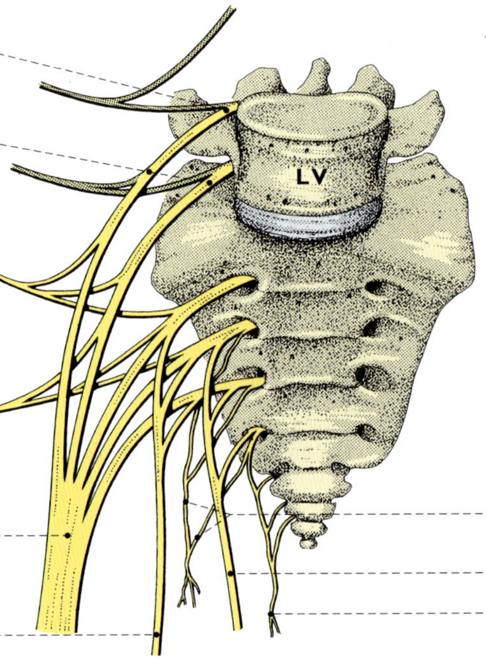

226 Plexus lumbalis et Plexus sacralis [Plexus lumbosacralis]

226 Lumbar and sacral plexuses

Schematic representations, ventral aspect
a Lumbar plexus
b Sacral plexus

226 Lenden- und Kreuzbeingeflechte

Schematische Darstellungen, Ventralansicht
a Lendengeflecht
b Kreuzbeingeflecht

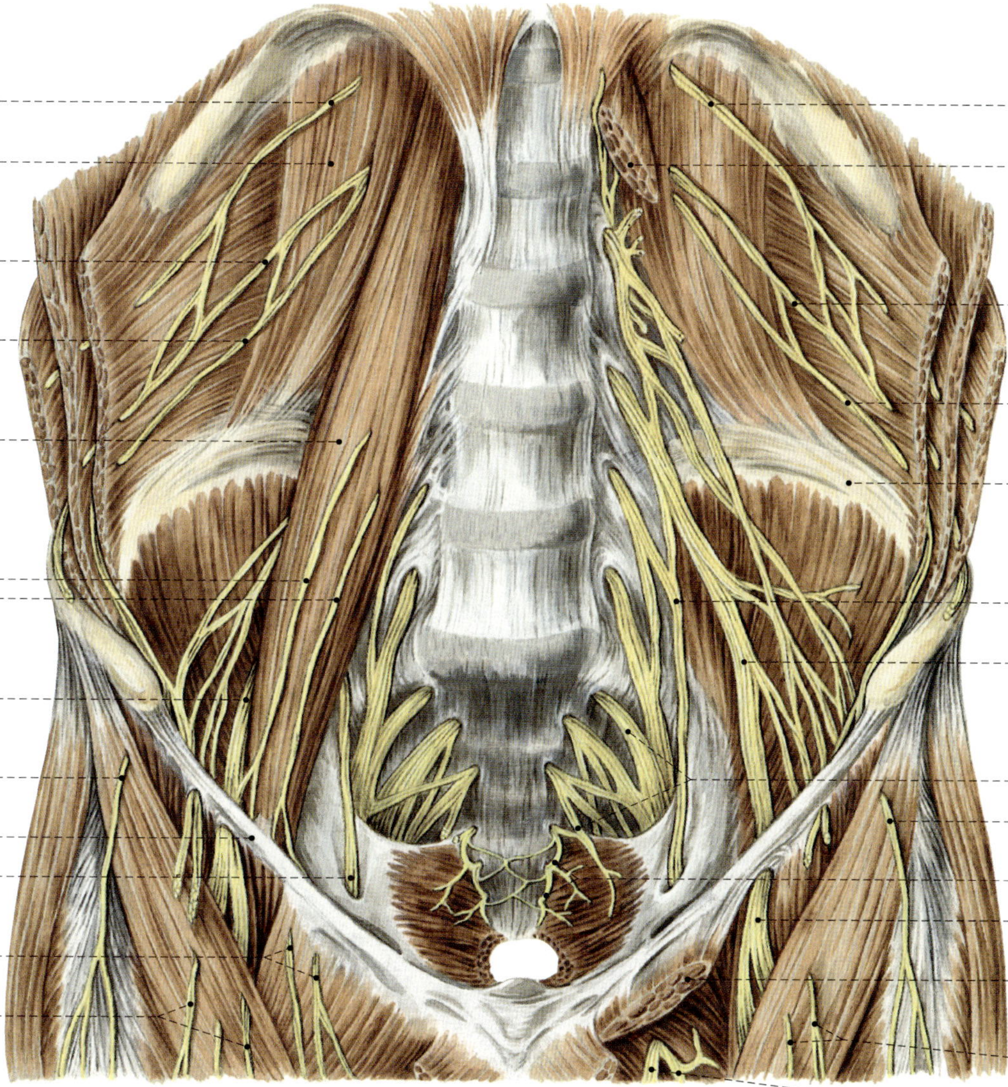

227 Plexus lumbosacralis

227 Lumbosacral plexus (40%)
On the left, the psoas major muscle was removed and the pectineus muscle cut near to its origin. Ventral aspect

227 Lendenkreuzbeingeflecht (40%)
Links ist der Musculus psoas major abgetragen, der Musculus pectineus ursprungsnah geschnitten. Ventralansicht

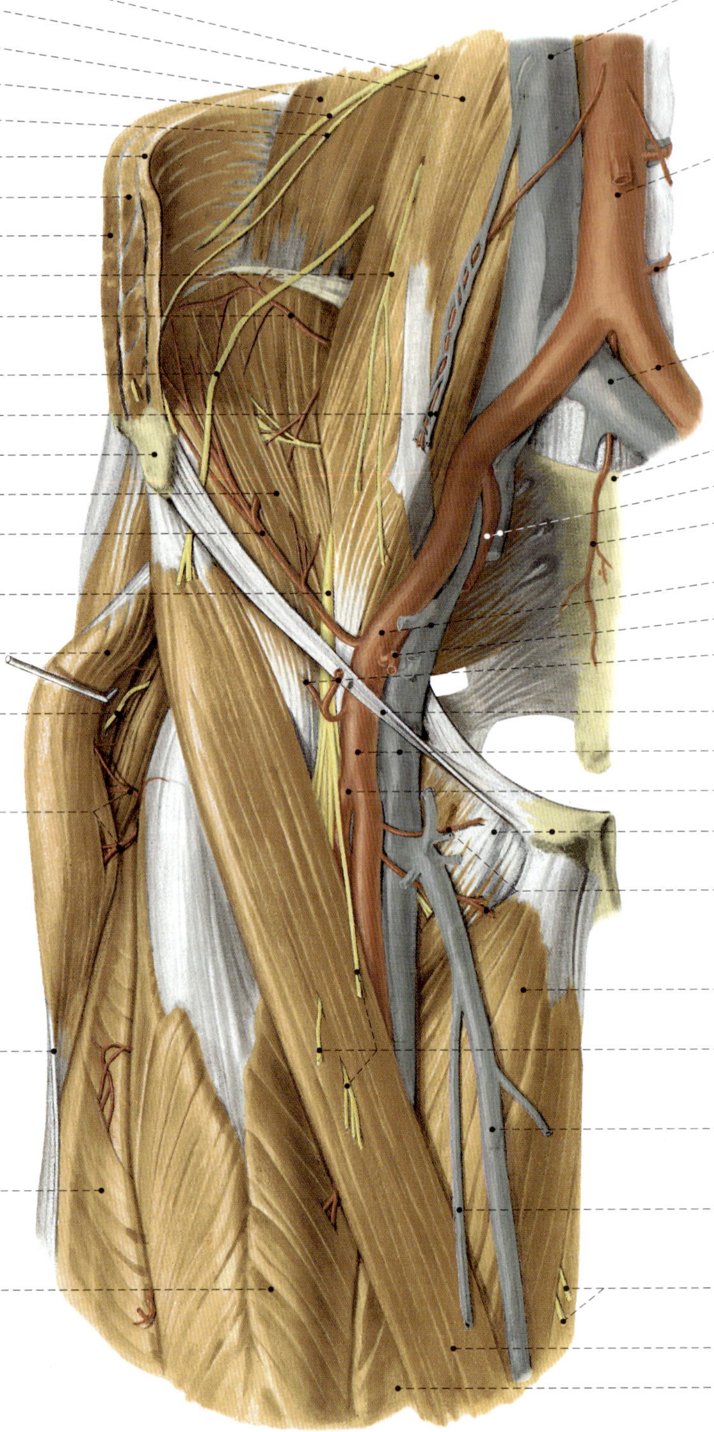

228 Abdomen et Femur

228 Blood vessels and nerves of the posterior abdominal wall and the thigh of a male (50%)
Ventral aspect

228 Blutgefässe und Nerven der hinteren Bauchwand und des Oberschenkels eines Mannes (50%)
Ventralansicht

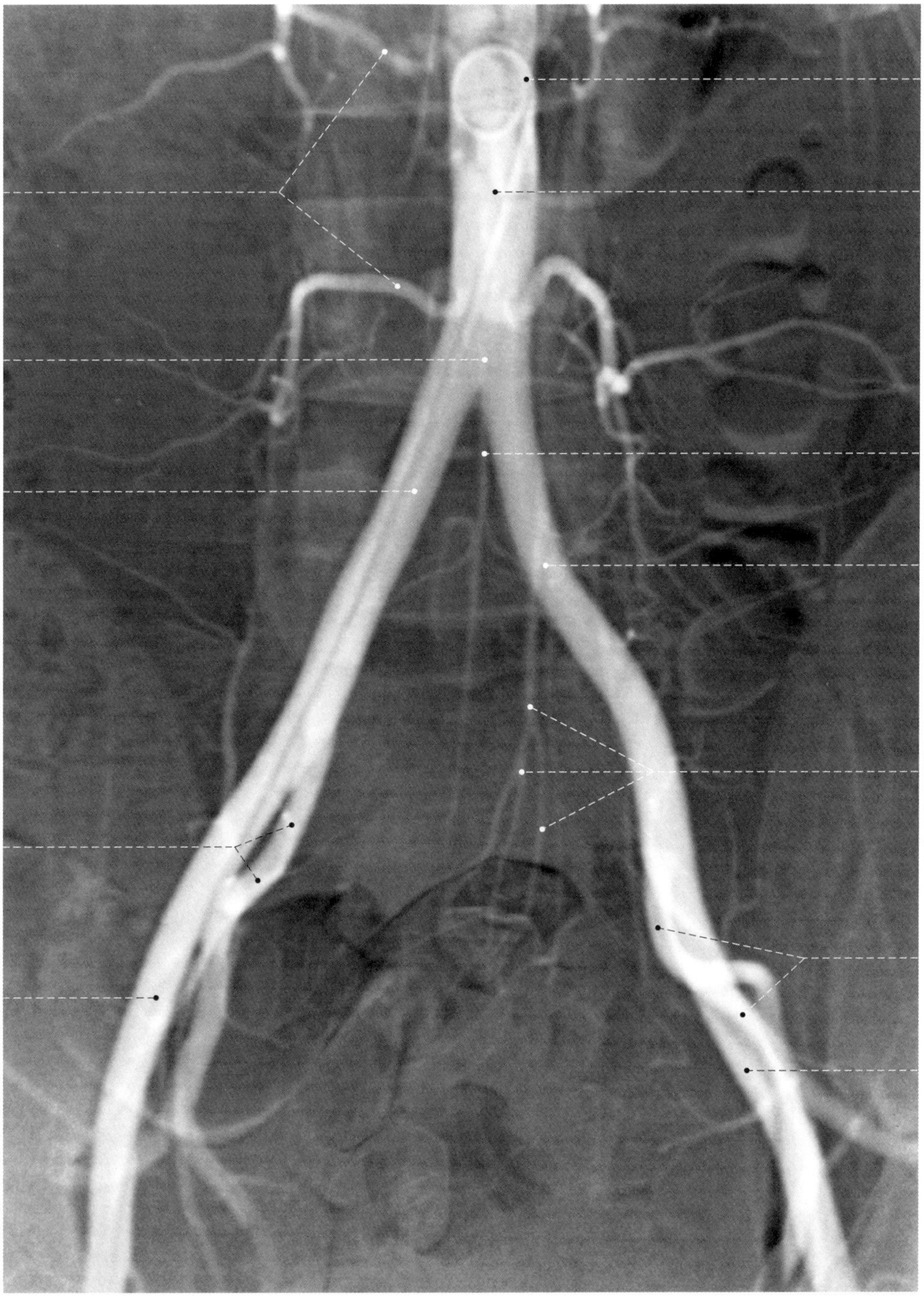

229 Abdomen

229 Large abdominal arteries (80%)

Arteriogram of the caudal abdominal aorta, the aortic bifurcation and the iliac arteries, posteroanterior radiograph

229 Grosse Baucharterien (80%)

Arteriogramm der unteren Bauchaorta, der Aortenbifurkation und der Beckenarterien, posteroanteriore Röntgenaufnahme

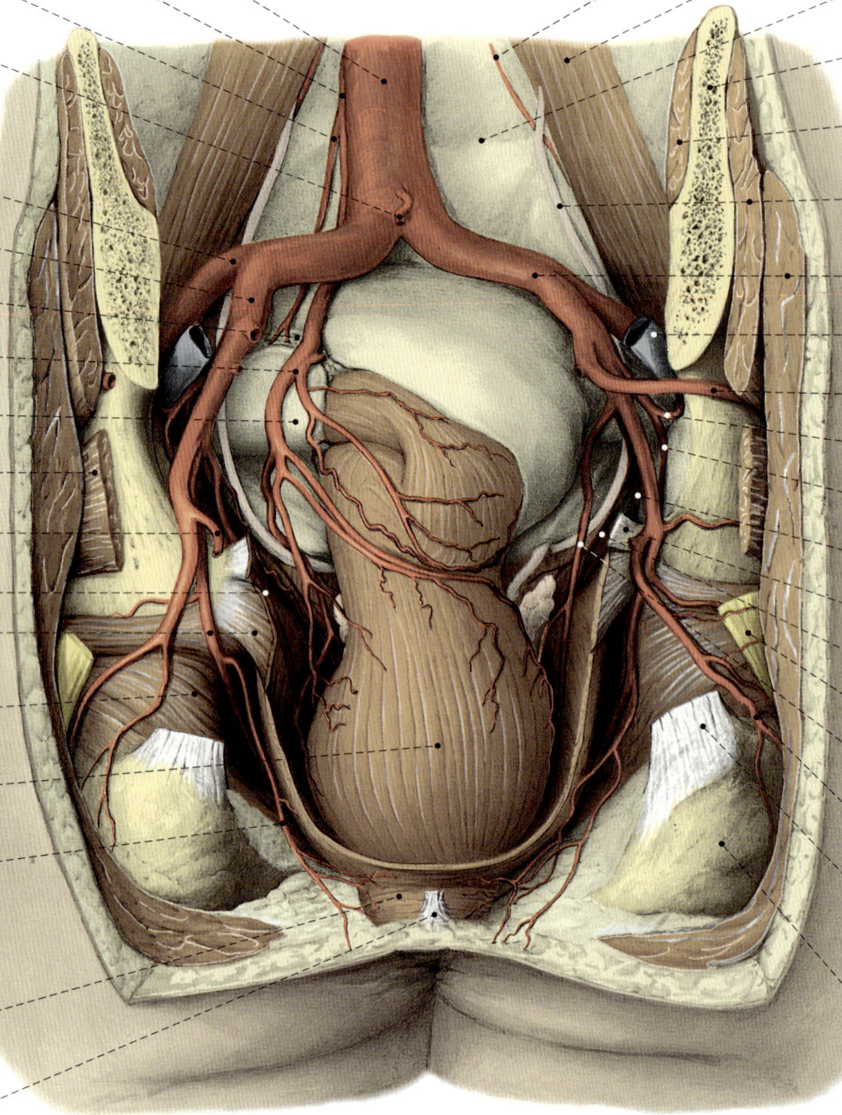

230 Rectum

230 Arterial supply of the rectum of a male (60%)

The sacrum was removed, the coccygeus muscle, the muscles of the gluteal region and the sacrospinous and sacrotuberous ligaments were cut off and partially excised. Dorsal aspect

230 Arterienversorgung des Mastdarms eines Mannes (60%)

Das Kreuzbein ist entfernt, der Musculus coccygeus, die Gesässmuskeln und die Ligamenta sacrospinale et sacrotuberale sind durchtrennt und teilweise abgetragen. Dorsalansicht

a

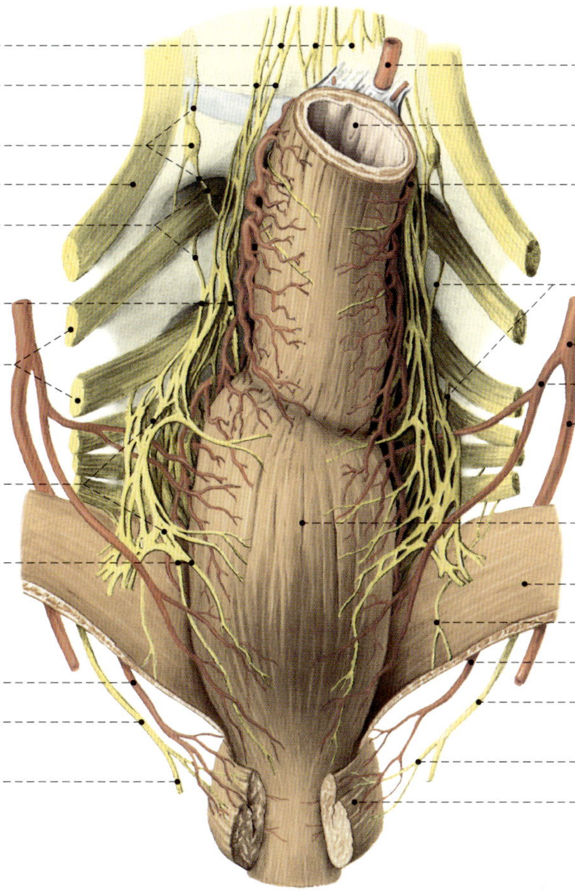

b

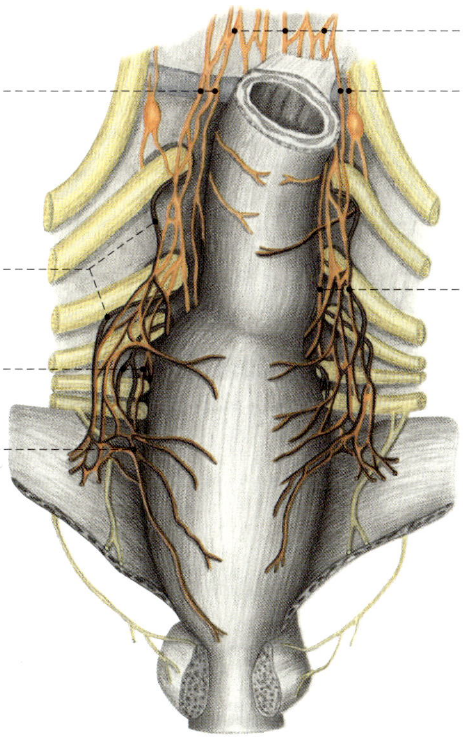

231 Rectum

231 Rectum (45%)

Ventral aspect

a Arteries and nerves of the rectum

b Explanatory drawing as to Fig. a. Sympathetic fibres are shown by **orange**, parasympathetic ones by **brown** color.

231 Mastdarm (45%)

Ventralansicht

a Arterien und Nerven des Mastdarms

b Erklärende Zeichnung zu Abb. a. Sympathische Fasern sind **orange**, parasympathische **braun** dargestellt.

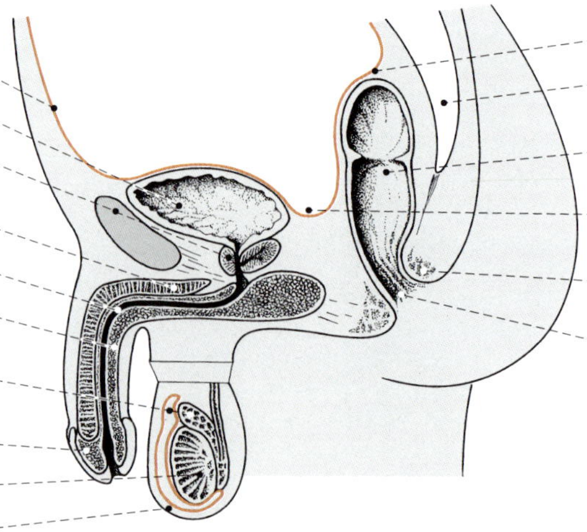

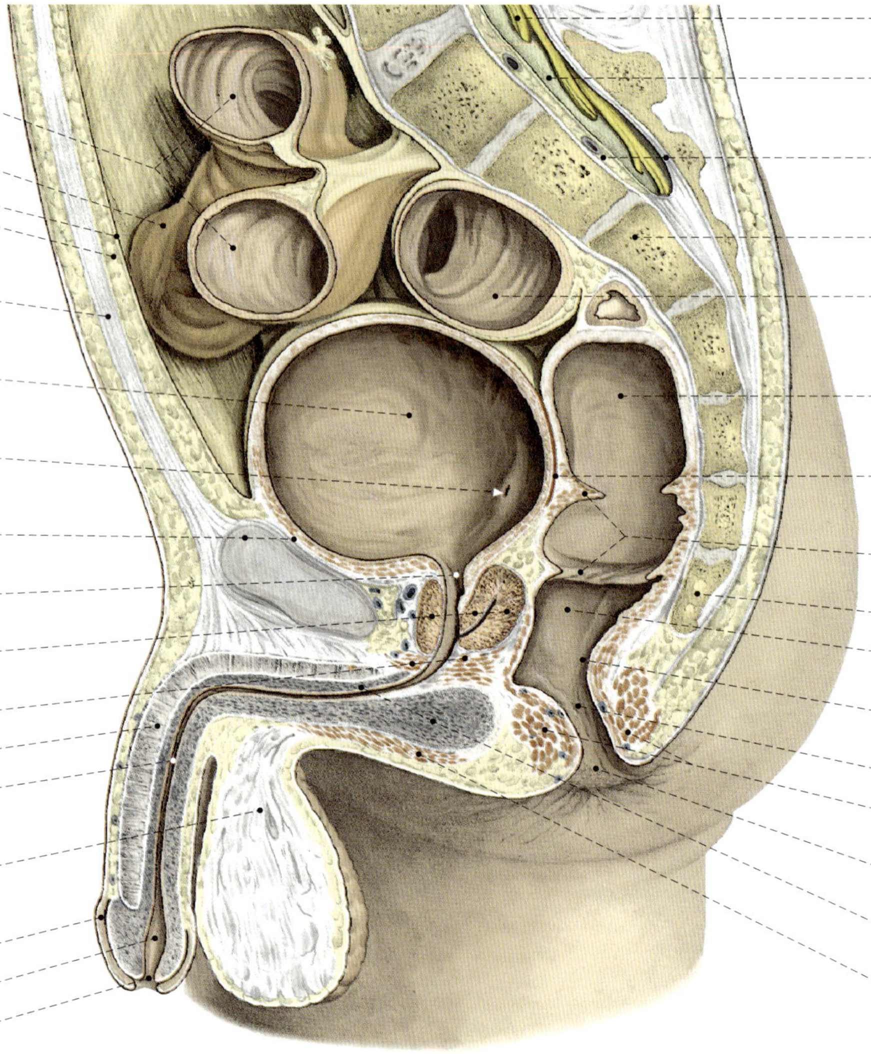

232 Pelvis et Apparatus urogenitalis

232 Male pelvis and urogenital apparatus

a Male pelvis, schematized median section, medial aspect
b Pelvic organs of an 18-year-old male, median section (55%), medial aspect of the right part (Anatomical Collection, Basel)

232 Männliches Becken mit Harn- und Geschlechtsapparat

a Männliches Becken, schematisierter Medianschnitt, Medialansicht
b Beckenorgane eines 18jährigen Mannes, Medianschnitt (55%), Medialansicht der rechten Hälfte (Anatomische Sammlung, Basel)

a

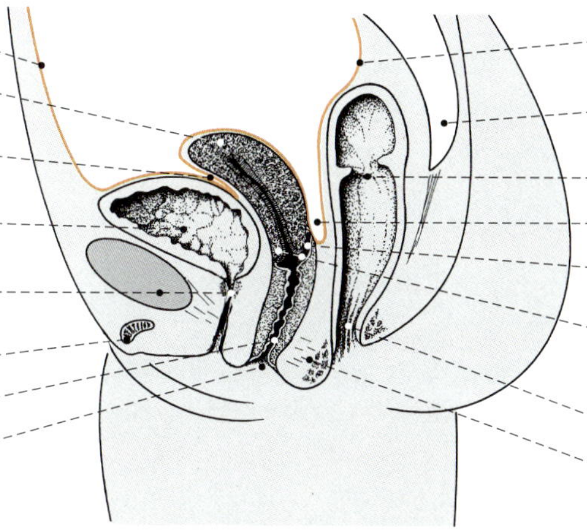

b

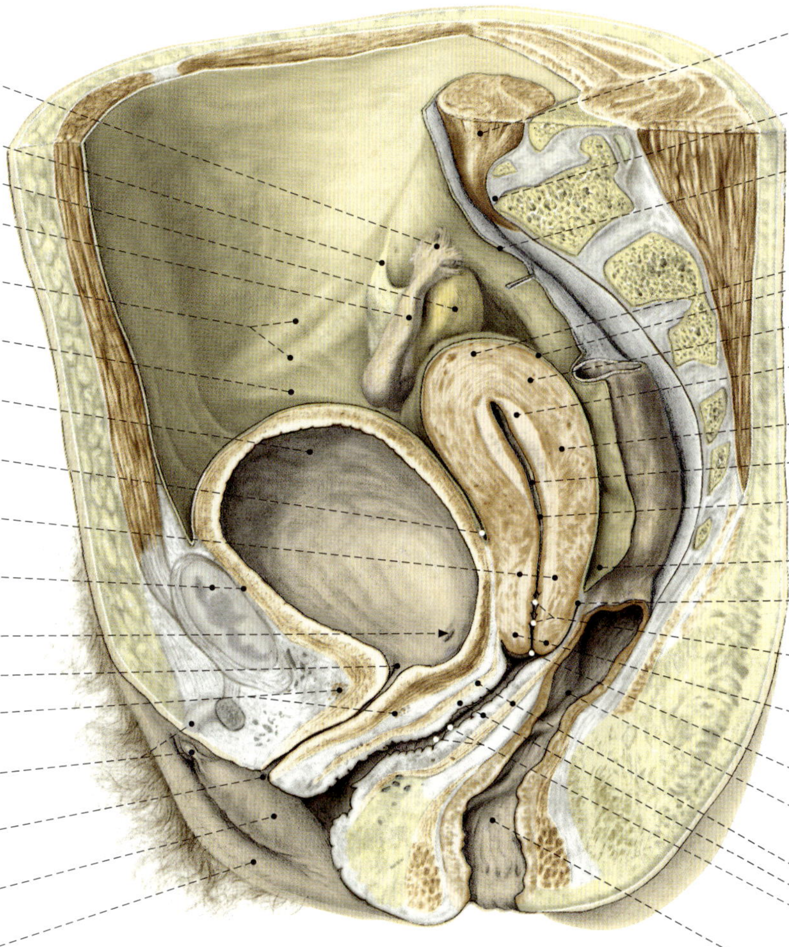

233 Pelvis et Apparatus urogenitalis

233 Female pelvis and urogenital apparatus

a Female pelvis, schematized median section, medial aspect

b Pelvic organs of a 23-year-old female, median section (55%), medial aspect of the right part (Anatomical Collection, Basel)

233 Weibliches Becken mit Harn- und Geschlechtsapparat

a Weibliches Becken, schematisierter Medianschnitt, Medialansicht

b Beckenorgane einer 23jährigen Frau, Medianschnitt (55%), Medialansicht der rechten Hälfte (Anatomische Sammlung, Basel)

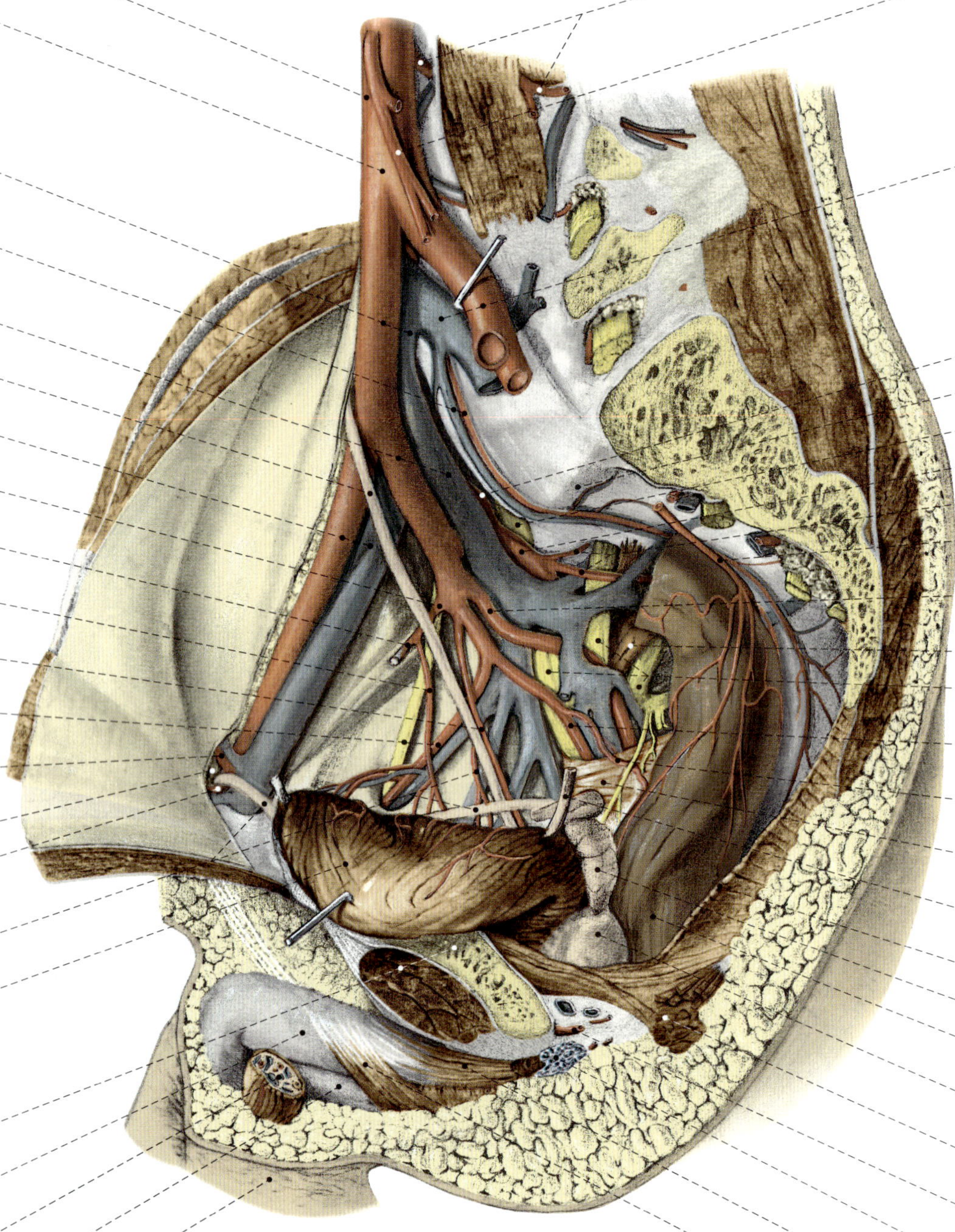

234 ⟨Arteriae, Venae et Nervi pelvis⟩

234 Blood vessels and nerves of the male pelvis (70%)
Sagittal section to the left of the median plane, medial aspect of the right part

234 Blutgefässe und Nerven des männlichen Beckens (70%)
Sagittalschnitt links der Medianebene, Medialansicht der rechten Hälfte

a

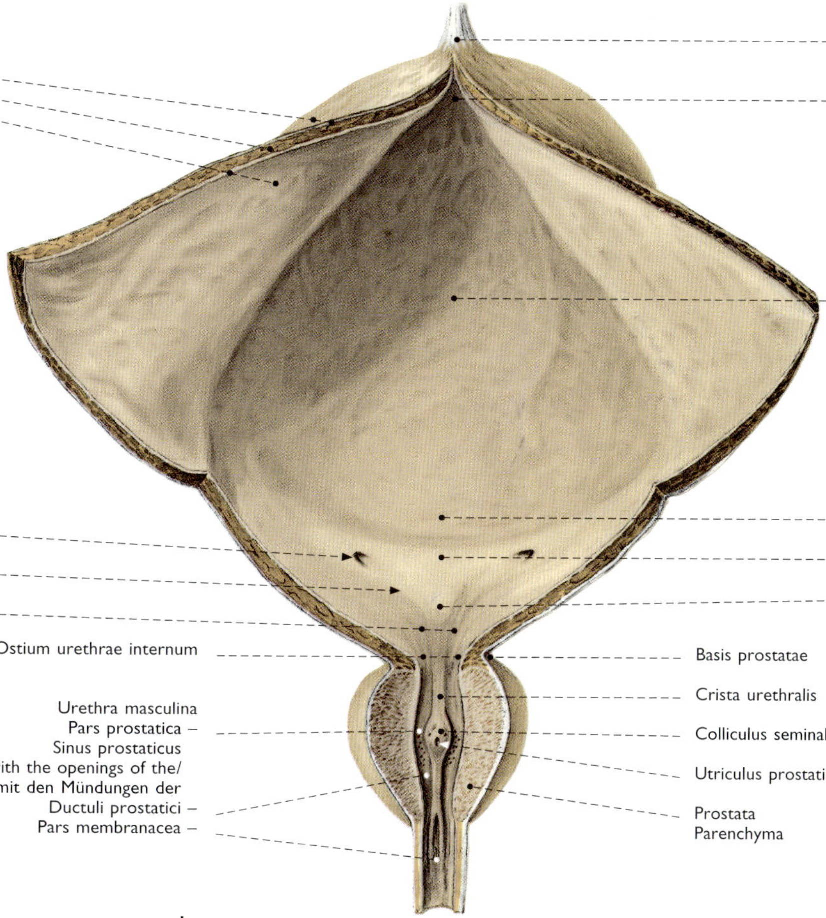

b

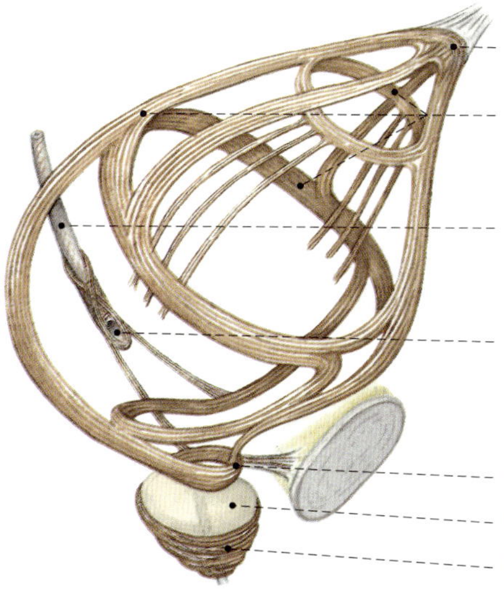

c

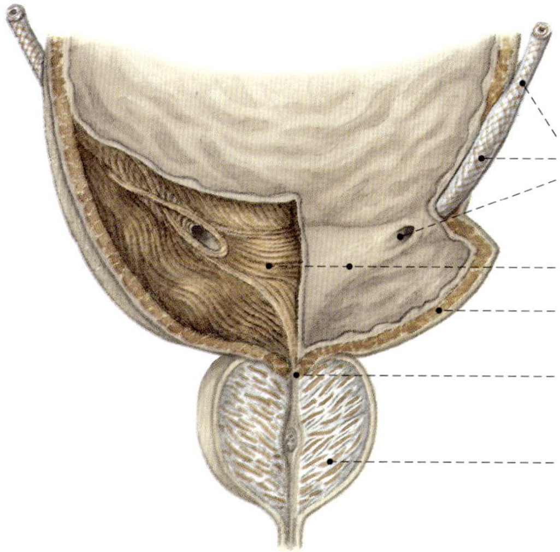

235 Vesica urinaria

235 Urinary bladder

a Urinary bladder and urethra of a male. The bladder and prostate were incised along the midsagittal plane and opened (80%). Ventral aspect

b Arrangement of the muscles in the bladder wall (according to Ferner, 1975) (60%), schematic representation, right lateral aspect

c Ureteral orifices and trigone of the bladder (according to Ferner, 1975, and Leonhardt, 1987). On the left step-cut in the bladder wall to show the intramural part of the ureter, on the right removal of the mucosa of the bladder base to demonstrate the muscle arrangement in the trigone (80%), ventral aspect

235 Harnblase

a Harnblase und Harnröhre eines Mannes. Die Harnblase und die Prostata sind median gespalten und aufgeklappt (80%). Ventralansicht

b Anordnung der Muskulatur in der Harnblasenwand (nach Ferner, 1975) (60%), schematische Darstellung, rechte Seitenansicht

c Ureterostien und Harnblasendreieck (nach Ferner, 1975, und Leonhardt, 1987). Links Stufenschnitt durch die Harnblasenwand zur Darstellung der intramuralen Strecke des Harnleiters, rechts Entfernung der Schleimhaut des Harnblasenbodens zur Darstellung der Muskelanordnung im Harnblasendreieck (80%), Ventralansicht

a

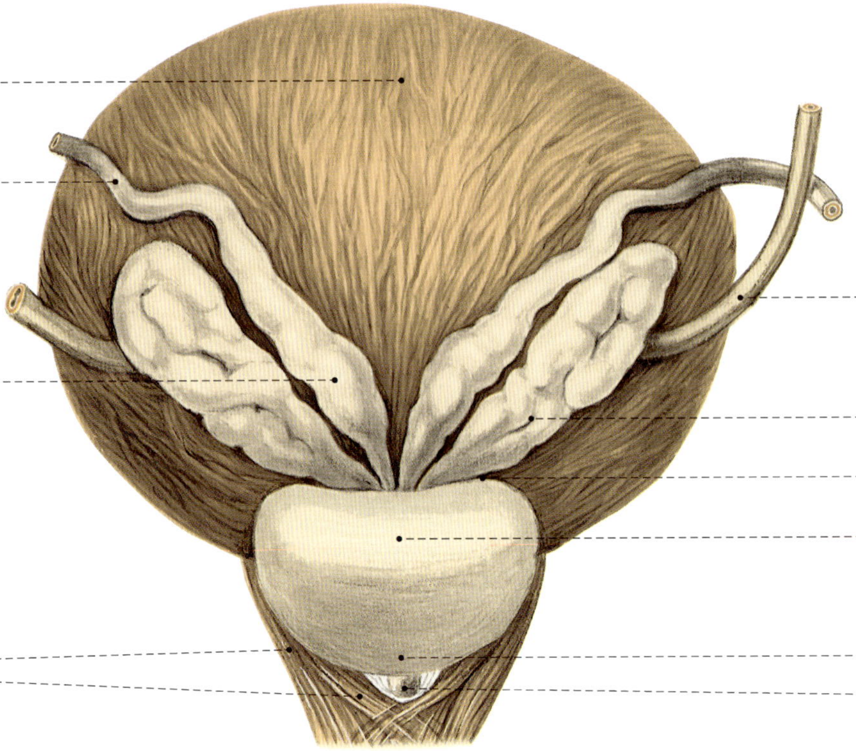

b

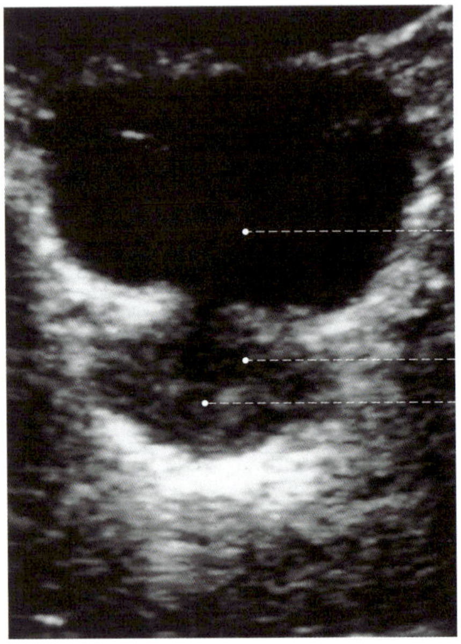

c

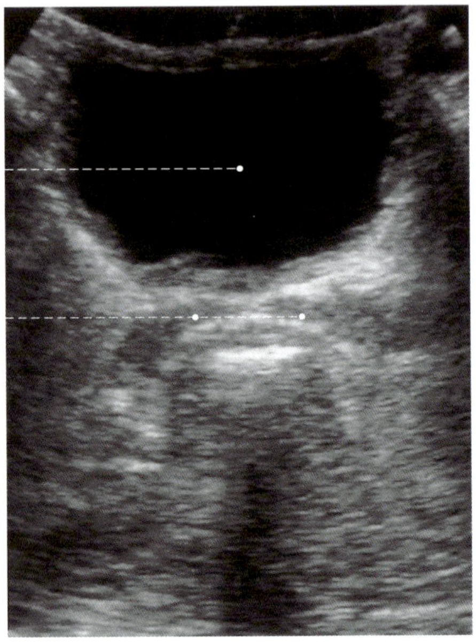

236 Vesica urinaria, Ductus deferens, Vesicula seminalis et Prostata

236 Urinary bladder, deferent duct, seminal vesicle and prostate

a Dorsal aspect (100%)
b, c Ultrasound images (80%), transverse sections above the symphysis through
b the filled urinary bladder and the prostate
c the filled urinary bladder and the seminal vesicles

236 Harnblase, Samenleiter, Samenbläschen und Vorsteherdrüse

a Dorsalansicht (100%)
b, c Ultraschallbilder (80%), Querschnitte oberhalb der Symphyse durch
b die gefüllte Harnblase und die Prostata
c die gefüllte Harnblase und die Samenbläschen

a

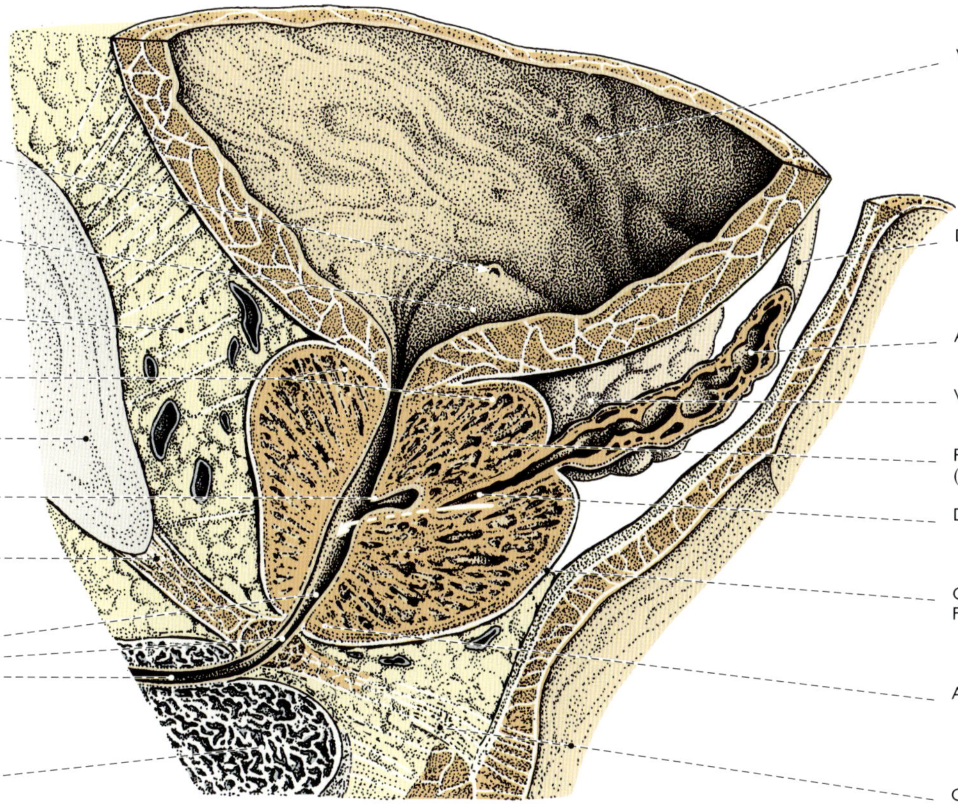

b

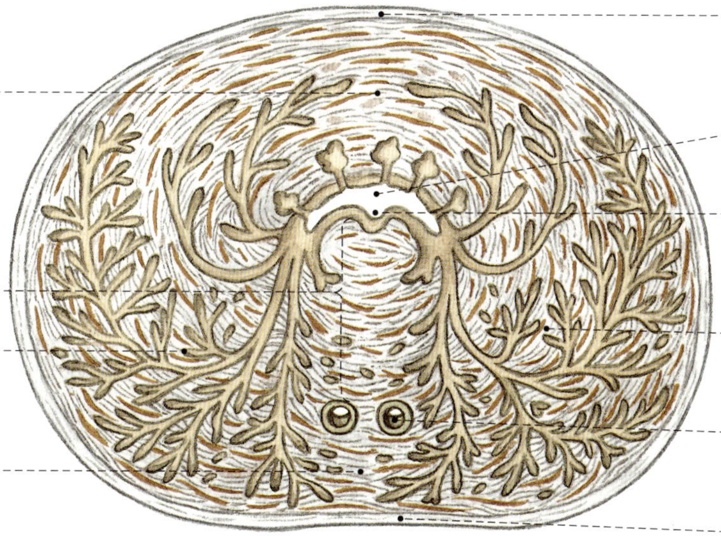

237 Vesica urinaria, Ductus deferens, Vesicula seminalis et Prostata

237 Urinary bladder, deferent duct, seminal vesicle and prostate

a Median section, medial aspect of the right half (120%)

b Transverse section through the prostate (250%), schematic representation, superior aspect

237 Harnblase, Samenleiter, Samenbläschen und Vorsteherdrüse

a Medianschnitt, Medialansicht der rechten Hälfte (120%)

b Querschnitt durch die Vorsteherdrüse (250%), schematische Darstellung, Kranialansicht

a

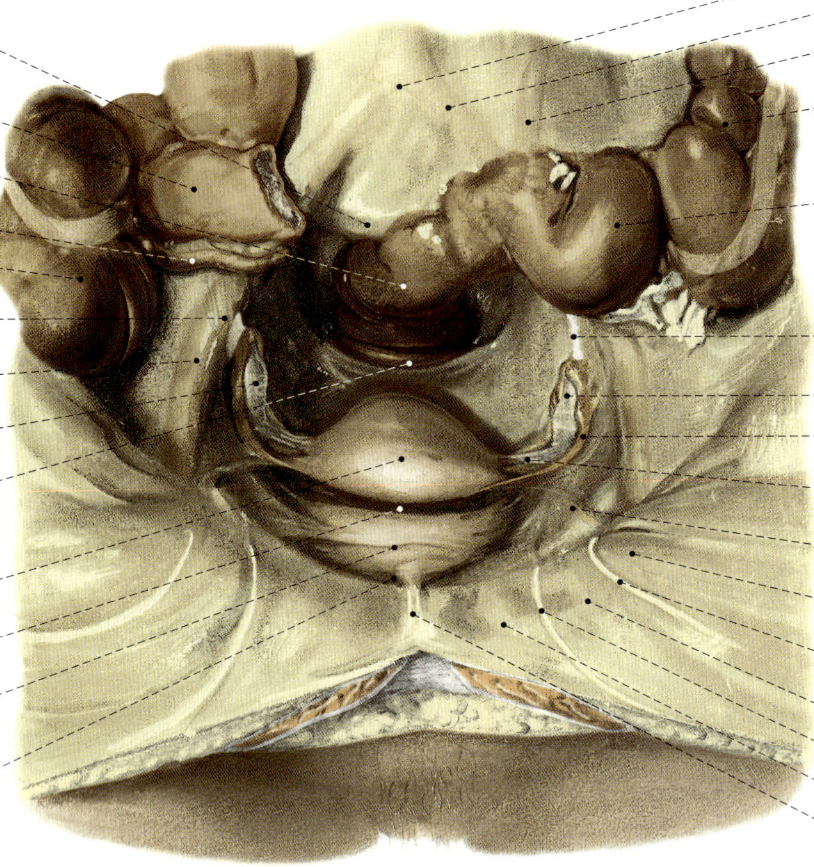

b

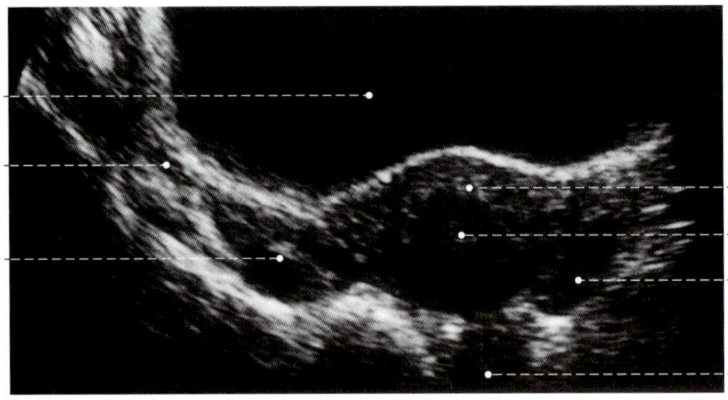

238 ⟨Organa pelvis et Paries anterior abdominis⟩

238 Pelvic organs and anterior abdominal wall

a Pelvic organs of a female in situ. The ventral abdominal wall was opened by a median incision and the small intestine nearly completely removed (45%). Cranioventral aspect

b Ultrasound image of uterus and ovaries (60%), transverse section above the symphysis

238 Beckenorgane und vordere Bauchwand

a Beckenorgane einer Frau in situ. Die vordere Bauchwand ist durch einen Medianschnitt eröffnet und der Dünndarm fast vollständig entfernt (45%). Kranioventralansicht

b Ultraschallbild von Gebärmutter und Eierstöcken (60%), Querschnitt oberhalb der Symphyse

a

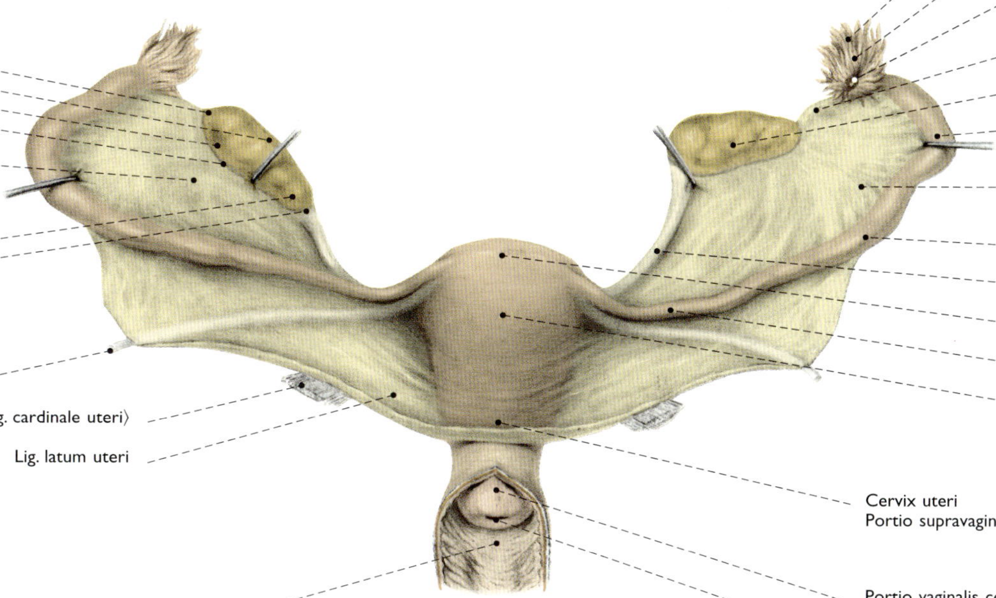

b

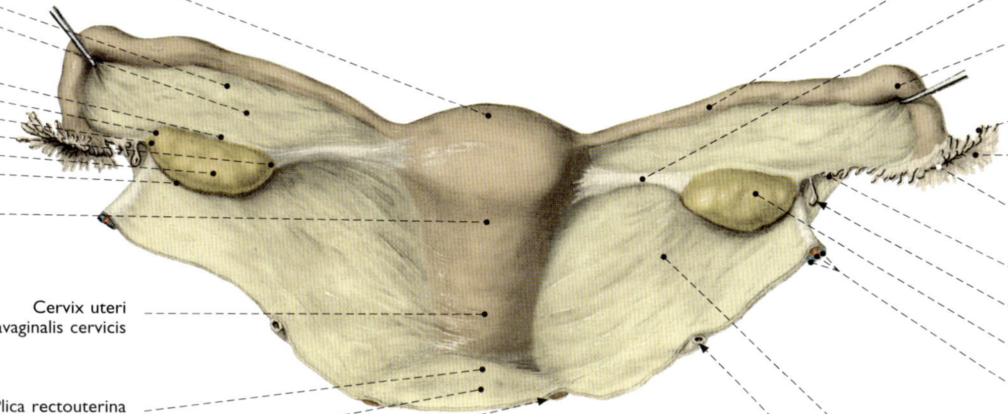

239 Organa genitalia feminina interna

239 Internal genitalia of a young woman (60%)

The ovaries and uterine tubes are drawn apart. The broad ligament was cut off from its parietal attachment. In Fig. a the anterior vaginal wall was opened in front.

a Ventral aspect
b Dorsal aspect

239 Innere Geschlechtsorgane einer jungen Frau (60%)

Eierstöcke und Eileiter sind durch Haken auseinandergezogen, das Ligamentum latum ist am Parietalansatz durchtrennt. In Abb. a ist die vordere Vaginalwand ventromedian eröffnet.

a Ventralansicht
b Dorsalansicht

a

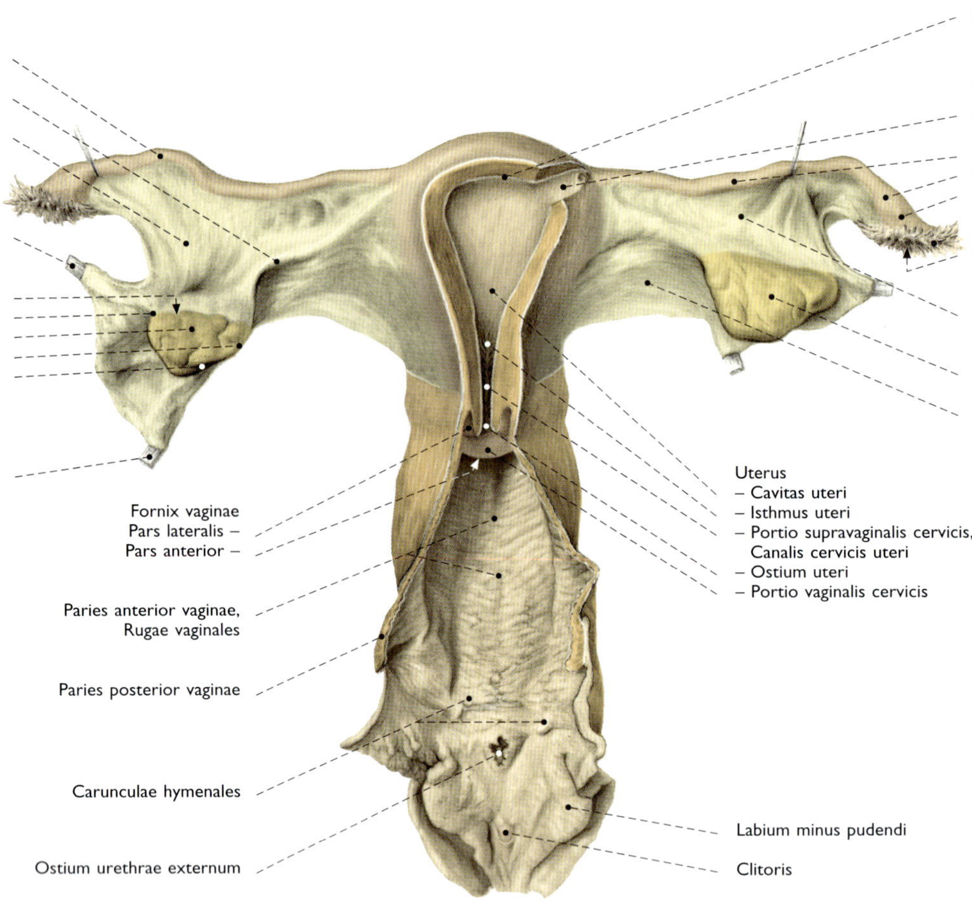

b

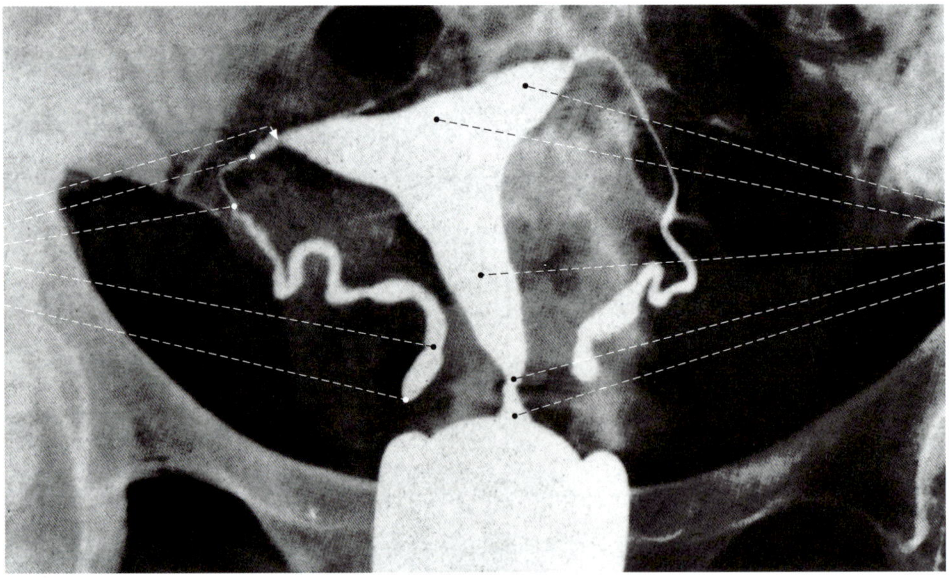

240 Organa genitalia feminina interna

240 Internal female genitalia

a A triangular window was cut in the dorsal uterine wall. The dorsal vaginal wall was opened by a median incision (60%). Dorsal aspect

b Radiograph of the uterus and the uterine tubes (hysterosalpingography). Slight displacement of the internal female genitalia to the right (70%)

240 Innere weibliche Geschlechtsorgane

a Dreiseitiger Fensterschnitt in der hinteren Uteruswand. Die hintere Vaginalwand ist median durchtrennt und aufgeklappt (60%). Dorsalansicht

b Röntgenaufnahme von Uterus und Tuben nach Kontrastmittelfüllung (Hysterosalpingographie). Leichte Rechtsneigung der inneren weiblichen Geschlechtsorgane (70%)

a

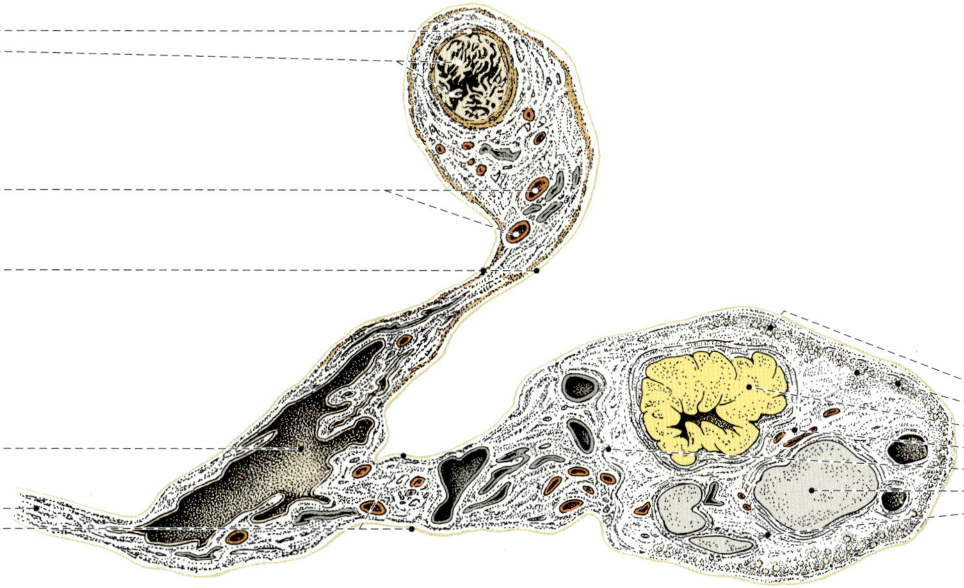

b

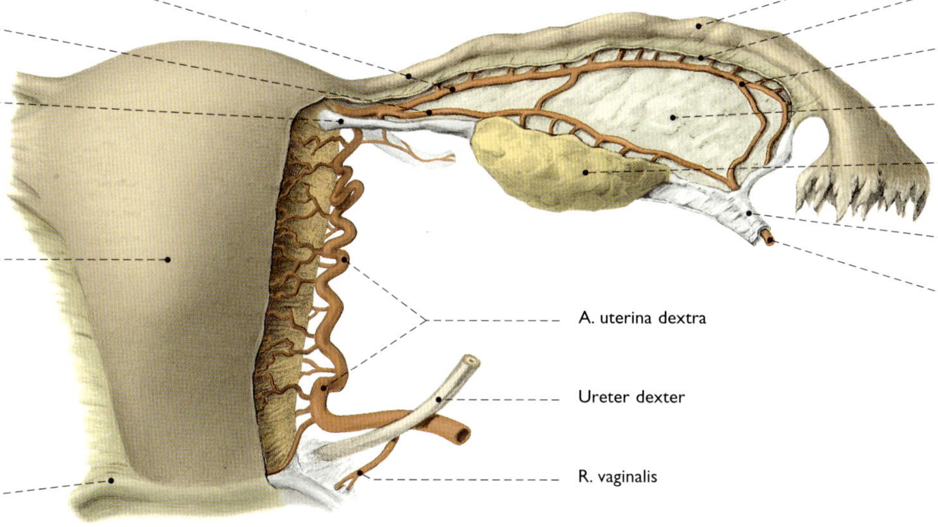

241 Organa genitalia feminina interna

241 Internal female genitalia

a Sagittal section through the ovary and the uterine tube (230%)
b Blood supply of the uterus, the uterine tubes and the ovary (70%), dorsal aspect

241 Innere weibliche Geschlechtsorgane

a Sagittalschnitt durch Ovar und Tube (230%)
b Arterien von Uterus, Tuben und Ovar (70%), Dorsalansicht

a

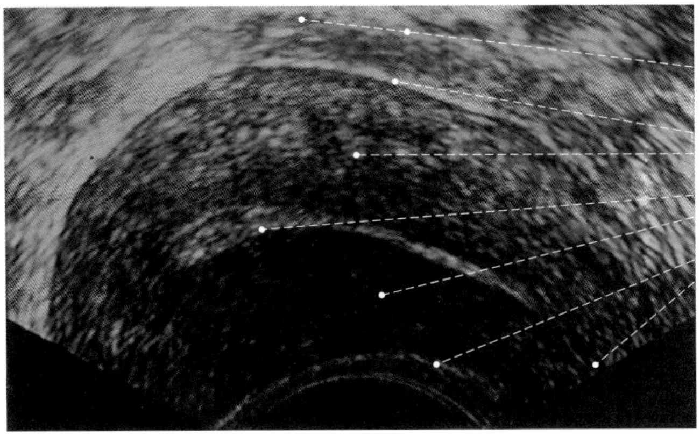

b

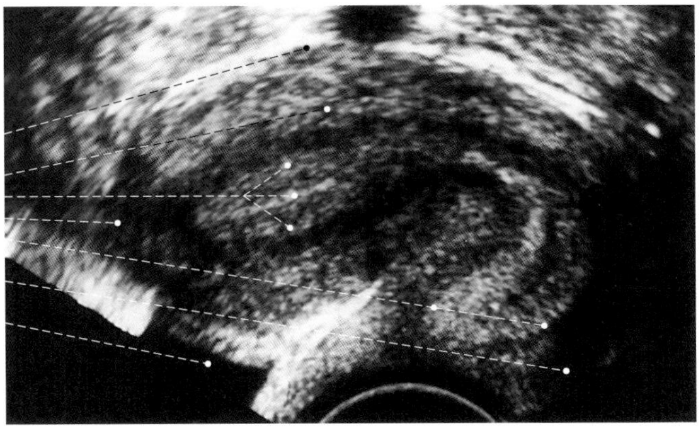

c

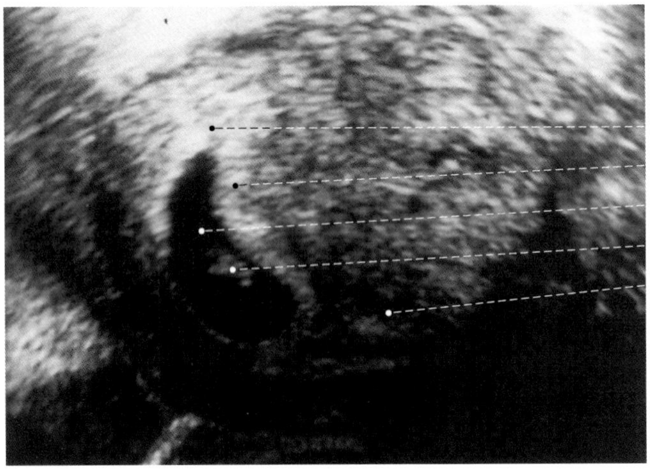

d

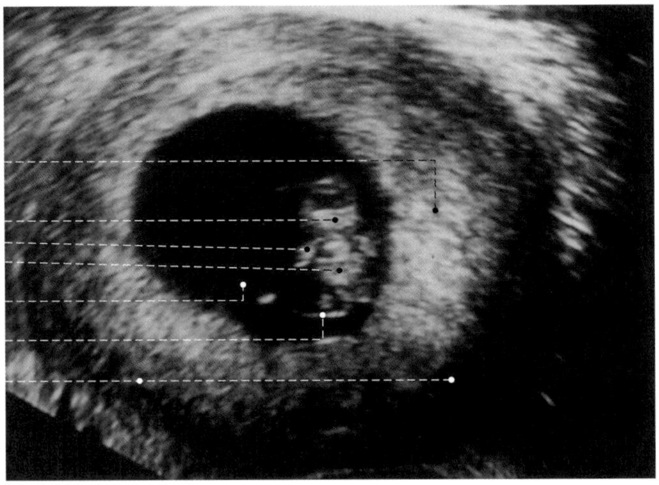

242 Organa genitalia feminina interna

242 Internal female genitalia

Ultrasound images of the uterus

- a Longitudinal section through the corpus uteri during early proliferative stage (100%)
- b Longitudinal section through corpus and cervix uteri during secretory stage (70%)
- c, d Sections through the uterus during early pregnancy (100%)
- c in the seventh week of pregnancy (fifth week after conception)
- d in the ninth week of pregnancy (seventh week after conception)

242 Innere weibliche Geschlechtsorgane

Ultraschallbilder der Gebärmutter

- a Längsschnitt durch den Gebärmutterkörper während der frühen Proliferationsphase (100%)
- b Längsschnitt durch Gebärmutterkörper und Gebärmutterhals während der Sekretionsphase (70%)
- c, d Schnitt durch die Gebärmutter während der Frühschwangerschaft (100%)
- c in der siebten Schwangerschaftswoche (fünfte Woche nach Befruchtung)
- d in der neunten Schwangerschaftswoche (siebte Woche nach Befruchtung)

a

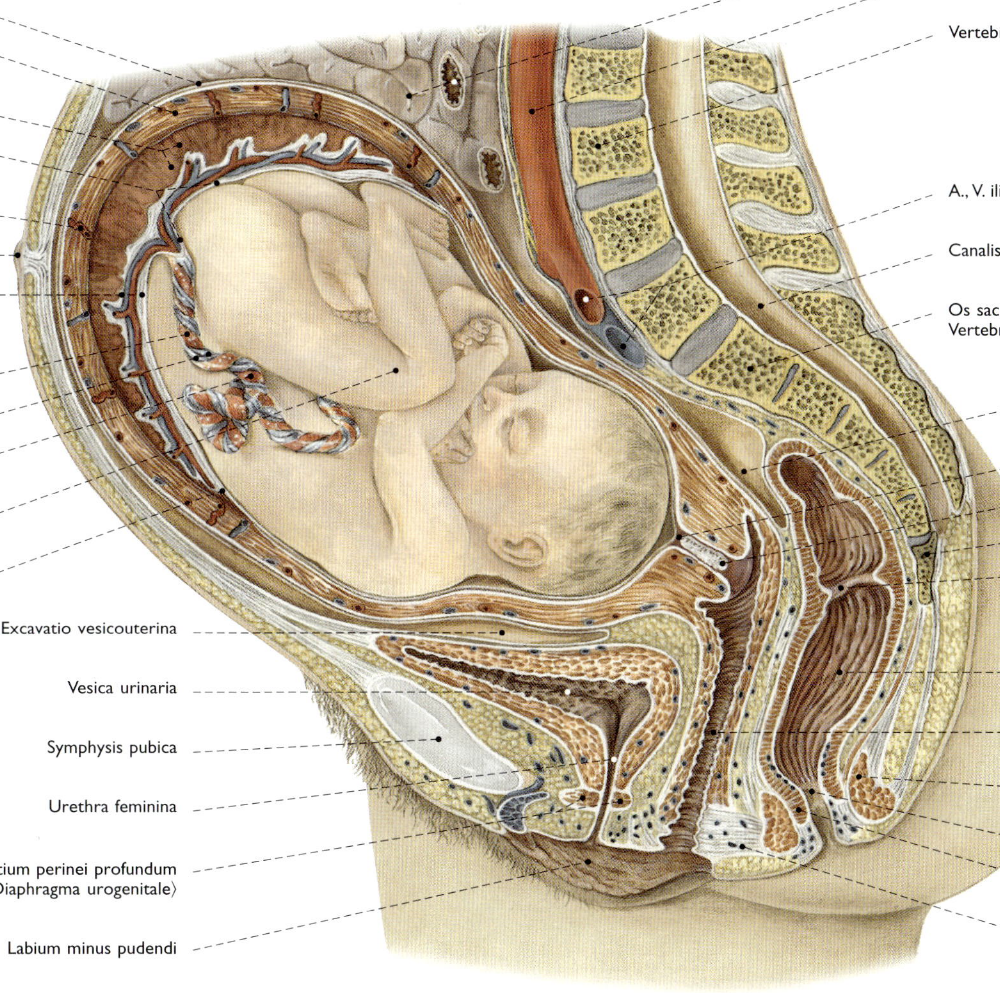

b

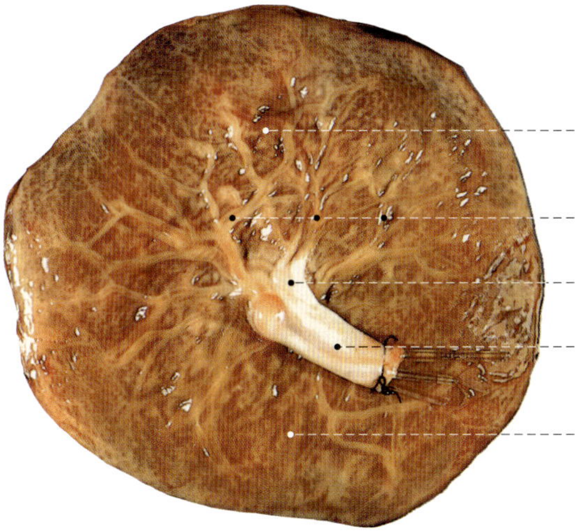

c

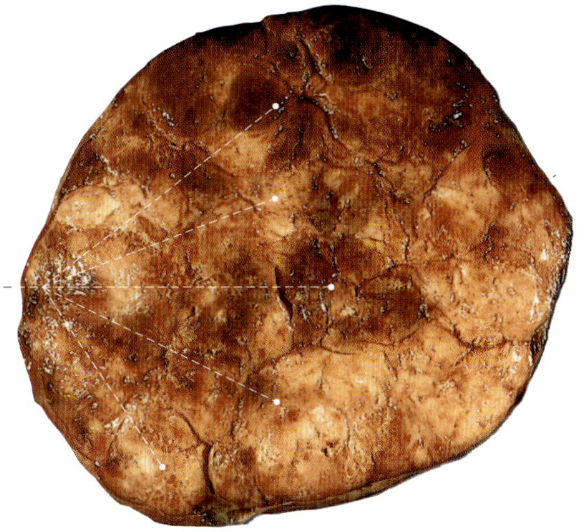

243 Organa genitalia feminina interna

243 Internal female genitalia and placenta (40%)

- a Female pelvis with an uterus advanced in pregnancy, medial aspect of a median section (fetus and umbilical cord were left intact)
- b, c Recently delivered placenta
- b Fetal surface with umbilical cord
- c Maternal surface

243 Innere weibliche Geschlechtsorgane und Plazenta (40%)

- a Weibliches Becken mit hochschwangerer Gebärmutter, Medialansicht eines medianen Sagittalschnitts (Fetus und Nabelschnur sind intakt dargestellt)
- b, c Geborene Plazenta
- b Kindliche Seite mit Nabelschnur
- c Mütterliche Seite

a

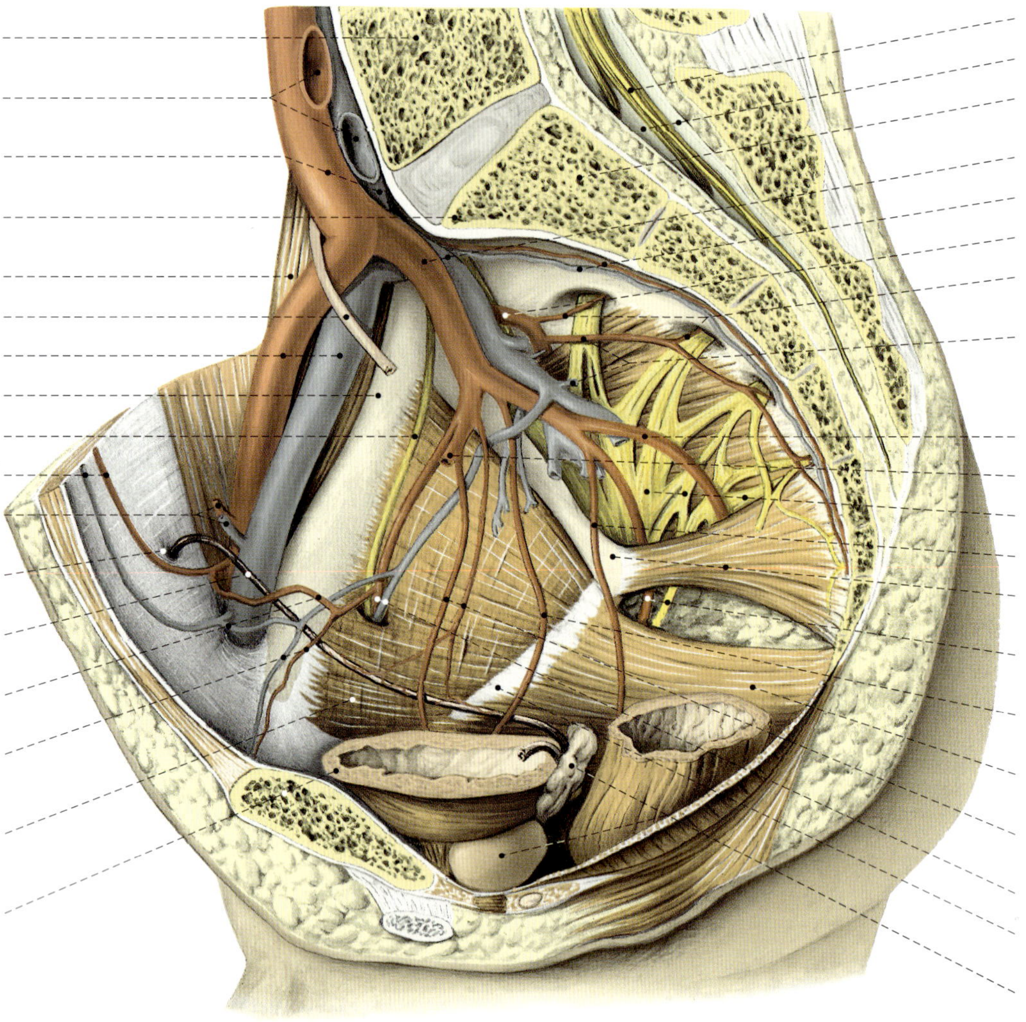

b

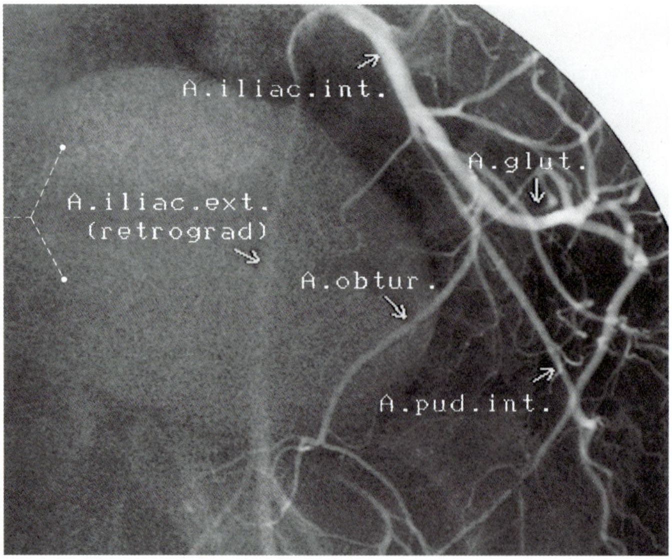

244 ⟨Arteriae, Venae et Organa pelvis⟩

244 Blood vessels and pelvic organs

(70%)

a Sagittal section slightly to the left of the median plane through a male pelvis, medial aspect of the right half

b Arteriogram of the right internal iliac artery and its branches in a male, medial aspect

244 Blutgefässe und Beckenorgane

(70%)

a Sagittalschnitt etwas links der Medianebene durch ein männliches Becken, Medialansicht der rechten Hälfte

b Arteriogramm der rechten inneren Beckenarterie und ihrer Äste bei einem Mann, Medialansicht

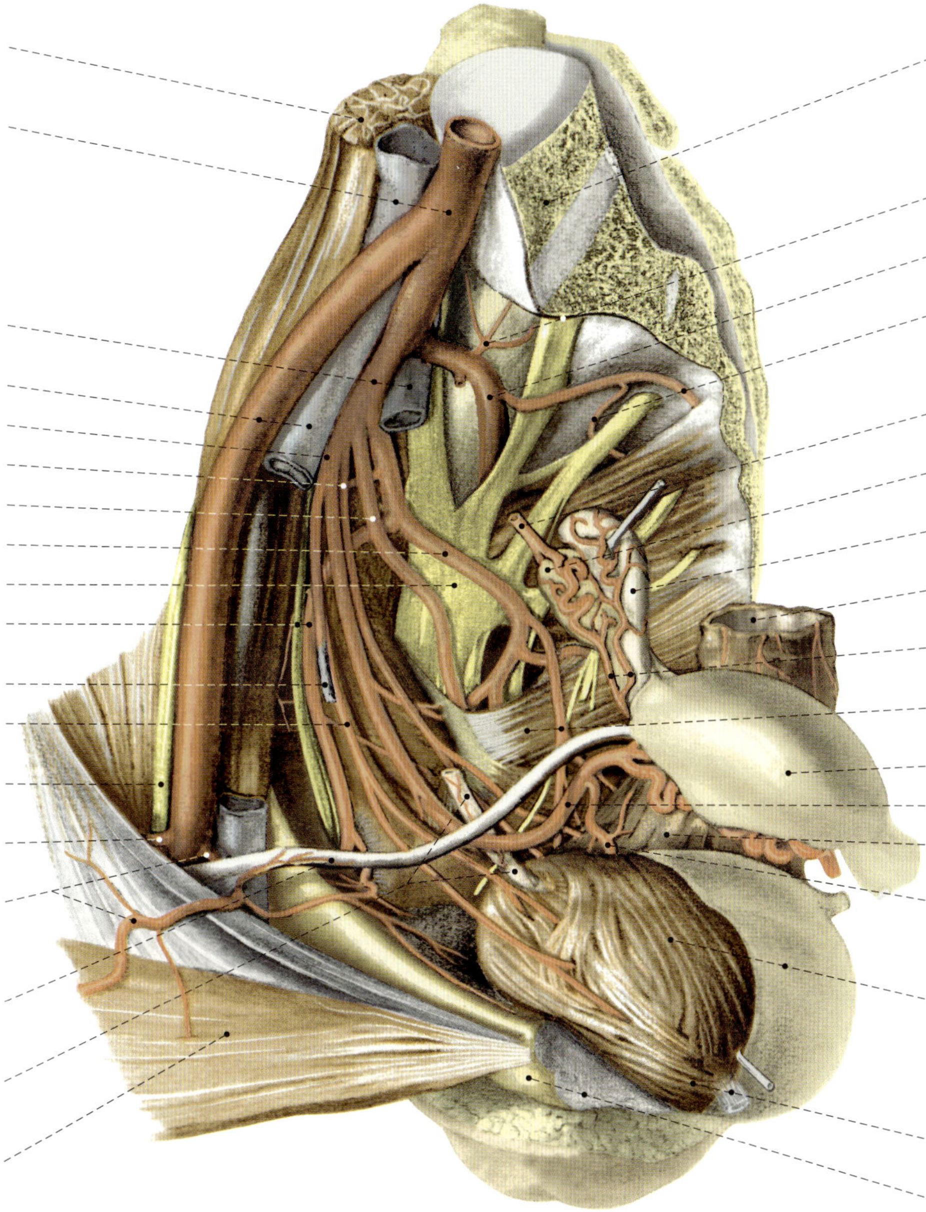

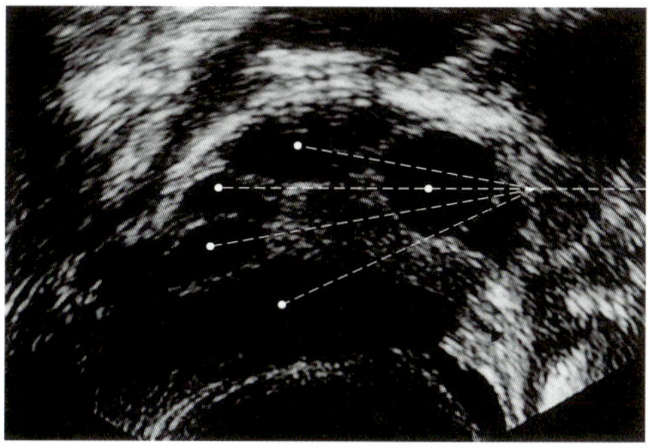

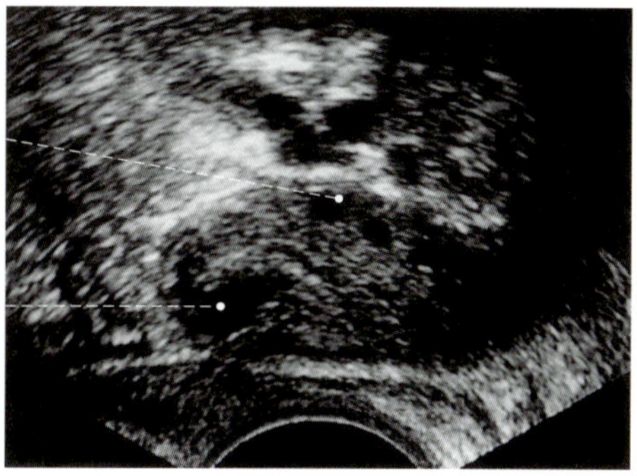

245 ⟨Organa, Arteriae et Nervi pelvis⟩

245 Organs, arteries and nerves of the female pelvis

a Median section. The unpaired viscera were left intact (70%). Oblique aspect from the left, above and anterior

b, c Ultrasound images of the ovaries of a female in the phase of sexual maturity (100%)

245 Organe, Arterien und Nerven des weiblichen Beckens

a Medianschnitt. Die unpaaren Beckenorgane sind nicht durchschnitten (70%). Schrägansicht von links, oben und vorne

b, c Ultraschallbilder der Eierstöcke einer geschlechtsreifen Frau (100%)

a

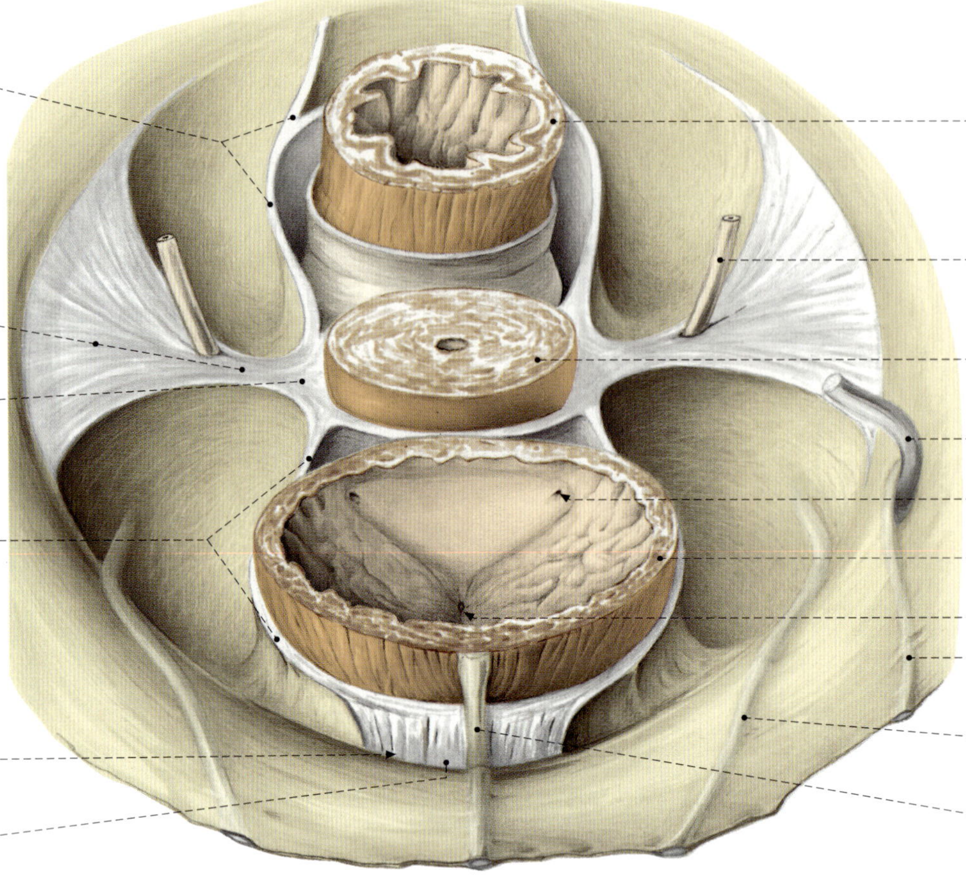

b c

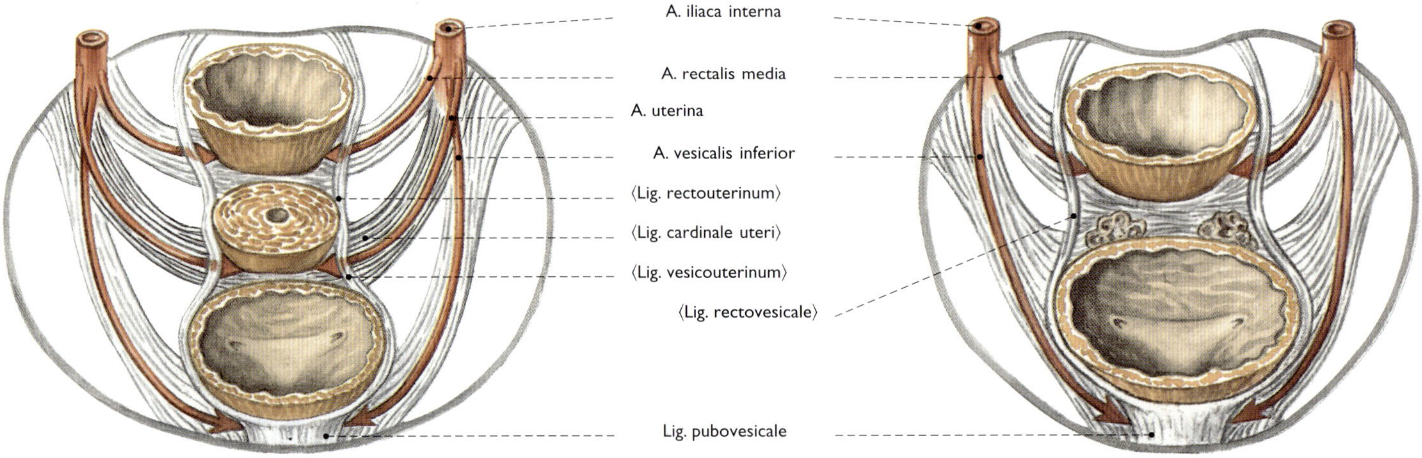

246 Pelvis

246 Organs and connective tissue of the lesser pelvis

Schematic representations, cranioventral aspect

- a Connective tissue in the lesser pelvis of a female (70%)
- b, c Sagittal and frontal connective tissue strands (according to Lierse, 1984) (40%)
- b in the lesser pelvis of a female
- c in the lesser pelvis of a male

The arrows indicate the course of the main supply vessels and nerves.

246 Organe und Bindegewebe des kleinen Beckens

Schematische Darstellungen, Kranioventralansicht

- a Bindegewebe im kleinen Becken einer Frau (70%)
- b, c Sagittale und frontale Bindegewebsgurtung (nach Lierse, 1984) (40%)
- b im weiblichen kleinen Becken
- c im männlichen kleinen Becken

Die Pfeile geben den hauptsächlichen Verlauf der Gefäss-Nerven-Strassen an.

a

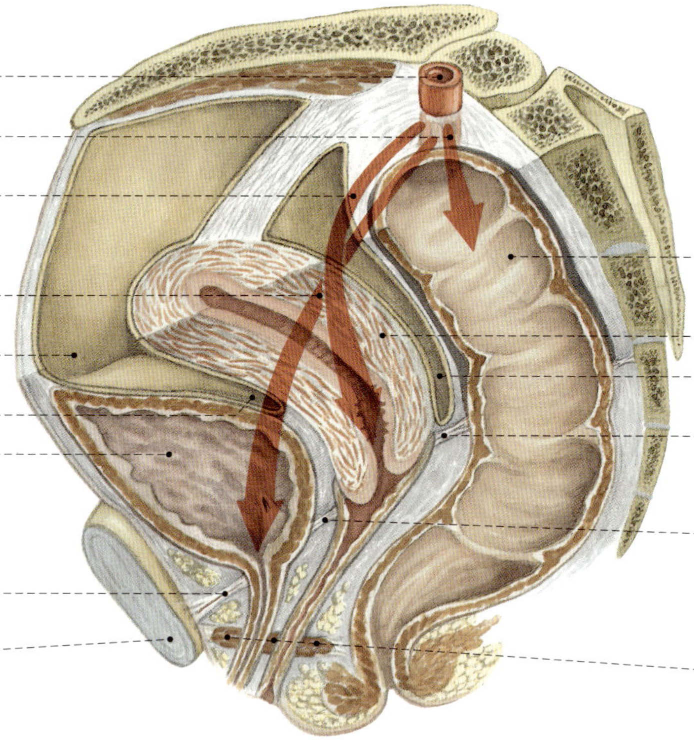

b

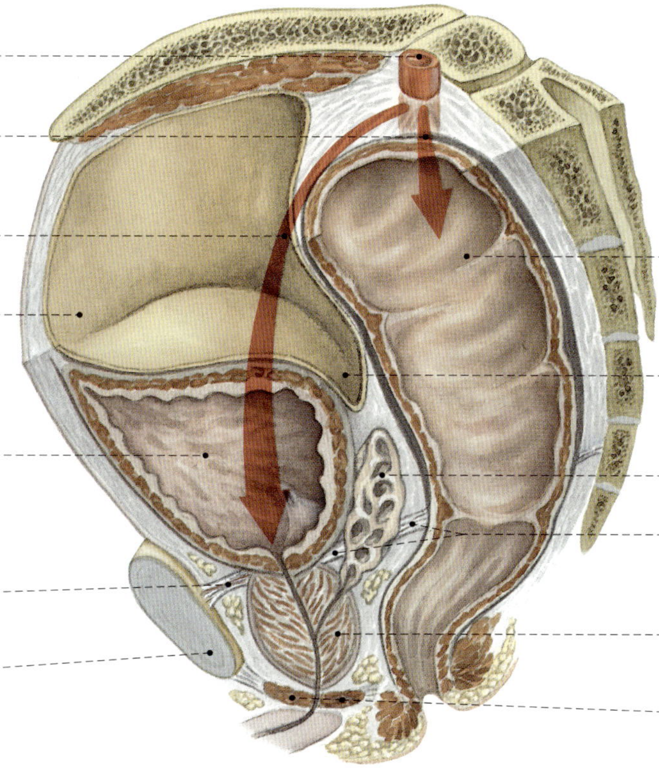

247 Pelvis

247 Organs and connective tissue of the lesser pelvis (50%)

Schematic representations (according to Lierse, 1984), midsagittal sections, medial aspect of the right halfs

a Female lesser pelvis
b Male lesser pelvis

The arrows indicate the course of the main supply vessels and nerves.

247 Organe und Bindegewebe des kleinen Beckens (50%)

Schematische Darstellungen (nach Lierse, 1984), Mediansagittalschnitte, Medialansicht der rechten Hälften

a Weibliches kleines Becken
c Männliches kleines Becken

Die Pfeile geben den Verlauf der hauptsächlichen Gefäss-Nerven-Strassen an.

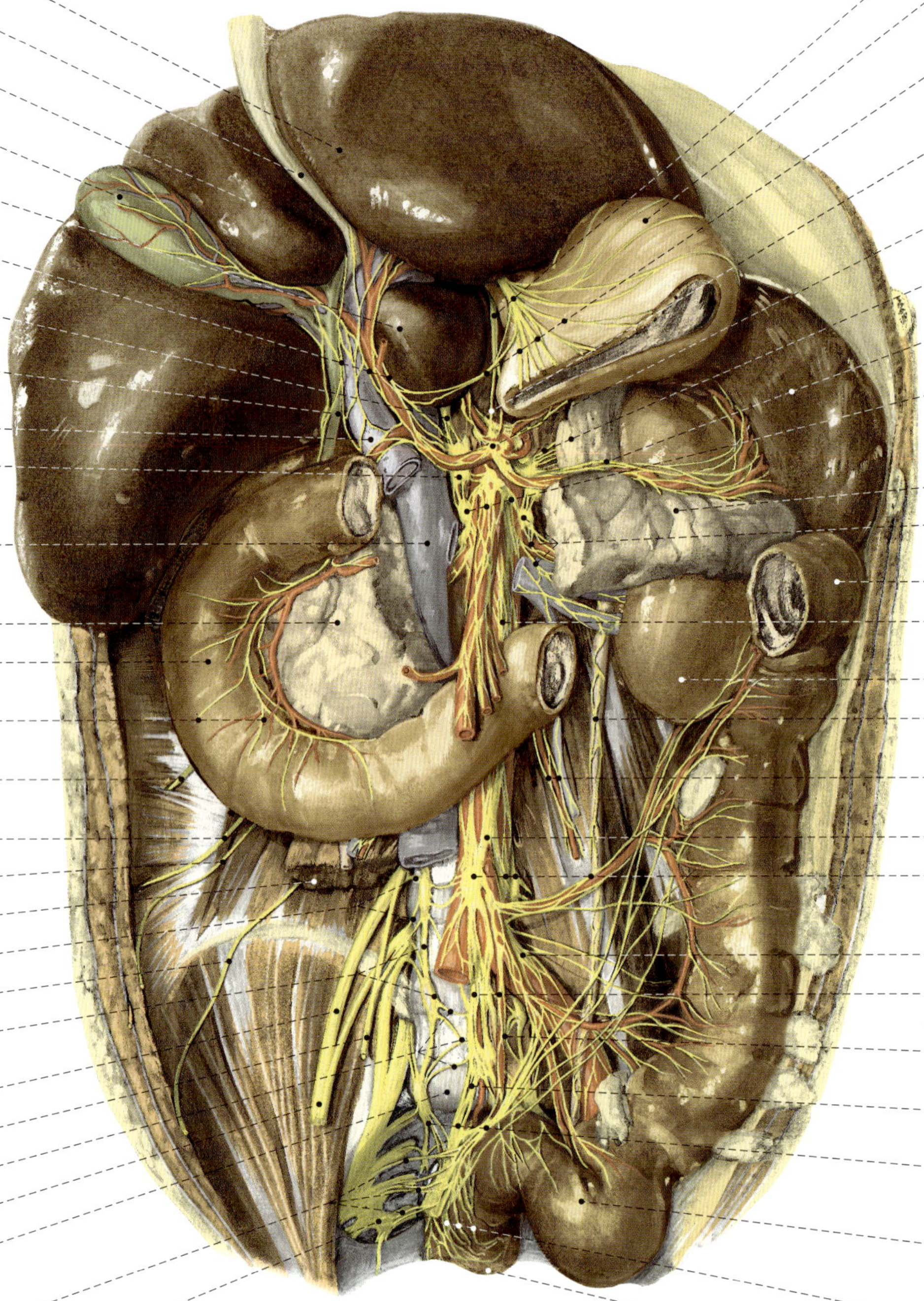

248 Spatium retroperitoneale

248 Autonomic nervous system in the retroperitoneal space (40%)
Ventral aspect

248 Autonomes Nervensystem im Retroperitonealraum (40%)
Ventralansicht

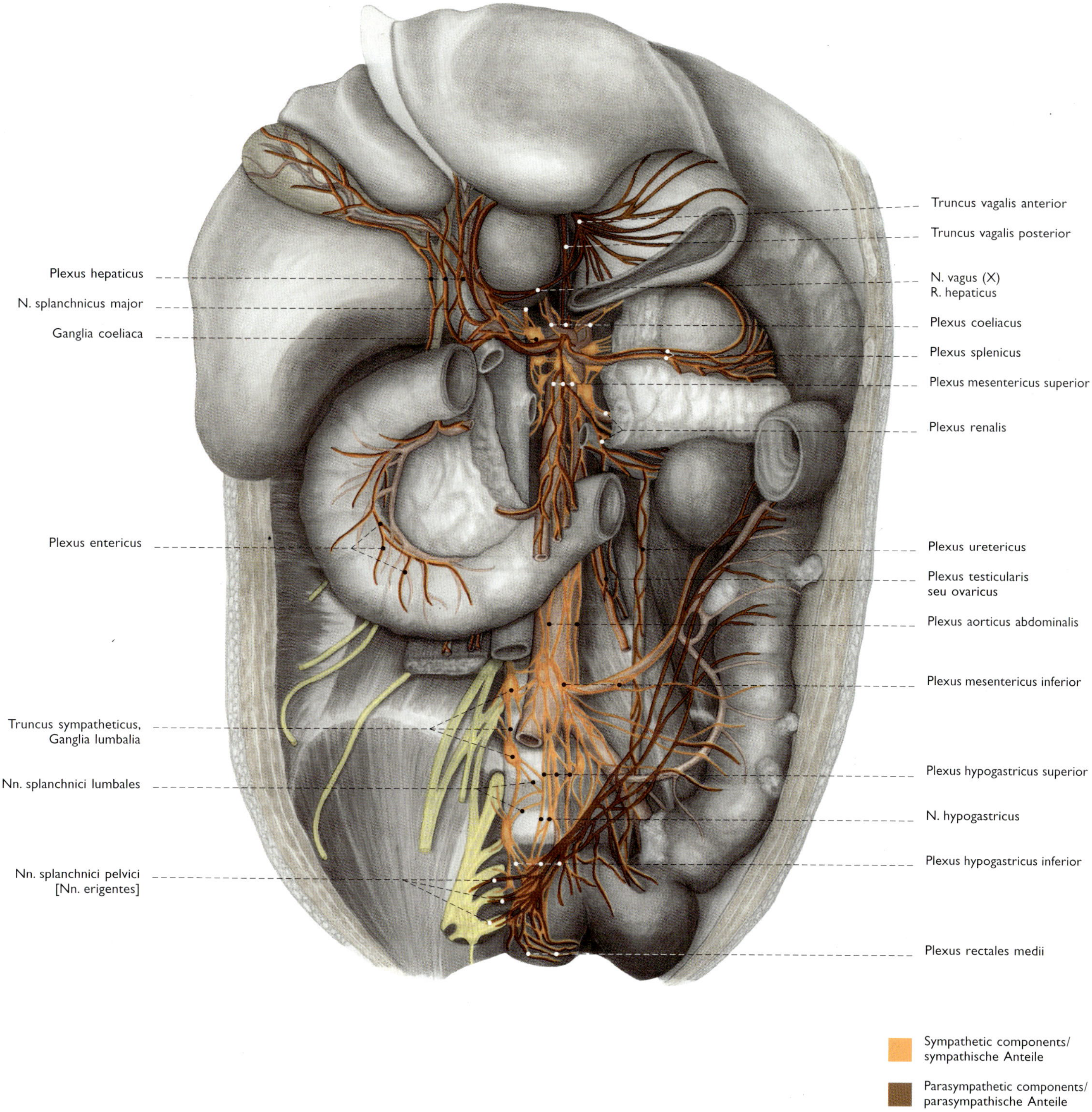

249 Spatium retroperitoneale

249 Autonomic nervous system in the retroperitoneal space (40%)

Explanatory drawing as to Fig. 248. Sympathetic fibres are shown by **orange**, parasympathetic ones by **brown** color. Ventral aspect

249 Autonomes Nervensystem im Retroperitonealraum (40%)

Erklärende Zeichnung zu Abb. 248. Sympathische Fasern sind **orange**, parasympathische **braun** dargestellt. Ventralansicht

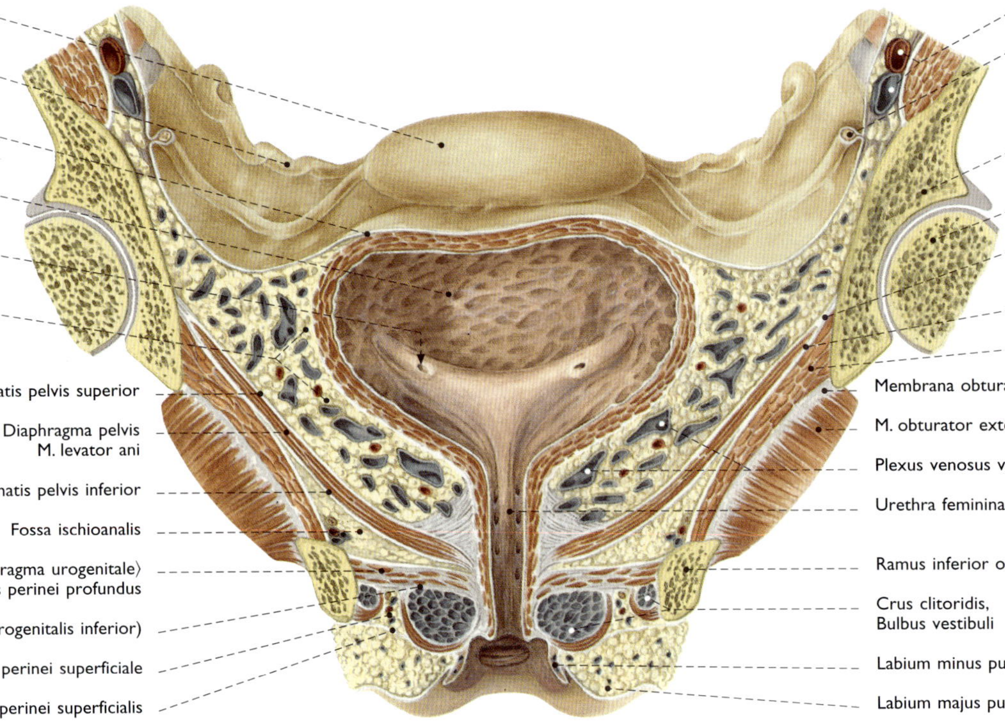

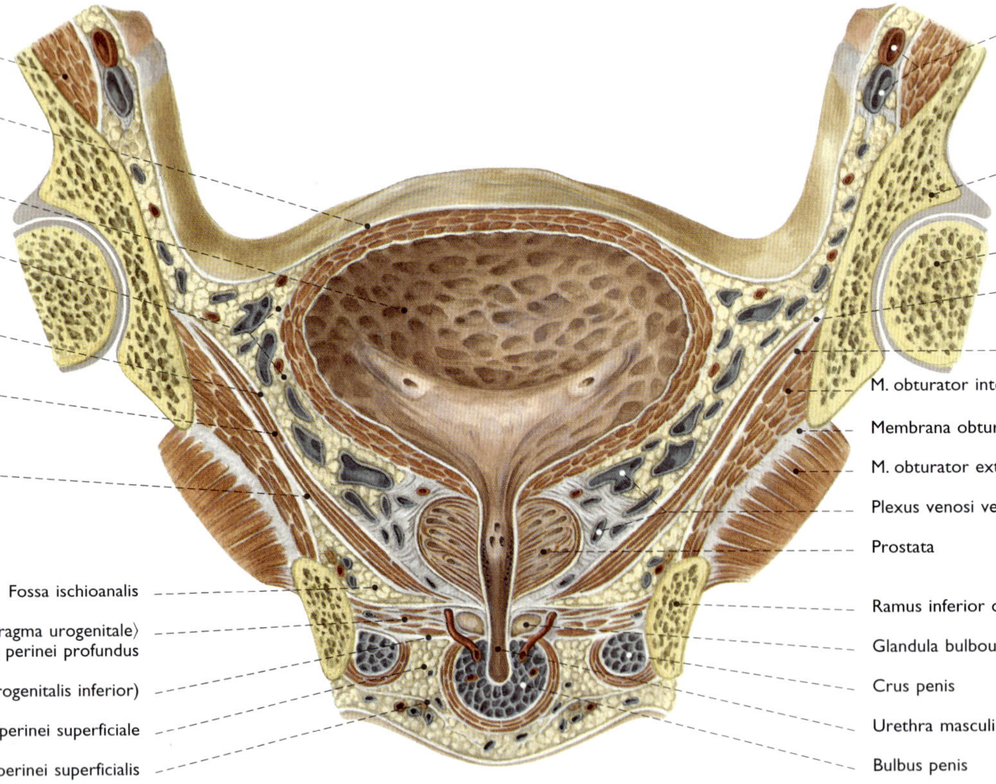

250 Pelvis

250 Female and male pelvis (60%)

Ventral aspect
Coronal sections through the pelvis in the plane of the urethra and urinary bladder
a of a female
b of a male

250 Weibliches und männliches Becken (60%)

Ventralansicht
Frontalschnitte durch das Becken in der Ebene der Harnröhre und der Harnblase
a einer Frau
b eines Mannes

a

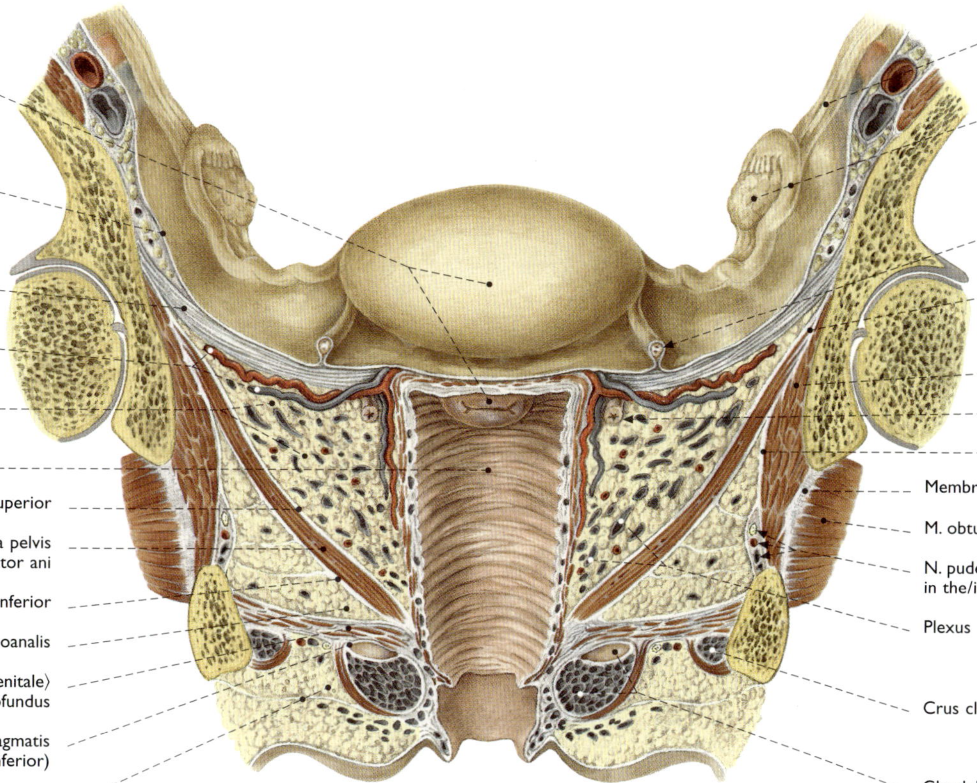

b

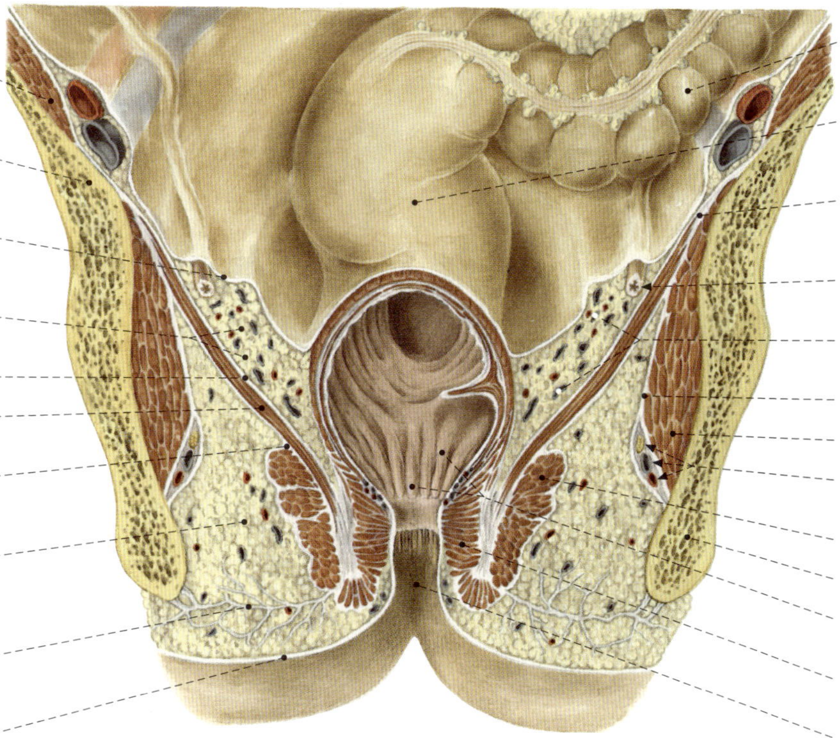

251 Pelvis

251 Female and male pelvis

Ventral aspect
Coronal sections through the pelvis
a in the plane of the vagina of a female (60%)
b in the plane of the rectum and anus (50%)

251 Weibliches und männliches Becken

Ventralansicht
Frontalschnitte durch das Becken
a in der Ebene der Vagina einer Frau (60%)
b in der Ebene von Mastdarm und After (50%)

a

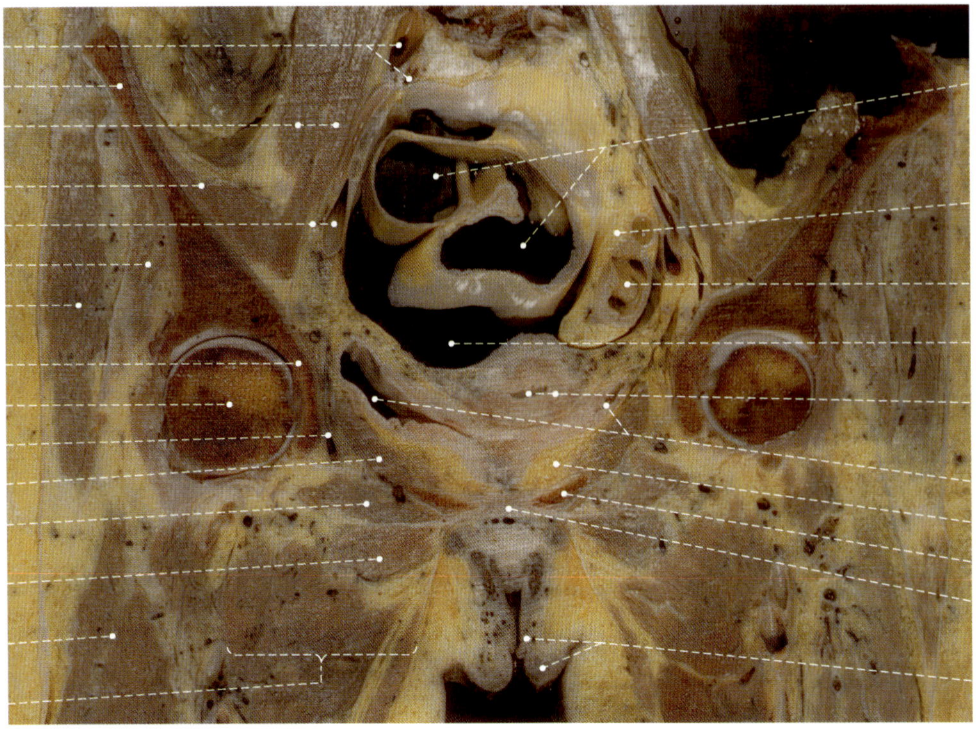

b

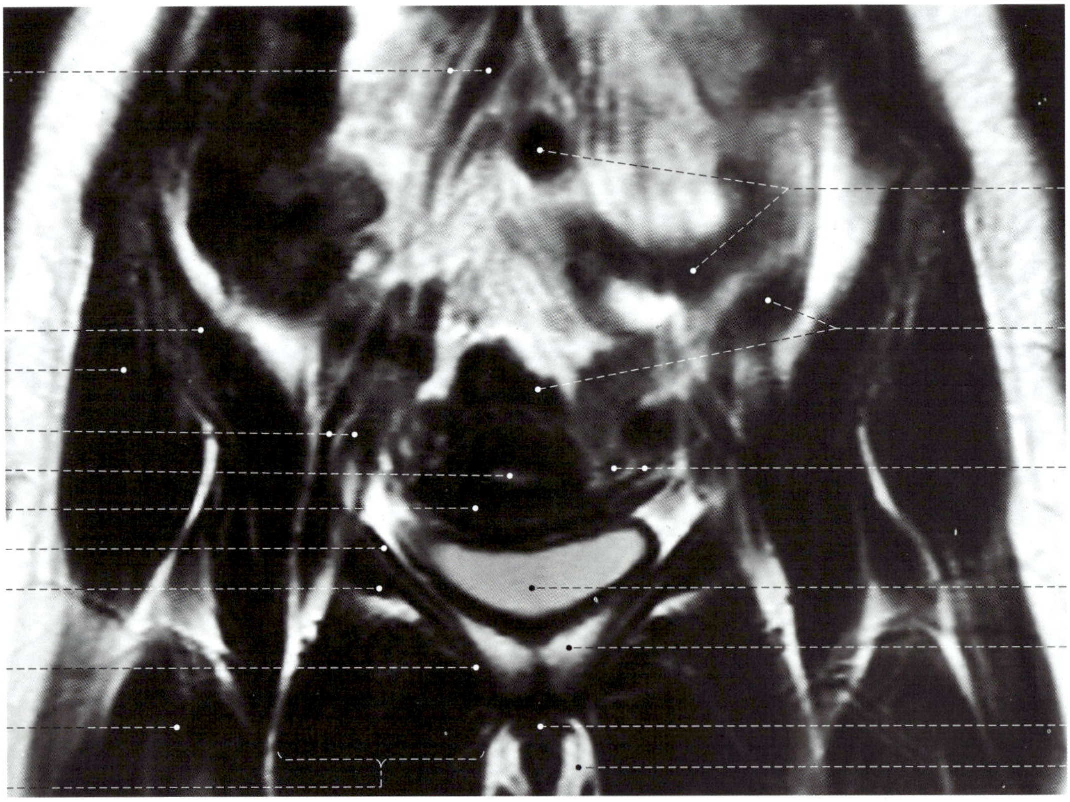

252 Pelvis

252 Pelvis of a female (40%)

Coronal sections through ventral parts of the lesser pelvis of a female just behind the symphysis, ventral aspect

a Anatomical section
b Magnetic resonance image (MRI, T_2-weighted)

252 Weibliches Becken (40%)

Frontalschnitte durch ventrale Abschnitte des kleinen Beckens einer Frau kurz hinter der Symphyse, Ventralansicht

a Anatomischer Schnitt
b Kernspintomogramm (MRI, T_2-gewichtet)

a

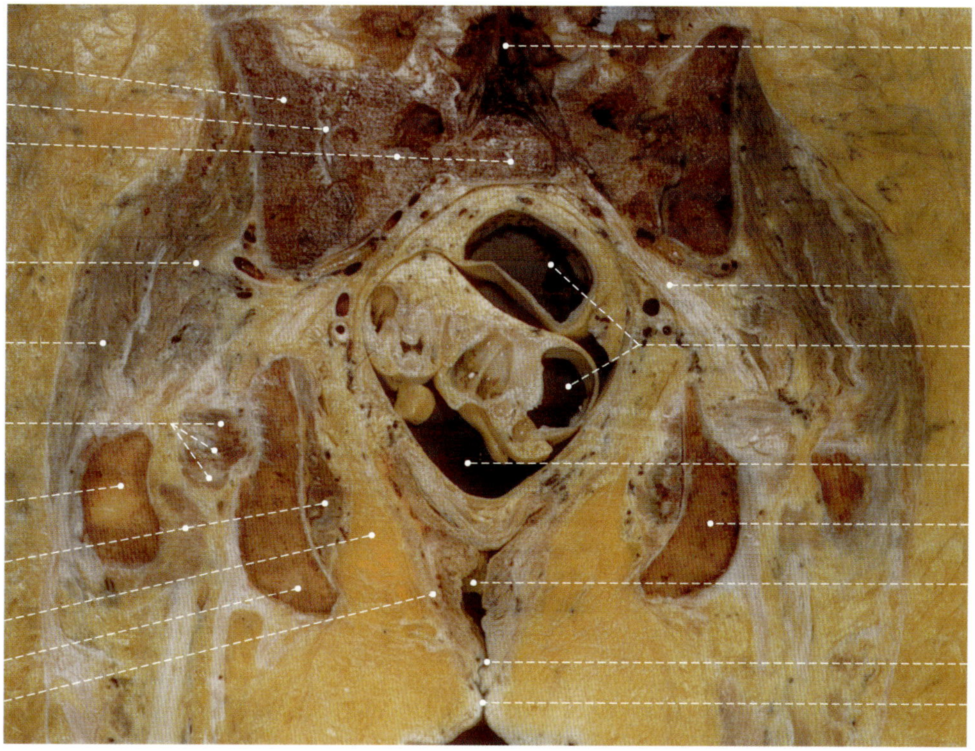

b

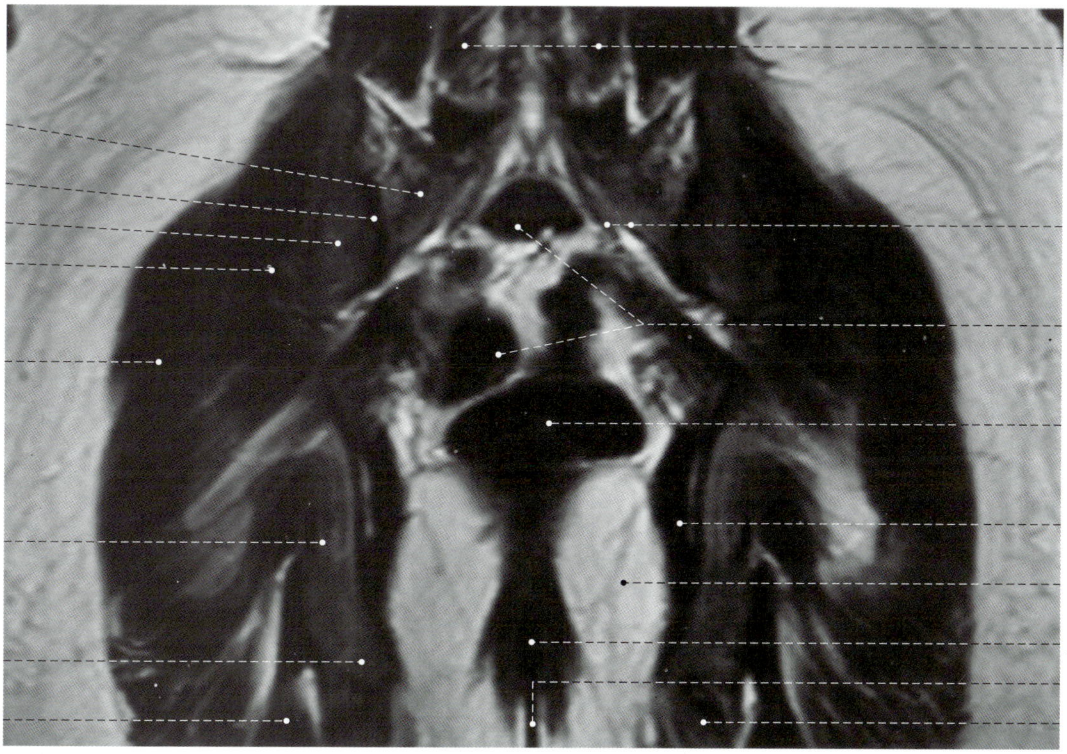

253 Pelvis

253 Pelvis of a female (40%)

Coronal sections through the dorsal parts of the lesser pelvis of a female in the plane of the anus, ventral aspect

a Anatomical section

b Magnetic resonance image (MRI, T_2-weighted)

253 Weibliches Becken (40%)

Frontalschnitte durch dorsale Abschnitte des kleinen Beckens einer Frau in der Ebene des Afters, Ventralansicht

a Anatomischer Schnitt

b Kernspintomogramm (MRI, T_2-gewichtet)

a

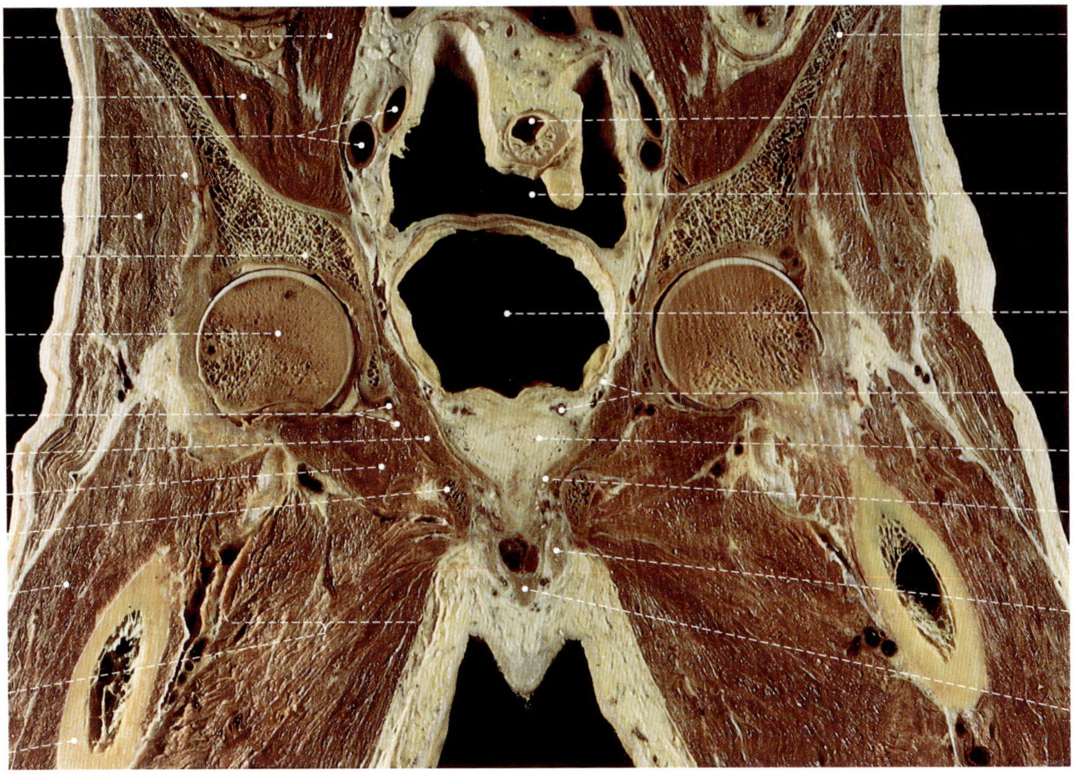

b

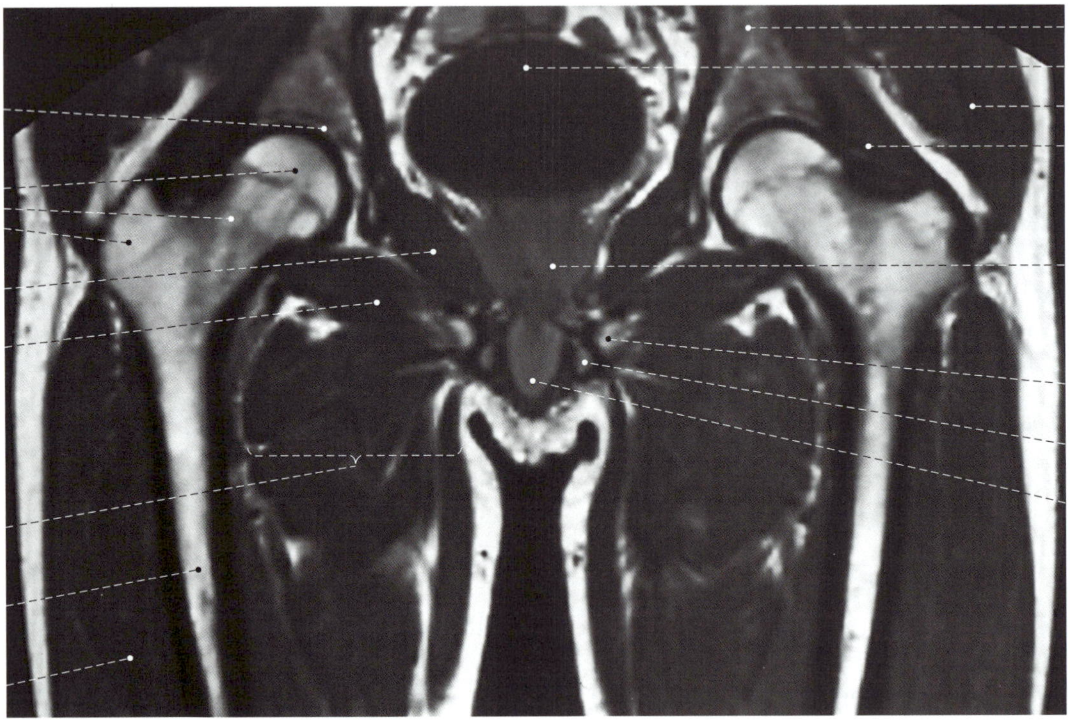

254 Pelvis

254 Pelvis of a male (40%)

Coronal sections through ventral parts of the lesser pelvis of a male just behind the symphysis, ventral aspect

a Anatomical section
b Magnetic resonance image (MRI, T_1-weighted)

254 Männliches Becken (40%)

Frontalschnitte durch ventrale Abschnitte des kleinen Beckens eines Mannes kurz hinter der Symphyse, Ventralansicht

a Anatomischer Schnitt
b Kernspintomogramm (MRI, T_1-gewichtet)

a

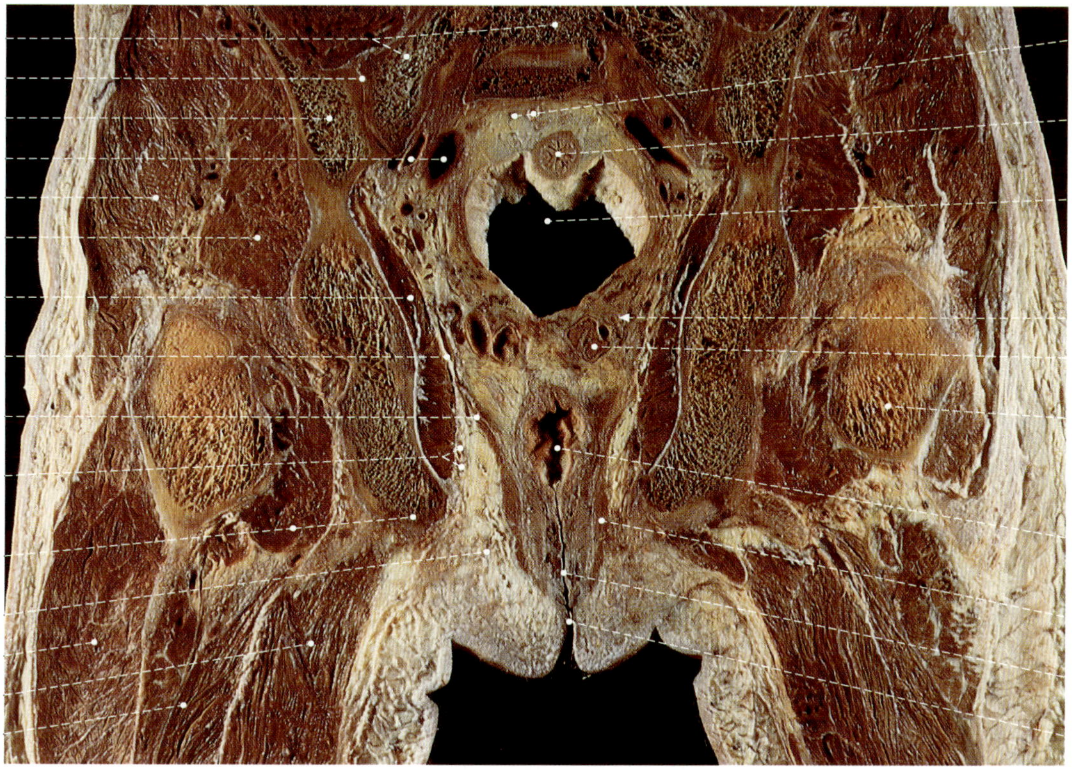

b

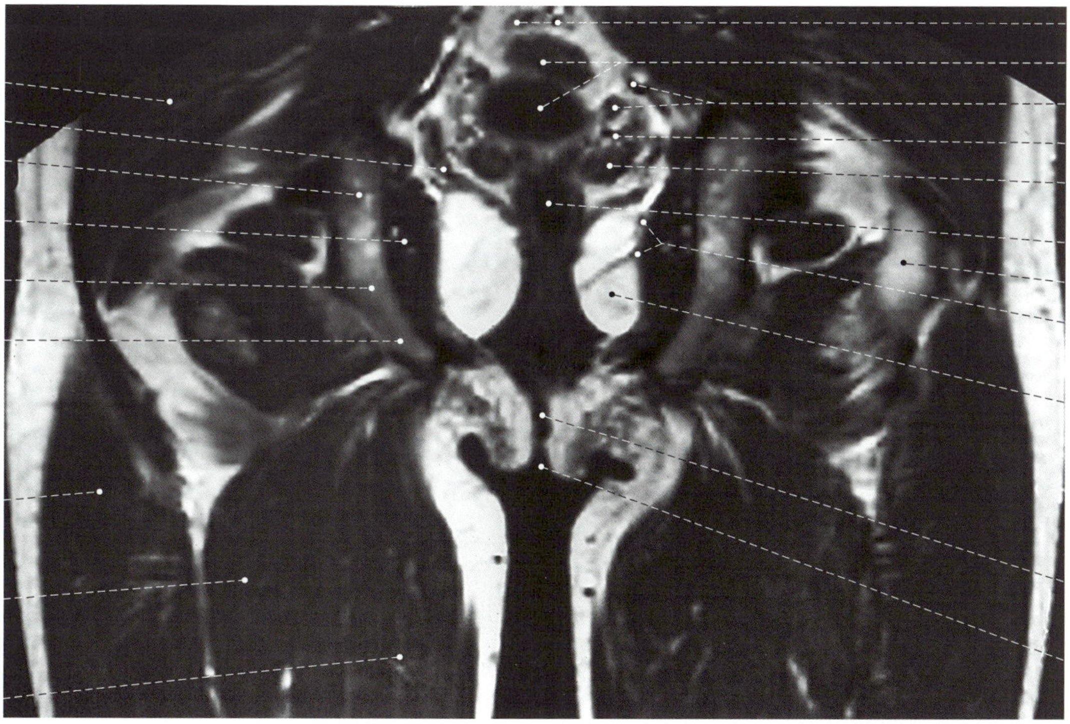

255 Pelvis

255 Pelvis of a male (40%)

Coronal sections through dorsal parts of the lesser pelvis of a male in the plane of the anus, ventral aspect

a Anatomical section

b Magnetic resonance image (MRI, T_1-weighted)

255 Männliches Becken (40%)

Frontalschnitte durch dorsale Abschnitte des kleinen Beckens eines Mannes in der Ebene des Afters, Ventralansicht

a Anatomischer Schnitt

b Kernspintomogramm (MRI, T_1-gewichtet)

a

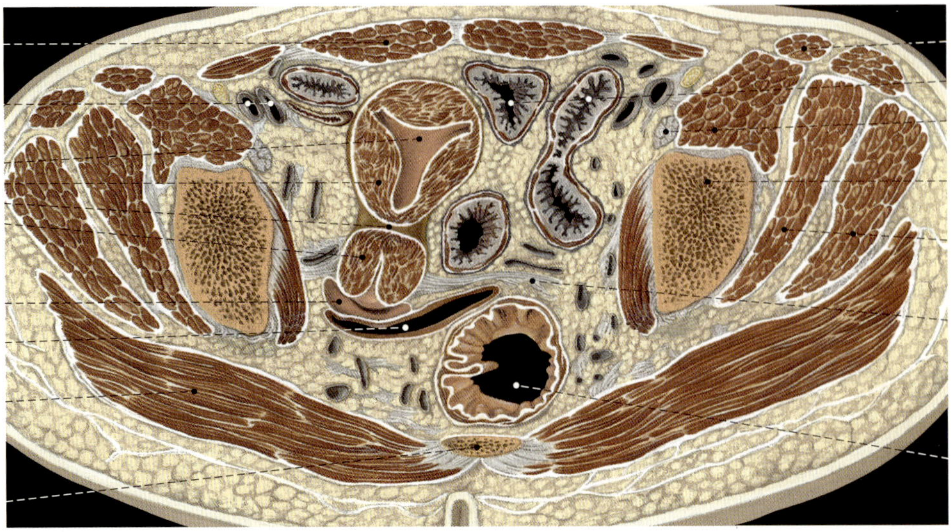

b

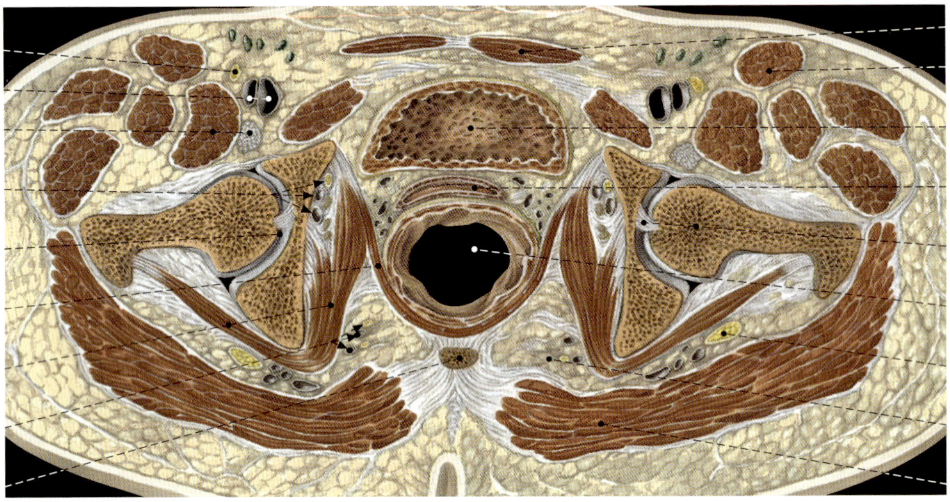

c

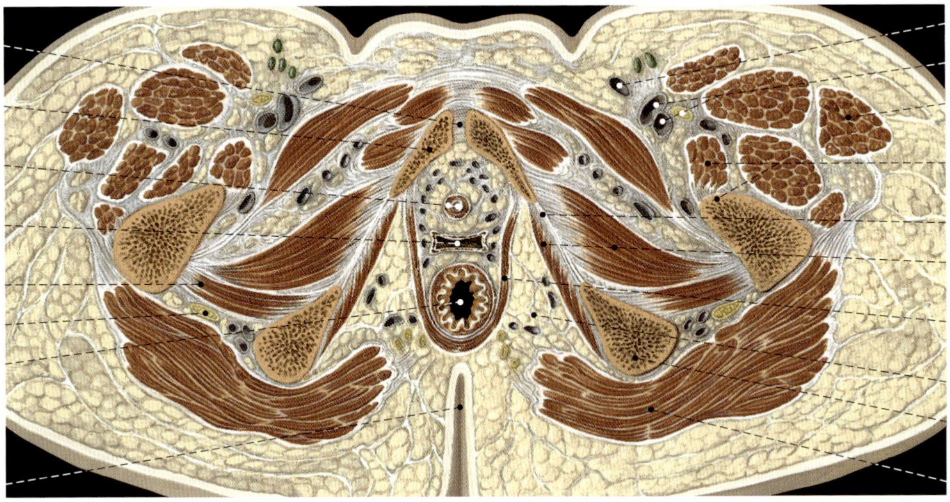

256 Pelvis

256 Pelvis of a female (40%)

Inferior aspect
Transverse anatomical sections (painted) through
a cranial
b middle
c caudal
horizontal planes of the lesser pelvis of a female. Slight displacement of the uterus to the right in Fig. a

256 Weibliches Becken (40%)

Kaudalansicht
Anatomische Querschnitte (gezeichnet) durch
a kraniale
b mittlere
c kaudale
Horizontalebenen des kleinen Beckens einer Frau. Leichte Rechtsverlagerung der Gebärmutter in Abb. a

a

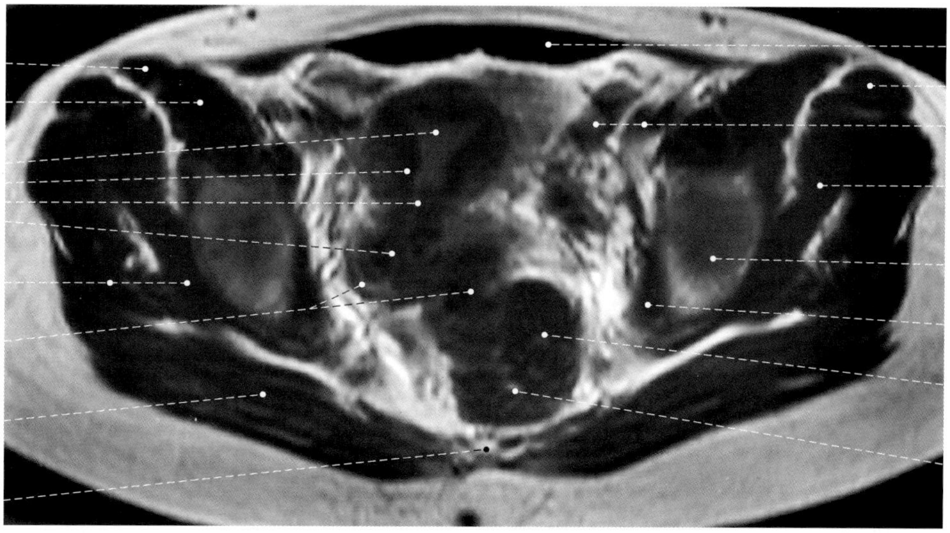

b

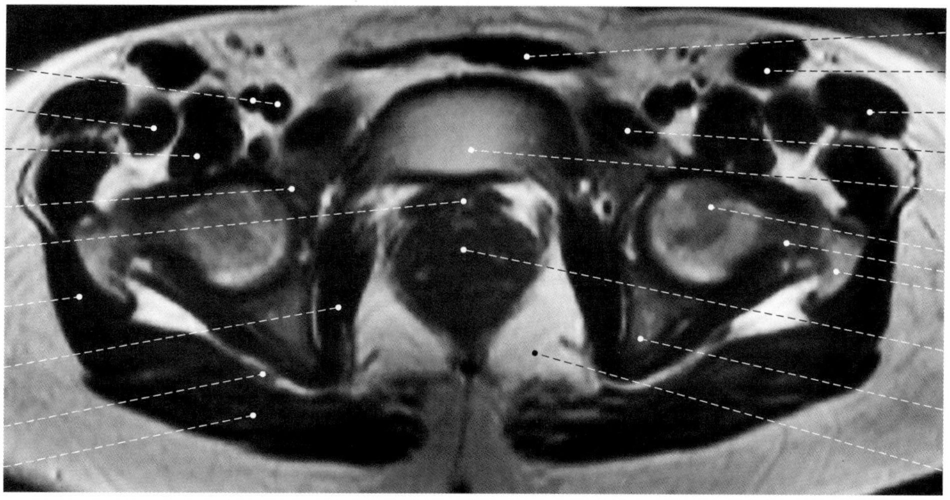

c

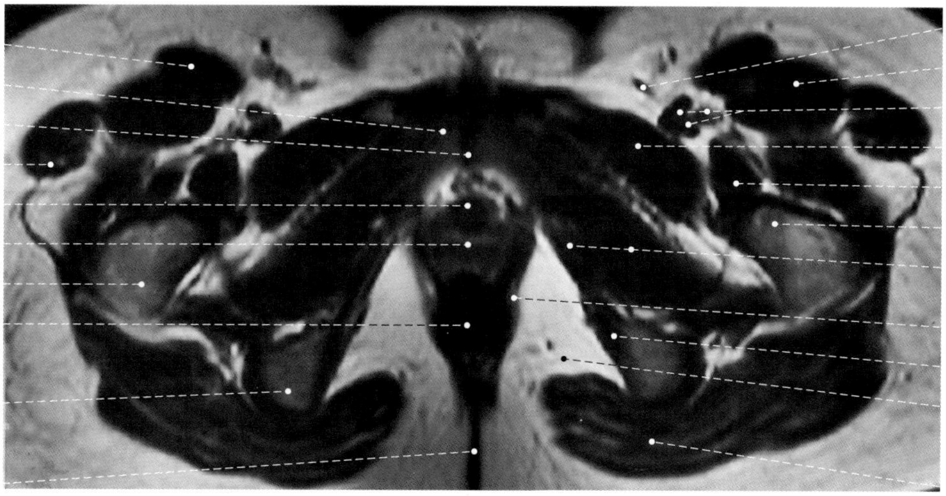

257 Pelvis

257 Pelvis of a female (40%)

Inferior aspect
Transverse magnetic resonance images (MRI, T_2-weighted) through
a cranial
b middle
c caudal
horizontal planes of the lesser pelvis of a female. Slight displacement of the uterus to the right in Fig. a

257 Weibliches Becken (40%)

Kaudalansicht
Horizontale Kernspintomogramme (MRI, T_2-gewichtet) durch
a kraniale
b mittlere
c kaudale
Horizontalebenen des kleinen Beckens einer Frau. Leichte Rechtsverlagerung der Gebärmutter in Abb. a

a

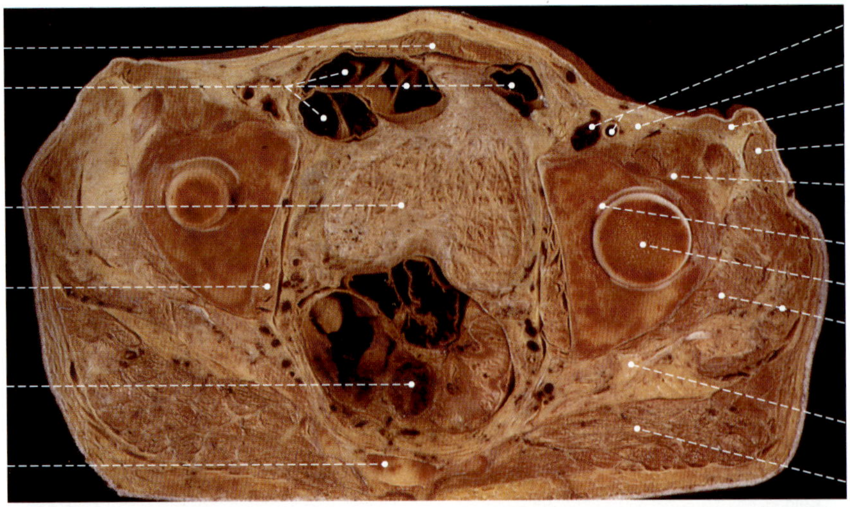

b

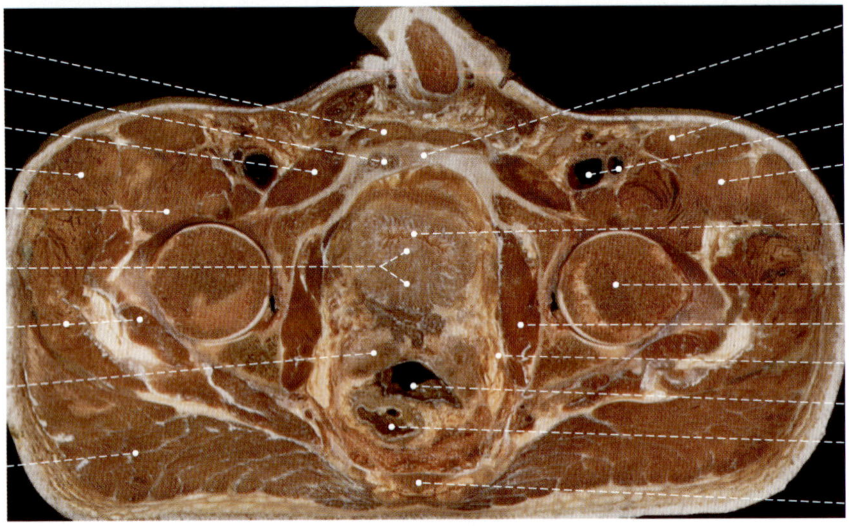

c

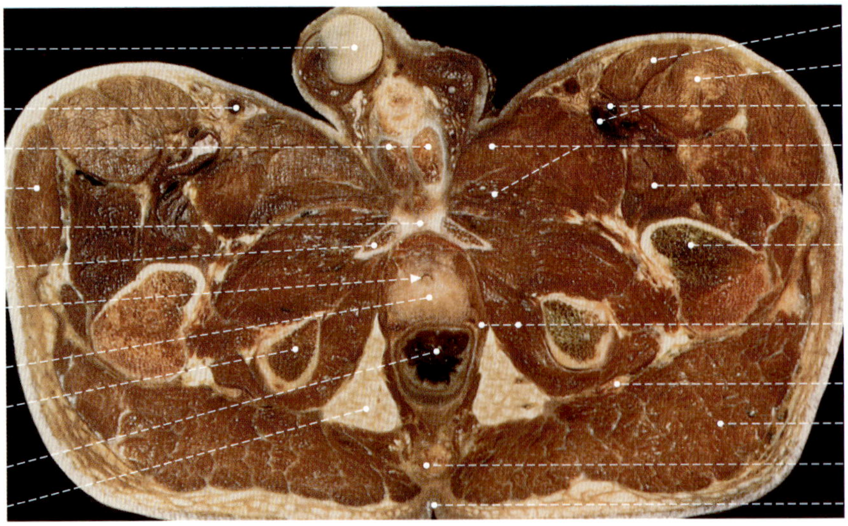

258 Pelvis

258 Pelvis of a male (40%)

Transverse anatomical sections through
a cranial
b middle
c caudal
horizontal planes of the lesser pelvis of a male
a, c Inferior aspect
b Superior aspect

258 Männliches Becken (40%)

Anatomische Querschnitte durch
a kraniale
b mittlere
c kaudale
Horizontalebenen des kleinen Beckens eines Mannes
a, c Kaudalansicht
b Kranialansicht

a

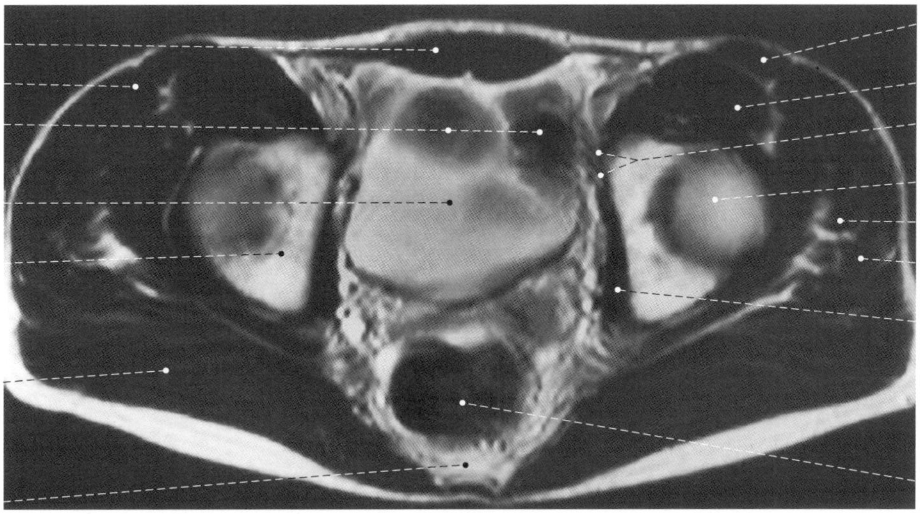

b

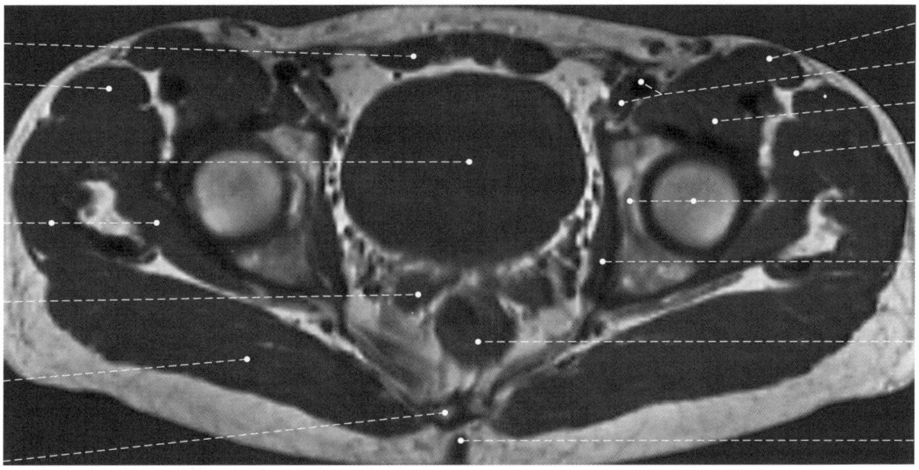

c

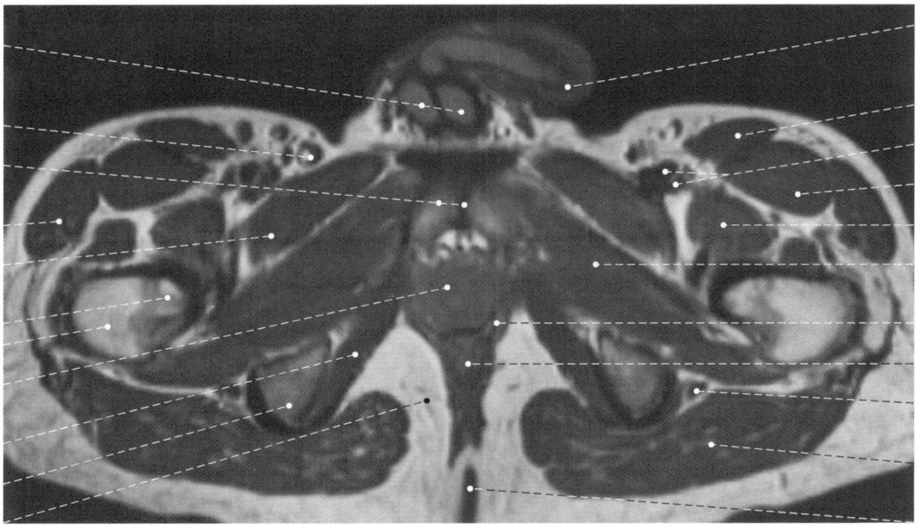

259 Pelvis

259 Pelvis of a male (40%)

Inferior aspect
Transverse magnetic resonance images
(MRI, T_2- or T_1-weighted, respectively) through
a cranial
b middle
c caudal
horizontal planes of the lesser pelvis of a male

259 Männliches Becken (40%)

Kaudalansicht
Horizontale Kernspintomogramme
(MRI, T_2- bzw. T_1-gewichtet) durch
a kraniale
b mittlere
c kaudale
Horizontalebenen des kleinen Beckens eines Mannes

a

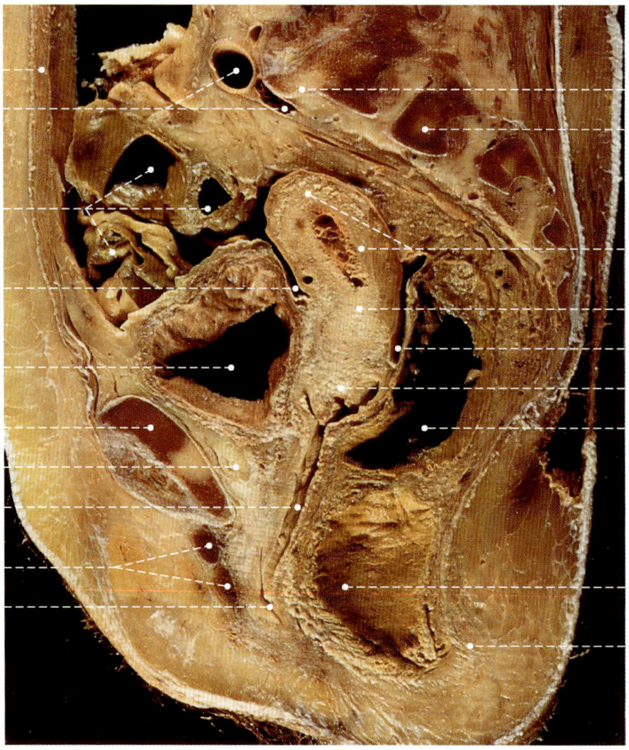

b

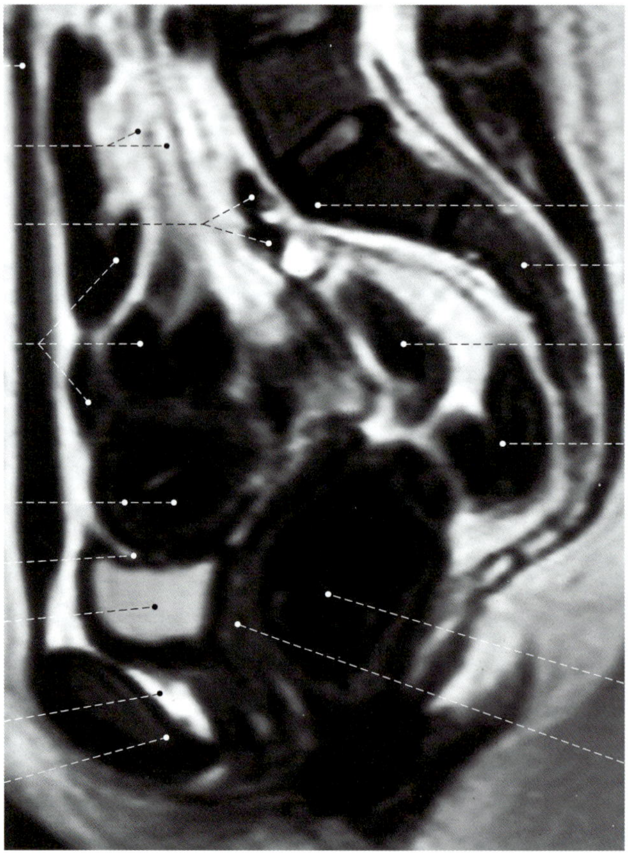

260 Pelvis

260 Pelvis of a female (45%)

Midsagittal sections

a Anatomical section, medial aspect of the right half of the body

b Magnetic resonance image (MRI, T_2-weighted)

260 Weibliches Becken (45%)

Mediansagittale Schnitte

a Anatomischer Schnitt, Medialansicht der rechten Körperhälfte

b Kernspintomogramm (MRI, T_2-gewichtet)

a

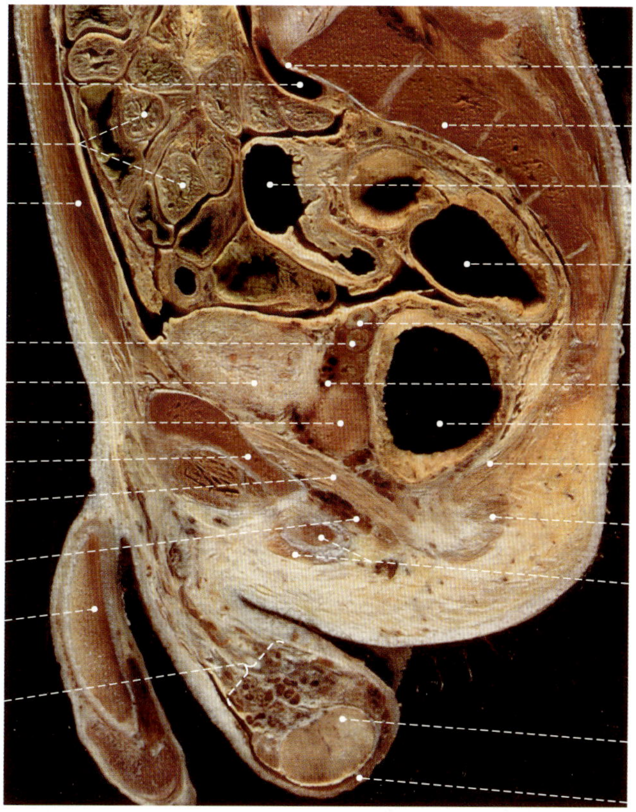

b

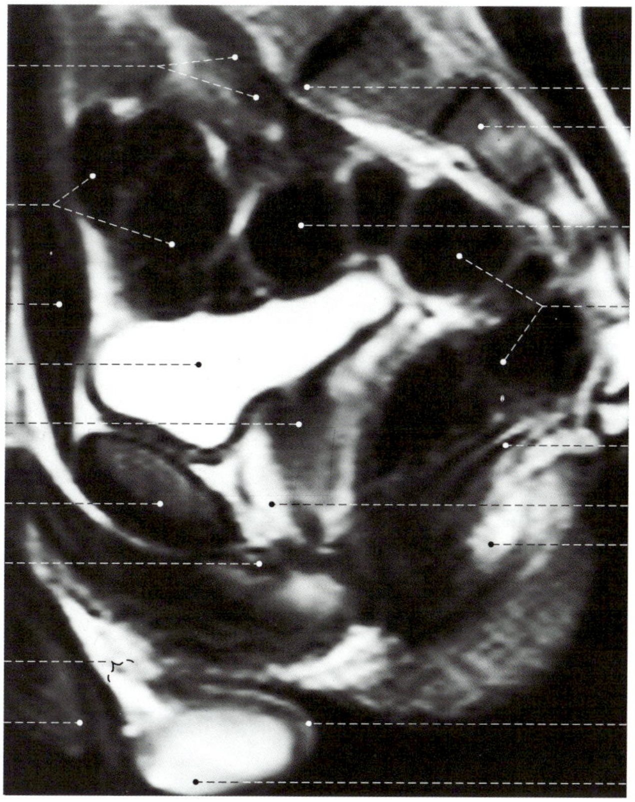

261 Pelvis

261 Pelvis of a male (45%)

Midsagittal sections

a Anatomical section, medial aspect of the right half of the body

b Magnetic resonance image (MRI, T_2-weighted)

261 Männliches Becken (45%)

Mediansagittale Schnitte

a Anatomischer Schnitt, Medialansicht der rechten Körperhälfte

b Kernspintomogramm (MRI, T_2-gewichtet)

Diaphragma pelvis et Organa genitalia externa

Pelvic Floor and External Genitalia

Beckenboden und äussere Geschlechtsorgane

a

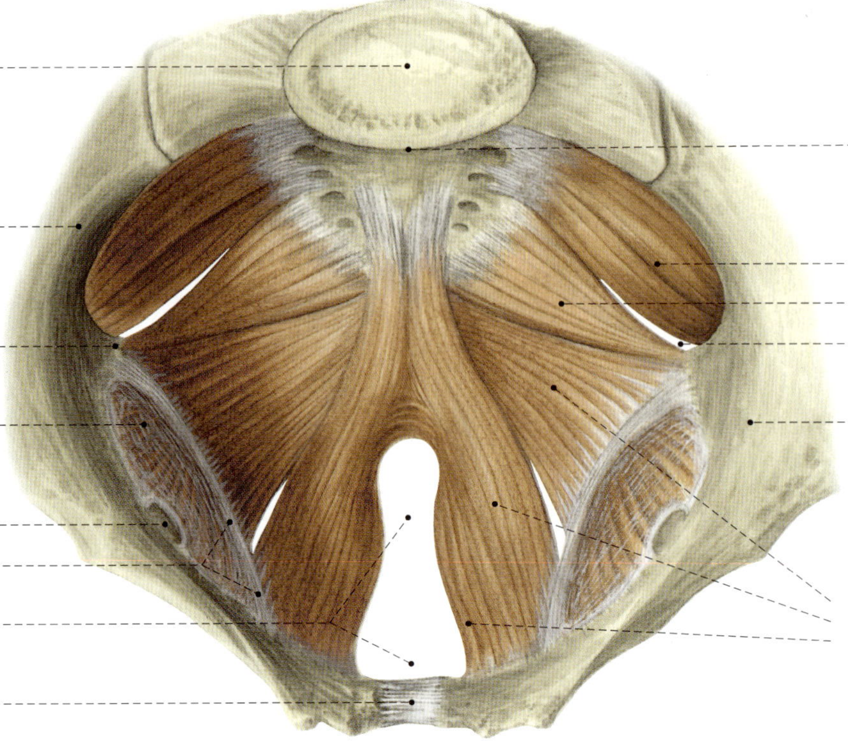

b

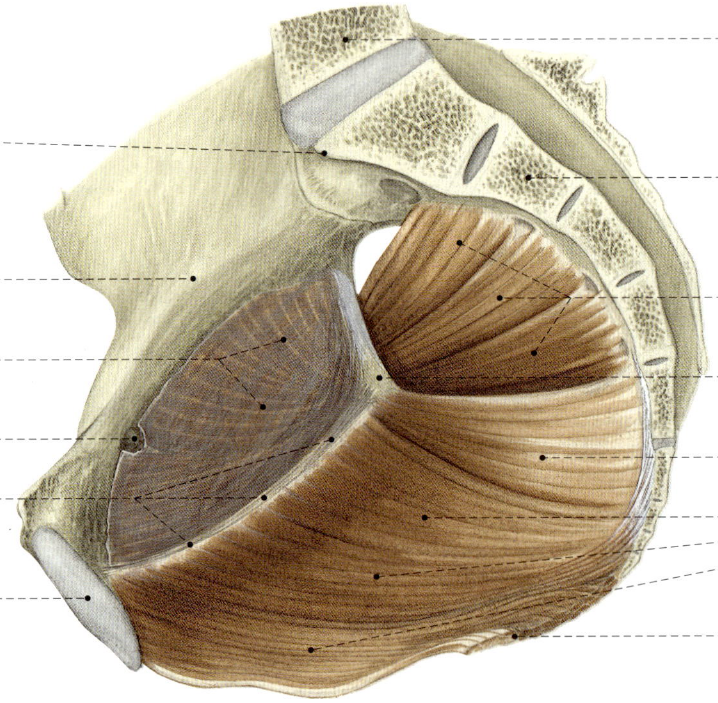

264 Diaphragma pelvis

264 Pelvic floor (60%)

Muscles of the pelvic floor (pelvic diaphragm)

a Superior aspect
b Medial aspect of the right half of the pelvis

264 Beckenboden (60%)

Beckenbodenmuskulatur (Diaphragma pelvis)

a Kranialansicht
b Medialansicht der rechten Beckenhälfte

a

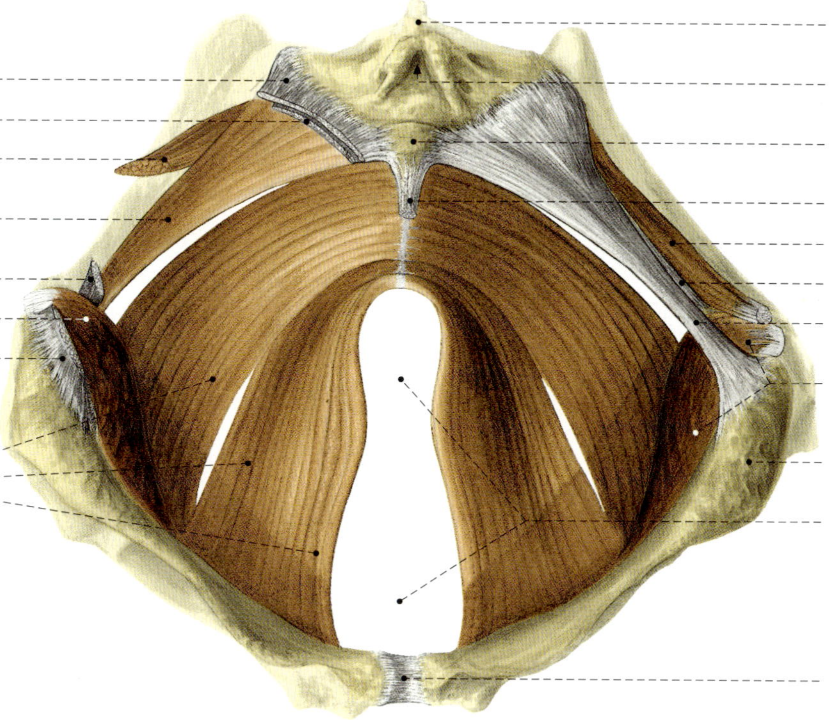

b

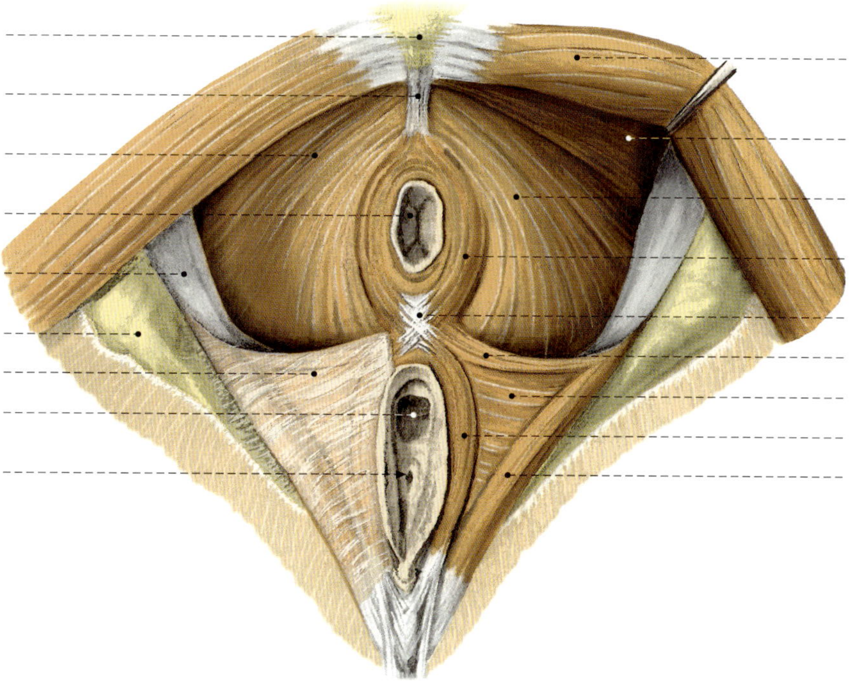

265 Diaphragma pelvis et ⟨Diaphragma urogenitale⟩

265 Pelvic floor and perineum of a female (60%)

Inferior aspect

a Muscles of the pelvic floor (pelvic diaphragm)
b Muscles of the pelvic floor and the perineum

265 Beckenboden und Damm einer Frau (60%)

Kaudalansicht

a Beckenbodenmuskulatur (Diaphragma pelvis)
b Beckenboden- und Dammuskulatur

a

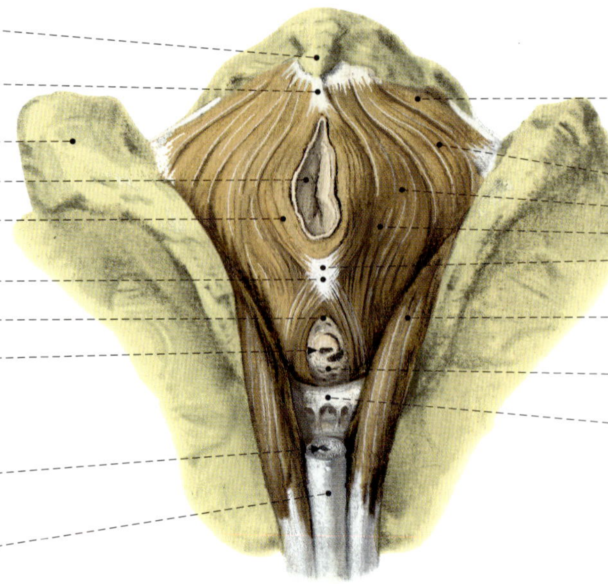

b

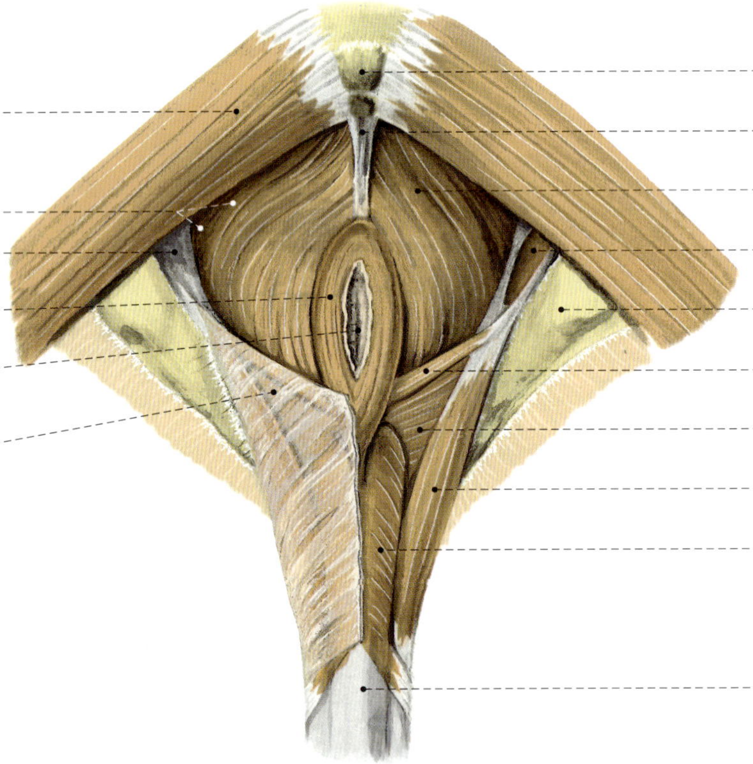

266 Diaphragma pelvis et ⟨Diaphragma urogenitale⟩

266 Pelvic floor and perineum of a male (60%)

Inferior aspect

a Muscles of the pelvic floor (pelvic diaphragm)
b Muscles of the pelvic floor and the perineum

266 Beckenboden und Damm eines Mannes (60%)

Kaudalansicht

a Beckenbodenmuskulatur (Diaphragma pelvis)
b Beckenboden- und Dammuskulatur

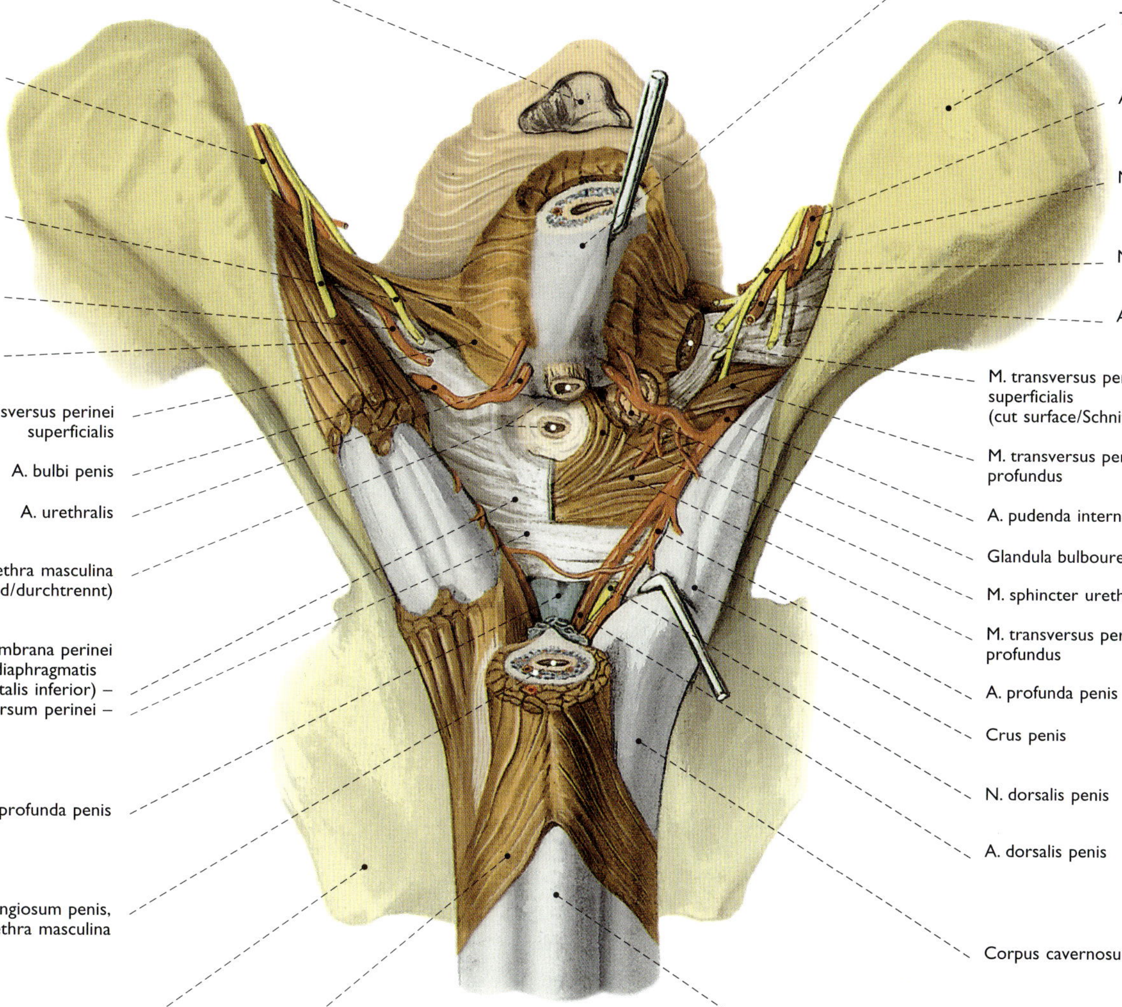

267 Penis et ⟨Diaphragma urogenitale⟩

267 Penis and urogenital diaphragm
(90%)
Inferior aspect
The corpus spongiosum penis was transected and the bulb of penis retracted.

267 Penis und Diaphragma urogenitale
(90%)
Kaudalansicht
Das Corpus spongiosum penis ist durchtrennt, der Bulbus penis zurückgeklappt.

a

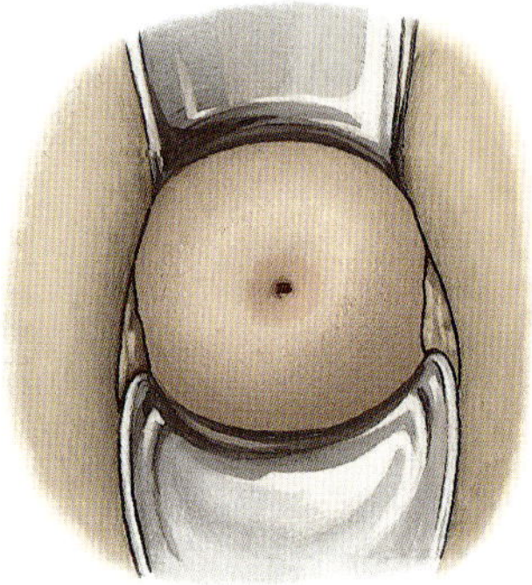

b

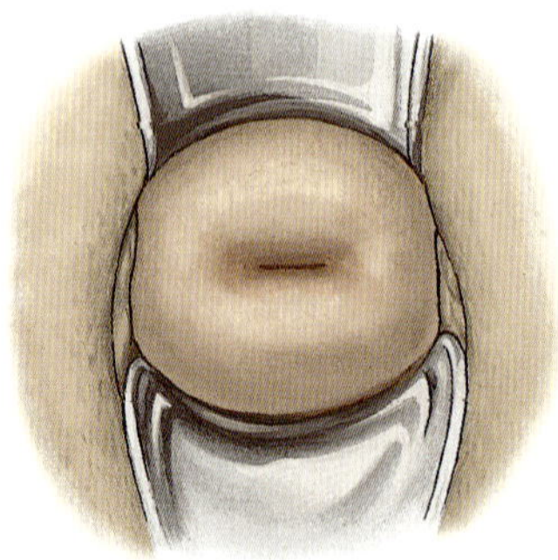

c

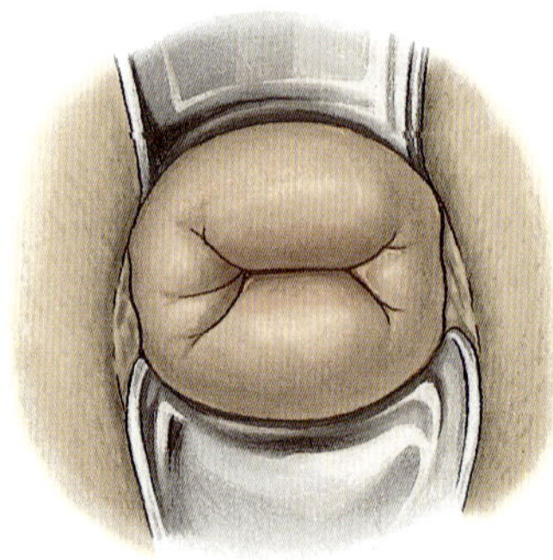

d

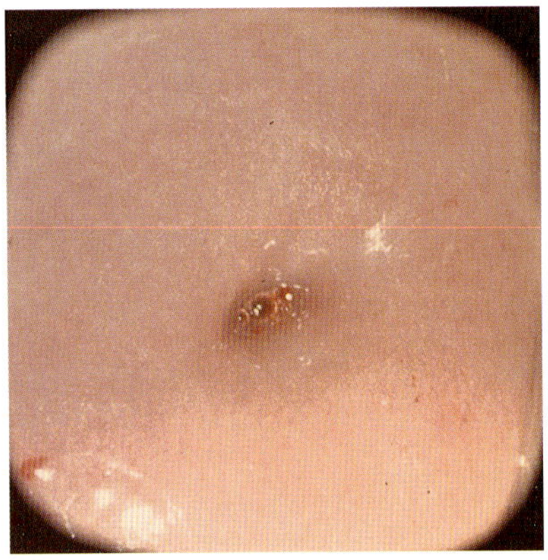

e

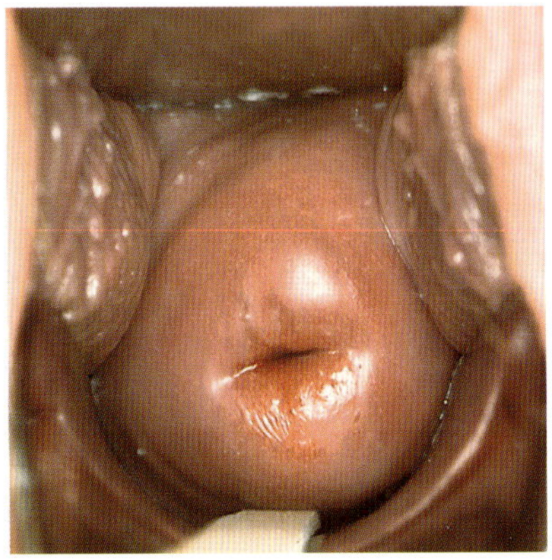

f

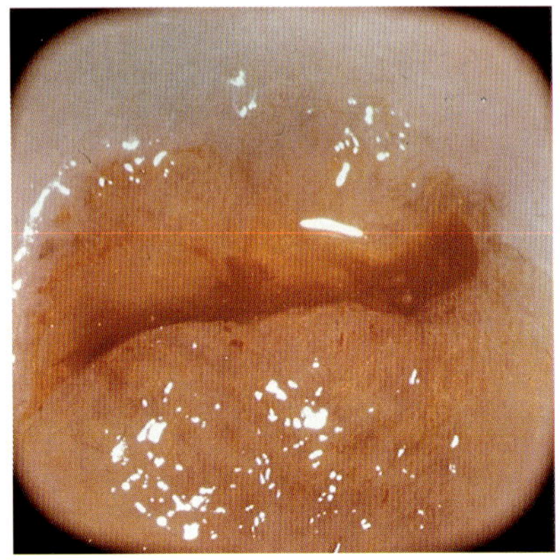

g

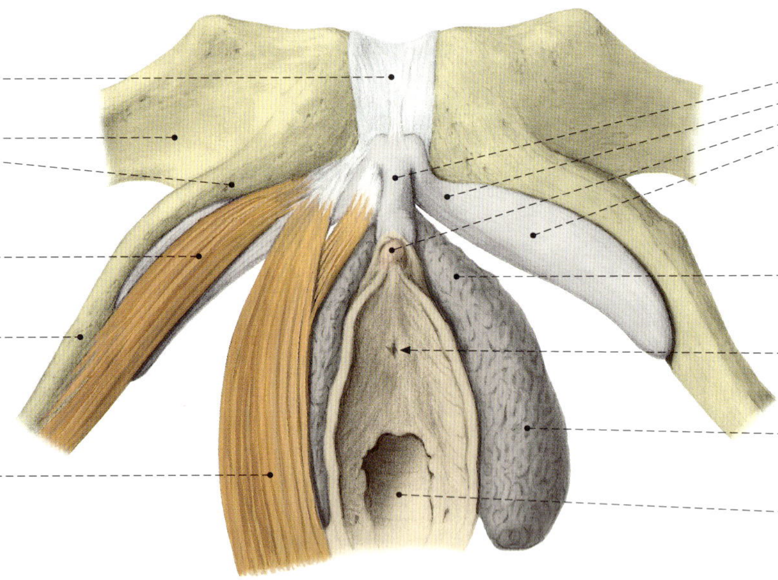

268 Portio vaginalis cervicis uteri et Clitoris

268 Vaginal part of the cervix uteri and clitoris

a–f Vaginal aspect of the external os uteri of a
a, b, d, e nulliparous woman (a, b, e, 100%; d, 200%)
c, f multiparous woman (c, 100%; f, 200%)
d–f Original photographs
g Erectile bodies of the clitoris (100%), inferior aspect

268 Portio vaginalis der Cervix uteri und Klitoris

a–f Vaginalansicht des äusseren Muttermunds einer
a, b, d, e Nullipara (a, b, e: 100%; d: 200%)
c, f Multipara (c: 100%; f: 200%)
d–f Originalphotographien
g Schwellkörper der Klitoris (100%), Kaudalansicht

a

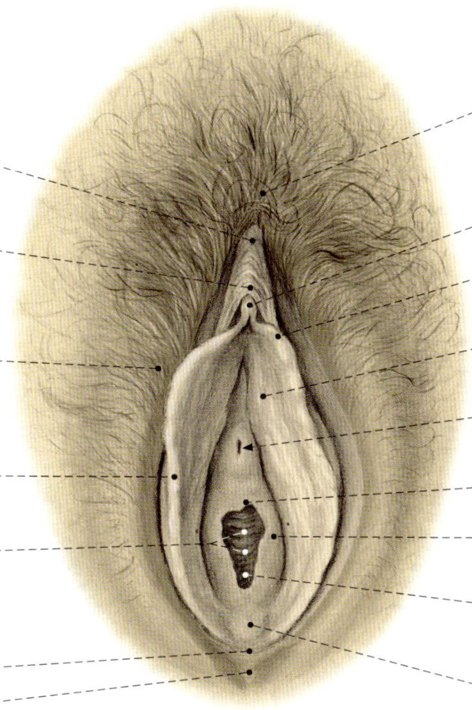

b

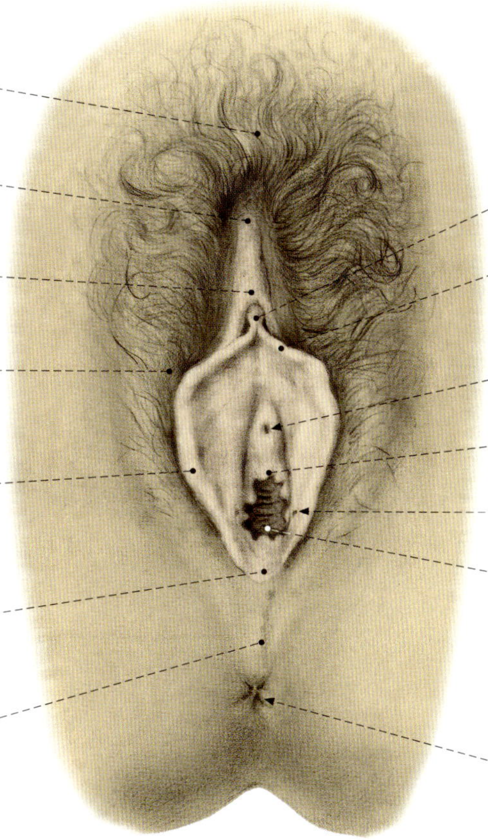

269 Organa genitalia feminina externa

269 External genitals of a female (80%)

Inferior aspect
- a Genitals of a virgin
- b Genitals after defloration

269 Äussere Geschlechtsorgane einer Frau (80%)

Kaudalansicht
- a Geschlechtsorgane einer Jungfrau
- b Geschlechtsorgane nach Defloration

a

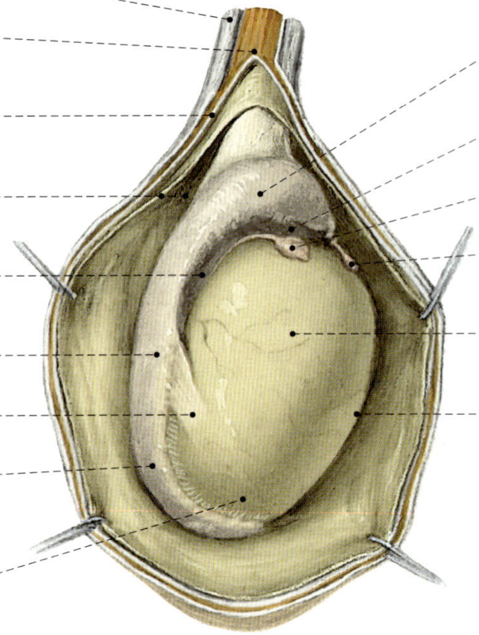

b

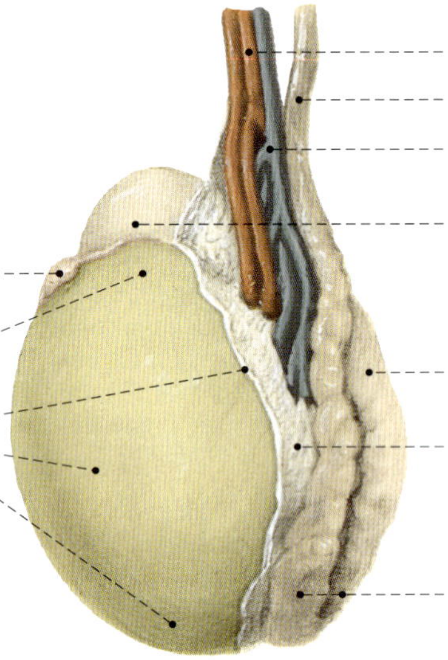

c

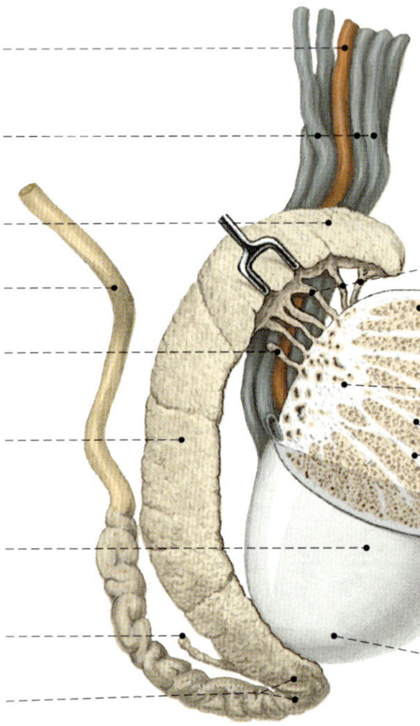

270 Testis et Epididymis

270 Testis and epididymis (100%)

a, b The coverings of the right testis were opened (a) and removed (b).
a Right lateral aspect
b Medial aspect
c The lateral superior quadrant of the right testis was excised by a rectangular incision. Right lateral aspect

270 Hoden und Nebenhoden (100%)

a, b Die Hüllen des rechten Hodens sind eröffnet (a) bzw. entfernt (b).
a Rechte Seitenansicht
b Medialansicht
c Der seitliche obere Quadrant des rechten Hodens ist herausgeschnitten. Rechte Seitenansicht

a

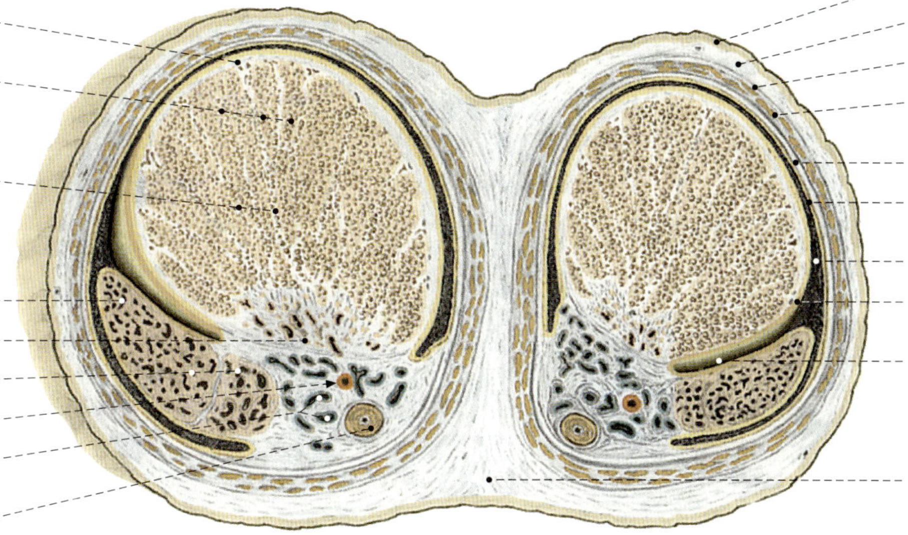

b

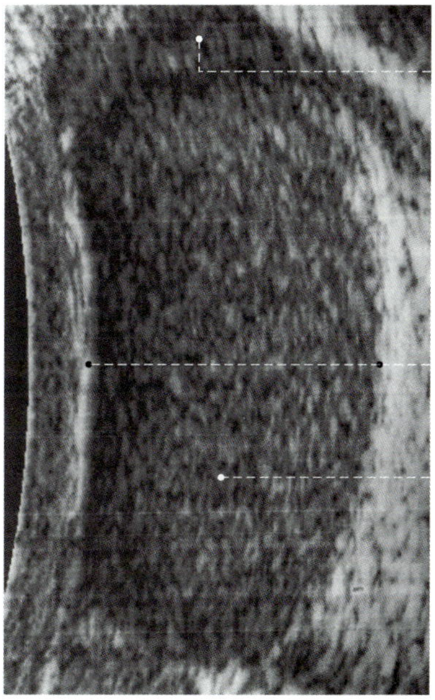

c

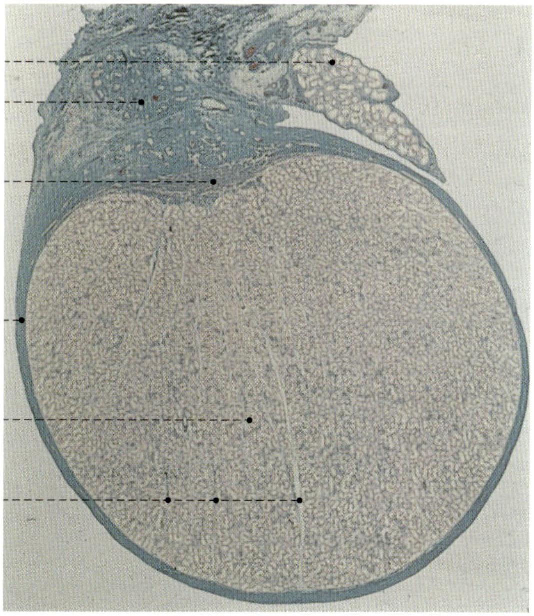

271 Scrotum et Testis

271 Scrotum and testis

a Transverse section through the scrotum and its contents (150%)
b Longitudinal ultrasound image of the testis (150%)
c Histological picture of the testis and the head of the epididymis sectioned in a transverse plane (260%)

271 Hodensack und Hoden

a Horizontaler Querschnitt durch den Hodensack und seine Inhaltsgebilde (150%)
b Ultraschallbild des Hodens im Längsschnitt (150%)
c Histologisches Bild von Hoden und Nebenhodenkopf im Querschnitt (260%)

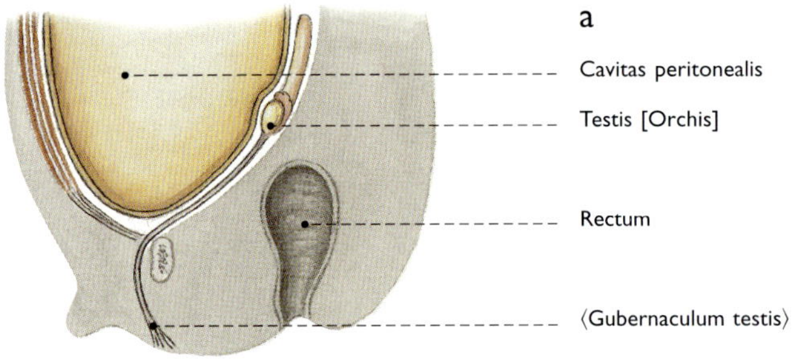

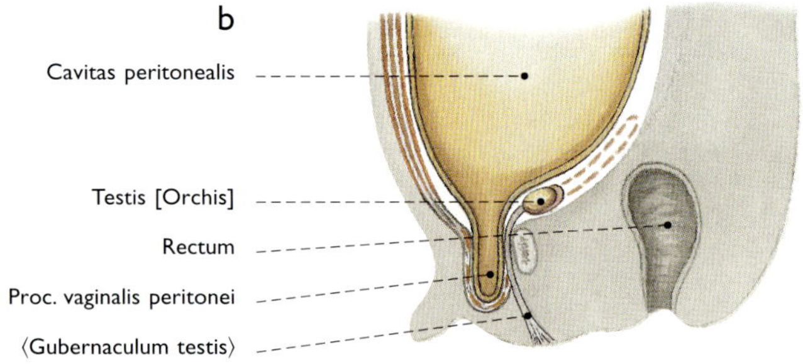

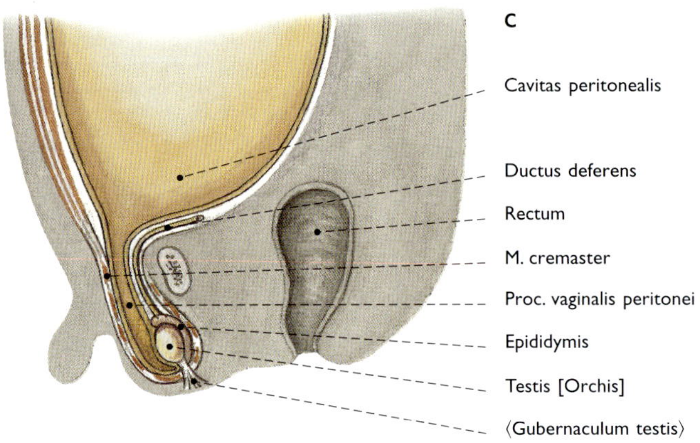

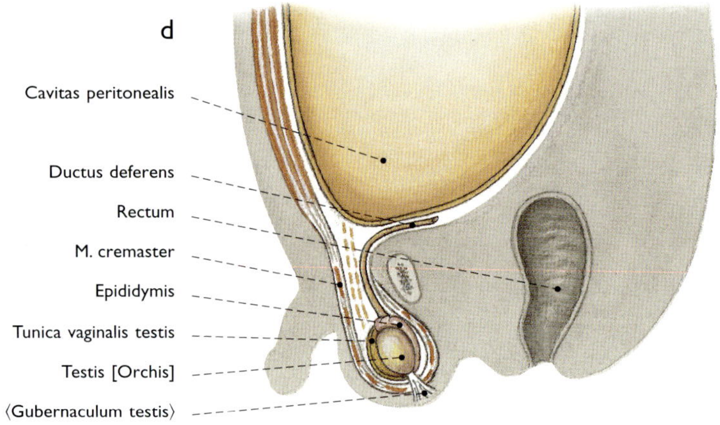

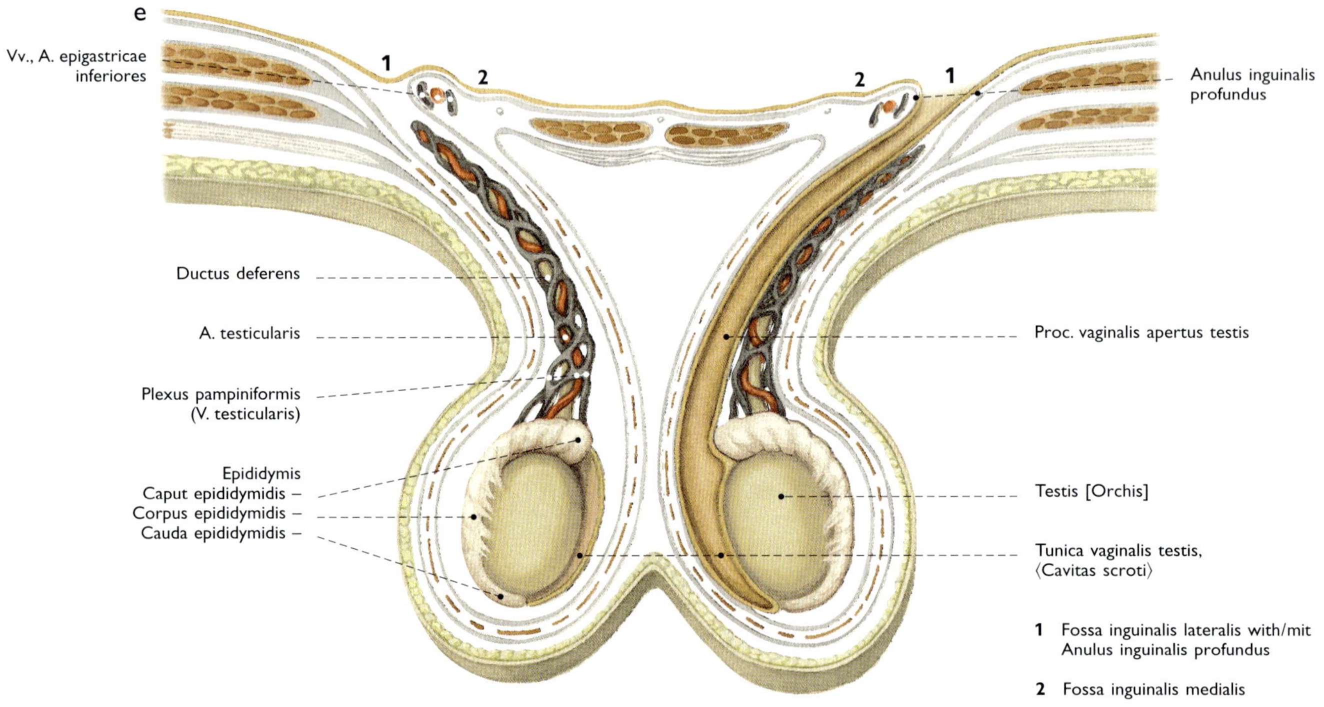

1 Fossa inguinalis lateralis with/mit Anulus inguinalis profundus

2 Fossa inguinalis medialis

272 Descensus testis et Regio inguinalis

272 Prenatal descent of the testis and inguinal region after birth

a–d Diverse stages of the prenatal descent of the testis. Medial aspect of the right half of paramedian sagittal sections

e Inguinal region of a male. On the right, normal situation. On the left, persistent processus vaginalis. Schematized section through the anterior abdominal wall on the level of the inguinal canal and through the scrotum (according to Benninghoff, 1985), ventral aspect

272 Vorgeburtlicher (pränataler) Abstieg des Hodens und Leistenregion nach der Geburt

a–d Verschiedene Stadien des vorgeburtlichen Abstiegs des Hodens. Medialansicht der rechten Hälfte von paramedianen Sagittalschnitten

e Leistenregion eines Mannes. Rechts, normale Situation. Links, offener persistierender Processus vaginalis. Schematisierter Querschnitt durch die vordere Bauchwand in Höhe des Leistenkanals und durch den Hodensack (nach Benninghoff, 1985), Ventralansicht

a

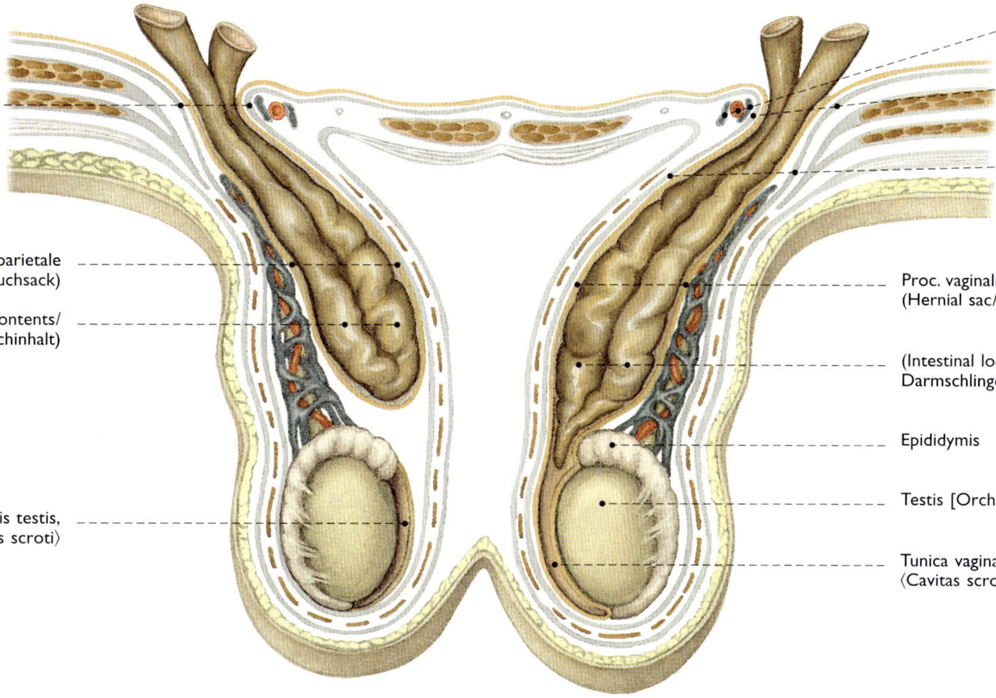

b

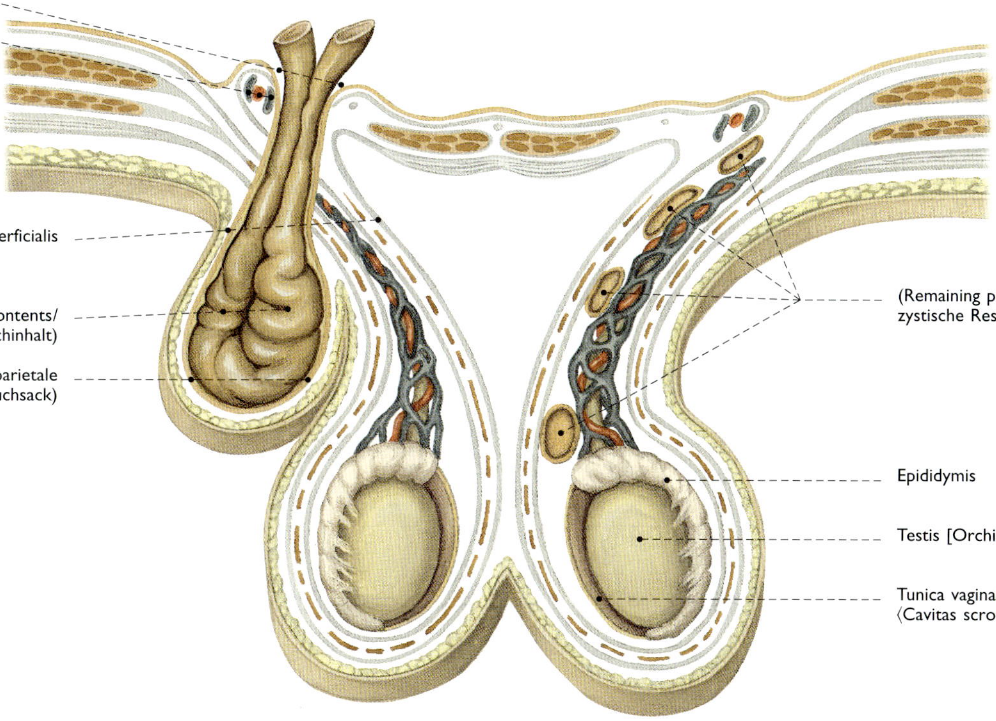

273 Regio inguinalis

273 Inguinal region of a male

Schematized sections through the anterior abdominal wall on the level of the inguinal canal and through the scrotum (according to Benninghoff, 1985), ventral aspect

a Lateral indirect inguinal herniae descending along the inguinal canal. The inner hernial ring is the deep inguinal ring. On the left, persistent processus vaginalis and complete congenital hernia

b On the right, medial direct inguinal hernia. The inner hernial ring is the medial inguinal fossa. On the left, persistent cysts in the spermatic cord as remainders of the processus vaginalis

273 Leistenregion eines Mannes

Schematisierte Querschnitte durch die vordere Bauchwand in Höhe des Leistenkanals und durch den Hodensack (nach Benninghoff, 1985), Ventralansicht

a Laterale indirekte Leistenhernien abgestiegen durch den Leistenkanal. Die innere Bruchpforte ist der innere Leistenring. Links: offener persistierender Processus vaginalis mit einem angeborenen indirekten Leistenbruch

b Rechts: medialer direkter Leistenbruch. Die innere Bruchpforte ist die Fossa inguinalis medialis. Links: persistierende Zysten im Samenstrang als Reste des Processus vaginalis

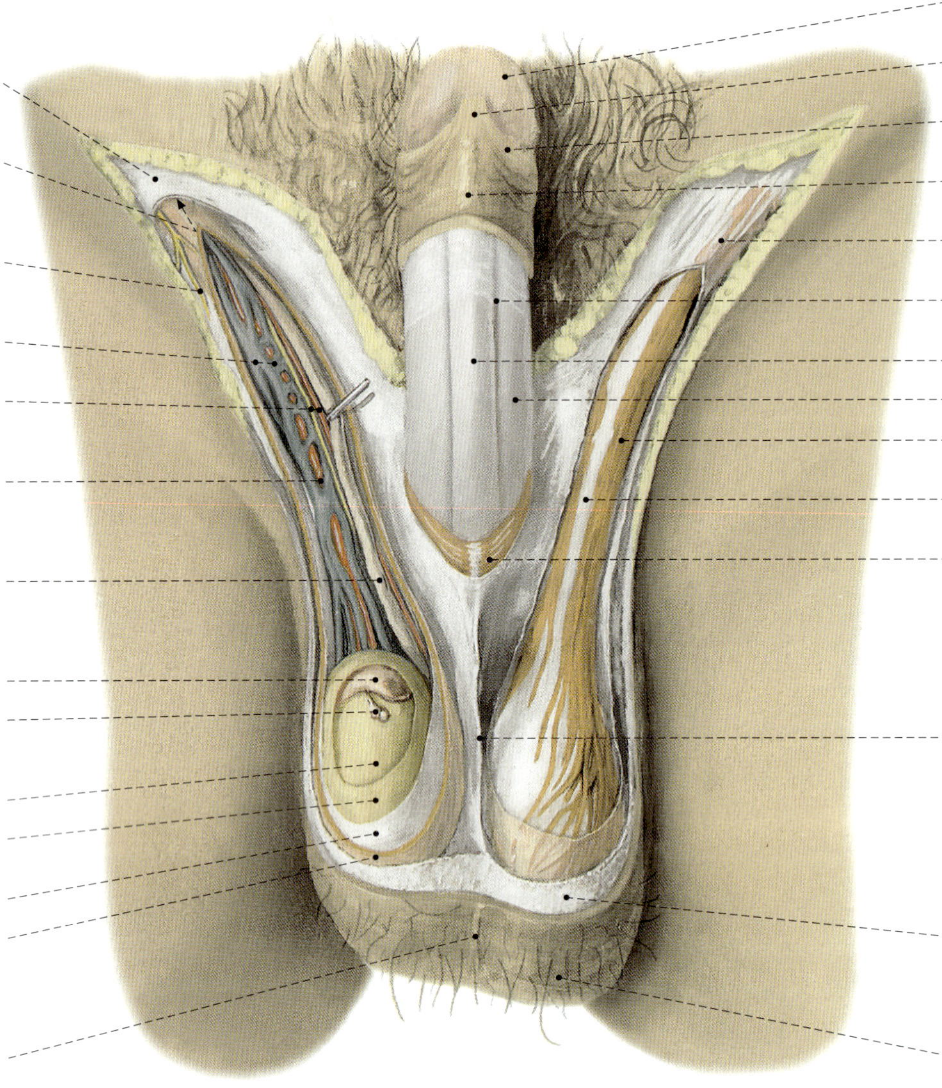

274 Organa genitalia masculina externa

274 External genitals of a male (70%)

The penis was turned upwards, the skin and the fascia penis profunda were largely removed from the shaft of the penis. The skin of the scrotum was excised in the region of the spermatic cords. On the right side the structures of the spermatic cord are exposed by opening the external and internal spermatic fasciae, the interposed cremasteric fascia and the cremasteric muscle. On the left side the cremaster muscle is shown after having removed the external spermatic and cremasteric fasciae. Ventral aspect

274 Äussere Geschlechtsorgane eines Mannes (70%)

Der Penis ist hochgeklappt, Haut und Fascia penis profunda sind an der Unterseite des Penis weitgehend entfernt. Die Haut an der Vorderseite des Hodensacks und der Samenstränge ist abgetragen. Rechts ist der Inhalt des Samenstranges nach Spaltung der Fasciae spermatica externa, cremasterica et spermatica interna dargestellt, links ist der Musculus cremaster nach Abtragung der Fasciae spermatica externa et cremasterica gezeigt. Ventralansicht

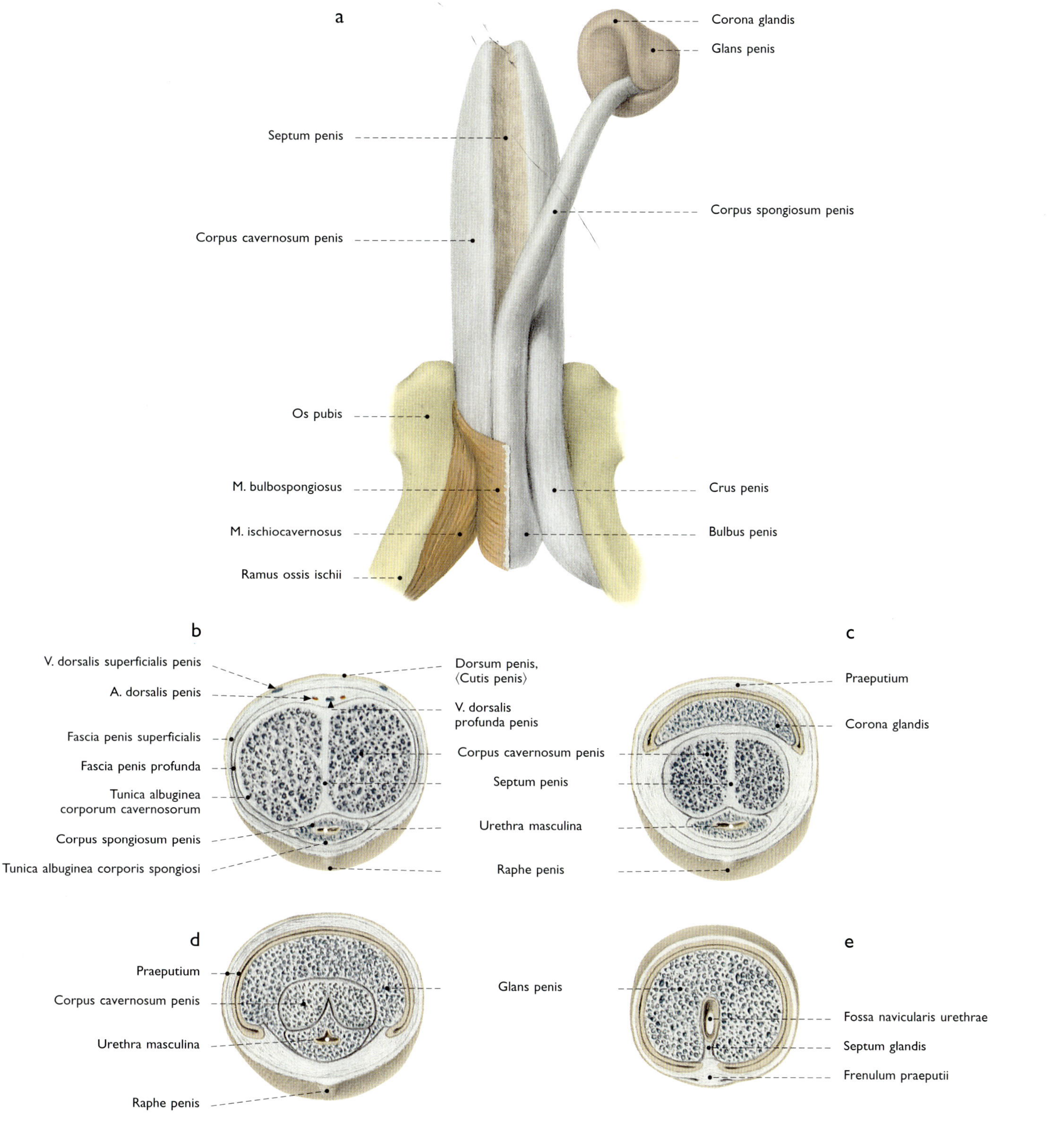

275 Penis

275 Penis (80%)

- a Erectile bodies of the penis. The glans penis and the distal part of the corpus spongiosum were detached from the corpora cavernosa penis and displaced to the left. Inferior aspect
- b–e Distal aspect of transverse sections
- b through the body of the penis
- c at the level of the collum glandis
- d through the posterior part of the glans penis
- e through the anterior part of the glans penis

275 Penis (80%)

- a Schwellkörper des Penis. Die Glans penis und der distale Teil des Corpus spongiosum sind von den Corpora cavernosa penis abgelöst und nach links abgehoben. Kaudalansicht
- b–e Distalansicht von Transversalschnitten
- b durch den Penisschaft
- c in Höhe des Collum glandis
- d durch den proximalen Bereich der Glans penis
- e durch den distalen Bereich der Glans penis

a

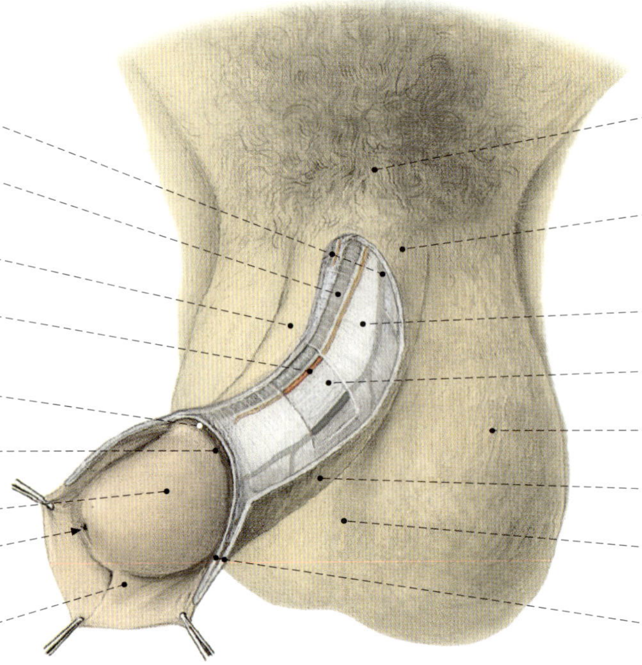

b

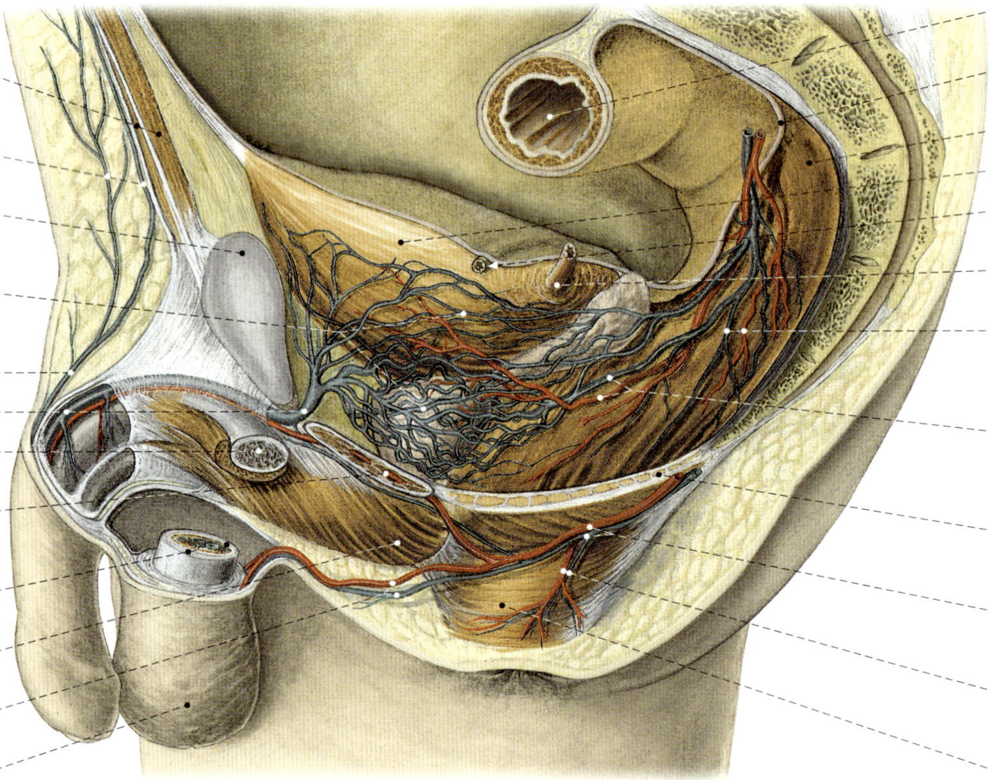

276 Organa genitalia masculina externa

276 External genitals of a male

a The preputial sac was opened and a longitudinal strip of skin removed from the left side of the prepuce and the shaft of the penis. A small rectangular window was cut into the superficial fascia of the penis (80%). Ventral aspect

b Drainage of superficial and deep dorsal veins of the penis (60%). Left paramedian section through the lesser pelvis, medial aspect of the right half

276 Äussere Geschlechtsorgane eines Mannes

a Auf der linken oberen Seite von Penisschaft und Vorhaut sind ein Längsstreifen der Haut und ein kleines rechteckiges Stück der oberflächlichen Penisfaszie abgetragen (80%). Ventralansicht

b Abfluss der oberflächlichen und tiefen dorsalen Penisvenen (60%). Linker Paramedianschnitt durch das kleine Becken, Medialansicht der rechten Hälfte

a

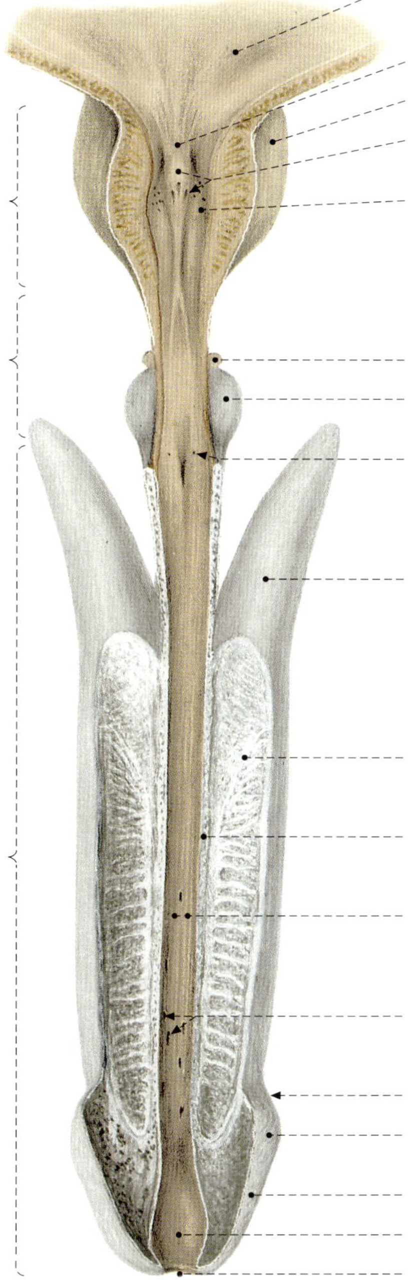

b

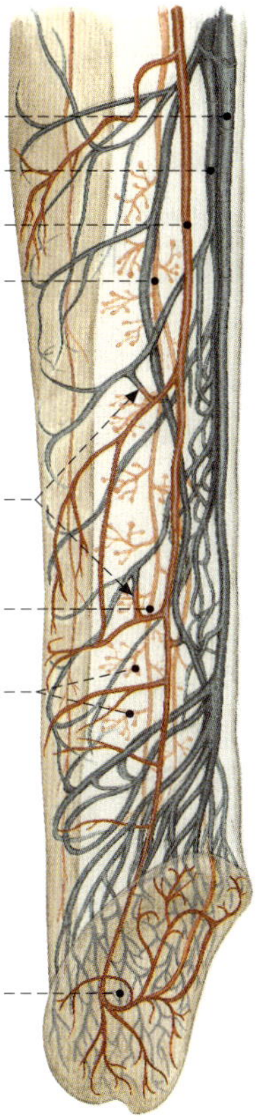

277 Urethra masculina et Penis

277 Urethra of a male and arteries of the penis (80%)

- a The urethra was opened by a median section from the internal to the external urethral orifices. The cut surfaces were turned outwards. Ventral aspect
- b Arteries of the penis including the helicine arteries (according to Ferner, 1975, and Lierse, 1984), schematic representation, right lateral aspect

277 Männliche Harnröhre und Penisarterien (80%)

- a Die Harnröhre ist durch einen Längsschnitt von der inneren bis zur äusseren Harnröhrenöffnung gespalten, die Schnittflächen sind auseinandergezogen. Ventralansicht
- b Arterien des Penis einschliesslich der Arteriae helicinae (nach Ferner, 1975, und Lierse, 1984), schematische Darstellung, rechte Seitenansicht

a

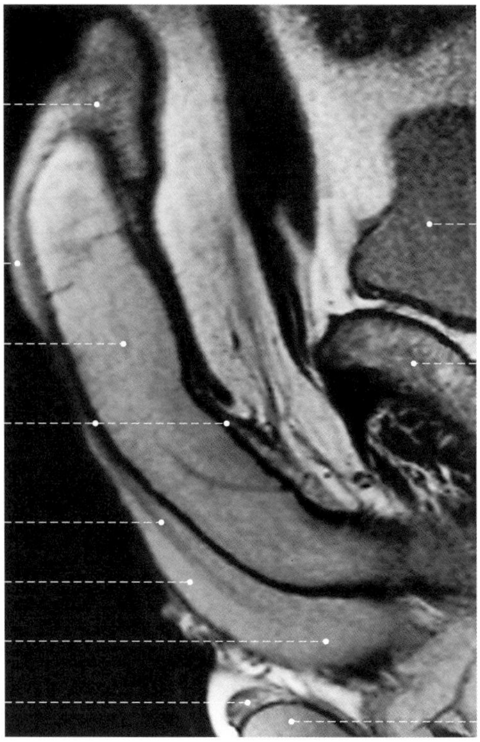

b

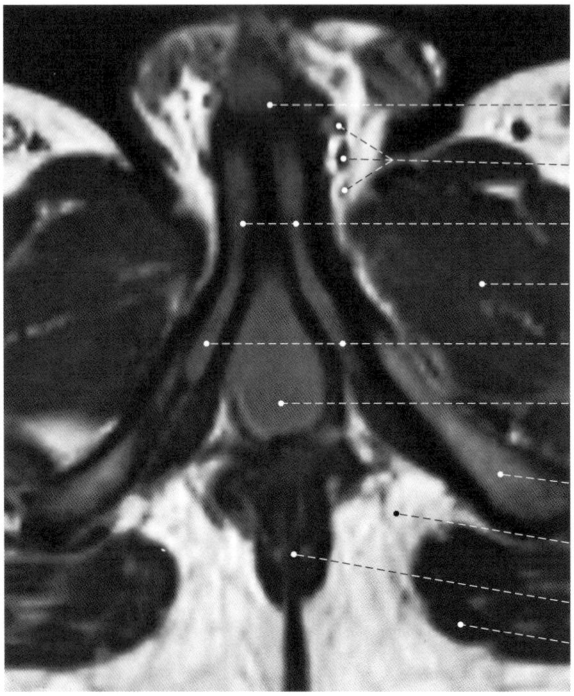

c

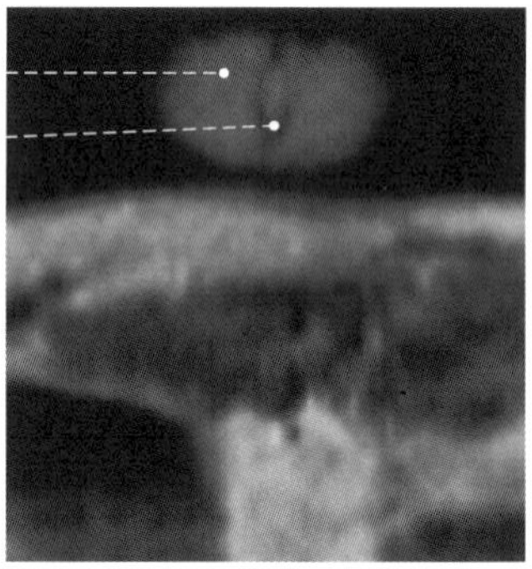

d

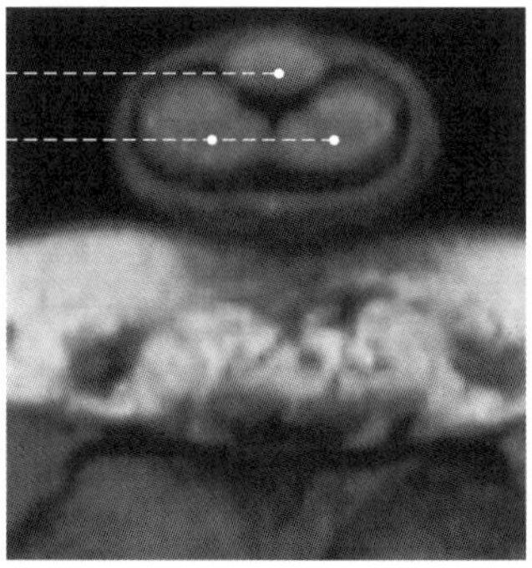

278 Penis et Urethra masculina

278 Penis and urethra of a male

a Sagittal magnetic resonance image (MRI, T_2-weighted) of an erected penis lying on the anterior abdominal wall (80%), medial aspect of the right half
b–d Transverse magnetic resonance images (MRI, T_1-weighted), inferior aspect,
b through the root of the penis, the anus, and the ischiopubic rami (50%)
c, d through the penis put on the anterior abdominal wall (90%)
c distally through the glans penis
d proximally through the body of the penis

278 Penis und männliche Harnröhre

a Sagittales Kernspintomogramm (MRI, T_2-gewichtet) eines erigierten, auf der vorderen Bauchwand liegenden Penis (80%), Medialansicht der rechten Hälfte
b–d Horizontale Kernspintomogramme (MRI, T_1-gewichtet), Kaudalansicht,
b durch die Peniswurzel, den Anus und die unteren Schambeinäste (50%)
c, d durch den auf die vordere Bauchwand gelegten Penis (90%)
c distal durch die Glans penis
d proximal durch den Penisschaft

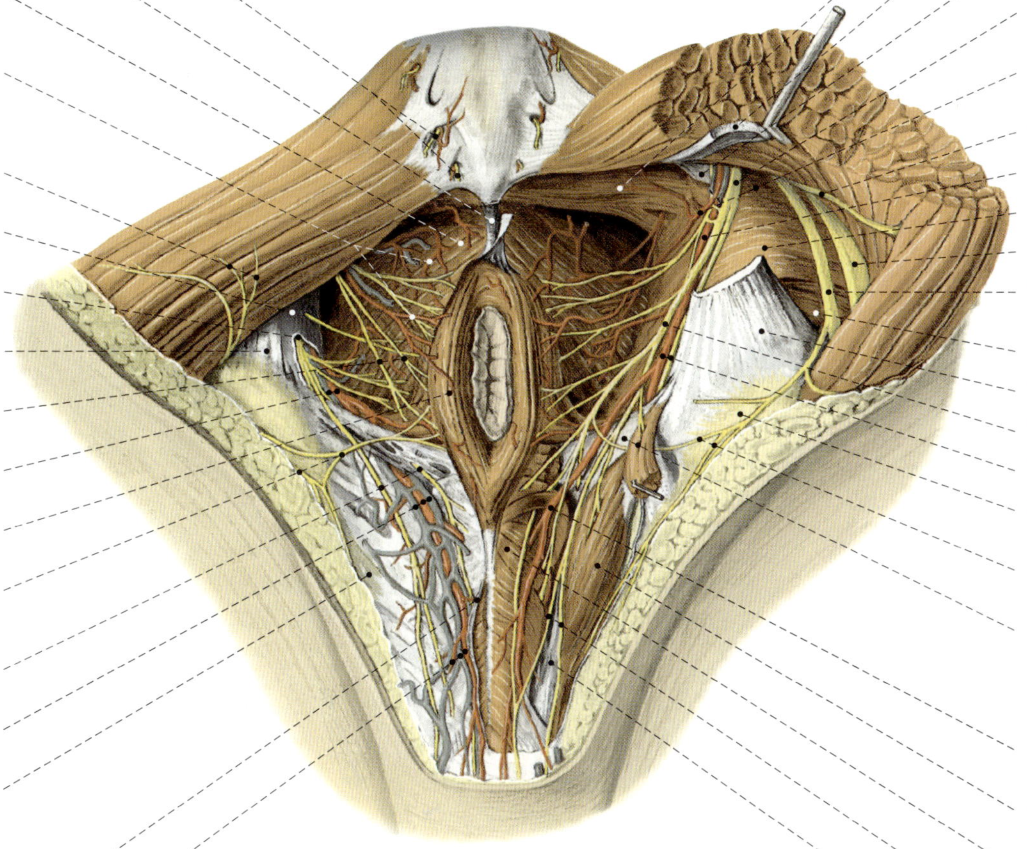

279 Regio perinealis

279 Blood vessels and nerves of the perineal region of a male (70%)
Inferior aspect

279 Blutgefässe und Nerven der Dammregion eines Mannes (70%)
Kaudalansicht

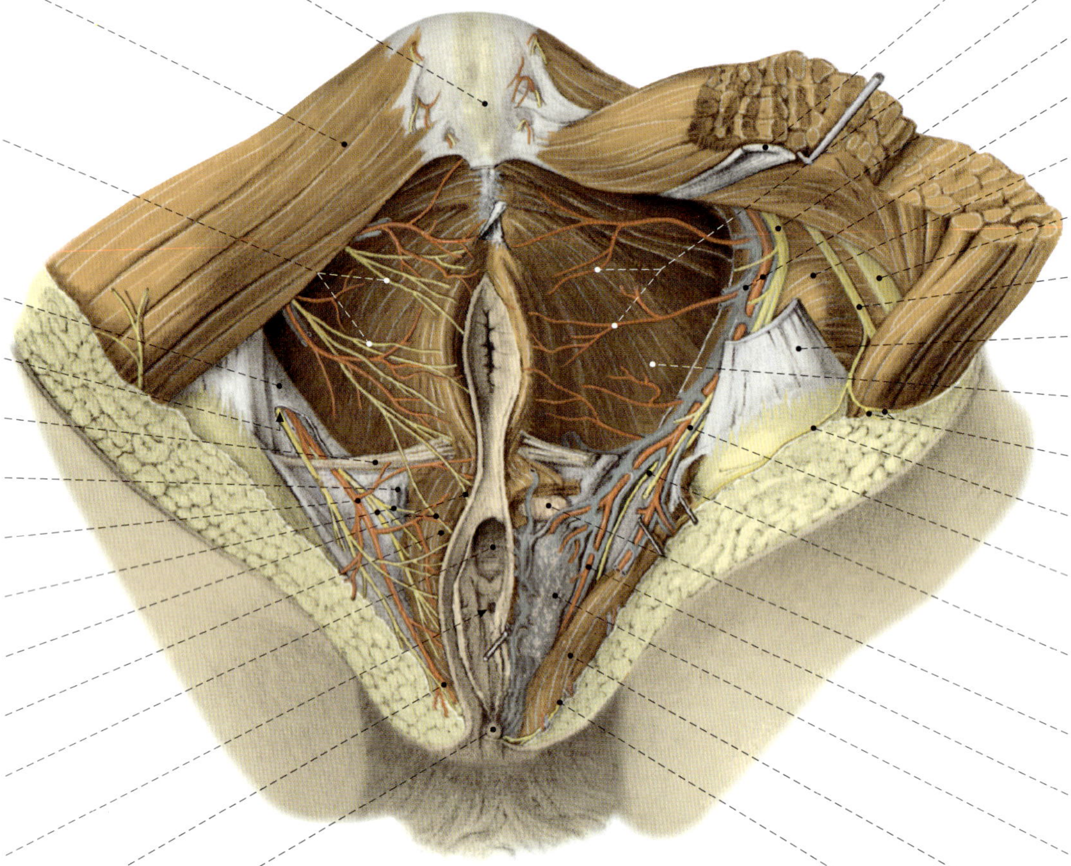

280 Regio perinealis

280 Blood vessels and nerves of the perineal region of a female (70%)
Inferior aspect

280 Blutgefässe und Nerven der Dammregion einer Frau (70%)
Kaudalansicht

a

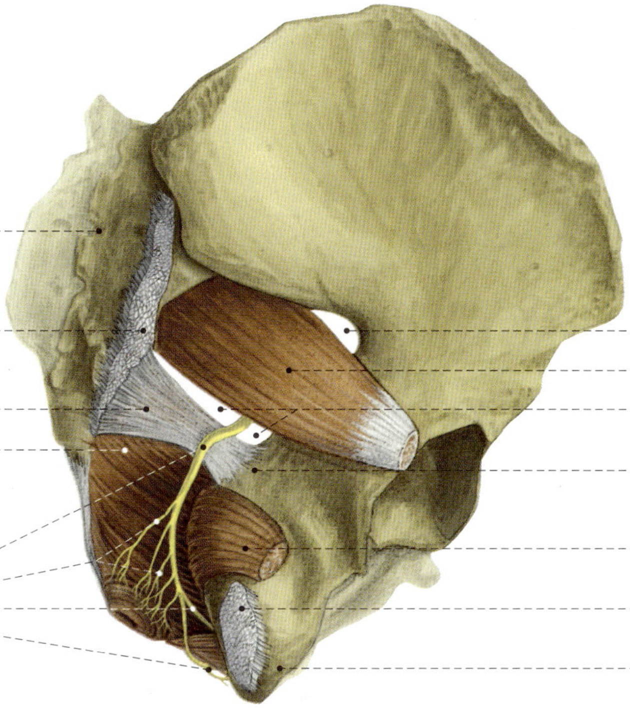

b

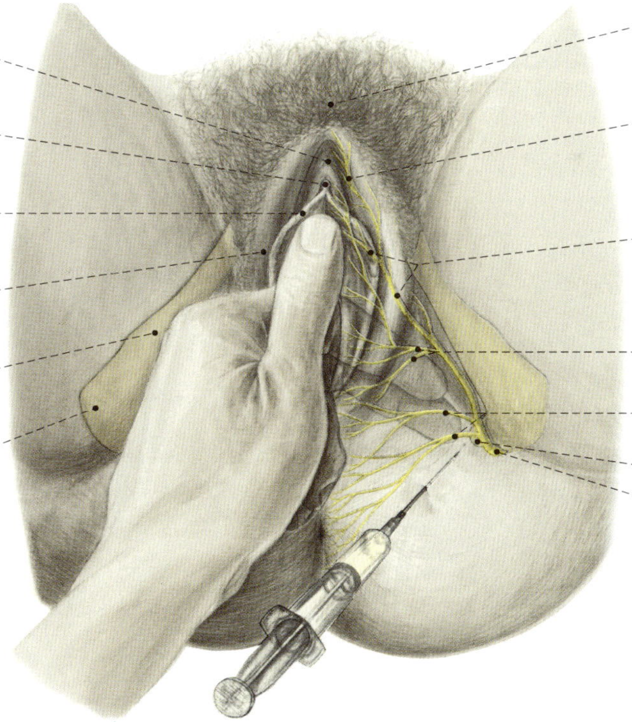

281 Regio perinealis

281 Nerves of the perineal region of a female (40%)

a Course of the right pudendal nerve, right dorsolateral aspect
b Pudendal anesthesia on the left side, schematic representation, inferior aspect

281 Nerven der Dammregion einer Frau (40%)

a Verlauf des rechten Nervus pudendus, rechte dorsolaterale Ansicht
b Linksseitige Pudendusanästhesie, schematische Darstellung, Kaudalansicht

Systema nervosum centrale

Central Nervous System

Zentralnervensystem

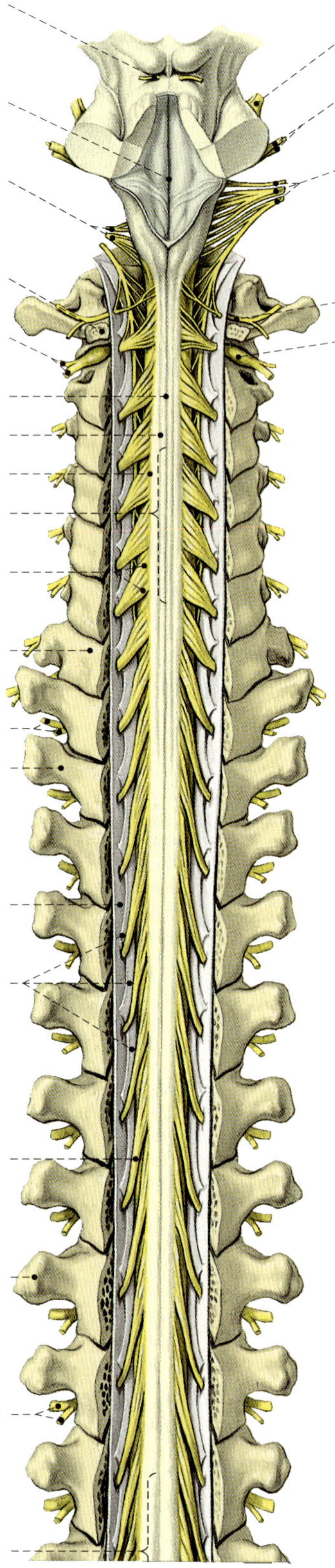

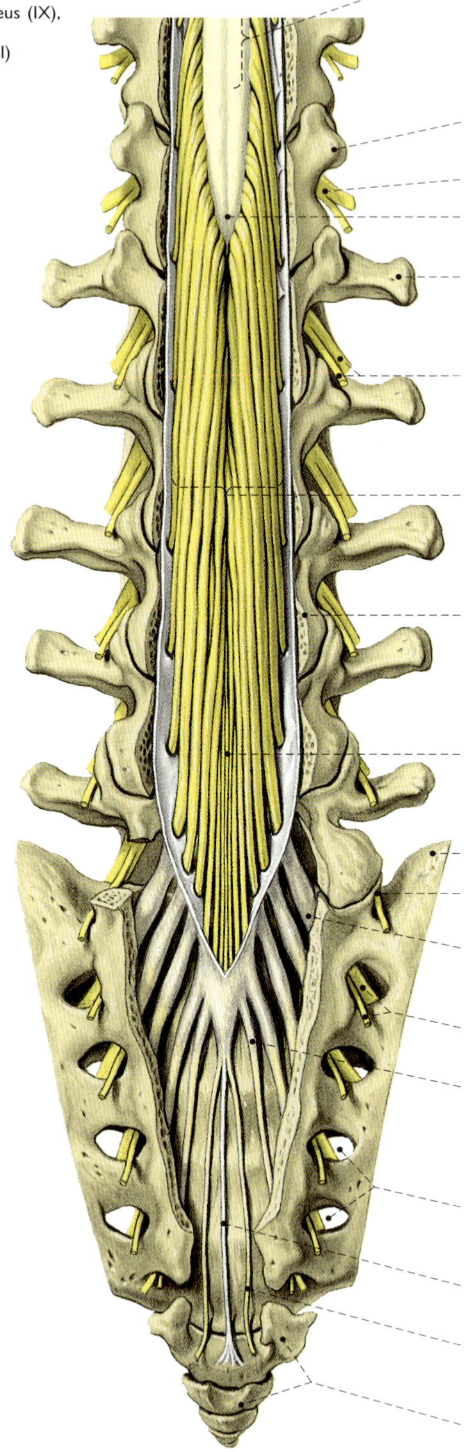

284 Medulla spinalis et Dura mater spinalis

284 Spinal cord and spinal dura mater in the vertebral canal (50%)

The vertebral canal was opened dorsally.
Dorsal aspect
a Spinal cord
b Cauda equina

284 Rückenmark und harte Rückenmarkhaut im Wirbelkanal (50%)

Der Wirbelkanal ist dorsal eröffnet.
Dorsalansicht
a Rückenmark
b Cauda equina

a

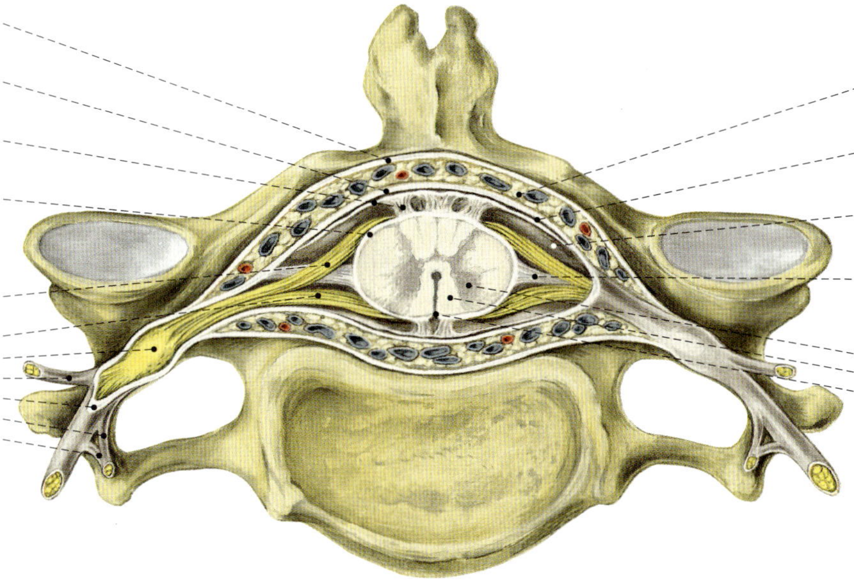

b

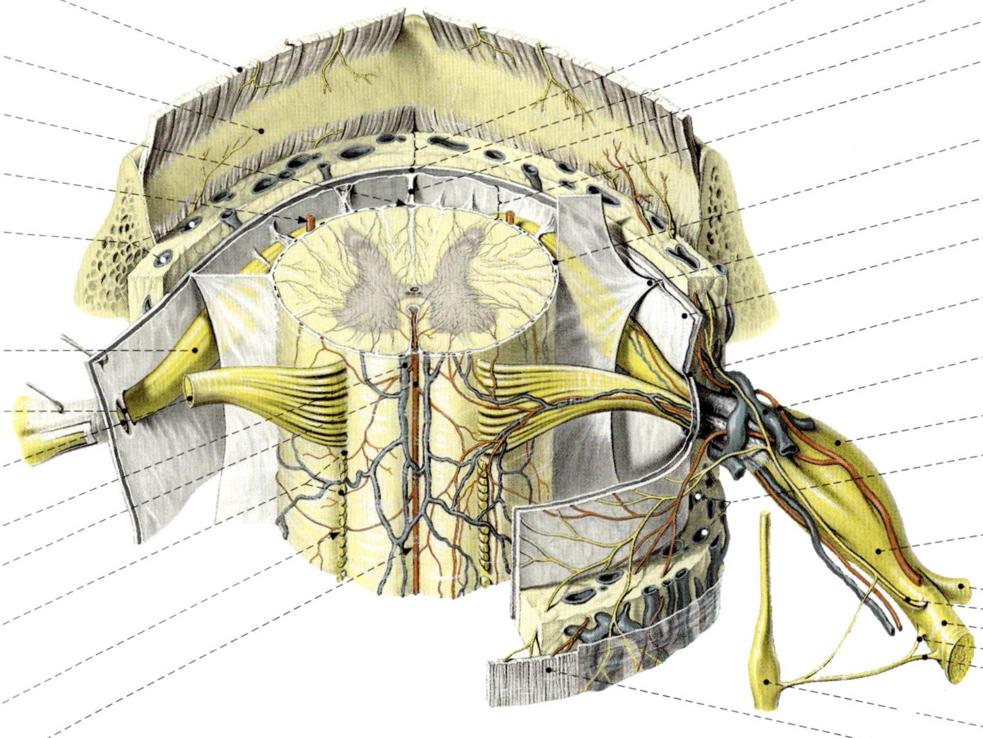

285 Medulla spinalis, Meninges et Radices nervi spinalis

285 Spinal cord, meninges and spinal roots

Transverse sections of the spinal cord and the spinal roots

- a Section through the vertebral canal of the cervical spine (230%), superior aspect
- b Schematized three-dimensional representation after removal of the vertebral bodies and the transverse processes (400%), ventrocranial aspect

285 Rückenmark, Hirnhäute und Spinalnervenwurzeln

Querschnitte durch Rückenmark und Spinalnervenwurzeln

- a Schnitt durch den Wirbelkanal der Halswirbelsäule (230%), Kranialansicht
- b Schematisierte dreidimensionale Darstellung nach Entfernung der Wirbelkörper und der Querfortsätze (400%), Ventrokranialansicht

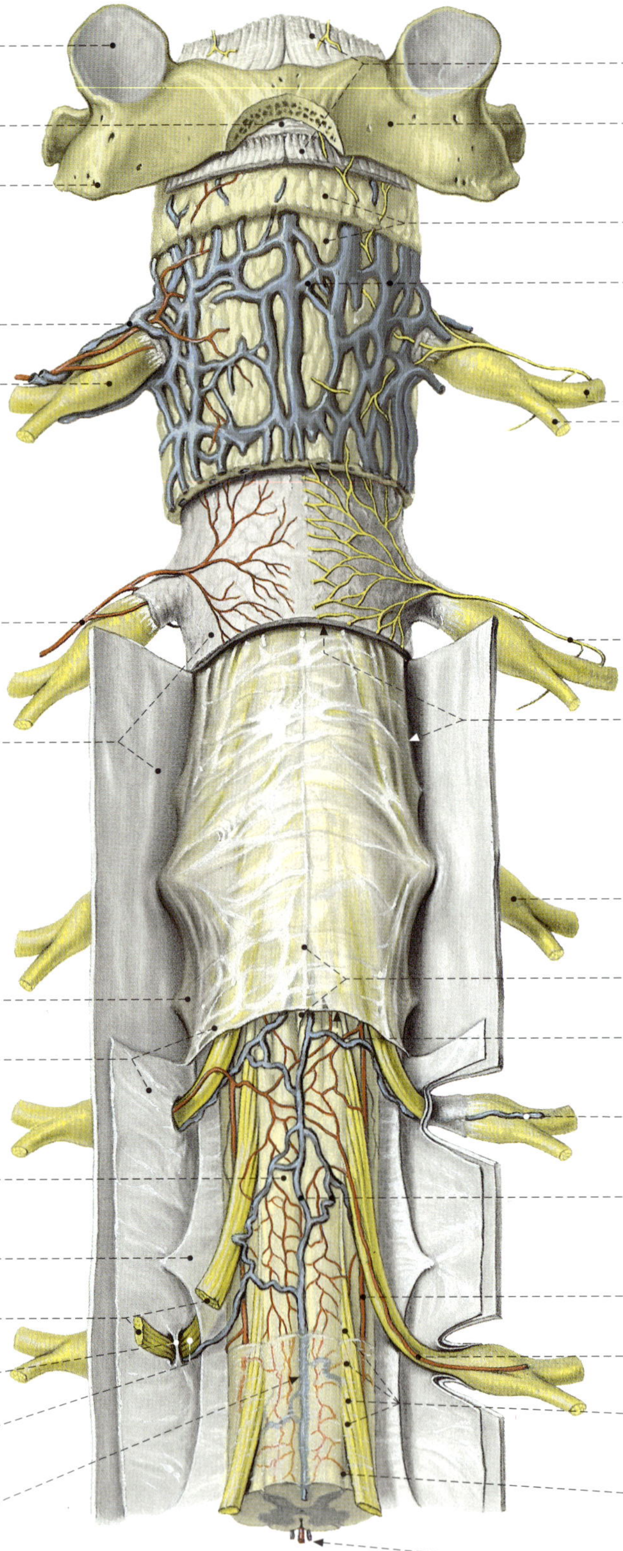

286 Medulla spinalis, Meninges et Nervi spinales

286 Spinal cord, meninges and spinal nerves

The investing structures of the spinal cord and the blood vessels are demonstrated in layers (150%). Dorsal aspect

286 Rückenmark, Hirnhäute und Spinalnerven

Die Hüllen des Rückenmarks und die Blutgefässe sind schichtweise dargestellt (150%). Dorsalansicht

a

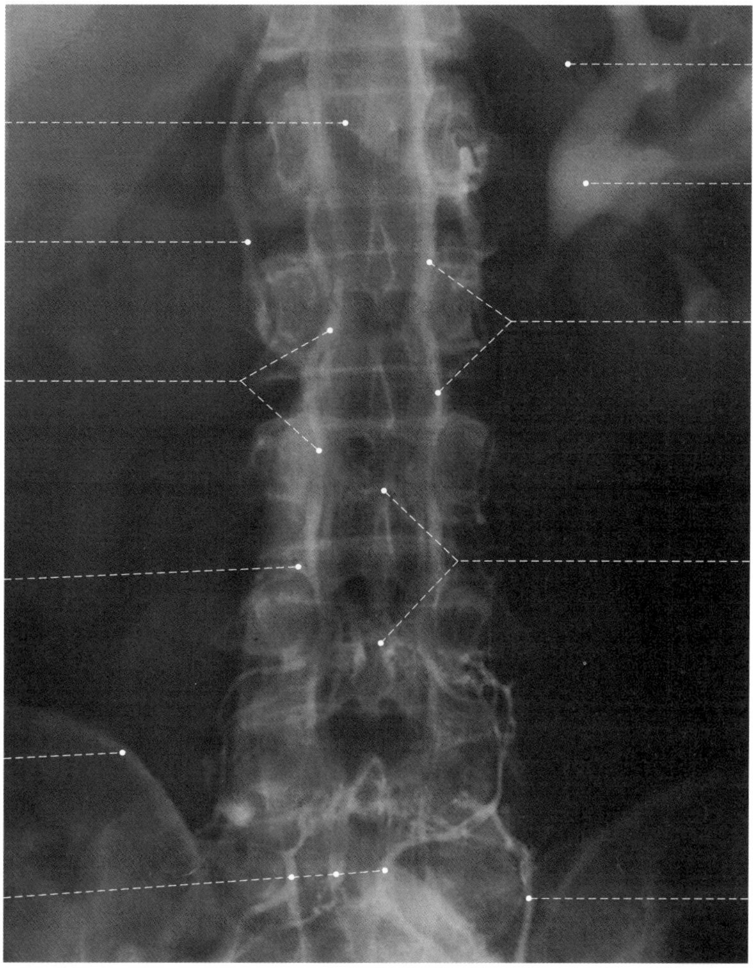

b

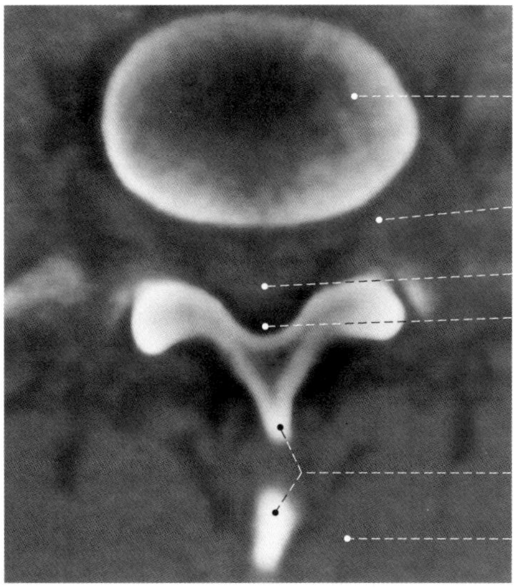

c

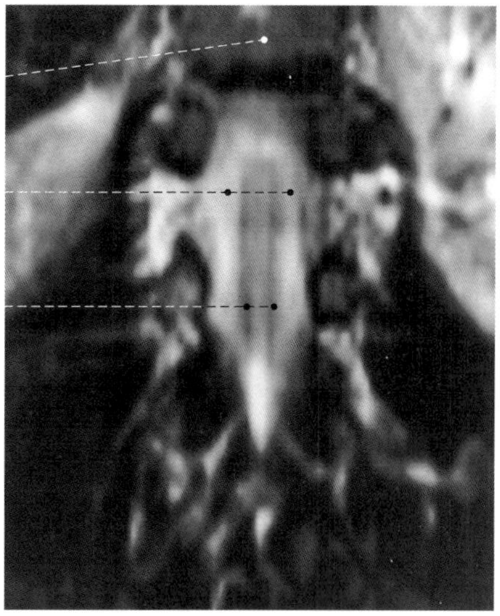

287 Medulla spinalis

287 Veins of the vertebral column and cauda equina

a Lumbar venogram and filling of the internal vertebral venous plexus, posteroanterior radiograph (60%)
b Transverse computed tomogram (CT) through the fifth lumbar vertebra and the cauda equina (80%), inferior aspect
c Coronal magnetic resonance image (MRI, T_2-weighted) of the lumbar spine (60%)

287 Venen der Wirbelsäule und Cauda equina

a Venogramm der aufsteigenden Venae lumbales und Füllung des Plexus venosus vertebralis internus, posteroanteriore Röntgenaufnahme (60%)
b Horizontales Computertomogramm (CT) durch den fünften Lendenwirbel und die Cauda equina (80%), Kaudalansicht
c Frontales Kernspintomogramm (MRI, T_2-gewichtet) durch die Lendenwirbelsäule (60%)

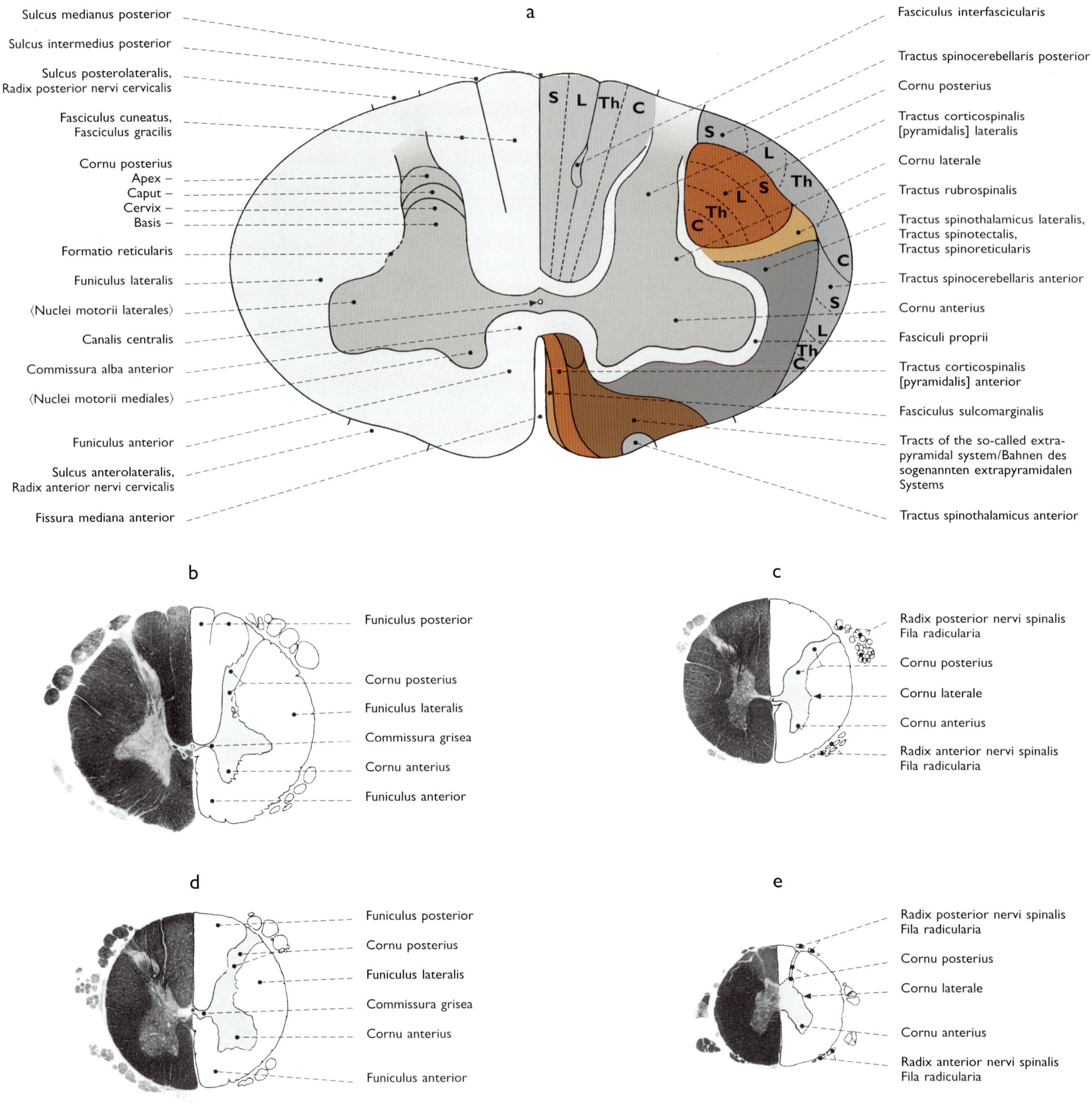

288 Medulla spinalis

288 Spinal cord

a Schematic transverse section at the endcervical level. Nerve cell groups are shown on the left side, tracts on the right side of the drawing. Descending tracts are given by brown and red, ascending ones by blue, and intersegmental tracts by white color (1000%).

b–e Transverse sections of the spinal cord (400%) at the levels of the

b cervical part
c thoracic part
d lumbar part
e sacral part

288 Rückenmark

a Schematischer Transversalschnitt des kaudalen Halsmarks. Auf der linken Bildseite sind Zellgruppen, rechts Fasersysteme dargestellt. Absteigende Bahnen sind braun und rot, aufsteigende Bahnen blau, Grundbündel weisslich koloriert (1000%).

b–e Transversalschnitte des Rückenmarks (400%) in Höhe des

b Halsmarks
c Brustmarks
d Lendenmarks
e Sakralmarks

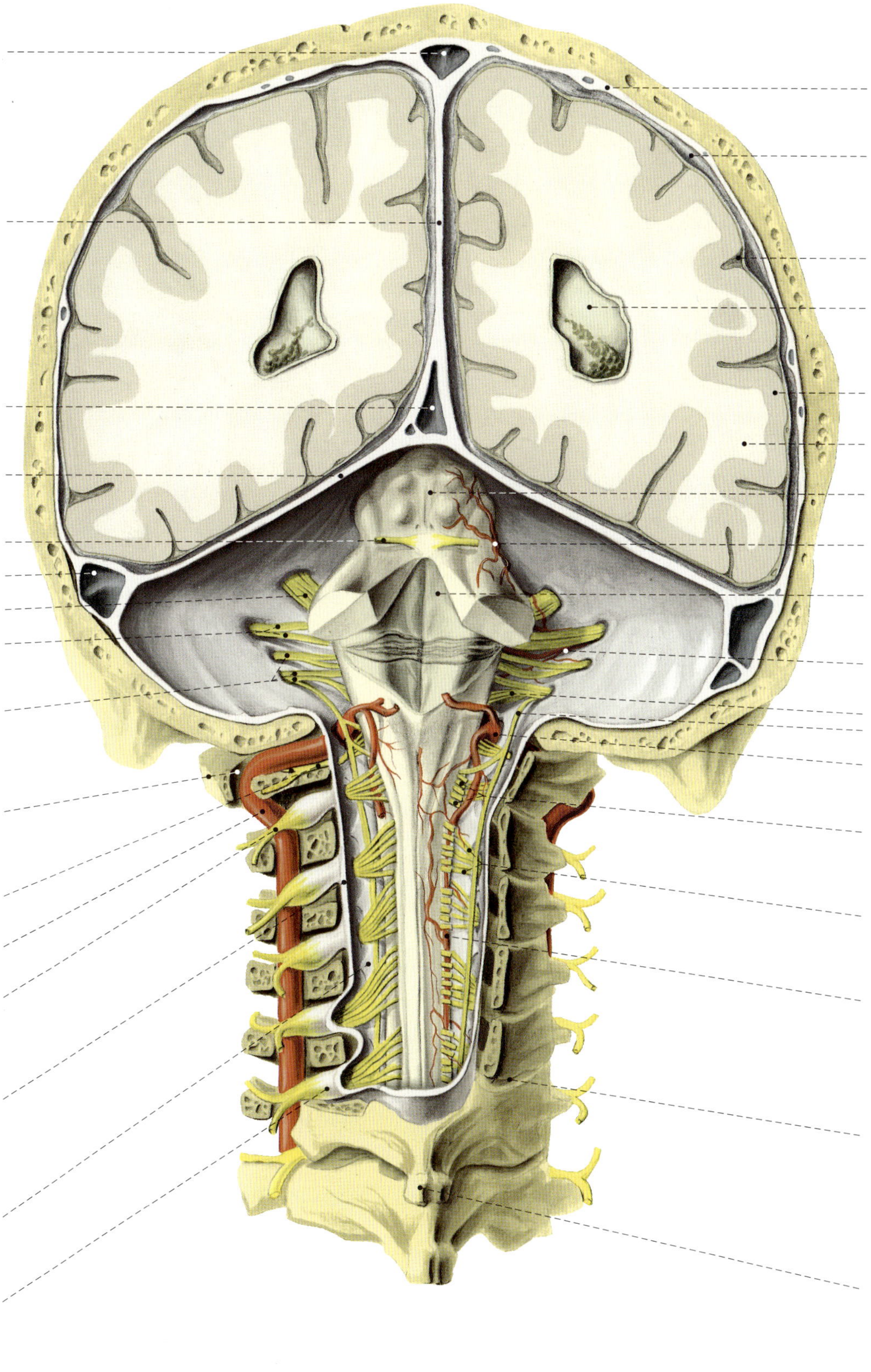

289 Truncus encephali et Dura mater

289 Brain stem and dura mater

(75%)

Coronal section through the head and neck.
The neural arches were removed.
Dorsal aspect

289 Hirnstamm und harte Hirnhaut

(75%)

Frontalschnitt durch Kopf und Hals.
Der Wirbelkanal ist von dorsal eröffnet.
Dorsalansicht

a

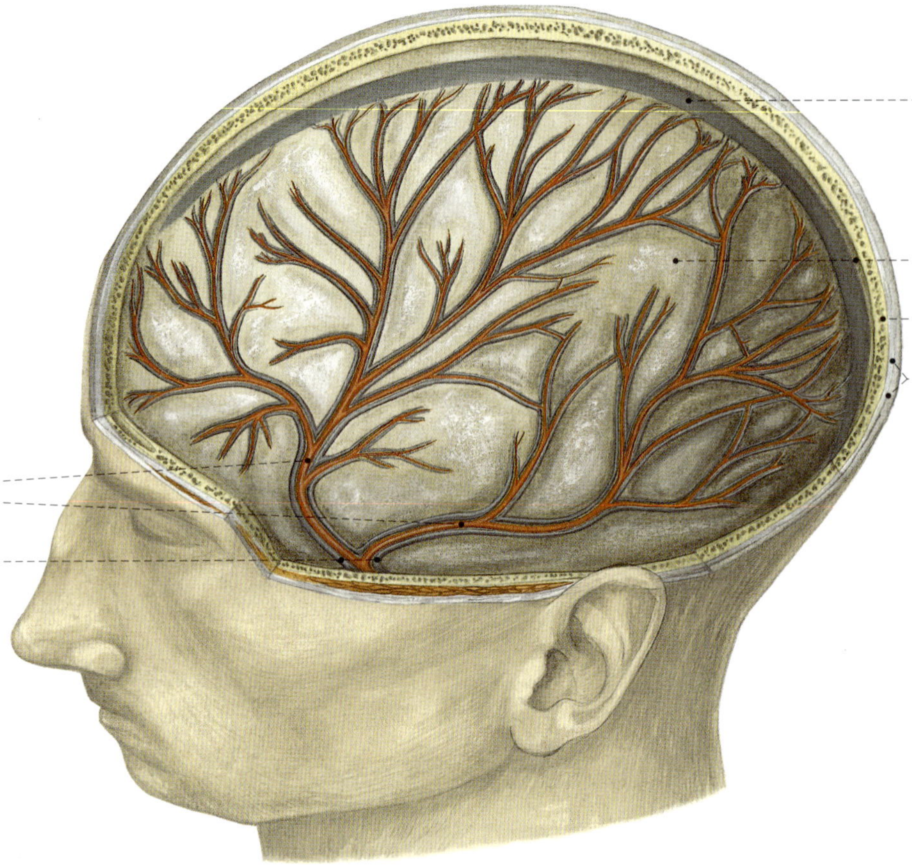

b

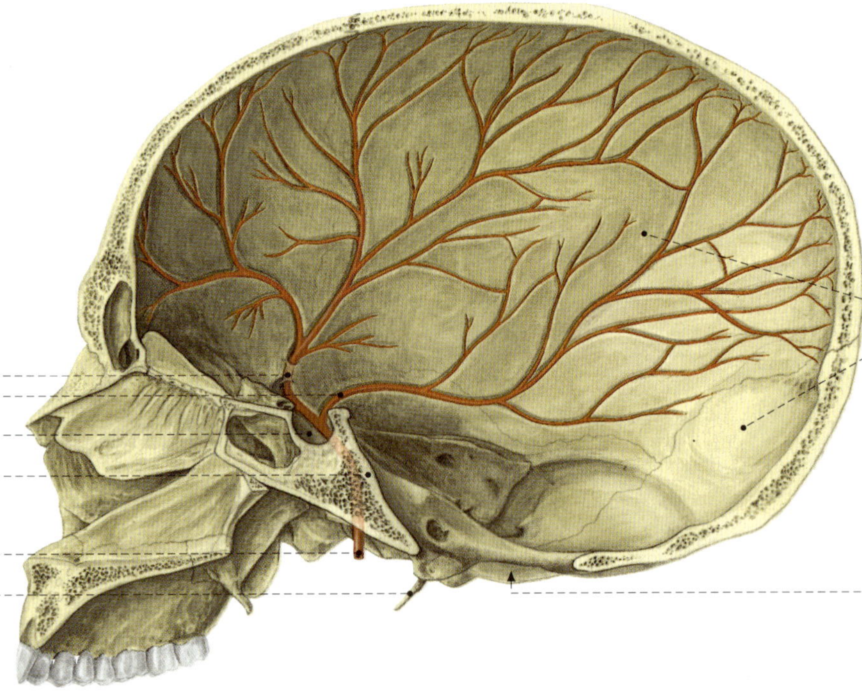

290 Arteriae meningeae

290 Meningeal arteries (50%)

a Meningeal arteries and veins apposed externally to the dura mater. The bones of the lateral skull were removed. Left parietolateral aspect

b Meningeal arteries lying in osseous sulci of the lateral skull, medial aspect of the right half

290 Hirnhautarterien (50%)

a Hirnhautarterien und -venen, der harten Hirnhaut aussen aufliegend. Die seitlichen Schädelknochen sind entfernt. Ansicht von links seitlich oben

b Hirnhautarterien in Knochenfurchen des seitlichen Schädeldachs verlaufend, Medialansicht der rechten Hälfte

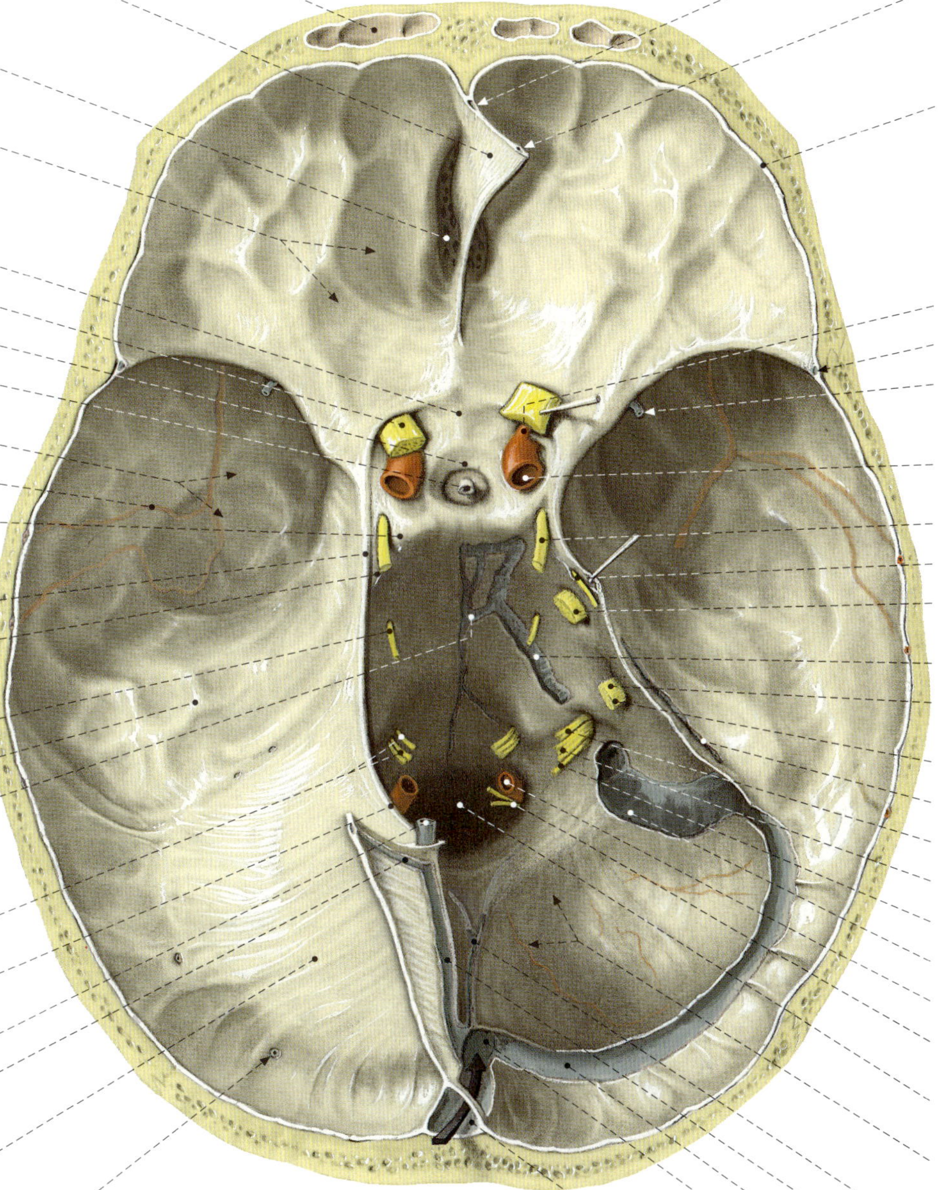

291 Dura mater cranialis et Sinus durae matris

291 Cerebral dura mater and dural venous sinuses (80%)

The right half of the tentorium cerebelli was removed, as well as the falx cerebri except for its anterior and posterior attachments. Some of the dural sinuses were opened. Superior aspect

291 Harte Hirnhaut und venöse Blutleiter (80%)

Die rechte Hälfte des Tentorium cerebelli ist entfernt, ebenso die Falx cerebri bis auf ihren vorderen und hinteren Ansatz. Die Sinus durae matris sind teilweise eröffnet. Parietalansicht

a

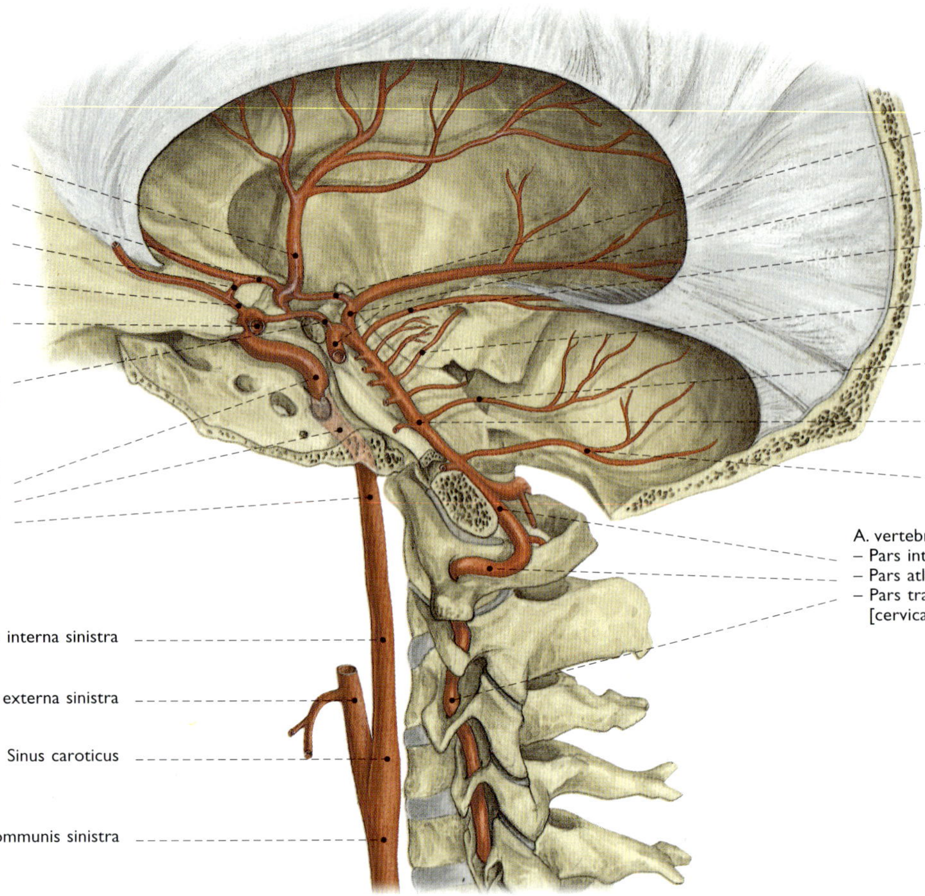

b

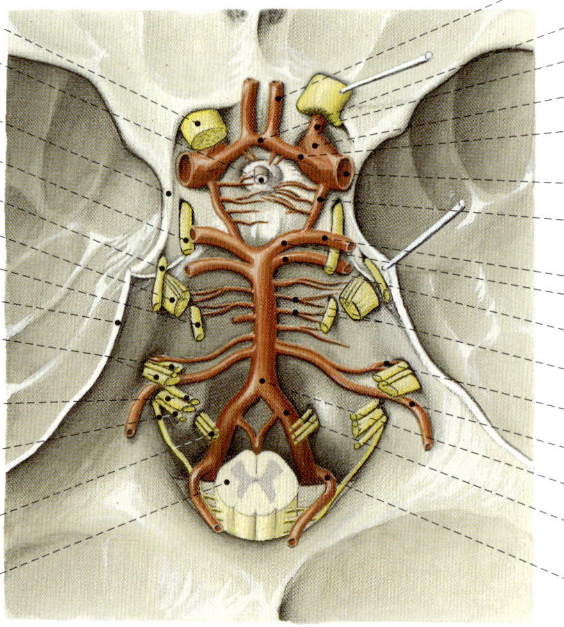

292 Dura mater et Circulus arteriosus cerebri

292 Dura mater and arterial circle of the cerebrum

a Internal carotid and vertebral arteries, arterial circle of the cerebrum (60%), sagittal section to the left of the median plane, parietomedial aspect of the right half
b Arterial circle of the cerebrum (90%), parietal aspect

292 Harte Hirnhaut und Circulus arteriosus cerebri

a Innere Kopfschlagader, Wirbelschlagader und Circulus arteriosus cerebri (60%), Sagittalschnitt links der Medianebene, Parietomedialansicht der rechten Hälfte
b Circulus arteriosus cerebri (90%), Parietalansicht

a

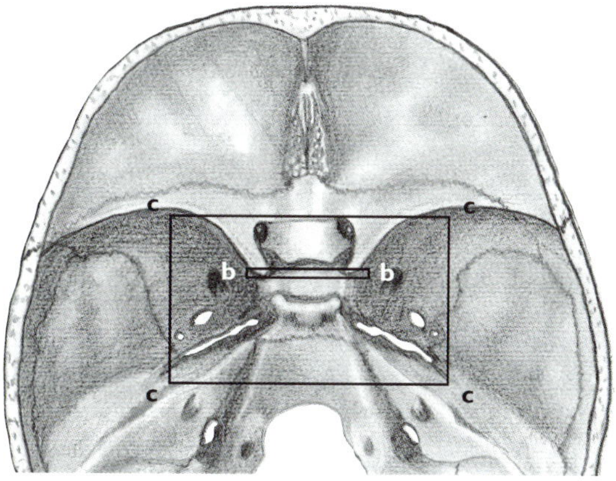

b

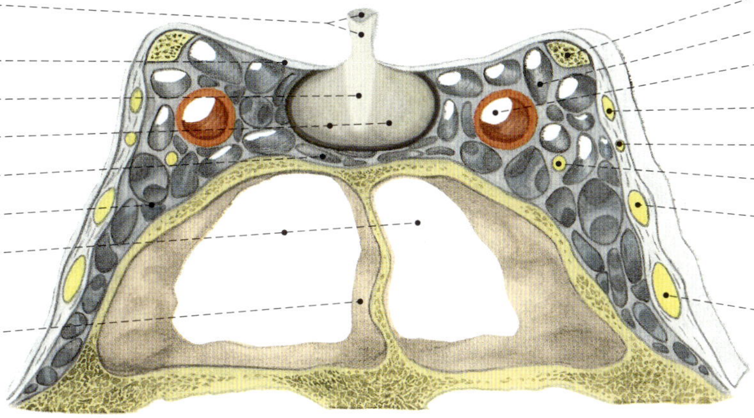

c

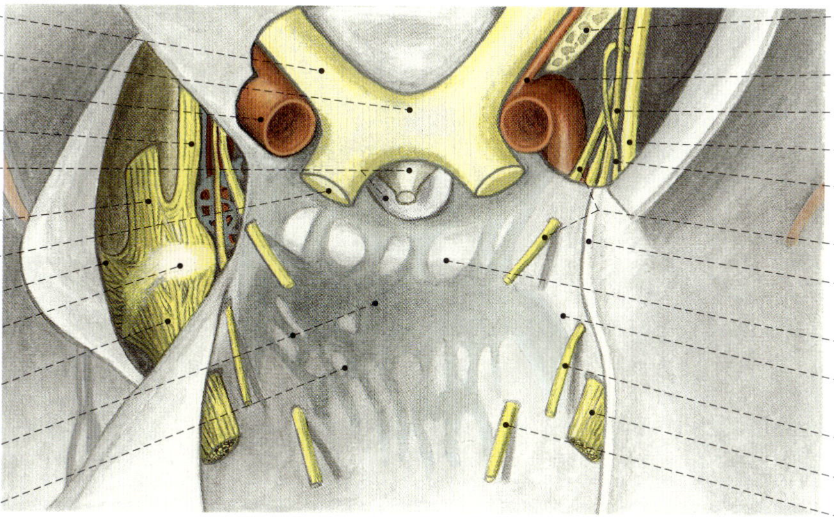

293 Sinus cavernosus

293 Cavernous sinus

a Internal surface of the cranial base. The section planes of Figs. b and c are indicated.
b Coronal section through the cavernous sinus, the hypophysis, and the sphenoidal sinuses (300%)
c View of central areas of the base of the skull. On the left the dura mater is turned upwards, the cavernous sinus opened, and the trigeminal ganglion exposed (200%). Parietal aspect

293 Sinus cavernosus

a Innere Schädelbasis mit Angabe der Schnittebenen von Abb. b und c
b Frontalschnitt durch den Sinus cavernosus, die Hypophyse und die Keilbeinhöhlen (300%)
c Aufsicht auf den zentralen Schädelbasisbereich. Links ist die harte Hirnhaut hochgehoben, der Sinus cavernosus eröffnet und das Trigeminusganglion freigelegt (200%). Parietalansicht

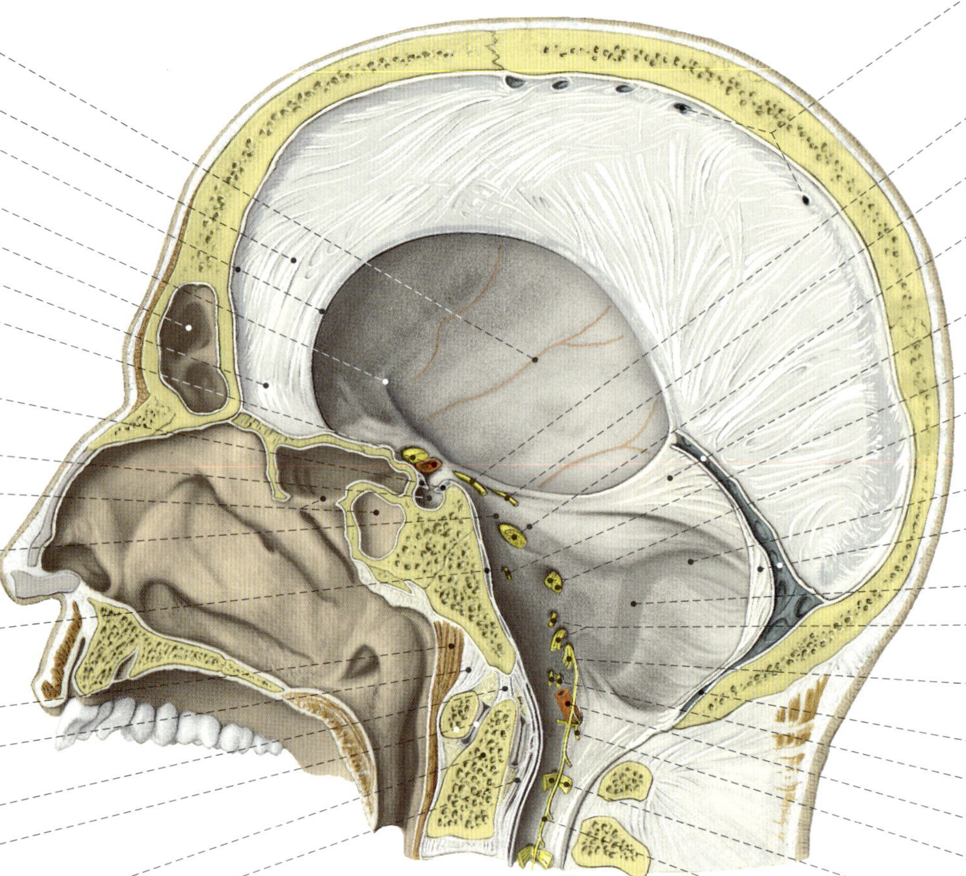

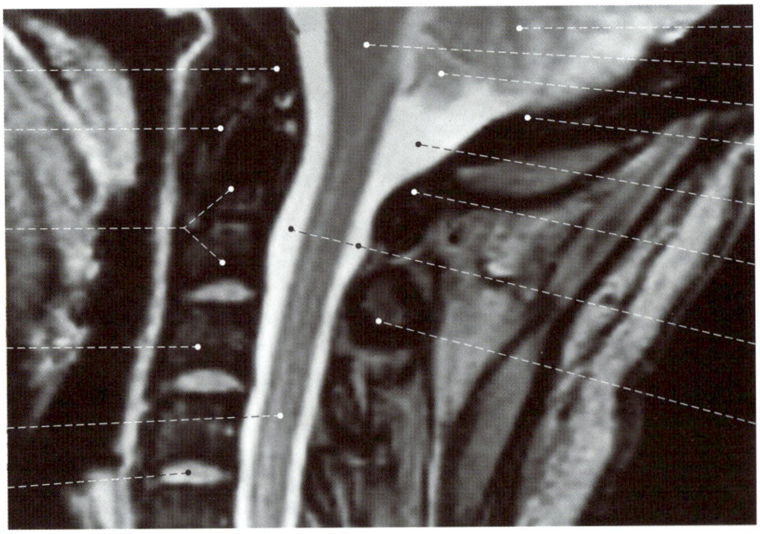

294 Cavitas cranii et Meninges

294 Cranial cavity and meninges

a Cranial cavity sectioned slightly to the left of the median plane. The nasal septum was removed (50%). Medial aspect of the right half

b Sagittal magnetic resonance image (MRI, T_2-weighted) of the brain stem, the cerebellomedullary cistern and adjacent structures (70%)

294 Schädelhöhle und Hirnhäute

a Paramedianschnitt durch die Schädelhöhle etwas links der Medianebene. Die Nasenscheidewand ist entfernt (50%). Medialansicht der rechten Hälfte

b Sagittales Kernspintomogramm (MRI, T_2-gewichtet) durch den Hirnstamm, die Cisterna cerebellomedullaris und benachbarte Strukturen (70%)

a

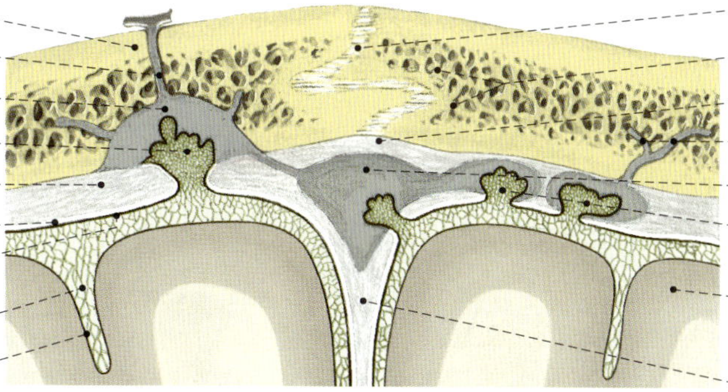

b

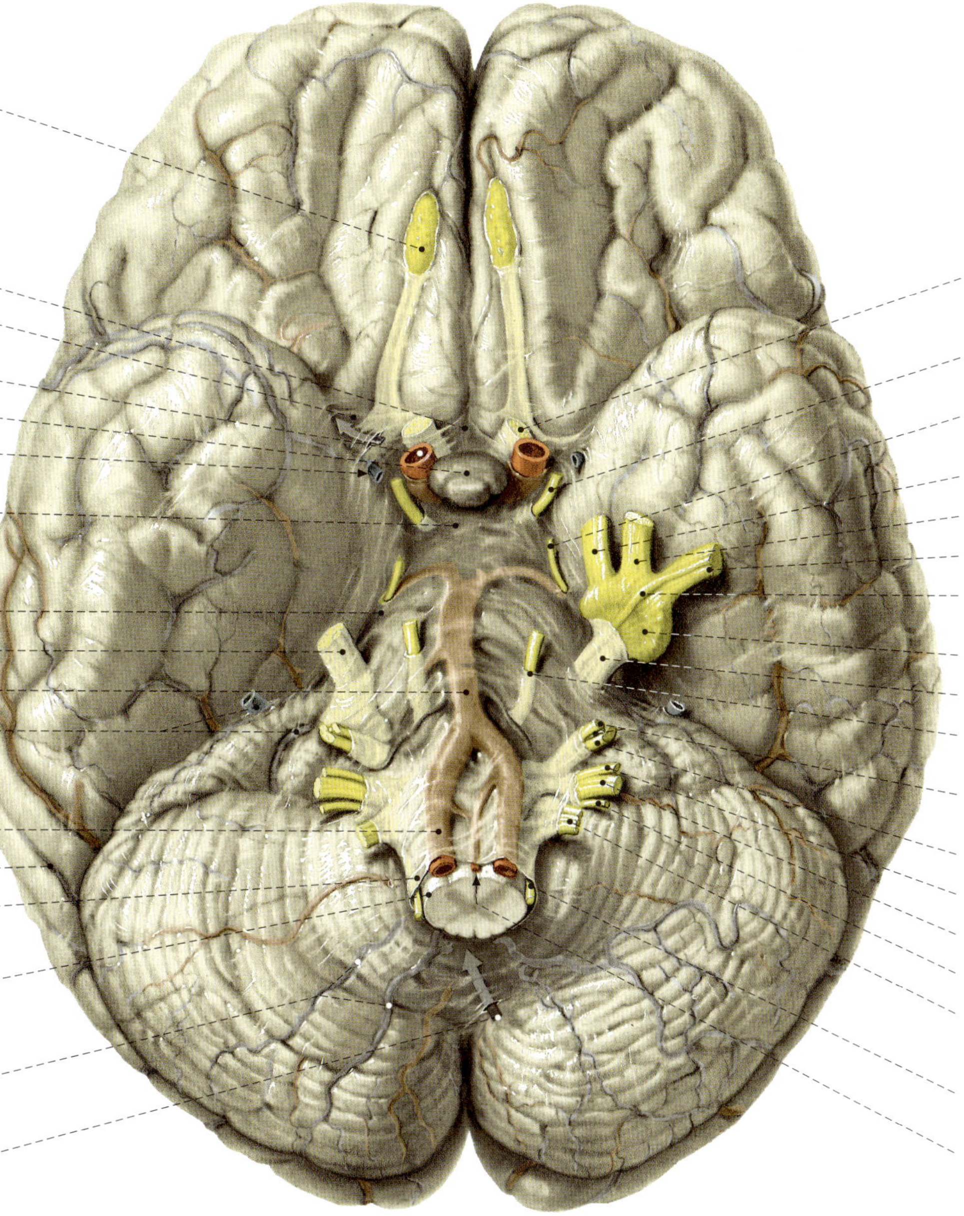

295 Cerebrum et Meninges

295 Brain and meninges (100%)

a Meninges on the cerebral vault, schematic coronal section
b Brain with leptomeninges (100%), inferior aspect

295 Gehirn und Hirnhäute (100%)

a Hirnhäute am Schädeldach, schematischer Frontalschnitt
b Gehirn mit weichen Hirnhäuten (100%), Basalansicht

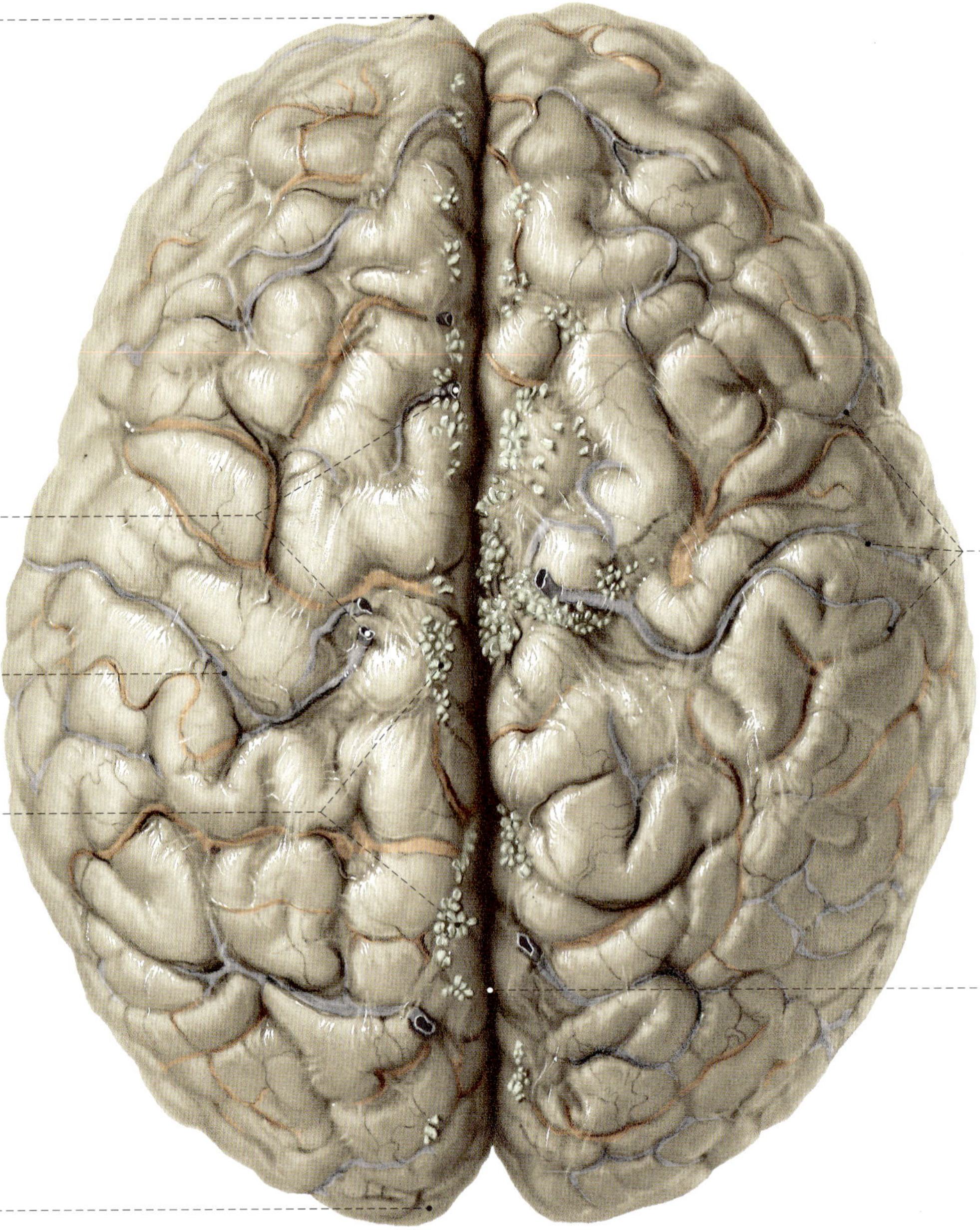

296 Cerebrum et Meninges

296 Brain with leptomeninges
(100%)
Superior aspect

296 Gehirn mit weichen Hirnhäuten
(100%)
Parietalansicht

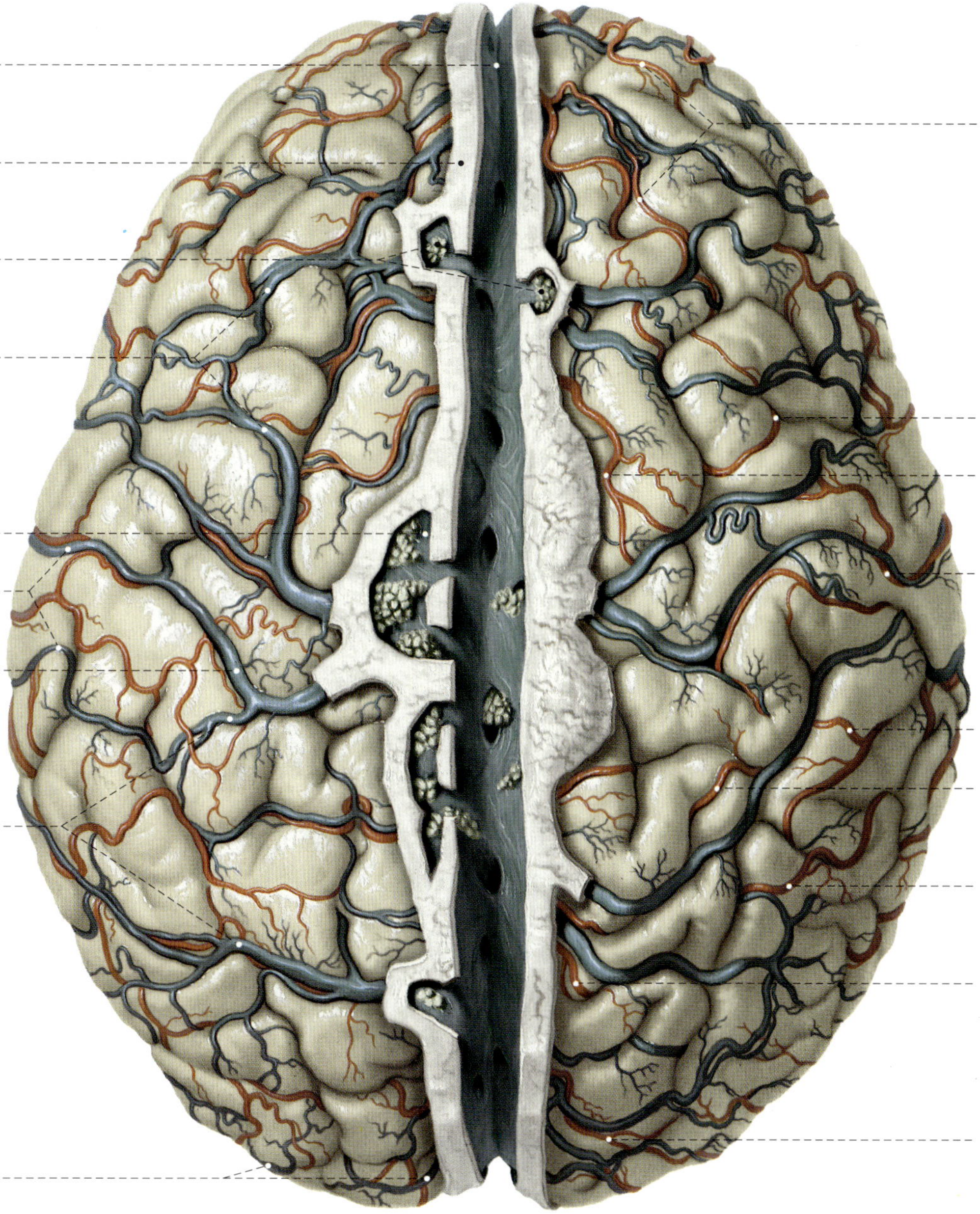

297 Arteriae cerebrales et Venae superficiales cerebri

297 Superficial arteries and veins of the cerebral hemispheres (100%)
Superior aspect

297 Oberflächliche Arterien und Venen des Gehirns (100%)
Parietalansicht

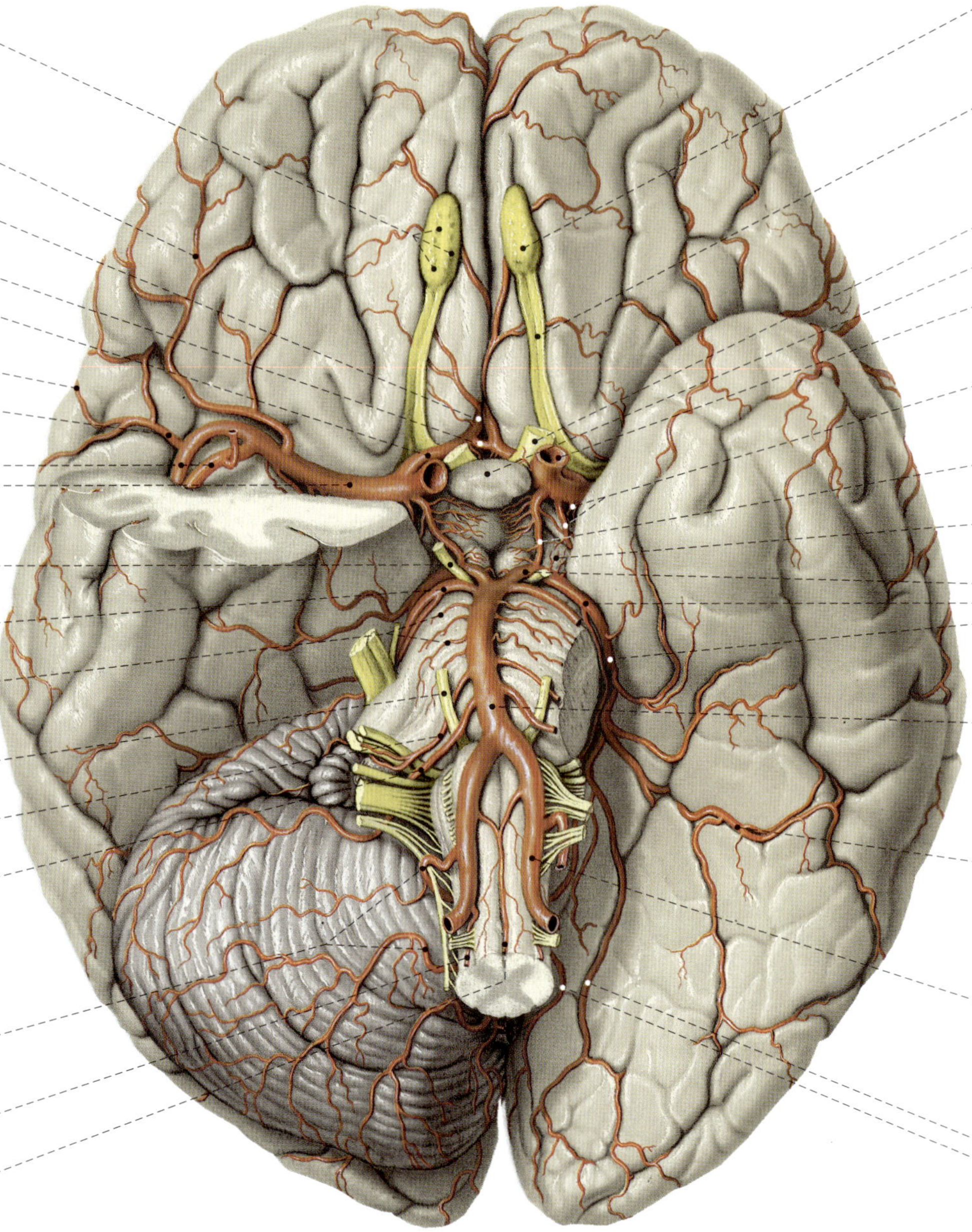

298 Arteriae cerebrales et Arteriae cerebelli

298 Arteries of the brain (100%)
Basal aspect

298 Arterien des Gehirns (100%)
Basalansicht

a

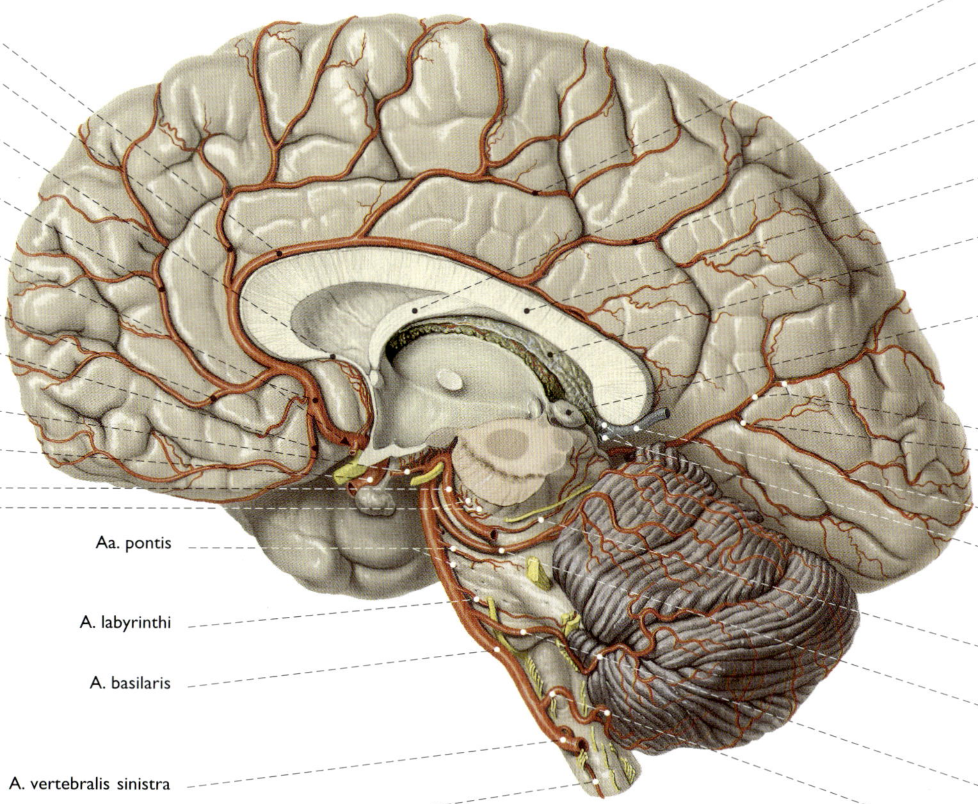

b

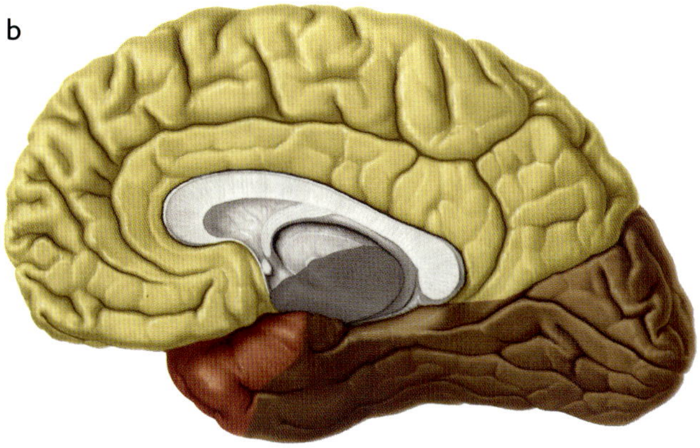

c

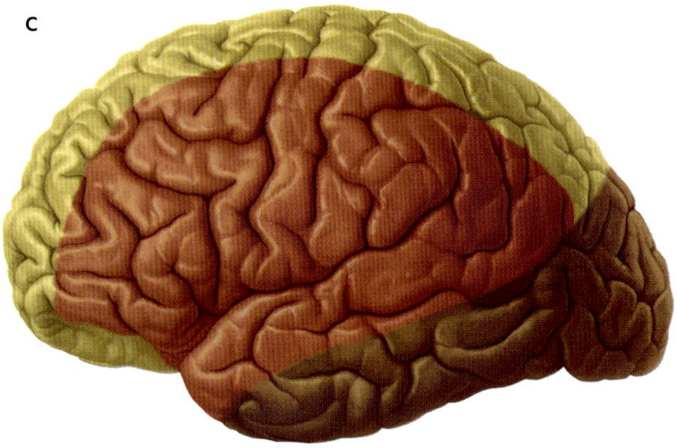

A. cerebri anterior A. cerebri media A. cerebri posterior

299 Arteriae cerebrales

299 Arteries of the brain

- a Medial aspect of the right cerebral hemisphere and lateral aspect of the left cerebellar hemisphere (80%)
- b, c Circulation areas of the anterior (**yellow**), middle (**red**), and posterior (**brown**) cerebral arteries at the cerebral cortex (50%)
- b Medial aspect
- c Lateral aspect

299 Arterien des Gehirns

- a Medialansicht der rechten Grosshirnhemisphäre und Seitenansicht der linken Kleinhirnhemisphäre (80%)
- b, c Versorgungsgebiete der Arteriae cerebri anterior (**gelb**), media (**rot**) et posterior (**braun**) an der Grosshirnrinde (50%)
- b Medialansicht
- c Seitenansicht

a

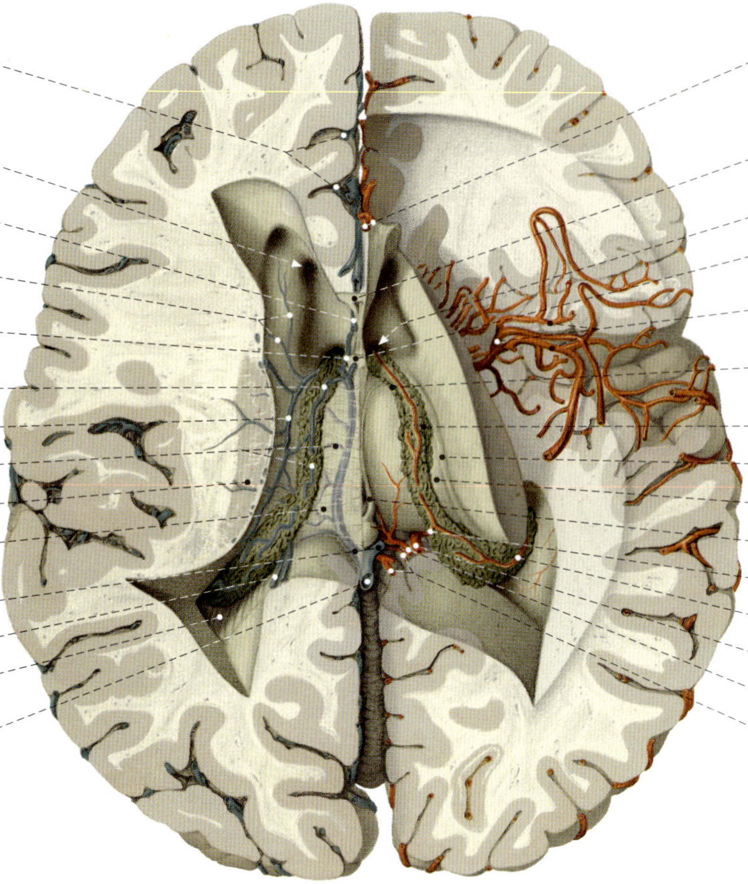

b

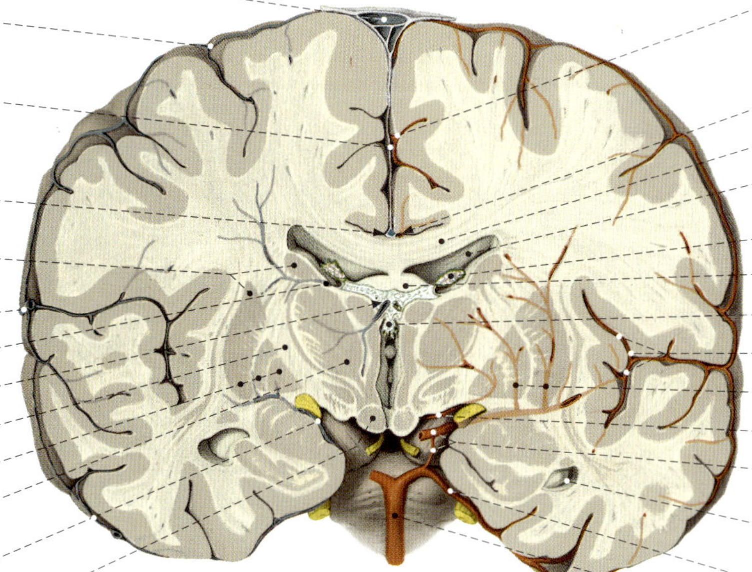

300 Arteriae cerebrales et Venae cerebri

300 Blood vessels of the brain (70%)

a Horizontal section through the upper part of the lateral ventricle. The corpus callosum, the fornix and parts of the hemispheres were removed. Superior aspect

b Coronal section at the level of the mamillary bodies, frontal aspect

300 Blutgefässe des Gehirns (70%)

a Horizontalschnitt mit Präparation der Seitenventrikel. Balken, Crura fornicis und Teile der Hemisphären sind entfernt. Parietalansicht

b Frontalschnitt in Höhe der Corpora mamillaria, Frontalansicht

a

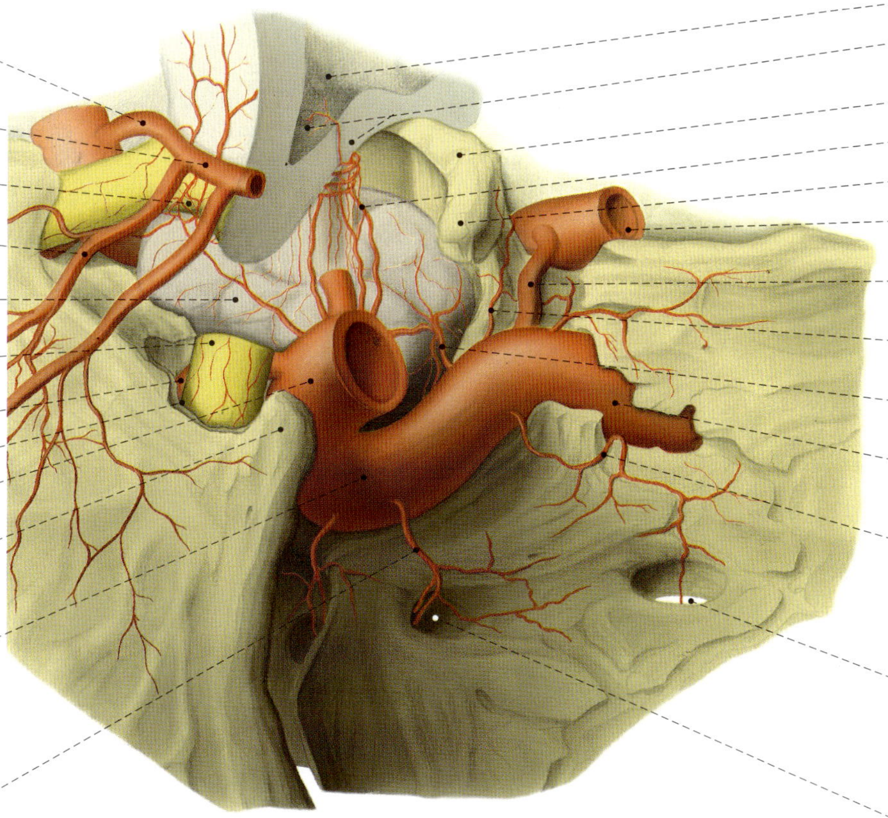

b

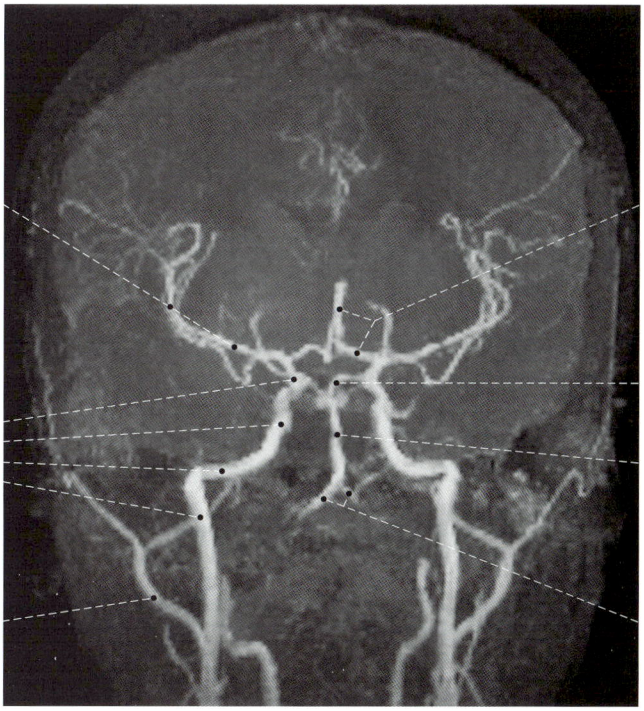

301 Arteriae cerebrales

301 Arteries of the brain

a Hypophysial arteries (300%), left superior aspect
b Coronal magnetic resonance angiogram (MRA) of the cerebral arteries (50%), frontal aspect

301 Arterien des Gehirns

a Arterien der Hypophyse (300%), linke Parietalansicht
b Frontales Kernspinangiogramm (MRA) der Gehirnarterien (50%), Frontalansicht

a

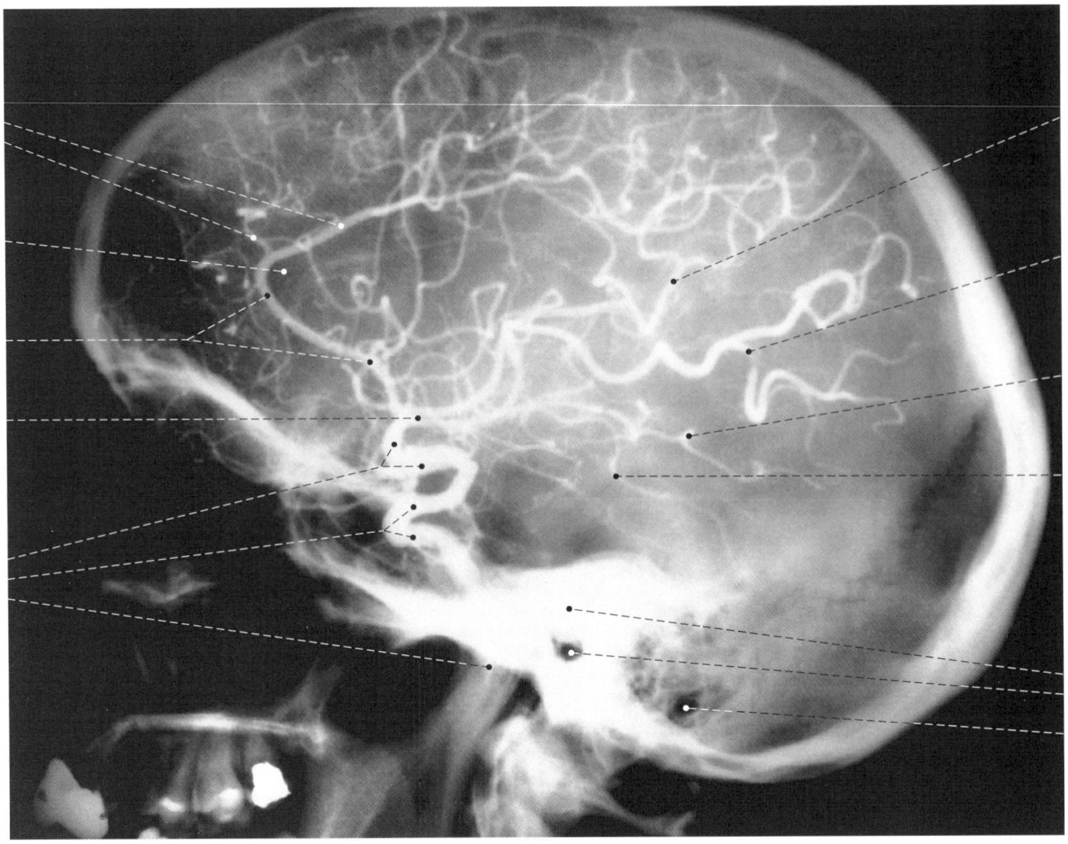

b

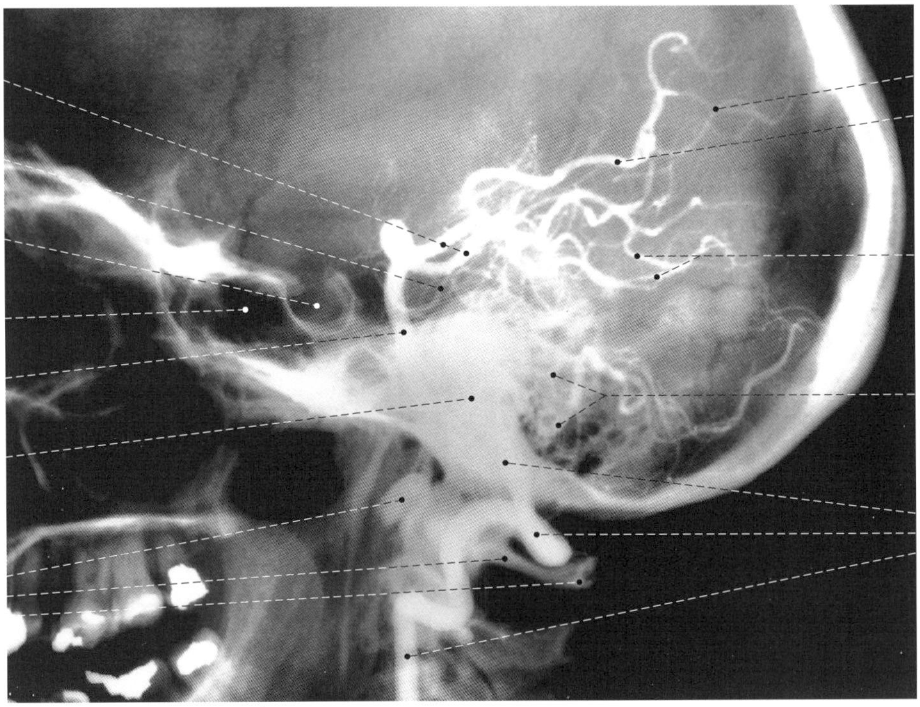

302 Arteriae cerebrales

302 Arteries of the brain (75%)

Lateral radiographs

a Selective arteriogram of the internal carotid artery

b Selective arteriogram of the vertebral artery

302 Arterien des Gehirns (75%)

Seitliche Röntgenaufnahmen

a Selektives Arteriogramm der inneren Kopfschlagader (Arteria carotis interna)

b Selektives Arteriogramm der Wirbelschlagader (Arteria vertebralis)

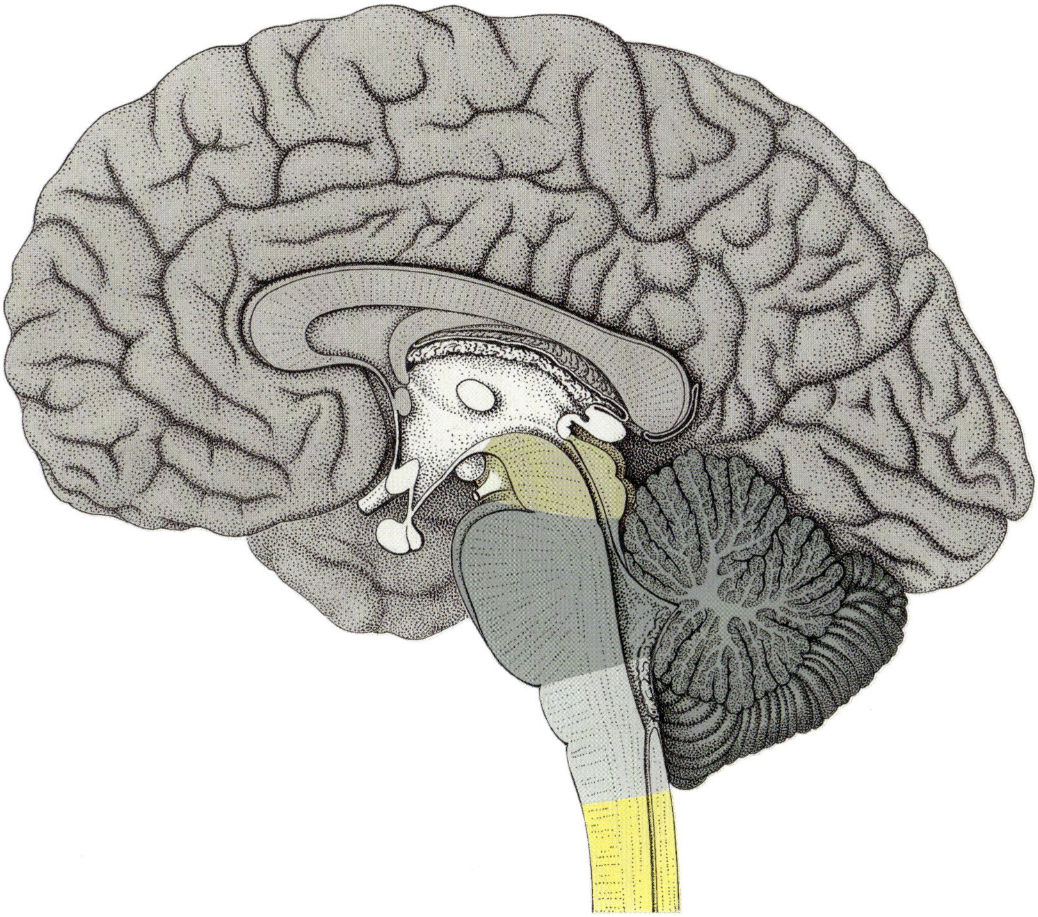

- Encephalon
  - Prosencephalon
    - Telencephalon [Cerebrum]
    - Diencephalon
  - Truncus encephali
    - Mesencephalon
    - Rhombencephalon
      - Pons and/und Cerebellum [Metencephalon]
      - Medulla oblongata [Myelencephalon]
- Medulla spinalis

303 Systema nervosum centrale

303 Division of the central nervous system

Schematic representation of the right hemisphere, medial aspect of a median section

303 Gliederung des zentralen Nervensystems

Schematische Darstellung der rechten Hemisphäre, Medialansicht eines Medianschnitts

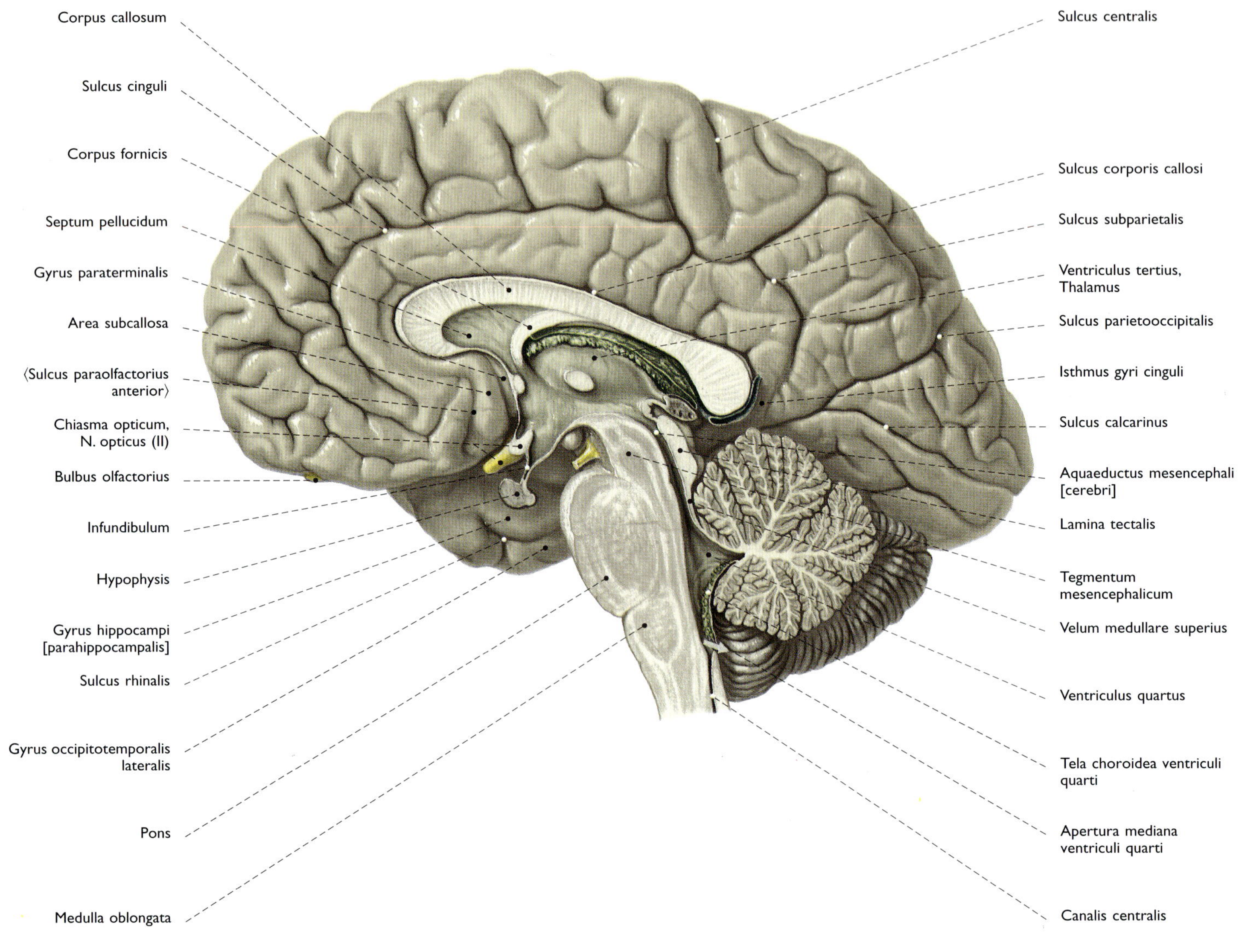

304 Encephalon

304 Brain (80%)

Median section, medial aspect of the right hemisphere

304 Gehirn (80%)

Medianschnitt, Medialansicht der rechten Hemisphäre

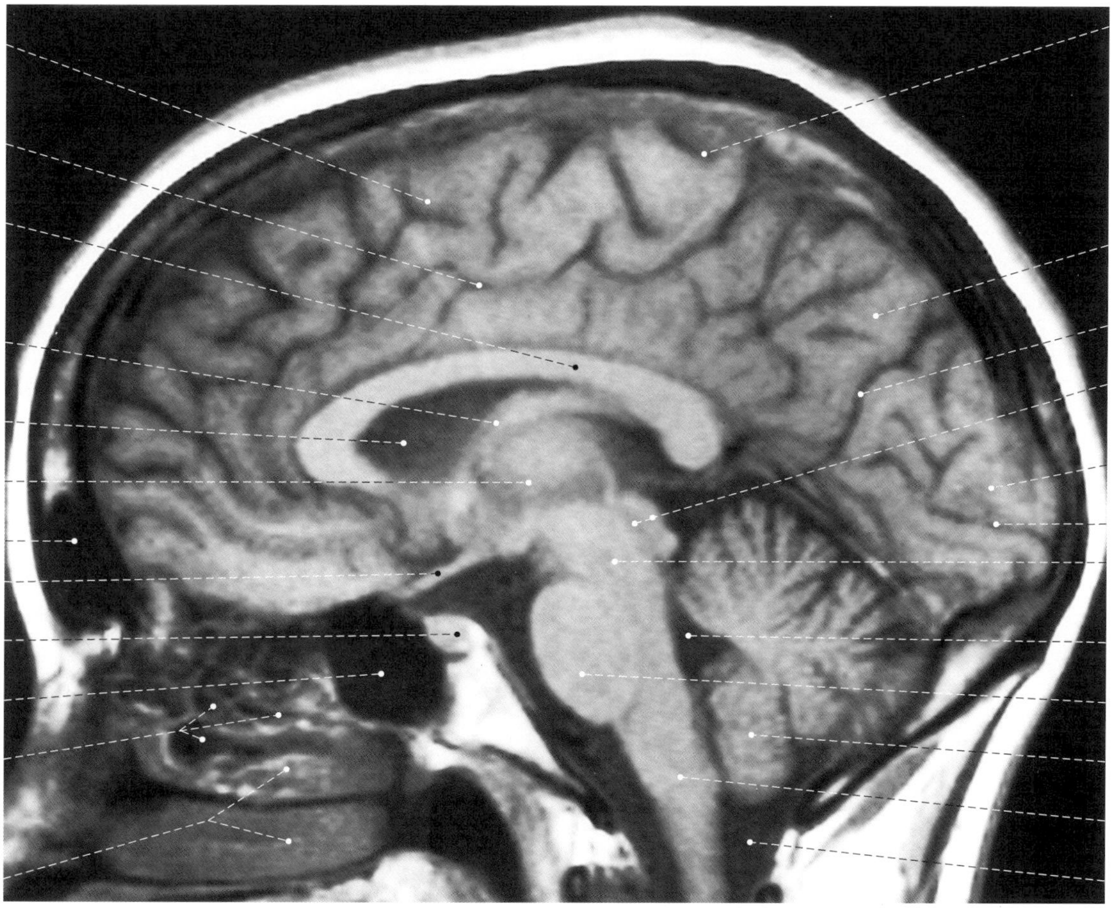

305 Encephalon

305 Brain (80%)

Sagittal, paramedian magnetic resonance image (MRI, T_1-weighted)

305 Gehirn (80%)

Sagittales, paramedianes Kernspintomogramm (MRI, T_1-gewichtet)

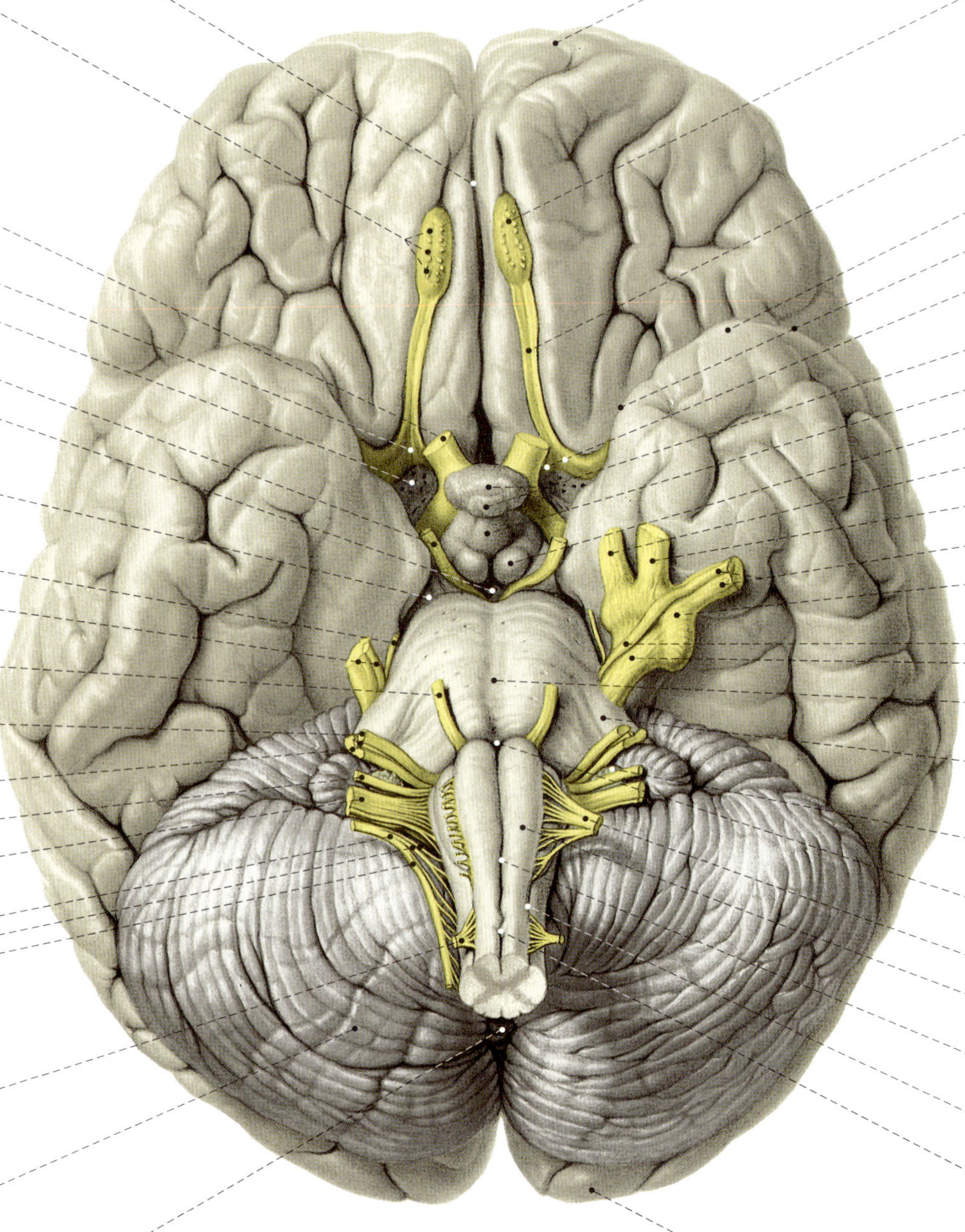

306 Truncus encephali, Facies inferior hemisphaerii et Nervi craniales

306 Brain and cranial nerves (100%)
Inferior aspect

306 Gehirn und Hirnnerven (100%)
Basalansicht

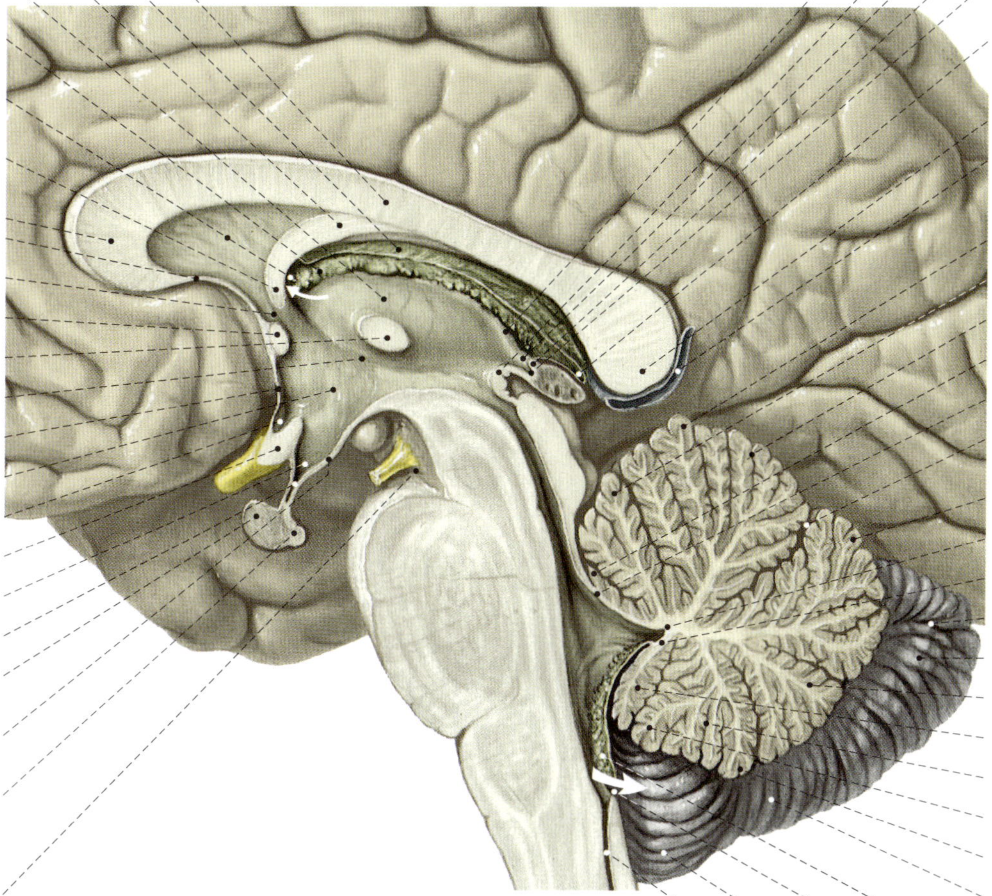

307 Truncus encephali, Hypophysis, Ventriculi tertius et quartus

307 Brain stem, third and fourth ventricles (110%)

Median section, medial aspect of the right hemisphere

307 Hirnstamm, dritter und vierter Ventrikel (110%)

Medianschnitt, Medialansicht der rechten Hemisphäre

a

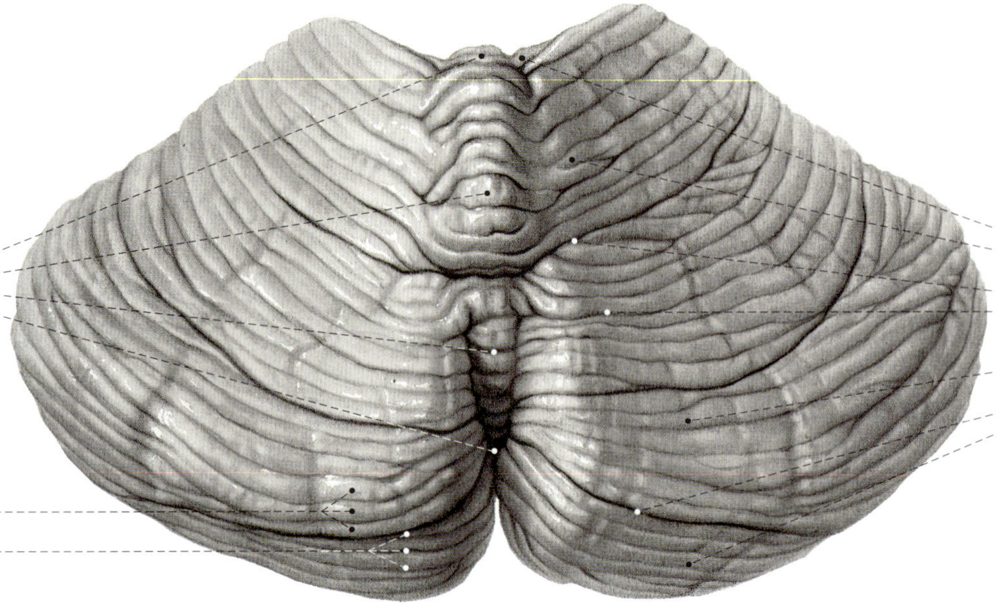

b

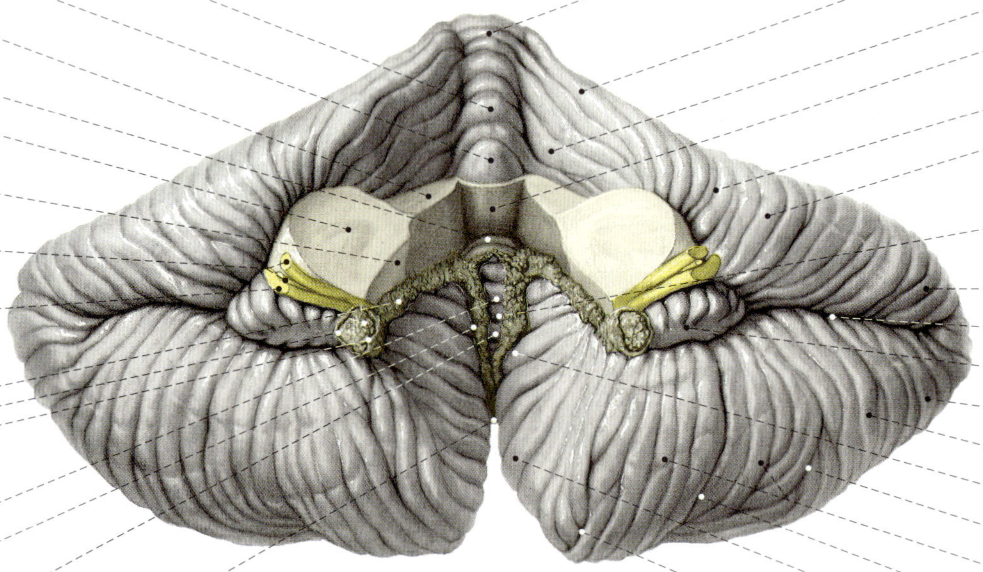

308 Cerebellum

308 Cerebellum (120%)

a Occipitoparietal aspect
b The cerebellar peduncles were transected. Ventral aspect

308 Kleinhirn (120%)

a Okzipitoparietalansicht
b Die Kleinhirnstiele sind durchtrennt. Ventralansicht

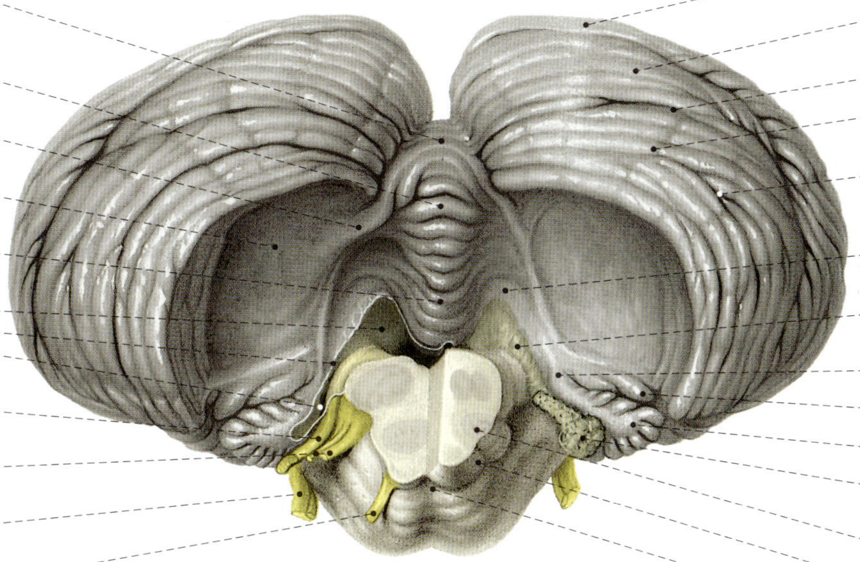

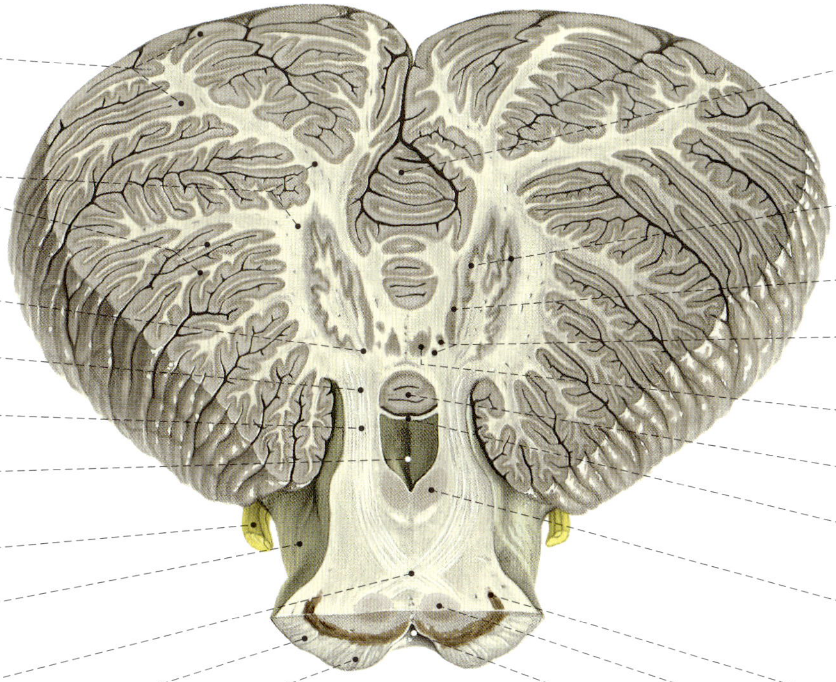

309 Cerebellum

309 Cerebellum (100%)

a The cerebellar tonsils were removed. Inferior aspect

b Section through cerebellum and midbrain, superior aspect

309 Kleinhirn (100%)

a Die Kleinhirntonsillen sind entfernt. Basalansicht

b Schnitt durch Kleinhirn und Mittelhirn, Parietalansicht

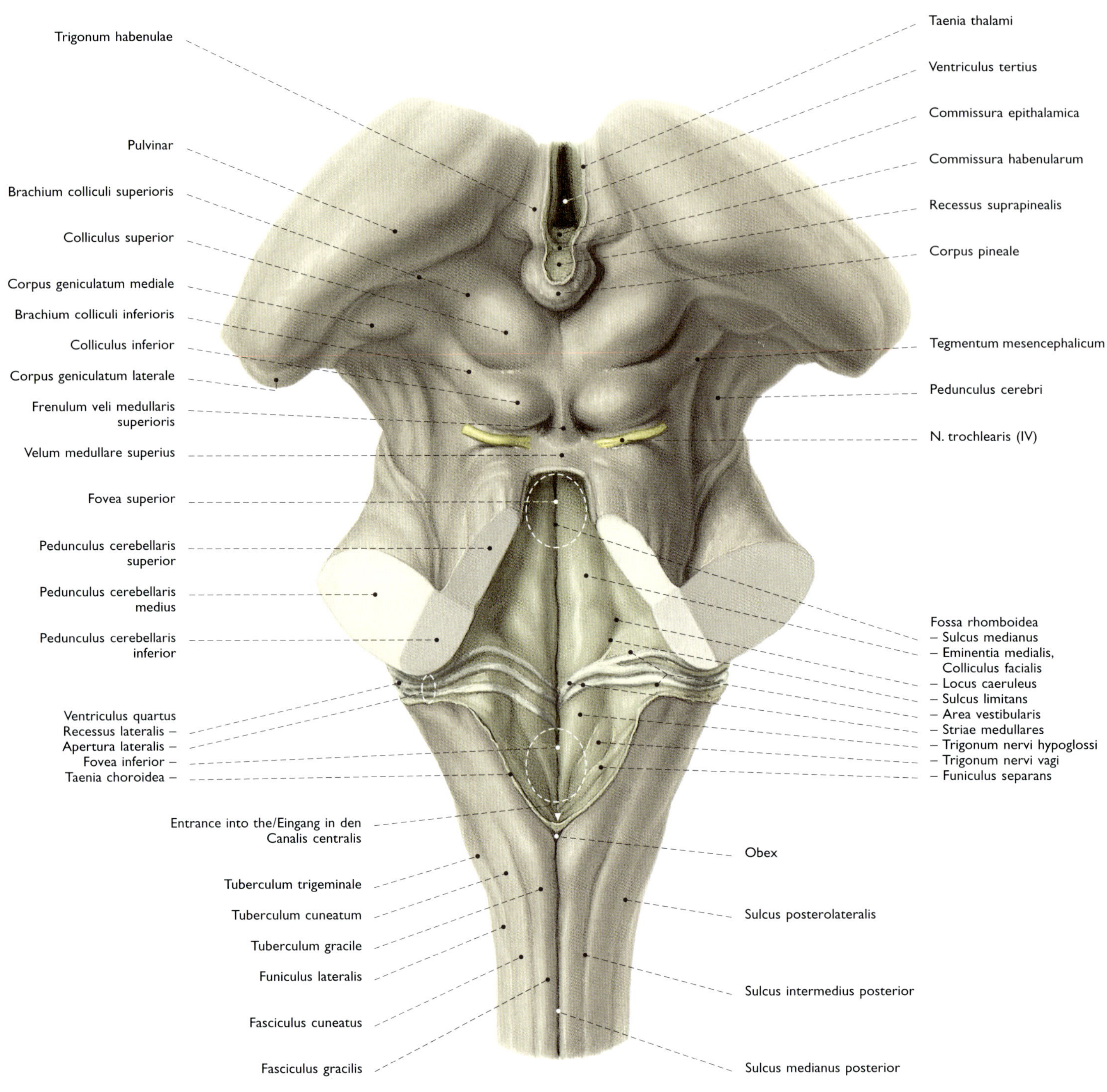

310 Truncus encephali et Ventriculus quartus

310 Brain stem and fourth ventricle (180%)
Dorsal aspect

310 Hirnstamm und vierter Ventrikel (180%)
Okzipitalansicht

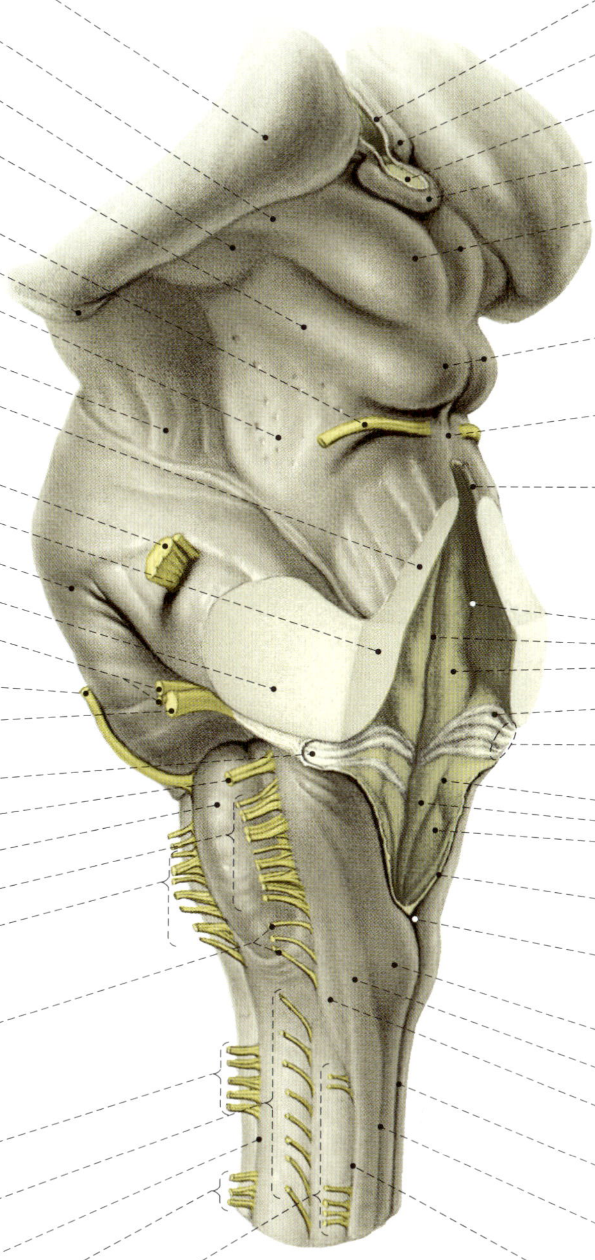

311 Truncus encephali et Ventriculus quartus

311 Brain stem and fourth ventricle
(180%)
Left dorsolateral aspect

311 Hirnstamm und vierter Ventrikel
(180%)
Linke Dorsolateralansicht

a

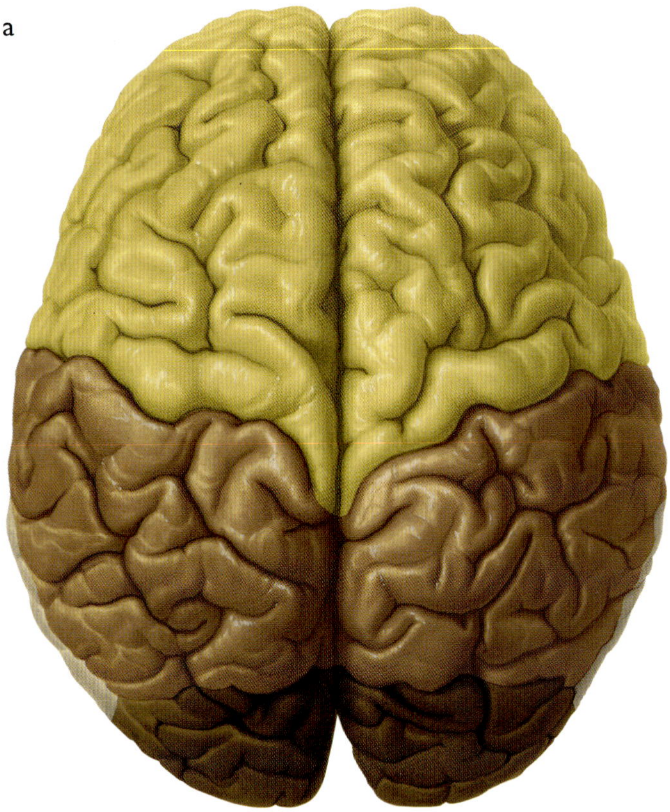

Lobus frontalis

Lobus parietalis

Lobus occipitalis

Lobus temporalis

⟨Lobus limbicus⟩

b

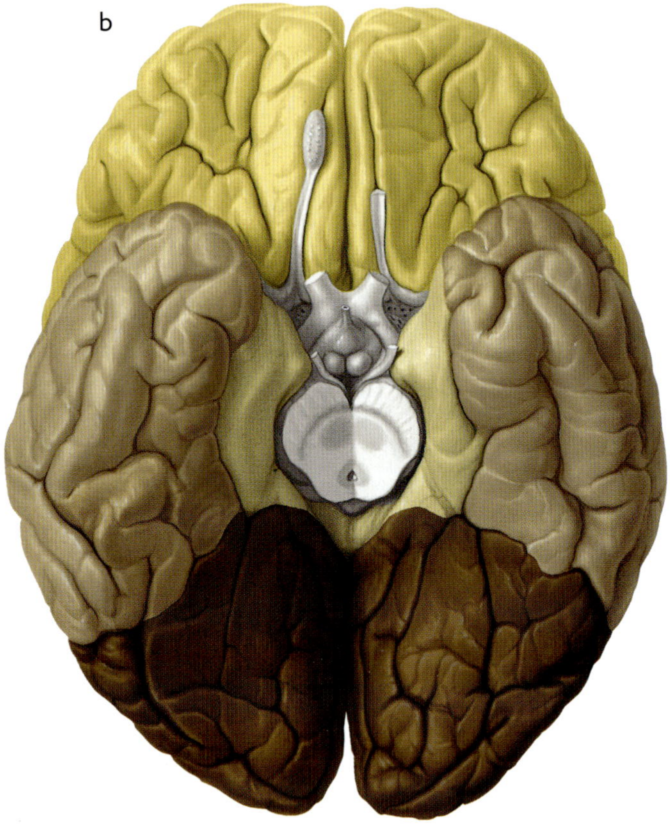

312 Lobi cerebrales

312 Cerebral lobes (70%)

The diverse cerebral lobes are indicated by different colors.

a Parietal aspect

b Inferior (basal) aspect

312 Lappen des Grosshirns (70%)

Die verschiedenen Grosshirnlappen sind durch unterschiedliche Farben markiert.

a Parietalansicht

b Basalansicht

a

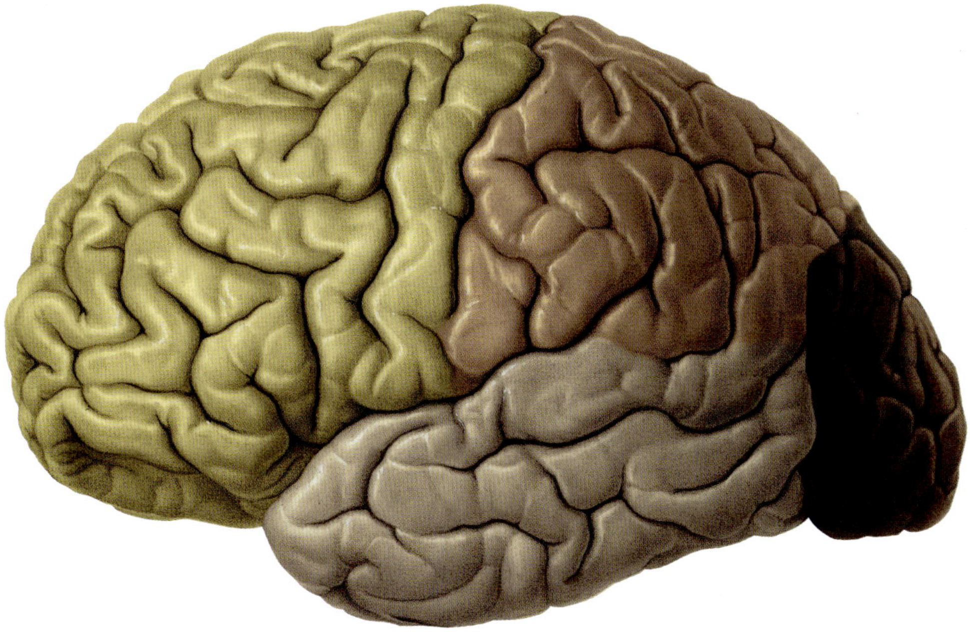

Lobus frontalis

Lobus parietalis

Lobus occipitalis

Lobus temporalis

⟨Lobus limbicus⟩

b

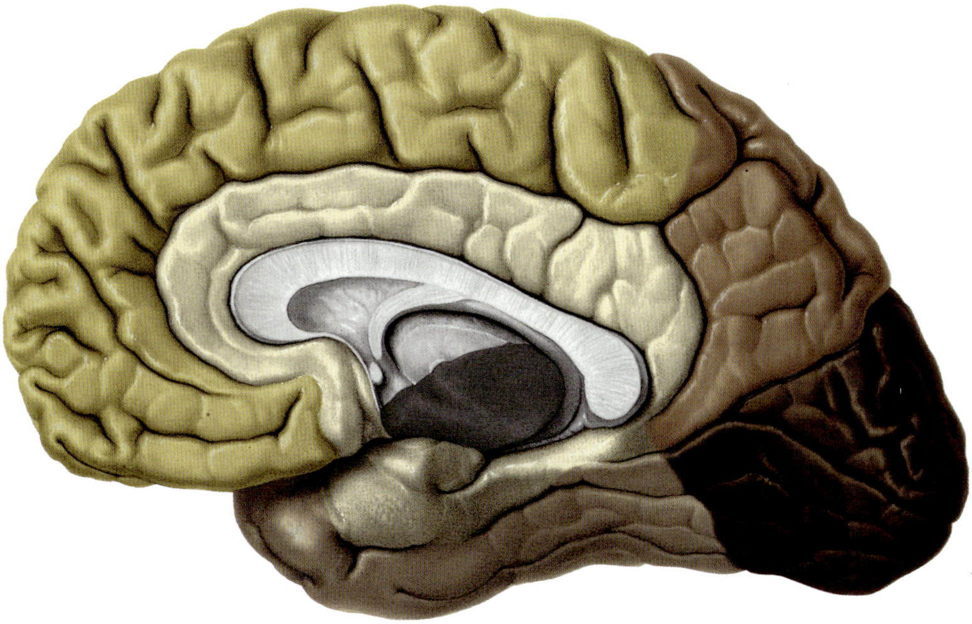

313 Lobi cerebrales

313 Cerebral lobes (80%)

The diverse cerebral lobes are indicated by different colors.

a Left lateral aspect of the left hemisphere
b Medial aspect of the right hemisphere

313 Lappen des Grosshirns (80%)

Die verschiedenen Grosshirnlappen sind durch unterschiedliche Farben markiert.

a Linke Seitenansicht der linken Hemisphäre
b Medialansicht der rechten Hemisphäre

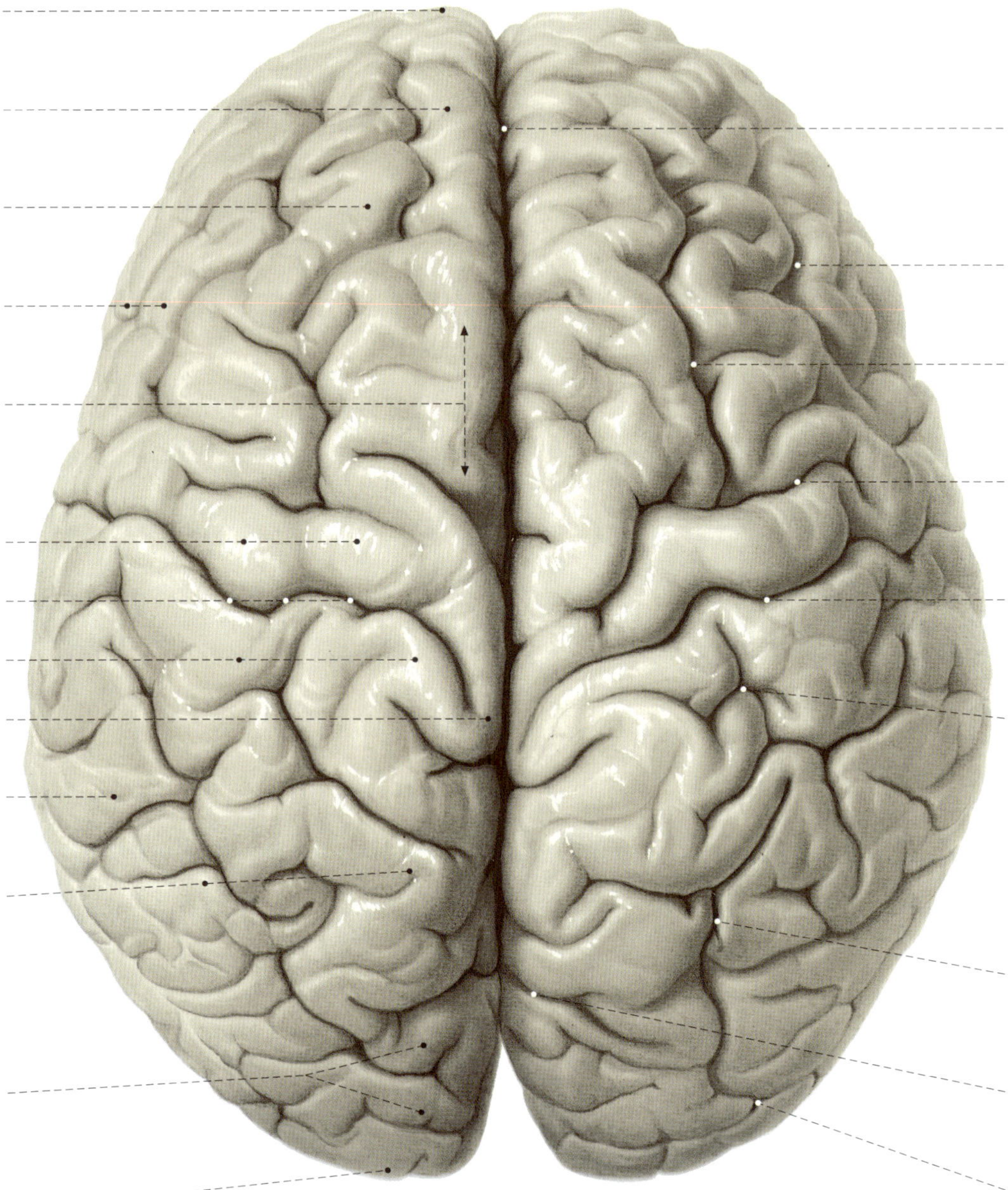

314 Facies superolateralis hemisphaerii

314 Cerebral hemispheres (100%)
Parietal aspect

314 Grosshirnhemisphären (100%)
Parietalansicht

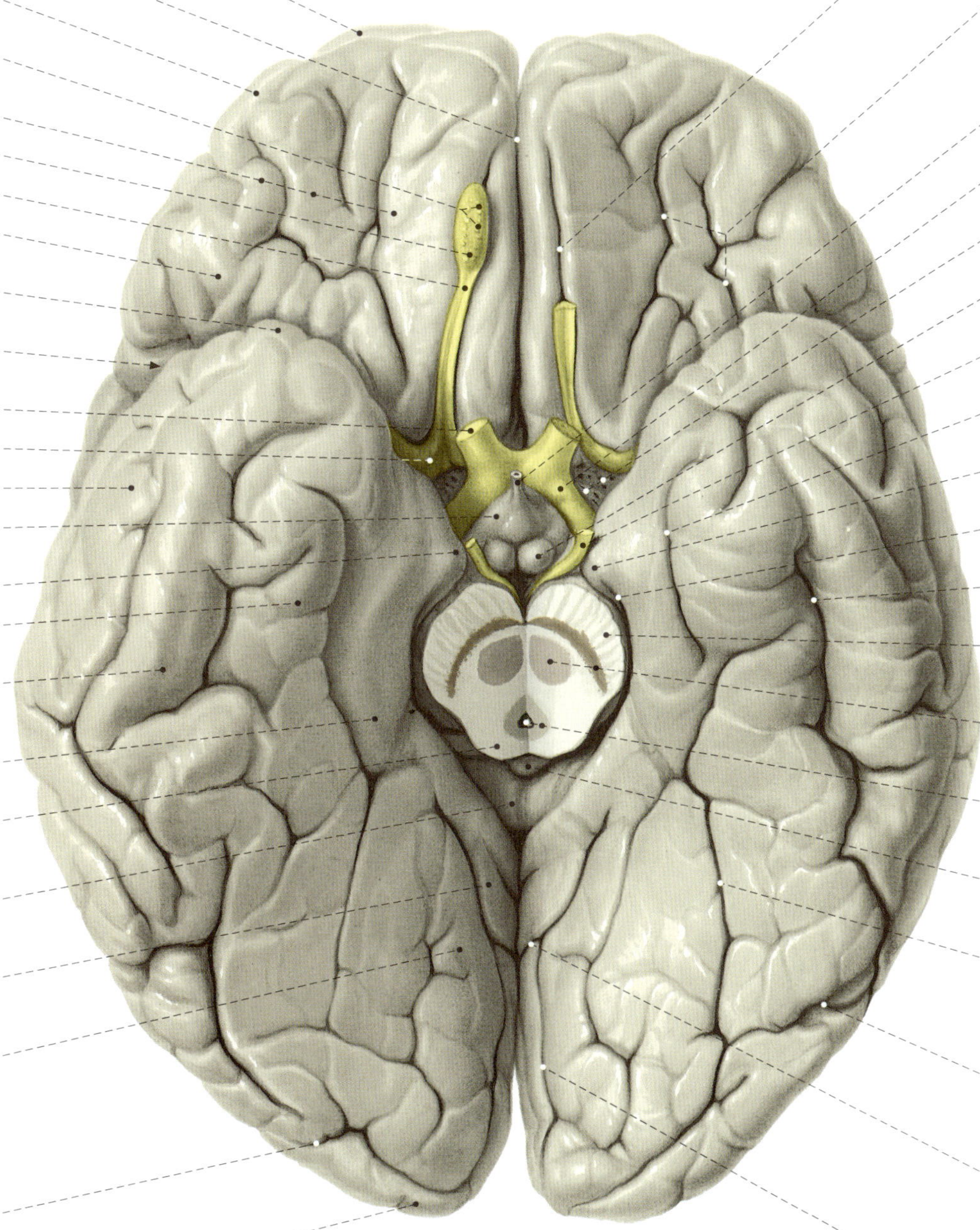

315 Facies inferior hemisphaerii

315 Cerebral hemispheres (100%)
The mesencephalon was transected.
Inferior (basal) aspect

315 Grosshirnhemisphären (100%)
Das Mittelhirn ist durchtrennt.
Basalansicht

a

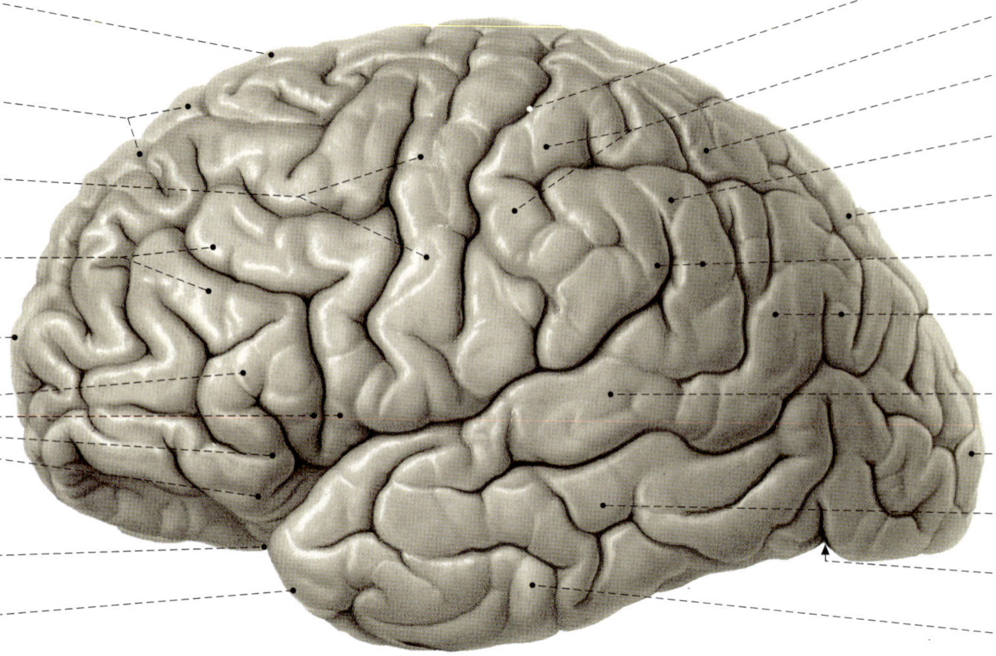

b

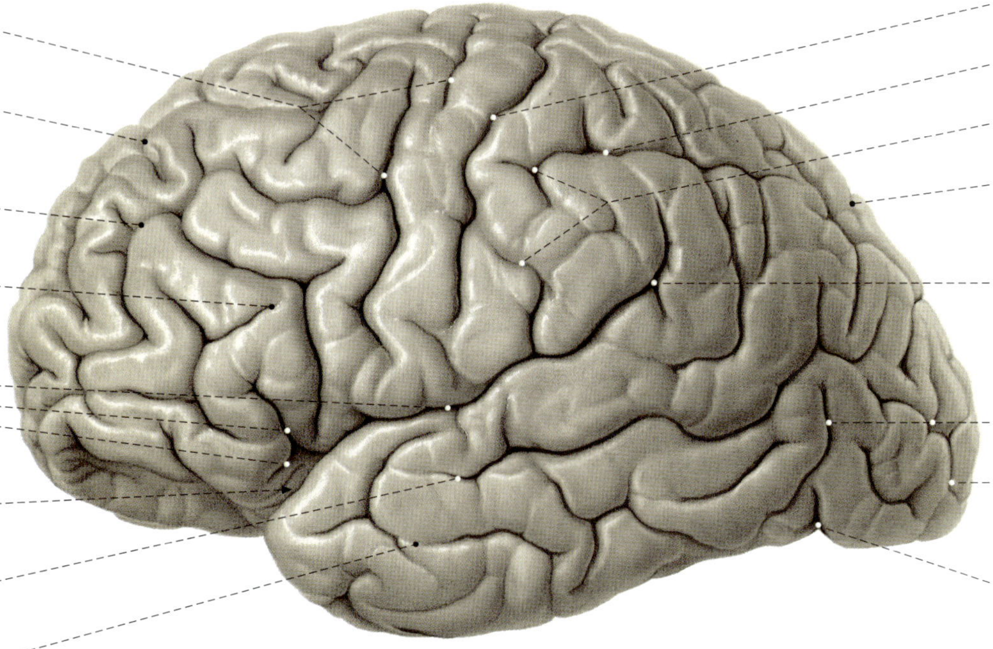

316 Facies superolateralis hemisphaerii

316 Left cerebral hemisphere (80%)
Left lateral aspect
Lettering of the
a gyri
b sulci

316 Linke Grosshirnhemisphäre (80%)
Linke Seitenansicht
Beschriftung der
a Gyri
b Sulci

a

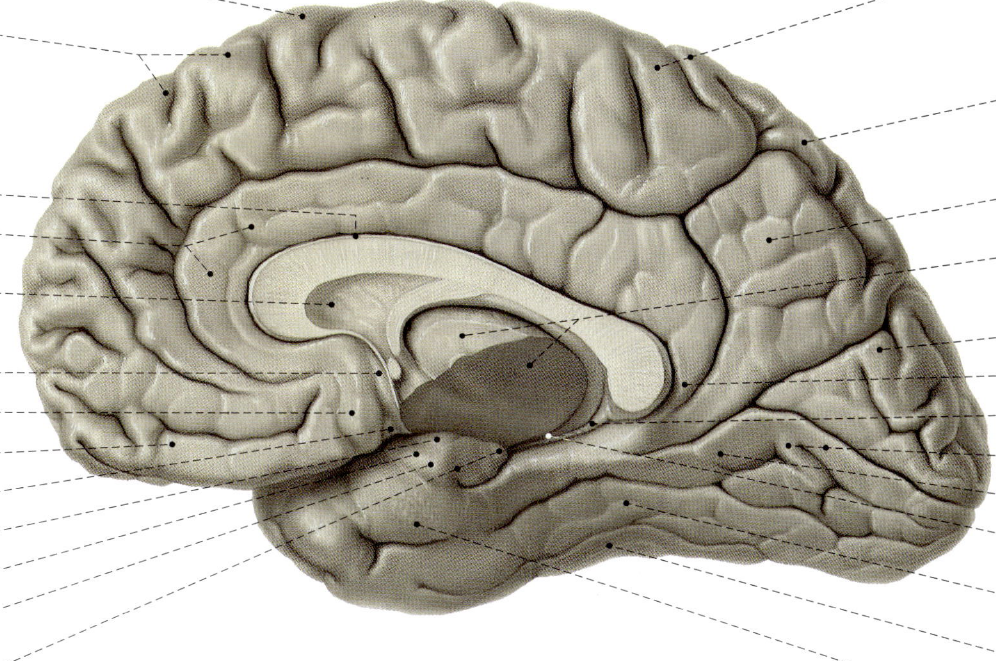

b

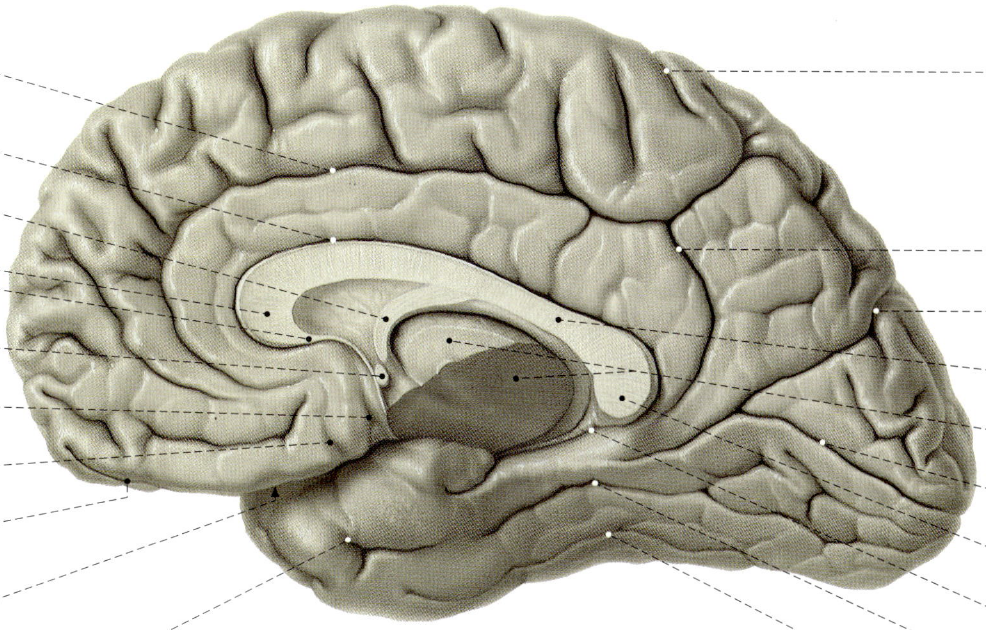

317 Facies medialis et inferior hemisphaerii

317 Right cerebral hemisphere (80%)

Medial aspect of a median section. The corpus callosum and the diencephalon were transected. Lettering of the

a gyri
b sulci

317 Rechte Grosshirnhemisphäre (80%)

Medialansicht eines Medianschnittes. Der Balken und das Zwischenhirn sind durchtrennt. Beschriftung der

a Gyri
b Sulci

a

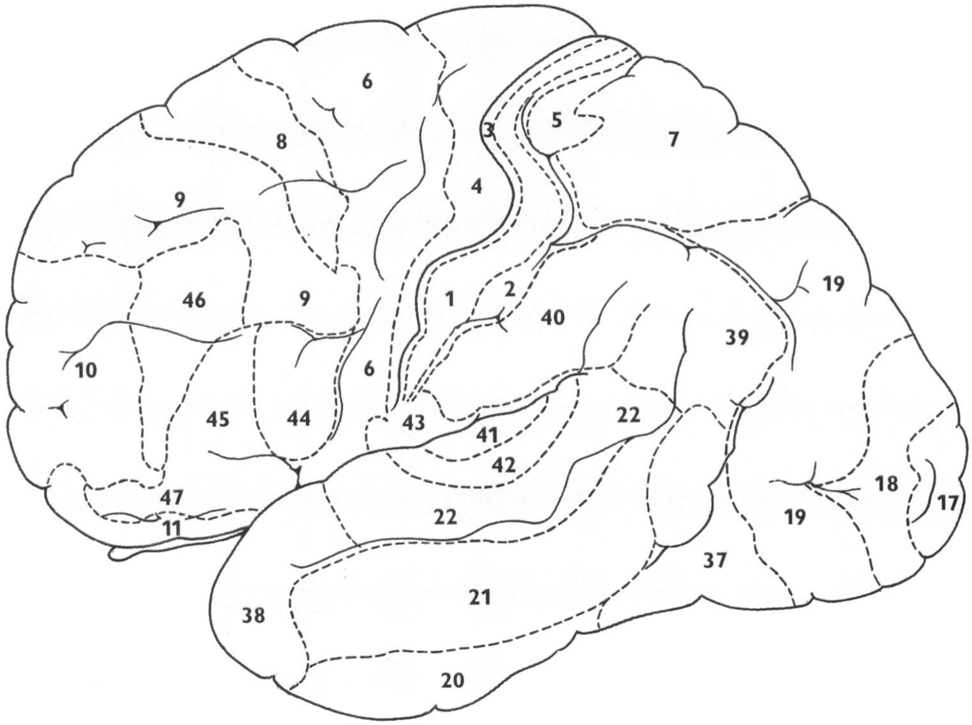

b

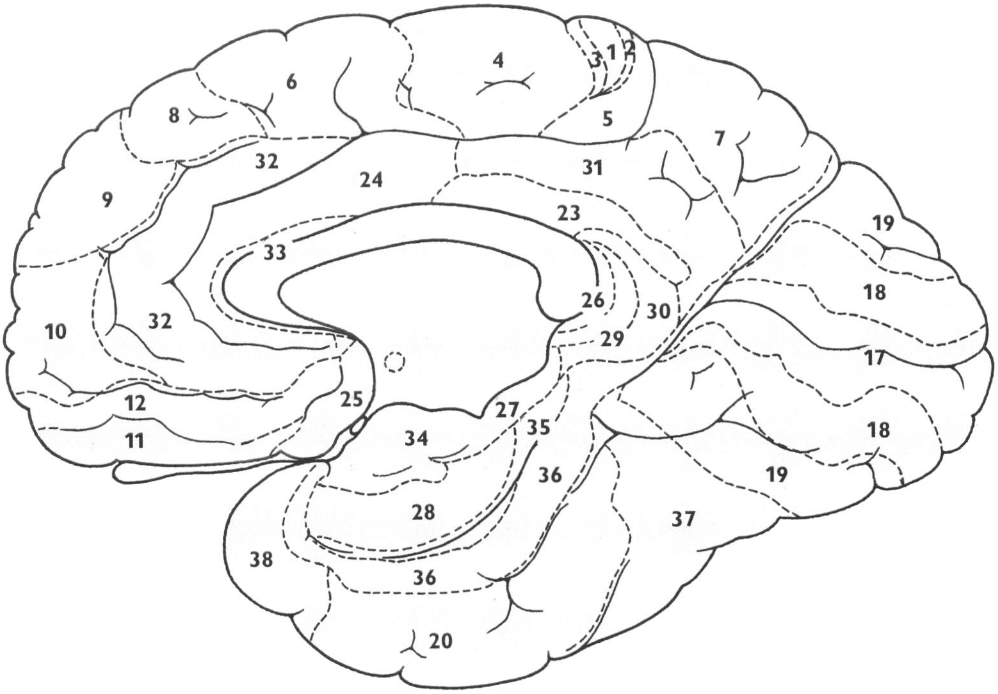

318 Cortex cerebralis

318 Cerebral cortex (80%)

Brodmann's areas 1–47

a Left hemisphere, lateral aspect

b Right hemisphere, medial aspect

318 Grosshirnrinde (80%)

Brodmannsche Rindenfelder 1–47

a Linke Hemisphäre, Seitenansicht

b Rechte Hemisphäre, Medialansicht

a

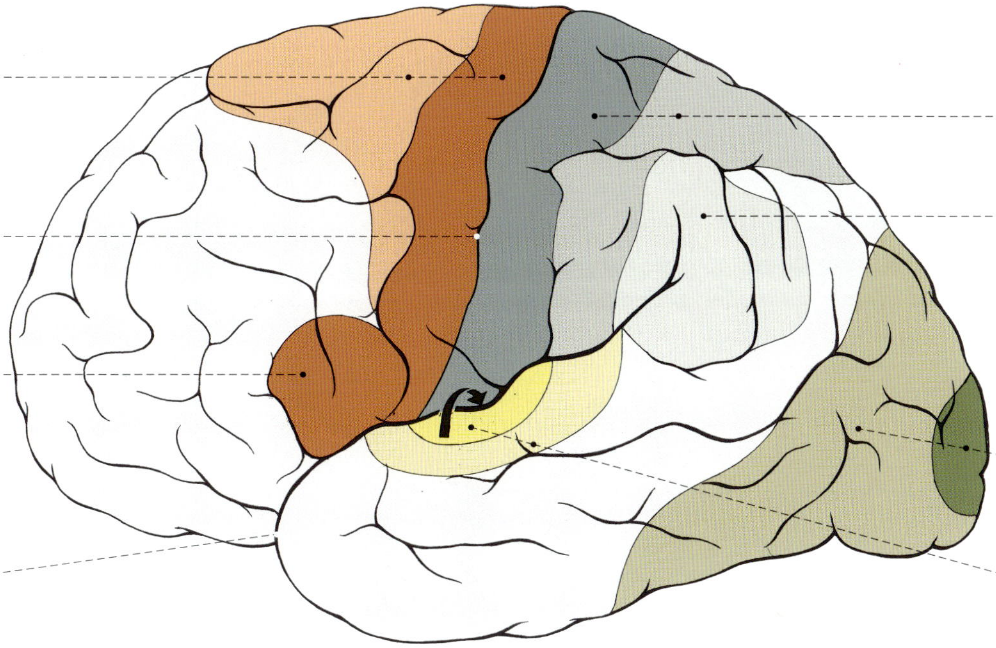

b

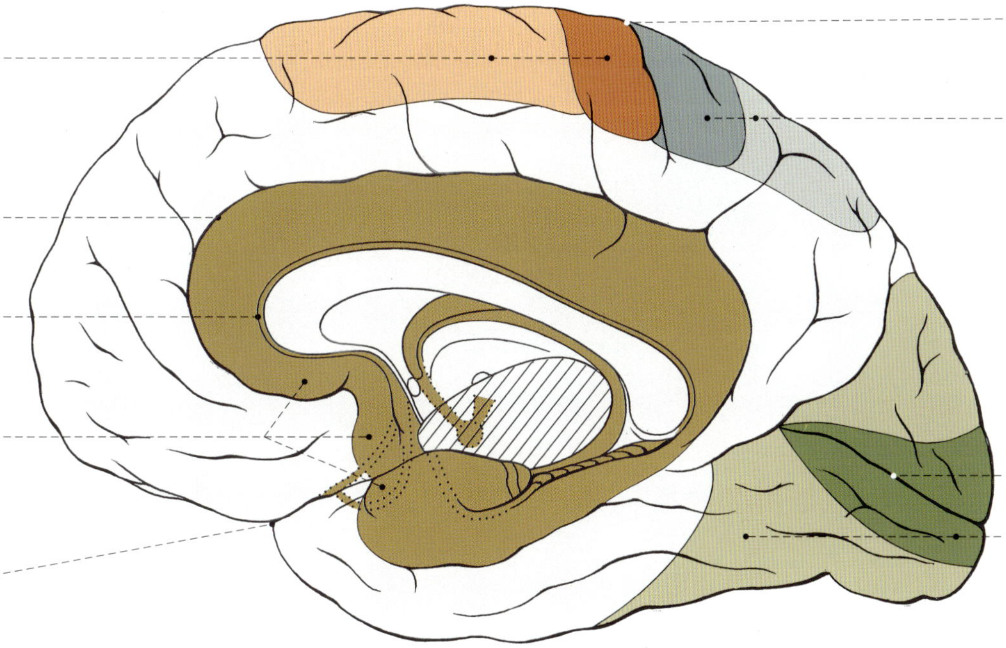

319 Cortex cerebralis

319 Cerebral cortex (80%)

Main first (dark) and second (light) functional areas

a Left hemisphere, lateral aspect
b Right hemisphere, medial aspect

319 Grosshirnrinde (80%)

Primäre (dunkel) und sekundäre (hell) funktionelle Rindenfelder

a Linke Hemisphäre, Seitenansicht
b Rechte Hemisphäre, Medialansicht

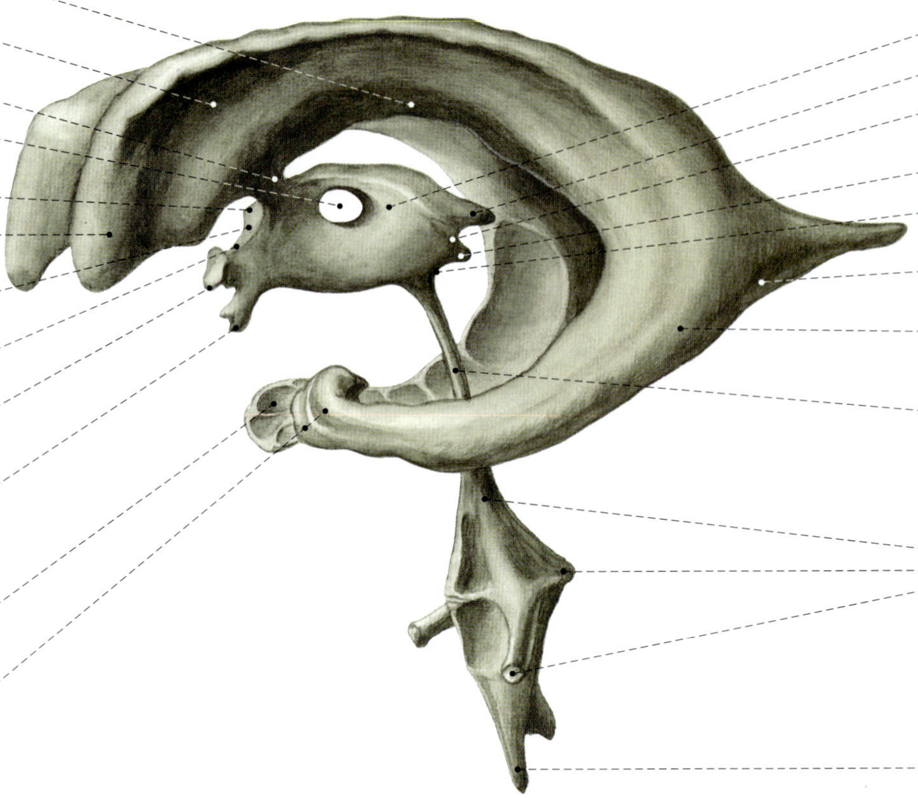

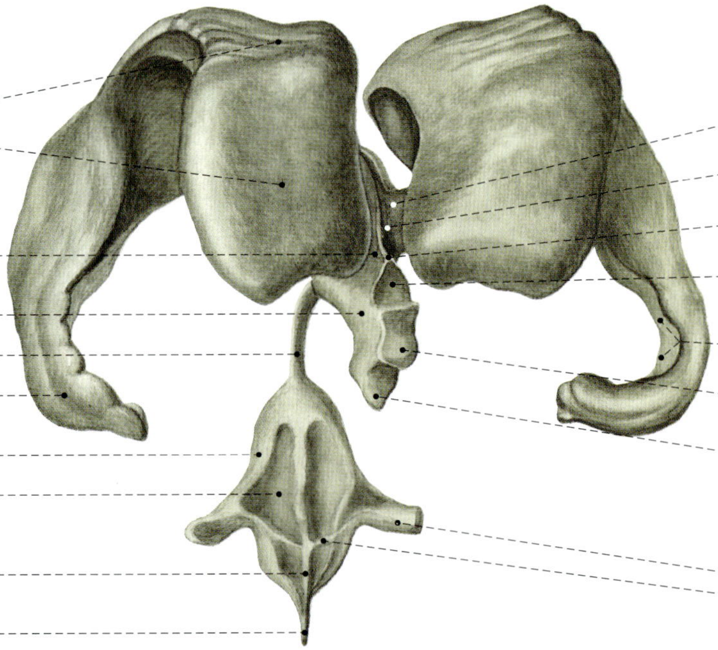

320 Ventriculi encephali

320 Ventricular system of the brain

Cast of the ventricular system (120%)

a Left lateral aspect

b Frontal aspect

320 Ventrikel des Gehirns

Ausguss des Ventrikelsystems (120%)

a Linke Seitenansicht

b Frontalansicht

a

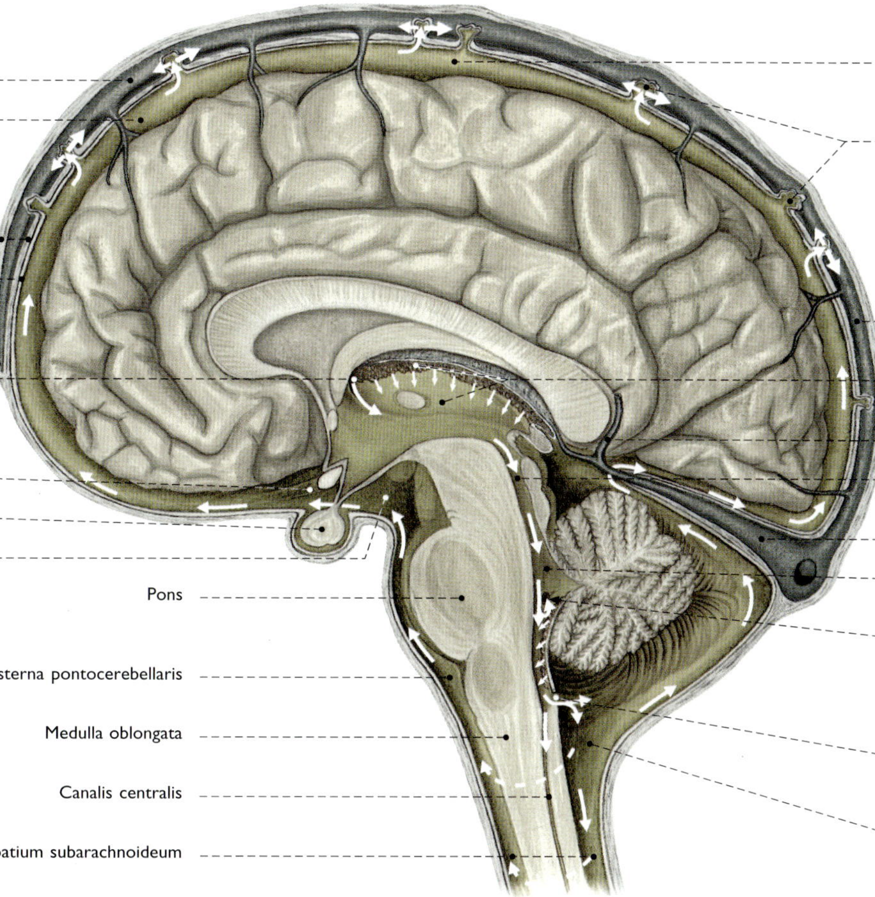

b

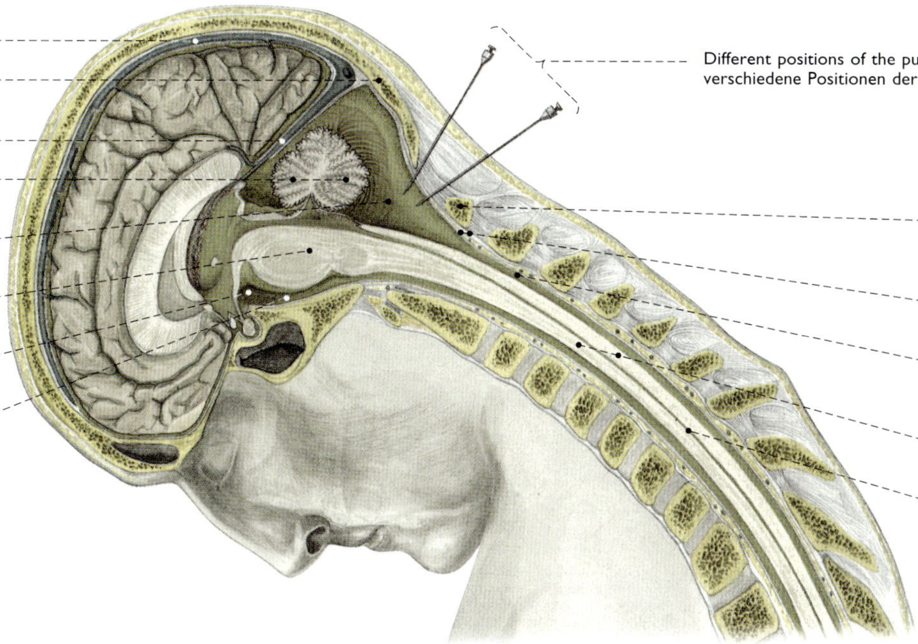

321 Spatium subarachnoideum, Cisternae subarachnoideae et Liquor cerebrospinalis

321 Subarachnoid space, subarachnoid cisterns and cerebrospinal fluid

a Circulation of the cerebrospinal fluid (60%)
b Puncture of the cerebellomedullary cistern (suboccipital puncture) (35%)

321 Subarachnoidalraum, Subarachnoidalzisternen und Liquor cerebrospinalis

a Zirkulation des Liquor cerebrospinalis (60%)
b Punktion der Cisterna cerebellomedullaris (Subokzipitalpunktion) (35%)

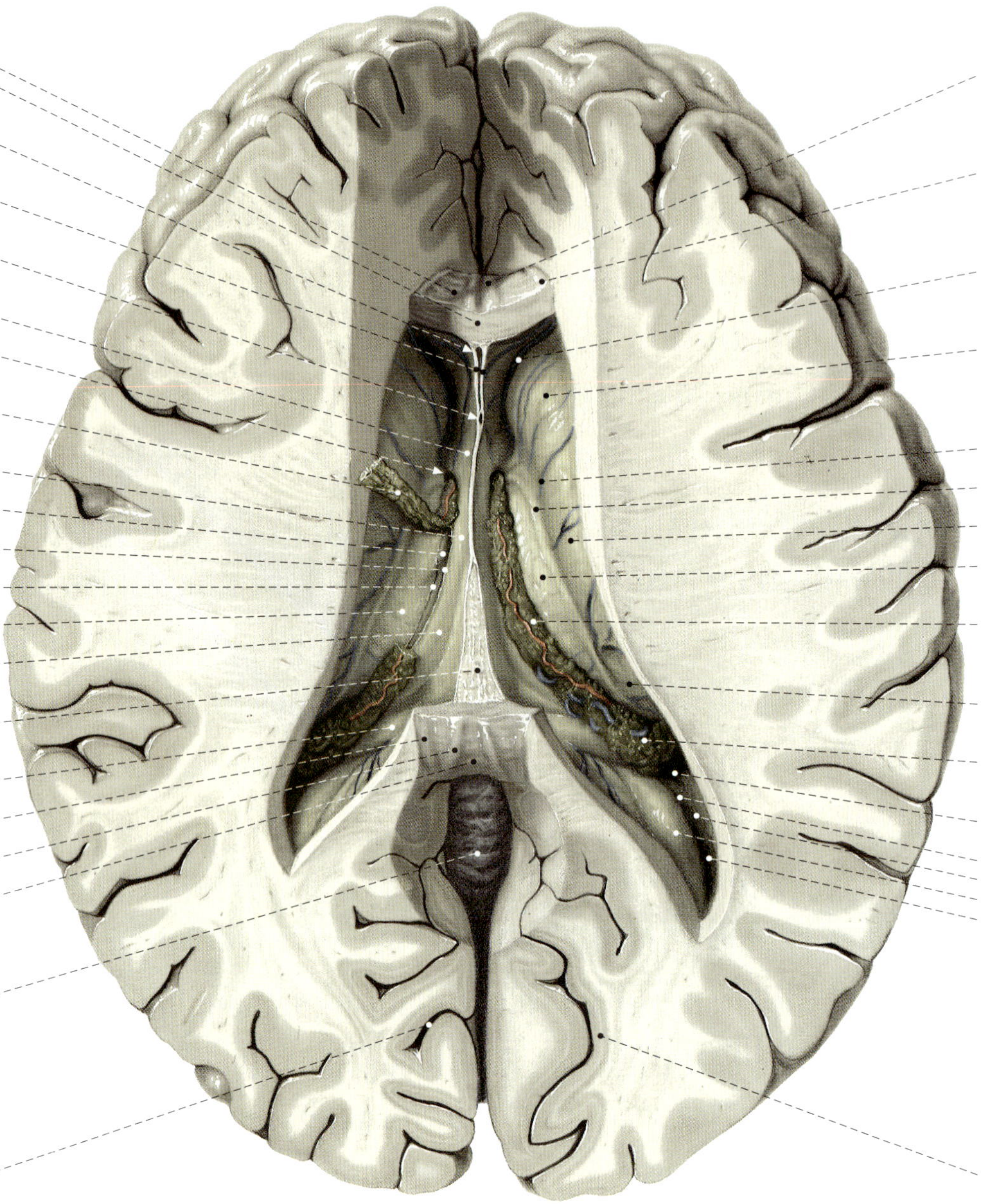

322 Ventriculi laterales cerebri

322 Lateral ventricles of the brain
(100%)
The lateral ventricles were opened from above, the body of the corpus callosum was removed. Superior aspect of a horizontal section

322 Seitenventrikel des Gehirns
(100%)
Die Seitenventrikel sind von oben eröffnet, der Balken ist abgetragen. Parietalansicht eines Horizontalschnittes

a

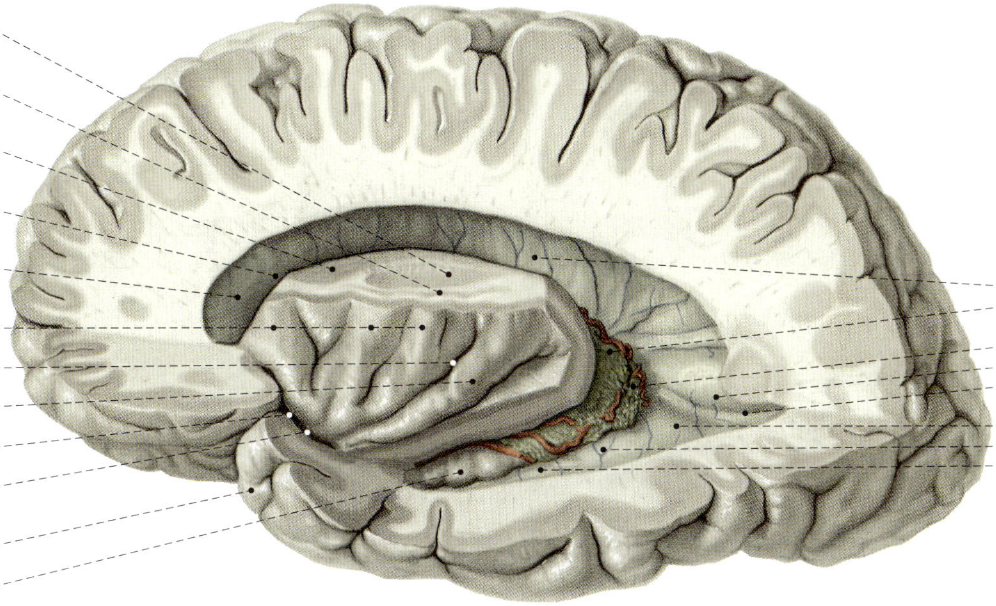

b

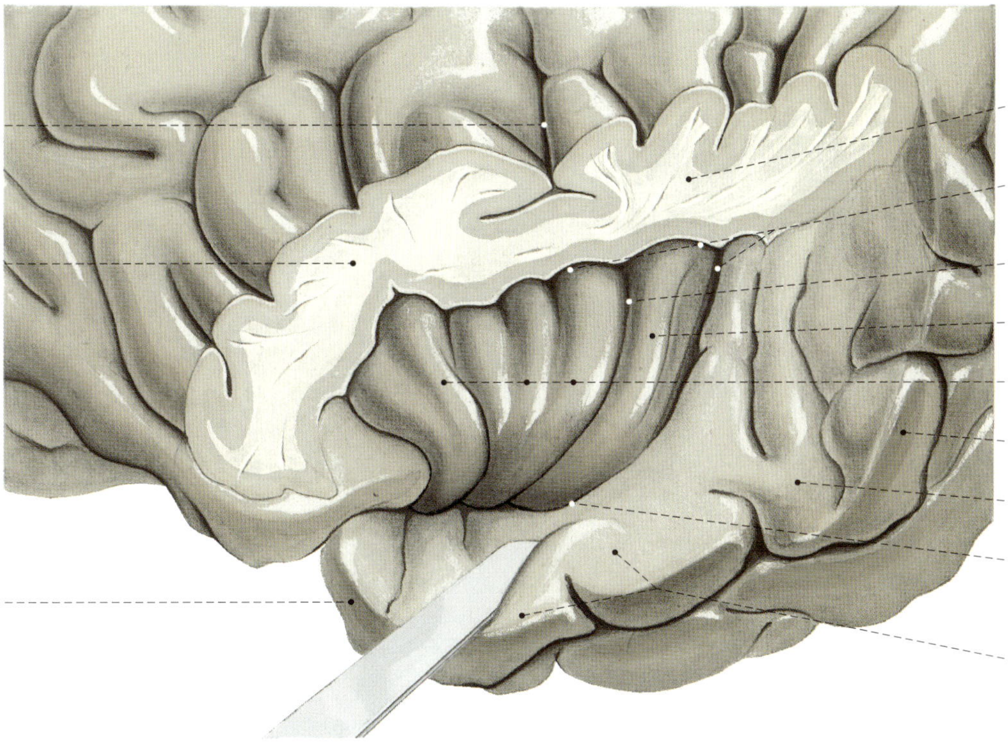

323 Ventriculus lateralis et Lobus insularis

323 Lateral ventricle and insular lobe

a The lateral ventricle of the left hemisphere was opened (80%). Left lateral aspect

b Parts of the frontal, parietal and temporal lobes of the left hemisphere were removed (100%). Left lateral aspect

323 Seitenventrikel und Insel

a Der Seitenventrikel der linken Hemisphäre ist eröffnet (80%). Linke Seitenansicht

b Teile von Stirn-, Scheitel- und Schläfenlappen der linken Hemisphäre sind entfernt (100%). Linke Seitenansicht

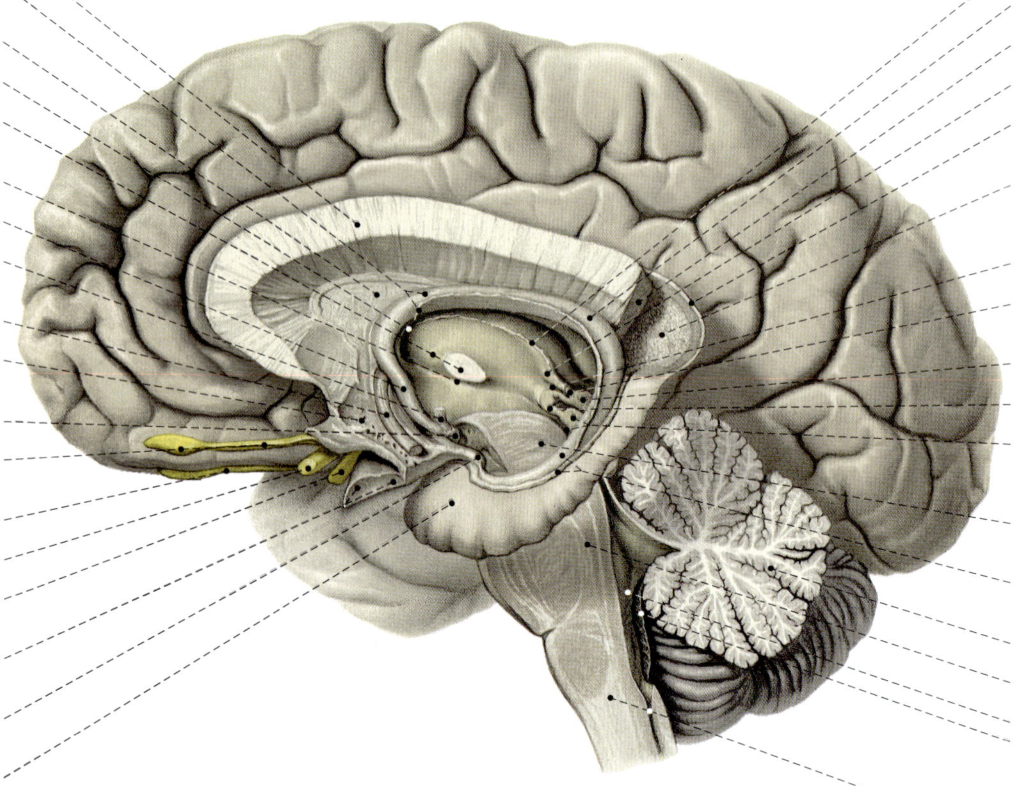

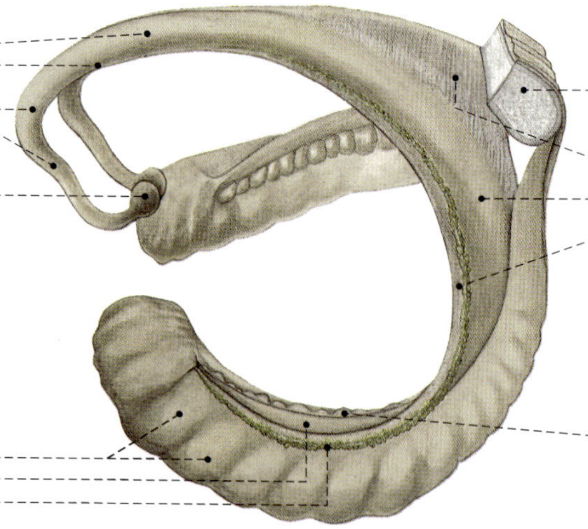

324 Fornix et Hippocampus

324 Fornix and hippocampus

a Dissection of the left cerebral hemisphere. The right hemisphere is intact, the brain stem was sectioned in the median plane (80%). Left lateral aspect

b Fornices and hippocampi of both sides (120%), three-dimensional schematic representation, left parietolateral aspect

324 Fornix und Hippocampus

a Präparation der linken Seite. Die rechte Hemisphäre ist intakt, der Hirnstamm median durchschnitten (80%). Linke Seitenansicht

b Fornices und Hippocampi beider Seiten (120%), räumliche schematische Darstellung, Ansicht von links seitlich oben (linke Parietolateralansicht)

a

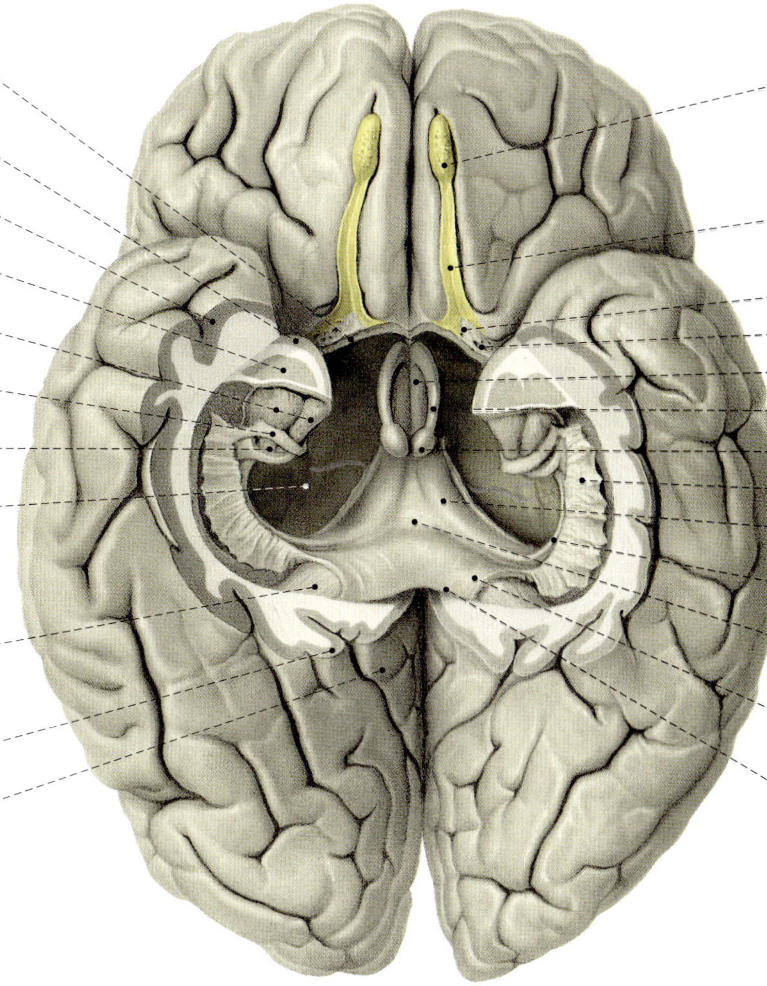

b

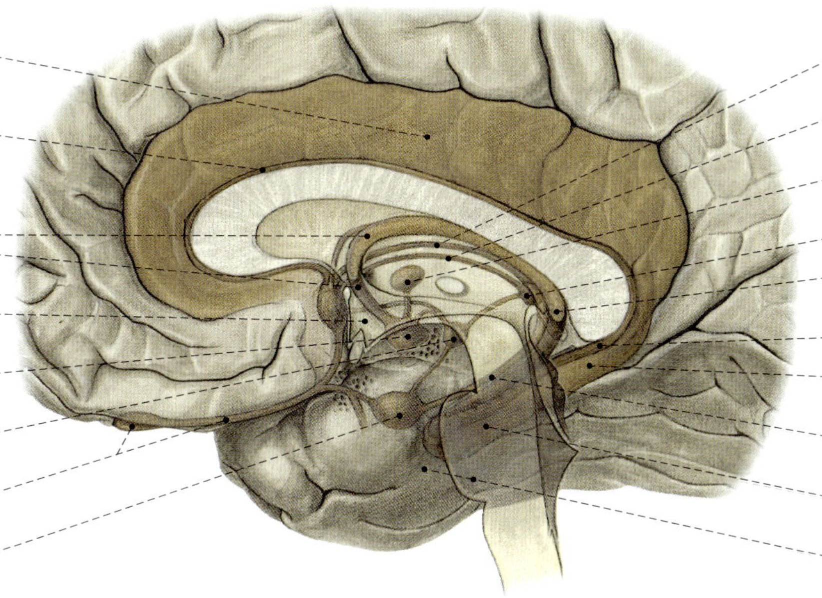

325 Fornix

325 Fornix and limbic system

a The undersurface of the brain was partly dissected in order to show the fornix and some other elements of the limbic system (80%). Inferior (basal) aspect

b The structures of the limbic system are emphasized by brown color (100%). Medial aspect of the right hemisphere

325 Fornix und limbisches System

a Die Unterfläche des Gehirns ist teilweise abgetragen, die Fornix und einige andere Anteile des limbischen Systems sind dargestellt (80%). Basalansicht

b Die Strukturen des limbischen Systems sind in brauner Farbe hervorgehoben (100%). Medialansicht der rechten Hemisphäre

a

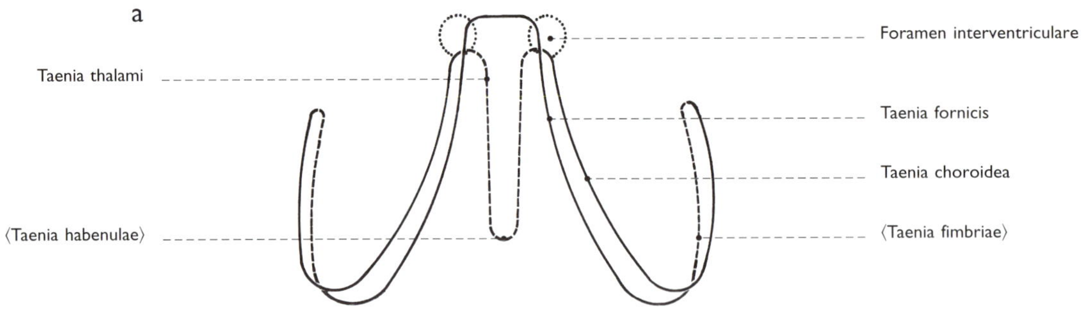

b

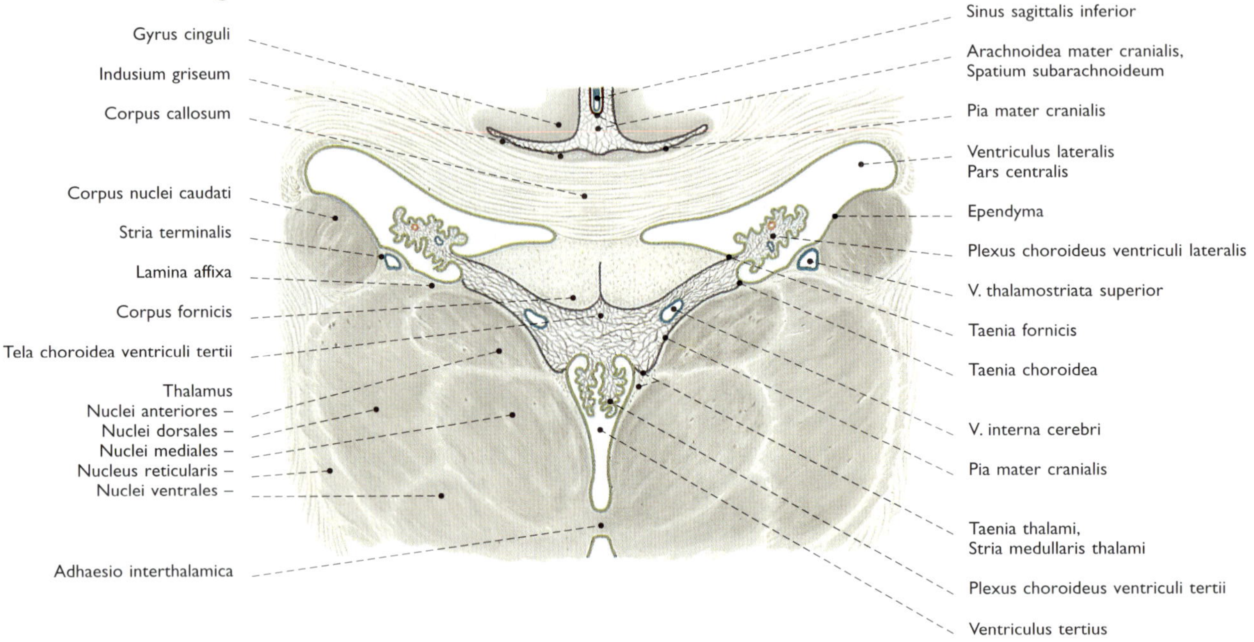

c

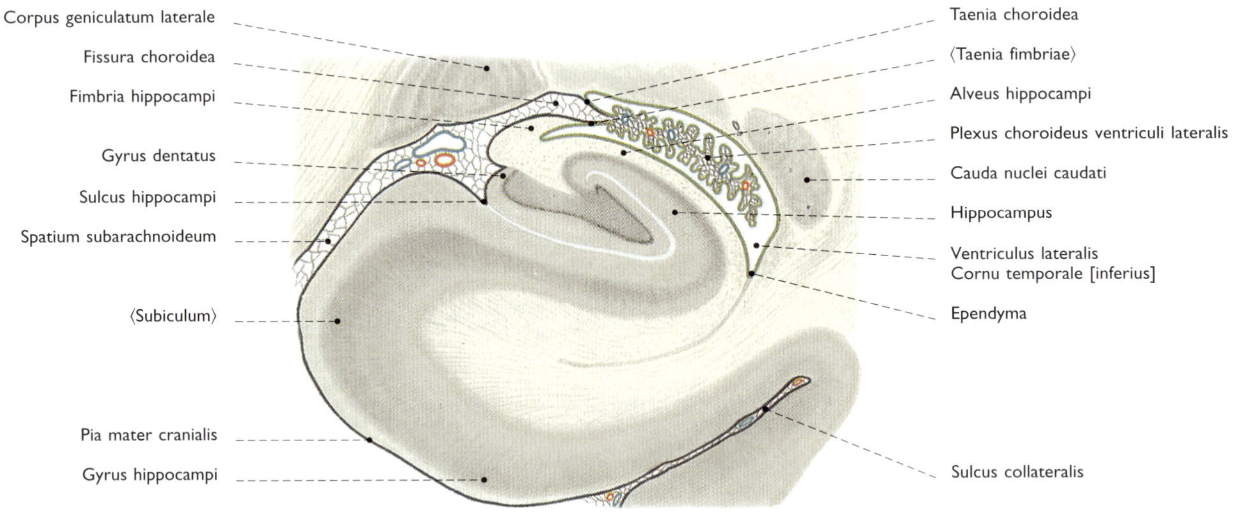

326 ⟨Plexus choroidei prosencephali⟩

326 Choroid plexuses of the forebrain

Schematic representations

- a Line of tearing of the choroid plexuses of both sides, parietal aspect
- b, c Occipital aspect of coronal sections through
  - b the central part of the lateral ventricles and the third ventricle (200%)
  - c the temporal horn of the right lateral ventricle (300%)

326 Adergeflechte des Vorderhirns

Schematische Darstellungen

- a Abrisslinie der Adergeflechte beider Seiten, Parietalansicht
- b, c Okzipitalansicht von Frontalschnitten durch
  - b die Pars centralis der Seitenventrikel und den dritten Ventrikel (200%)
  - c das Unterhorn des rechten Seitenventrikels (300%)

a

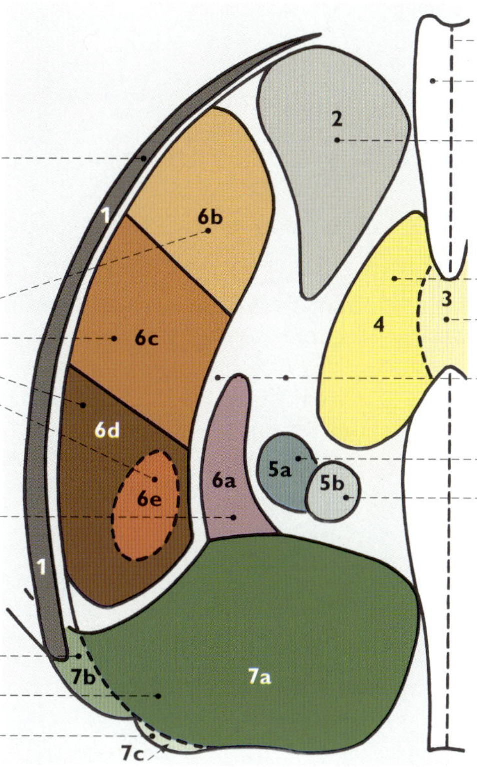

b

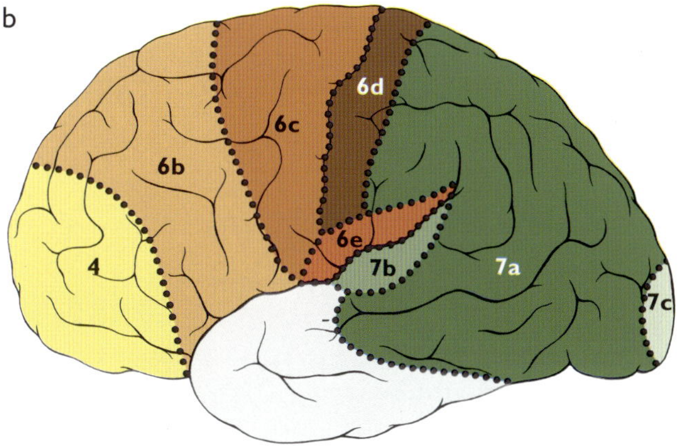

c

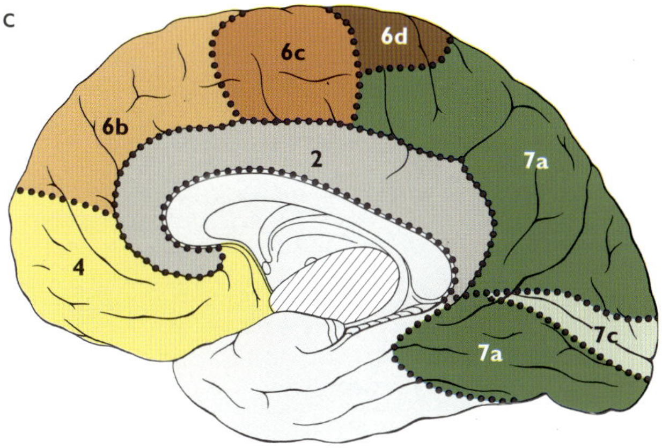

327 Thalamus

327 Thalamus

The thalamic nuclei and their projection onto the cerebral cortex. Identical numbers indicate nuclei and their functionally connected cortical areas. The nuclei 1, 3 and 6a seem to be without specific projection to the cerebral cortex.

a Main nuclear masses of the left thalamus, schematic representation, superior aspect of a horizontal section

b Fields of thalamic projection onto the superolateral surface of the left cerebral hemisphere, left lateral aspect

c Fields of thalamic projection onto the medial surface of the right cerebral hemisphere, medial aspect

327 Thalamus

Projektion der Thalamuskerne auf die Grosshirnrinde. Die Zahlen bezeichnen die mit den Kerngebieten korrespondierenden Rindenfelder. Für die Kerne 1, 3 und 6a ist bisher keine spezifische Rindenprojektion bekannt.

a Hauptkerngebiete des linken Thalamus, schematische Darstellung, Parietalansicht eines Horizontalschnittes

b Projektionsfelder der Thalamuskerne auf die Kraniolateralfläche der linken Grosshirnrinde, linke Seitenansicht

c Projektionsfelder der Thalamuskerne auf die Medialfläche der rechten Grosshirnrinde, Medialansicht

a

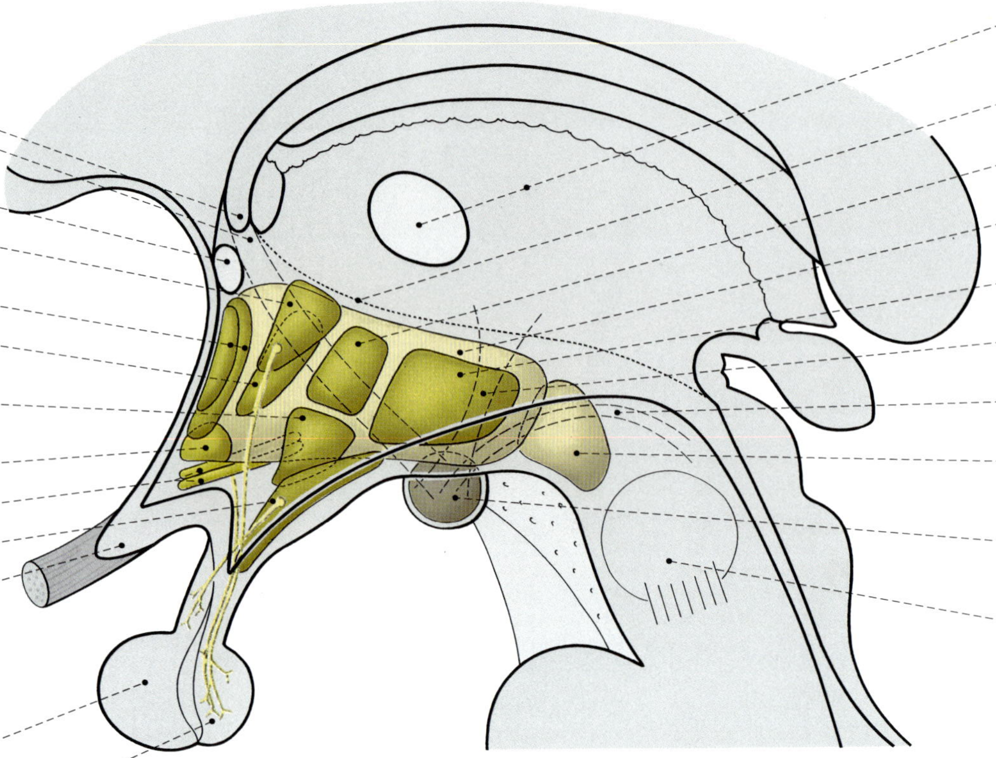

b

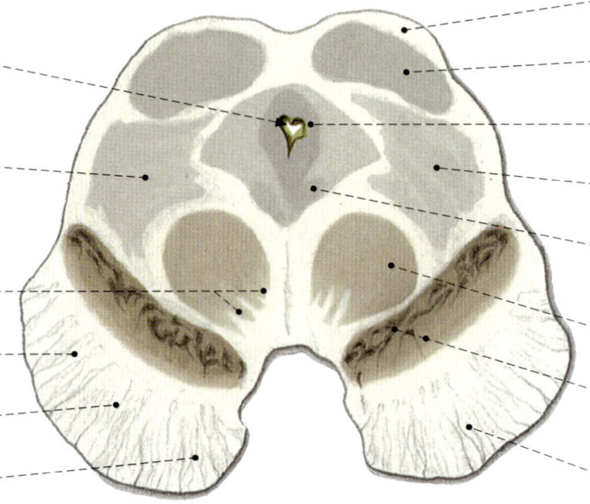

328 Hypothalamus et Mesencephalon

328 Hypothalamus and midbrain (250%)

a Nuclei of the hypothalamus, schematic representation, medial aspect
b Oblique section through the midbrain at the level of the superior colliculi

328 Hypothalamus und Mittelhirn (250%)

a Kerne des Hypothalamus, schematische Darstellung, Medialansicht
b Schrägschnitt durch das Mittelhirn in Höhe der Colliculi superiores

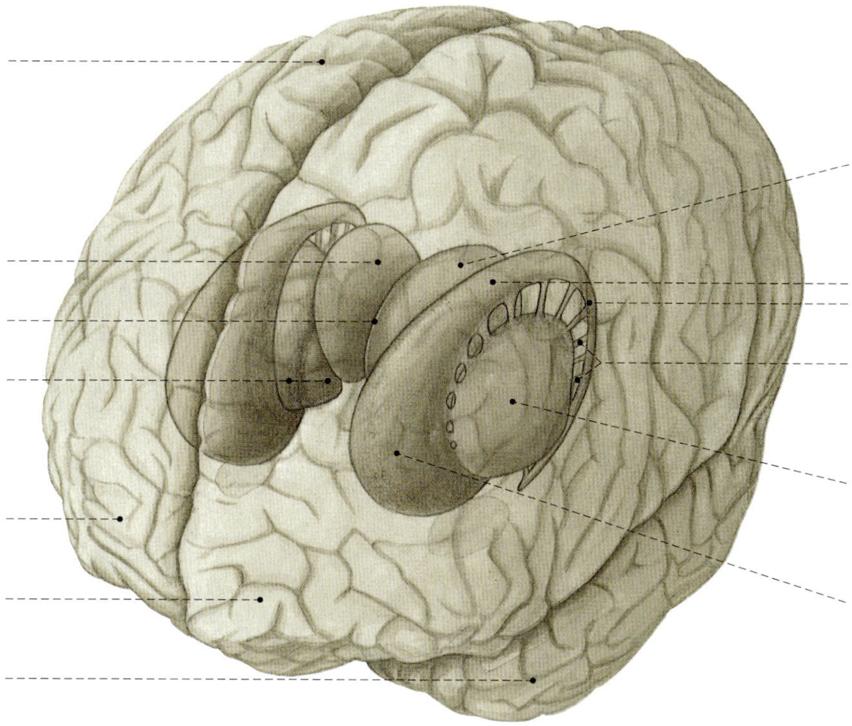

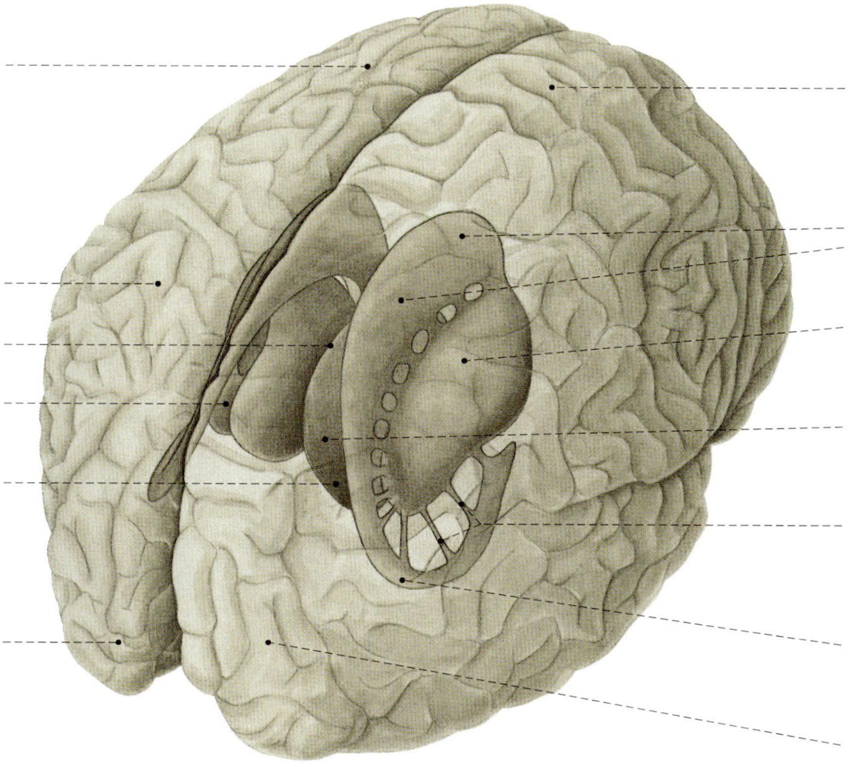

329 ⟨Nuclei prosencephali⟩

329 Nuclei of the forebrain (70%)

Thalamus, striate body (= caudate nucleus and putamen), and lentiform nucleus (= putamen and globi pallidi) are shown in the interior of the two cerebral hemispheres. Schematized three-dimensional representations

a Left lateral anterior superior aspect (left frontoparietolateral aspect)

b Right lateral posterior superior aspect (right occipitoparietolateral aspect)

329 Kerne des Vorderhirns (70%)

Thalamus, Corpus striatum (= Nucleus caudatus und Putamen) und Nucleus lentiformis (= Putamen und Globi pallidi) im Inneren der beiden Grosshirnhemisphären, schematisierte dreidimensionale Darstellungen

a Ansicht von links seitlich vorne oben (linke Frontoparietolateralansicht)

b Ansicht von rechts seitlich hinten oben (rechte Okzipitoparietolateralansicht)

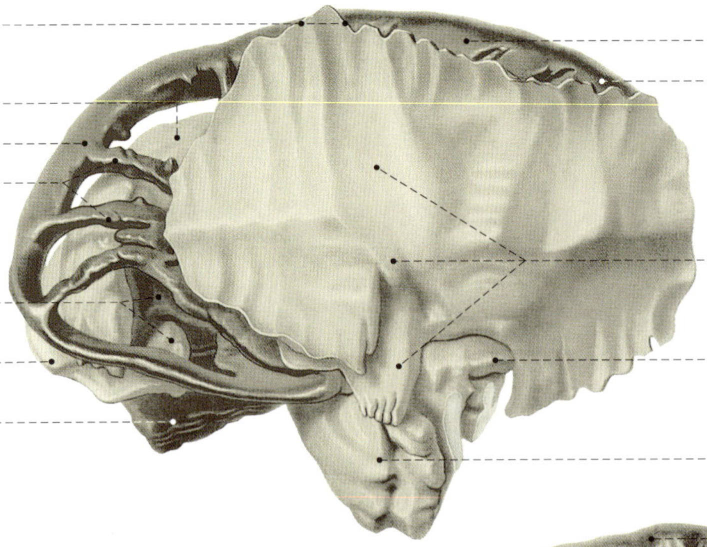

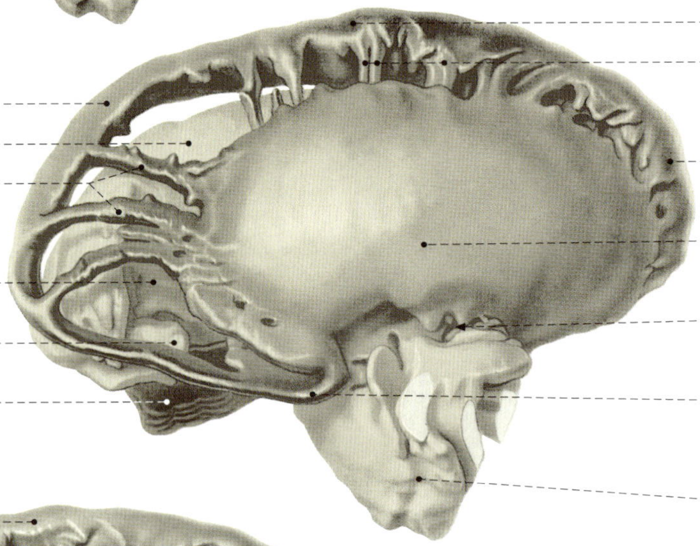

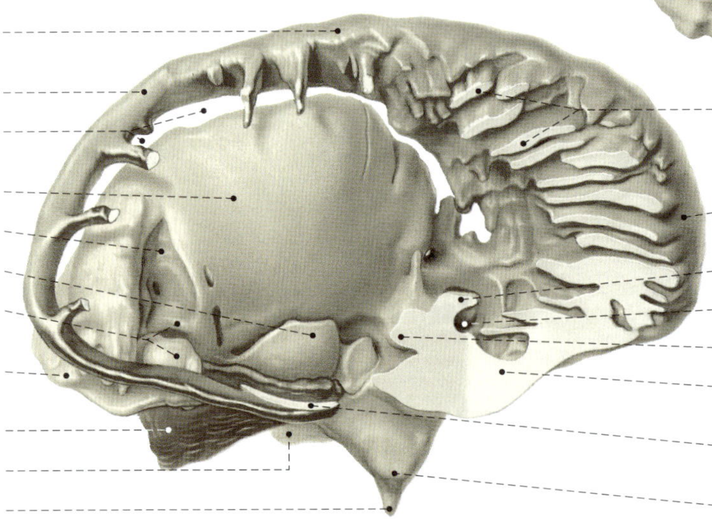

330 ⟨Nuclei prosencephali et mesencephali⟩

330 Nuclei of the forebrain and the midbrain

Collapsible model of reconstruction of the nuclei of the right hemisphere (Anatomical Collection, Basel) (150%), right lateral aspect

a Claustrum and caudate nucleus
b Putamen and caudate nucleus after removal of the claustrum
c Thalamus and caudate nucleus after removal of claustrum, lentiform nucleus, and amygdaloid body

330 Kerne des Vorder- und Mittelhirns

Zerlegbares Rekonstruktionsmodell der Kerne der rechten Hirnhälfte (Anatomische Sammlung, Basel) (150%), rechte Seitenansicht

a Claustrum und Nucleus caudatus
b Putamen und Nucleus caudatus nach Entfernung des Claustrum
c Thalamus und Nucleus caudatus nach Entfernung von Claustrum, Nucleus lentiformis und Corpus amygdaloideum

a

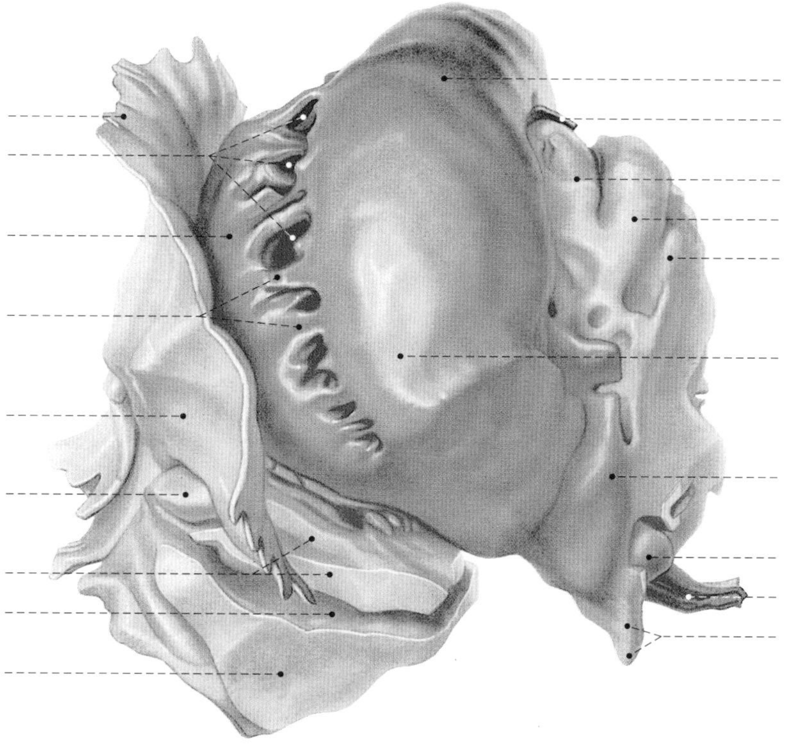

b

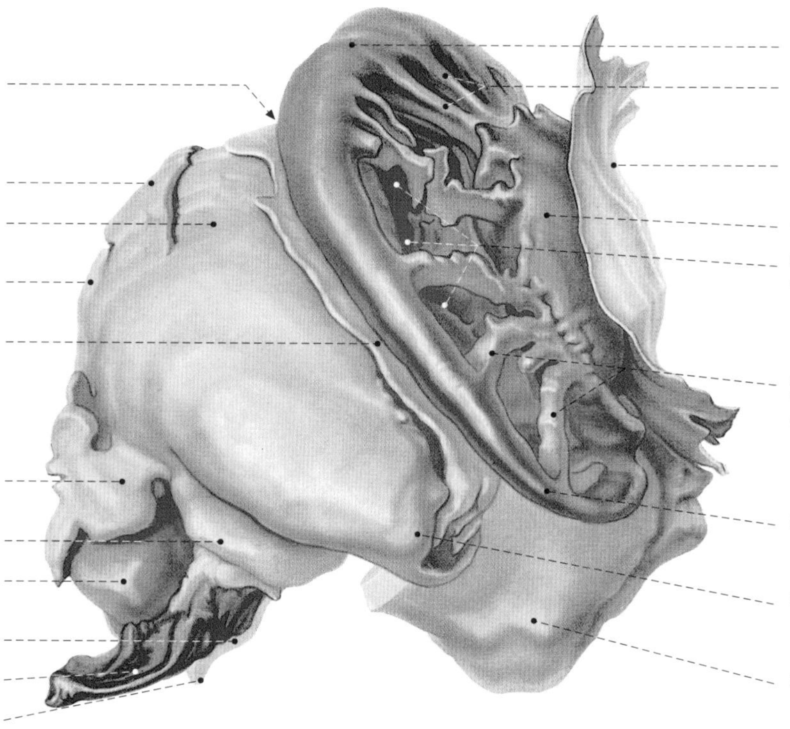

331 ⟨Nuclei prosencephali et mesencephali⟩

331 Nuclei of the forebrain and the midbrain

Collapsible model of reconstruction of the nuclei of the right hemisphere (Anatomical Collection, Basel) (200%)

a Frontal aspect
b Occipital aspect

331 Kerne des Vorder- und Mittelhirns

Zerlegbares Rekonstruktionsmodell der Vorder- und Mittelhirnkerne der rechten Hirnhälfte (Anatomische Sammlung, Basel) (200%)

a Frontalansicht
b Okzipitalansicht

a

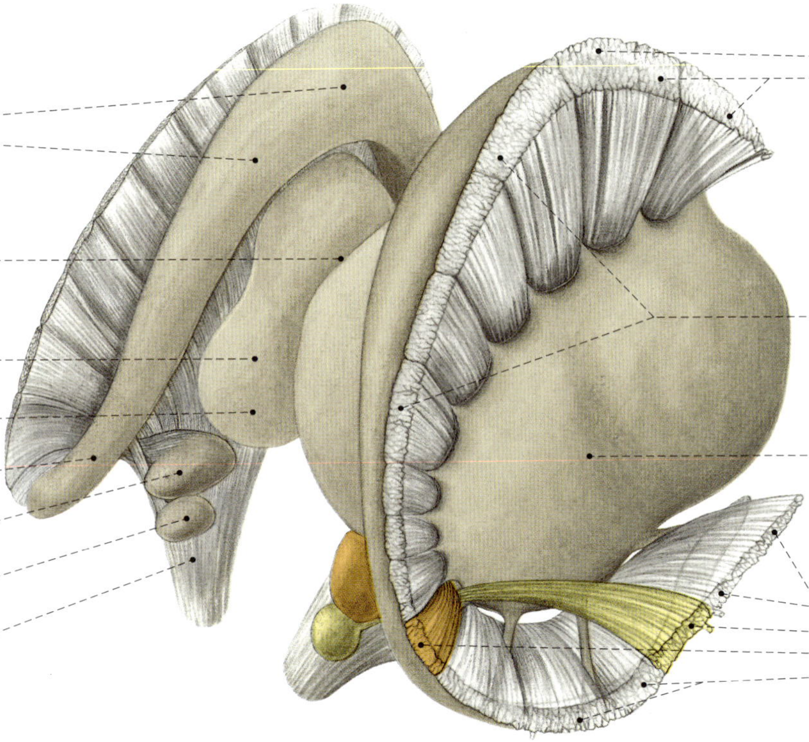

b

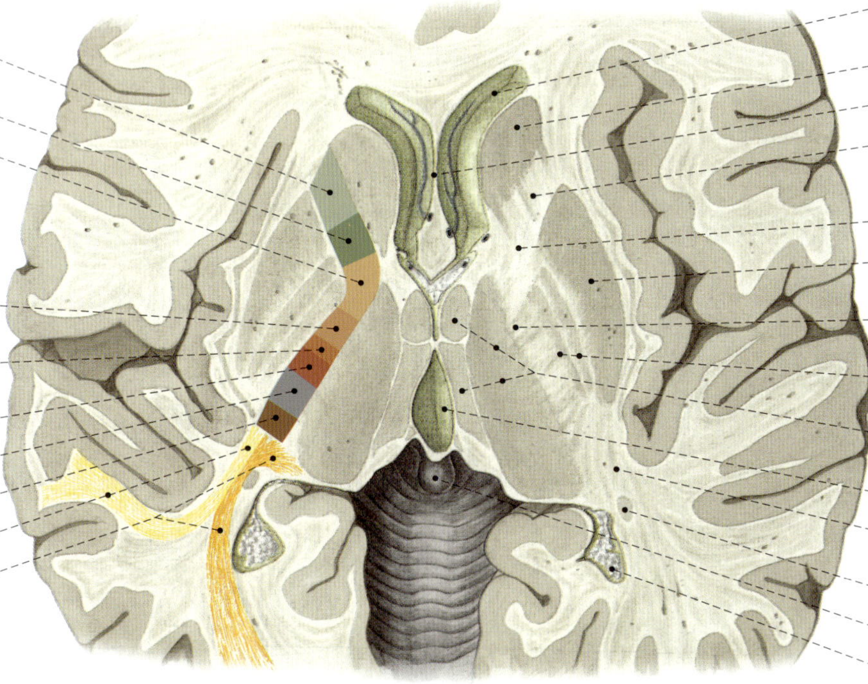

332 ⟨Nuclei prosencephali⟩ et Capsula interna

332 Nuclei of the forebrain and internal capsule

a Thalamus, striate body (= caudate nucleus and putamen), and lentiform nucleus (= putamen and globi pallidi) with the internal capsule (250%), right occipitolateral aspect

b Horizontal section through the internal capsule and adjacent nuclei. On the left side, the main components of the internal capsule are marked by different colors (90%). Parietal aspect

332 Kerne des Vorderhirns und Capsula interna

a Thalamus, Corpus striatum (= Nucleus caudatus und Putamen) und Nucleus lentiformis (= Putamen und Globi pallidi) mit Capsula interna (250%), rechte Okzipitolateralansicht

b Horizontalschnitt durch die innere Kapsel und benachbarte Kerne. Links sind die Hauptbestandteile der inneren Kapsel durch verschiedene Farben hervorgehoben (90%). Parietalansicht

a

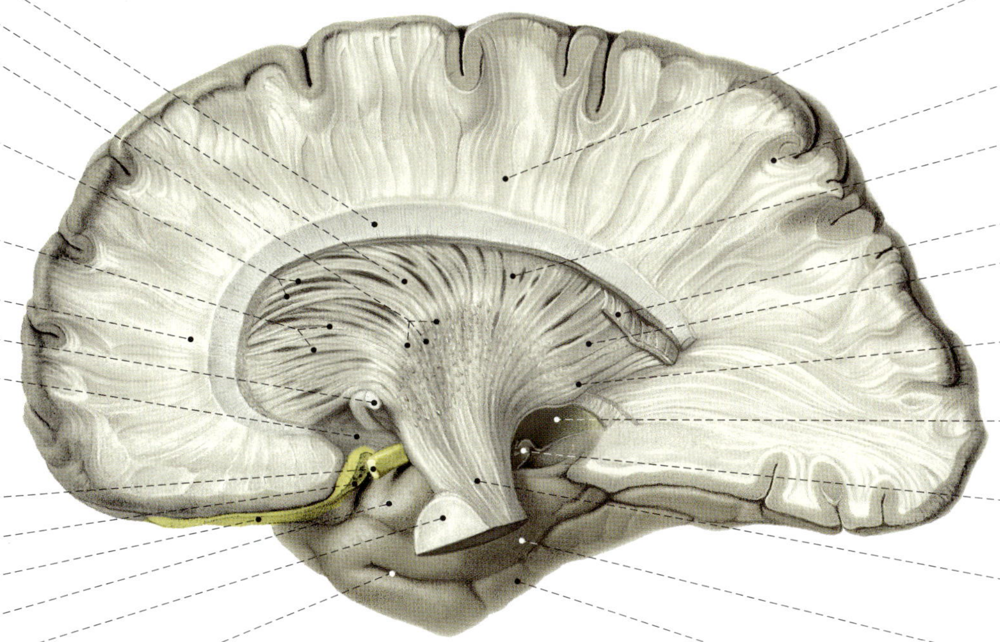

b

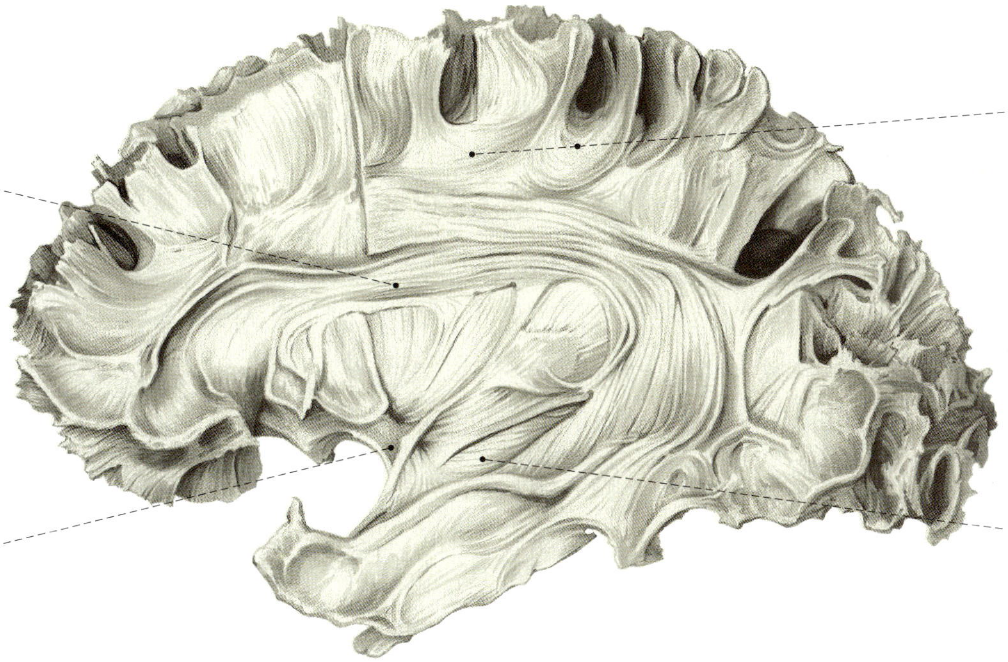

333 Neurofibrae et Fasciculi cerebrales

333 Nerve fascicles of the cerebrum

(80%)

a Internal capsule with the corona radiata of the right hemisphere. The caudate nucleus and the thalamus were removed. Medial aspect of the right half
b Association tracts of the left hemisphere, left lateral aspect

333 Fasersysteme des Gehirns

(80%)

a Capsula interna mit Corona radiata der rechten Hemisphäre. Nucleus caudatus und Thalamus sind entfernt. Medialansicht der rechten Hälfte
b Assoziationsbahnen der linken Hemisphäre, linke Seitenansicht

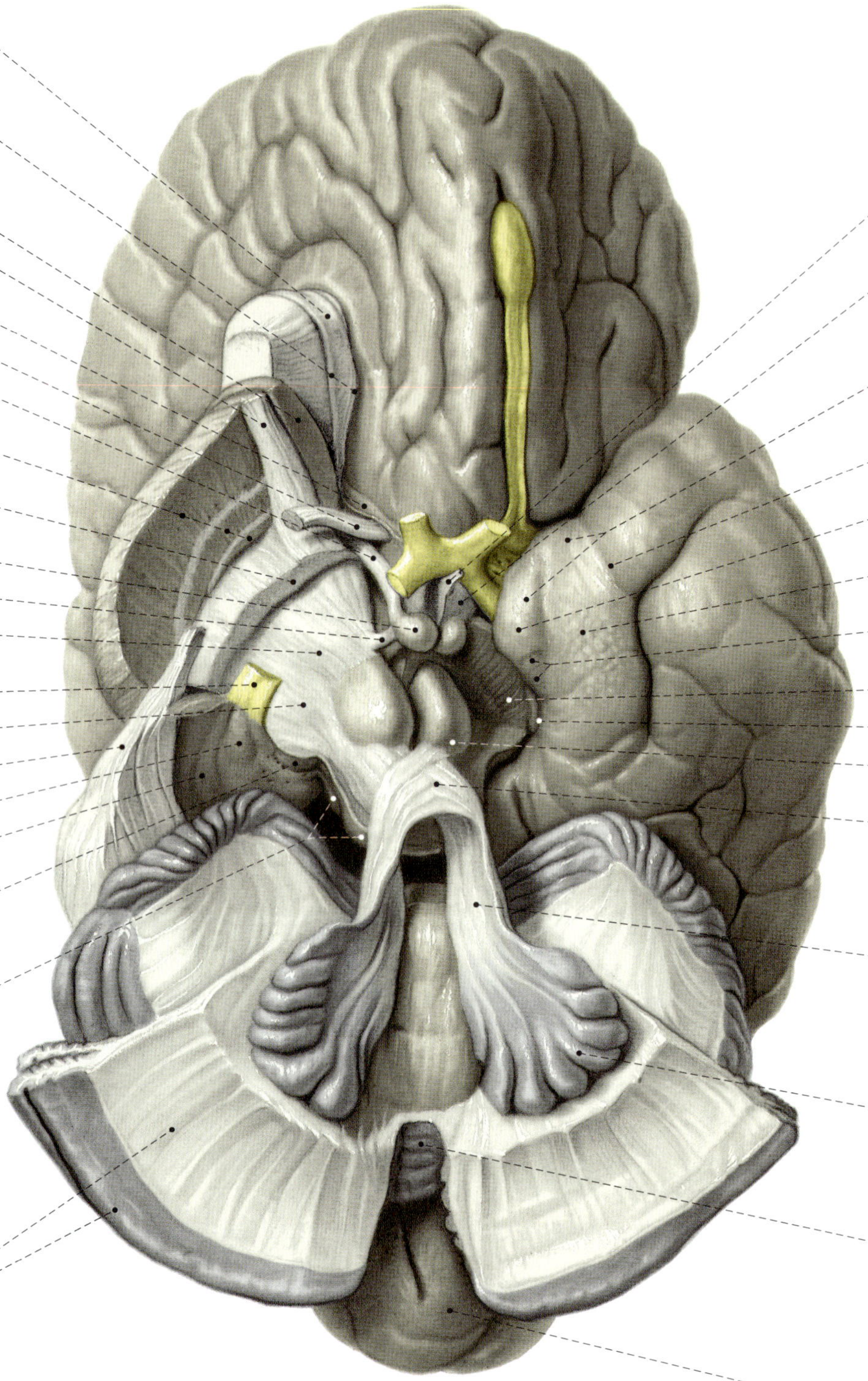

334 Pedunculi cerebellares superiores

334 Cerebellorubral connections
(100%)
Dissection revealing the cerebellorubral fibers (decussation of Stilling) and several fiber systems in the mesencephalic and diencephalic regions, inferior (basal) aspect

334 Zerebellorubrale Verbindungen
(100%)
Faserpräparat der dentatorubralen Verbindung (Stillingsche Schere) und verschiedener Fasersysteme des Mittel- und Zwischenhirns, Basalansicht

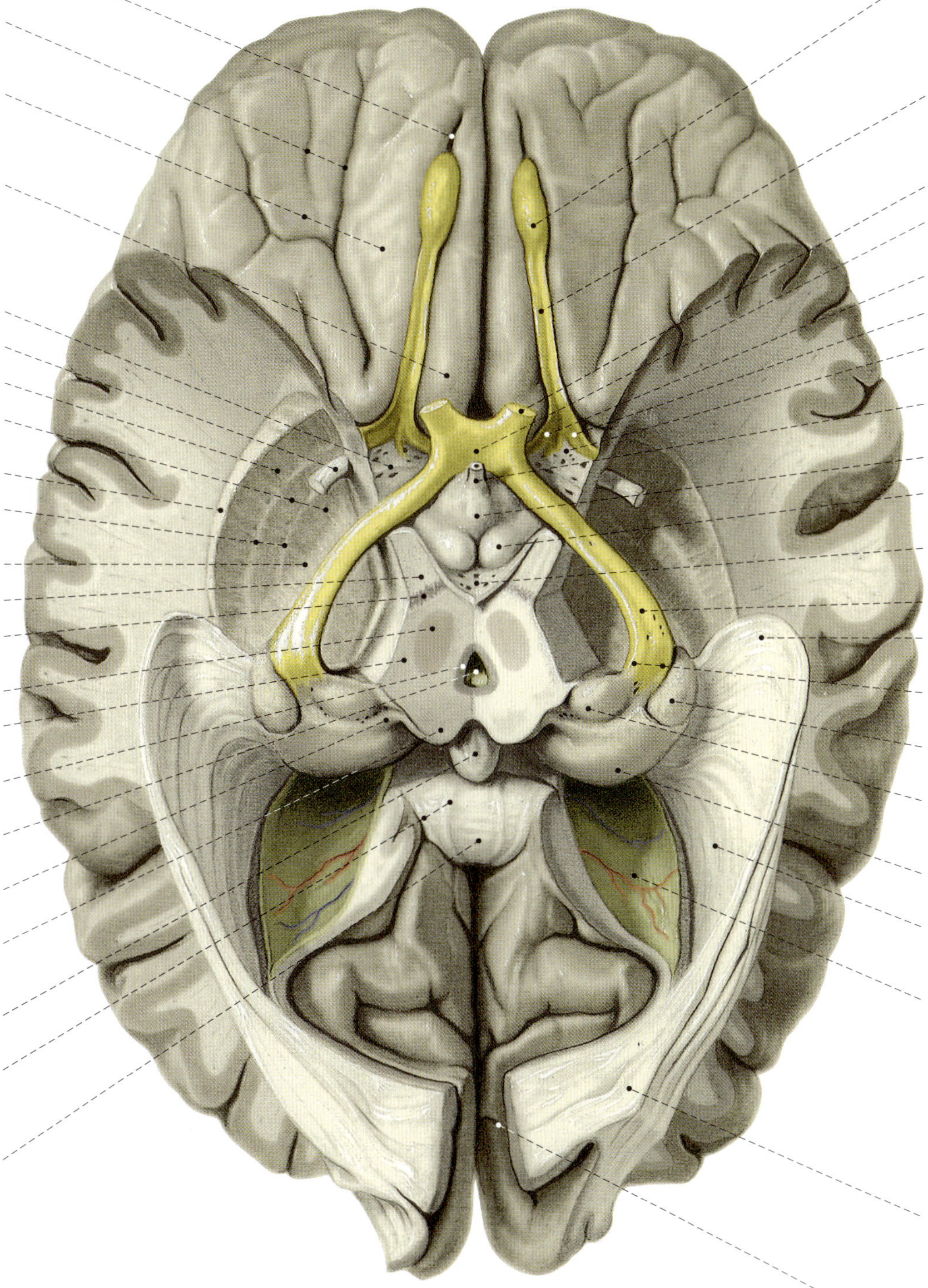

335 Tractus opticus et Radiatio optica

335 Visual pathway (100%)

Dissection showing the optic radiation. The fibers pass from the chiasma to the visual cortex. Inferior (basal) aspect

335 Sehbahn (100%)

Faserpräparat der Sehbahn vom Chiasma opticum zur Sehrinde, Basalansicht

a

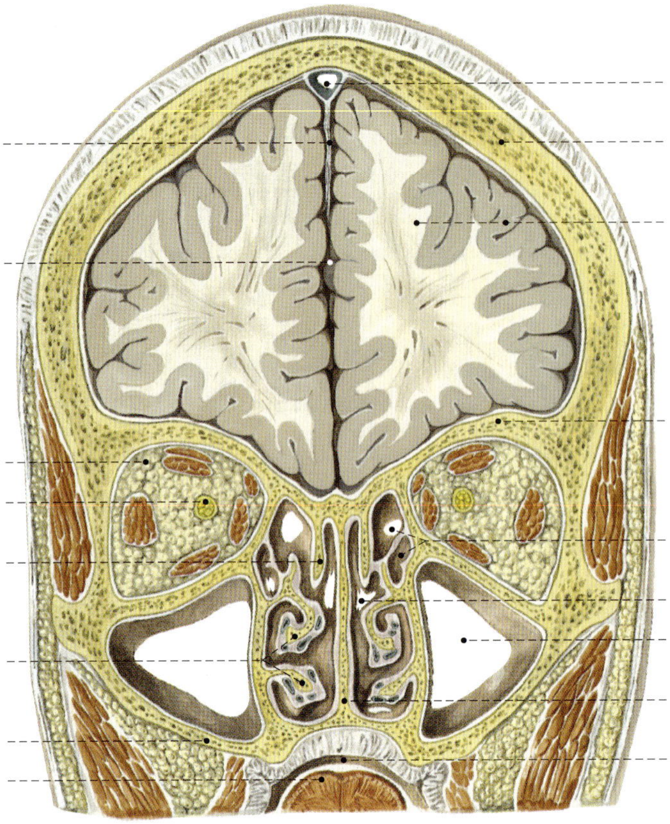

b

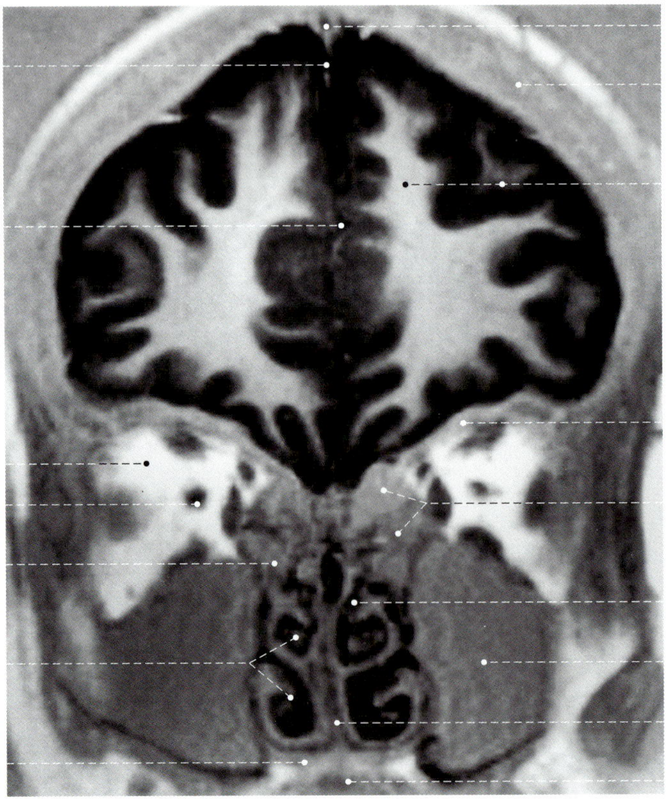

336 Encephalon

336 Brain (75%)

Coronal sections through the frontal cerebral lobes, adjacent skull bones, and the orbital and nasal cavities, frontal aspect

a Anatomical section (painted)
b Magnetic resonance image (MRI, T_1-weighted)

336 Gehirn (75%)

Frontalschnitte durch die Stirnlappen des Gehirns, die umgebenden Schädelknochen sowie die Augen- und Nasenhöhlen, Frontalansicht

a Anatomischer Schnitt (gezeichnet)
b Kernspintomogramm (MRI, T_1-gewichtet)

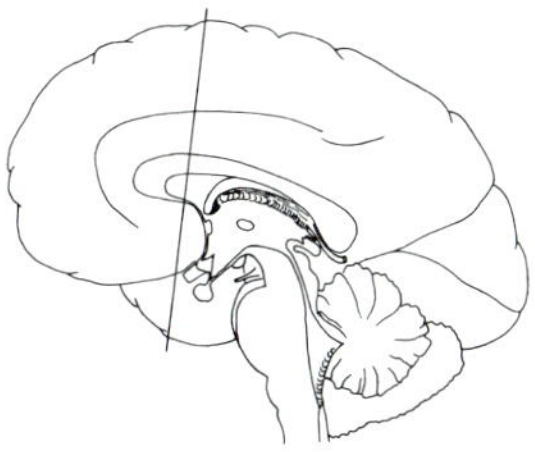

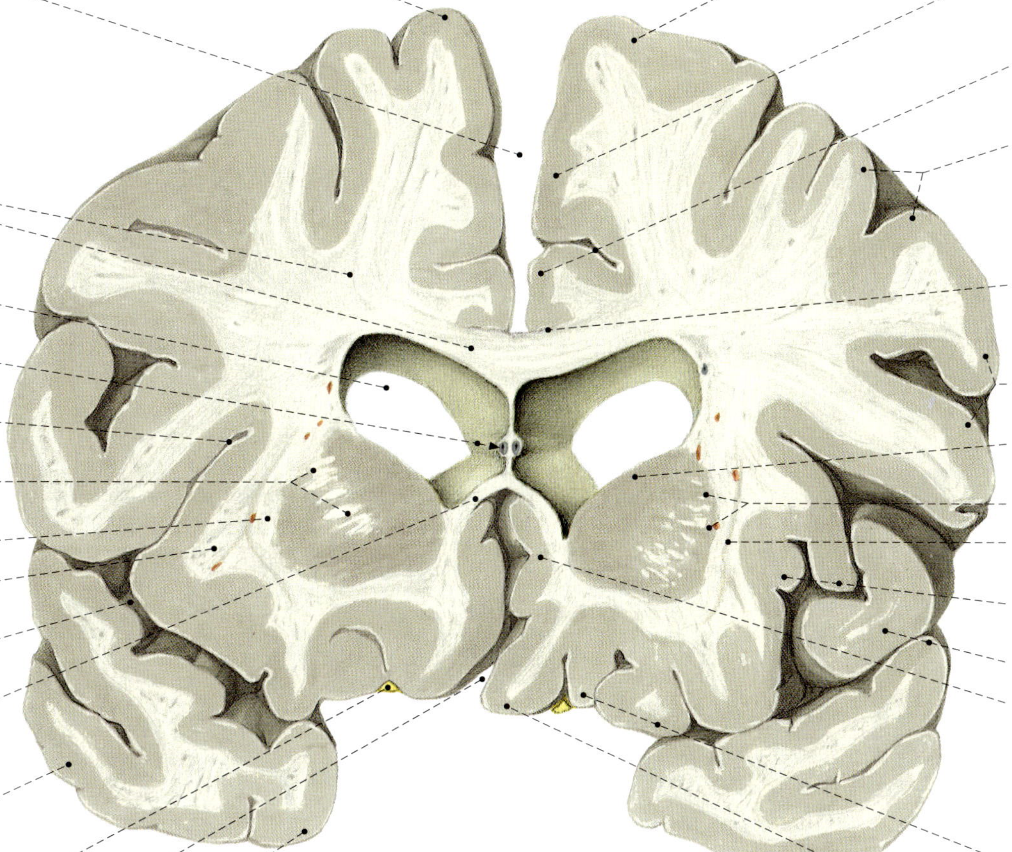

337 Encephalon

337 Brain (100%)

Coronal section in the plane of the anterior horn of the lateral ventricles, occipital aspect

337 Gehirn (100%)

Frontalschnitt in der Ebene des Vorderhorns der Seitenventrikel, Okzipitalansicht

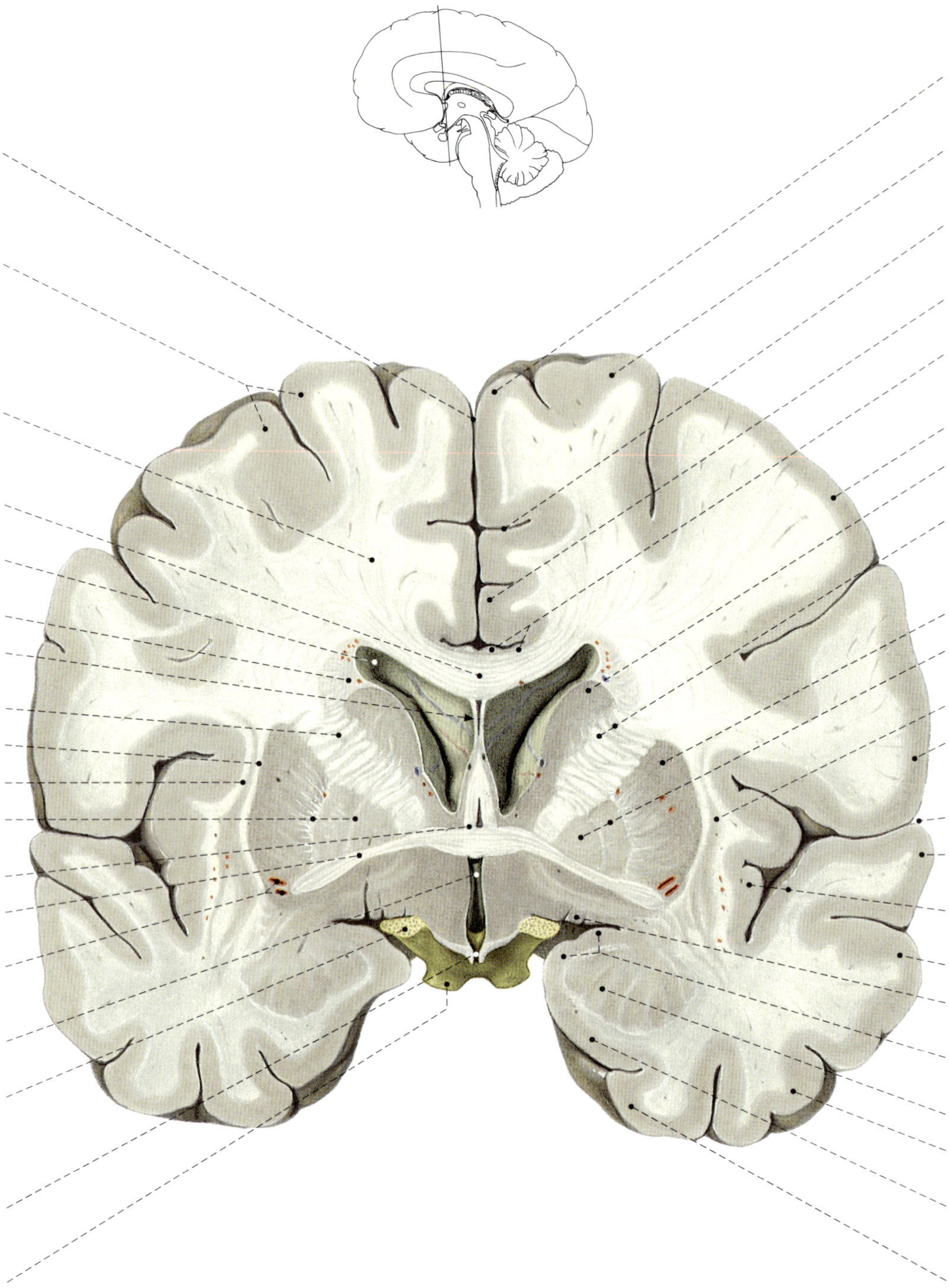

338 Encephalon

338 Brain (100%)

Coronal section in the plane of the anterior commissure, occipital aspect

338 Gehirn (100%)

Frontalschnitt in der Ebene der Commissura anterior, Okzipitalansicht

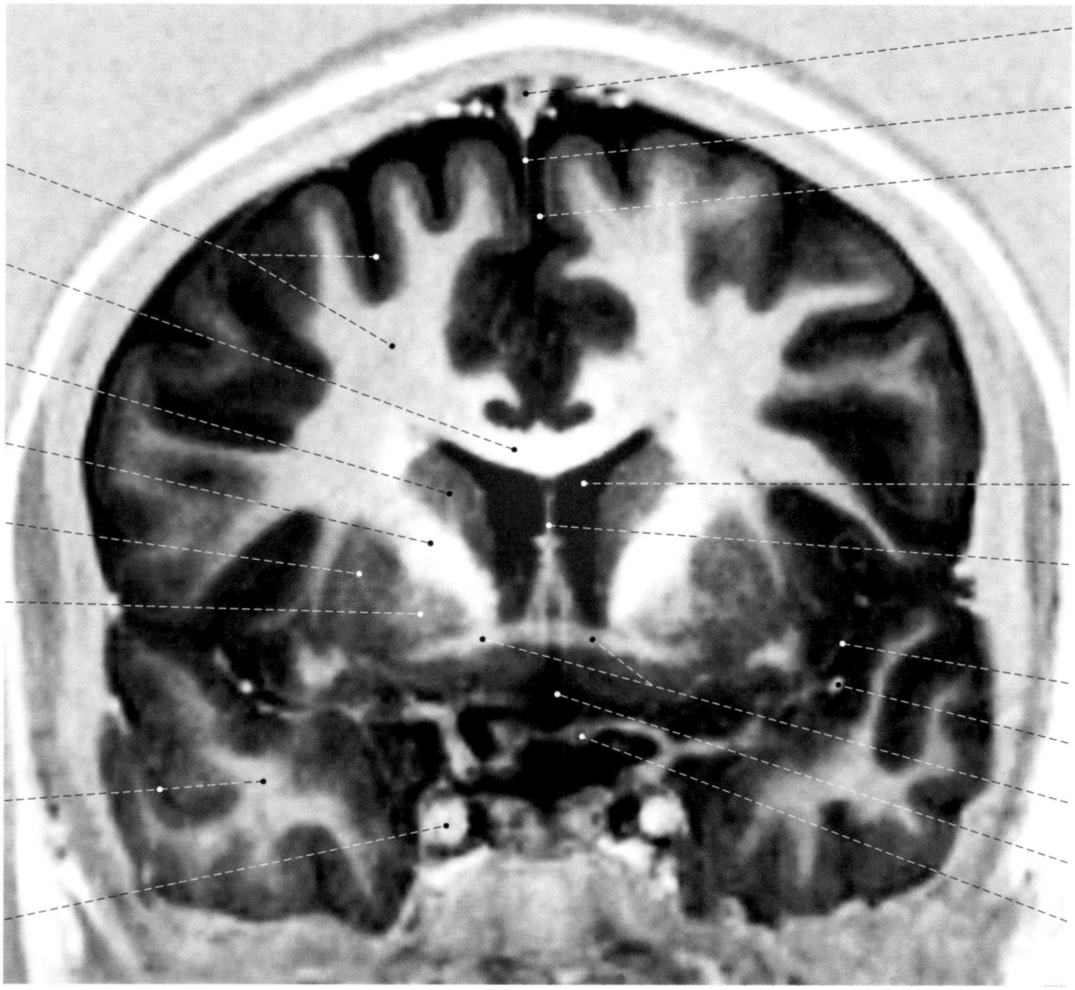

339 Encephalon

339 Brain (100%)

Coronal magnetic resonance image (MRI, T_1-weighted) in the plane of the anterior commissure, in correspondence to Fig. 338

339 Gehirn (100%)

Frontales Kernspintomogramm (MRI, T_1-gewichtet) in der Ebene der vorderen Kommissur, entsprechend Abb. 338

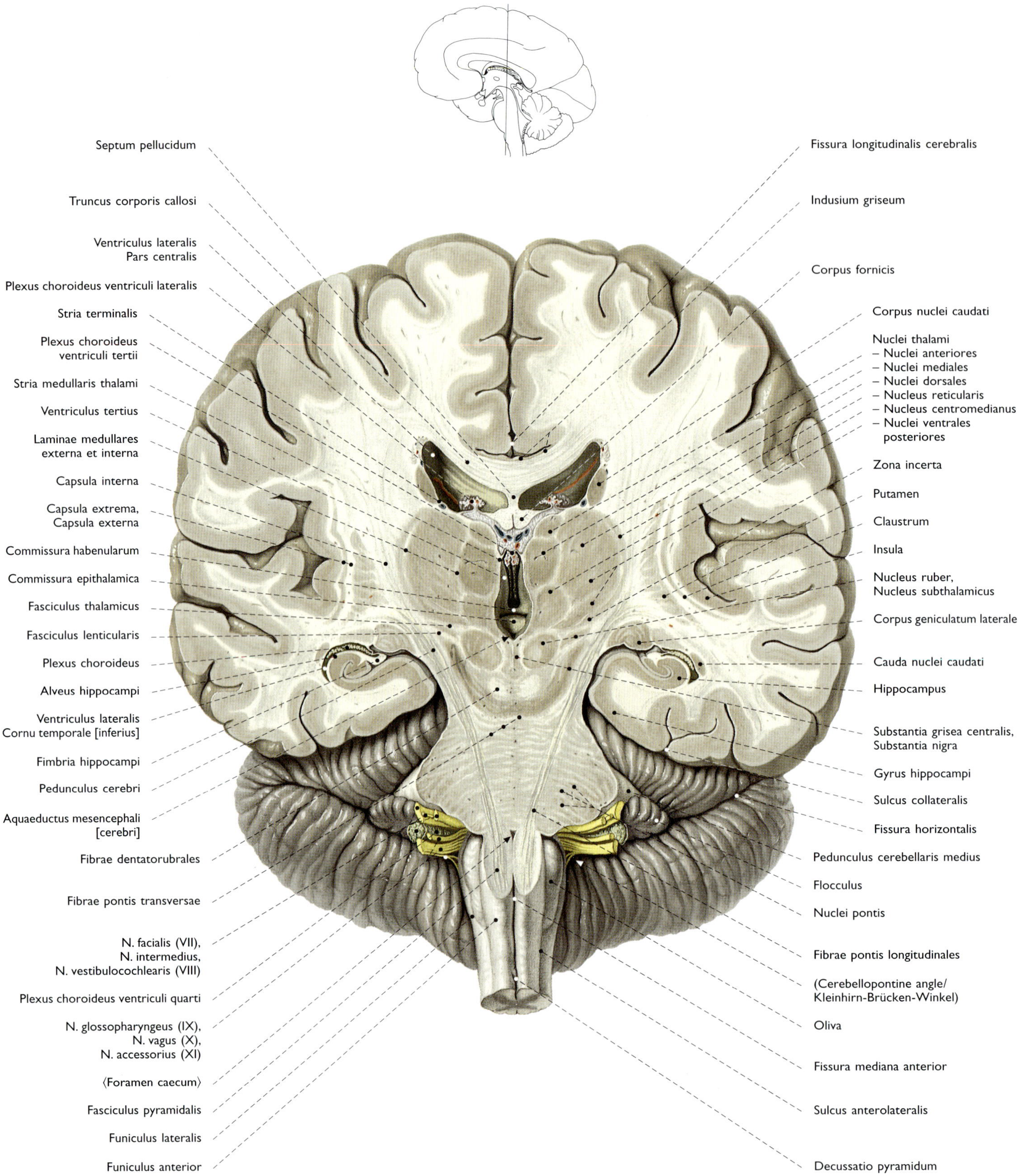

340 Encephalon

340 Brain (100%)
Coronal section in the plane of the cerebral peduncles, frontal aspect

340 Gehirn (100%)
Frontalschnitt in der Ebene der Grosshirnstiele, Frontalansicht

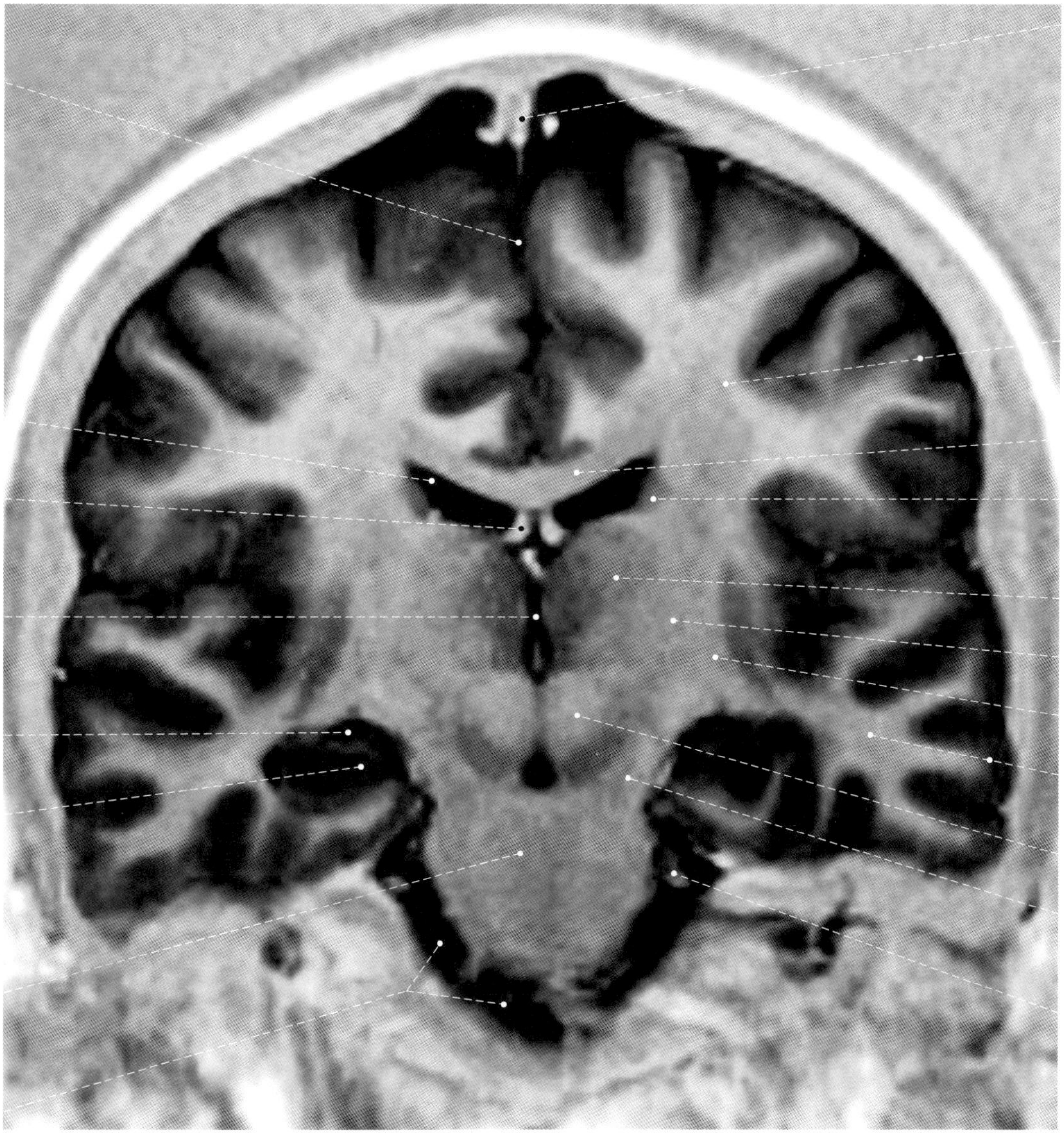

341 Encephalon

341 Brain (100%)

Coronal magnetic resonance image (MRI, T_1-weighted) in the plane of the cerebral peduncles, in correspondence to Fig. 340

341 Gehirn (100%)

Frontales Kernspintomogramm (MRI, T_1-gewichtet) in der Ebene der Grosshirnstiele, entsprechend Abb. 340

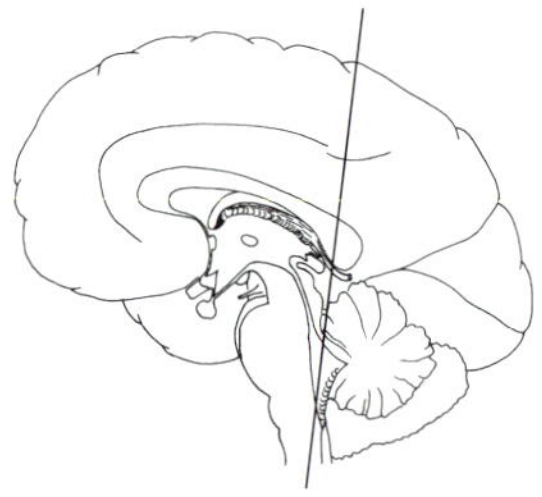

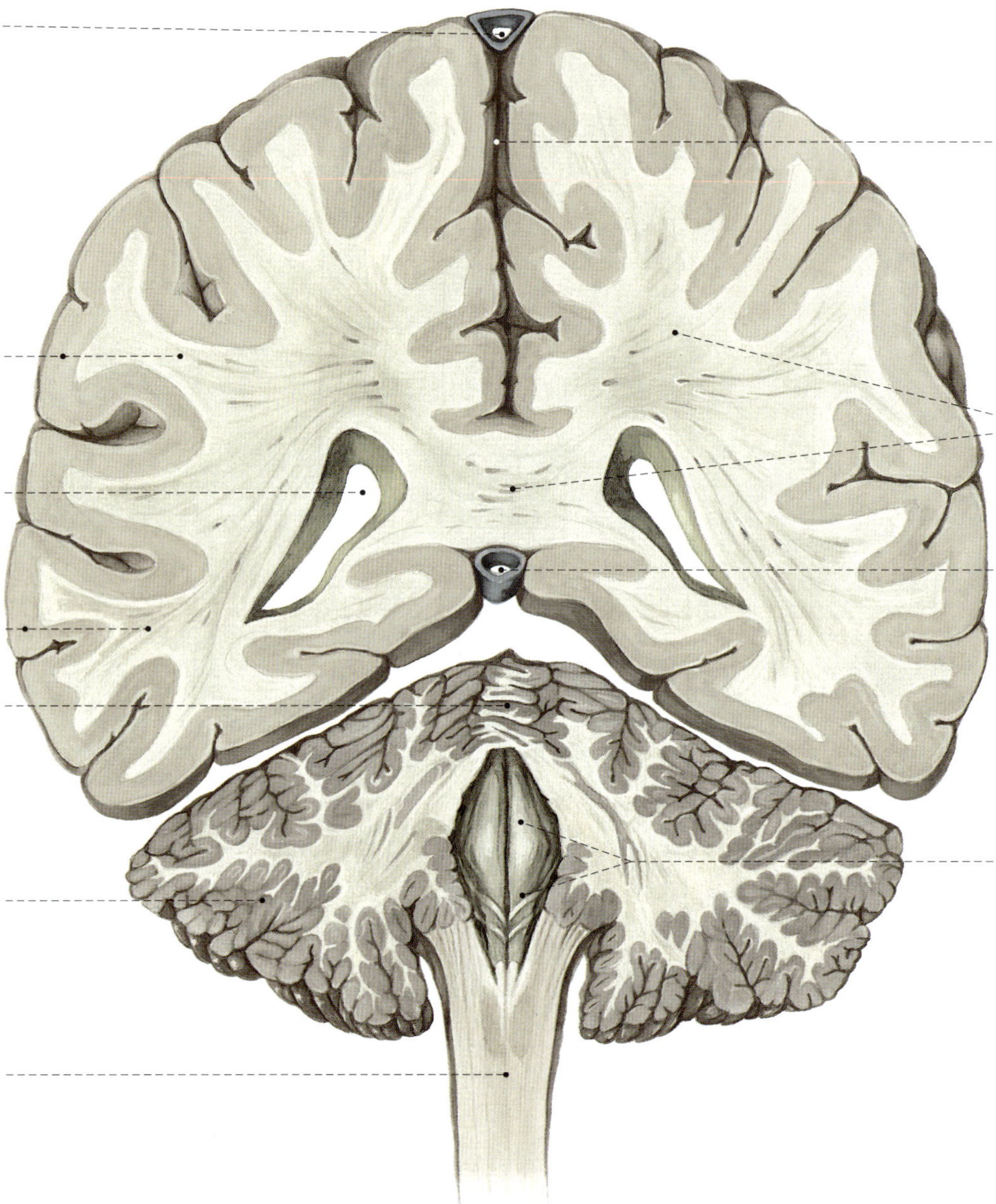

342 Encephalon

342 Brain (100%)

Coronal section in the plane of the posterior horn of the lateral ventricles and the fourth ventricle, occipital aspect

342 Gehirn (100%)

Frontalschnitt in der Ebene des Hinterhorns der Seitenventrikel und des vierten Ventrikels, Okzipitalansicht

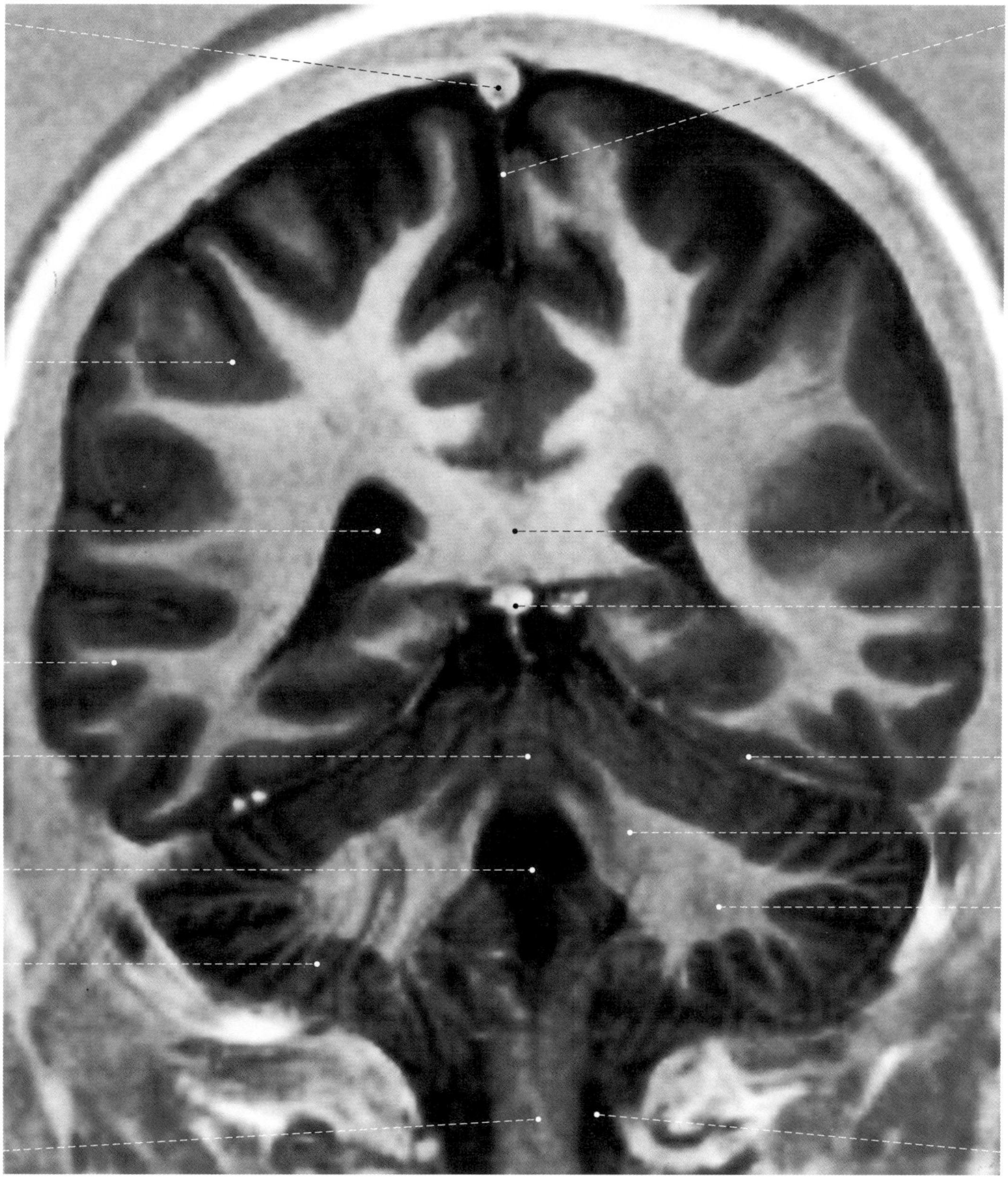

343 Encephalon

343 Brain (100%)

Coronal magnetic resonance image (MRI, T_1-weighted) in the plane of the posterior horn of the lateral ventricles and the fourth ventricle, in correspondence to Fig. 342

343 Gehirn (100%)

Frontales Kernspintomogramm (MRI, T_1-gewichtet) in der Ebene des Hinterhorns der Seitenventrikel und des vierten Ventrikels, entsprechend Abb. 342

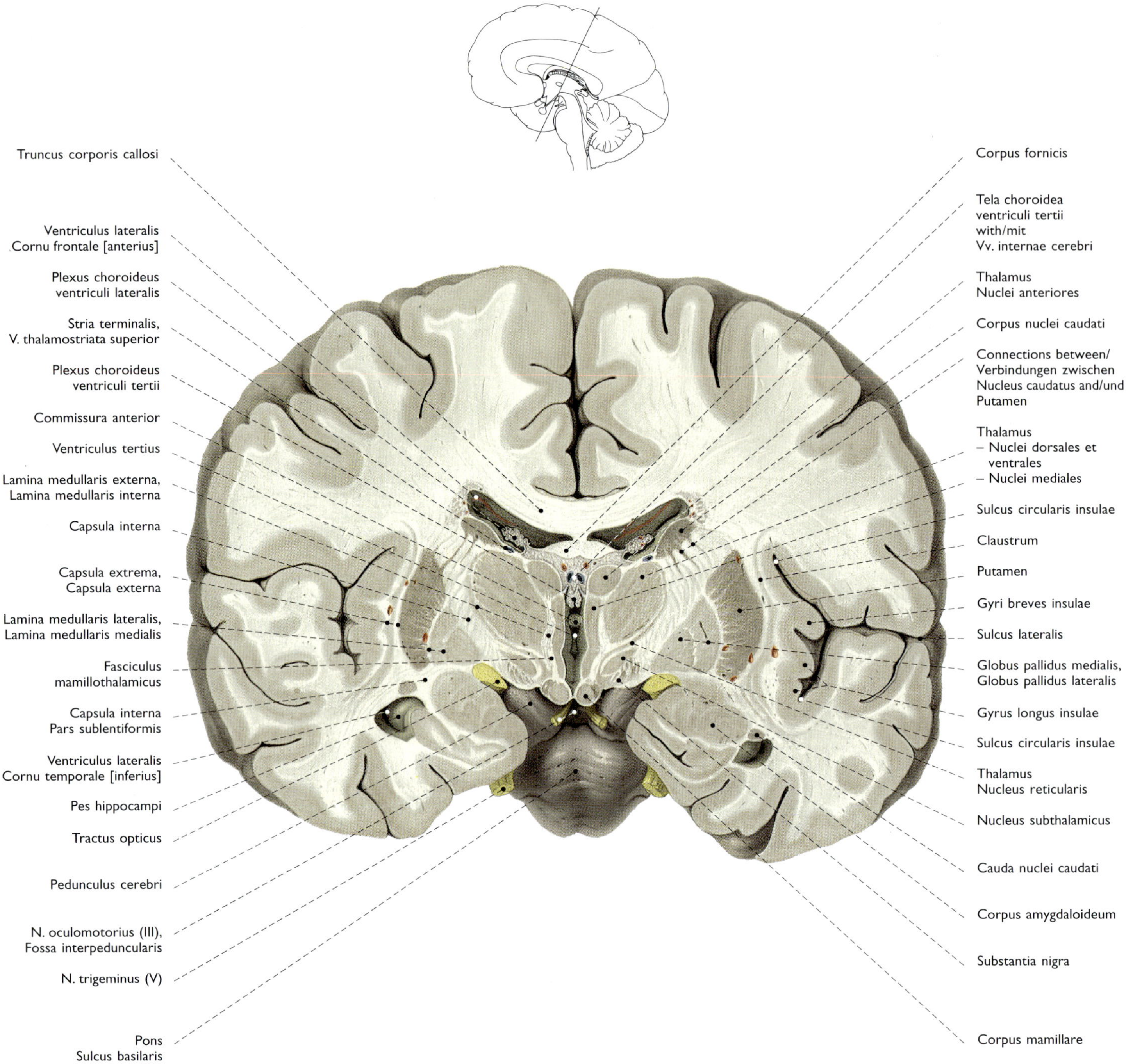

344 Encephalon

344 Brain (100%)

Coronal section in the plane of the mamillary bodies, frontal aspect

344 Gehirn (100%)

Frontalschnitt in der Ebene der Corpora mamillaria, Frontalansicht

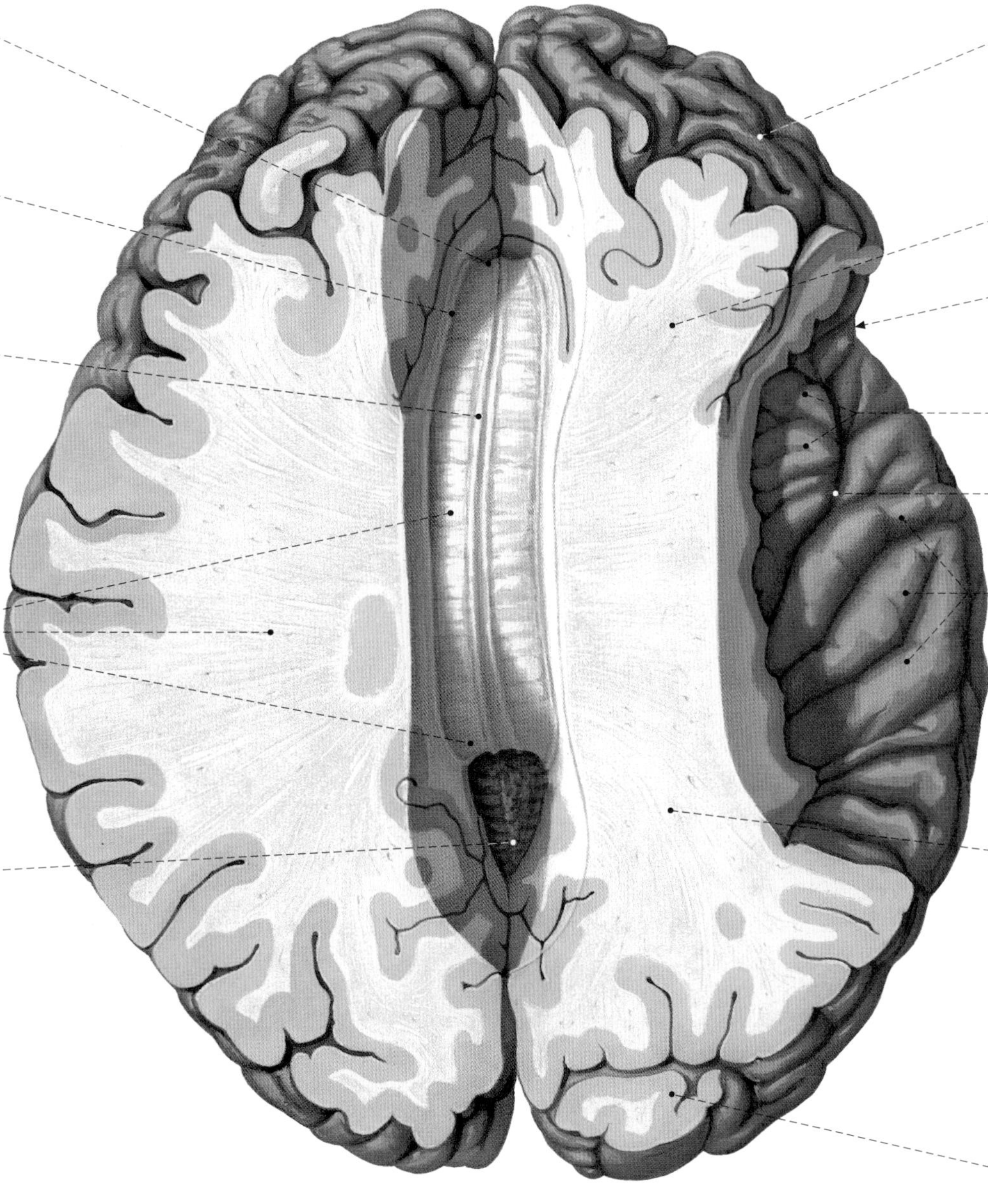

345 Encephalon

345 Brain (100%)

Horizontal section above the corpus callosum. The upper surface of the temporal lobe and the insula were exposed in the right hemisphere. Parietal aspect

345 Gehirn (100%)

Horizontalschnitt über dem Balken. Die obere Fläche des Schläfenlappens und die Insel sind an der rechten Hemisphäre freigelegt. Parietalansicht

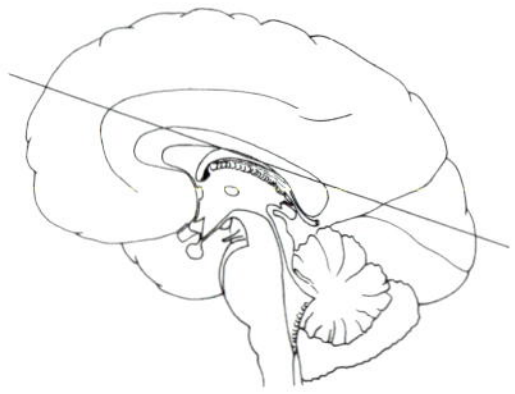

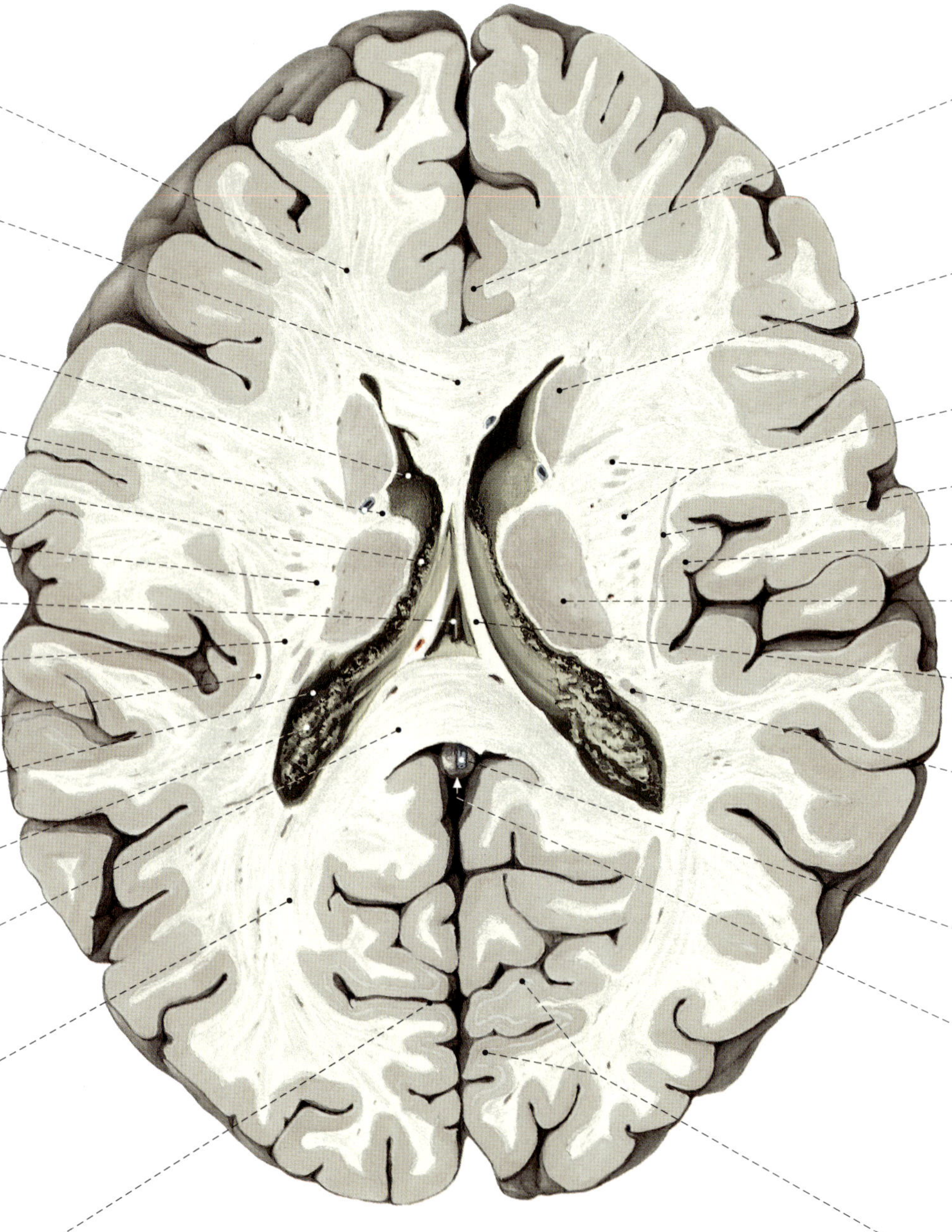

346 Encephalon

346 Brain (100%)

Transverse section at the level of the body of the fornix, and the anterior and posterior horns of the lateral ventricles, parietal aspect

346 Gehirn (100%)

Horizontalschnitt in Höhe des Corpus fornicis und der Vorder- und Hinterhörner der Seitenventrikel, Parietalansicht

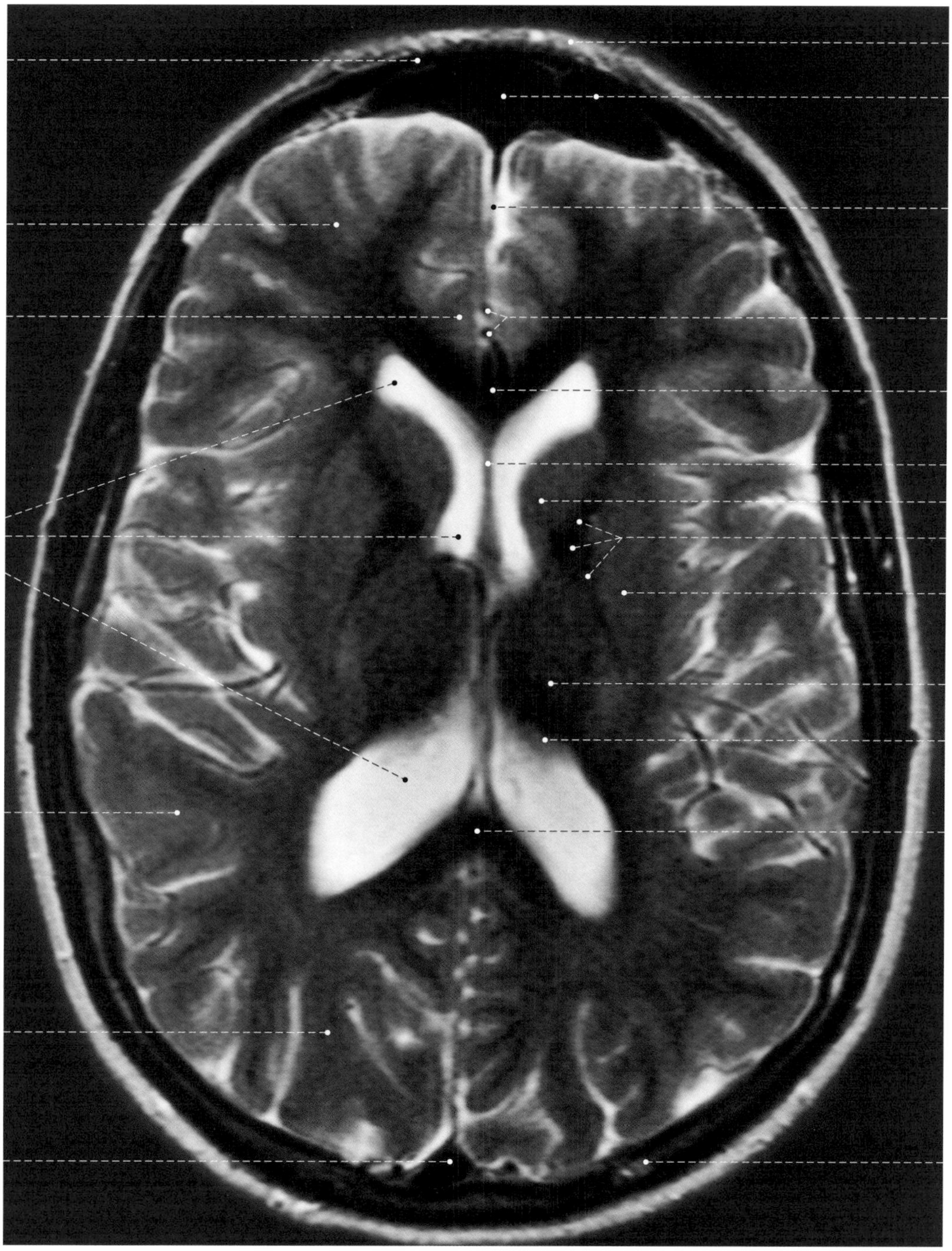

347 Encephalon

347 Brain (100%)

Transverse magnetic resonance image (MRI, T_2-weighted) at the level of the anterior and posterior horns of the lateral ventricles, in correspondence to Fig. 346

347 Gehirn (100%)

Horizontales Kernspintomogramm (MRI, T_2-gewichtet) in Höhe der Vorder- und Hinterhörner der Seitenventrikel, entsprechend Abb. 346

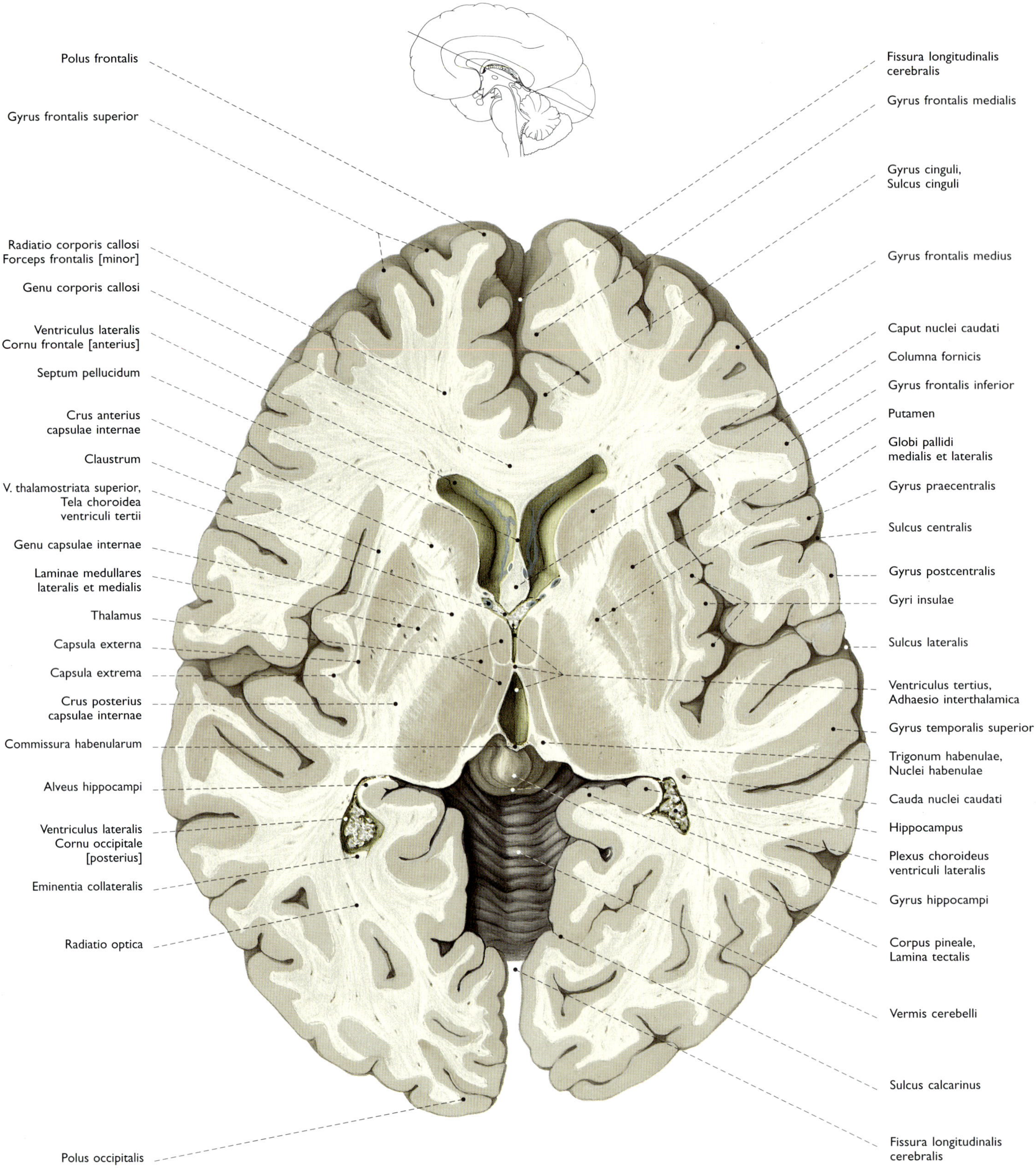

348 Encephalon

348 Brain (100%)

Transverse section at the level of the third ventricle, the anterior and posterior horns of the lateral ventricles, and the pineal body, parietal aspect

348 Gehirn (100%)

Horizontalschnitt in Höhe des dritten Ventrikels, der Vorder- und Hinterhörner der Seitenventrikel und der Zirbeldrüse, Parietalansicht

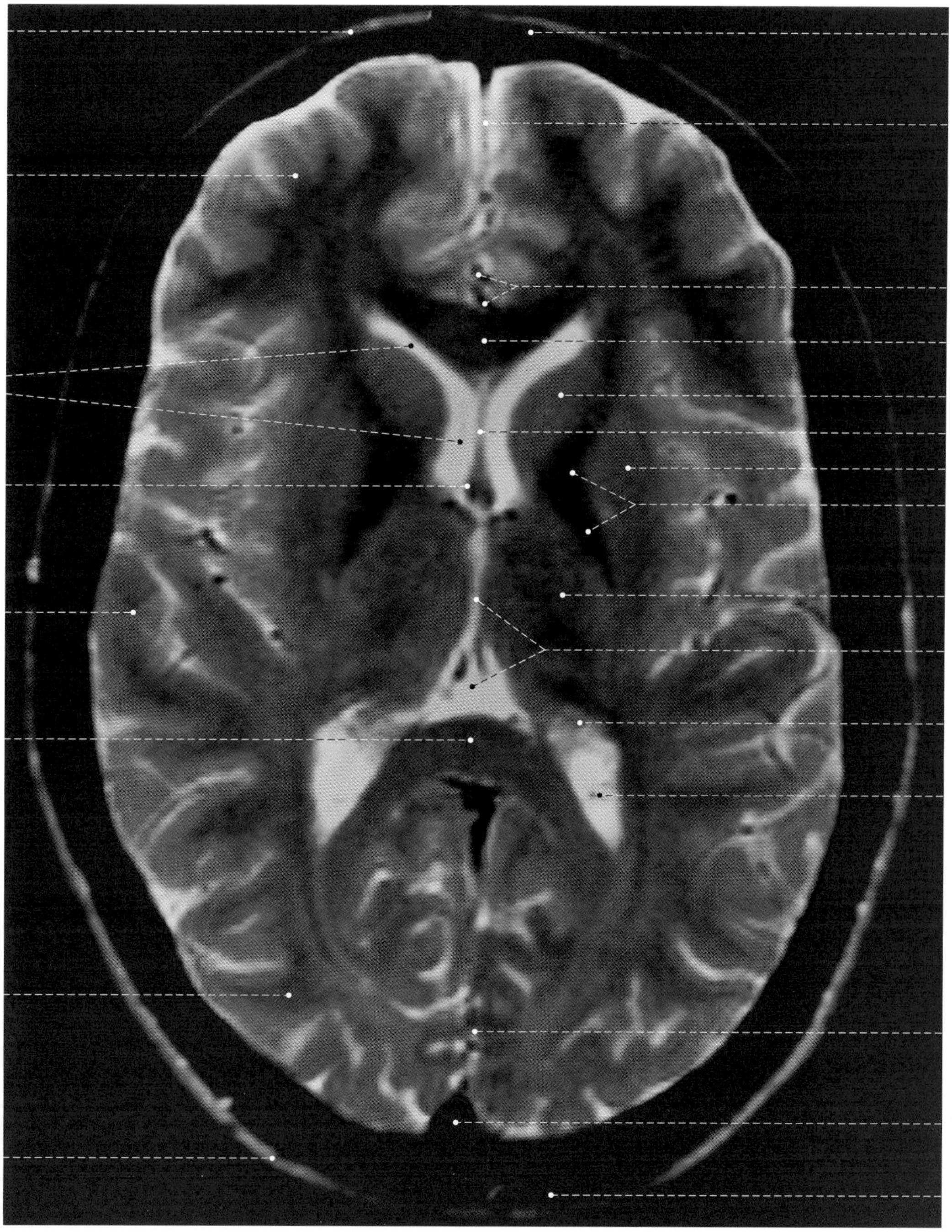

349 Encephalon

349 Brain (100%)

Transverse magnetic resonance image (MRI, T_2-weighted) at the level of the third ventricle, and the anterior and posterior horns of the lateral ventricles, in correspondence to Fig. 348

349 Gehirn (100%)

Horizontales Kernspintomogramm (MRI, T_2-gewichtet) in Höhe des dritten Ventrikels und der Vorder- und Hinterhörner der Seitenventrikel, entsprechend Abb. 348

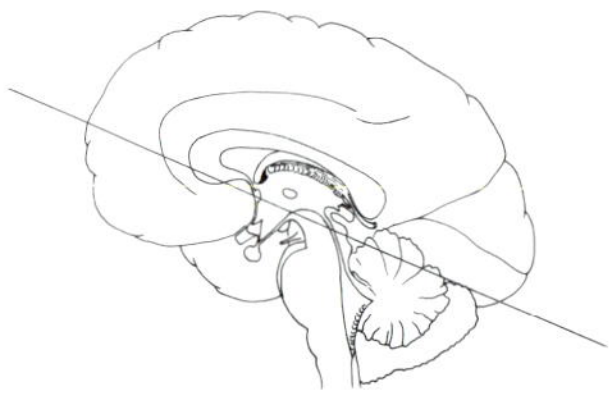

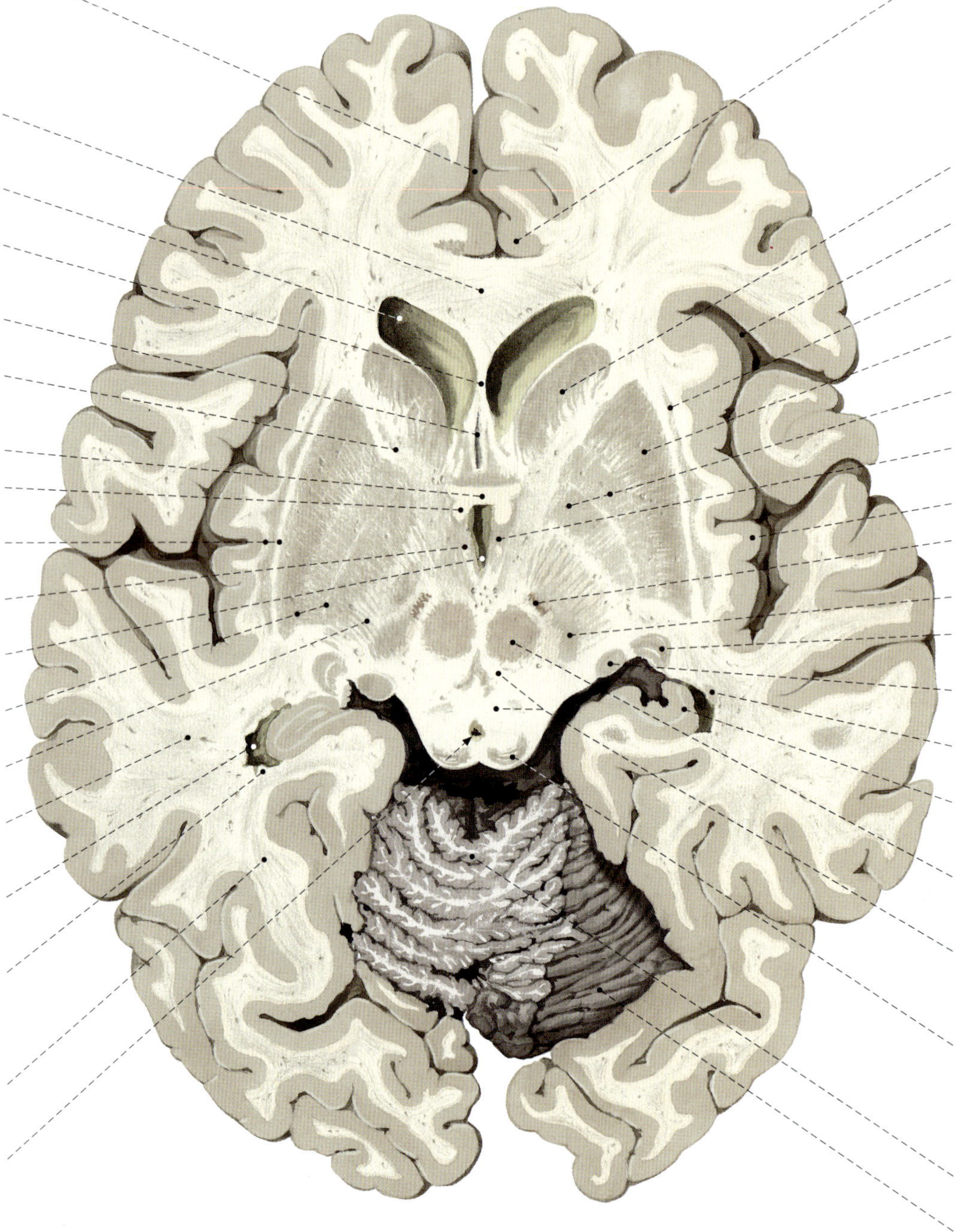

350 Encephalon

350 Brain (100%)

Transverse section at the level of the anterior commissure, superior aspect

350 Gehirn (100%)

Horizontalschnitt in Höhe der Commissura anterior, Parietalansicht

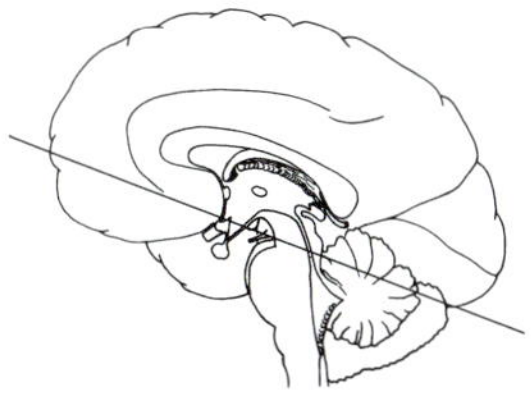

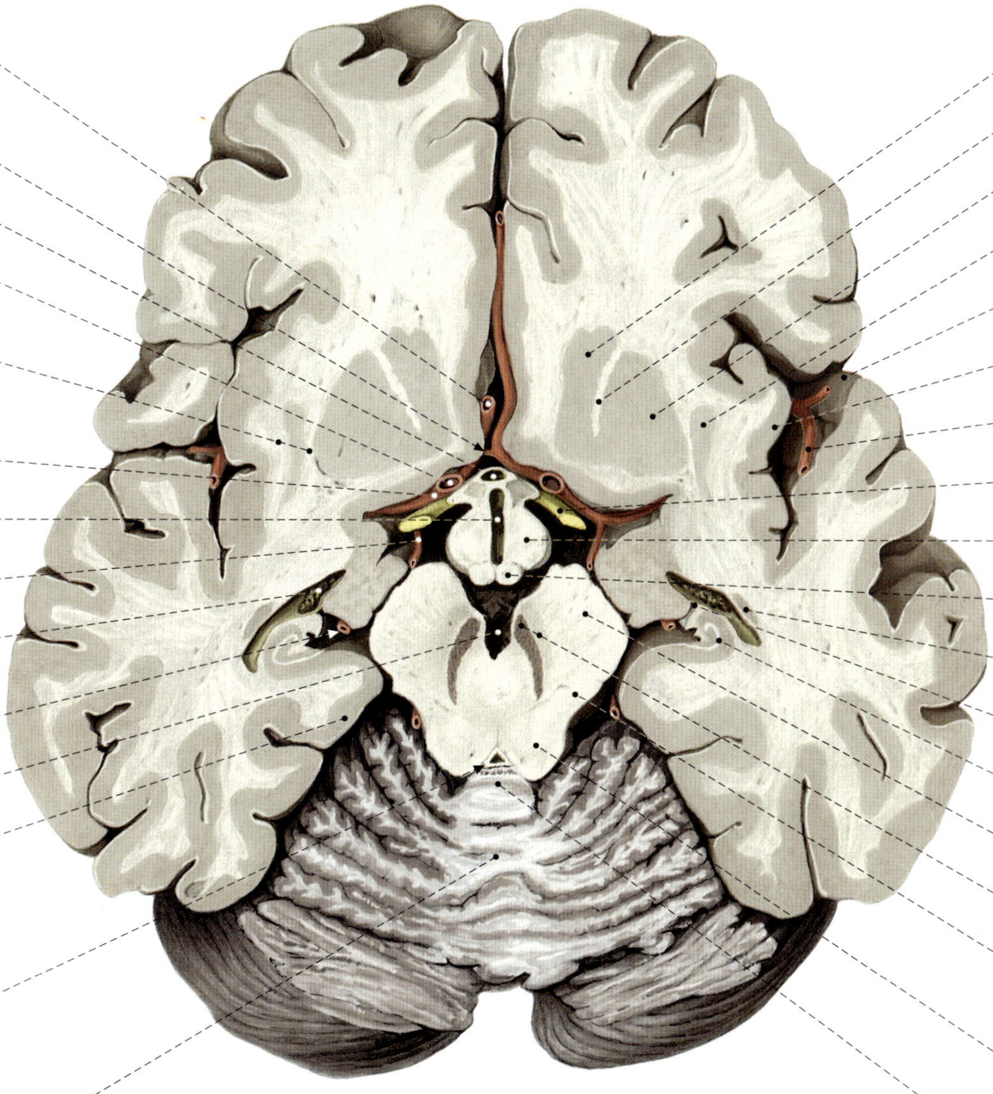

351 Encephalon

351 Brain (100%)

Transverse section at the level of the midbrain and the mamillary bodies, superior aspect

351 Gehirn (100%)

Horizontalschnitt in Höhe des Mittelhirns und der Corpora mamillaria, Parietalansicht

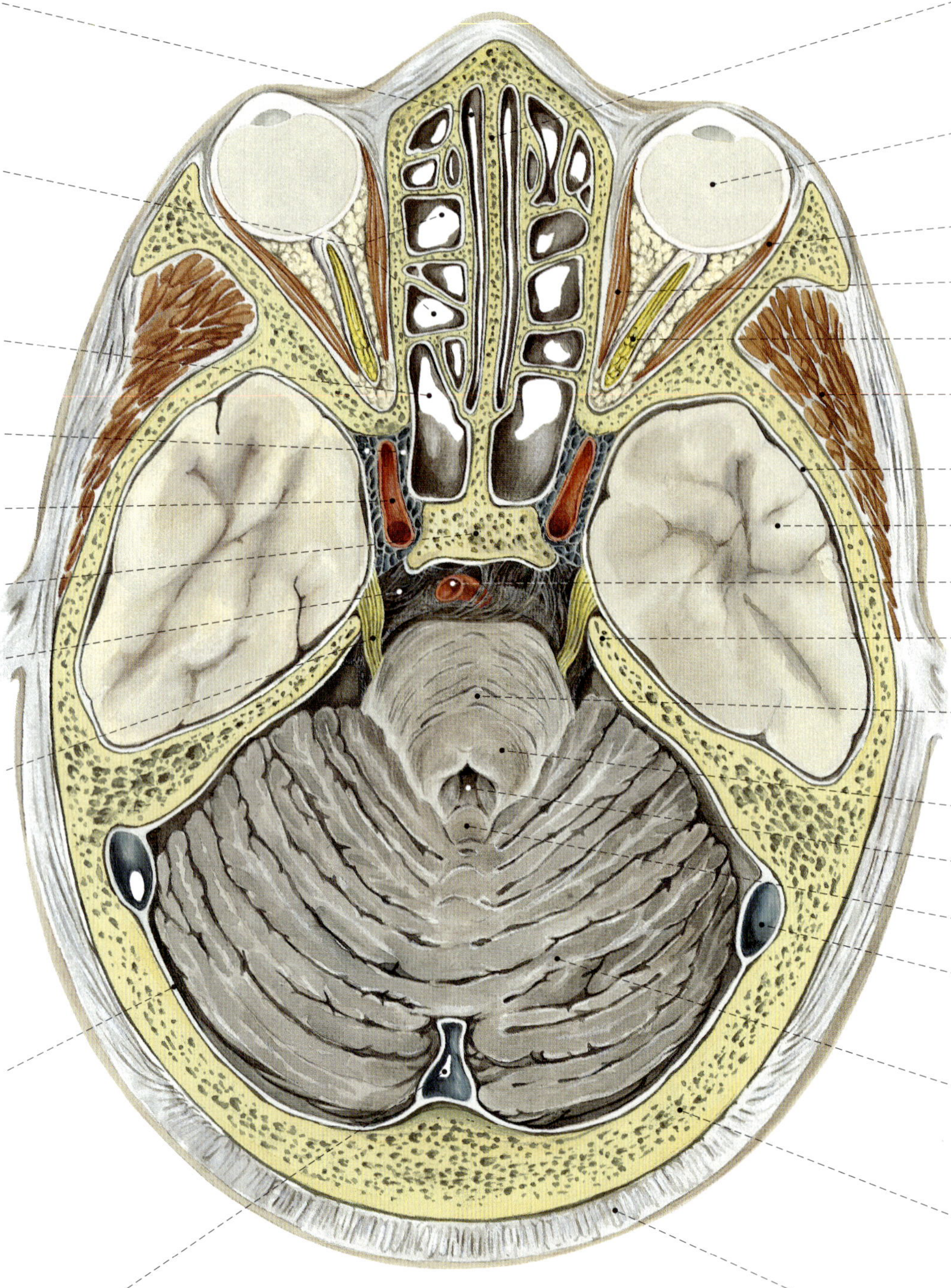

352 Encephalon

352 Brain (100%)

Transverse section through the head at the level of the optic nerves, the dorsum sellae, the pons, and the cerebral aqueduct, superior aspect

352 Gehirn (100%)

Horizontalschnitt durch den Kopf in Höhe der Sehnerven, des Dorsum sellae, der Brücke und des Aquaeductus mesencephali, Parietalansicht

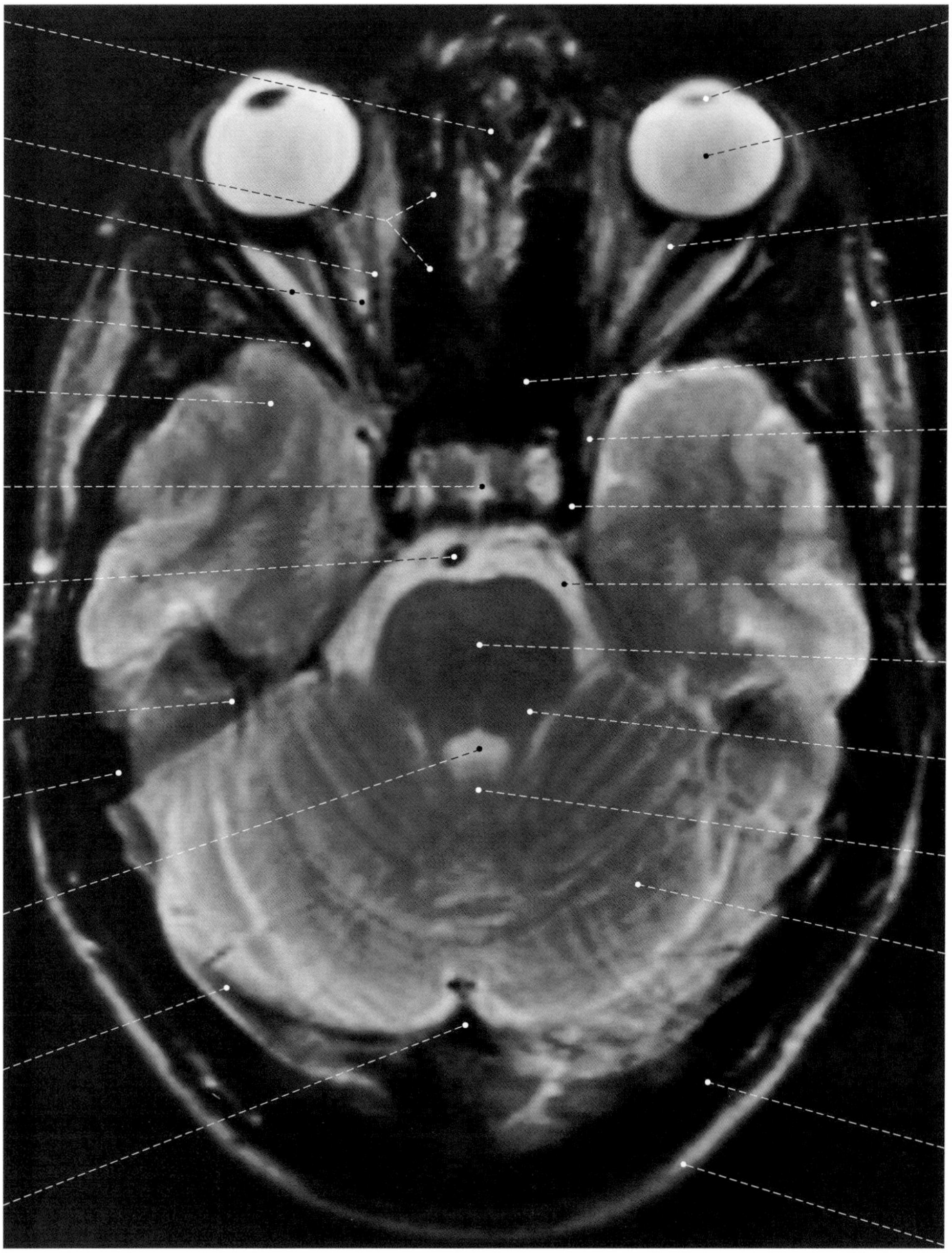

353 Encephalon

353 Brain (100%)
Transverse magnetic resonance image (MRI, T_2-weighted) at the level of the optic nerves, the dorsum sellae, the pons, and the cerebral aqueduct, in correspondence to Fig. 352, inferior aspect

353 Gehirn (100%)
Horizontales Kernspintomogramm (MRI, T_2-gewichtet) in Höhe der Sehnerven, des Dorsum sellae, der Brücke und des Aquaeductus mesencephali, entsprechend Abb. 352, Kaudalansicht

a

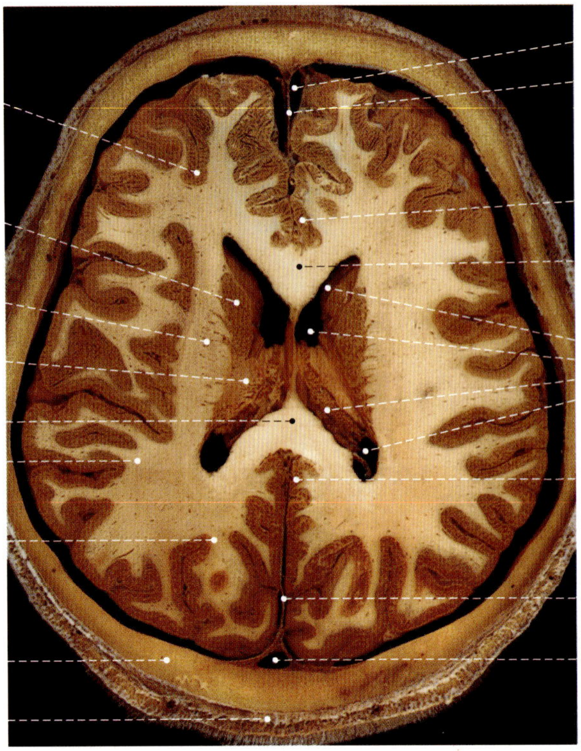

b

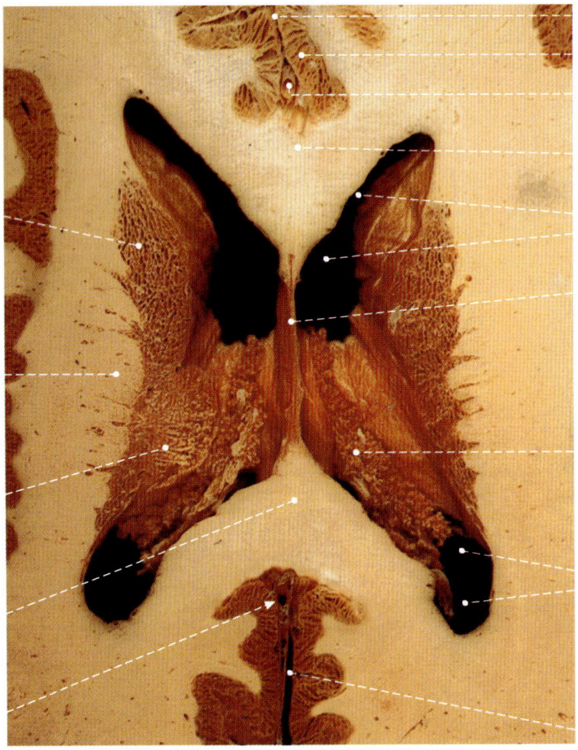

354 Encephalon

354 Brain

Anatomical transverse section of the head and brain at a superior level through the central part and the anterior and posterior horns of the lateral ventricles, parietal aspect

a Cross section through the whole head (50%)

b Magnification of central parts of the brain (110%)

354 Gehirn

Kranial geführter anatomischer Horizontalschnitt durch Kopf und Gehirn in Höhe des Zentralteils sowie der Vorder- und Hinterhörner der Seitenventrikel, Parietalansicht

a Querschnitt durch den ganzen Kopf (50%)

b Ausschnittvergrösserung zentraler Gehirnregionen (110%)

a

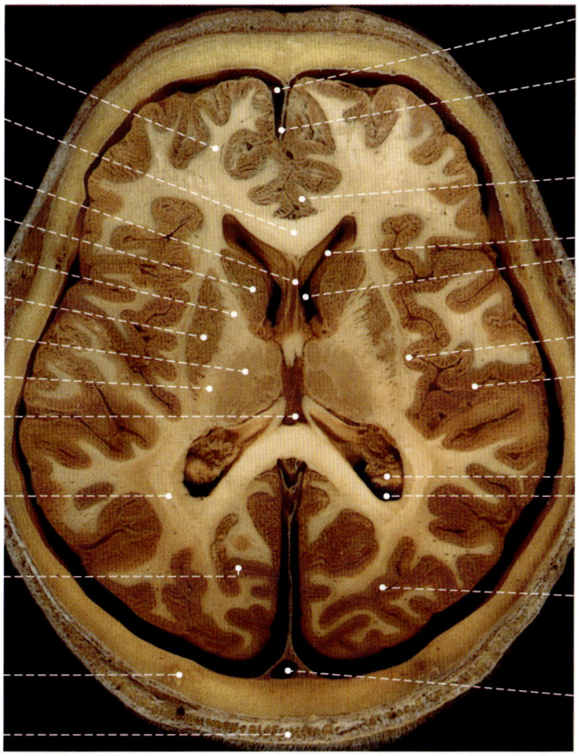

b

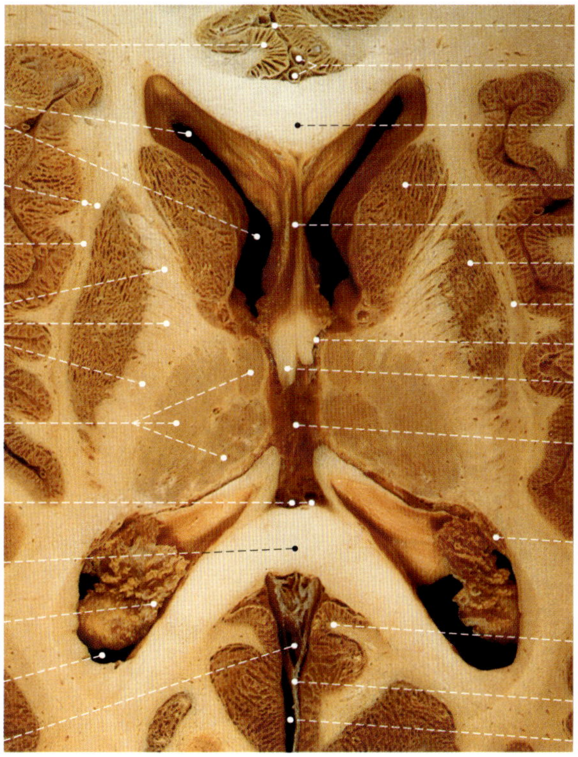

355 Encephalon

355 Brain

Anatomical transverse section (more caudal than in Fig. 354) through the head and brain at the level of the third ventricle, the cranial parts of the internal capsule, and the anterior and posterior horns of the lateral ventricles, parietal aspect

a Cross section through the whole head (50%)

b Magnification of central parts of the brain (110%)

355 Gehirn

Anatomischer Horizontalschnitt (kaudaler als Abb. 354) durch Kopf und Gehirn auf Höhe des dritten Ventrikels, der kranialen Abschnitte der inneren Kapsel sowie der Vorder- und Hinterhörner der Seitenventrikel, Parietalansicht

a Querschnitt durch den ganzen Kopf (50%)

b Ausschnittvergrösserung zentraler Gehirnregionen (110%)

a

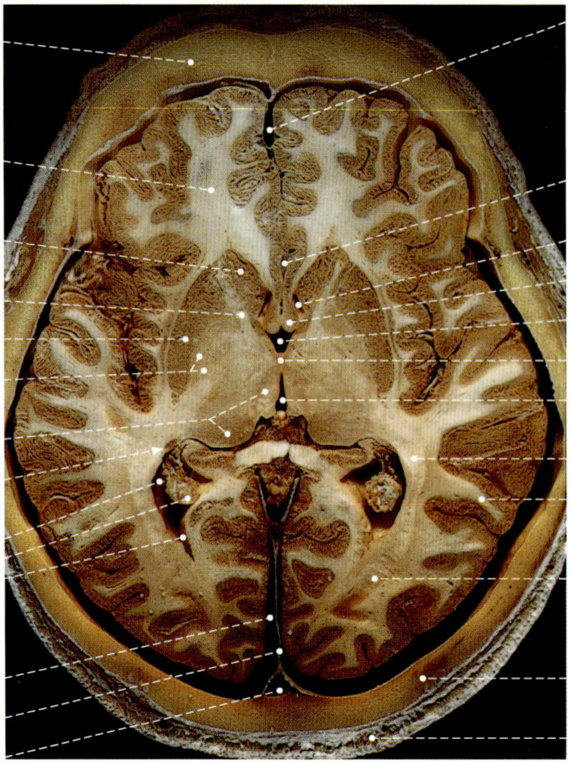

b

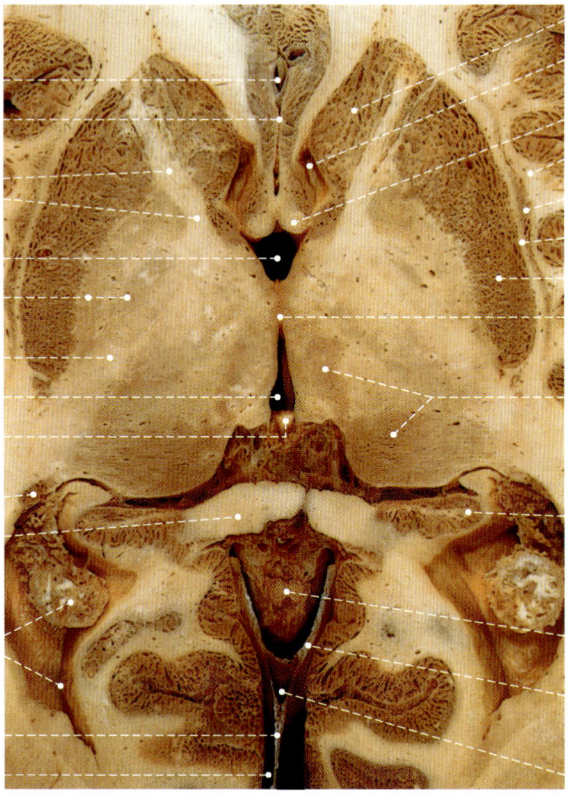

356 Encephalon

356 Brain

Anatomical transverse section (more caudal than in Fig. 355) through the head and brain at the level of the caudal parts of the third ventricle, the internal capsule, and the posterior horn of the lateral ventricles, parietal aspect

a Cross section through the whole head (50%)
b Magnification of central parts of the brain (110%)

356 Gehirn

Anatomischer Horizontalschnitt (kaudaler als Abb. 355) durch Kopf und Gehirn auf Höhe der kaudalen Abschnitte des dritten Ventrikels, der inneren Kapsel und des Hinterhorns der Seitenventrikel, Parietalansicht

a Querschnitt durch den ganzen Kopf (50%)
b Ausschnittvergrösserung zentraler Gehirnregionen (110%)

a

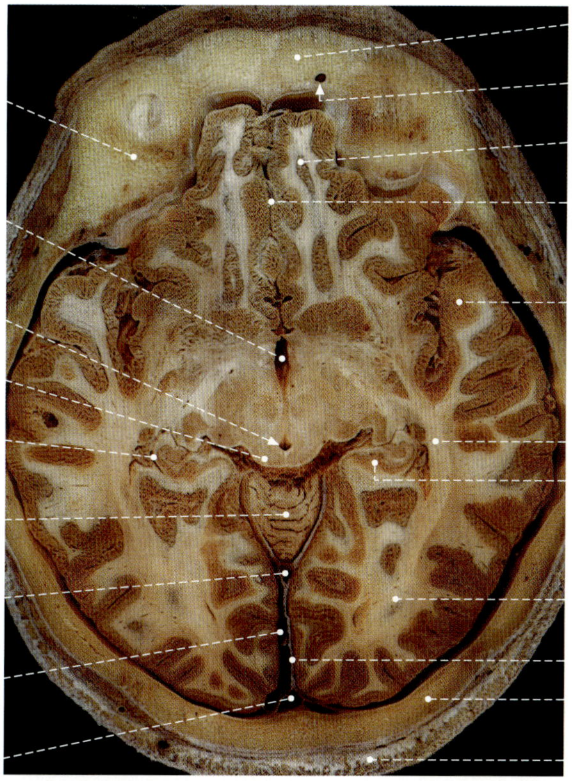

b

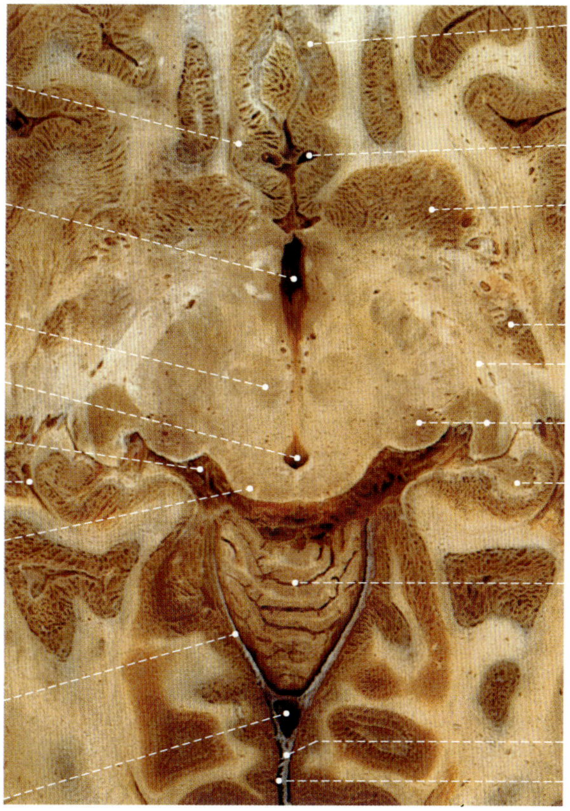

357 Encephalon

357 Brain

Anatomical transverse section (more caudal than in Fig. 356) through the head and brain at the level of the orbital roof, the cerebral aqueduct, and the vermis of the cerebellum, parietal aspect

a Cross section through the whole head (50%)
b Magnification of central parts of the brain (110%)

357 Gehirn

Anatomischer Horizontalschnitt (kaudaler als Abb. 356) durch Kopf und Gehirn auf Höhe des Orbitadachs, des Aquaeductus mesencephali und des Kleinhirnwurms, Parietalansicht

a Querschnitt durch den ganzen Kopf (50%)
b Ausschnittvergrösserung zentraler Gehirnregionen (110%)

a

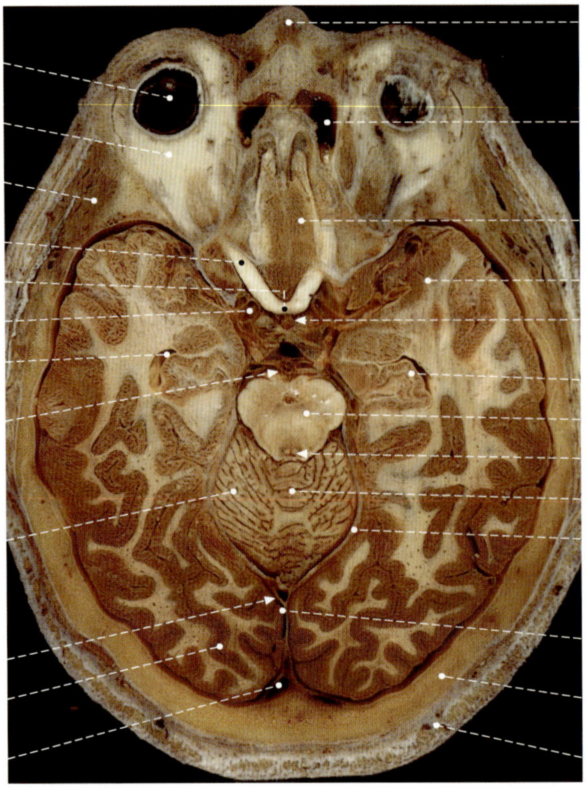

b

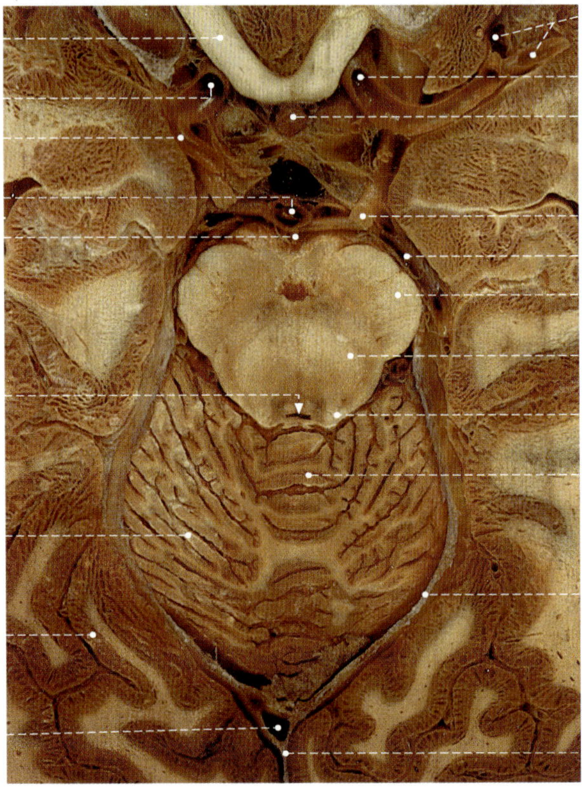

358 Encephalon

358 Brain

Anatomical transverse section (more caudal than in Fig. 357) through the head and brain at the level of the optic chiasma, the cerebral aqueduct, and cranial parts of the cerebellum, parietal aspect

a Cross section through the whole head (50%)
b Magnification of central parts of the brain (110%)

358 Gehirn

Anatomischer Horizontalschnitt (kaudaler als Abb. 357) durch Kopf und Gehirn auf Höhe der Sehnervenkreuzung, des Aquaeductus mesencephali und kranialer Kleinhirnabschnitte, Parietalansicht

a Querschnitt durch den ganzen Kopf (50%)
b Ausschnittvergrösserung (110%)

a

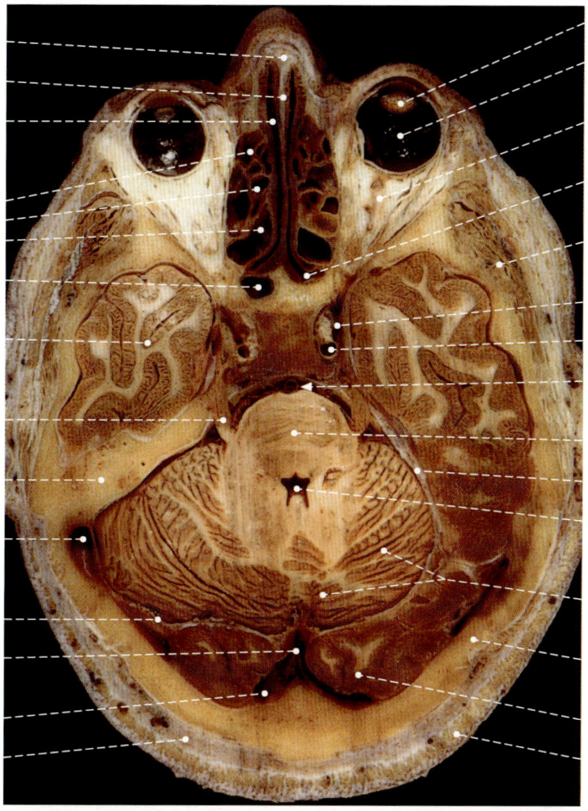

b

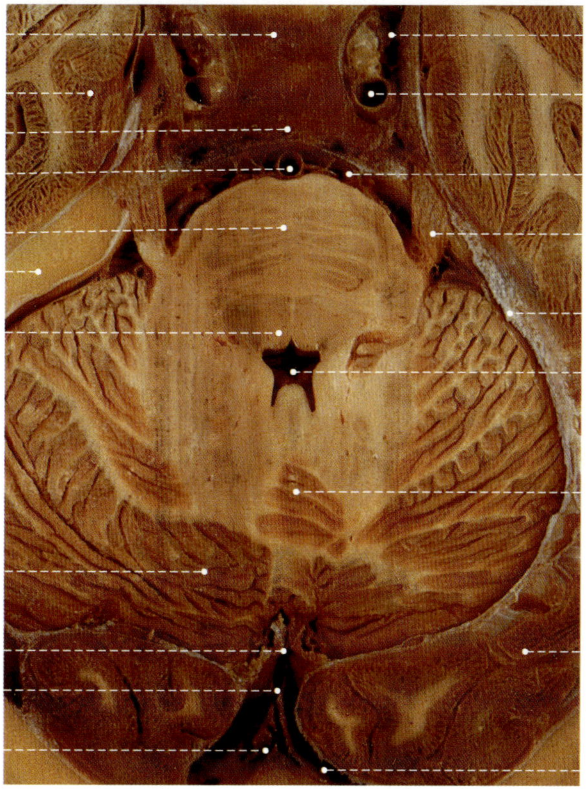

359 Encephalon

359 Brain

Anatomical transverse section (more caudal than in Fig. 358) through the head and brain at the level of the optic nerves, the cavernous sinus, the pons, the fourth ventricle, and the cerebellum, parietal aspect

a Cross section through the whole head (50%)
b Magnification of central parts of the brain (110%)

359 Gehirn

Anatomischer Horizontalschnitt (kaudaler als Abb. 358) durch Kopf und Gehirn auf Höhe der Sehnerven, des Sinus cavernosus, der Brücke, des vierten Ventrikels und des Kleinhirns, Parietalansicht

a Querschnitt durch den ganzen Kopf (50%)
b Ausschnittvergrösserung (110%)

a

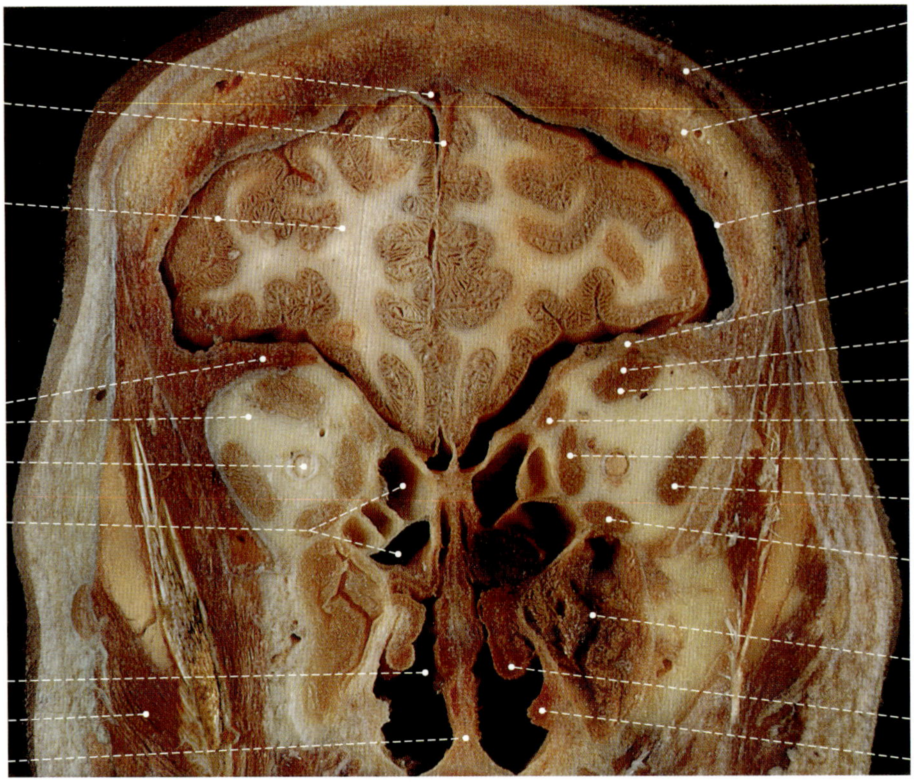

b

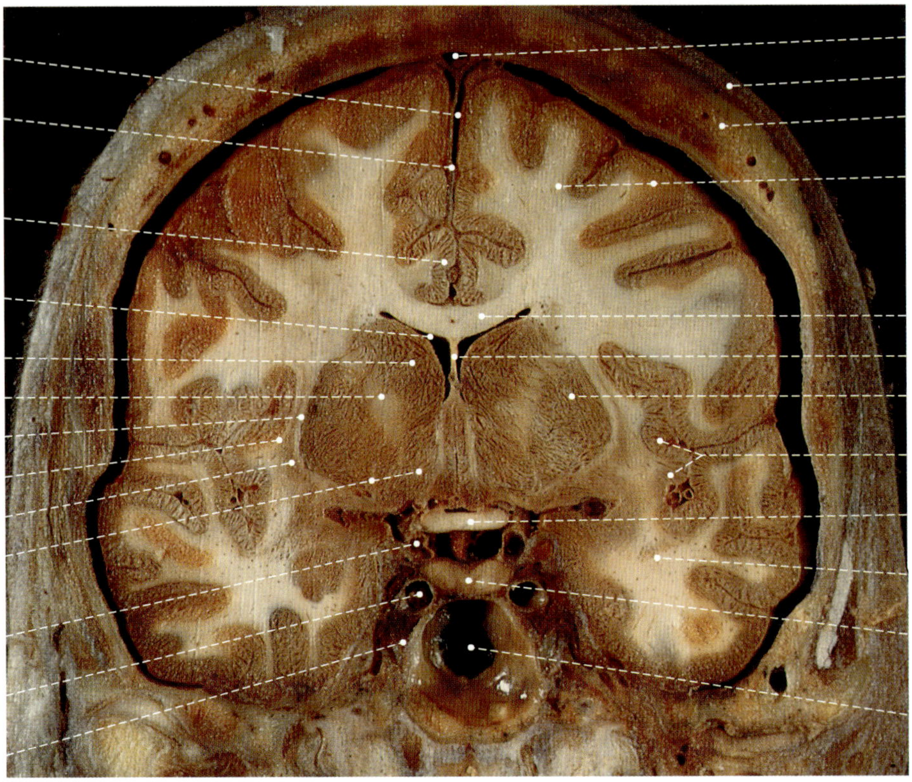

360 Encephalon

360 Brain (80%)

Occipital aspect
Coronal anatomical sections of the head and brain (80%)

- a in a plane through the frontal cerebral lobes and the orbits behind the eyeballs
- b in a plane through the optic chiasma, the anterior horn of the lateral ventricles, and the head of the caudate nuclei

360 Gehirn (80%)

Okzipitalansicht
Frontale anatomische Schnitte von Kopf und Gehirn (80%)

- a in einer Ebene durch die Frontallappen des Gehirns und die Augenhöhlen hinter den Augäpfeln
- b in einer Ebene durch die Sehnervenkreuzung, das Vorderhorn der Seitenventrikel und den Kopf der Nuclei caudati

a

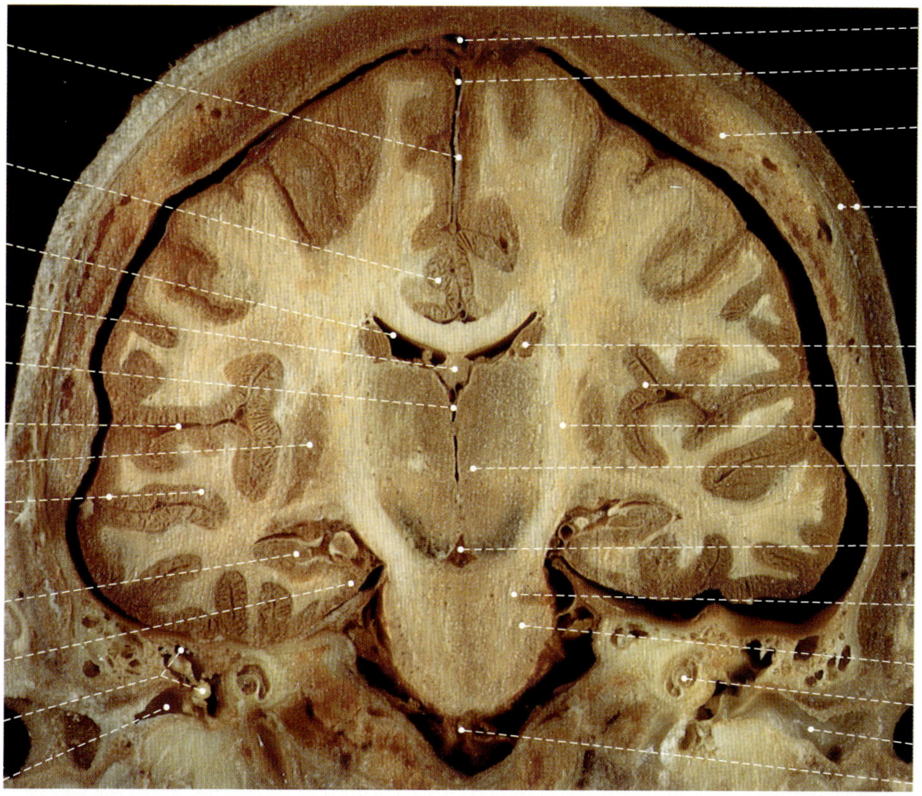

b

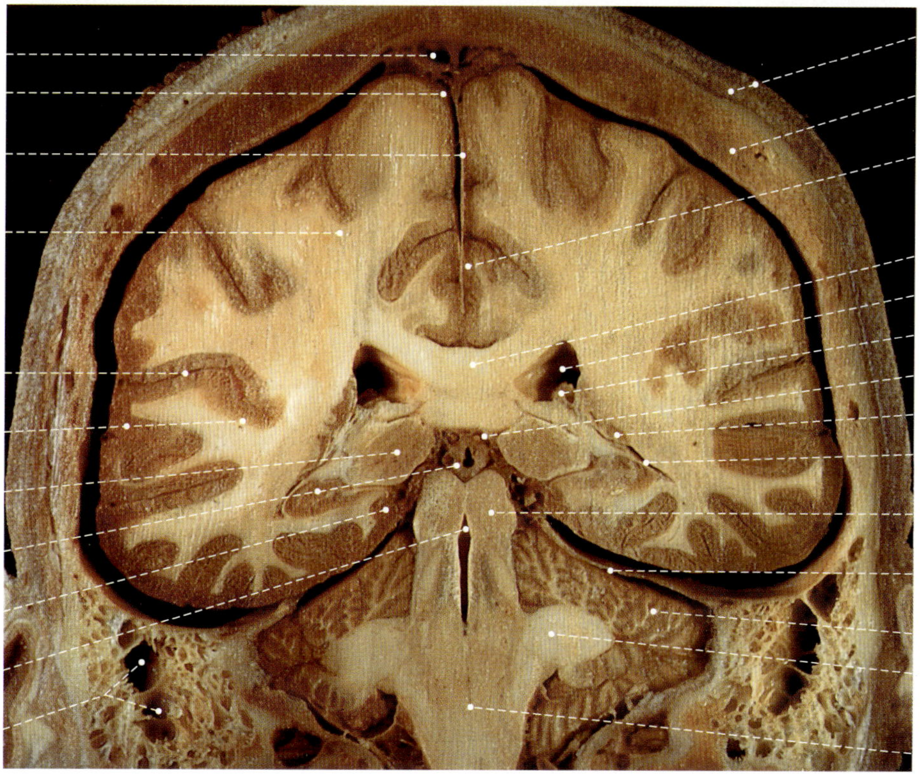

361 Encephalon

361 Brain (80%)

Occipital aspect
Coronal anatomical sections of the head and brain (80%)

a in a plane through the central part of the lateral ventricles, the third ventricle, and the pons
b in a plane through the pineal body, the cerebral aqueduct, and the middle cerebellar peduncles

361 Gehirn (80%)

Okzipitalansicht
Frontale anatomische Schnitte von Kopf und Gehirn (80%)

a in einer Ebene durch den Zentralteil der Seitenventrikel, den dritten Ventrikel und die Brücke
b in einer Ebene durch das Corpus pineale, den Aquaeductus mesencephali und die mittleren Kleinhirnstiele

a

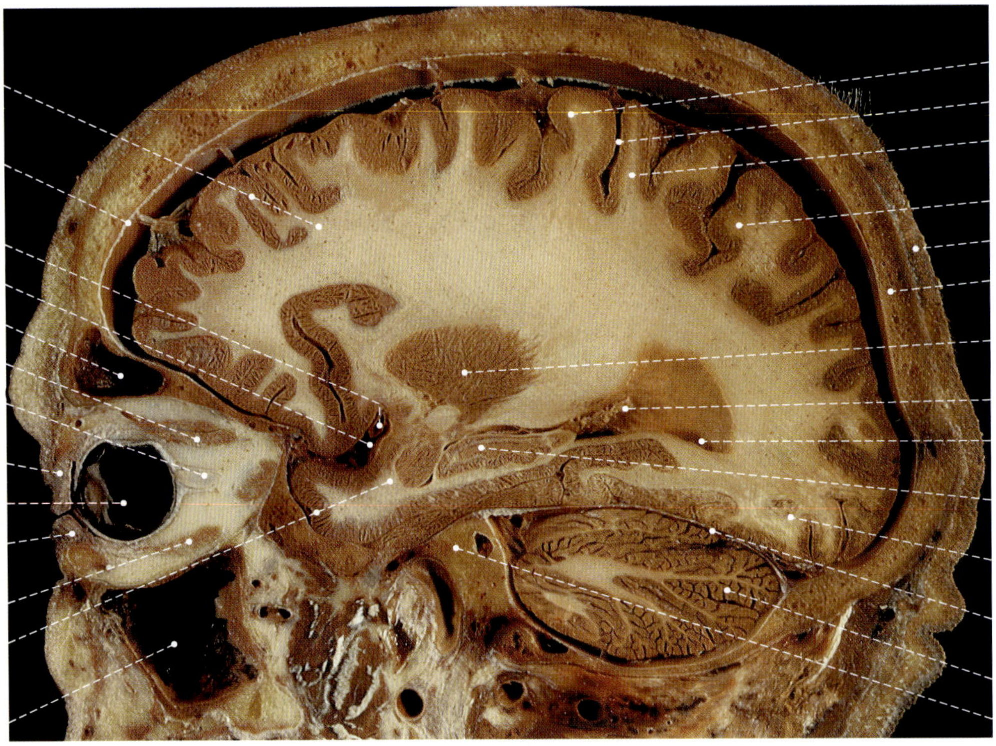

b

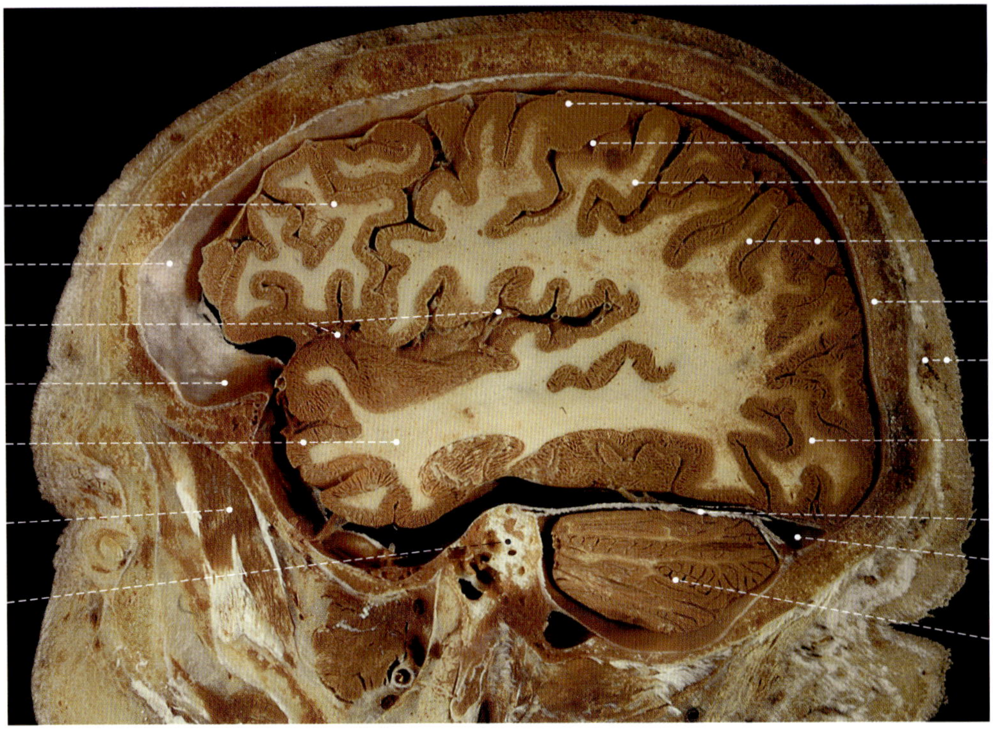

362 Encephalon

362 Brain (70%)

Sagittal anatomical sections through the right half of head and brain

a in a paramedian plane through the right eyeball, the putamen, and the inferior horn of the right lateral ventricle

b in a more lateral plane through the lateral angle of the right orbit, the lateral cerebral sulcus, and the lateral parts of the cerebellum
Medial aspect of the lateral segments

362 Gehirn (70%)

Sagittale anatomische Schnitte durch die rechte Hälfte von Kopf und Gehirn

a in einer paramedianen Ebene durch den rechten Augapfel, das Putamen und das Unterhorn des rechten Seitenventrikels

b in einer seitlicheren Ebene durch den lateralen Winkel der rechten Augenhöhle, den Sulcus lateralis cerebri und die seitlichen Kleinhirnabschnitte
Medialansicht der seitlichen Schnittblöcke

Organum visus et Orbita

Visual Organ and Orbital Cavity

Sehorgan und Augenhöhle

a

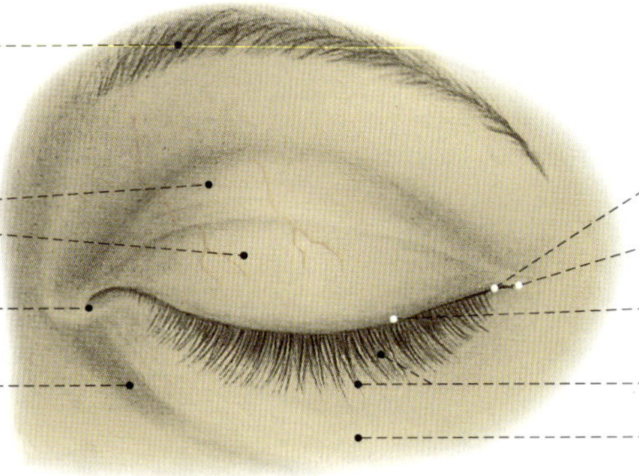

b

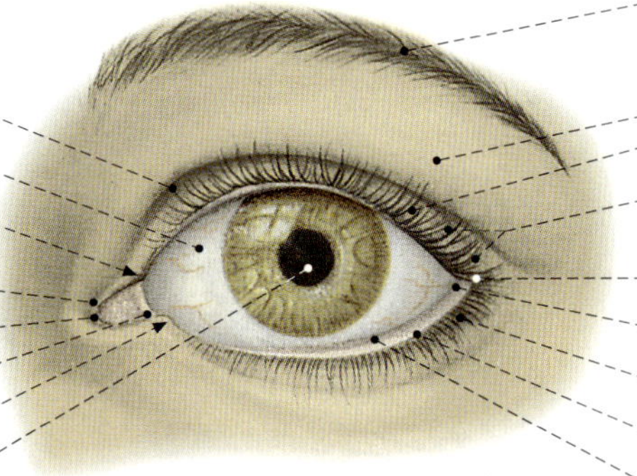

c

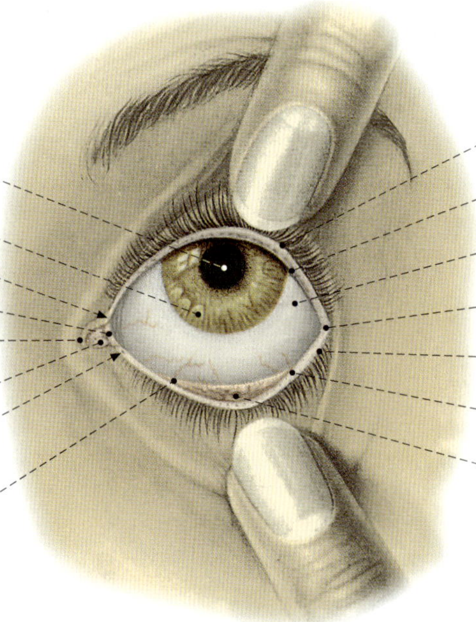

364 Oculus

364 Eye (110%)

Frontal aspect

a Lids of the left eye closed
b Lids of the left eye open
c Upper lid of the left eye drawn upwards and lower lid slightly everted

364 Auge (110%)

Frontalansicht

a Lider des linken Auges geschlossen
b Lider des linken Auges offen
c Oberlid des linken Auges nach oben, Unterlid nach unten gezogen und leicht ektropioniert

a

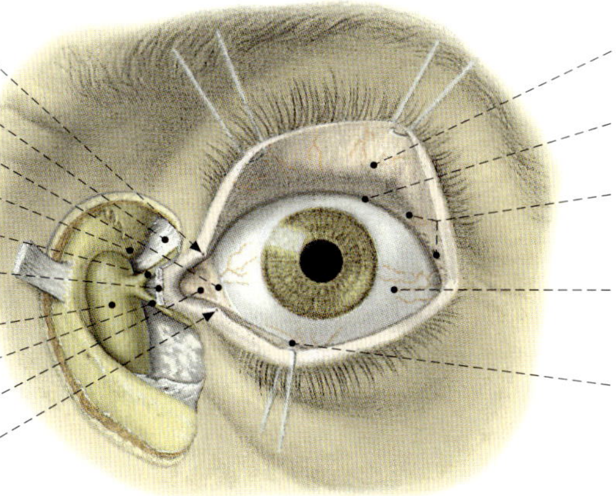

b

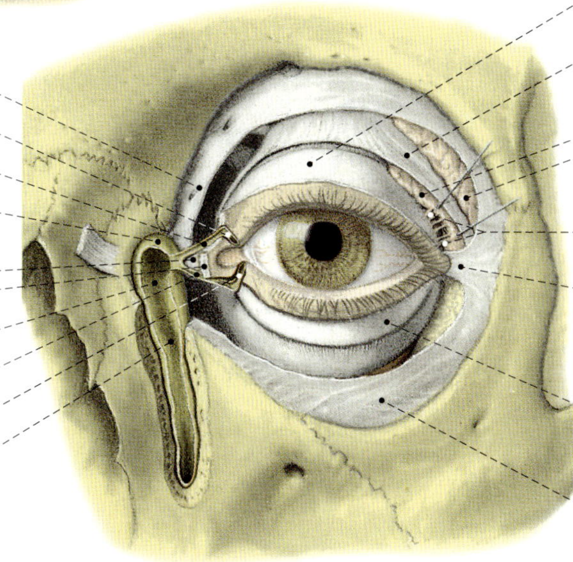

c

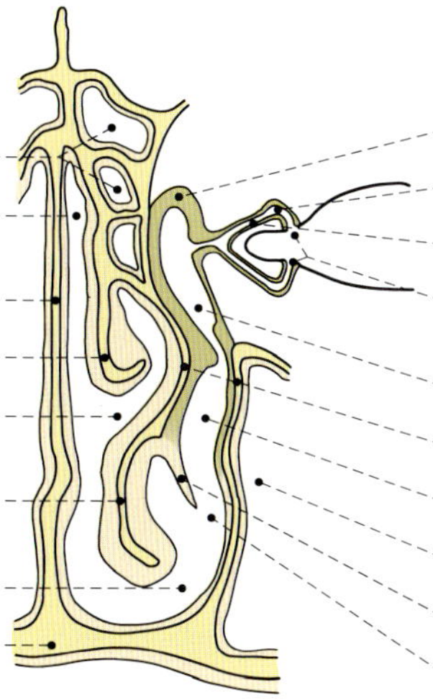

365 Apparatus lacrimalis

365 Lacrimal apparatus

Frontal aspect

a The lacrimal apparatus of the left eye was exposed. The upper lid was pulled upwards by sutures, while the lower lid was drawn slightly downwards (90%).

b Left orbit and its contents after removal of the muscles of facial expression. The tendon of the levator palpebrae superioris muscle was divided by a bow-shaped incision. The lacrimal apparatus was exposed and opened (90%).

c Schematized coronal section through the lacrimal ducts, nasolacrimal duct and nasal cavity of the left side (100%)

365 Tränenapparat

Frontalansicht

a Der Tränenapparat des linken Auges ist freigelegt. Das Oberlid ist durch Fadenschlingen gefasst und nach oben, das Unterlid leicht nach unten gezogen (90%).

b Linke Orbita und Inhaltsgebilde nach Entfernung der mimischen Muskulatur und bogenförmiger Durchtrennung des Musculus levator palpebrae superioris. Der Tränenapparat ist freigelegt (90%).

c Übersicht über die Tränenwege des linken Auges, schematischer Frontalschnitt durch Tränenröhrchen, Tränennasengang und Nasenhöhle (100%)

a

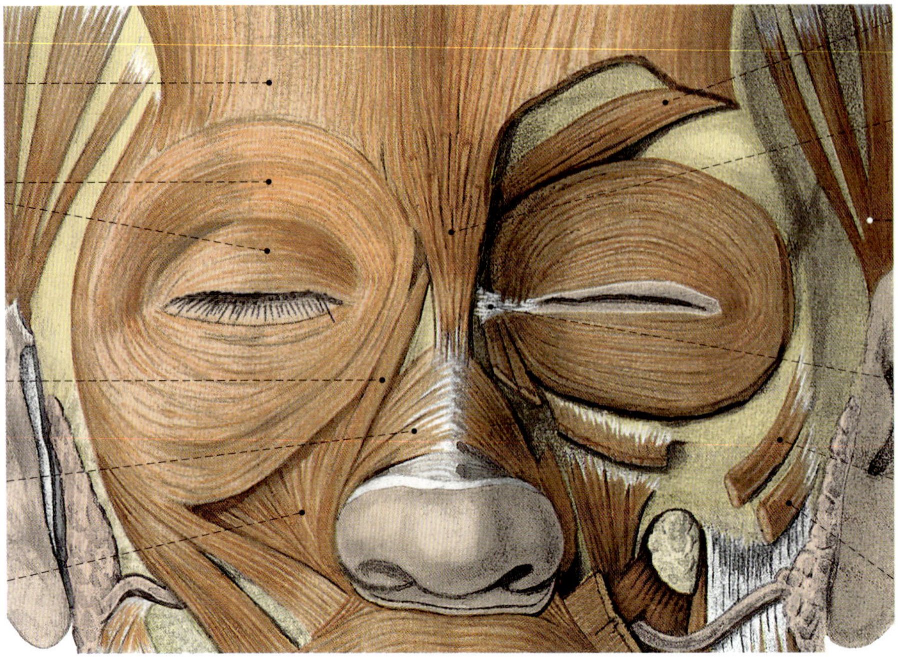

b

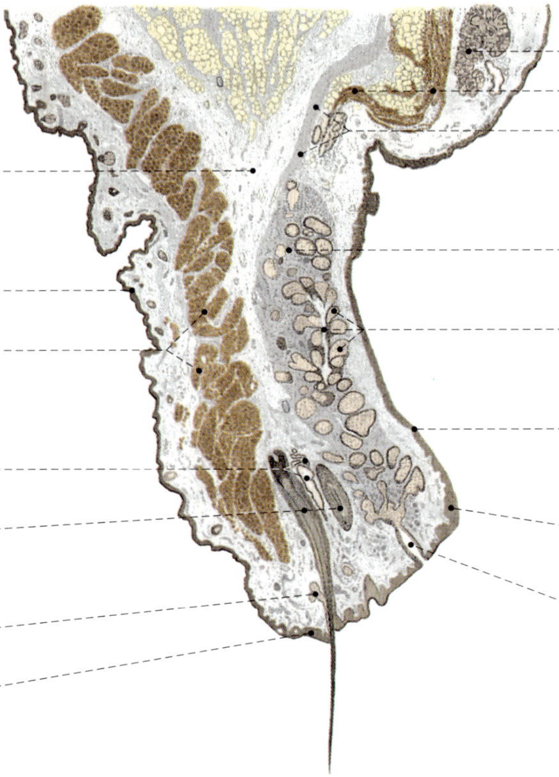

366 Regio orbitalis

366 Orbital region

a Muscles of facial expression around the eyes (75%), frontal aspect
b Sagittal section through the upper eyelid (700%), medial aspect

366 Augenregion

a Mimische Muskeln im Bereich der oberen Gesichtshälfte (75%), Frontalansicht
b Sagittalschnitt durch das Oberlid (700%), Medialansicht

a

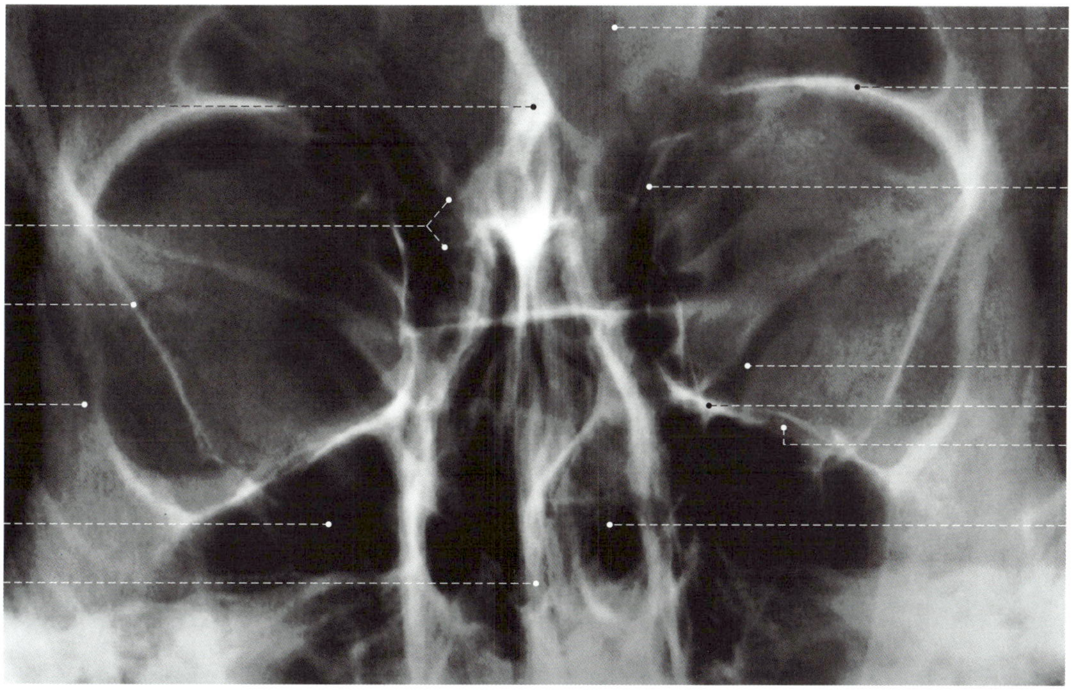

b

367 Orbita

367 Orbital cavity (110%)

a Posteroanterior radiograph of both orbits
b Left osseous orbit, frontal aspect
Individual bones are indicated by different colors.

367 Augenhöhle (110%)

a Posteroanteriore Röntgenaufnahme beider Augenhöhlen
b Linke knöcherne Orbita, Frontalansicht
Die verschiedenen Knochen sind durch unterschiedliche Farben dargestellt.

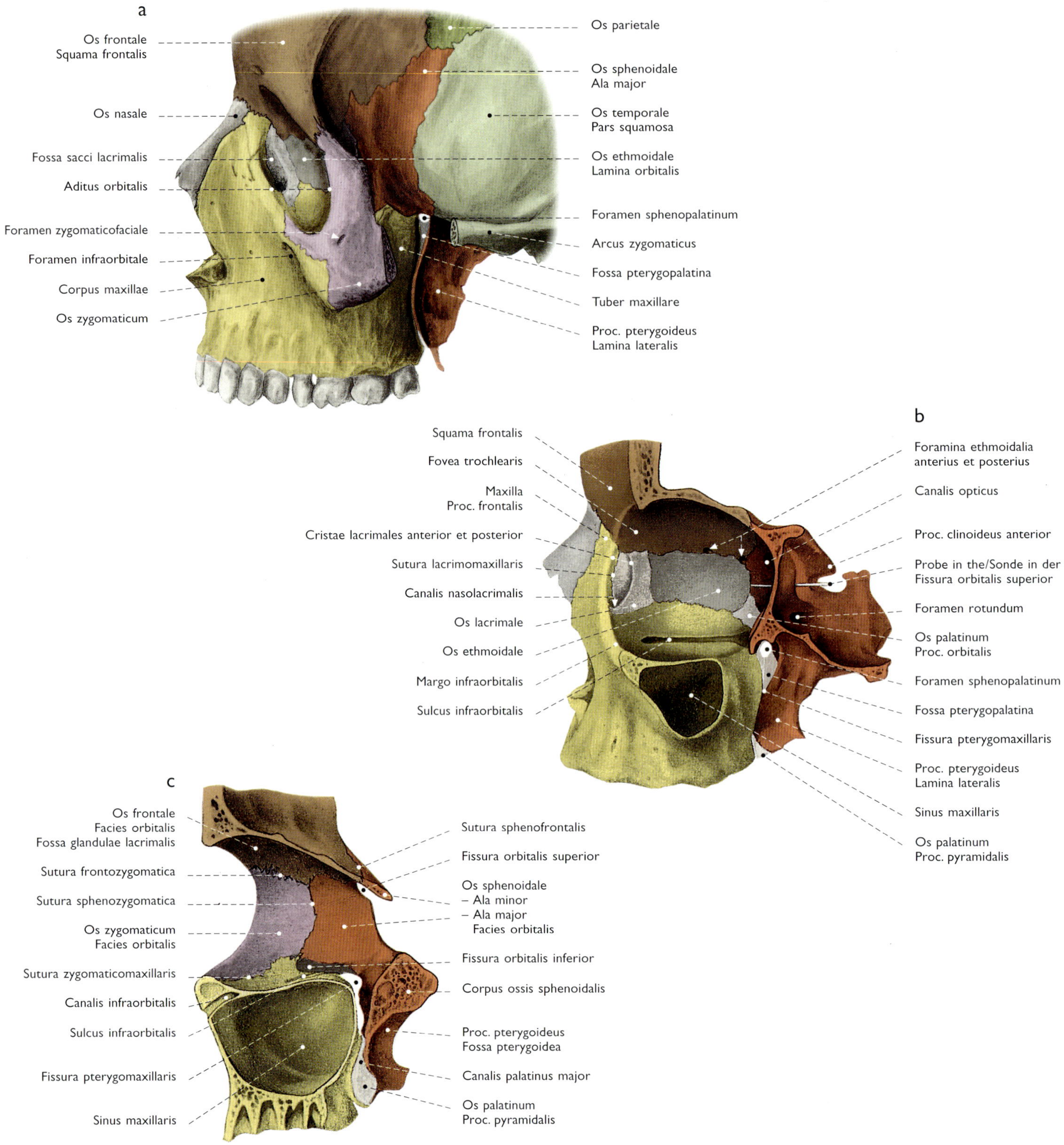

368 Orbita

368 Walls of the orbital cavity (65%)

Individual bones are indicated by different colors.

a Lateral wall of the left orbit and adjacent bones, left lateral aspect

b Medial wall of the left orbit and adjacent bones, left lateral aspect

c Lateral wall of the right orbit and adjacent bones, medial aspect

368 Wände der Augenhöhle (65%)

Die verschiedenen Knochen sind in unterschiedlichen Farben dargestellt.

a Seitenwand (Paries lateralis) der linken Augenhöhle und angrenzende Knochen, linke Seitenansicht

b Mediale Wand (Paries medialis) der linken Augenhöhle und angrenzende Knochen, linke Seitenansicht

c Seitenwand (Paries lateralis) der rechten Augenhöhle und angrenzende Knochen, Medialansicht

a

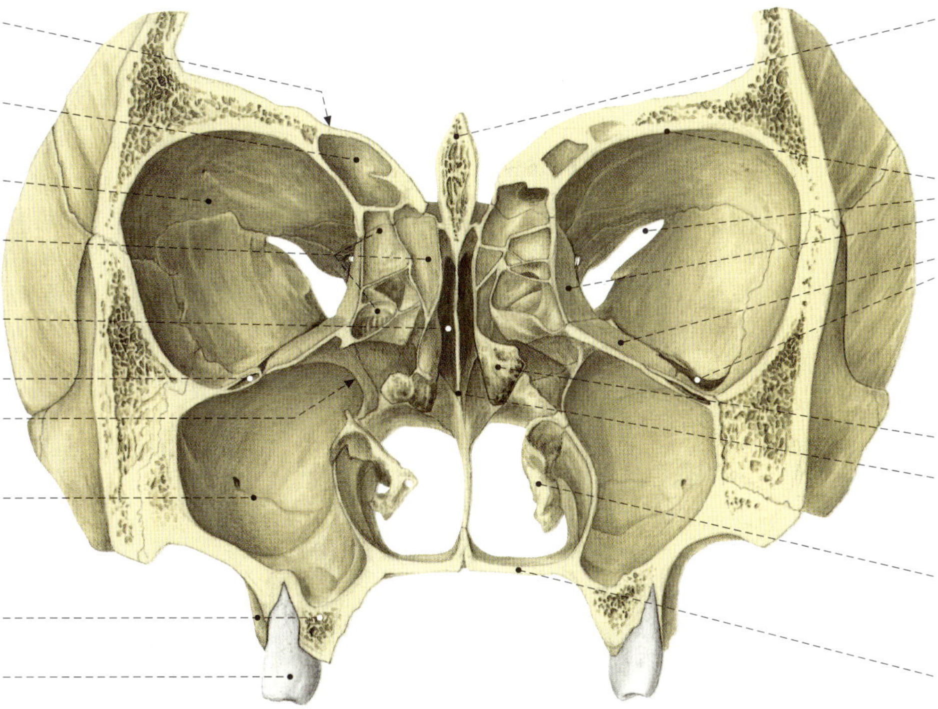

b

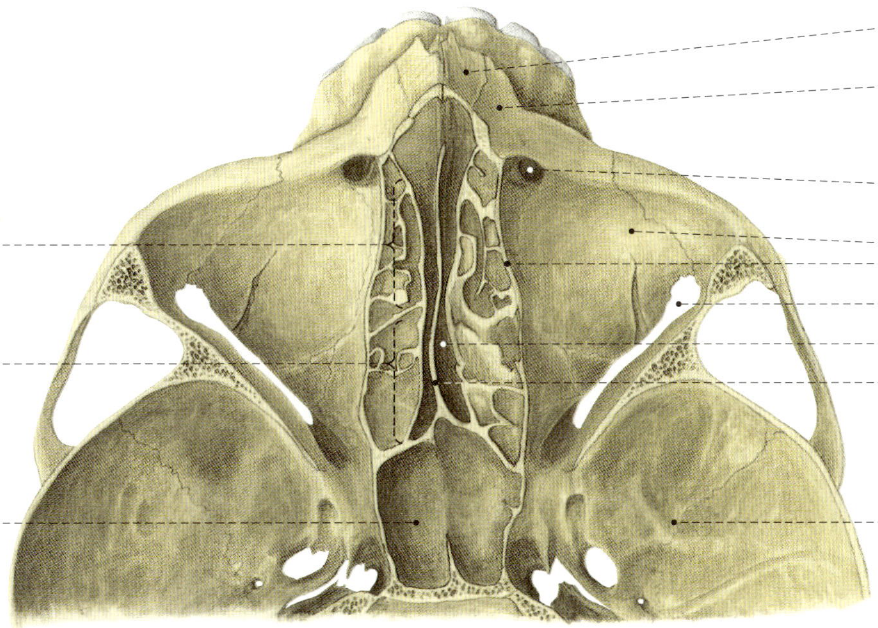

369 Orbita

369 Orbital cavity (85%)

Sections through the osseous walls of the orbits, the nasal cavity and paranasal sinuses

a Coronal section through the viscerocranium, frontal aspect

b Transverse section through the middle parts of the orbits, the nasal cavity, the ethmoidal and sphenoidal sinuses, superior aspect

369 Augenhöhle (85%)

Schnitte durch die knöchernen Wände der Augenhöhlen, der Nasenhaupt- und Nasennebenhöhlen

a Frontalschnitt durch den Gesichtsschädel, Frontalansicht

b Horizontalschnitt durch die mittleren Abschnitte der Augenhöhlen, die Nasenhöhle sowie die Sieb- bein- und Keilbeinhöhlen, Parietalansicht

a

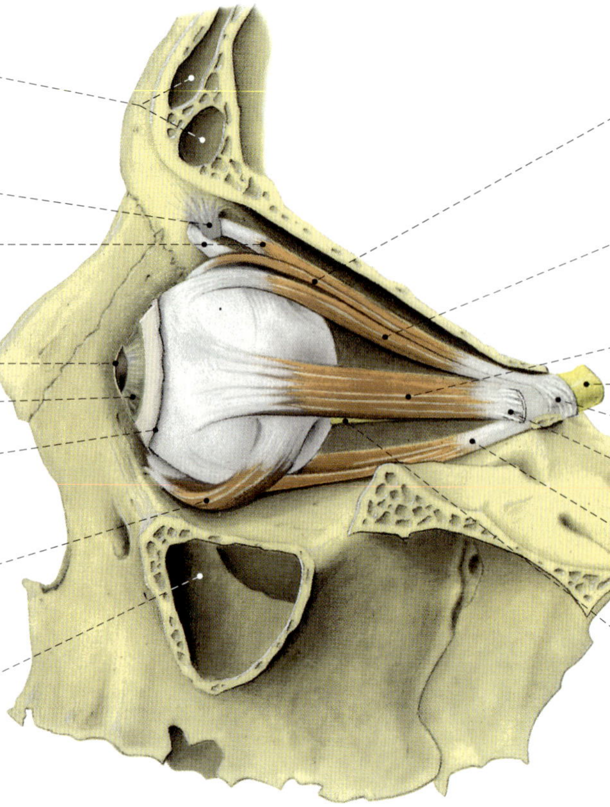

b

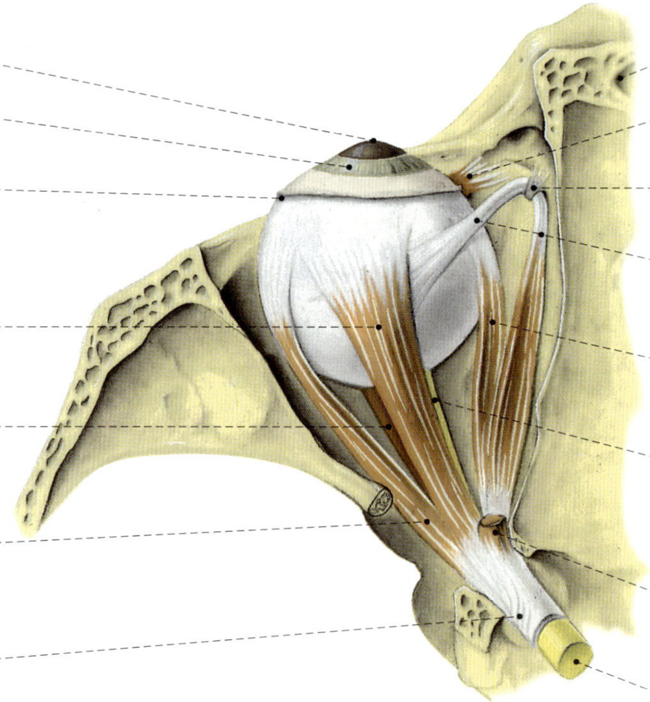

370 Musculi bulbi oculi

370 Extrinsic muscles of the left eyeball (100%)

- a The lateral wall of the left orbit was removed. Left lateral aspect of the left eyeball
- b The orbital roof was removed. Superior aspect of the left eyeball

370 Äussere Muskeln des linken Augapfels (100%)

- a Die laterale Wand der linken Augenhöhle ist abgetragen. Linke Seitenansicht des linken Augapfels
- b Das Dach der Augenhöhle ist abgetragen. Parietalansicht des linken Augapfels

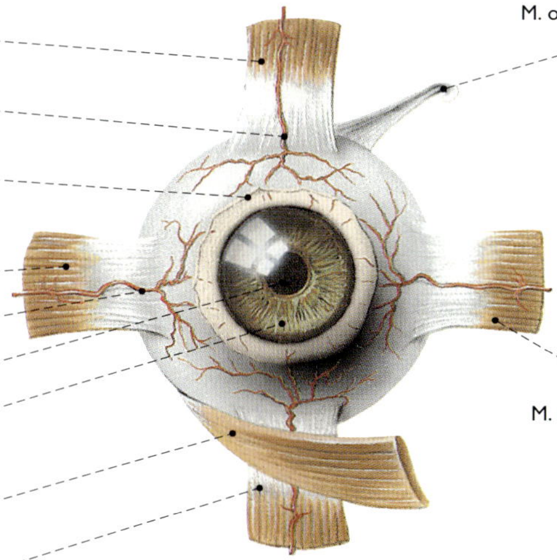

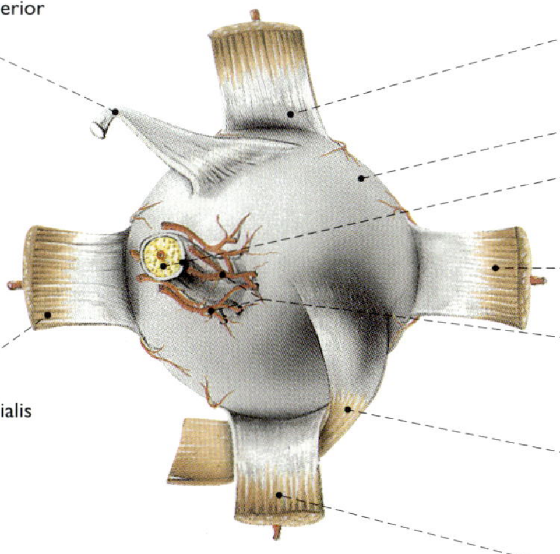

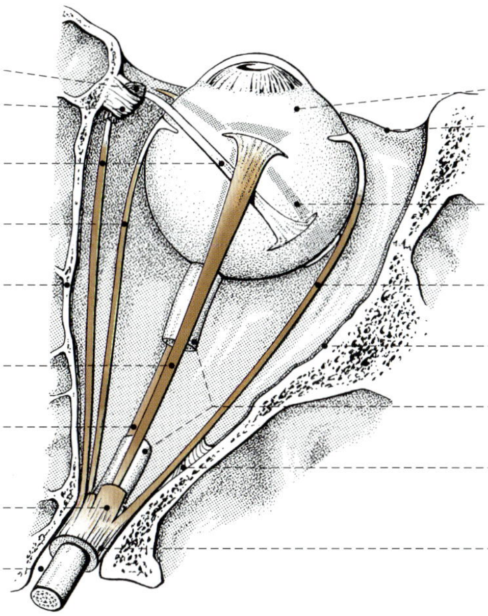

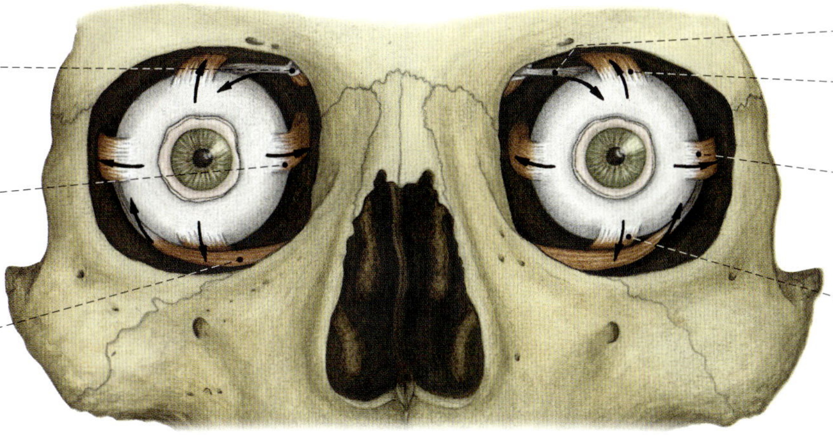

371 Musculi bulbi oculi

371 Extrinsic muscles of the eyeball

a Frontal aspect of the right eyeball (125%)
b Dorsal aspect of the right eyeball (125%)
c Superior aspect of the right orbit (100%), schematic representation
d Frontal aspect of the osseous orbits and both eyeballs with muscle insertions (80%). The arrows indicate the ocular movements after muscle contraction.

371 Äussere Muskeln des Augapfels

a Frontalansicht des rechten Augapfels (125%)
b Dorsalansicht des rechten Augapfels (125%)
c Parietalansicht der rechten Augenhöhle (100%), schematische Darstellung
d Frontalansicht der knöchernen Augenhöhlen und der beiden Augäpfel mit Muskelansätzen (80%). Die Pfeile geben die Augenbewegungen nach Muskelkontraktion an.

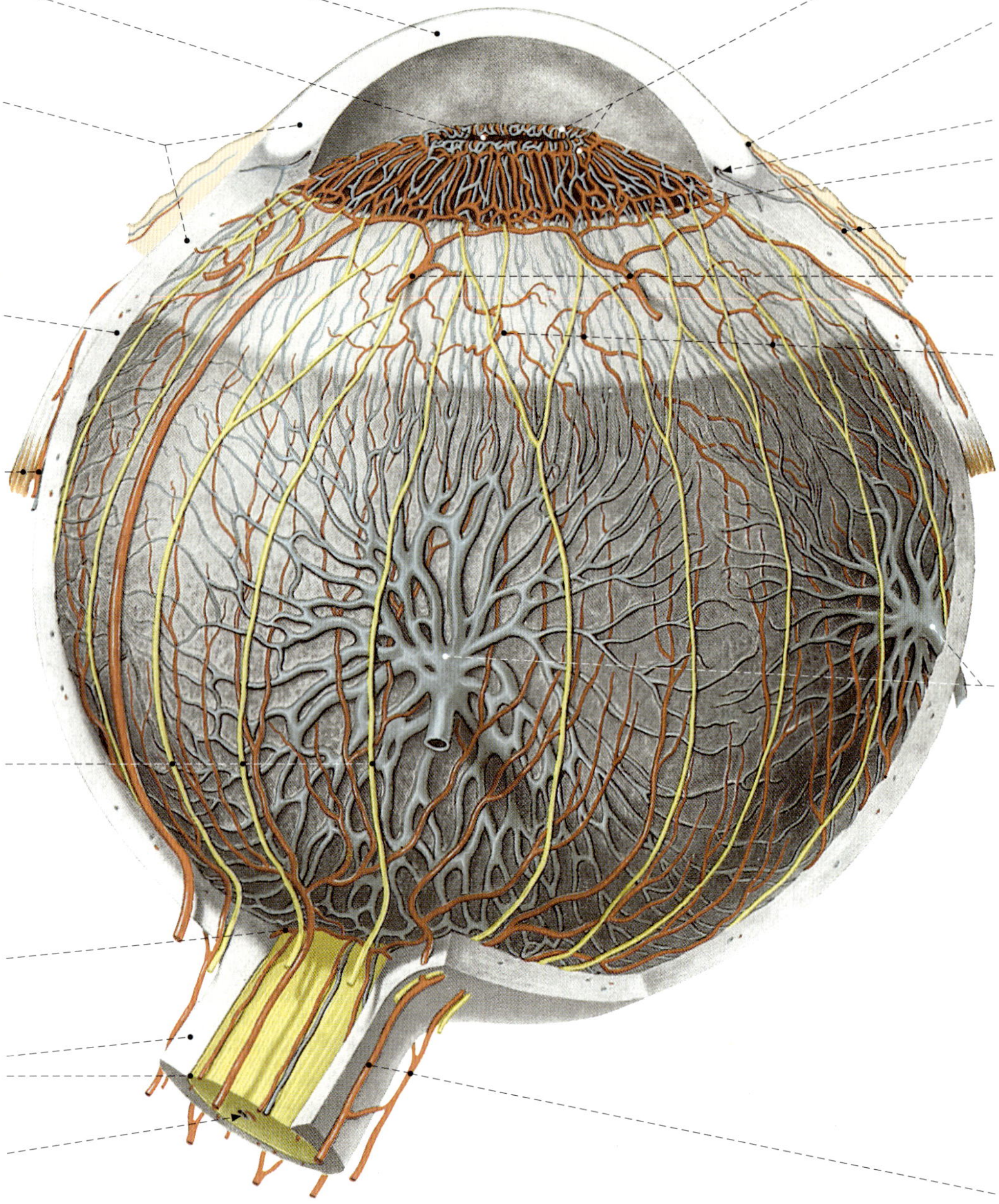

372 ⟨Arteriae et Venae bulbi oculi⟩

372 Blood vessels of the eyeball (500%)
The upper part of the fibrous tunic of the right eyeball was removed nearly completely.
Superior aspect

372 Blutgefässe des Augapfels (500%)
Die obere Hälfte der Tunica fibrosa des rechten Augapfels ist fast vollständig entfernt.
Parietalansicht

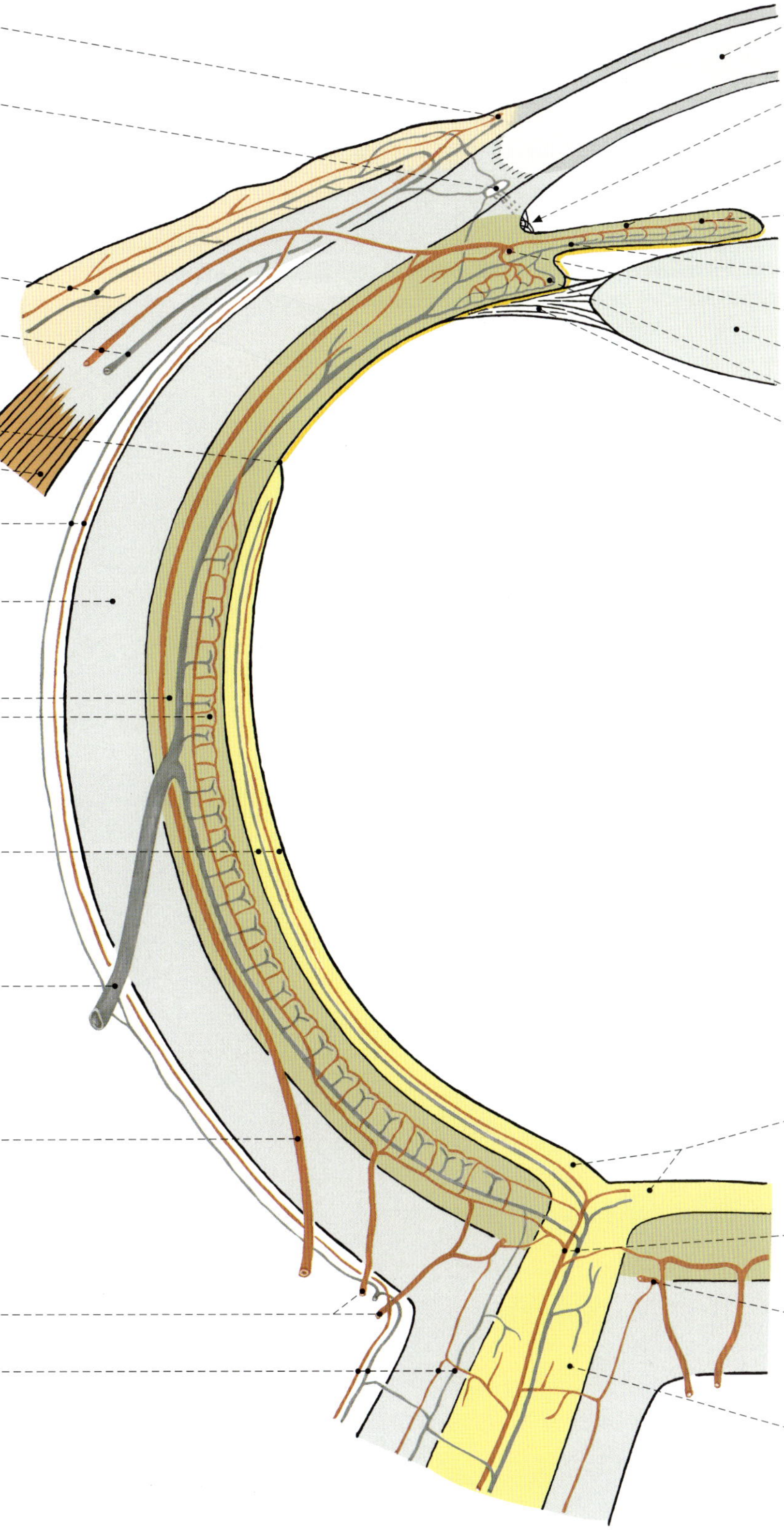

373 ⟨Arteriae et Venae bulbi oculi⟩

373 Blood vessels of the eyeball (700%)

Transverse section through the right eyeball incised along the equatorial plane and schematic representation of the blood vessels in the nasal half of the eyeball, superior aspect

373 Blutgefässe des Augapfels (700%)

Horizontalschnitt durch den rechten Augapfel in der Äquatorialebene und schematische Darstellung der Blutgefässe in der nasalen Hälfte des Augapfels, Parietalansicht

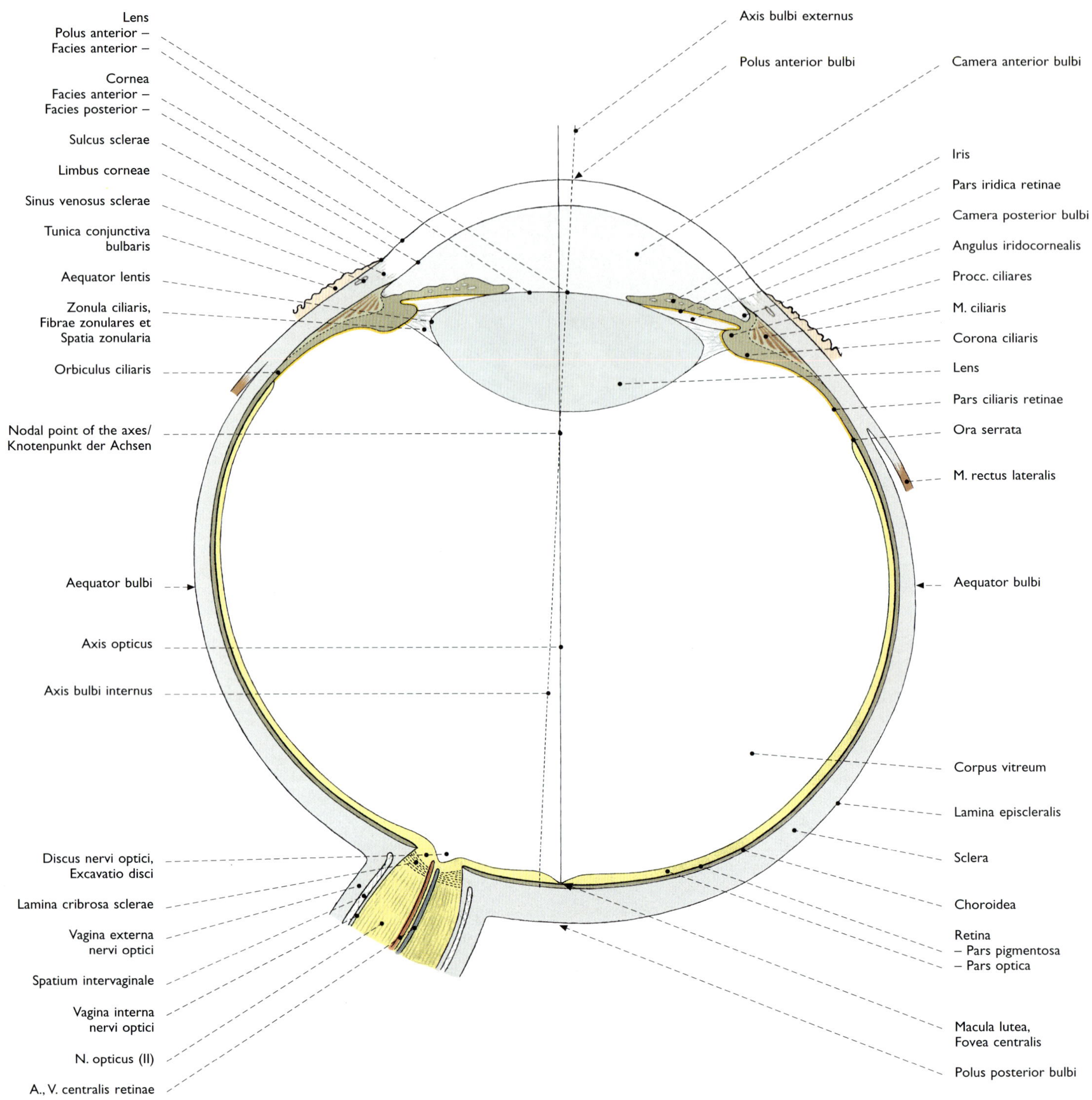

374 Bulbus oculi

374 Eyeball (500%)

Schematized transverse section through the right eyeball of an adult with normal vision, parietal aspect

374 Augapfel (500%)

Schematisierter Horizontalschnitt durch den rechten Augapfel eines normalsichtigen Erwachsenen, Parietalansicht

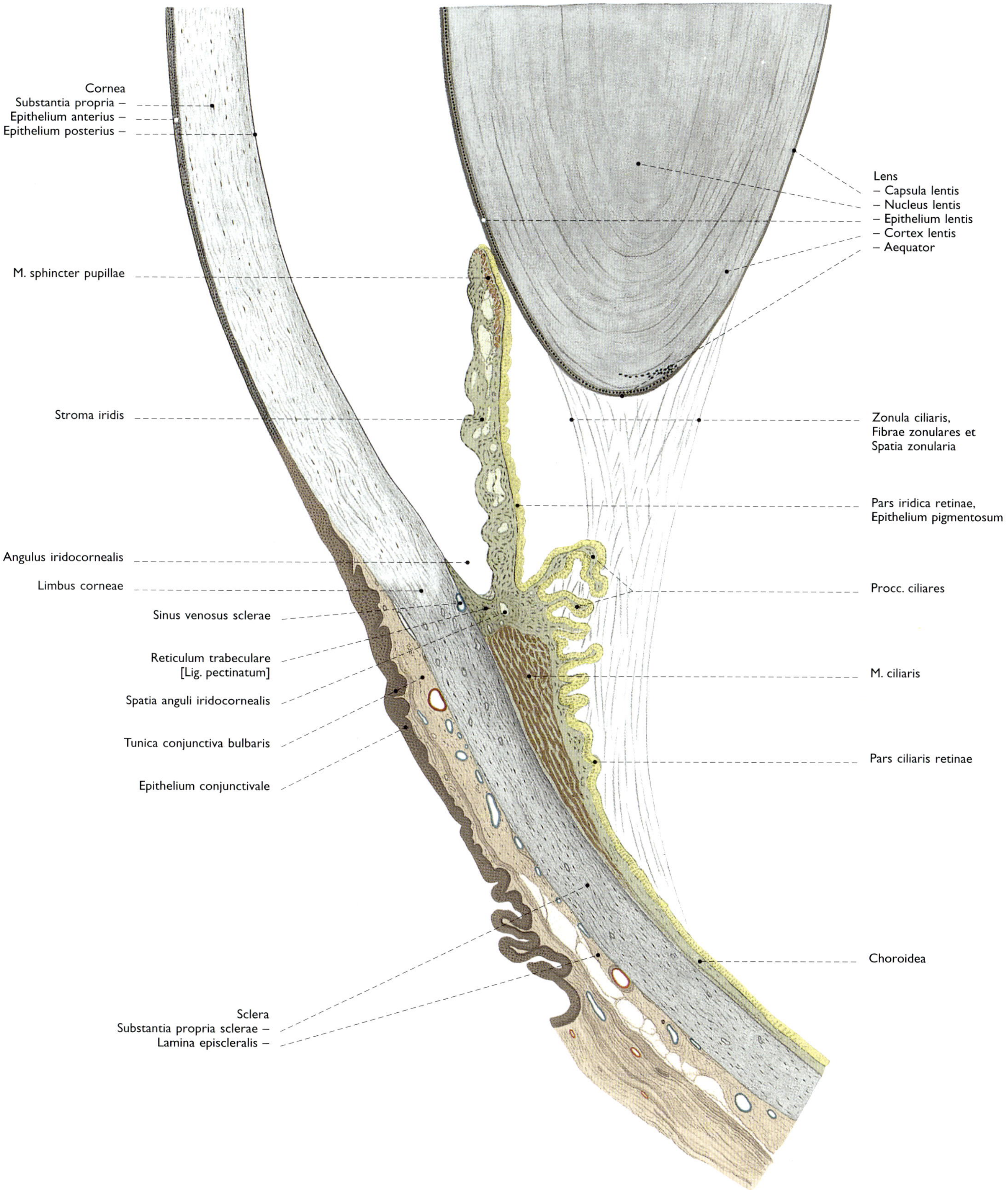

375 Bulbus oculi

375 Eyeball

Meridional section through the anterior part of the eyeball (2000%)

375 Augapfel

Meridionalschnitt durch den vorderen Teil des Augapfels (2000%)

a

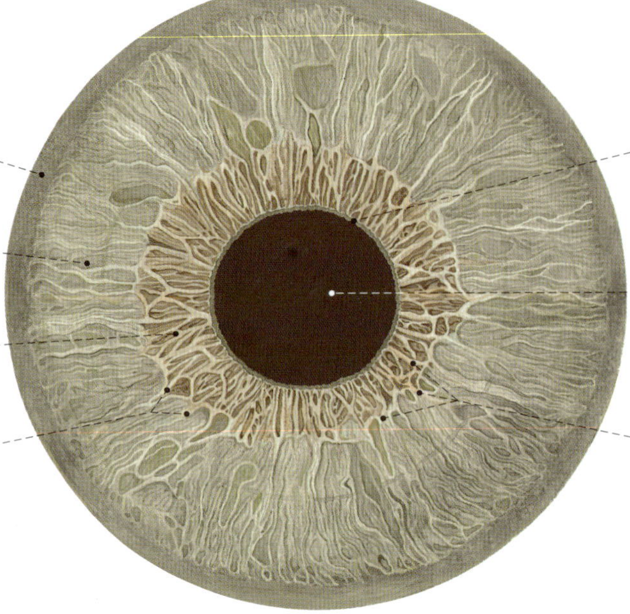

b

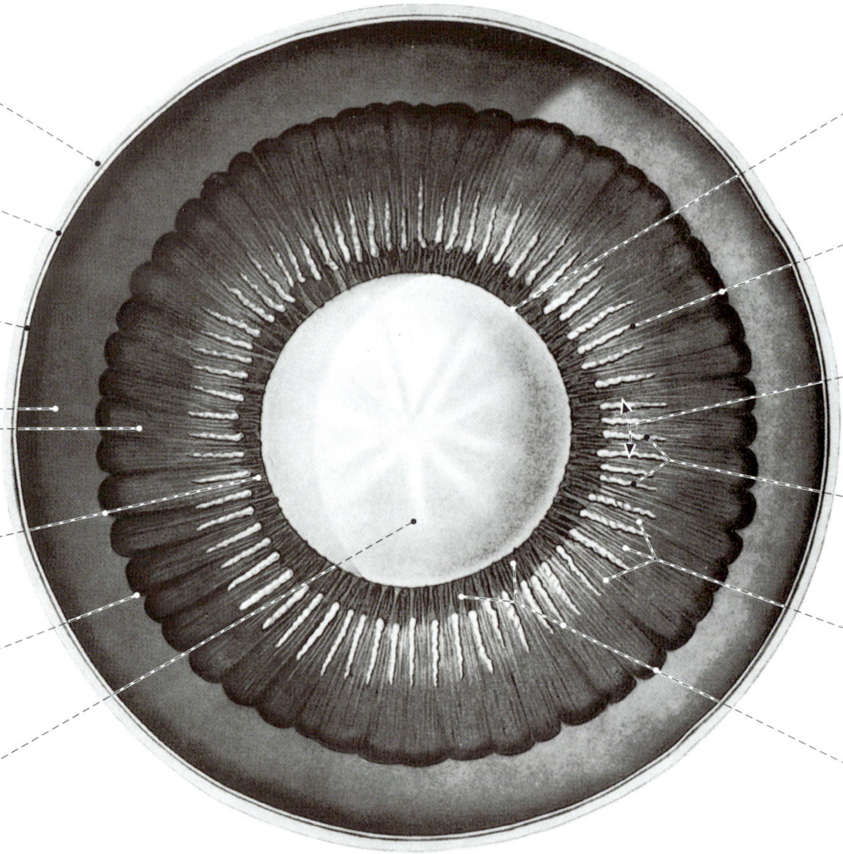

376 Iris et Corpus ciliare

376 Iris and ciliary body

a Ventral aspect after removal of the cornea (700%)

b Dorsal aspect after removal of the vitreous body (400%)

376 Regenbogenhaut und Ziliarkörper

a Ventralansicht nach Ablösung der Hornhaut (700%)

b Dorsalansicht nach Entfernung des Glaskörpers (400%)

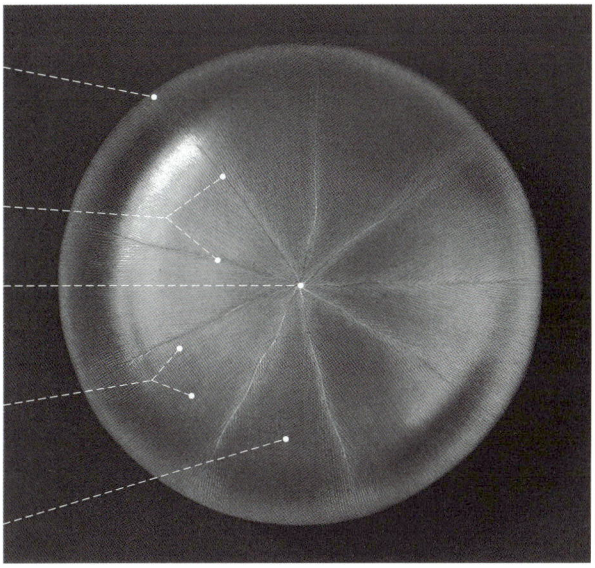

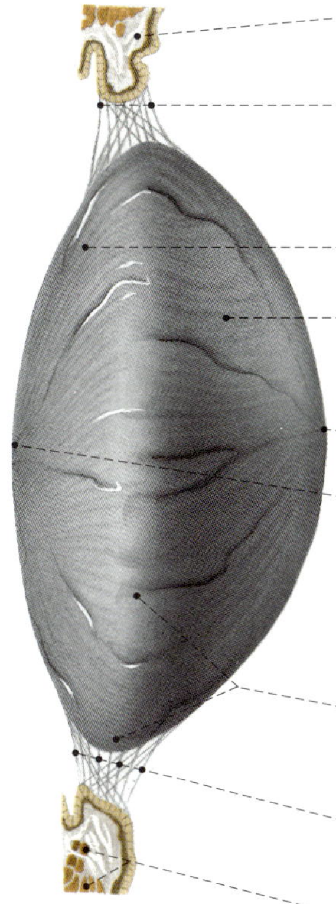

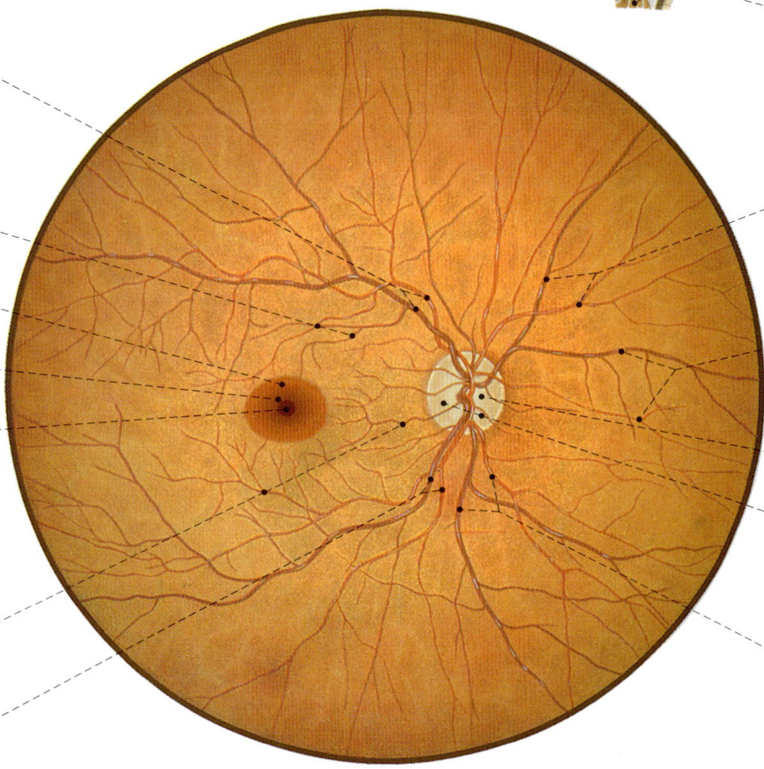

377 Lens et Retina

377 Crystalline lens and retina (600%)

a Crystalline lens, the lens star showing several rays, frontal aspect
b Crystalline lens, lateral aspect
c Retina of the right eye, ophthalmoscopic view

377 Linse und Netzhaut (600%)

a Linse mit mehrstrahligem Linsenstern, Frontalansicht
b Linse, Seitenansicht
c Netzhaut des rechten Auges, Augenhintergrund im Ophthalmoskop

a

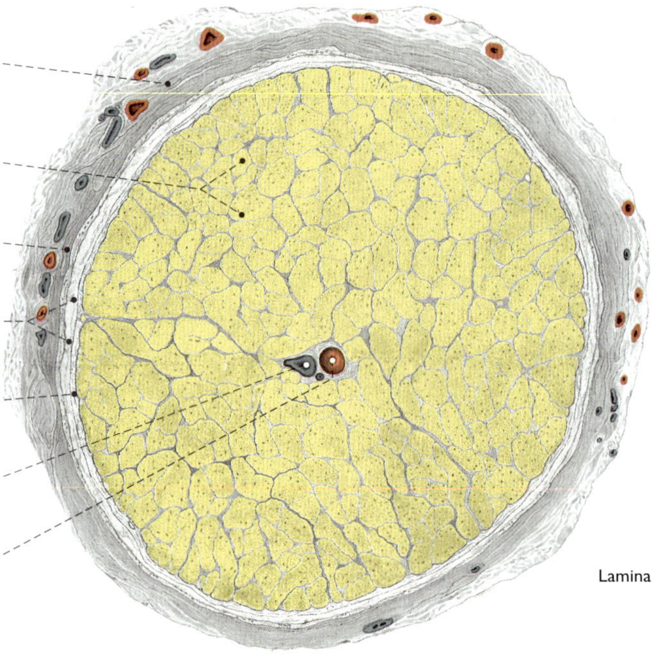

b

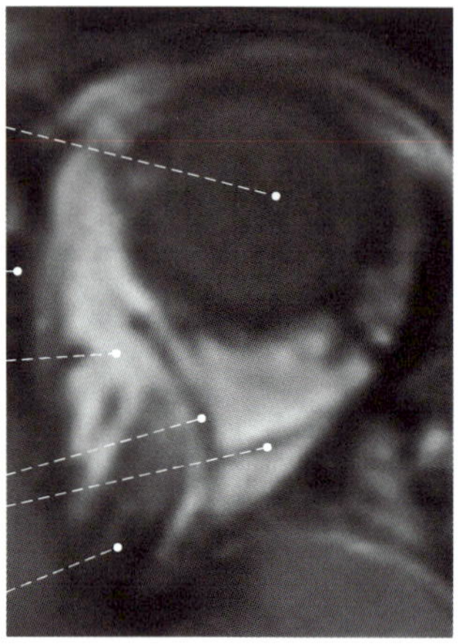

c

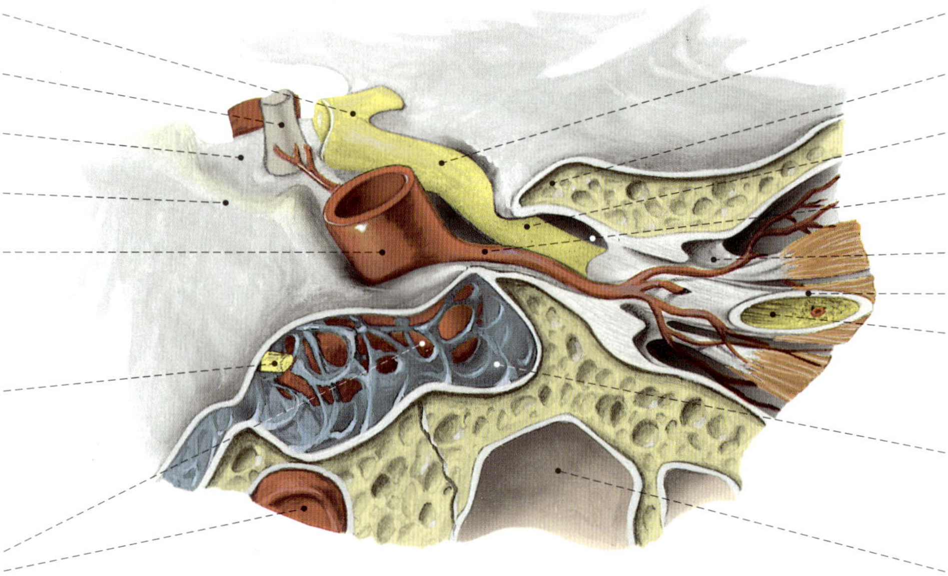

378 Nervus opticus et Arteria ophthalmica

378 Optic nerve and ophthalmic artery

a Optic nerve of the right eye, cross section close to the eyeball (1 600%), frontal aspect
b Optic nerve, eyeball and ophthalmic artery, transverse magnetic resonance image (MRI, T_1-weighted) (100%)
c Oblique section through the optic canal of the right orbit (250%), right lateral aspect

378 Sehnerv und Arteria ophthalmica

a Sehnerv des rechten Auges, Querschnitt in Bulbusnähe (1 600%), Frontalansicht
b Sehnerv, Augapfel und Arteria ophthalmica, horizontales Kernspintomogramm (MRI, T_1-gewichtet) (100%)
c Sehnerv und Arteria ophthalmica, schräger Schnitt durch den Canalis opticus der rechten Augenhöhle (250%), rechte Seitenansicht

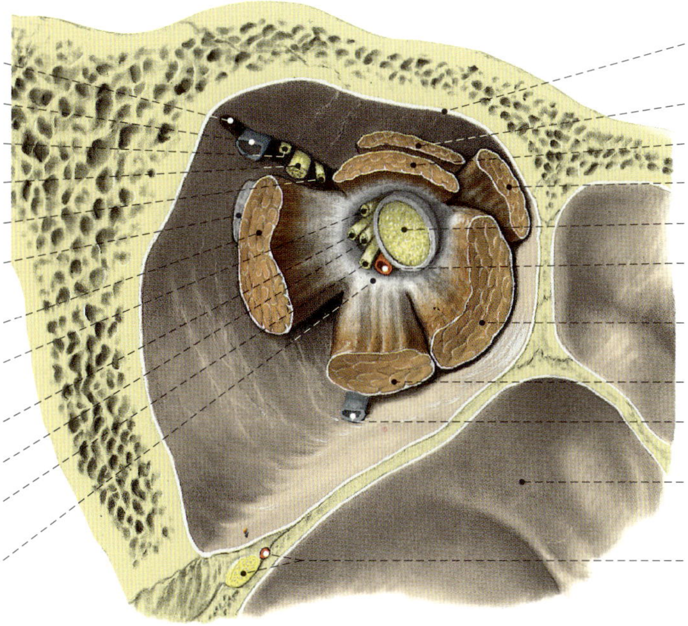

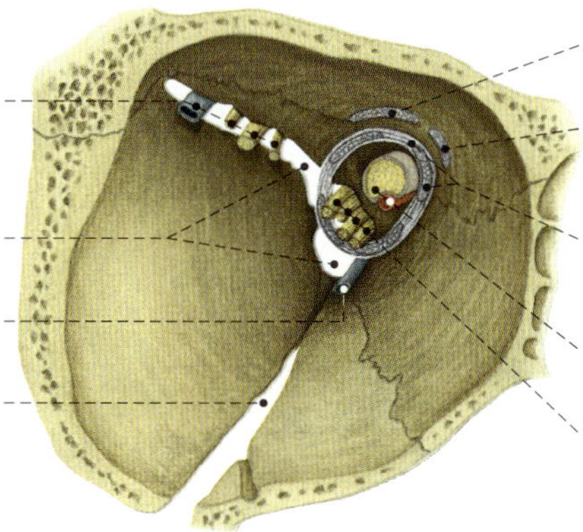

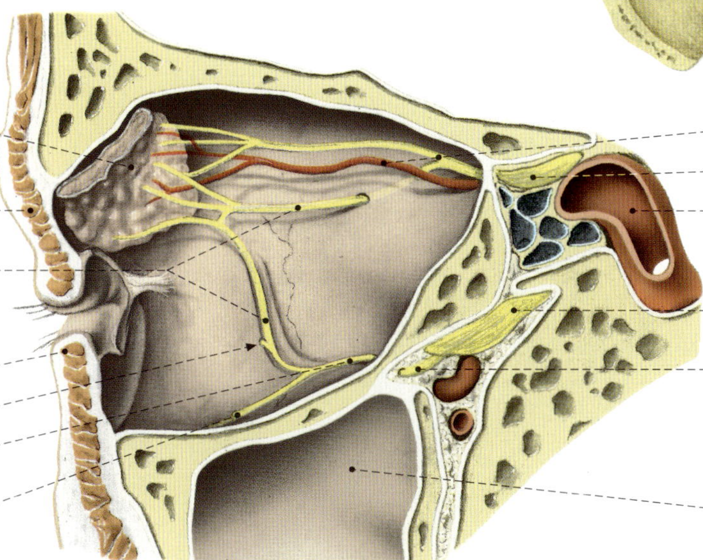

379 Orbita

379 Right orbital cavity

a Coronal section. The common tendinous ring and the origins of the extrinsic muscles of the eyeball are shown (150%). Frontal aspect

b Coronal section. The extrinsic muscles of the eyeball were transected close to their origins, the nerves and blood vessels close to their points of entry into or exit from the orbit (150%). Frontal aspect

c Sagittal section. The lateral wall of the right orbit is shown after removal of the eyeball (100%). Medial aspect

379 Rechte Augenhöhle

a Frontalschnitt. Darstellung des gemeinsamen Sehnenrings und des Ursprung der äusseren Augenmuskeln (150%). Frontalansicht

b Frontalschnitt. Die äusseren Augenmuskeln sind knapp nach ihrem Ursprung, die Gefässe und Nerven unmittelbar nach ihrem Eintritt in bzw. vor ihrem Austritt aus der Augenhöhle durchtrennt (150%). Frontalansicht

c Sagittalschnitt. Die seitliche Wand der rechten Orbita ist nach Entfernung des Augapfels dargestellt (100%). Medialansicht

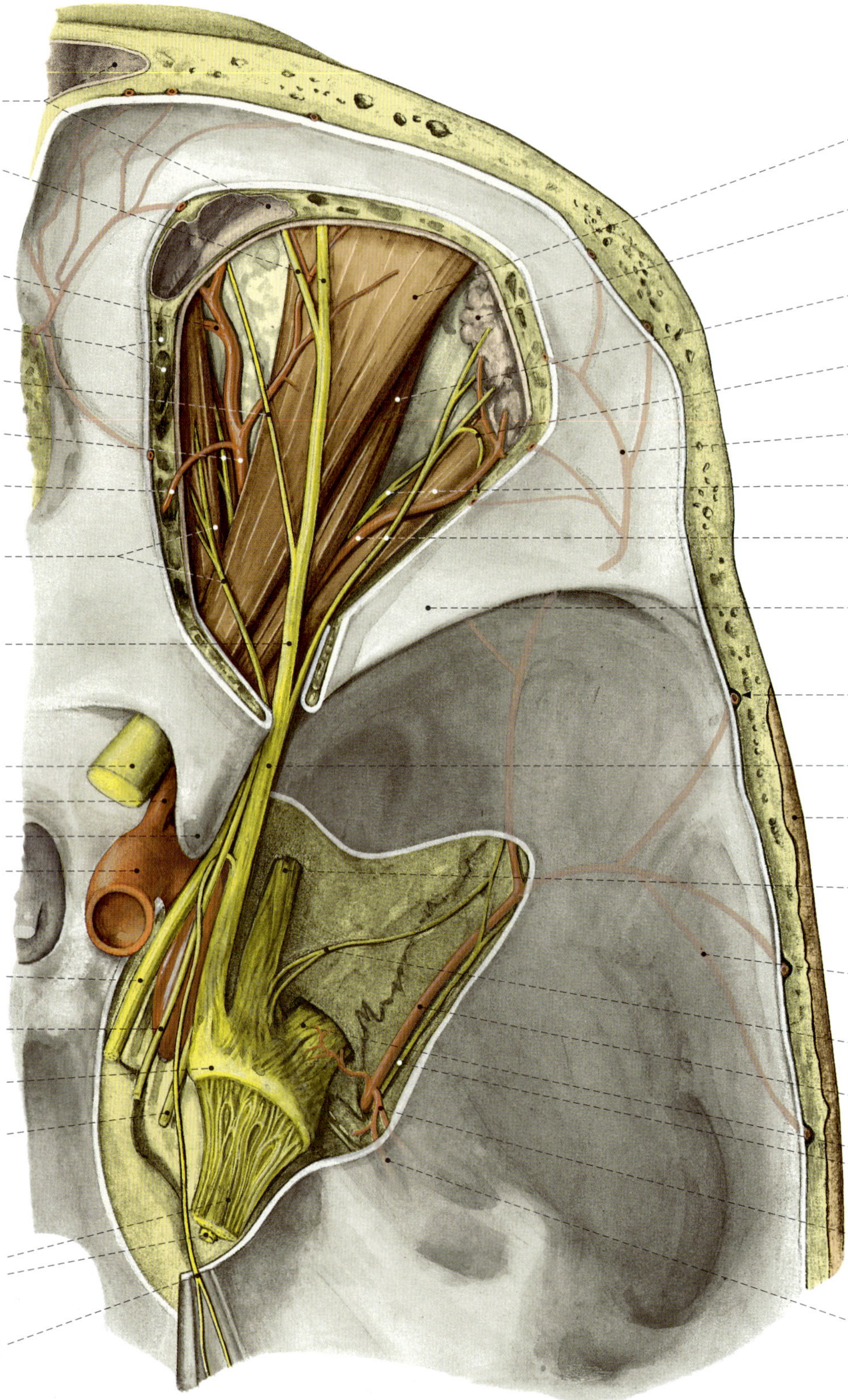

380 Orbita et Fossa cranii media

380 Right orbital cavity and middle cranial fossa (200%)

The orbital roof and the dura mater beyond the right orbital cavity were removed, the right cavernous sinus was opened. Superior aspect

380 Rechte Augenhöhle und mittlere Schädelgrube (200%)

Das Orbitadach und die harte Hirnhaut über der rechten Augenhöhle sind entfernt, der rechte Sinus cavernosus ist eröffnet. Parietalansicht

a

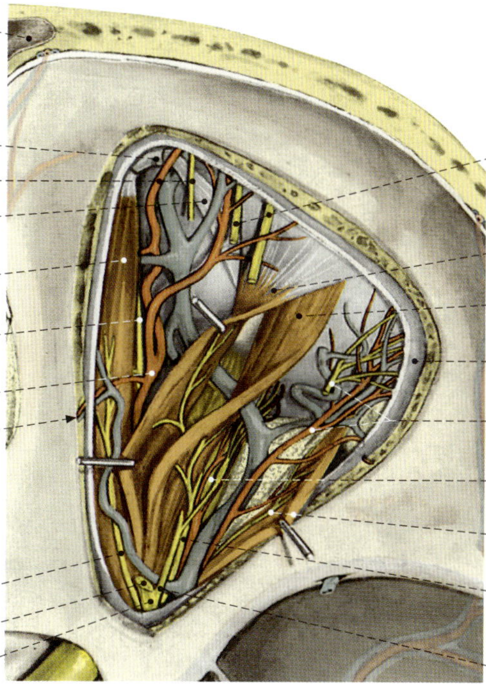

b

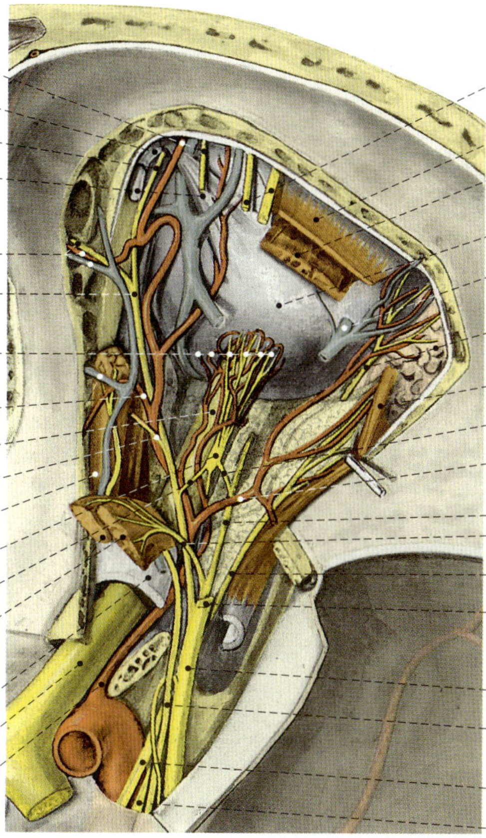

381 Orbita

381 Right orbital cavity (140%)

The orbital roof was removed. Superior aspect
a Superficial dissection
b Deeper dissection just above the optic nerve. The upper extrinsic muscles of the eyeball were transected and removed.

381 Rechte Augenhöhle (140%)

Das Orbitadach ist abgetragen. Parietalansicht
a Oberflächliche Schicht
b Tiefere Schicht etwas oberhalb des Nervus opticus. Die oberen äusseren Augenmuskeln sind durchtrennt und abgetragen.

a

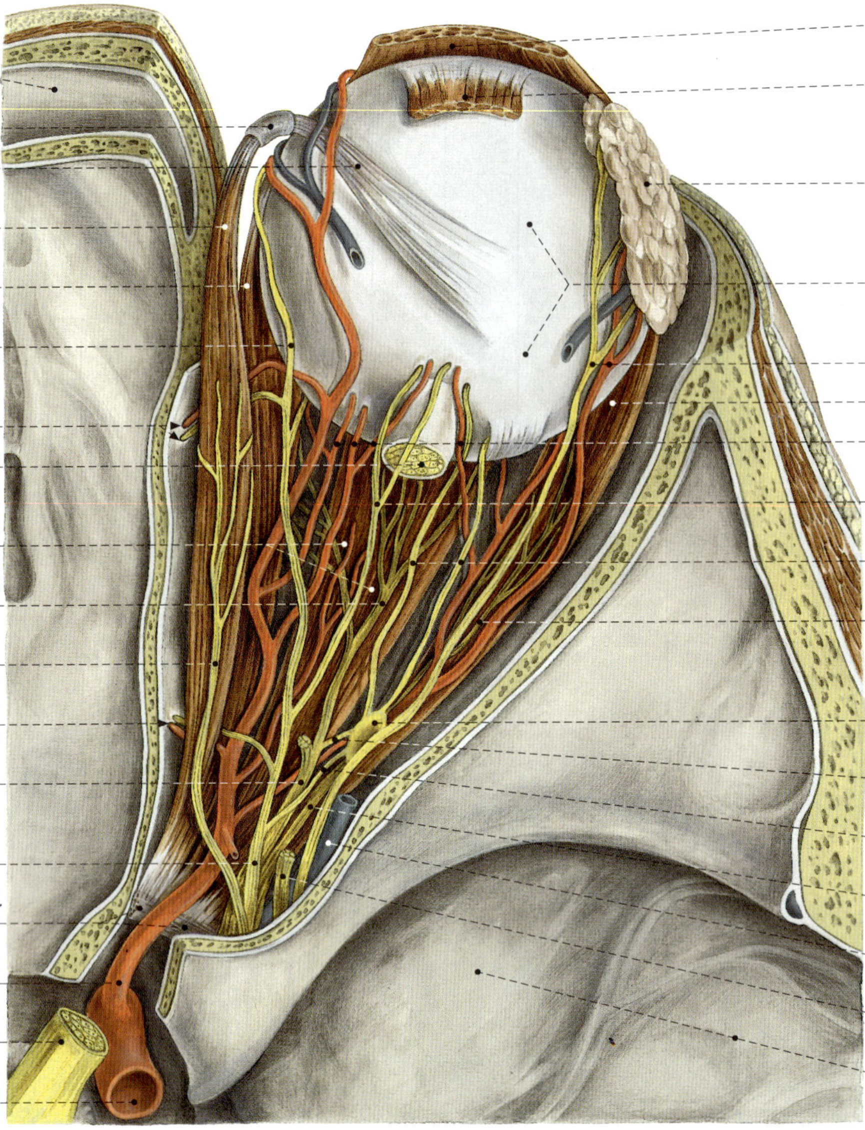

b

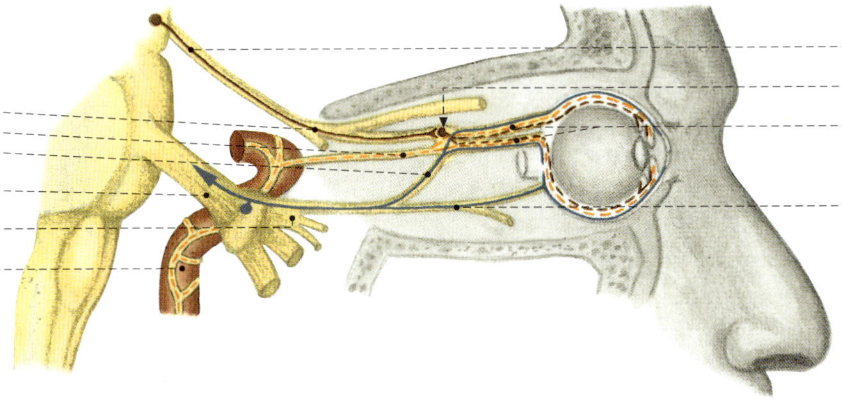

382 Orbita

382 Right orbital cavity

a Deep dissection. The orbital roof was completely removed, the upper extrinsic muscles of the eyeball and the optic nerve were transected and removed (150%). Superior aspect

b Ciliar ganglion and autonomic innervation of the eyeball (80%), schematic representation, right lateral aspect

382 Orbita

a Tiefe Schicht. Das Orbitadach ist vollständig abgetragen, die oberen äusseren Augenmuskeln und der Sehnerv sind durchtrennt und entfernt (150%). Parietalansicht

b Ganglion ciliare und vegetative Innervation des Augapfels (80%), schematische Darstellung, rechte Seitenansicht

a

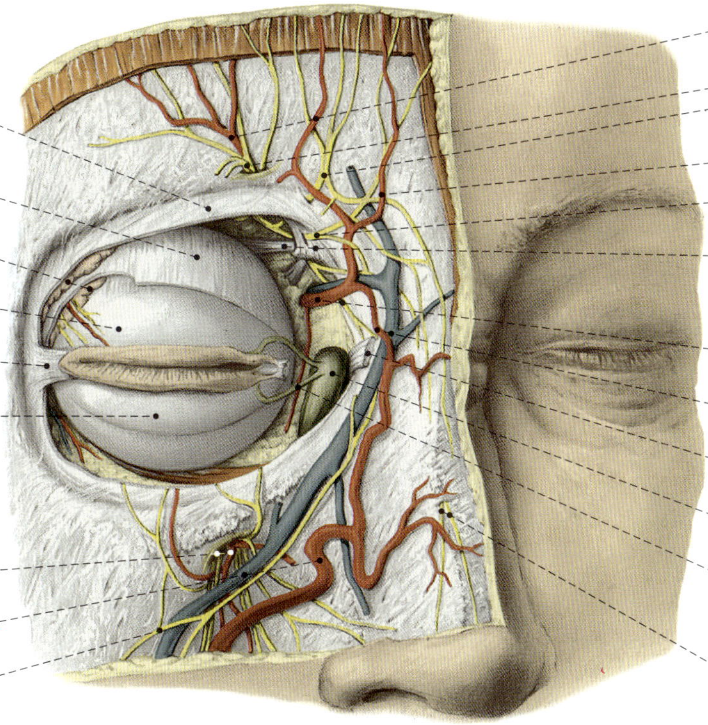

b

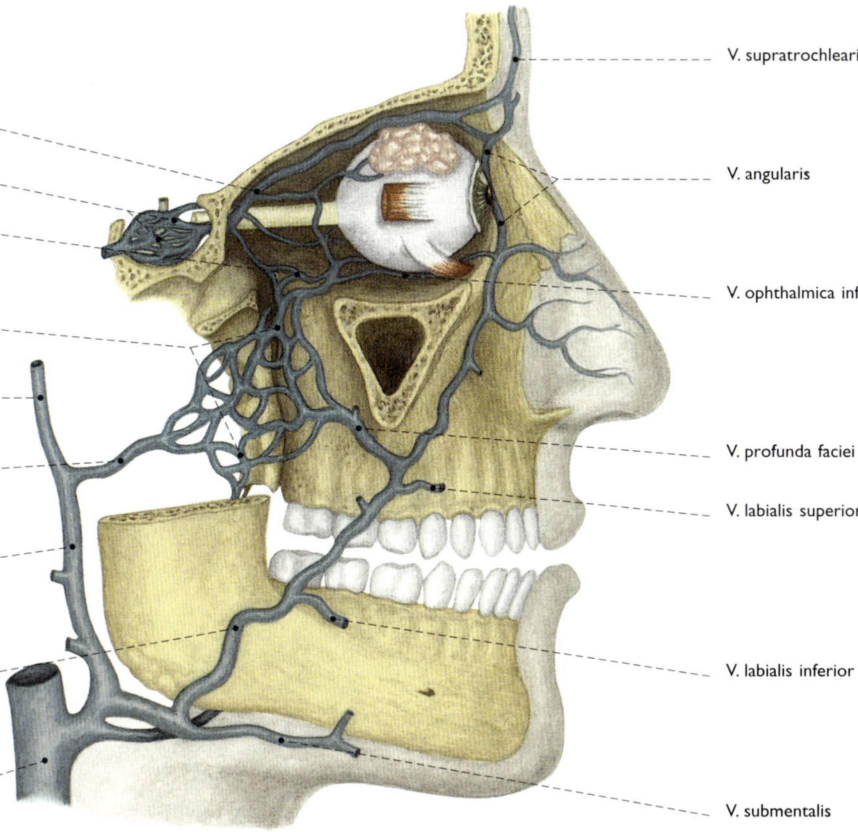

383 Regio orbitalis

383 Right orbital region

a Gross anatomy of the right orbital region (90%)
b Drainage of the orbital veins. The lateral wall of the right orbit and parts of the mandibular ramus were removed, the right maxillary sinus is opened (80%). Right lateral aspect

383 Rechte Augenregion

a Topographie der rechten Augenregion (90%)
b Abfluss der Orbitavenen. Die Seitenwand der rechten Augenhöhle und Teile des Unterkieferastes sind entfernt, die rechte Kieferhöhle ist eröffnet (80%). Rechte Seitenansicht

a

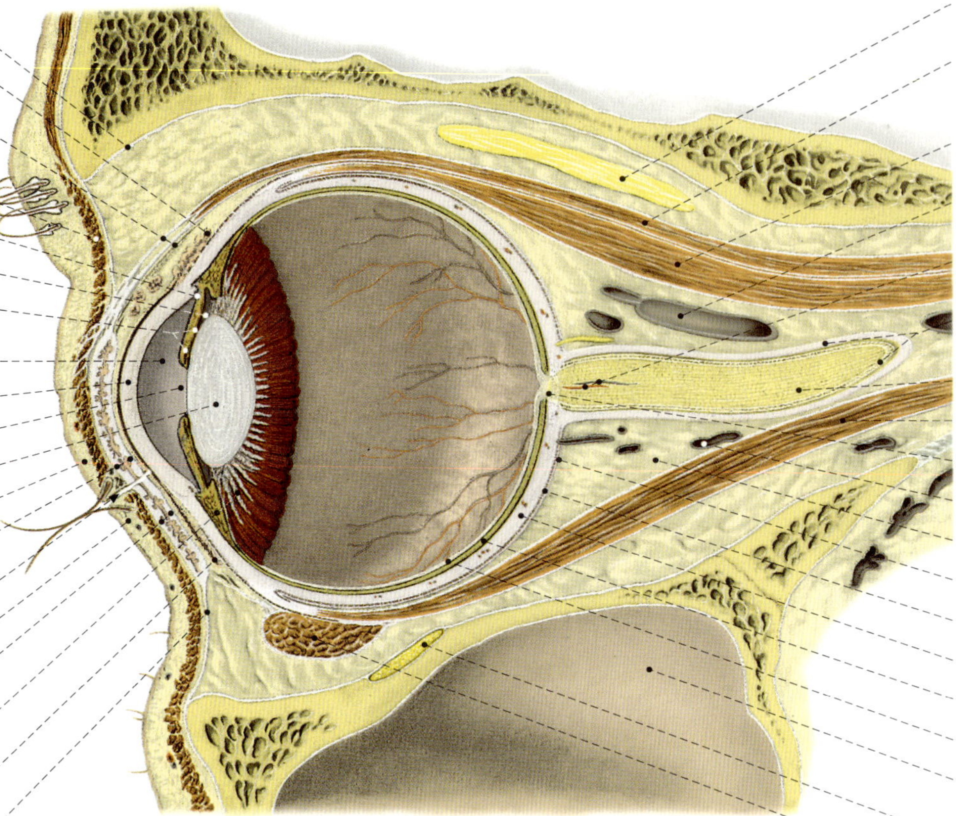

b

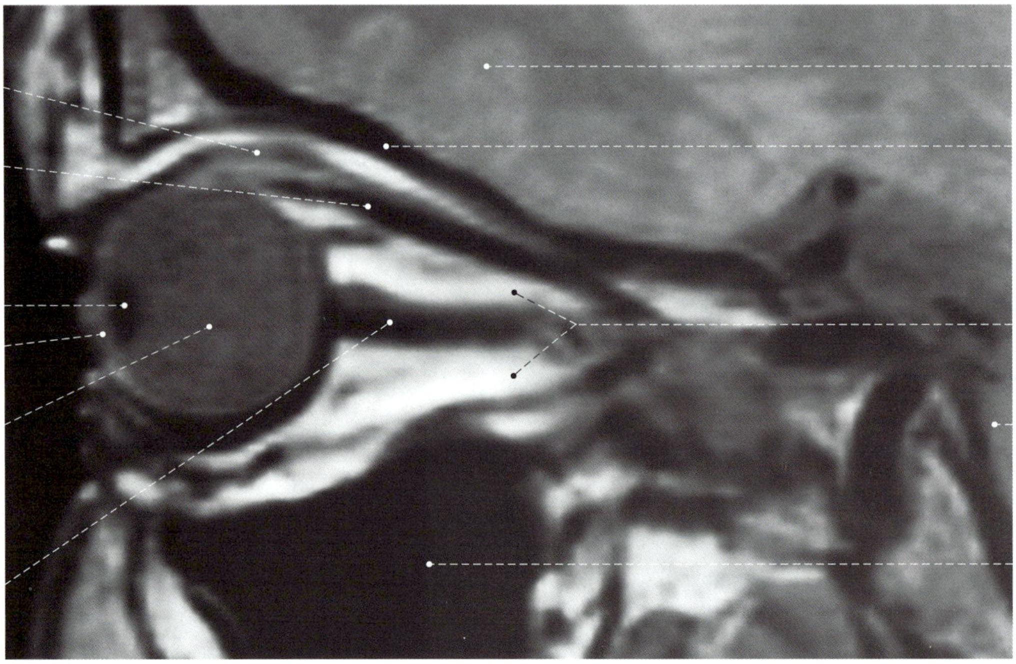

384 Orbita

384 Orbital cavity

Oblique vertical sections along the axis of the optic nerve through the right orbit, medial aspect

a Anatomical section (200%)

b Magnetic resonance image (MRI, T_1-weighted) (140%)

384 Augenhöhle

Schräge Vertikalschnitte entlang der Sehnervenachse durch die rechte Augenhöhle, Medialansicht

a Anatomischer Schnitt (200%)

b Kernspintomogramm (MRI, T_1-gewichtet) (140%)

a

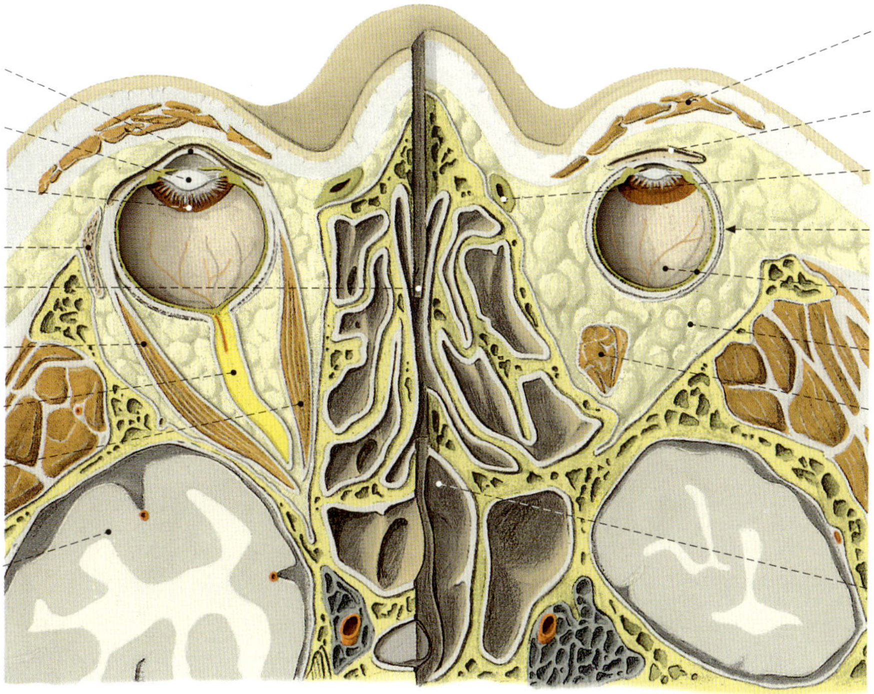

b

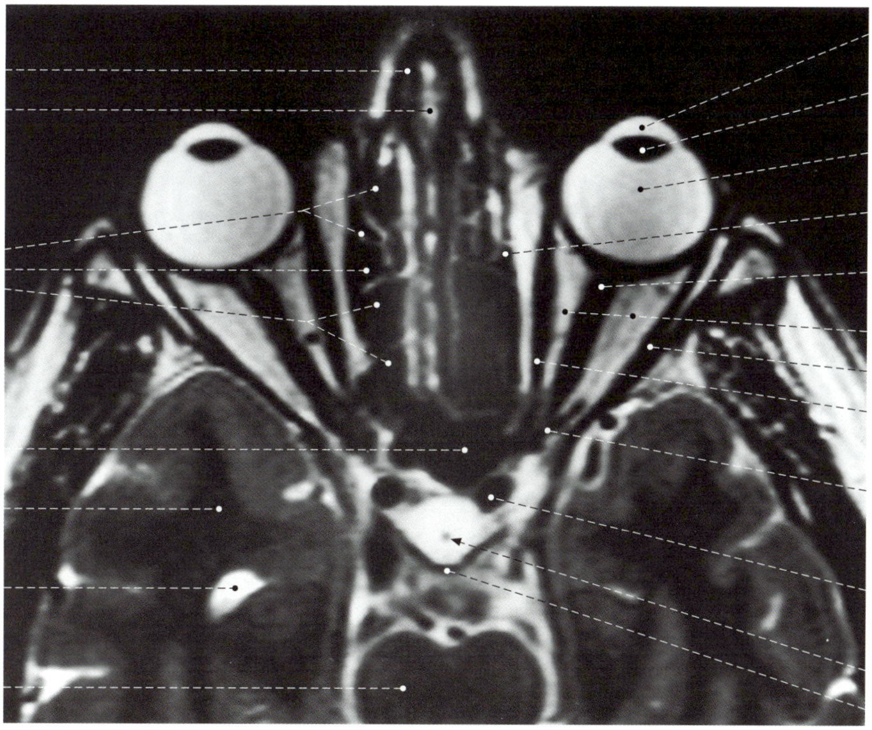

385 Orbita

385 Orbital cavity 90%)

a Stepped anatomical transverse section through both orbits and eyeballs, superior aspect

b Transverse magnetic resonance image (MRI, T_2-weighted) through both orbits, eyeballs and optic nerves, inferior aspect

385 Augenhöhle (90%)

a Gestufter anatomischer Horizontalschnitt durch beide Augenhöhlen und die Augäpfel, Parietalansicht

b Horizontales Kernspintomogramm (MRI, T_2-gewichtet) durch beide Augenhöhlen, Augäpfel und Sehnerven, Kaudalansicht

a

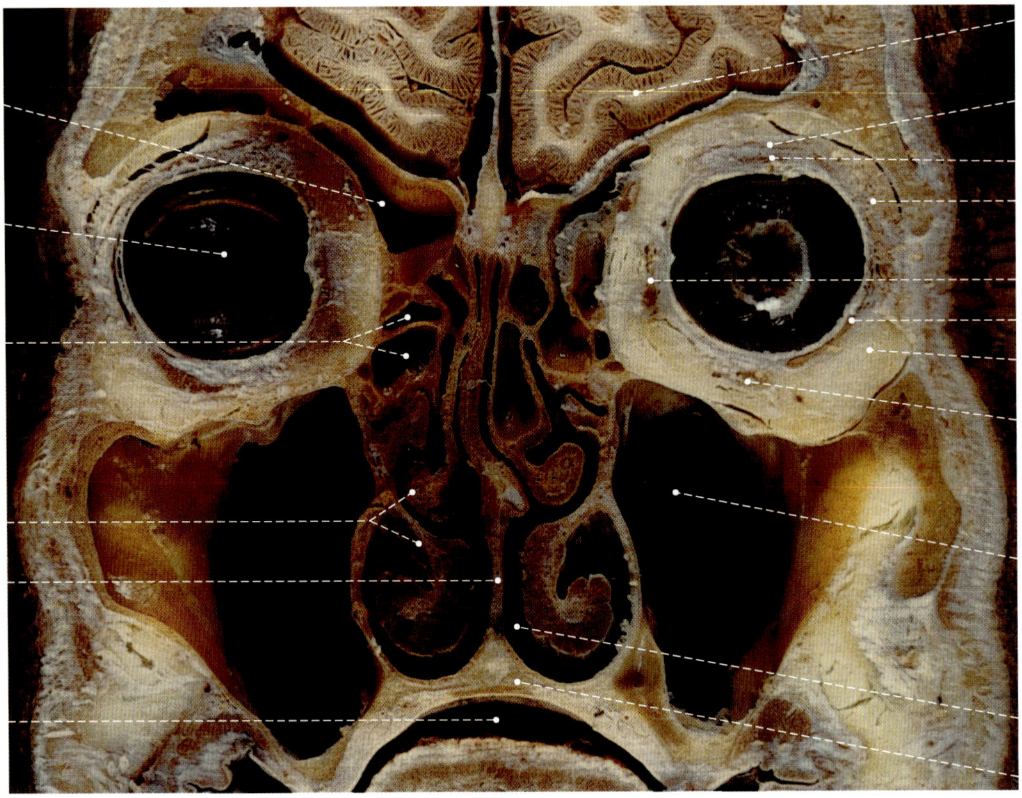

b

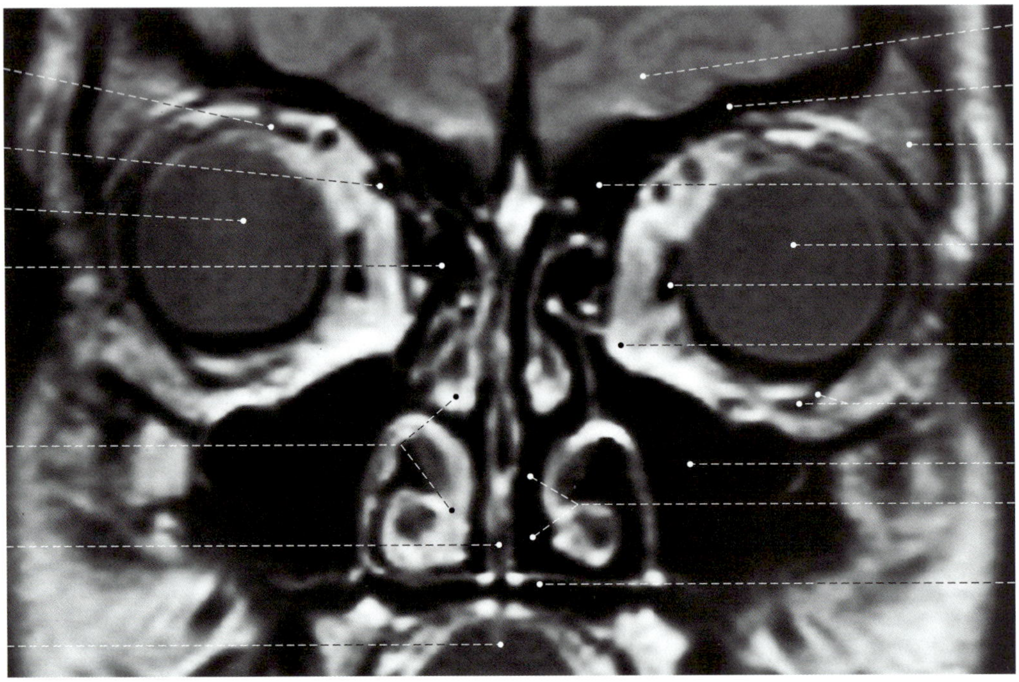

386 Orbita

386 Orbital cavity (100%)

Coronal sections through ventral parts of both orbits, the nasal cavity and paranasal sinuses, frontal aspect

a Anatomical section
b Magnetic resonance image (MRI, T_1-weighted)

386 Augenhöhle (100%)

Frontalschnitte durch ventrale Abschnitte beider Augenhöhlen sowie der Nasenhaupt- und Nasennebenhöhlen, Frontalansicht

a Anatomischer Schnitt
b Kernspintomogramm (MRI, T_1-gewichtet)

a

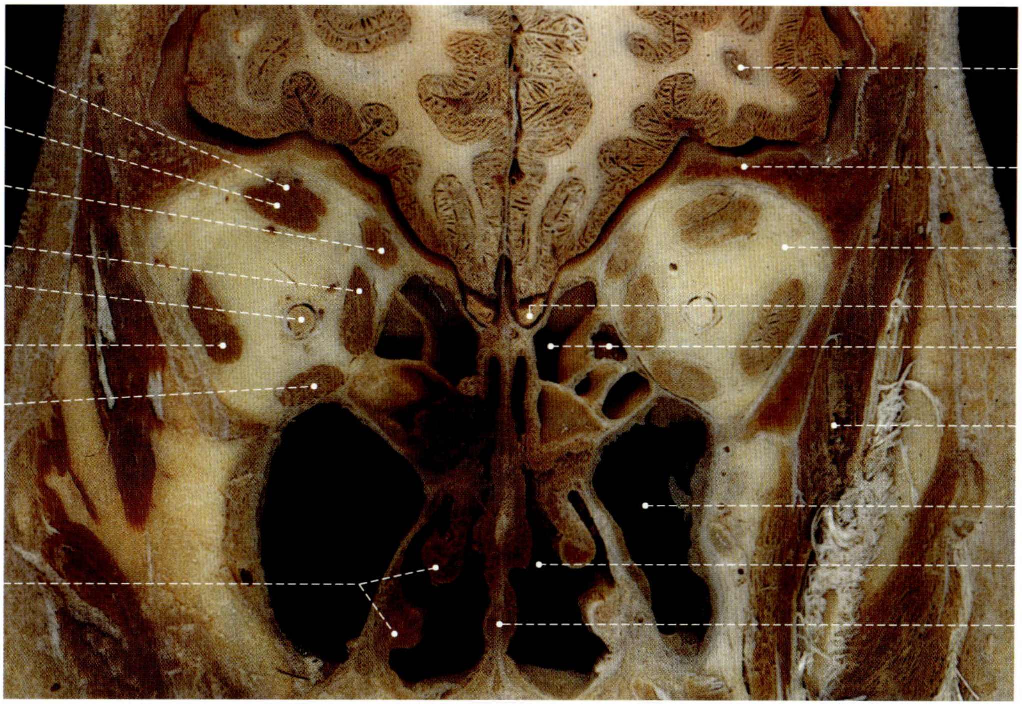

b

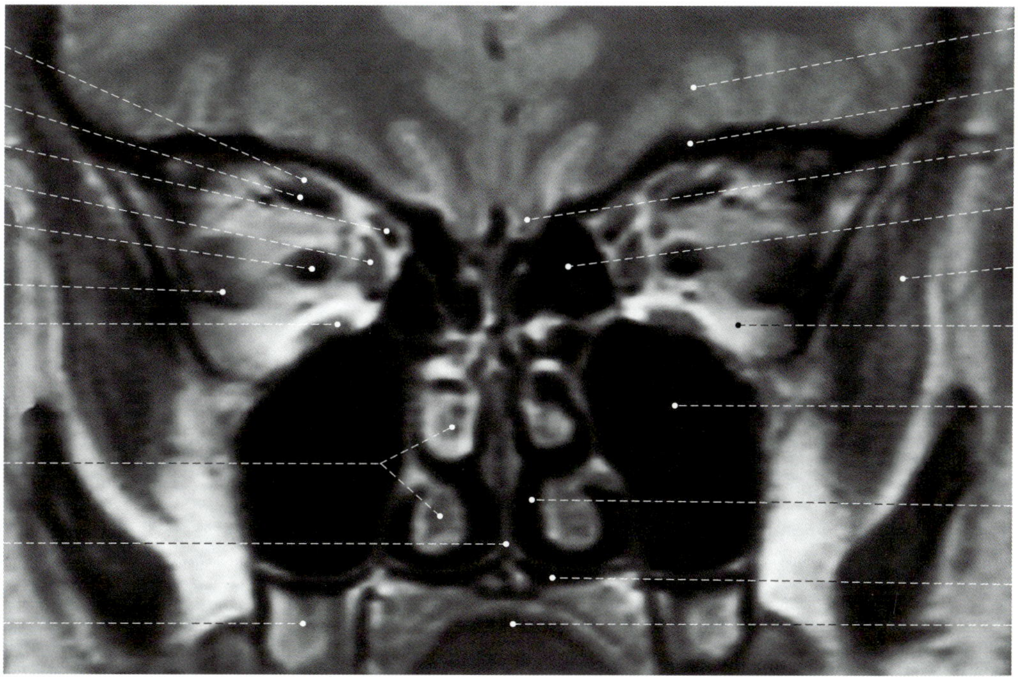

387 Orbita

387 Orbital cavity (100%)

Coronal sections through dorsal parts of both orbits, the nasal cavity and paranasal sinuses, frontal aspect

a Anatomical section
b Magnetic resonance image (MRI, T_1-weighted)

387 Augenhöhle (100%)

Frontalschnitte durch dorsale Abschnitte beider Augenhöhlen sowie der Nasenhaupt- und Nasennebenhöhlen, Frontalansicht

a Anatomischer Schnitt
b Kernspintomogramm (MRI, T_1-gewichtet)

Organum vestibulocochleare

Vestibulocochlear Organ

Gehör- und Gleichgewichtsorgan

a

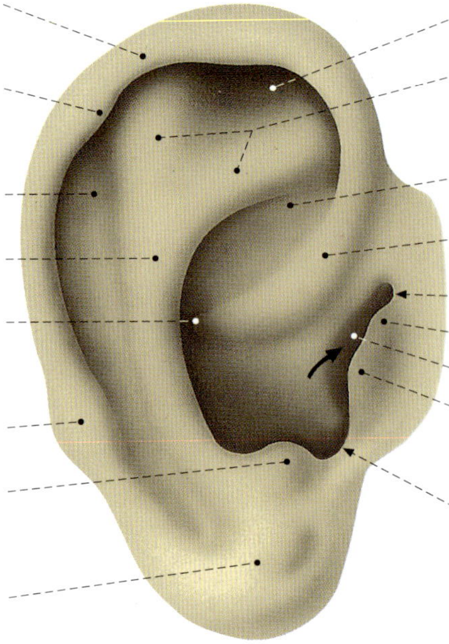

b

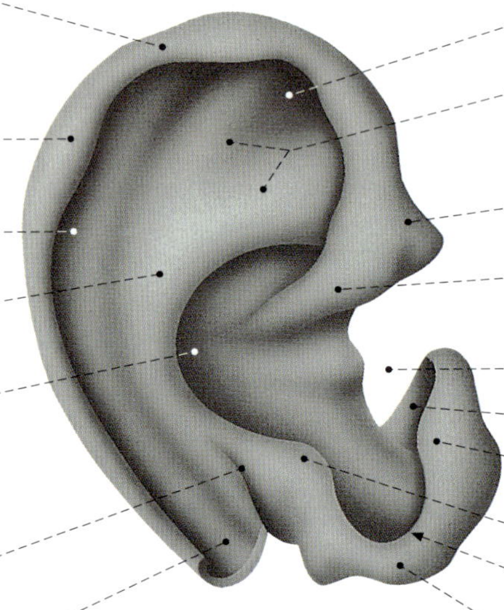

390 Auricula

390 External ear (110%)

Right external ear, lateral aspect

a Auricle
b Cartilage of the auricle

390 Äusseres Ohr (110%)

Rechtes äusseres Ohr, Seitenansicht

a Ohrmuschel
b Ohrknorpel, Cartilago auricularis

a

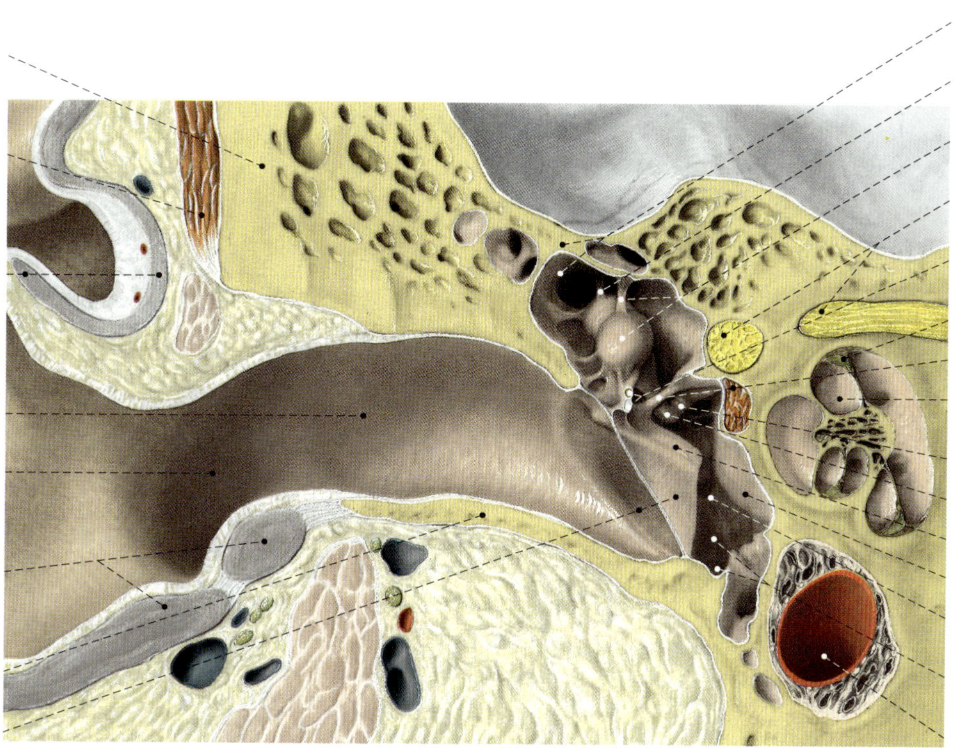

b

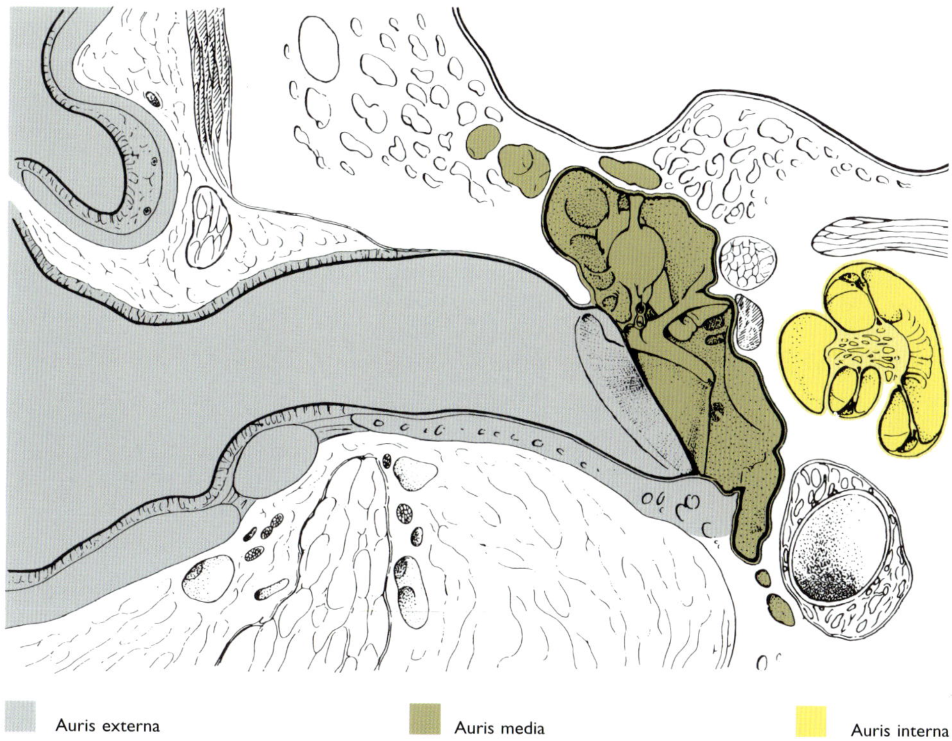

391 Organum vestibulocochleare

391 Organ of hearing (300%)

a, b Coronal section through the external, middle and internal ear of the right side, frontal aspect

b The parts of the external, middle and internal ear are indicated by different colors.

391 Gehörorgan (300%)

a, b Frontalschnitt durch den rechten äusseren Gehörgang, die rechte Paukenhöhle und das Labyrinthorgan, Frontalansicht

b Die Anteile des äusseren, des Mittel- und Innenohrs sind in verschiedenen Farben dargestellt.

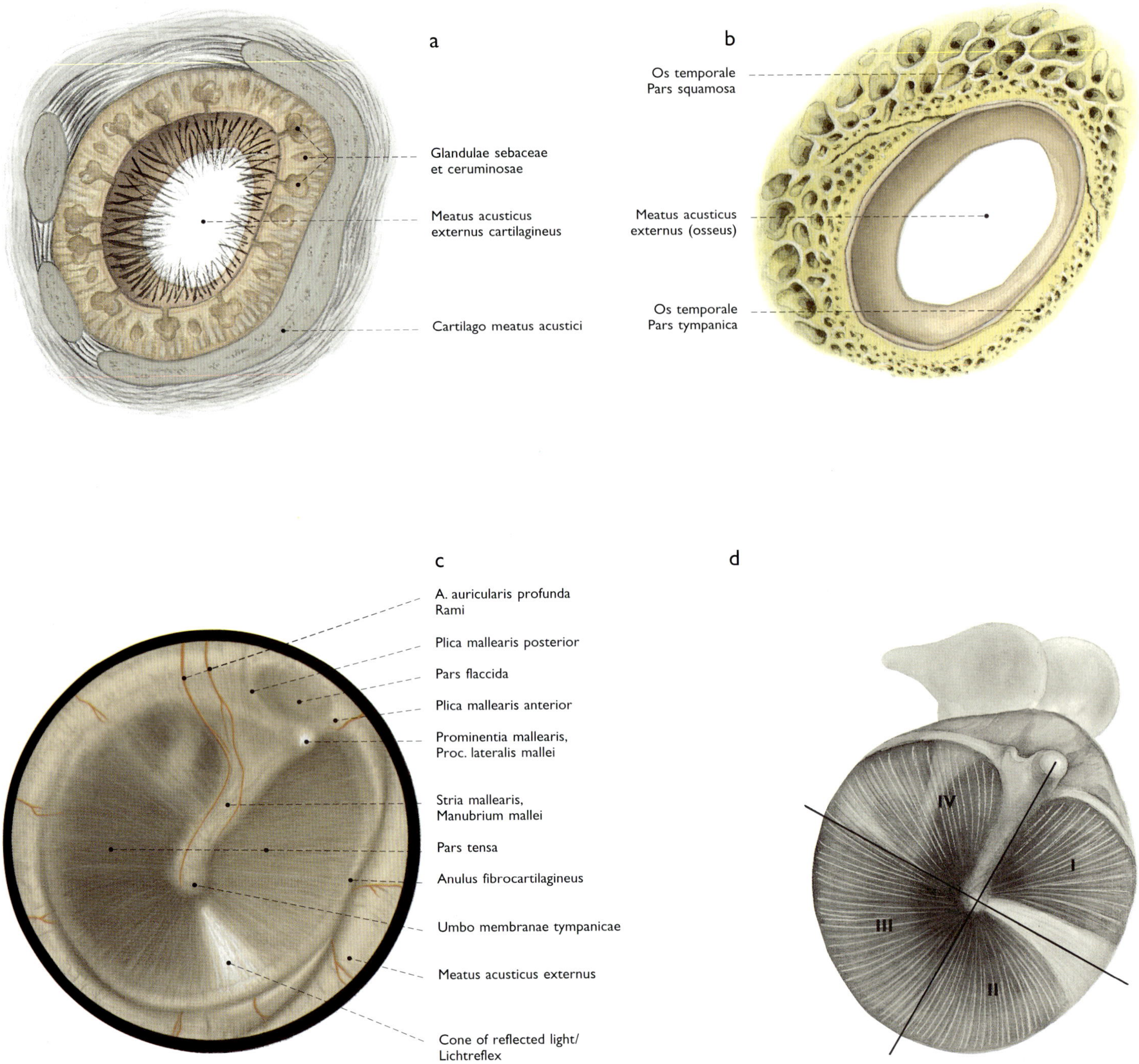

392 Meatus acusticus externus et Membrana tympanica

392 External acoustic meatus and tympanic membrane of the right ear

Lateral aspect

a, b Cross sections through the
a cartilaginous and
b non-cartilaginous parts of the right external acoustic meatus (500%)
c View of the right tympanic membrane (eardrum) as seen through an otoscope in a living subject (700%)
d Division of the right eardrum into quadrants, schematic representation

392 Äusserer Gehörgang und Trommelfell des rechten Ohres

Seitenansicht

a, b Querschnitte durch den
a knorpeligen und
b nichtknorpeligen Teil des rechten äusseren Gehörgangs (500%)
c Vitalaspekt des rechten Trommelfells bei Betrachtung mit dem Ohrenspiegel (700%)
d Quadrantengliederung des rechten Trommelfells, schematische Darstellung

a

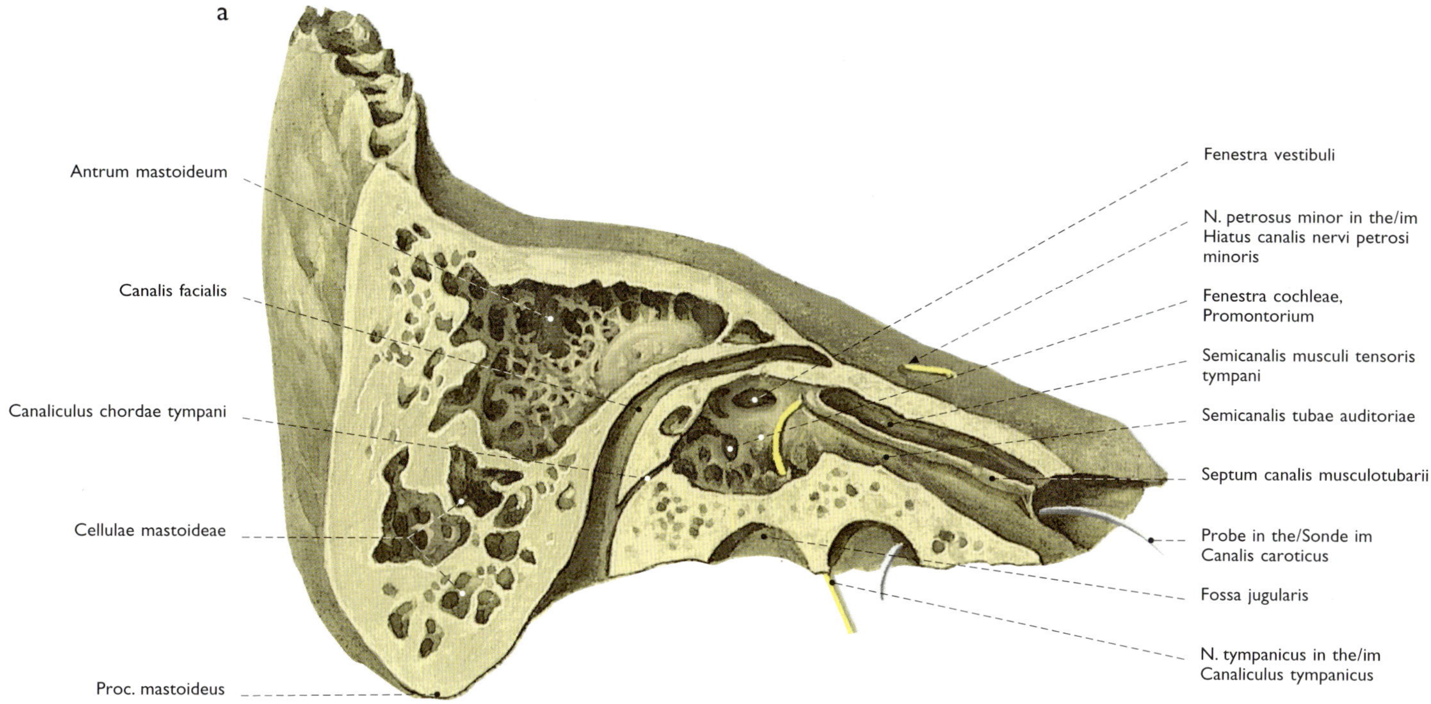

b

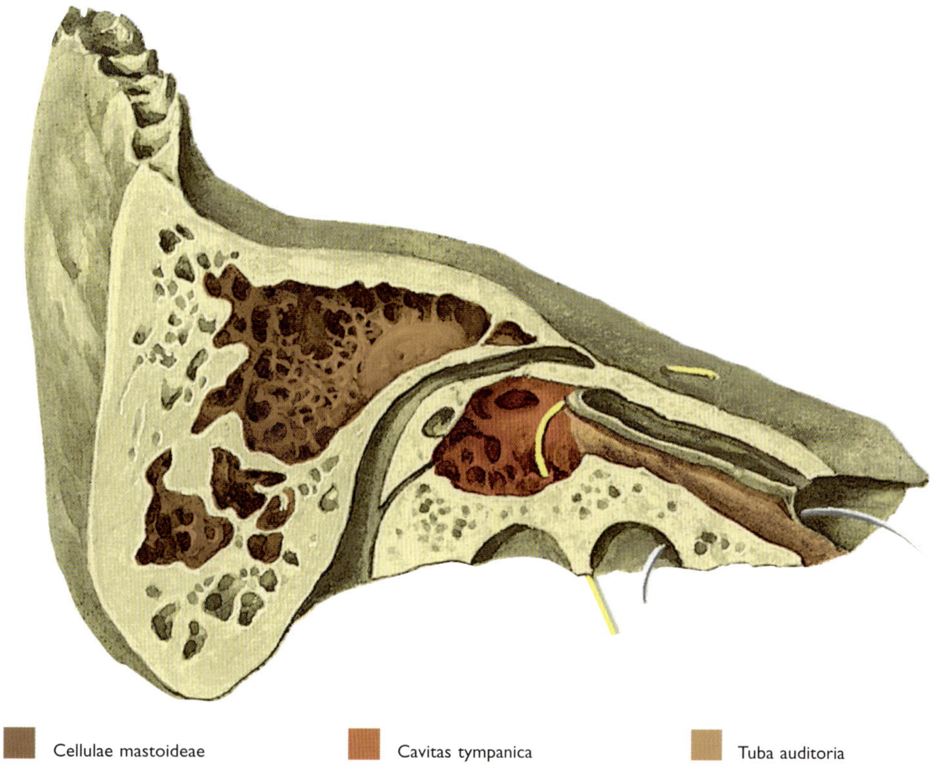

393 Os temporale

393 Right temporal bone

a, b Vertical section through the petrous bone parallel to its long axis. The tympanic cavity and the facial nerve canal were opened (100%). Frontolateral aspect

b The three parts of the middle ear are indicated by different colors (mastoid air cells **dark-brown**, tympanic cavity **red**, auditory tube **light-brown**).

393 Rechtes Schläfenbein

a, b Vertikalschnitt parallel zur Felsenbeinlängsachse. Die Paukenhöhle und der Fazialiskanal sind eröffnet (100%). Frontolateralansicht

b Die drei Anteile des Mittelohrs sind in verschiedenen Farben dargestellt (Mastoidzellen **dunkelbraun**, Paukenhöhle **rot**, Ohrtrompete **hellbraun**).

a

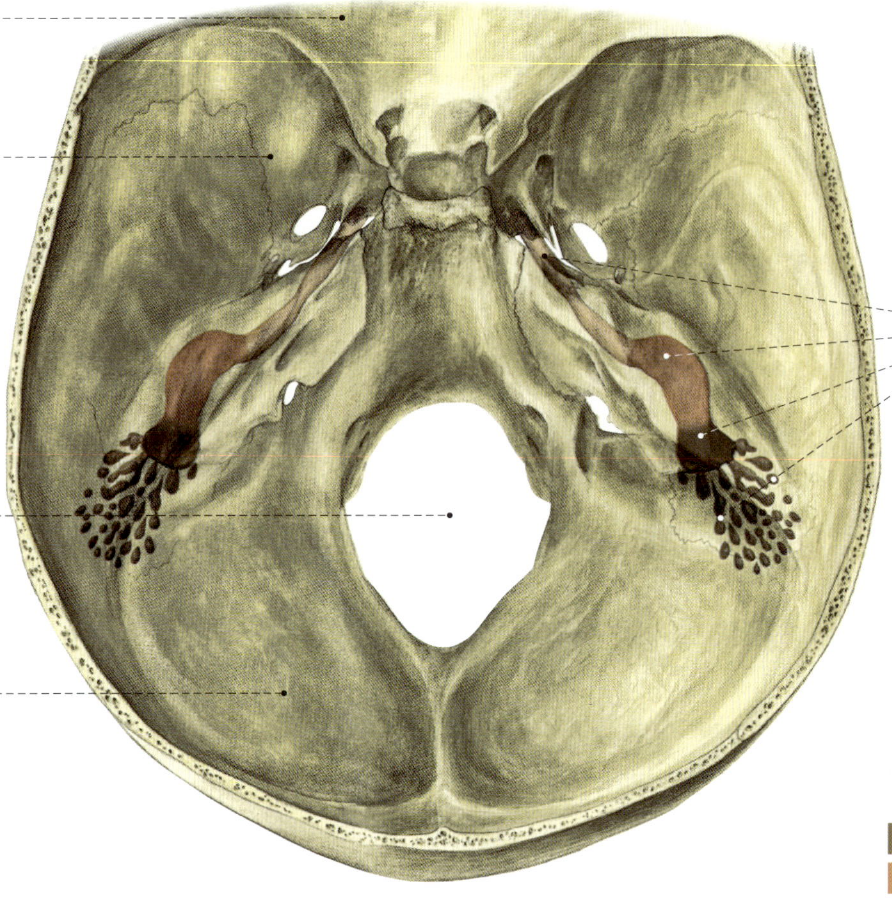

b

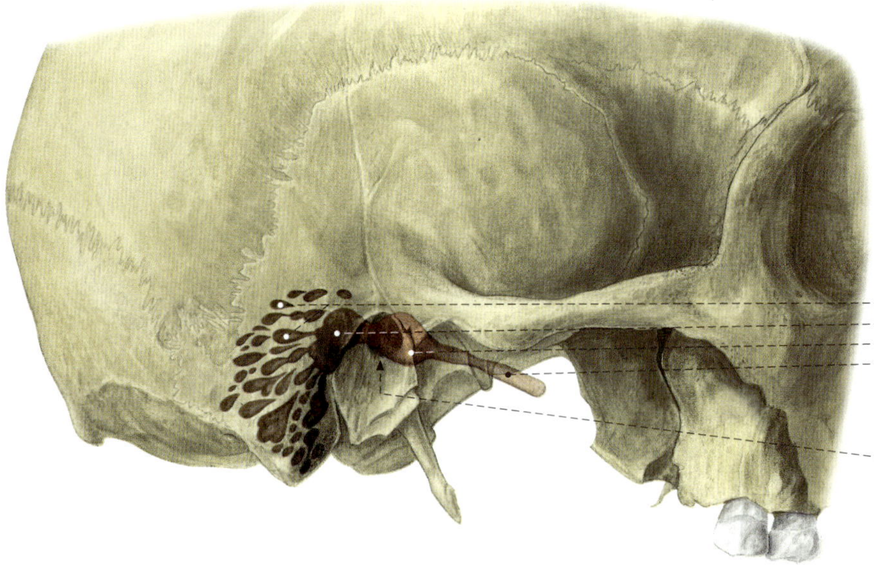

394 Os temporale

394 Temporal bone (80%)

Projection of the middle ear

a onto the internal surface of the cranial base, parietal aspect

b onto the lateral external surface of the skull, right lateral aspect

The three parts of the middle ear are indicated by different colors (mastoid air cells **dark-brown**, tympanic cavity **red**, auditory tube **light-brown**).

394 Schläfenbein (80%)

Projektion des Mittelohrs

a auf die innere Schädelbasis, Parietalansicht

b auf die seitliche Aussenfläche des Schädels, rechte Seitenansicht

Die drei Anteile des Mittelohrs sind in verschiedenen Farben dargestellt (Mastoidzellen **dunkelbraun**, Paukenhöhle **rot**, Ohrtrompete **hellbraun**).

a

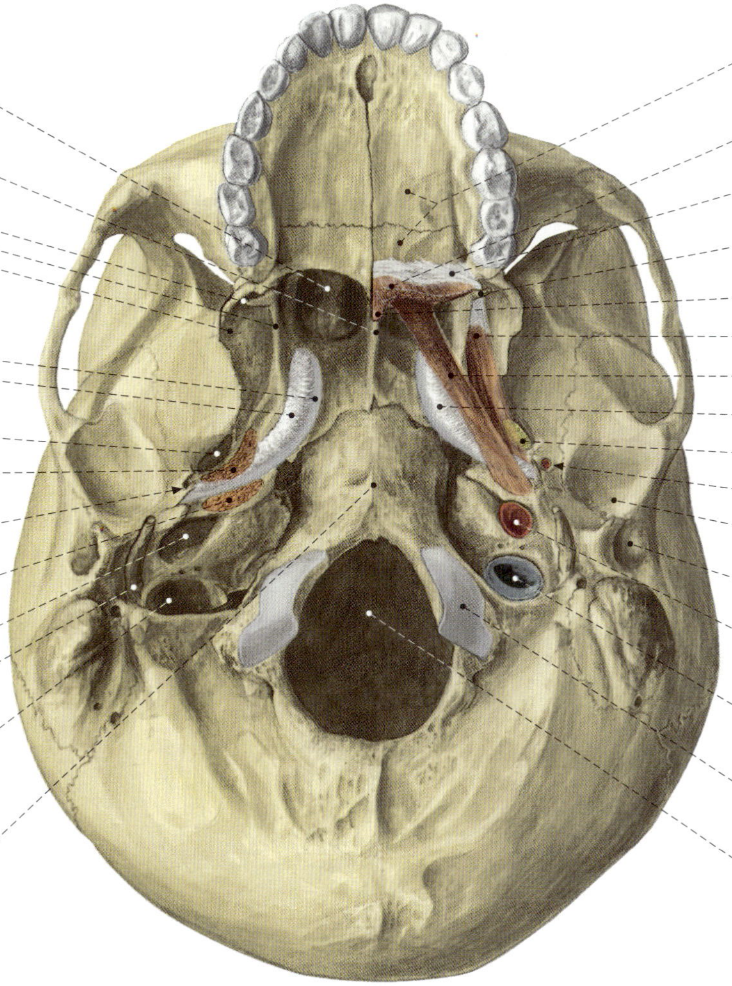

b

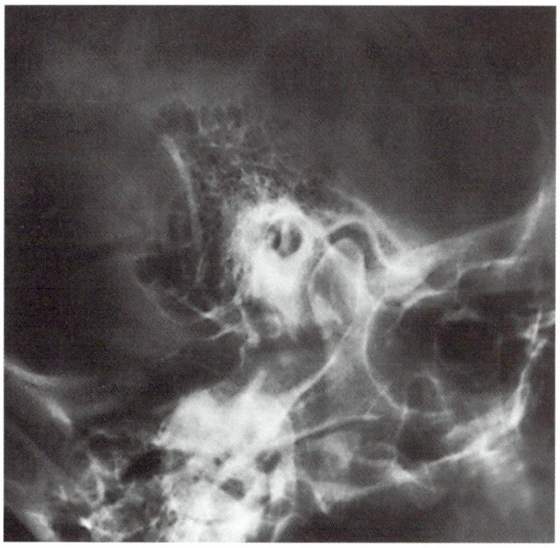

c

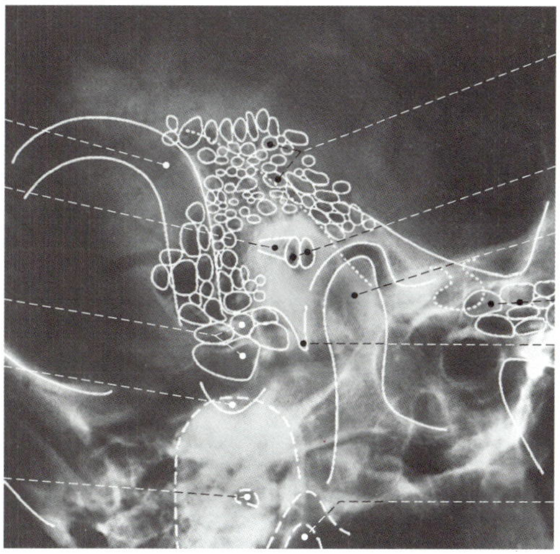

395 Os temporale

395 Temporal bone (100%)

a External surface of the cranial base with auditory tubes. The muscles of the soft palate are shown on the left, the areas of attachment of the levator veli palatini and tensor veli palatini on the right half of the skull (70%). Norma basilaris (inferior aspect)
b Parietolateral radiograph of the petrous bones according to Schüller
c Explanatory drawing as to Fig. b

395 Schläfenbein (100%)

a Äussere Schädelbasis mit Ohrtrompeten (Tubae auditoriae). Die Muskeln des weichen Gaumens sind an der linken, die Ursprungsfelder der Musculi levator et tensor veli palatini an der rechten Schädelhälfte eingezeichnet (70%). Norma basilaris (Kaudalansicht)
b Parietolaterale Röntgenaufnahme der Felsenbeine nach Schüller
c Erklärende Zeichnung zu Abb. b

a

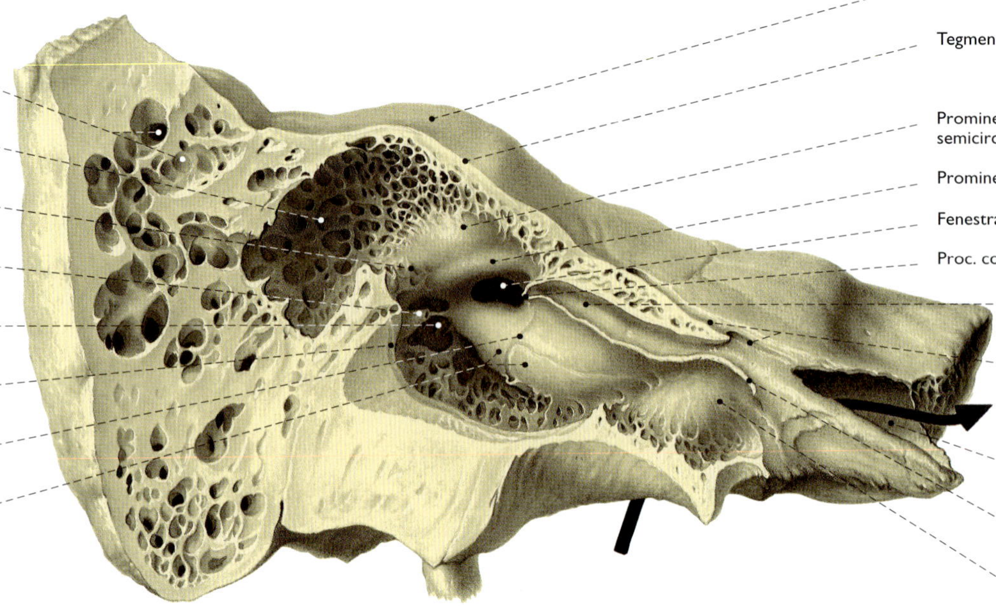

b

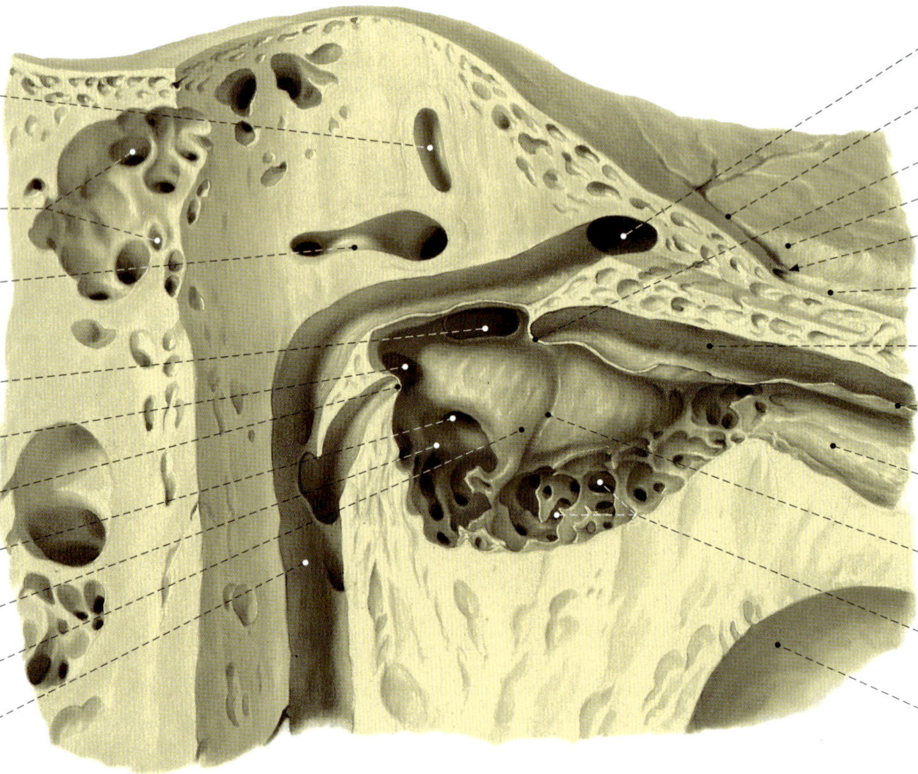

396 Cavitas tympanica

396 Right tympanic cavity

Frontolateral aspect of the medial labyrinthine wall of the tympanic cavity. The fenestra vestibuli was opened by removing the stapes. The mucous membrane was taken away.

a Vertical section through the petrous bone parallel to its longitudinal axis (200%)

b Section through the tympanic cavity parallel to its medial wall. The facial nerve canal was opened and the facial nerve removed (500%).

396 Rechte Paukenhöhle

Frontolateralansicht der medialen Wand (Paries labyrinthicus) der Paukenhöhle. Das ovale Fenster (Fenestra vestibuli) ist durch Entfernung des Steigbügels eröffnet, die Schleimhaut entfernt.

a Vertikalschnitt durch das Felsenbein parallel zu seiner Längsachse (200%)

b Schnitt durch die Paukenhöhle parallel zu ihrer medialen Wand. Der Fazialiskanal ist eröffnet, der Nervus facialis entfernt (500%).

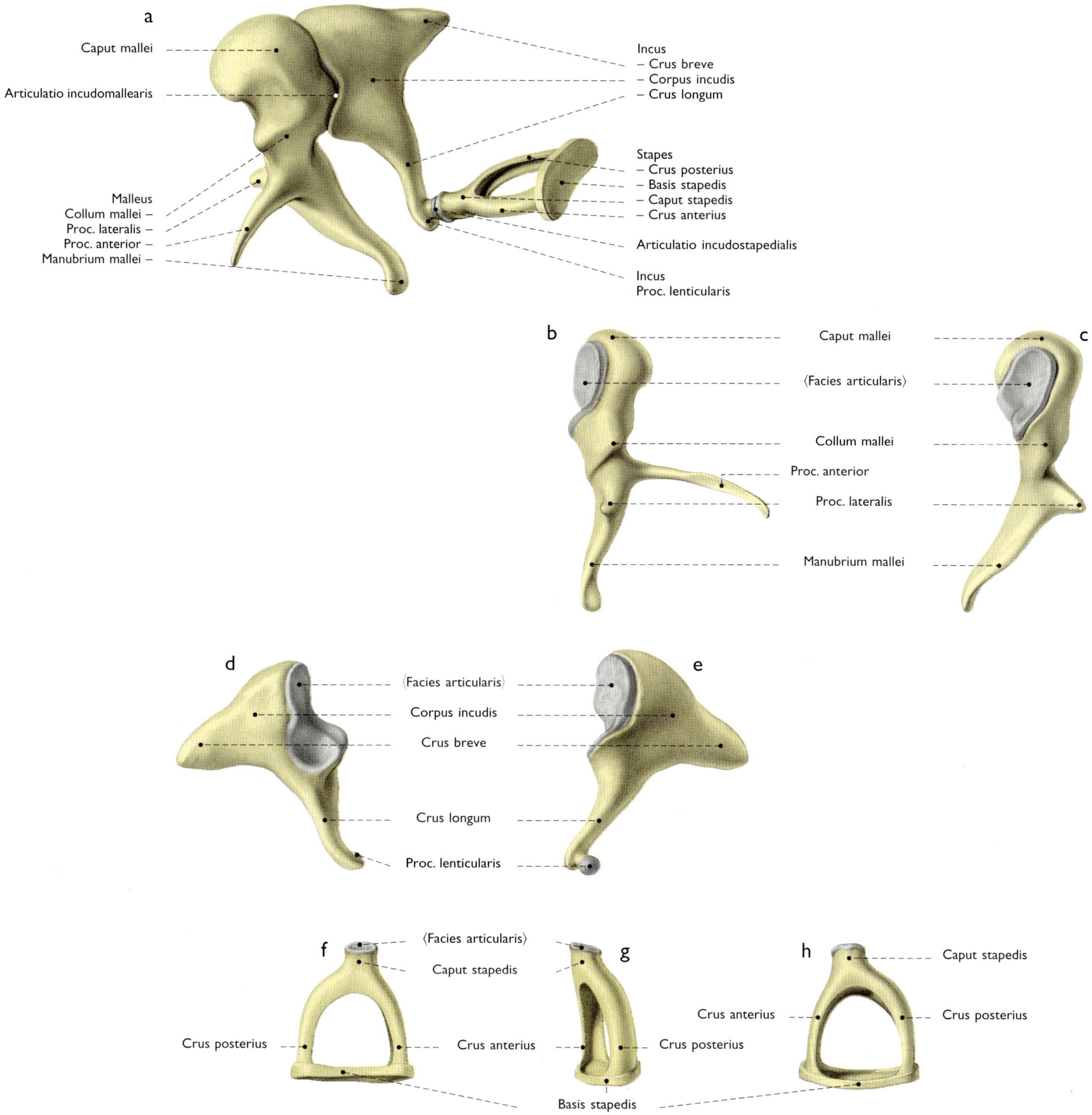

397 Ossicula auditoria

397 Auditory ossicles

- a Right auditory ossicles and articulations in their natural position (750%), superior medial aspect
- b, c Malleus (650%)
  - b Lateral aspect
  - c Dorsal aspect
- d, e Incus (650%)
  - d Lateral aspect
  - e Medial aspect
- f–h Stapes (650%)
  - f Inferior aspect
  - g Dorsal aspect
  - h Superior aspect

397 Gehörknöchelchen

- a Gehörknöchelchen und Gelenke der rechten Paukenhöhle in natürlicher Lage (750%), Kraniomedialansicht
- b, c Hammer (650%)
  - b Seitenansicht
  - c Dorsalansicht
- d, e Amboss (650%)
  - d Seitenansicht
  - e Medialansicht
- f–h Steigbügel (650%)
  - f Kaudalansicht
  - g Dorsalansicht
  - h Kranialansicht

a

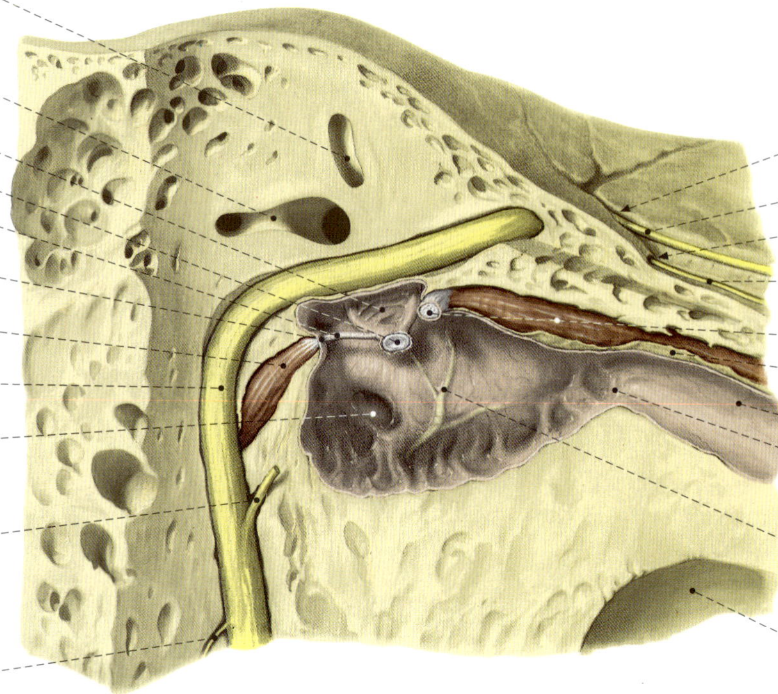

b

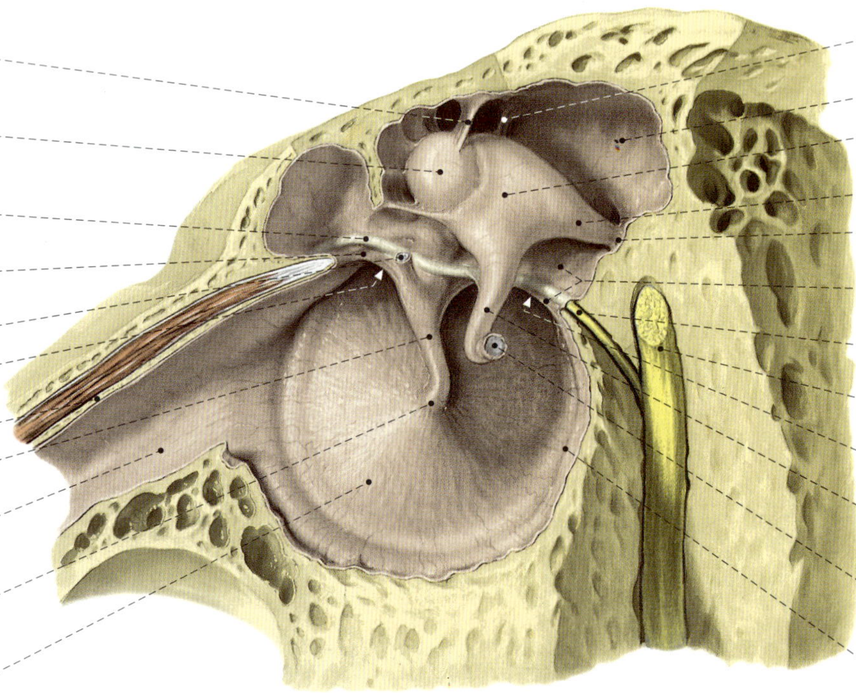

398 Cavitas tympanica

398 Right tympanic cavity (450%)

a Section through the tympanic cavity parallel to its medial wall, frontolateral aspect of the medial labyrinthine wall
b Section through the tympanic cavity parallel to the lateral wall, medial aspect of the lateral membranous wall

398 Rechte Paukenhöhle (450%)

a Schnitt durch die Paukenhöhle parallel zur medialen Wand, Frontolateralansicht der medialen Wand (Paries labyrinthicus)
b Schnitt durch die Paukenhöhle parallel zur lateralen Wand, Medialansicht der lateralen Wand (Paries membranaceus)

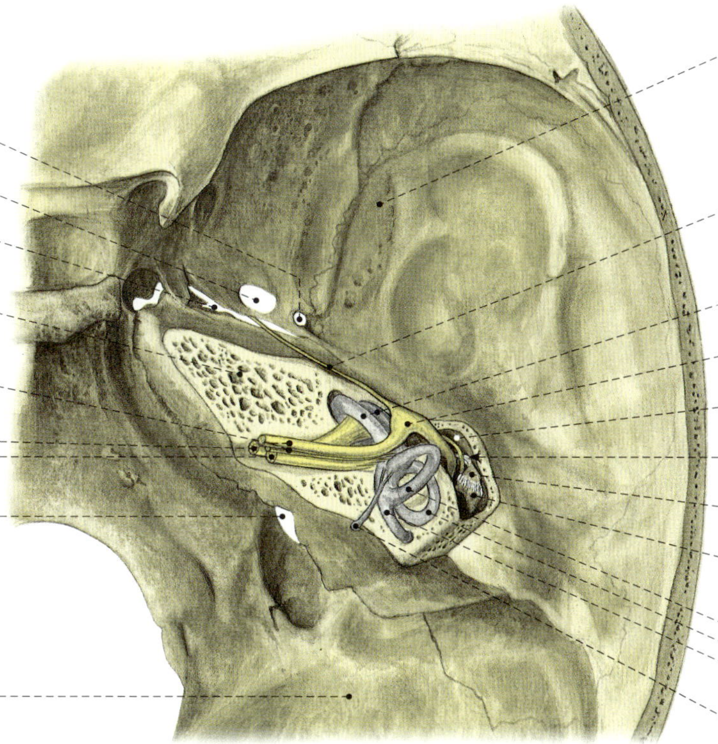

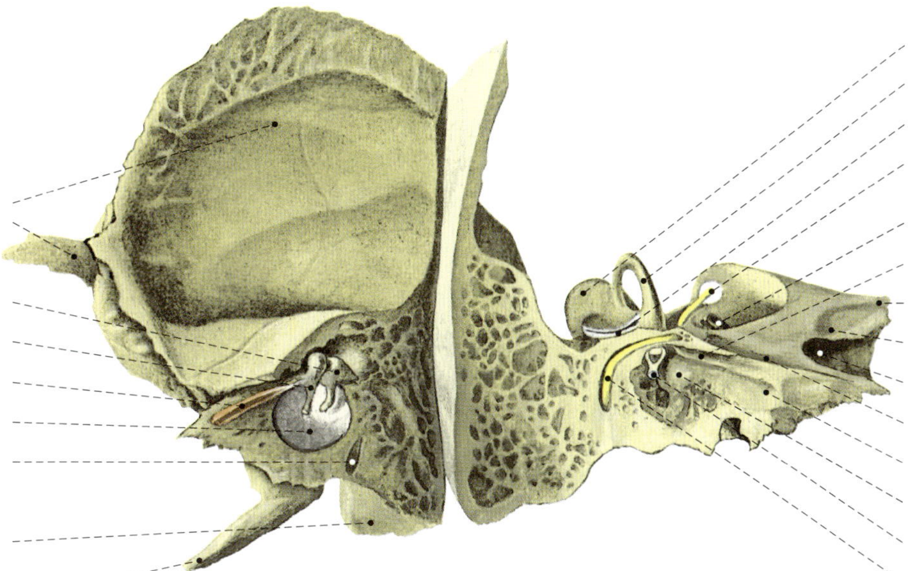

399 Os temporale et Canalis facialis

399 Temporal bone and facial nerve canal (100%)

a Transverse section through the right petrous bone in the plane of the internal acoustic meatus. Exposure of the cochlear duct, the semicircular ducts, and the facial and vestibulocochlear nerves. The facial nerve canal and the tympanic cavity were opened. Superior aspect

b Vertical section through the middle ear, the facial nerve canal, and the mastoid air cells of the right petrous bone. The lateral part was turned outwards. The semicircular canals and the cochlea were exposed. Frontolateral aspect

399 Schläfenbein und Fazialiskanal (100%)

a Horizontalschnitt durch das rechte Felsenbein in der Ebene des inneren Gehörgangs. Darstellung der Schnecke, der Bogengänge und der Nervi facialis et vestibulocochlearis. Der Fazialiskanal und die Paukenhöhle sind eröffnet. Parietalansicht

b Vertikalschnitt durch die Paukenhöhle, den Fazialiskanal und die Zellen des Warzenfortsatzes des rechten Felsenbeins. Der seitliche Schnittblock ist nach aussen geklappt. Die Canales semicirculares und die Schnecke sind freipräpariert. Frontolateralansicht

a

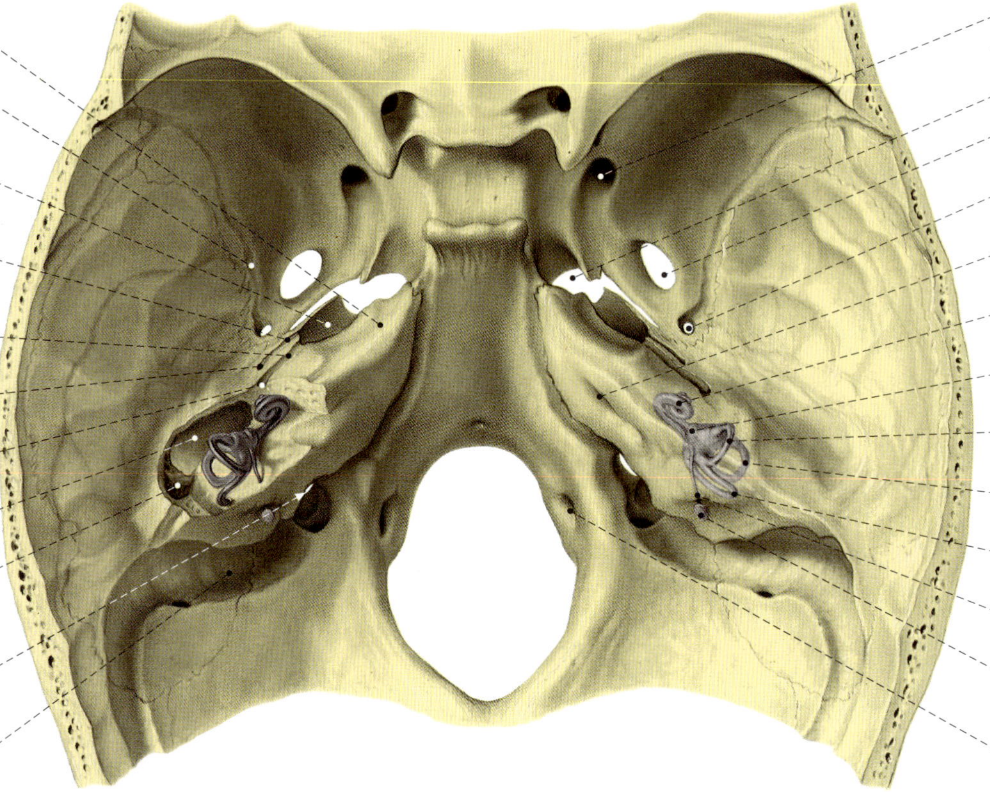

b

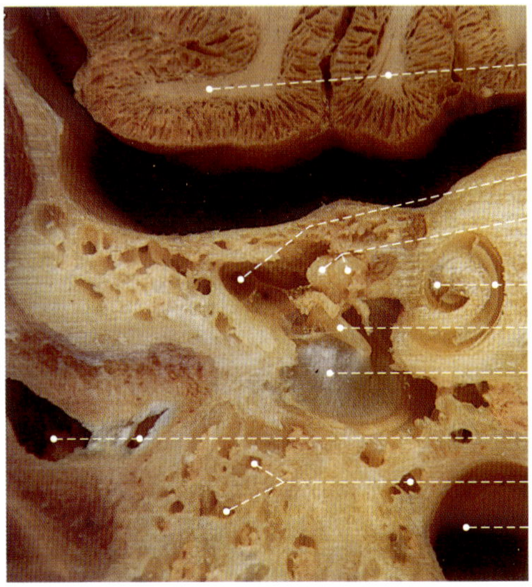

c

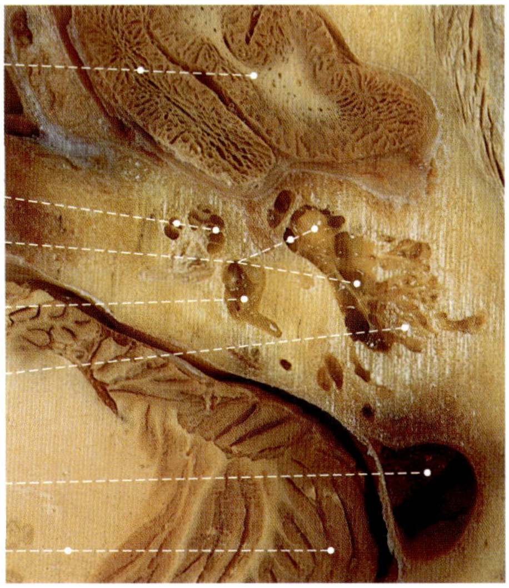

400 Labyrinthus osseus

400 Bony labyrinth

a Internal surface of the cranial base (100%). On the right side, the bony labyrinth is shown in projection onto the superior surface of the petrous temporal bone, on the left a cast of the bony labyrinth in its natural position.

b, c Anatomical sections through the petrous bone, the tympanic cavity, and the cochlea

b Coronal section through the right petrous bone (180%), frontal aspect

c Transverse section through the right petrous bone (130%), parietal aspect

400 Knöchernes Labyrinth

a Innere Schädelbasis (100%). Rechts ist die Projektion des knöchernen Labyrinths auf die Innenfläche des Felsenbeins dargestellt, links ein Ausguss des knöchernen Labyrinths in seiner natürlichen Lage.

b, c Anatomische Schnitte durch Felsenbein, Paukenhöhle und Schnecke

b Frontalschnitt durch das rechte Felsenbein (180%), Frontalansicht

c Horizontalschnitt durch das rechte Felsenbein (130%), Parietalansicht

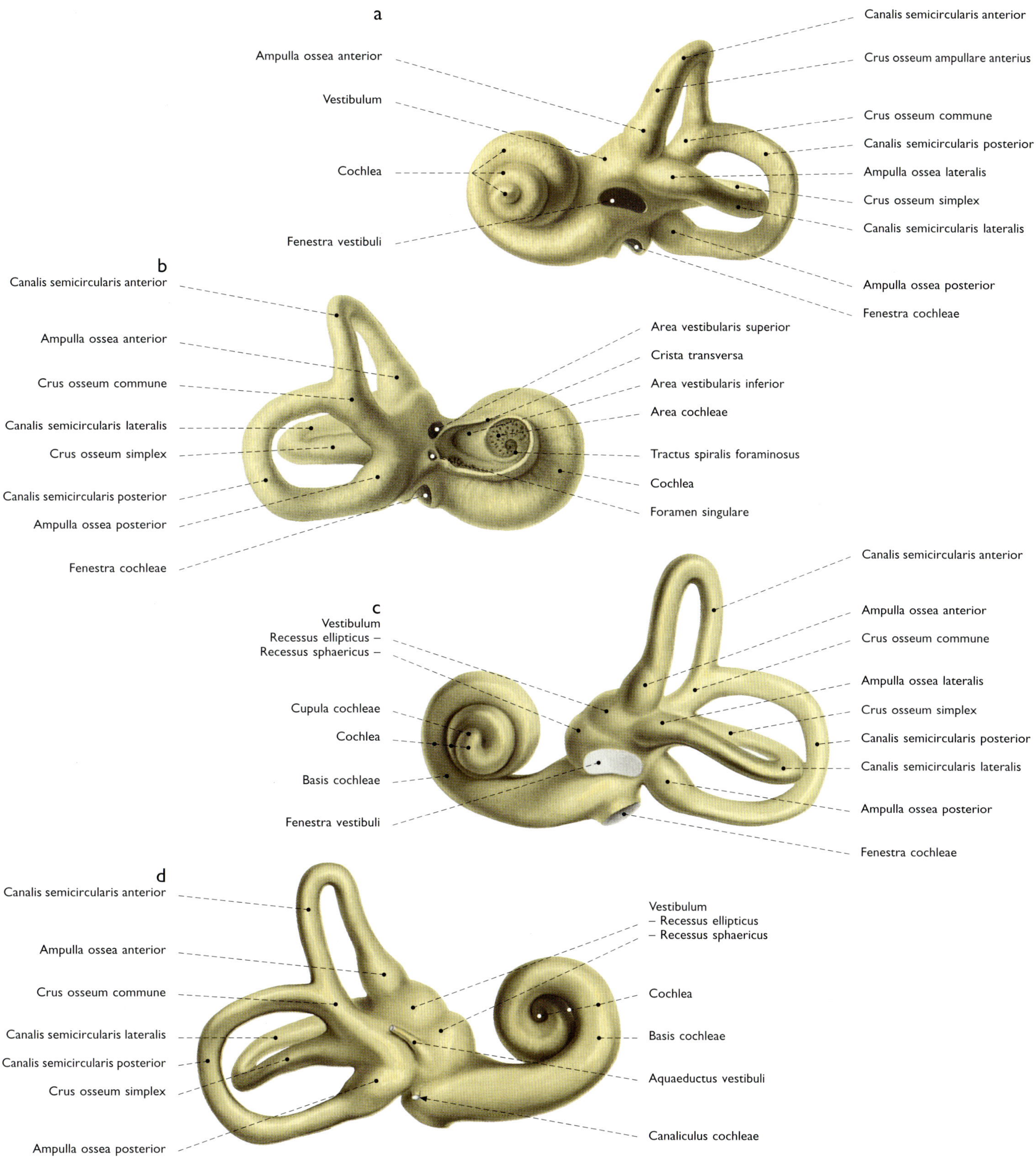

401 Labyrinthus osseus

401 Bony labyrinth

a, b Left labyrinth of a newborn child (900%)
c, d Left labyrinth of an adult (400%), cast preparation
a, c Frontolateral aspect
b, d Dorsomedial aspect

401 Knöchernes Labyrinth

a, b Linkes Labyrinth eines Neugeborenen (900%)
c, d Linkes Labyrinth eines Erwachsenen (400%), Ausgusspräparat
a, c Frontolateralansicht
b, d Dorsomedialansicht

a

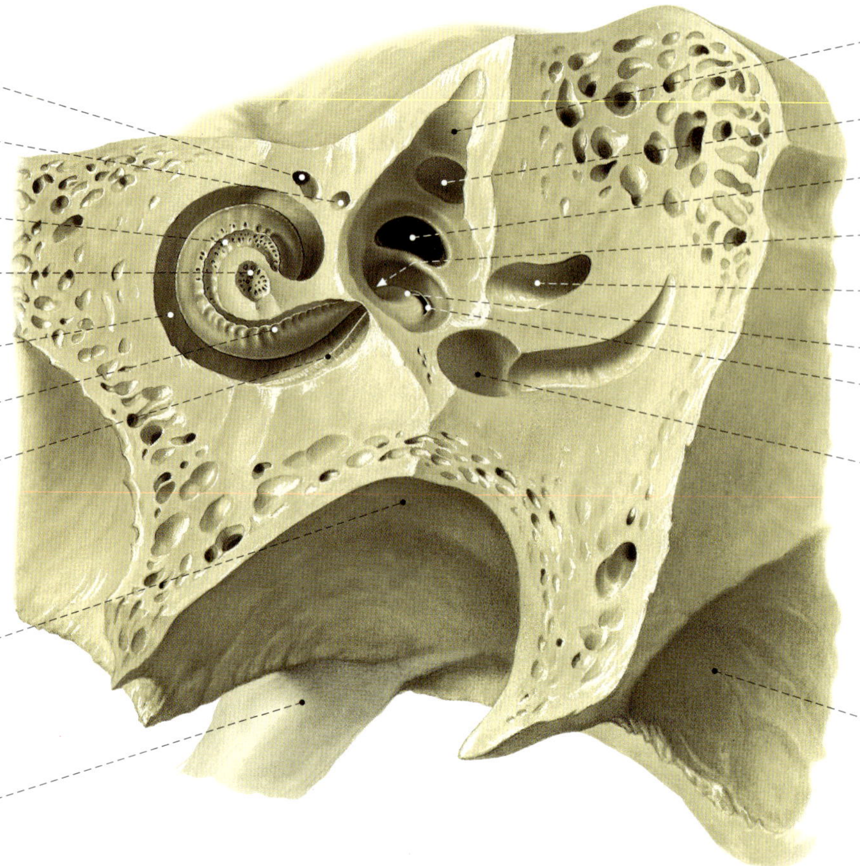

b

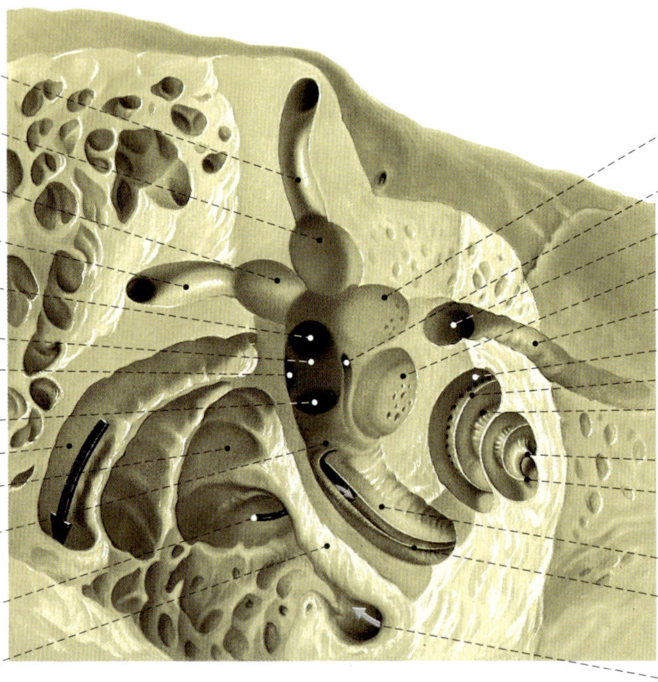

402 Labyrinthus osseus

402 Bony labyrinth (350%)

Vestibule, cochlea, and parts of the semicircular canals of the right petrous temporal bone,

a revealed and opened by dissection of the petrous bone from medial, dorsomedial aspect

b exposed and opened by dissection of the petrous bone from lateral, frontolateral aspect

402 Knöchernes Labyrinth (350%)

Vestibulum, Schnecke und Teile des Bogengangsystems des rechten Felsenbeins,

a dargestellt und eröffnet durch Aufmeisselung des Felsenbeins von medial, Dorsomedialansicht

b freigelegt und eröffnet durch Präparation des Felsenbeins von lateral, Frontolateralansicht

a

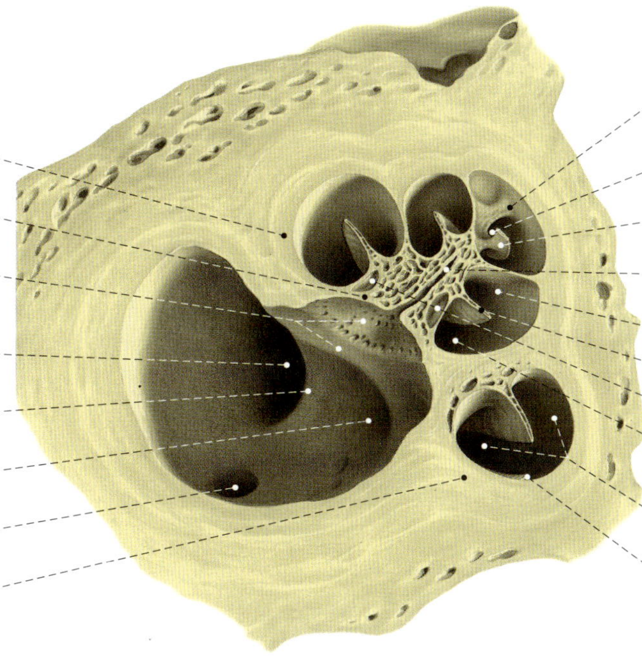

b

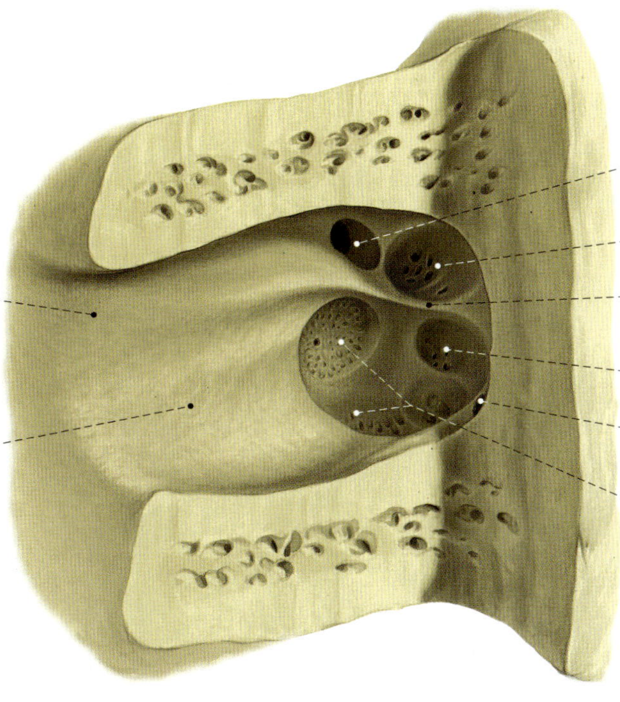

403 Auris interna

403 Internal ear (500%)

a The cochlea and the fundus of the internal acoustic meatus were opened by an axial bisection through the right cochlea. Parietal aspect

b Internal acoustic meatus after removal of the dorsomedial wall of the right internal acoustic meatus, medial aspect

403 Innenohr (500%)

a Die Schnecke und der Boden des inneren Gehörgangs sind durch einen axialen Halbierungsschnitt durch die rechte Schnecke eröffnet. Parietalansicht

b Innerer Gehörgang nach Entfernung der dorsomedialen Wand des rechten inneren Gehörgangs, Medialansicht

a

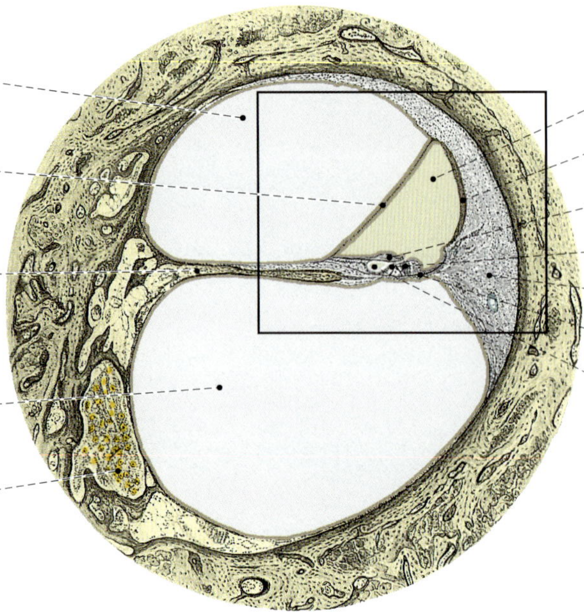

b

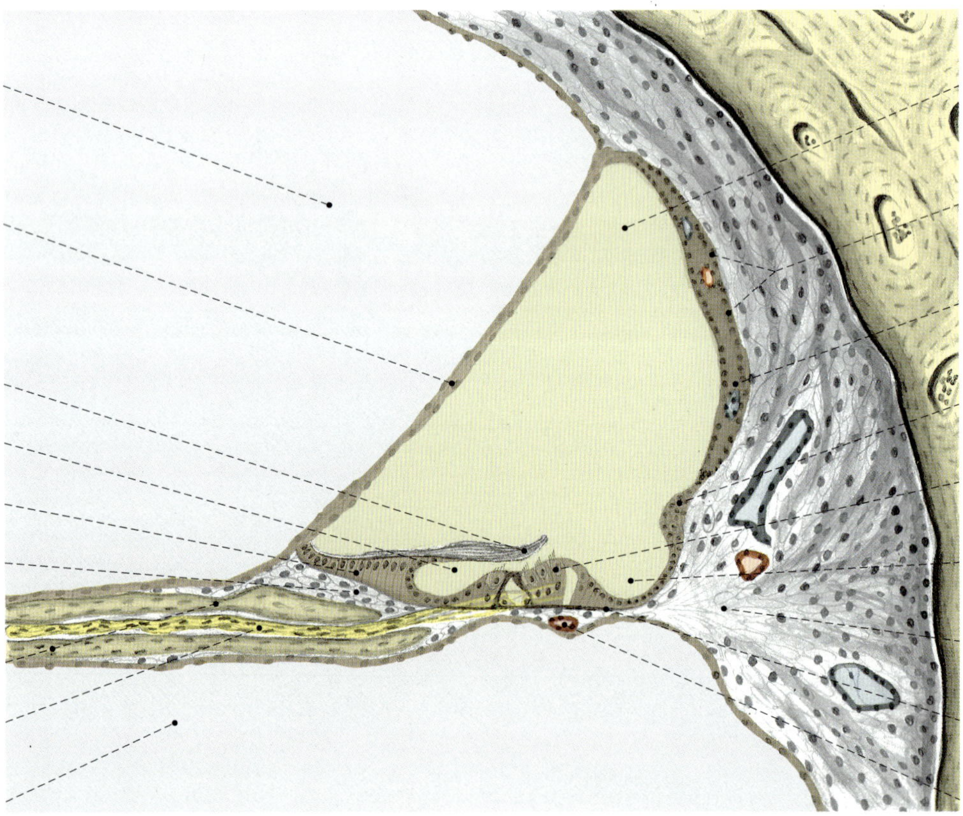

404 Cochlea, Ductus cochlearis et Organum spirale

404 Cochlea, cochlear duct and organ of Corti

a Transverse section through one of the coils of the cochlea (2000%)

b Enlargement of an area similar to the rectangle in Fig. a showing a cross section through the cochlear duct (8000%)

404 Schnecke, Schneckengang und Gehörorgan

a Querschnitt durch eine Windung der Schnecke (2000%)

b Ausschnittvergrösserung aus Abb. a, Querschnitt durch die häutige Schnecke (8000%)

a

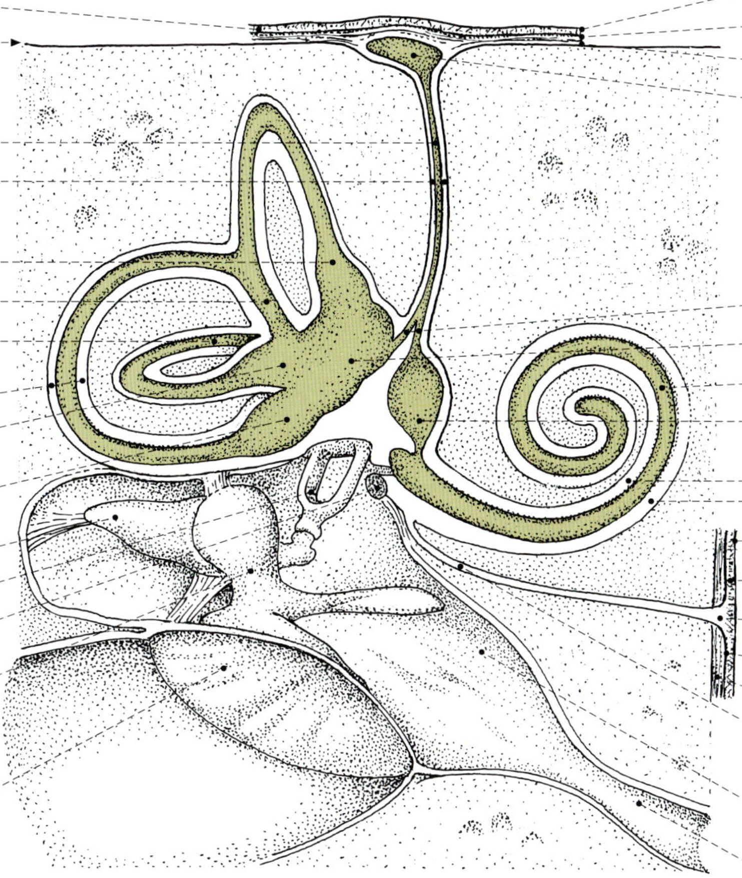

b

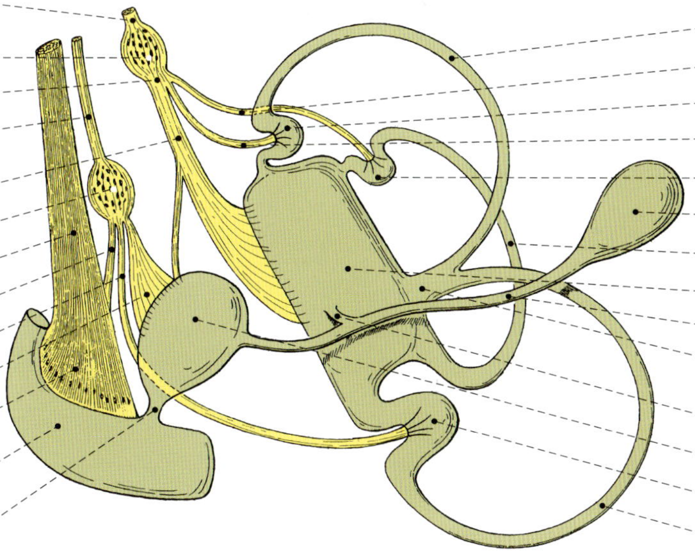

405 Labyrinthus membranaceus et Nervus vestibulocochlearis

405 Membraneous labyrinth and vestibulocochlear nerve

Schematic representations

a Membraneous labyrinth. The membraneous labyrinth filled with endolymph is represented in light-green color, the surrounding perilymphatic spaces are given in white color.

b Ramification of the vestibulocochlear nerve (N. VIII)

405 Häutiges Labyrinth und Nervus vestibulocochlearis

Schematische Darstellungen

a Häutiges Labyrinth. Die Endolymphräume des häutigen Labyrinths sind grünlich dargestellt, die umgebenden Perilymphräume weiss belassen.

b Verzweigung des Nervus vestibulocochlearis (N. VIII)

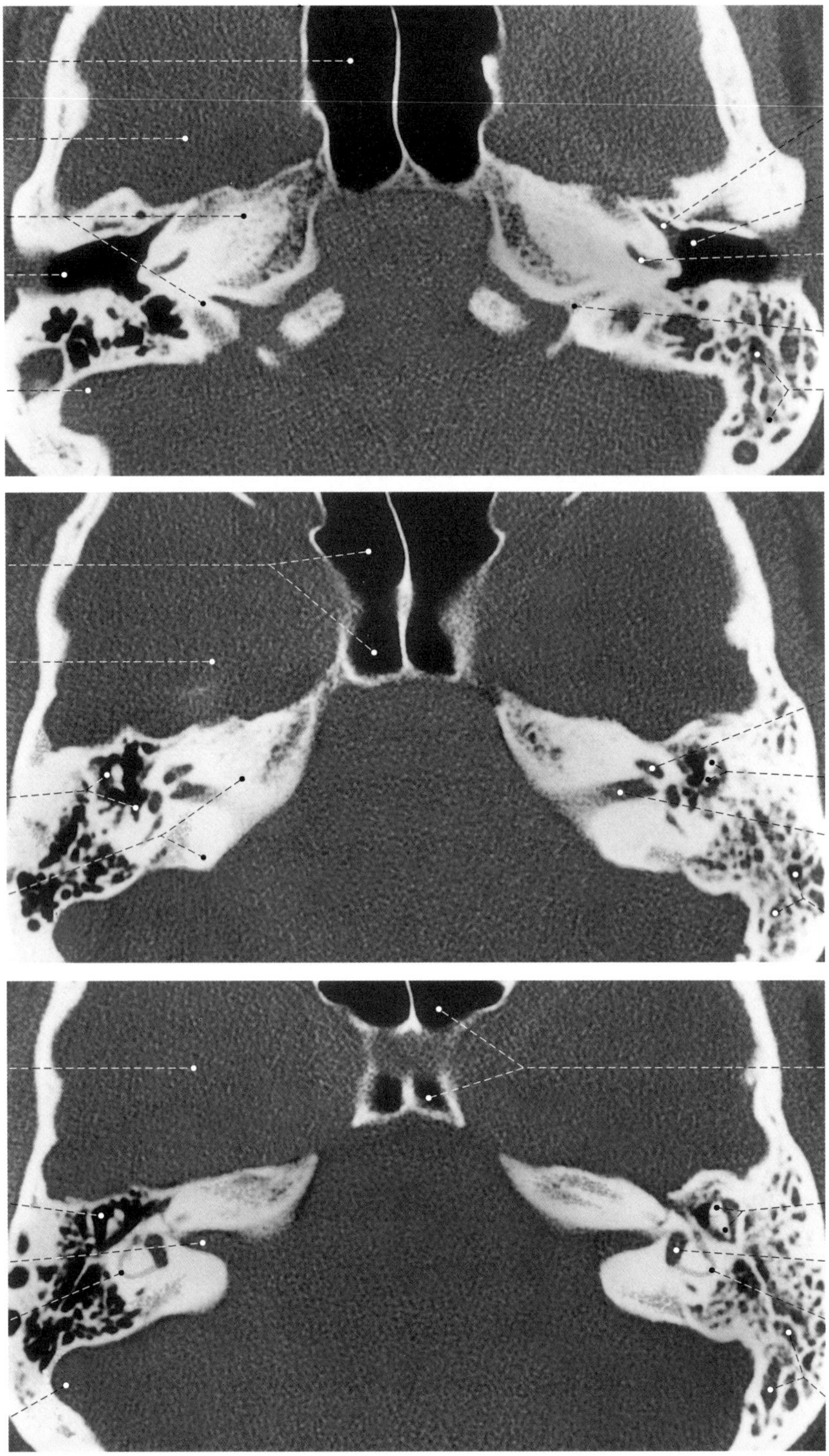

406 Os temporale

406 Petrous part of the temporal bone (85%)

Horizontal computed tomograms (CT) through caudal parts of the petrous bone, inferior aspect

- a Section through the external acoustic meatus, the auditory tube, and the base of the cochlea
- b Section (a little more cranial than Fig. a) through the tympanic cavity, the base of the cochlea, and the internal acoustic meatus
- c Section (a little more cranial than Fig. b) through the tympanic cavity, the lateral semicircular duct, the vestibule, and the internal acoustic meatus

406 Felsenbein des Schläfenbeins (85%)

Horizontale Computertomogramme (CT) durch kaudale Bereiche des Felsenbeins, Kaudalansicht

- a Schnitt durch den äusseren Gehörgang, die Ohrtrompete und die Schneckenbasis
- b Schnitt (etwas kranialer als Abb. a) durch die Paukenhöhle, die Schneckenbasis und den inneren Gehörgang
- c Schnitt (etwas kranialer als Abb. b) durch die Paukenhöhle, den seitlichen Bogengang, das Vestibulum und den inneren Gehörgang

a

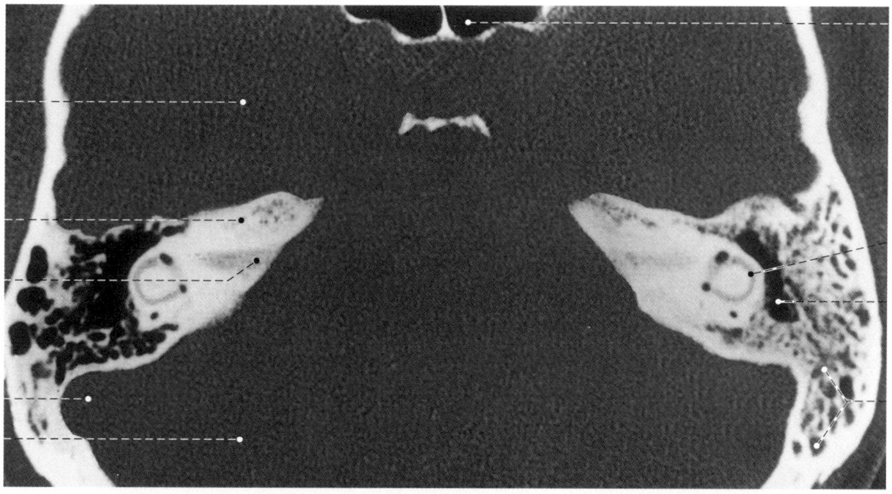

b

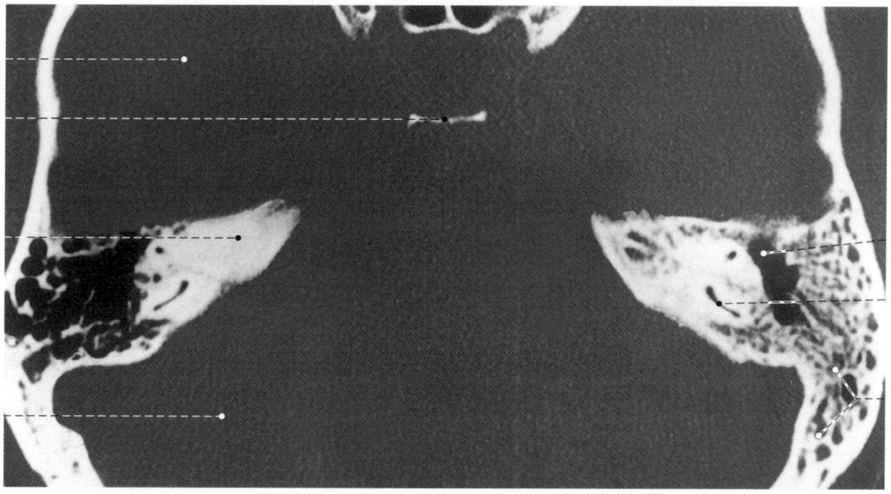

c

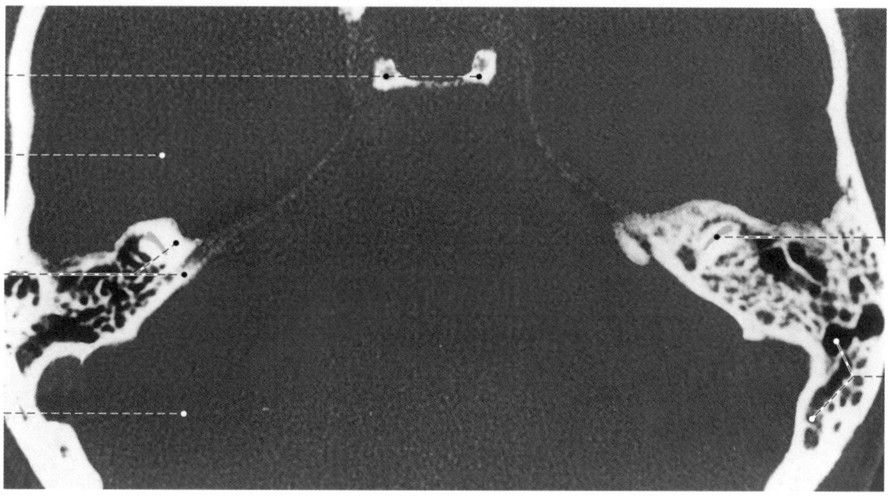

407 Os temporale

407 Petrous part of the temporal bone (85%)

Horizontal computed tomograms (CT) through cranial parts of the petrous bone, inferior aspect

a Section through the epitympanic recessus, the lateral semicircular duct and mastoid air cells
b Section (a little more cranial than Fig. a) through the epitympanic recessus, the posterior semicircular duct and mastoid air cells
c Section (a little more cranial than Fig. b) through the anterior semicircular duct and mastoid air cells

407 Felsenbein des Schläfenbeins (85%)

Horizontale Computertomogramme (CT) durch kraniale Bereiche des Felsenbeins, Kaudalansicht

a Schnitt durch den Recessus epitympanicus, den seitlichen Bogengang und die Mastoidzellen
b Schnitt (etwas kranialer als Abb. a) durch den Recessus epitympanicus, den hinteren Bogengang und die Mastoidzellen
c Schnitt (etwas kranialer als Abb. b) durch den vorderen Bogengang und die Mastoidzellen

a

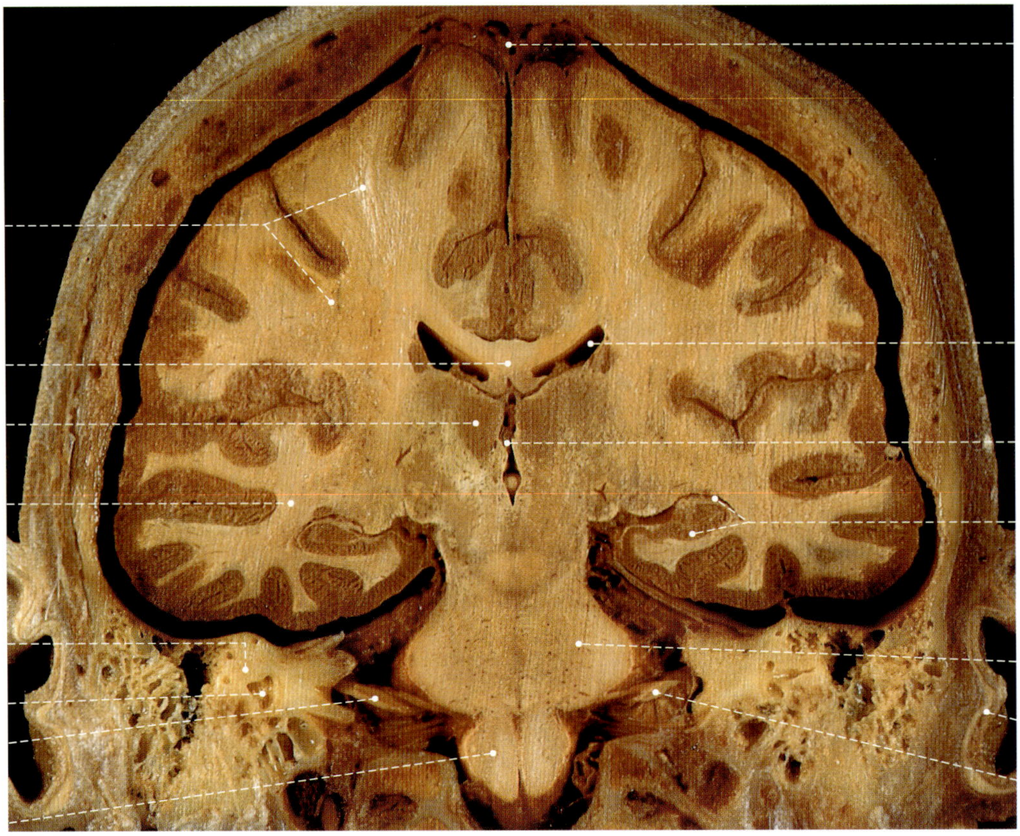

b

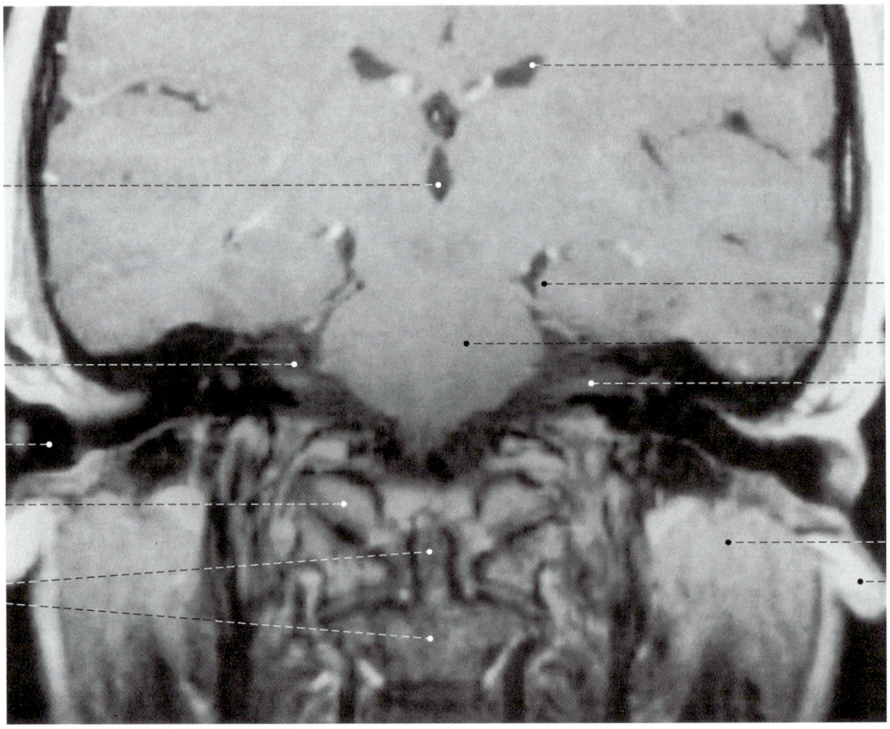

408 Os temporale et Nervus vestibulocochlearis

408 Temporal bone and vestibulocochlear nerve (100%)

Coronal sections through the parietal cerebral lobes, the petrous bones, and the pons

a Anatomical section, occipital aspect
b Computed tomogram (CT), frontal aspect

408 Schläfenbein und Nervus vestibulocochlearis (100%)

Frontalschnitte durch die Parietallappen des Grosshirns, die Felsenbeine und die Brücke

a Anatomischer Schnitt, Okzipitalansicht
b Computertomogramm (CT), Frontalansicht

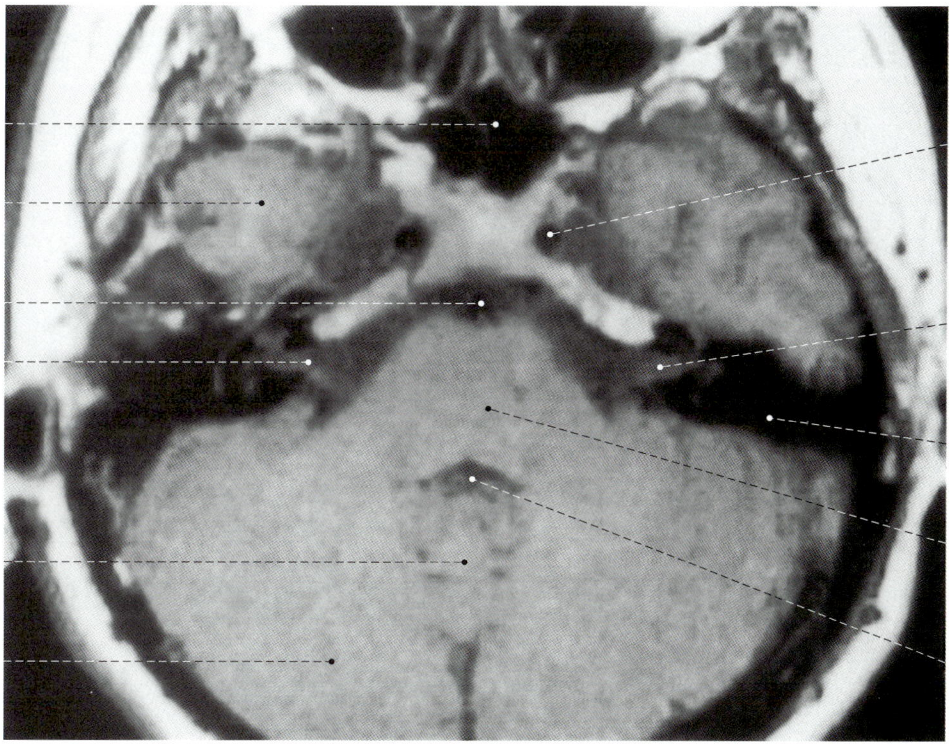

409 Os temporale et Nervus vestibulocochlearis

409 Temporal bone and vestibulocochlear nerve (100%)

Transverse sections through the temporal cerebral lobes, the petrous bones, the pons, and the cerebellum, inferior aspect

a Anatomical section
b Computed tomogram (CT)

409 Schläfenbein und Nervus vestibulocochlearis (100%)

Horizontalschnitte durch die Temporallappen des Grosshirns, die Felsenbeine, die Brücke und das Kleinhirn, Kaudalansicht

a Anatomischer Schnitt
b Computertomogramm (CT)

Index rerum

Subject Index

Sachwortverzeichnis

Volume numbering is given in bold print, followed by page numbers.
Brackets in the text of illustrations were omitted in the subject index.

Die Bandnummern sind fett gedruckt, gefolgt von den Seitenzahlen im jeweiligen Band.
Die in den Bildbeschriftungen stehenden Klammern sind im Sachwortverzeichnis weggelassen.

A

Abdomen **1**: 2, 75; **2**: 154–155, 162, 167, 172, 182, 190–193, 196–197, 199–202, 204–207, 218–225, 228–229
– arteries **2**: 218–219, 229
– lower **2**: 192–193, 202, 225
– – arteries **2**: 202
– lymph nodes **2**: 163
– – vessels **2**: 163
– Lymphgefässe **2**: 163
– Lymphknoten **2**: 163
– surface relief **1**: 62
– upper **2**: 182, 191, 199, 208, 222–223
– – arteries **2**: 196–198
– – – of retroperitoneal organs **2**: 203
– – autonomic nervous system **2**: 167
– – lymph nodes **2**: 198
– – – – of retroperitoneal organs **2**: 203
– – – vessels **2**: 198
– – – – of retroperitoneal organs **2**: 203
– – organs **2**: 154–155
– – retroperitoneal organs **2**: 199, 203
– – sympathetic nervous system **2**: 167
– – veins of retroperitoneal organs **2**: 203
– veins **2**: 218
Abduction **1**: 5
Abduktion **1**: 5
Abstieg, vorgeburtlicher, des Hodens **2**: 272
Acetabulum **1**: 172, 175–176, 191, 193, 249, 258; **2**: 205, 252, 254
– roof **1**: 177
Achselhöhle **1**: 72, 144
– Arterien **2**: 93, 149
– Lymphgefässe **1**: 72–74
– Lymphknoten **1**: 72–74
– Nerven **2**: 93, 149
– Venen **2**: 93, 149
Achsen, Bulbus **2**: 374
– Körper **1**: 5
Acromion **1**: 53, 63–64, 84–85, 87, 95–97, 105, 108, 136, 142–143, 149
Adduction **1**: 5
Adductor muscles, thigh **1**: 220, 221
Adduktion **1**: 5
Adduktoren, Oberschenkel **1**: 220–221
Adenohypophysis **2**: 293, 307, 328
Adergeflechte, Vorderhirn **2**: 326
Adhaesio interthalamica **2**: 307, 320, 324, 326–328, 348, 356
Aditus ad antrum mastoideum **2**: 396
– laryngis **2**: 46, 50–51, 101
– orbitalis **2**: 368
Adminiculum lineae albae **1**: 77
Aequator bulbi **2**: 374
– lentis **2**: 374–377
Agger nasi **2**: 55
Air cells, mastoid **2**: 393–394, 399, 407
Ala(-ae) cristae galli **2**: 22
– lobuli centralis cerebelli **2**: 308
– major ossis sphenoidalis **2**: 4, 6, 8, 10, 12, 17, 19, 31, 58, 60, 63, 367–368
– minor ossis sphenoidalis **2**: 5, 10, 12, 19, 58, 60, 367–368, 380
Ala(-ae)
– nasi **2**: 44
– ossis ilii **1**: 173–176, 191, 220–221, 225; **2**: 205, 225, 252, 254
– – sacri → Ala sacralis
– sacralis **1**: 34–35
– vomeris **2**: 23
Alveolus(-i) dentalis(-es) mandibulae **2**: 25
– – maxillae **2**: 24
– pulmonis **2**: 115
Alveus hippocampi **2**: 324, 326, 340, 348
Amboss **2**: 397
Amnion **2**: 243
Ampulla canaliculi lacrimalis inferioris **2**: 365
– – – superioris **2**: 365
– ductus deferentis **1**: 17, 77; **2**: 236–237
– duodeni **2**: 173–176, 194, 223
– membranacea anterior **2**: 405
– – lateralis **2**: 405
– – posterior **2**: 405
– ossea anterior **2**: 401–402
– – lateralis **2**: 401–402
– – posterior **2**: 401–402
– recti **1**: 17; **2**: 170, 179, 230–232, 243, 253, 255, 260–261
– tubae uterinae **1**: 17; **2**: 239–240
Anastomosis(-es) arteriovenosae penis **2**: 277
Anesthesia, pudendal **2**: 281
Angiography, coronary **2**: 146–147
Angle, cerebellopontine **2**: 340
– neck-shaft, of thigh **1**: 179
Angulus acromialis **1**: 84–85
– acromii → Angulus acromialis
– costae **1**: 42, 44
– frontalis ossis parietalis **2**: 22
– inferior scapulae **1**: 84–85, 108, 111
– infrasternalis **1**: 41
– iridocornealis **2**: 373–375, 384
– mandibulae **2**: 5, 7, 25, 31
– mastoideus ossis parietalis **2**: 22
– occipitalis ossis parietalis **2**: 22
– oculi lateralis **2**: 364
– – medialis **2**: 364
– oris **2**: 38, 44
– sphenoidalis ossis parietalis **2**: 22
– sterni **1**: 45; **2**: 161
– subpubicus **1**: 172, 175
– superior scapulae **1**: 84–85, 108
– venosus **1**: 15
Ankle bone **1**: 187
– joint **1**: 188, 206, 208–214
– synovial sheaths **1**: 241
Ansa cervicalis **1**: 112; **2**: 34–35, 87, 93–94, 97–100
– subclavia **2**: 96, 100, 167
Antebrachium **1**: 2, 105, 150–154, 156–159
Anterior **1**: 5
Anteversion **1**: 5
Antihelix **2**: 390
Antitragus **2**: 390
Antrum mastoideum **2**: 393–396, 400
– pyloricum **2**: 173–176
Anulus fibrocartilagineus membranae tympanicae **2**: 392, 398
– fibrosus dexter cordis **2**: 135
– – disci intervertebralis **1**: 39, 47–48
– – sinister cordis **2**: 133, 135
– inguinalis profundus **1**: 67, 77; **2**: 272–273
– – superficialis **1**: 66–67, 218; **2**: 273–274
– iridis major **2**: 376
– – minor **2**: 376
– lymphaticus cardiae **2**: 198
– tendineus communis **2**: 370–371, 378–379, 381–382
– tympanicus **2**: 17
– umbilicalis **1**: 63
Anus **1**: 17, 19, 261; **2**: 232, 251, 253, 255, 269, 278
Aorta **2**: 105–107, 133
– abdominalis → Pars abdominalis aortae
– ascendens → Pars ascendens aortae
– descendens → Pars descendens aortae
– thoracica → Pars thoracica aortae
Aortenbifurkation **2**: 229
Apertura externa aquaeductus vestibuli **2**: 20
– – canaliculi cochleae **2**: 20, 405
– – – tympanici **2**: 20
– – canalis carotici **2**: 8, 20
– interna canalis carotici **2**: 20, 399–400
– lateralis ventriculi quarti **2**: 308, 310–311, 321
– mediana ventriculi quarti **2**: 304, 307–308, 321
– pelvis superior **1**: 172
– sinus frontalis **2**: 21, 59, 62
– – sphenoidalis **2**: 11, 19, 55–56
– thoracis inferior **1**: 41
– – superior **1**: 41
Aperture(s), nasal, posterior **2**: 23
Apex capitis fibulae **1**: 182–183
– cartilaginis arytaenoideae **2**: 64, 69
– cordis **2**: 124–125, 132–133, 141–142, 159
– cornus posterioris medullae spinalis **2**: 288
– dentis axis **1**: 30
– linguae **2**: 39, 46
– nasi **2**: 44
– ossis sacri **1**: 34–35
– partis petrosae ossis temporalis **2**: 20, 352, 359, 399
– patellae **1**: 181
– prostatae **2**: 236–237
– pulmonalis → Apex pulmonis
– pulmonis **2**: 72, 153
– – dextri **2**: 111–112, 114, 126
– – sinistri **2**: 111–112, 126
– radicis dentis **2**: 40
– vesicae urinariae **2**: 235, 238, 245
Aponeurosis(-es) bicipitalis → Aponeurosis musculi bicipitis brachii
– dorsales digitorum manus **1**: 124–126, 130, 165, 167
– – – pedis **1**: 239
– epicranialis → Galea aponeurotica
– linguae → Aponeurosis lingualis
– lingualis **2**: 46–47
– musculi bicipitis brachii **1**: 106, 114, 136, 140–141, 151–153

Aponeurosis(-es) musculi
– – obliqui externi abdominis **1**: 64, 66; **2**: 274
– – trapezii **1**: 53
– palatina **2**: 395
– palmar **1**: 120
– palmaris **1**: 114–116, 120, 160, 166
– plantar **1**: 237
– plantaris **1**: 188, 210–211, 213, 237–238, 279
Apparatus alimentarius → Systema alimentarium
– lacrimal **2**: 365
– lacrimalis **2**: 365
– respiratorius → Systema respiratorium
– urogenital, female **2**: 233
– – male **2**: 232
– urogenitalis **2**: 232–233
Appendix(-ices) epididymidis **2**: 270
– epiploicae **2**: 177, 179
– fibrosa hepatis **2**: 154, 180–181, 191
– omentales → Appendices epiploicae
– testis **1**: 17; **2**: 270, 274
– vermiform **2**: 177–178, 195
– – different positions **2**: 178
– vermiformis **1**: 16, 18–19; **2**: 172, 177–178, 193, 195, 202–204, 238
– – verschiedene Positionen **2**: 178
– vesiculosa epoophori **2**: 239
Aquaeductus cerebri → Aquaeductus mesencephali
– cochleae **2**: 405
– mesencephali **2**: 304–305, 315, 320–321, 324, 328, 335, 340, 350–353, 357–358, 361
– vestibuli **2**: 401–402, 405
Aqueduct, cerebral **2**: 352–353, 357–358, 361
Arachnoidea mater cranialis **2**: 295, 321, 326, 405
– – encephali → Arachnoidea mater cranialis
– – spinalis **2**: 73, 285–286, 295, 321
Arbor bronchialis **2**: 114–115, 119–121
Arch, alveolar, mandible **2**: 25
– – maxilla **2**: 24
– lateral, foot **1**: 189
– longitudinal, foot **1**: 189
– medial, foot **1**: 189
– transverse, foot **1**: 189
Arcus alveolaris mandibulae **2**: 25
– – maxillae **2**: 24
– anterior atlantis **1**: 30, 36–37, 40, 51, 60; **2**: 37, 52, 294, 302
– aortae **1**: 12–14, 18–19, 43; **2**: 78–79, 97, 104–107, 110, 118, 122–131, 139, 142–144, 154–155, 161–162, 164–165, 172
– cartilaginis cricoideae **2**: 52, 64–66, 68, 70, 108
– costalis **1**: 40–42, 62–63; **2**: 107, 150
– ductus thoracici **1**: 15, 76; **2**: 149, 163
– iliopectineus **1**: 190, 219, 249
– inguinalis → Ligamentum inguinale
– palatoglossus **2**: 38–39, 50, 52
– palatopharyngeus **2**: 38–39, 45, 50, 52
– palmaris profundus **1**: 138, 162–163
– – superficialis **1**: 138, 161, 163
– plantaris profundus **1**: 250, 279–281
– posterior atlantis **1**: 30–31, 36–37, 49–51, 55, 58–60; **2**: 284, 294, 302, 321
– pubicus → Arcus pubis
– pubis **1**: 172
– superciliaris **2**: 21
– tendineus musculi levatoris ani **2**: 244, 250–251, 255, 264
– – – solei **1**: 268
– venosus dorsalis pedis **1**: 245, 270, 278
– – jugularis **2**: 33, 97
– vertebrae **1**: 29–30, 33, 60; **2**: 209, 284–286
– – axis **1**: 30
– zygomaticus **2**: 6, 8, 31, 48, 58, 368
Area(-ae) cochleae **2**: 401–403
– gastricae **2**: 174
– hypothalamica lateralis **2**: 328
– intercondylaris anterior tibiae **1**: 182
– – posterior tibiae **1**: 182
– nervi facialis meatus acustici interni **2**: 403
– nuda faciei diaphragmaticae hepatis **2**: 180–181, 206–207
– striata **2**: 322, 346
– subcallosa **2**: 304, 317, 357
– tegmentalis ventralis **2**: 328
– vestibularis fossae rhomboideae **2**: 310
– – inferior meatus acustici interni **2**: 401, 403
– – superior meatus acustici interni **2**: 401–403
Area(s), acoustic **2**: 319
– autonomous, nerves of lower limb **1**: 242–243
– – – – upper limb **1**: 132–133
– functional, first, of cerebral cortex **2**: 319
– – second, of cerebral cortex **2**: 319
– motor **2**: 319
– somatosensory **2**: 319
– speech, motor **2**: 319
– – sensory **2**: 319
– visual **2**: 319
Areola mammae **1**: 70–71
Arm, arteries **1**: 138–139
– cutaneous innervation **1**: 132–133
– Dermatome **1**: 132–134, 244
– dermatomes **1**: 132–134, 244
– Hautinnervation **1**: 132–133
– Hautnerven **1**: 132–133
– Oberflächenrelief **1**: 105
– segmental innervation **1**: 132–134
– segmentale Innervation **1**: 132–134
– surface relief **1**: 105
– upper **1**: 144–149
– – arteries **1**: 140–143
– – muscles **1**: 106–111
– – nerves **1**: 140–143
– – subcutaneous nerves **1**: 135–136
– – – veins **1**: 135–137
Armarterien **1**: 138–139
Armgeflecht **2**: 94
Arteria(-ae) abdominis **2**: 198, 203
– adrenalis inferior/media/superior → Arteria suprarenalis inferior/media/superior
– alveolaris inferior **2**: 76–77, 85–89
– – superior posterior **2**: 76, 86–87
– angularis **2**: 76, 85–87, 383
– appendicularis **2**: 202–203
– arcuata **1**: 250, 271, 278, 281
– auricularis posterior **2**: 76–77, 85–86, 91–92, 98–99
– – profunda **2**: 77, 88–89, 392
– axillaris **1**: 138–141, 143, 148; **2**: 79, 93–94, 149
– azygos vaginae → Rami vaginales arteriae uterinae
– basilaris **2**: 292, 295, 298–302, 352–353, 358–359, 361, 409
– brachialis **1**: 138–143, 145–147, 152–153, 156; **2**: 149
– buccalis **2**: 76, 86–87
– bulbi oculi **2**: 372–373
– – penis **2**: 267
– callosomarginalis **2**: 297, 299–300, 302, 347, 355
– canalis pterygoidei **2**: 88–89
– capitis **2**: 76–79
– carotis communis **1**: 12, 14; **2**: 33, 36, 68, 71, 74–75, 77–79, 93, 96–97, 99–100, 105–107, 118, 122–125, 130–131, 139, 142–143, 149, 156, 162–163, 165, 167, 292
– – externa **2**: 34, 73, 76–79, 81, 86–89, 96–97, 99–100, 292, 301
Arteria(-ae) carotis
– – interna **2**: 61–63, 72–73, 76–79, 86–89, 95–97, 100, 166, 291–295, 298–299, 301–302, 339, 351–353, 358–360, 378–380, 382, 385, 391, 395, 409
– caudae pancreatis **2**: 188
– centralis(-es) anterolaterales **2**: 298, 300, 338
– – longa **2**: 300
– – posterolaterales **2**: 300
– – retinae **2**: 58, 372–374, 378, 384
– cerebelli **2**: 298
– cerebrales **2**: 297–302
– cerebri anterior **2**: 292, 298–302, 347, 349, 351, 354–358
– – – circulation area **2**: 299
– – – Versorgungsgebiet **2**: 299
– – media **2**: 292, 298–302, 339, 351, 358, 362
– – – circulation area **2**: 299
– – – Versorgungsgebiet **2**: 299
– – posterior **2**: 292, 298–302, 351, 358
– – – circulation area **2**: 299
– – – Versorgungsgebiet **2**: 299
– cervicalis ascendens **2**: 79, 93, 100, 118, 149
– – profunda **1**: 60; **2**: 78–79, 118
– – superficialis **2**: 91–92, 98, 100, 149
– choroidea anterior **2**: 300
– ciliares anteriores **2**: 371–373
– – posteriores **2**: 382
– – – breves **2**: 371–373, 381
– – – longae **2**: 372, 373
– circumflexa anterior humeri **1**: 138–141, 143
– – femoris lateralis **1**: 250–253; **2**: 228
– – – medialis **1**: 250–253, 256
– – humeri anterior/posterior → Arteria circumflexa anterior/posterior humeri
– – iliaca profunda **1**: 251–253; **2**: 228, 234, 244–245
– – – superficialis **1**: 75, 248, 251, 253; **2**: 228
– – posterior humeri **1**: 138–140, 143
– – scapulae **1**: 135, 139, 143
– colica dextra **2**: 200–203
– – media **2**: 200–203
– – sinistra **2**: 202, 219
– collateralis media **1**: 138, 143
– – radialis **1**: 138–139, 142–143, 152–154
– – ulnaris inferior **1**: 138, 140–141, 152–153
– – – superior **1**: 138–143, 152–153
– colli **2**: 78–79
– comitans nervi ischiadici **1**: 256
– – – mediani **1**: 153
– communicans anterior **2**: 292, 298–299, 301, 351
– – posterior **2**: 292, 298–301, 351, 358
– conjunctivales anteriores **2**: 372–373
– coronaria(-ae) **2**: 142–143
– – dextra **2**: 104, 124, 129, 132, 134–135, 140–144, 146, 148, 162, 167
– – sinistra **2**: 107, 128–130, 133–135, 141–142, 144, 147–148, 162
– cystica **2**: 196–197, 200
– descendens genicularis **1**: 248, 250, 252, 269
– – genus → Arteria descendens genicularis
– digitales dorsales manus **1**: 165, 167
– – – pedis **1**: 271, 278, 281
– – palmares communes **1**: 160–163, 165
– – – propriae **1**: 160–163, 165, 167
– – plantares communes **1**: 279–280
– – – propriae **1**: 279–280
– dorsalis clitoridis **2**: 280
– – nasi **2**: 76
– – pedis **1**: 250, 271, 278, 280–281
– – penis **2**: 234, 267, 275–277, 279
– ductus deferentis **2**: 234, 244, 274
– encephali → Arteriae cerebrales

Arteria(-ae)
– epigastrica inferior **1**: 17, 67, 75, 77, 252–253; **2**: 196, 200, 228, 234, 244–246, 272–273
– – superficialis **1**: 75, 248, 252–253; **2**: 228
– – superior **1**: 75–76
– episcleralis **2**: 372–373
– ethmoidalis anterior **2**: 77, 88, 90, 380–382
– – posterior **2**: 77, 88, 90, 380–382
– facialis **2**: 44, 76–77, 79, 81, 85–87, 92, 98–100, 383
– femoralis **1**: 14, 64–65, 75, 219, 248–253, 259–260, 262–263, 269; **2**: 163, 228, 256–259
– fibularis **1**: 250, 266–268, 271, 273–274, 278
– frontobasalis lateralis **2**: 298
– – medialis **2**: 299
– gastrica(-ae) breves **2**: 198
– – dextra **2**: 196, 198
– – posterior **2**: 188, 199, 219
– – sinistra **2**: 167, 196–200, 203–204, 219
– gastroduodenalis **2**: 196–198, 200, 203
– gastroepiploica dextra/sinistra → Arteria gastroomentalis dextra/sinistra
– gastroomentalis dextra **2**: 196–198, 200, 202–203
– – sinistra **2**: 196–198
– glutaea inferior/superior→ Arteria glutaealis inferior/superior
– glutaealis inferior **1**: 256–257; **2**: 230, 234, 244–245
– – superior **1**: 256–257; **2**: 230, 234, 244–245
– gyri angularis **2**: 297, 302
– helicinae penis **2**: 277
– hepatica communis **2**: 167, 196–200, 219
– – propria **2**: 181, 183–185, 196–199, 203
– hypophysialis inferior **2**: 301
– – superior **2**: 301
– ileales **2**: 188, 200–204, 219
– ileocolica **2**: 200–203, 219
– iliaca communis **1**: 12–14, 194; **2**: 162–163, 202, 218–219, 225, 228–230, 234, 238, 243–245, 252, 254, 260–261
– – externa **1**: 12–14, 75, 77, 250–253; **2**: 162–163, 220–221, 228–230, 233–234, 238, 244–245, 250, 252, 256, 258–259
– – interna **1**: 12–14, 251; **2**: 162–163, 220–221, 228–231, 234, 244–247, 253, 255
– iliolumbalis **2**: 228, 230, 234, 245
– inferior anterior cerebelli **2**: 292, 298–299, 302
– – lateralis genus **1**: 250, 267, 269, 271
– – medialis genus **1**: 250, 267–269, 271
– – posterior cerebelli **2**: 292, 298–299, 302
– infraorbitalis **2**: 58, 76, 85–87, 379, 383
– insularis(-es) **2**: 300, 339
– intercapitales manus **1**: 161–162, 165
– intercostalis(-es) posterior(-iores) **1**: 14, 76; **2**: 96, 106, 162–163, 167, 285–286
– – suprema **2**: 79, 118
– interossea anterior **1**: 138, 153–155, 157
– – communis **1**: 138–139, 153, 155
– – posterior **1**: 138, 153–155, 157, 165
– – recurrens **1**: 138, 154–155
– jejunales **2**: 188, 200–204, 219
– labialis inferior **2**: 76, 85–86
– – superior **2**: 76, 85–87
– labyrinthi **2**: 77, 289, 292, 298–299
– lacrimalis **2**: 77, 378–382
– laryngea superior **2**: 97–99, 101
– lienalis → Arteria splenica
– ligamenti teretis uteri **2**: 245
– lingualis **2**: 76–77, 79, 86–87, 96–100
– lumbalis(-es) **1**: 14; **2**: 162, 228–229, 234
– malleolaris anterior lateralis **1**: 271, 278
– – – medialis **1**: 271, 278
– masseterica **2**: 76
– maxillaris **2**: 76–77, 86–89, 99

Arteria(-ae)
– media genus **1**: 267
– membri inferioris **1**: 250–251, 269
– – superioris **1**: 138–139
– meningea(-ae) **2**: 290
– – anterior **2**: 77
– – media **2**: 76–77, 86–89, 290–291, 294, 380, 395
– – posterior **2**: 77, 100
– mesencephalicae **2**: 289, 298–299
– mesenterica inferior **1**: 12, 14; **2**: 162, 199, 202, 211, 218–219, 230–231, 234
– – superior **1**: 12, 14; **2**: 105, 162, 167, 188, 198–203, 207, 211, 213–214, 218–219, 223–224
– metacarpales dorsales **1**: 165
– – palmares **1**: 162–163
– metatarsales dorsales **1**: 271, 278, 281
– – plantares **1**: 279–280
– nasalis externa → Arteria dorsalis nasi
– – posterior lateralis **2**: 77, 88–89
– – – septi **2**: 77
– nutriens fibulae **1**: 268
– – humeri **1**: 143
– – tibiae **1**: 268
– obturatoria **1**: 193, 252–253; **2**: 230, 234, 244–245, 252, 254, 256
– occipitalis **1**: 60; **2**: 76–77, 85–86, 91–93, 98–100
– – lateralis **2**: 298
– – medialis **2**: 298–299
– ophthalmica **2**: 58, 76–77, 86, 88–89, 291–293, 301, 378–383, 385
– ovarica **1**: 14; **2**: 162–163, 215, 238–239, 241, 245
– palatina(-ae) ascendens **2**: 76–77, 99
– – descendens **2**: 76–77, 87–89
– – major **2**: 88–89
– – minores **2**: 89
– pancreatica dorsalis **2**: 188
– – magna **2**: 188
– pancreaticoduodenalis inferior **2**: 188, 198, 200, 219
– – superior **2**: 197, 219
– – – anterior **2**: 198, 200, 203
– – – posterior **2**: 188, 198, 203
– paracentralis **2**: 297, 299
– parietalis posterior **2**: 297
– parietooccipitalis **2**: 297
– pedis **1**: 281
– pelvis **2**: 234, 244–245
– perforantes **1**: 250–251, 253, 256, 262–263
– pericallosa **2**: 302
– perinealis **2**: 267, 276, 279–280
– pharyngea ascendens **2**: 76–77, 79, 99–100
– phrenica inferior **1**: 14; **2**: 162, 196, 200, 203
– plantaris lateralis **1**: 214, 250, 279–281
– – medialis **1**: 214, 250, 279–281
– – profunda **1**: 250, 271, 278, 280
– polaris inferior renis **2**: 215
– – superior renis **2**: 162
– pontis **2**: 292, 298–299
– poplitea **1**: 245, 250, 255–256, 261, 264–265, 267–269, 272
– praecunealis **2**: 297, 299
– princeps pollicis **1**: 162–163, 166
– profunda brachii **1**: 138–143, 145–146
– – femoris **1**: 250–253, 260, 262; **2**: 228
– – linguae **2**: 47
– – penis **2**: 267, 277
– pudenda(-ae) externae **1**: 248, 252–253; **2**: 228
– – interna **1**: 256–257; **2**: 230–231, 234, 244–245, 251, 255–256, 267, 276, 279–280
– pulmonalis **2**: 96, 115

Arteria(-ae) pulmonalis
– – dextra **1**: 13; **2**: 104, 110–111, 113, 119, 123–125, 131, 139, 142–143, 157, 161, 165
– – sinistra **1**: 13; **2**: 107, 111, 119, 122–125, 131, 139, 142–143, 155, 157, 165
– radialis **1**: 123, 138–141, 152–153, 155, 157–163, 165
– – indicis **1**: 162–163
– rectalis inferior **2**: 230–231, 276, 279–280
– – media **2**: 230–231, 234, 244–247, 276
– – superior **2**: 202, 229–231, 234
– recurrens radialis **1**: 138–139, 141, 152–153
– – tibialis anterior **1**: 250, 271
– – ulnaris **1**: 138–139, 143, 153
– renalis **1**: 12, 14; **2**: 162, 208, 210–212
– – dextra **2**: 203, 213–215, 218–219
– – sinistra **2**: 167, 199, 203, 213–215, 219
– sacralis(-es) lateralis(-es) **2**: 230, 234, 244–245
– – mediana **1**: 14; **2**: 162, 202, 218, 228–230, 234, 244, 255
– segmenti anterioris inferioris renalis **2**: 212, 214
– – – superioris renalis **2**: 212, 214
– – inferioris renalis **2**: 212, 214
– segmentorum renalium **2**: 215
– sigmoidea(-ae) **2**: 202, 219, 229–230
– sphenopalatina **2**: 76, 88–90
– spinalis anterior **2**: 285–286, 295, 298
– – anterolateralis **2**: 298
– – posterior **2**: 285–286, 289, 299
– spirales uteri **2**: 243
– splenica **2**: 167, 188–189, 196–200, 203, 207, 219, 221
– sternocleidomastoidea **2**: 97, 99
– stylomastoidea **2**: 76, 100
– subclavia **1**: 12, 14, 60, 64, 72, 75, 138–139; **2**: 35, 72, 78–79, 81, 92–93, 96–97, 100, 105–107, 111–112, 118, 122–125, 130, 139, 142–143, 156, 162–163, 165, 167
– sublingualis **2**: 47, 97–99
– submentalis **2**: 44, 76–77, 98–99
– subscapularis **1**: 140–141
– sulci centralis **2**: 297–298
– – postcentralis **2**: 297
– – praecentralis **2**: 297–298
– superior cerebelli **2**: 292, 298–299, 302
– – lateralis genus **1**: 250, 255–256, 267, 269, 271
– – medialis genus **1**: 250, 256, 267–269
– supraorbitalis **2**: 76–77, 85–87, 380–381, 383
– suprarenalis inferior **2**: 162, 215
– – media **1**: 14; **2**: 162
– – superior **2**: 162
– suprascapularis **1**: 140; **2**: 78–79, 92–93, 98, 100, 149
– supratrochlearis **2**: 76, 85–87, 381, 383
– suralis(-es) **1**: 250, 255–256, 266–267, 269
– tarsalis lateralis **1**: 250, 271, 278, 281
– – medialis **1**: 271, 278
– temporalis(-es) anterior **2**: 302
– – media **2**: 76, 302
– – posterior **2**: 302
– – profunda(-ae) **2**: 76, 87
– – superficialis **2**: 76–77, 81, 85–89, 98–99
– testicularis **1**: 67, 77; **2**: 162, 211, 215, 218, 228, 230, 270–272, 274
– thalamostriatae anterolaterales → Arteriae centrales anterolaterales
– thoracica interna **1**: 75–76, 139; **2**: 78–79, 97, 118, 122, 149, 154, 162
– – lateralis **1**: 72, 139; **2**: 93, 149
– thoracis **1**: 75–76
– thoracoacromialis **2**: 91, 93, 98, 149
– thoracodorsalis **1**: 72; **2**: 93, 149
– thyroidea inferior **2**: 79, 97, 100–101, 118, 149

Arteria(-ae) thyroidea
– – superior **2**: 74, 76–77, 79, 93, 97–101, 149
– tibialis anterior **1**: 250, 268–269, 271, 273–275, 278, 281
– – posterior **1**: 250, 266–268, 273–275, 277, 279–281
– transversa cervicis **1**: 60; **2**: 81, 92–93, 118, 149, 162
– – colli → Arteria transversa cervicis
– – faciei **2**: 76, 85–86, 98
– trunci corporis **1**: 14
– tympanica anterior **2**: 77, 87–89
– – superior **2**: 380
– ulnaris **1**: 123, 138–139, 152–153, 155, 157–163, 165
– umbilicalis(-es) **1**: 12–13; **2**: 230, 234, 243–245
– urethralis **2**: 267
– uterina **2**: 241, 245–247, 251
– vaginae nervi optici **2**: 373
– vaginalis **2**: 245
– vertebralis **1**: 14, 51, 60; **2**: 33, 68, 72–75, 77–79, 96–97, 100, 118, 162, 286, 289, 291–292, 294–295, 298–299, 301–302
– vesicalis inferior **2**: 230, 234, 244–247, 276
– – superior **2**: 234, 244–245
– zygomaticoorbitalis **2**: 76, 85, 98
Arterien, Achselhöhle **2**: 93, 149
– Augapfel **2**: 372–373
– Bauchwand **1**: 75
– Bein **1**: 251
– Brustkorb **1**: 76
– Brustwand **1**: 75; **2**: 93
– Dickdarm **2**: 202
– dorsale Rumpfwand **2**: 162
– Eierstock **2**: 241
– Eileiter **2**: 241
– Ellbogenregion **1**: 142–143
– Ellenbeuge **1**: 140–141
– Finger **1**: 165
– Fuss **1**: 271, 281
– Fussohle **1**: 279–280
– Fussrücken **1**: 278
– Gebärmutter **2**: 241
– Gehirn **2**: 298–302
– Gesäss **1**: 255–256
– Gesicht **2**: 86–90
– Hals **2**: 78–79, 93, 97–99, 118, 149, 165
– Hand **1**: 163
– Handrücken **1**: 165
– Herz **2**: 142–148
– – Variationen **2**: 144–145
– hintere Bauchwand **2**: 228
– Hirnhäute **2**: 290
– Hohlhand **1**: 160–162
– Hypophyse **2**: 301
– Kehlkopf **2**: 101
– kleines Becken **2**: 234, 244–245
– Knie **1**: 252
– Kniekehle **1**: 255–256, 267–268
– Kopf **2**: 76–79, 86–90
– Leber **2**: 183
– Lunge **2**: 119
– männliches Becken **2**: 234
– Mastdarm **2**: 230–231
– Mediastinum **2**: 149, 165
– Mesenterien **2**: 203
– Niere **2**: 214–215
– – obere Polarterie **2**: 214
– – untere Polarterie **2**: 214–215
– – Variationen **2**: 214
– Oberarm **1**: 140–143
– Oberbauch **2**: 196–198
– obere Gliedmasse **1**: 138–139
– oberflächliche, Gehirn **2**: 297
– – Gesicht **2**: 85
Arterien, oberflächliche
– – Hals **2**: 92
– – Kopf **2**: 85
– Oberschenkel **1**: 251–253, 255–256; **2**: 228
– Penis **2**: 277
– Rachenwand **2**: 100
– retroperitoneale Organe des Oberbauchs **2**: 203
– Rückenmark **2**: 286
– Schilddrüse **2**: 101
– Schulter **1**: 140, 142–143
– Spatium peripharyngeum **2**: 100
– Thorax **2**: 97, 118
– Unterarm **1**: 152–155
– Unterbauch **2**: 202
– untere Gliedmasse **1**: 250–251, 269
– Unterschenkel **1**: 267–268, 271
– vordere Bauchwand **1**: 77
– weibliches Becken **2**: 245
– Wirbelsäule **2**: 286
– Zunge **2**: 97
Arteriola macularis inferior **2**: 377
– – superior **2**: 377
– medialis retinae **2**: 377
– nasalis retinae inferior **2**: 377
– – – superior **2**: 377
– temporalis retinae inferior **2**: 377
– – – superior **2**: 377
Artery(-ies), acetabular **1**: 193
– anterior abdominal wall **1**: 77
– – elbow region **1**: 140–141
– arm **1**: 138–139
– axilla **2**: 93, 149
– back of foot **1**: 278
– – – hand **1**: 165
– brain **2**: 298–302
– carotid, internal **2**: 292, 302
– coronary, heart **2**: 142–148
– – – origin **2**: 135
– – – variations **2**: 144–145
– dorsal body wall **2**: 162
– eyeball **2**: 372–373
– face **2**: 86–90
– female pelvis **2**: 245
– fingers **1**: 165
– foot **1**: 271, 281
– forearm **1**: 152–155
– gluteal region **1**: 255–256
– hand **1**: 163
– head **2**: 78–79, 86–90
– heart **2**: 135, 142–143, 147–148
– helicine, penis **2**: 277
– hepatic **2**: 183
– hypophysial **2**: 301
– iliac **2**: 229
– knee **1**: 252
– large intestine **2**: 202
– larynx **2**: 101
– leg **1**: 267–268, 271
– lesser pelvis **2**: 234, 244–245
– liver **2**: 183
– lower abdomen **2**: 202
– – limb **1**: 250–251, 269
– male pelvis **2**: 234
– mediastinum **2**: 149, 165
– meningeal **2**: 290
– mesenterial, superior **2**: 200–201
– mesenteries **2**: 203
– neck **2**: 78–79, 93, 97–99, 118, 149, 165
– ophthalmic **2**: 378
– ovary **2**: 241
– palm of hand **1**: 160–162
– penis **2**: 277
– peripharyngeal space **2**: 100
– popliteal fossa **1**: 255–256, 267–268
Artery(-ies)
– posterior abdominal wall **2**: 228
– – elbow region **1**: 142–143
– pulmonary **2**: 119
– rectum **2**: 230–231
– renal **2**: 215
– – multiple **2**: 214
– – polar, lower **2**: 214–215
– – – upper **2**: 214
– – variations **2**: 214
– retroperitoneal organs of upper abdomen **2**: 203
– shoulder **1**: 140, 142–143
– skull **2**: 76–77
– sole of foot **1**: 279–280
– spinal cord **2**: 286
– superficial, brain **2**: 297
– – face **2**: 85
– – head **2**: 85
– – neck **2**: 92
– thigh **1**: 251–253, 255–256; **2**: 228
– thorax **1**: 76; **2**: 93, 97, 118
– thyroid gland **2**: 101
– tongue **2**: 97
– upper abdomen **2**: 196–198
– – arm **1**: 140–143
– – limb **1**: 138–139
– uterine tubes **2**: 241
– uterus **2**: 241
– ventral body wall **1**: 75
– vertebral **1**: 60; **2**: 292, 302
– – column **2**: 286
Articulatio(-ones) **1**: 9
– acromioclavicularis **1**: 82–83, 95–96, 111
– atlantoaxialis(-es) **1**: 49–51
– – lateralis **1**: 36, 49–51; **2**: 72
– – mediana **1**: 50–51; **2**: 294
– atlantooccipitalis **1**: 49–50; **2**: 72
– calcaneocuboidea **1**: 215
– capitis costae **1**: 47
– carpometacarpales **1**: 82–83, 102, 104
– costotransversaria **1**: 47; **2**: 75
– costovertebrales **1**: 46–47
– coxae **1**: 170–171, 173, 177, 192–194, 220, 225; **2**: 220–221, 256
– cricoarytaenoidea **2**: 65
– cricothyroidea **2**: 65–68
– cubiti **1**: 82–83, 90–91, 98, 100–101, 116, 147, 156
– cuboideonavicularis **1**: 215
– cuneonavicularis **1**: 215
– genus **1**: 170–171, 195–204
– glenohumeralis → Articulatio humeri
– humeri **1**: 82–83, 95–97, 109–110
– humeroradialis **1**: 82–83, 90–91, 100
– humeroulnaris **1**: 82–83, 90–91, 100–101
– iliofemoralis → Articulatio coxae
– incudomallearis **2**: 397
– incudostapedialis **2**: 397
– interphalangeae manus/pedis → Articulationes interphalangeales manus/pedis
– interphalangeales manus **1**: 82–83, 104, 124, 165
– – pedis **1**: 206, 213
– manus **1**: 102, 104
– mediocarpalis **1**: 104
– metacarpophalangeae → Articulationes metacarpophalangeales
– metacarpophalangealis(-es) **1**: 82–83, 102, 104, 124
– metatarsophalangeae → Articulationes metatarsophalangeales
– metatarsophalangealis(-es) **1**: 206
– ossis pisiformis **1**: 104
– pedis **1**: 170–171, 206, 208–213, 215

Articulatio(-ones)
– radiocarpalis **1**: 82–83, 104, 159
– radioulnaris distalis **1**: 82–83, 99, 102, 104
– – proximalis **1**: 82–83, 90–91, 99–101
– sacroiliaca **1**: 38, 170–175, 178, 194; **2**: 194, 220, 225, 253, 255
– sternoclavicularis **1**: 45, 62, 82
– sternocostales **1**: 45
– subtalaris **1**: 208–211, 214–215
– synovialis **1**: 10
– talocalcanea → Articulatio subtalaris
– talocalcaneonavicularis **1**: 207–211, 214–215
– talocruralis **1**: 170–171, 188, 206, 208–211, 214, 234
– talonavicularis **1**: 188
– tarsometatarsales **1**: 207, 215
– temporomandibularis **2**: 11, 26–27, 31
– tibiofibularis(-es) **1**: 170, 205, 272
– zygapophysialis(-es) **1**: 36, 48, 51, 55; **2**: 75, 289
Association tracts, brain **2**: 333
Assoziationsbahnen, Gehirn **2**: 333
Athletiker **1**: 6
Atlantoaxialgelenke **1**: 49–51
Atlantookzipitalgelenk **1**: 49–50
Atlas [C I] **1**: 28, 30–31, 36–37, 40, 49–51, 55, 58–60; **2**: 7, 52, 72–73, 79, 284, 289, 294, 302, 321, 408
– cruciform ligament **1**: 50
– Kreuzband **1**: 50
Atmungssystem **1**: 16; **2**: 52–53
Atrium dextrum cordis **1**: 12–13; **2**: 104, 122, 124–126, 128–132, 137, 139–143, 154, 158–161, 167
– left, of heart **2**: 133
– meatus nasi medii **2**: 55
– right, of heart **2**: 132, 140
– sinistrum cordis **1**: 12–13; **2**: 104, 107, 122, 125, 127, 131, 133, 136–137, 139–141, 143, 155, 158, 161
Augapfel **2**: 62, 362, 370–371, 374–375, 384–386
– autonome Innervation **2**: 382
– parasympathische Innervation **2**: 382
– sympathische Innervation **2**: 382
– vegetative Innervation **2**: 382
Augapfelarterien **2**: 372–373
Augapfelvenen **2**: 372–373
Auge **2**: 364–366
Augenhintergrund **2**: 377
– Spiegelbild **2**: 377
Augenhöhle **2**: 58, 63, 336, 360, 362, 367–371, 378–387
Augenhöhlenvenen **2**: 383
Augenlider **2**: 364–366
Augenmuskeln, äussere **2**: 370–371, 379
– – gemeinsamer Sehnenring **2**: 379
Augenregion **2**: 366, 383
Auricle **2**: 390
– cartilage **2**: 390
Auricula **2**: 390, 408
– dextra cordis **2**: 124, 128–129, 132, 139–142, 158, 166
– sinistra cordis **2**: 122, 124–126, 129–130, 132–133, 139–143, 154, 158, 166
Auris externa **2**: 391, 394
– interna **2**: 391, 403
– media **2**: 391, 394
Ausatmung **1**: 40–41
Ausscheidungsurogramm **2**: 217
Aussenrotation **1**: 5
Autonomgebiete, Nerven des Arms **1**: 132–133
– – – Beins **1**: 242–243
Axes, longitudinal **1**: 5
– of bulb **2**: 374
– sagittal **1**: 5
– transversal **1**: 5
Axes
– vertical **1**: 5
Axilla **1**: 72–73, 144
– arteries **2**: 93
– lymph nodes **1**: 72–74
– – vessels **1**: 72–74
– Lymphgefässe **1**: 72–74
– Lymphknoten **1**: 72–74
– nerves **2**: 93
– veins **2**: 93
Axis [C II] **1**: 30, 36–37, 40, 49–51, 53; **2**: 55, 72, 294, 408
– bulbi externus **2**: 374
– – internus **2**: 374
– externus/internus bulbi oculi → Axis bulbi externus/internus
– opticus **2**: 374
– pelvis **1**: 178

B

Back, genuine muscles **1**: 56–59
– muscles **1**: 52–60; **2**: 37
– of foot, arteries **1**: 278
– – – fascia **1**: 230
– – – muscles **1**: 231–232, 236
– – – nerves **1**: 278
– – – subcutaneous nerves **1**: 278
– – – – veins **1**: 278
– – – veins **1**: 278
– – hand **1**: 165
– – – arteries **1**: 165
– – – muscles **1**: 125
– – – nerves **1**: 165
– – – subcutaneous nerves **1**: 164
– – – – veins **1**: 164
– – – veins **1**: 164
– surface relief **1**: 52
Balken, Gehirn **2**: 345
Bänder, Beckengürtel **1**: 190–191
– Fuss **1**: 206, 208, 212–213
– Kehlkopf **2**: 66
– Kiefergelenk **2**: 26–27
Bandhaft **1**: 9
Bandscheibe **1**: 48
Bandscheibenvorfall **1**: 48
Base, cranial → Base of skull
– fibrous, heart **2**: 135
– of skull **2**: 8–11, 16, 395, 400
– – – external surface **2**: 395
– – – internal surface **2**: 293, 400
– – – muscle attachments **2**: 37, 395
– – urinary bladder **2**: 235
Basis(-es) cartilaginis arytaenoideae **2**: 64
– cochleae **2**: 401, 403, 406
– cornus posterioris medullae spinalis **2**: 288
– cranii **2**: 16
– – externa **2**: 8–9, 395
– – interna **2**: 10–11, 400
– mandibulae **2**: 25, 41, 99
– metacarpalis(-es) **1**: 94, 103, 123, 129
– metatarsalis(-es) **1**: 186, 209–210
– modioli cochleae **2**: 403
– nasi **2**: 44
– ossis metacarpi/metatarsi → Basis metacarpalis/metatarsalis
– – sacri **1**: 34–35, 172, 174; **2**: 264
– patellae **1**: 181
– phalangis manus **1**: 94
– – pedis **1**: 186
– prostatae **2**: 235–237
– pulmonalis → Basis pulmonis
– pulmonis dextri **2**: 111–112
– – sinistri **2**: 111–112
– pyramidis renalis **2**: 212
Basis(-es)
– stapedis **2**: 397
Bauch **2**: 182, 220–225
– Lymphgefässe **2**: 163
– Lymphknoten **2**: 163
– Oberflächenrelief **1**: 62
Baucharterien **2**: 218–219, 229
Bauchfellhöhle **2**: 170–171
– fetale **2**: 170
Bauchmuskeln **1**: 62–66, 228
Bauchorgane **1**: 18–21; **2**: 170–172, 190–193
Bauchspeicheldrüse **2**: 188, 198, 206
Bauchvenen **2**: 218
Bauchwand, Arterien **1**: 75
– hintere **2**: 206–207, 228
– – Arterien **2**: 228
– – Nerven **2**: 228
– – Venen **2**: 228
– Muskeln **1**: 62–66
– Venen **1**: 75
– vordere **1**: 77, 79; **2**: 238, 272–273
– – Arterien **1**: 77
– – Venen **1**: 77
Becken, kleines **2**: 235, 246–247, 276
– – Arterien **2**: 234, 244–245
– – männliches **2**: 232, 234, 236–237, 244, 246–247, 250–251, 254–255, 258–259, 261
– – – Bindegewebe **2**: 246–247
– – – Organe **2**: 246–247
– – Venen **2**: 244
– – weibliches **2**: 233, 238–247, 250–253, 256–257, 260
– – – Arterien **2**: 245
– – – Bindegewebe **2**: 246–247
– – – Nerven **2**: 245
– – – Organe **2**: 245–247
– knöchernes **1**: 172–175, 178
– – kindliches **1**: 177
– – männliches **1**: 172, 174
– – weibliches **1**: 172, 174–175, 178
– Lymphgefässe **2**: 163
– Lymphknoten **2**: 163
– männliches **2**: 232, 234, 244
– Muskelanheftungen **1**: 228
– weibliches **2**: 233, 243
Beckenarterien **2**: 229, 234, 244–245
Beckenausgang **1**: 178
Beckenboden **2**: 264–266
Beckenbodenmuskeln **2**: 264–266
Beckeneingang **1**: 178
Beckenenge **1**: 178
Beckengürtel, Bänder **1**: 190–191
– Gelenke **1**: 190–191
Beckenmasse **1**: 178
Beckenneigung **1**: 178
Beckennerven **2**: 234
Beckenorgane **2**: 232–233, 238
– männliche **2**: 232, 234, 236–237, 244, 246–247
– weibliche **2**: 233, 238–243, 245–247
Beckenvenen **2**: 234, 244–245
Beckenweite **1**: 178
Bein, Dermatome **1**: 134, 242–244
– Hautinnervation **1**: 242–243
– Hautnerven **1**: 242–243
– Lymphgefässe **1**: 245
– Lymphknoten **1**: 245
– Oberflächenrelief **1**: 216–217
– segmentale Innervation **1**: 134, 242–243
Beinarterien **1**: 250–251, 269
Beinvenen **1**: 246–247
– oberflächliche **1**: 246, 248
– perforierende **1**: 246
– subkutane **1**: 246, 248
– tiefe **1**: 246
Bewegungsrichtungen **1**: 5

Bifurcatio aortae **1**: 14; **2**: 162, 218–219, 229, 234, 238
– tracheae **2**: 104, 106, 108, 110, 155
Bifurcation, aortic **2**: 229
– tracheal **2**: 155
Bile ducts **2**: 183
– – extrahepatic **2**: 186–187
– – intrahepatic **2**: 183
Bindegewebe, männliches kleines Becken **2**: 246–247
– weibliches kleines Becken **2**: 246–247
Binnenräume, Herz **2**: 139
Bladder, urinary **2**: 235–237
– – base **2**: 235
– – muscles **2**: 235
– – trigone **2**: 235
– – wall **2**: 235
Blasenpfeiler **2**: 246
Blinddarm **2**: 177–178, 195
Blutkreislauf **1**: 12–14
Blutleiter, venöse, Gehirn **2**: 291
Body(-ies), ciliary **2**: 376
– erectile, clitoris **2**: 268
– – penis **2**: 275
– fornix **2**: 346
– mamillary **2**: 300, 344, 351
– penis **2**: 275, 278
– pineal **2**: 348, 361
– striate **2**: 329, 332
– types **1**: 6
– wall, dorsal, arteries **2**: 162
– – – veins **2**: 162
– – ventral, arteries **1**: 75
– – – veins **1**: 75
Bogengang, hinterer **2**: 407
– seitlicher **2**: 406–407
– vorderer **2**: 407
Bogengänge **2**: 399, 402, 406–407
Bone(s), carpal **1**: 92–94
– compact **1**: 7
– development **1**: 8
– ethmoid **2**: 22
– foot **1**: 208–211
– frontal **2**: 21
– hyoid **2**: 27, 35, 47
– – muscle attachments **2**: 34–35
– interparietal **2**: 3
– lateral arch of foot **1**: 189
– lower leg **1**: 182–183
– medial arch of foot **1**: 189
– occipital **2**: 18
– parietal **2**: 22
– petrous **2**: 393, 395–396, 399–400, 402, 406–409
– scan **1**: 3
– sphenoid **2**: 19
– spongy **1**: 7
– sutural **2**: 3
– temporal **2**: 20, 393–395, 399–400, 402, 406–409
Borders, lobes of lungs **2**: 152
– lungs **2**: 152
– pleura **2**: 152
Boyd's perforating veins **1**: 246
Boyd-Venen **1**: 246
Brachium **1**: 2, 105, 135–137, 140–149
– colliculi inferioris **2**: 310–311
– – superioris **2**: 310–311, 335
Brachyzephalie **2**: 2
Brain **2**: 295–296, 304–306, 336–362
– arteries **2**: 298–302
– lateral ventricles **2**: 300, 322, 326
– stem **2**: 289, 294, 307, 310–311, 324
– superficial arteries **2**: 297
– – veins **2**: 297
Brain
– veins **2**: 300
– ventricles **2**: 320
Breast **1**: 70–74
Breitgesicht **2**: 2
Breitschädel **2**: 2
Brodmann-Felder der Grosshirnrinde **2**: 318
Brodmann's areas of cerebral cortex **2**: 318
Bronchialbaum **2**: 114–115, 119–121
– Verzweigung **2**: 114
Bronchien **2**: 108–110
– Lymphgefässe **2**: 110
– Lymphknoten **2**: 110
Bronchographie **2**: 120–121
Bronchography **2**: 120–121
Bronchus(-i) **2**: 108–110
– cardiacus dexter → Bronchus segmentalis basalis medialis dexter
– lingularis inferior **2**: 108–109, 114
– – superior **2**: 108–109, 114
– lobaris inferior dexter **2**: 108–110, 114, 120–121
– – – sinister **2**: 108–109, 114, 127, 164
– – medius dexter **2**: 108–110, 114, 120–121
– – superior dexter **2**: 108–110, 114, 120–121
– – – sinister **2**: 108–109, 114, 165
– lymph nodes **2**: 110
– – vessels **2**: 110
– principalis dexter **1**: 16; **2**: 105–106, 108–111, 113–114, 119–121, 155, 157, 161
– – sinister **1**: 16; **2**: 96, 105–111, 114, 119–121, 127, 155, 157
– segmentalis **2**: 115
– – anterior dexter **2**: 108–109, 114, 120–121
– – – sinister **2**: 108–109, 114
– – apicalis dexter **2**: 108–109, 113–114, 120–121
– – apicoposterior sinister **2**: 108–109, 114
– – basalis anterior dexter **2**: 108–109, 114, 120–121
– – – – sinister **2**: 108–109, 114
– – – lateralis dexter **2**: 108–109, 114, 120–121
– – – – sinister **2**: 108–109, 114
– – – medialis dexter **2**: 108–109, 114, 120–121
– – – posterior dexter **2**: 108–109, 114, 120–121
– – – – sinister **2**: 108–109, 114
– – lateralis dexter **2**: 108–109, 114, 120–121
– – medialis dexter **2**: 108–109, 114, 120–121
– – posterior dexter **2**: 108–109, 114, 120–121
– – superior dexter **2**: 109, 114, 120–121
– – – sinister **2**: 109, 114
Bruchinhalt **1**: 63, 67; **2**: 273
Bruchpforte **1**: 63; **2**: 273
– äussere **1**: 249
– innere **1**: 249
Bruchsack **1**: 63, 67, 249; **2**: 273
Brücke **2**: 352–353, 359, 361, 408–409
Brust, Lymphgefässe **2**: 163
– Lymphknoten **2**: 163
– Oberflächenrelief **1**: 62
– weibliche **1**: 70–74
Brustbein **1**: 41, 45
Brustbeinrippengelenke **1**: 45
Brustdrüse **1**: 70–74
– Lymphabfluss **1**: 72–74
Brustkorb **1**: 40–43, 76, 82–83
– Arterien **1**: 76
– Ausatmungsphase **1**: 40–41
– Einatmungsphase **1**: 40–41
– Lymphgefässe **1**: 76
– Muskelanheftungen **1**: 68
– Venen **1**: 76
Brustmark **2**: 288
Brustmuskeln **1**: 62–64
– autochthone **1**: 68
Brustorgane **1**: 18–21; **2**: 126–127, 150–151, 154–161, 172
Brustraum, autonomes Nervensystem **2**: 167
– oberer, Arterien **2**: 97
– – autonomes Nervensystem **2**: 95–96
– – Nerven **2**: 97
– – sympathisches Nervensystem **2**: 95–96, 166
– – Venen **2**: 97
– sympathisches Nervensystem **2**: 167
Brustwand **1**: 68, 72–74, 76, 79
– Arterien **1**: 75; **2**: 93
– Lymphgefässe **1**: 72–74
– Lymphknoten **1**: 72–74
– Muskeln **1**: 62–64
– Nerven **2**: 93
– Venen **1**: 75; **2**: 93
Brustwarze **1**: 70
Brustwirbel **1**: 29, 32, 47
Brustwirbelsäule **1**: 42, 46–47
Bulb of penis **2**: 267
Bulbus aortae **1**: 14; **2**: 130, 161–162
– cornus occipitalis ventriculi lateralis **2**: 320
– – posterioris ventriculi lateralis → Bulbus cornus occipitalis ventriculi lateralis
– oculi **2**: 9, 11, 59, 61–62, 352–353, 358–359, 362, 371, 374–375, 378, 381–382, 385–386
– olfactorius **1**: 22; **2**: 295, 298, 304, 306, 315, 324–325, 335
– penis **1**: 17, 259–260; **2**: 170, 250, 254, 267, 275, 277–278
– superior venae jugularis **2**: 49, 100
– vestibuli **2**: 250–251, 268–269, 280
Bulla ethmoidalis **2**: 56–57
Bursa apicis dentis axis **1**: 51
– bicipitoradialis **1**: 116
– infrapatellaris profunda **1**: 195
– omentalis **2**: 170–171, 191, 204
– subacromialis **1**: 96
– subcutanea praepatellaris **1**: 218, 230, 232
– subdeltoidea **1**: 96, 109
– subfascialis praepatellaris **1**: 200
– subtendinea musculi gastrocnemii medialis **1**: 233
– – – subscapularis **1**: 96
– suprapatellaris **1**: 10, 195
– synovialis **1**: 51
– trochanterica musculi glutaei maximi **1**: 223–224, 256
– – – – medii **1**: 224

C

Caecum **1**: 18–20, 194; **2**: 172, 177–178, 190, 192–196, 202–205, 225, 238
Calcaneus **1**: 184–188, 207–211, 214–215, 277
Calcar avis **2**: 322–323
Calf bone **1**: 182–183
Calix(-ices) renales **1**: 17; **2**: 217
– – majores **2**: 212, 216–217
– – minores **2**: 212, 216–217
Calvaria **1**: 87; **2**: 3, 290, 336, 347, 349, 354–362, 409
Camera anterior bulbi oculi **2**: 374, 384–385
– posterior bulbi oculi **2**: 374, 384
Canal(s), facial **2**: 89, 393, 396, 399
– incisive **2**: 56
– inguinal **1**: 66–67; **2**: 272–273
– mandibular **2**: 86
– optic **2**: 378
– semicircular **2**: 399, 402
– vertebral **2**: 284–285, 289

Canaliculus chordae tympani **2**: 393
– cochleae **2**: 401
– lacrimalis inferior **2**: 365, 383
– – superior **2**: 365
– tympanicus **2**: 20, 393, 402
Canalis(-es) adductorius **1**: 219, 252–253
– analis **1**: 16–17; **2**: 179, 232–233, 237, 243, 253, 255, 265–267
– arteriae vertebralis **1**: 31
– caroticus **2**: 8–9, 11, 20, 393, 395–396, 398, 400
– carpi **1**: 103, 121–124, 166
– centralis medullae spinalis **2**: 288, 304, 307, 310, 320–321, 324
– cervicis uteri **2**: 233, 240
– condylaris **1**: 50; **2**: 10, 18
– facialis **2**: 89, 393, 396, 399, 402
– hypoglossi **1**: 49–50; **2**: 9–10, 12, 18, 27, 400
– incisivus **2**: 12, 24, 45, 55–57, 90
– infraorbitalis **2**: 59, 61, 367–369
– inguinalis **1**: 17; **2**: 234, 274
– longitudinalis modioli cochleae **2**: 403
– mandibulae **2**: 41
– musculotubarius **2**: 11, 20
– nasolacrimalis **2**: 57, 365, 368
– nervi facialis → Canalis facialis
– – hypoglossi → Canalis hypoglossi
– obturatorius **1**: 190–191, 193, 225, 249; **2**: 264
– opticus **2**: 4, 10, 19, 58, 301, 367–368, 371, 378
– palatinus major **2**: 368
– pterygoideus **2**: 19, 23
– pudendalis **2**: 251, 255, 279–280
– pyloricus **2**: 173–176
– radicis dentis **2**: 40
– sacralis **1**: 22, 34–35, 40, 174, 194; **2**: 225, 244
– semicircularis(-es) **2**: 399
– – anterior **2**: 396, 398–402, 407
– – lateralis **2**: 396, 398–402, 406–407
– – posterior **2**: 399–401, 407, 409
– spiralis cochleae **2**: 403
– – modioli **2**: 402–404
– vertebralis **1**: 22, 39–40, 46, 60; **2**: 74, 158, 160–161, 243, 253, 289
Capitulum humeri **1**: 86, 90–91, 100–101, 149
Capsula adiposa perirenalis → Capsula adiposa renalis
Capsula adiposa renalis **2**: 208, 210, 213, 220–222, 224
– articularis **1**: 10–11
– – articulationis coxae **1**: 220
– – – cricoarytaenoideae **2**: 65
– – – cricothyroideae **2**: 65–67
– – – cubiti **1**: 98, 116
– – – genus **1**: 198, 200
– – – humeri **1**: 95–96, 109–110
– – – interphalangealis manus **1**: 104
– – – temporomandibularis **2**: 26, 31
– externa **2**: 335, 337–338, 340, 344, 346, 348, 350–351, 355–356, 360
– extrema **2**: 337–338, 340, 344, 346, 348, 350–351, 355–356, 360
– fibrosa renis **2**: 212
– glandulae suprarenalis **2**: 216
– interna **2**: 300, 330–335, 337–341, 344, 346–351, 354–356, 360–361
– lentis **2**: 375
– prostatica **2**: 237
Capsule, articular, of temporomandibular joint **2**: 26
– internal **2**: 332–333, 355–356
Caput **1**: 2; **2**: 13, 52–53
– breve musculi bicipitis brachii **1**: 64, 106–107, 110, 112, 141
Caput breve musculi bicipitis
– – – – femoris **1**: 199, 221–222, 224, 226, 229, 232–234, 255–256, 261, 263, 267–268, 271
– commune musculorum flexorum **1**: 115–116, 153
– cornus posterioris medullae spinalis **2**: 288
– costae **1**: 41, 44
– embryonis **2**: 242
– epididymidis **2**: 170, 270–272, 278
– femoris **1**: 173, 177, 179–180, 193–194, 251, 258; **2**: 195, 205, 220–221, 250, 252, 254, 256–259
– fibulae **1**: 181–183, 196–197, 199, 202–203, 205, 216, 220–222, 226, 231–233, 247, 255–256, 266–268, 271–272, 276–277
– fibulare → Caput fibulae
– humerale musculi flexoris carpi ulnaris **1**: 113
– – – pronatoris teretis **1**: 113, 115–116
– humeri **1**: 73, 85–87, 96–97, 148–149
– humeroulnare musculi flexoris digitorum superficialis **1**: 113, 115, 117
– laterale musculi flexoris hallucis brevis **1**: 240
– – – gastrocnemii **1**: 181, 195, 200, 203–204, 222, 224, 226, 229, 232–234, 265, 267–268, 272, 276–277
– – – tricipitis brachii **1**: 105, 108–109, 111, 113, 118–119, 143, 145–146, 148–149
– longum musculi bicipitis brachii **1**: 64, 95–97, 106–107, 109–112, 141, 144
– – – – femoris **1**: 199, 221–224, 226–227, 232–234, 255–256, 258, 261–263, 267–268
– – – tricipitis brachii **1**: 95, 105–112, 141–146, 148–149
– mallei **2**: 391, 397–400
– mandibulae **2**: 6, 11, 25–26, 63, 395
– mediale musculi flexoris hallucis brevis **1**: 240
– – – gastrocnemii **1**: 195, 200, 203–204, 221–222, 224, 229, 231, 233–234, 252, 265, 267–268, 272, 276–277
– – – tricipitis brachii **1**: 105–107, 109–110, 113–116, 118–119, 141–143, 145–146, 148–149, 152
– metacarpale **1**: 94, 123, 125, 129
– metatarsale **1**: 186
– musculi **1**: 10
– nuclei caudati **2**: 320, 322–323, 329–334, 337–339, 346–351, 354–357, 360
– obliquum musculi adductoris hallucis **1**: 238, 240
– – – – pollicis **1**: 122, 124
– ossis metacarpi/metatarsi → Caput metacarpale/metatarsale
– pancreatis **2**: 188, 198–200, 202–203, 223–224, 248
– phalangis digiti manus **1**: 94
– – – pedis **1**: 186
– profundum musculi flexoris pollicis brevis **1**: 123–124, 167
– – – tricipitis brachii → Caput mediale musculi tricipitis brachii
– radiale musculi flexoris digitorum superficialis **1**: 115, 117
– radii **1**: 88, 90–91, 99–101
– stapedis **2**: 391, 397–398
– superficiale musculi flexoris pollicis brevis **1**: 103, 121, 123–124, 166–167
– tali **1**: 186–188, 210, 215
– transversum musculi adductoris hallucis **1**: 238, 240
– – – – pollicis **1**: 121–122, 124
– ulnae **1**: 89, 92, 94, 116
– ulnare musculi flexoris carpi ulnaris **1**: 117
– – – pronatoris teretis **1**: 115–117
Carina tracheae **2**: 155
Carpalia → Ossa carpi
Carpus **1**: 92–94, 103–105, 123
Cartilage(s), arytenoid **2**: 64
– auricle **2**: 390
– cricoid **2**: 64
– epiglottic **2**: 64
– laryngeal **2**: 64–65
– thyroid **2**: 64
Cartilago(-ines) alaris major nasi **2**: 54
– articularis **1**: 10–11, 202–203
– arytaenoidea **2**: 64–66, 68–69
– auricularis **2**: 390–391
– corniculata **2**: 64–66
– costalis **1**: 41, 45
– cricoidea **2**: 33, 36, 64–65, 67, 106, 110
– epiglottica **2**: 64–66, 68, 70
– epiphysialis **1**: 8–9
– laryngeales **2**: 65
– meatus acustici **2**: 390–392
– nasi accessoria **2**: 54
– – lateralis **2**: 54
– septi nasi **2**: 13, 54
– thyroidea **1**: 18, 20–21; **2**: 35–36, 47–48, 51–52, 64–68, 70, 101, 106, 110, 149, 164
– tracheales **2**: 47–48, 50, 65–66, 68, 70, 105–106, 108–109
– triticea **2**: 65–67
– tubae auditoriae **2**: 395
Caruncula(-ae) hymenales **2**: 240, 269
– lacrimalis **2**: 364–365
– sublingualis **2**: 38, 45
Cauda epididymidis **2**: 270, 272
– equina **1**: 22, 39, 48, 55, 194; **2**: 225, 244, 284, 287
– helicis **2**: 390
– nuclei caudati **2**: 322, 326, 329–332, 340, 344, 346, 348, 350–351, 355–356, 361
– pancreatis **2**: 188, 199–200, 203, 206–207, 248
Caudalis **1**: 5
Cavitas amniotica **2**: 242
– articularis **1**: 10
– – articulationis humeri **1**: 96
– conchalis **2**: 390
– cranii **1**: 22, 40; **2**: 294
– dentis **2**: 40
– glenoidalis scapulae **1**: 84–85, 87, 96–97
– infraglottica **2**: 68
– nasi **1**: 16, 51; **2**: 5, 9, 11, 41, 55–63, 336, 352–353, 359–360, 369, 385–387
– orbitalis → Orbita
– oris **2**: 38–39
– – propria **1**: 16; **2**: 45, 52, 55, 58, 61, 336, 386–387
– pericardialis **2**: 104, 123, 129, 155, 159
– peritonealis **2**: 154, 170–171, 252–255, 272
– pharyngis **1**: 16; **2**: 38, 50–51
– pleuralis **2**: 153–154, 156–157, 159–160, 164
– pulparis → Cavitas dentis
– scroti **2**: 170, 232, 271–273
– thoracis **2**: 153
– trigeminalis **2**: 294
– tympanica **2**: 11, 361, 391, 393–394, 396, 398–400, 402, 405–407
– uteri **1**: 17; **2**: 233, 238, 240, 242, 252, 256–257
Cavity(-ies), cranial **2**: 294
– heart **2**: 139
– nasal **2**: 55–63, 336, 365, 369, 386–387
– – bony lateral wall **2**: 55–57
– – lateral wall **2**: 55–57, 88–90
– – – – innervation **2**: 88
– – mucous membrane **2**: 55–57
– oral **2**: 38–39
– orbital **2**: 58, 336, 367–369, 379–382, 384–387
– pericardial **2**: 123

Cavity(-ies)
– peritoneal **2**: 170–171
– – adult **2**: 170
– – fetal **2**: 170
– pharyngeal **2**: 50–51
– pleural **2**: 153
– thoracic **2**: 153
– tympanic **2**: 393–394, 396–400, 406
– – lateral membranous wall **2**: 398
– – medial labyrinthine wall **2**: 396, 398
Cavum scroti → Cavitas scroti
– septi pellucidi **2**: 322, 338
– trigeminale → Cavitas trigeminalis
– tympani → Cavitas tympanica
Cecum **2**: 177–178, 195
Celiacography **2**: 197
Cellula(-ae) ethmoidales → Sinus ethmoidales
– mastoideae **2**: 7, 9, 11, 302, 361, 393–396, 400, 406–407
– pneumaticae ossis temporalis **2**: 395
– tympanicae **2**: 396
Cementum dentis **2**: 40
Centrum perinei → Centrum tendineum perinei
– tendineum diaphragmatis **1**: 69; **2**: 218
– – perinei **1**: 17; **2**: 237, 243, 265–266
Cerebellum **1**: 22, 37; **2**: 13, 294, 299, 303, 305–309, 321, 324, 334, 342–343, 350–353, 357–359, 361–362, 409
Cerebrum **1**: 22; **2**: 13, 61–63, 295–297, 299, 303, 312–319, 324–325, 327, 336–362
– arterial circle **2**: 292
– frontal lobe **2**: 312–313, 360
– gyri **2**: 316–317
– limbic lobe **2**: 312–313
– nerve fascicles **2**: 333
– occipital lobe **2**: 312–313
– parietal lobe **2**: 312–313, 408
– sulci **2**: 316–317
– temporal lobe **2**: 312–313, 409
Cervix cornus posterioris medullae spinalis **2**: 288
– dentis **2**: 40
– uteri **1**: 17; **2**: 233, 239, 242, 246, 251, 256–257, 260, 268
– – Portio vaginalis **2**: 268
– – vaginal part **2**: 268
– vesicae urinariae **2**: 235, 258
Chest wall **1**: 68, 72–73, 75–76, 79
– – lymph nodes **1**: 72–74
– – – vessels **1**: 72–74
Chiasma crurale **1**: 234
– opticum **1**: 22; **2**: 13, 60, 293, 301, 304–305, 307, 328, 335, 358, 360, 378
– plantare **1**: 238
– tendinum digitorum manus **1**: 124
Choana(-ae) **2**: 17, 23, 45, 50–51, 58, 63, 395
Choanen **2**: 23
Cholangiopancreatography, retrograde, endoscopic **2**: 188
Cholangiopankreatikographie, endoskopisch-retrograde **2**: 188
Cholecystocholangiography **2**: 186
Cholezystcholangiographie **2**: 186
Chorda(-ae) arteriae umbilicalis **2**: 246
– obliqua membranae interosseae antebrachii **1**: 98–99
– tendineae cordis **2**: 132–134, 140
– – falsae cordis **2**: 132–134
– tympani **2**: 84, 87–89, 391, 398
– urachi **2**: 246
Chorion frondosum **2**: 242
Choroidea **2**: 373–376, 384
Cilium(-ia) **2**: 364, 366
Cingulum membri inferioris **1**: 172–175, 177
– – superioris **1**: 84, 87, 112
Cingulum
– pectorale → Cingulum membri superioris
– pelvicum → Cingulum membri inferioris
Circle, arterial, of cerebrum **2**: 292
Circulation of cerebrospinal fluid **2**: 321
Circulus arteriosus cerebri **2**: 292
– – iridis major **2**: 372–373
– – – minor **2**: 372–373
– vasculosus nervi optici **2**: 372–373
Circumferentia articularis radii **1**: 88, 99
– – ulnae **1**: 89, 92, 94
Cistern(s), cerebellomedullary **2**: 294
– – puncture **2**: 321
– subarachnoid **2**: 321
Cisterna(-ae) ambiens **2**: 295, 357
– cerebellomedullaris **2**: 294–295, 305, 321, 343
– – Punktion **2**: 321
– chiasmatica **2**: 295, 321
– chyli **1**: 15; **2**: 163
– fossae lateralis cerebri **2**: 295
– interpeduncularis **2**: 295, 321, 358
– magna → Cisterna cerebellomedullaris
– pontocerebellaris **2**: 321, 341, 352–353, 359
– subarachnoideae **2**: 321
– trigeminalis **2**: 295
Claustrum **2**: 323, 330–331, 337–338, 340, 344, 346, 348, 350–351, 355–356, 360
Clavicle **1**: 84, 112
Clavicula **1**: 18, 20–21, 43, 45, 62–63, 68, 73, 75, 82, 84, 87, 95–97, 105–108, 110–111, 136–137, 140, 142–143, 149; **2**: 33–36, 92, 94, 96, 98, 126, 150, 172
Cleavage lines of skin **1**: 11
Clitoris **1**: 17; **2**: 233, 240, 250–252, 260, 268–269, 280–281
– erectile bodies **2**: 268
Clivus **2**: 10, 19, 290, 293–294
Coccyx **1**: 34–35, 38–39; → Os coccygis
Cochlea **2**: 361, 399–404, 406, 408–409
Cockett's perforating veins **1**: 246
Cockett-Venen **1**: 246
Colliculus cartilaginis arytaenoideae **2**: 64
– facialis **2**: 310–311, 320
– inferior **2**: 310–311
– seminalis **2**: 235, 237, 277
– superior **2**: 310–311, 324, 328, 335, 350, 357
Collum **1**: 2, 60; **2**: 52–53, 72–75, 91–93, 95–96, 98–99, 118, 166
– anatomicum humeri **1**: 86
– chirurgicum humeri **1**: 86
– costae **1**: 44
– dentis → Cervix dentis
– femoris **1**: 173, 179–180, 194; **2**: 254, 257, 259
– fibulae **1**: 183
– glandis penis **2**: 275–277
– mallei **2**: 397
– mandibulae **2**: 25–26
– radii **1**: 88, 90–91, 98, 100
– scapulae **1**: 84–85, 95, 97
– tali **1**: 187–188, 210, 215
– vesicae biliaris **2**: 186
Collum-Diaphysen-Winkel → Kollodiaphysenwinkel
Colon ascendens **1**: 16, 18–20; **2**: 171–172, 177–178, 190–192, 194–196, 200, 202–207, 220, 223–224
– descendens **1**: 16, 18–19, 21, 194; **2**: 154, 171–172, 182, 193, 196, 202, 204–208, 220, 223–225, 238
– rectum → Rectum
– sigmoideum **1**: 16, 18–19, 21, 194; **2**: 172, 193, 196, 202, 204–205, 220–221, 225, 230, 232, 238, 248, 251–255, 257, 260–261, 276
Colon
– transversum **1**: 16, 18, 20–21; **2**: 154, 170, 172, 179, 186, 190–194, 196, 200, 202–205, 208, 223–224
Column, vertebral **1**: 28, 30–39, 42–43, 46–51
– – arteries **2**: 286
– – veins **2**: 286–287
Columna(-ae) anales **2**: 179, 251
– fornicis **2**: 307, 317, 320, 322, 324–325, 328, 334, 338, 348–350, 355–356
– renales **2**: 212
– vertebralis **1**: 28–39, 42–43, 46–51; **2**: 161
Commissura alba anterior medullae spinalis **2**: 288
– anterior **2**: 307, 317, 320, 324, 328, 330, 333–335, 338–339, 344, 350
– epithalamica **2**: 307, 310, 320, 324, 340
– fornicis **2**: 322, 324–325
– grisea medullae spinalis **2**: 288
– habenularum **2**: 307, 310, 320, 324, 340, 348
– labiorum anterior vulvae **2**: 269
– – oris **2**: 38, 44
– – posterior vulvae **2**: 269
– lateralis palpebrarum → Commissura palpebralis lateralis
– medialis palpebrarum → Commissura palpebralis medialis
– palpebralis lateralis **2**: 364
– – medialis **2**: 364
– posterior → Commissura epithalamica
Commissure, anterior, of brain **2**: 338–339, 350
Compartments, fascial, of neck **2**: 33
Concha auricularis **2**: 390
– nasalis inferior **2**: 4–5, 8, 12, 14, 23, 50, 52, 55–59, 62, 88–89, 305, 336, 360, 365, 369, 386–387
– – media **2**: 5, 22–23, 50, 52, 55–59, 62–63, 88, 305, 336, 360, 365, 369, 386–387
– – superior **2**: 22, 50, 52, 55–56, 58, 88–89, 336
– – suprema **2**: 52
– sphenoidalis **2**: 19
Condylus humeri **1**: 86
– lateralis femoris **1**: 10, 179–181, 196–199, 202–204, 259–261, 265, 277
– – tibiae **1**: 10, 182–183, 197–200, 202–203, 226, 232, 276–277
– medialis femoris **1**: 179–181, 196–199, 202–204, 220, 259–261, 265, 277
– – tibiae **1**: 182–183, 197–200, 202–203, 276–277
– occipitalis **1**: 40, 60; **2**: 8, 18, 72, 395
Confluens sinuum **2**: 291, 294, 352–353, 359
Conjugata vera pelvis **1**: 178
Connections, cerebellorubral **2**: 334
Connexus intertendinei musculi extensoris digitorum **1**: 125, 130
Constriction, esophageal, inferior (diaphragmatic) **2**: 106–107
– – middle (aortic) **2**: 106
– – upper (cricopharyngeal) **2**: 53, 106
Contents, hernial **1**: 63, 67; **2**: 273
Contraction, pelvic **1**: 178
Conus arteriosus **2**: 132, 139, 142, 158
– elasticus **2**: 110
– medullaris **1**: 22; **2**: 284
Cor **2**: 107, 122–125, 127–148, 158–160, 166, 172
Cord, spermatic **2**: 274
– spinal **1**: 23; **2**: 284–286, 288
– – arteries **2**: 286
– – investing structures **2**: 286
– – veins **2**: 286–287
– umbilical **2**: 243
Corium → Dermis

Cornea **2**: 370, 372–375, 384–385
Cornu anterius medullae spinalis **1**: 23, 79; **2**: 288
– – menisci lateralis **1**: 201
– – – medialis **1**: 201
– – ventriculi lateralis → Cornu frontale ventriculi lateralis
– coccygeum **1**: 34–35
– frontale ventriculi lateralis **2**: 300, 302, 320, 322–323, 332, 337–339, 344, 346–350, 354–355, 360
– inferius cartilaginis thyroideae **2**: 64–65, 68
– – marginis falciformis hiatus sapheni **1**: 218
– – ventriculi lateralis → Cornu temporale ventriculi lateralis
– laterale medullae spinalis **1**: 79; **2**: 288
– majus ossis hyoidei **2**: 27, 35–36, 46–49, 65–68, 70, 100–101
– minus ossis hyoidei **2**: 27, 36, 46, 65–66, 70
– occipitale ventriculi lateralis **2**: 300, 320, 322–323, 332, 335, 342–343, 346–349, 354–356, 362
– posterius medullae spinalis **1**: 23, 79; **2**: 288
– – menisci lateralis **1**: 201
– – – medialis **1**: 201
– – ventriculi lateralis → Cornu occipitale ventriculi lateralis
– sacrale **1**: 34–35
– superius cartilaginis thyroideae **2**: 51, 64–67, 70, 101
– – marginis falciformis hiatus sapheni **1**: 218
– temporale ventriculi lateralis **2**: 300, 320, 322–323, 326, 333–334, 340–341, 344, 350, 357–358, 361, 385, 408
– uteri **2**: 240
Corona ciliaris **2**: 374, 376
– clinica dentis **2**: 40
– dentis **2**: 40
– glandis penis **2**: 275–277
– radiata **2**: 333
Corpus(-ora) adiposum buccae **2**: 28–29, 44
– – calcaneum **1**: 279
– – infrapatellare **1**: 10, 195, 198, 200, 202–204
– – laryngis **2**: 65
– – orbitae **2**: 58–59, 61, 336, 353, 358, 360, 378, 384–387
– – praeepiglotticum **2**: 66, 68
– albicans **2**: 241
– amygdaloideum **2**: 325, 330–331, 338, 344
– calcanei **1**: 187
– callosum **2**: 13, 299–300, 304–305, 307, 315, 317, 322, 324–326, 333, 337–350, 354–355, 361, 408
– cavernosum(-a) clitoridis **2**: 268
– – penis **1**: 17; **2**: 170, 232, 234, 258–259, 267, 274–275, 277–279
– ciliare **2**: 373, 376–377, 384–385
– claviculae **1**: 84
– clitoridis **2**: 233, 260, 268
– costae **1**: 44
– epididymidis **2**: 270–272
– femoris **1**: 177, 179–180, 197, 202–203, 247, 251, 258–260; **2**: 254
– fibulae **1**: 182–183, 197, 205, 209, 214, 277
– fornicis **2**: 299–300, 304–305, 307, 322, 324–326, 340–341, 344, 346, 361
– gastricum **2**: 154, 173–176, 190–191, 194, 200, 222–223, 248
– geniculatum laterale **2**: 310–311, 326, 330, 332, 334–335, 340, 350, 357
– – mediale **2**: 310–311, 330–332, 334–335, 350, 357
– humeri **1**: 86–87, 90–91, 96–97, 100, 109, 111, 149
– incudis **2**: 397–400
Corpus(-ora)
– luteum **2**: 241
– mamillare **2**: 300, 306, 315, 324–325, 330–331, 334–335, 344, 351
– mammae **1**: 70–72, 74
– mammillare → Corpus mamillare
– mandibulae **1**: 36; **2**: 4–7, 25, 27, 41, 48
– maxillae **2**: 4, 24, 48, 58, 367–368
– medullare cerebelli **2**: 309, 334
– metacarpale **1**: 94
– metatarsale **1**: 186
– nuclei caudati **2**: 300, 322, 326, 329–332, 340–341, 344, 361
– ossis hyoidei **2**: 27, 36, 47–48, 52, 65–66, 68, 70
– – ilii **1**: 175
– – ischii **1**: 172, 175–176
– – metacarpi/metatarsi → Corpus metacarpale/metatarsale
– – sphenoidalis **2**: 19, 55, 63, 359, 368
– pancreatis **2**: 186, 188, 203, 206
– penis **2**: 276
– perineale → Centrum tendineum perinei
– phalangis digiti manus **1**: 94
– – – pedis **1**: 186
– pineale **1**: 15; **2**: 299, 310–311, 315, 324, 332, 335, 346, 348, 356, 361
– radii **1**: 88, 90–91, 99–100
– spongiosum penis **1**: 17; **2**: 170, 232, 234, 237, 266–267, 274–275, 277–278
– sterni **1**: 40–41, 45; **2**: 107, 158–159, 161
– striatum **2**: 300, 329, 332
– tali **1**: 187, 215
– tibiae **1**: 182–183, 197, 202, 205, 208–209, 214, 276–277
– ulnae **1**: 89–91, 99–101
– unguis **1**: 11
– uteri **2**: 233, 238–239, 242, 252, 256–257, 260
– ventriculare → Corpus gastricum
– vertebrae **1**: 29–33, 35–40, 43, 46–48, 51, 55; **2**: 53, 68, 71–72, 74–75, 158–161, 182, 203, 205, 213–214, 220–221, 224, 244–245, 287
– – axis **1**: 30, 36–37, 40, 51; **2**: 72, 294, 408
– vesicae biliaris **2**: 154, 186
– – urinariae **2**: 232–233, 235, 245, 258–259
– vitreum **2**: 374, 384–385
Corpusculum(-a) lamellosa **1**: 161–162
Cortex cerebellaris → Cortex cerebelli
– cerebelli **2**: 309, 334
– cerebralis **2**: 295, 299, 318–319, 327
– cerebri → Cortex cerebralis
– glandulae suprarenalis **2**: 216
– lentis **2**: 375
– ovarii **2**: 241
– renalis **2**: 208, 210, 212, 218
– visual **2**: 322, 335
Corti-Organ **2**: 404
Corti's organ **2**: 404
Costa(-ae) **1**: 18–21, 29, 38, 40–47, 55, 68–69, 73–74, 82–83, 85, 87, 148, 219; **2**: 35, 37, 75, 93, 105, 107, 118, 126, 153, 156–157, 161–162, 164, 167, 172, 209, 211, 213–215, 217, 222, 226, 287
Cotyledo(-ones) **2**: 243
Coverings of testis **2**: 270
Coxa valga **1**: 179
– vara **1**: 179
Cranialis **1**: 5
Cranium **2**: 2–16
– neonati **2**: 17
Crista(-ae) arcuata cartilaginis arytaenoideae **2**: 64
– capitis costae **1**: 44
– colli costae **1**: 44
Crista(-ae)
– ethmoidalis maxillae **2**: 24, 57
– frontalis **2**: 3, 10, 21
– galli **2**: 5, 7, 10, 12–13, 22–23, 55–56, 58–60, 90, 294, 367, 369
– iliaca **1**: 53, 69, 172–176, 191, 194, 217–219, 223, 226, 255–257; **2**: 211, 218, 227, 287
– infratemporalis alae majoris ossis sphenoidalis **2**: 19, 31
– intertrochanterica **1**: 180, 192
– lacrimalis anterior **2**: 24, 367–368
– – posterior **2**: 368
– matricis unguis **1**: 11
– medialis fibulae **1**: 182–183
– musculi supinatoris **1**: 89
– nasalis maxillae **2**: 24
– obturatoria **1**: 176
– occipitalis externa **2**: 18
– – interna **2**: 10, 18
– pubica **1**: 174, 176
– sacralis intermedia **1**: 34–35
– – lateralis **1**: 34–35
– – medialis → Crista sacralis intermedia
– – mediana **1**: 34–35, 194; **2**: 265
– spiralis ductus cochlearis **2**: 404
– supraepicondylaris lateralis **1**: 86, 98
– – medialis **1**: 86, 98
– supraventricularis **2**: 132
– terminalis atrii dextri cordis **2**: 132
– transversa meatus acustici interni **2**: 401, 403
– tuberculi majoris **1**: 86
– – minoris **1**: 86
– urethralis **2**: 235, 277
Crus (Crura) **1**: 2, 216–217, 266–268, 270–277
– anterius capsulae internae **2**: 332–334, 337, 348, 350, 355–356
– – stapedis **2**: 397
– antihelicis **2**: 390
– breve incudis **2**: 397–398
– cerebri → Pars anterior pedunculi cerebri
– clitoridis **2**: 250–251, 260, 268
– dextrum fasciculi atrioventricularis **2**: 140
– – partis lumbalis diaphragmatis **1**: 69; **2**: 105–106, 155, 182, 196, 208, 220–223
– fornicis **2**: 322, 324–325
– helicis **2**: 390
– laterale anuli inguinalis superficialis **1**: 66
– – cartilaginis alaris majoris nasi **2**: 54
– longum incudis **2**: 391, 397–398
– mediale anuli inguinalis superficialis **1**: 66
– – cartilaginis alaris majoris nasi **2**: 54
– membranaceum commune ductus semicircularis **2**: 405
– – simplex ductus semicircularis **2**: 405
– osseum ampullare anterius canalis semicircularis **2**: 401
– – commune canalis semicircularis **2**: 401–402
– – simplex canalis semicircularis **2**: 401–402
– penis **1**: 259–260; **2**: 234, 250, 254, 261, 267, 275–278
– posterius capsulae internae **2**: 332–334, 348, 350, 355–357
– – stapedis **2**: 397
– sinistrum fasciculi atrioventricularis **2**: 140
– – partis lumbalis diaphragmatis **1**: 69; **2**: 105–106, 182, 189, 196, 200
Culmen **2**: 307–308
Cuneus **2**: 317
Cupula cochleae **2**: 401, 403
– pleurae **2**: 153
Curvatura gastrica major **2**: 173–176, 198
– – minor **2**: 173–176, 198
– ventricularis major/minor → Curvatura gastrica major/minor

Cuspis(-ides) anterior valvae atrioventricularis dextrae **2**: 132, 135
– – – – sinistrae **2**: 133–135, 140
– dentis **2**: 39
– posterior valvae atrioventricularis dextrae **2**: 132, 134–135
– – – – sinistrae **2**: 133, 135, 140
– septalis valvae atrioventricularis dextrae **2**: 132, 135, 140
Cutis **1**: 11, 65, 67, 70, 74; **2**: 71, 251, 347, 349, 359, 361–362
– penis **2**: 275–276
– scroti **2**: 271
Cymba conchalis **2**: 390
Cysts, persistent, in spermatic cord **2**: 273

D

Dach der Mundhöhle **2**: 39
Damm **2**: 265–266
– männlicher **2**: 266, 279
– – Arterien **2**: 279
– – Muskeln **2**: 267
– – Nerven **2**: 279
– – Venen **2**: 279
– weiblicher **2**: 265, 280–281
– – Arterien **2**: 280
– – Nerven **2**: 280–281
– – Venen **2**: 280
Dammarterien **2**: 279–280
Dammnerven **2**: 279–280
Dammregion, männliche **2**: 279
– weibliche **2**: 280
Dammuskeln **2**: 265–266
Dammvenen **2**: 279–280
Darmschlingen **2**: 242, 273
Declive **2**: 307–308
Decussatio pedunculorum cerebellarium superiorum **2**: 309, 334
– pyramidum **2**: 306, 340
Decussation of Stilling **2**: 334
Defloration **2**: 269
Dens (Dentes) axis **1**: 30–31, 36–37, 40, 50–51, 60; **2**: 5, 52, 72–73, 294, 408
– canini **2**: 39–41
– – decidui **2**: 42
– decidui **2**: 42
– incisivi **1**: 36; **2**: 39–41, 59
– – decidui **2**: 42
– molares **2**: 39–41, 58, 369, 387
– permanentes **2**: 5, 39–41
– – inferiores **2**: 61
– – superiores **2**: 61
– praemolares **2**: 39–41
– – decidui **2**: 42
– serotini **2**: 40–41
Dentinum **2**: 40
Dentition, deciduous **2**: 42–43
– permanent **2**: 39–41
Dermatome, Arm **1**: 132–134, 244
– Bein **1**: 134, 242–244
– Hals **2**: 82–83
– Kopf **2**: 82–83
– Thorax **1**: 134, 244
Dermatomes, head **2**: 82–83
– lower limb **1**: 134, 242–244
– neck **2**: 82–83
– thorax **1**: 134, 244
– upper limb **1**: 132–134, 244
Dermis **1**: 11
Descensus testis **2**: 272
Descent, prenatal, of testis **2**: 272
Desmodontium **2**: 40
Development, bones **1**: 8
– hip joint **1**: 177
Dextrocardiogram **2**: 119
Dextrogramm des Herzens **2**: 119
Diameter conjugata pelvis **1**: 178
– diagonalis pelvis **1**: 178
– externa pelvis **1**: 178
– mediana pelvis **1**: 178
– obliqua pelvis **1**: 178
– transversa pelvis **1**: 178
Diaphragm **1**: 69
– pelvic **2**: 264–266
– urogenital **2**: 267
Diaphragma **1**: 12–14, 16, 18–21, 25, 43, 68–69, 79; **2**: 104–107, 113, 122, 126–131, 150–151, 153–155, 160, 162, 164, 167, 172, 182, 185, 189, 191, 196, 208–209, 211, 218, 220–223
– pelvis **2**: 250–251, 264–266
– sellae **2**: 55, 293–294, 378
– urogenitale **2**: 233, 237, 243, 247, 250–251, 261, 265–267, 276, 280
Diaphysis **1**: 8
Diastole **2**: 138
Dickdarm **2**: 202–205
Dickdarmarterien **2**: 202
Diencephalon **2**: 303, 334
Digitus(-i) anularis manus **1**: 105
– manus **1**: 165
– medius manus **1**: 104–105, 124–125, 165, 167
– minimus manus **1**: 105
– pedis **1**: 184–185, 188, 216–217
– primus [I] manus → Pollex
– – – pedis → Hallux
– quartus [IV] manus → Digitus anularis manus
– quintus [V] manus → Digitus minimus manus
– secundus [II] manus → Index
– tertius [III] manus → Digitus medius manus
Dilatation, esophageal, epiphrenic **2**: 107
Diploë **2**: 3, 295
Directions of motion **1**: 5
Disc(s), prolapse **1**: 48
Discus(-i) articularis articulationis radioulnaris distalis **1**: 104
– – – sternoclavicularis **1**: 45
– – – temporomandibularis **2**: 26, 31
– interpubicus **1**: 172, 174, 190, 193
– intervertebralis(-es) **1**: 31, 36–39, 43, 47–49, 51; **2**: 72, 156, 158–159, 182, 221, 294
– nervi optici **2**: 373–374, 377, 384
Distalis **1**: 5
Diverticulum, pharyngoesophageal **2**: 49
Division of central nervous system **2**: 303
Dodd's perforating veins **1**: 246
Dodd-Venen **1**: 246
Dolichozephalie **2**: 2
Dorsalaponeurose der Finger **1**: 124–125
Dorsalis **1**: 5
Dorsum **1**: 2
– linguae **2**: 38, 46–47, 51, 101
– manus **1**: 164–165
– nasi **2**: 44
– of foot, arteries **1**: 278
– – – fascia **1**: 230
– – – muscles **1**: 231–232, 236
– – – nerves **1**: 278
– – – subcutaneous nerves **1**: 278
– – – – veins **1**: 278
– – – veins **1**: 278
– – hand, arteries **1**: 165
– – – muscles **1**: 125
– – – nerves **1**: 165
– – – subcutaneous nerves **1**: 164
– – – – veins **1**: 164
– pedis **1**: 216, 278, 281
– penis **2**: 275–276
– sellae **1**: 49; **2**: 7, 10, 19, 90, 293, 301, 352–353, 359, 378, 385, 407
Dreieck, pleurafreies, oberes **2**: 150
– – unteres **2**: 150
Druckspannungstrajektorien **1**: 7
Drüsen, endokrine **1**: 15
Duct(s), cochlear **2**: 399, 404
– deferent **2**: 236–237
– hepatic **2**: 183
– lacrimal **2**: 365
– nasolacrimal **2**: 365
– semicircular **2**: 399, 402, 406–407
– – anterior **2**: 407
– – lateral **2**: 406–407
– – posterior **2**: 407
Ductulus(-i) aberrans inferior epididymidis **2**: 270
– – superior epididymidis **2**: 270
– biliferi **2**: 183
– efferentes testis **2**: 270
– excretorii glandulae lacrimalis **2**: 365
– prostatici **2**: 235, 277
Ductus arteriosus **1**: 13
– biliaris → Ductus choledochus
– choledochus **2**: 181, 183, 186–188, 198–199, 223, 248
– cochlearis **2**: 391, 404–405
– cysticus **2**: 181, 186–188, 198
– deferens **1**: 17, 67, 77; **2**: 234, 236–237, 244, 255, 261, 270–272, 274, 276
– ejaculatorius **2**: 237, 277
– endolymphaticus **2**: 399–400, 405
– epoophorontis longitudinalis **2**: 241
– excretorius glandularum tarsalium **2**: 366
– glandulae bulbourethralis **2**: 277
– hepaticus(-i) **2**: 183, 187
– – accessorii **2**: 187
– – communis **2**: 185–188
– – dexter **2**: 181
– – sinister **2**: 183
– intralobares pancreatis **2**: 188
– lactiferus(-i) **1**: 70–71
– lymphaticus dexter **1**: 15, 76; **2**: 81
– nasolacrimalis **2**: 56, 365, 369, 385
– pancreaticus **2**: 187–188
– – accessorius **2**: 188
– parotideus **2**: 28, 30, 44, 48, 86–87, 98
– reuniens **2**: 405
– semicircularis anterior **2**: 399, 405
– – lateralis **2**: 399, 405
– – posterior **2**: 399, 405
– sublingualis(-es) minores **2**: 45
– – major **2**: 45
– submandibularis **2**: 45, 47, 87, 99
– thoracicus **1**: 15; **2**: 149, 155, 163, 167
– – dexter → Ductus lymphaticus dexter
– utriculosaccularis **2**: 405
– venosus **1**: 13
Dünndarm **2**: 176, 194–195, 203
Duodenum **1**: 16, 18–19; **2**: 155, 170–175, 187–188, 191, 194, 199–200, 203–204, 206–207, 223–224, 248
Dura mater **2**: 289–292
– – cerebral **2**: 289–291
– – cranialis **1**: 51; **2**: 88, 289–291, 295, 297, 321, 352, 360, 362, 405
– – encephali → Dura mater cranialis
– – spinal **2**: 284, 289
– – spinalis **2**: 73, 75, 284–286, 289, 321

E

Ear, external **2**: 390–391
– internal **2**: 391, 403
– middle **2**: 391, 393–394
Eardrum **2**: 392
– quadrants **2**: 392

Ebene(n) **1**: 5
– frontale **1**: 5
– paramediane **1**: 5
– quere **1**: 5
– sagittale **1**: 5
Echocardiography **2**: 136, 138
Echokardiographie **2**: 136, 138
Edge of pallium **2**: 316–317
Eierstockarterien **2**: 241
Eierstöcke **2**: 238–239, 241, 245
Eileiter **2**: 239–241
Eileiterarterien **2**: 241
Einatmung **1**: 40–41
Elbow joint **1**: 90–91, 98, 100–101, 147, 156
– region **1**: 106
– – anterior, arteries **1**: 140–141
– – – nerves **1**: 140–141
– – – subcutaneous nerves **1**: 151
– – – – veins **1**: 151
– – posterior, arteries **1**: 142–143
– – – nerves **1**: 142–143
Ellbogengelenk **1**: 90–91, 98, 100–101, 147, 156
Ellbogenregion **1**: 106
– Arterien **1**: 142–143
– Nerven **1**: 142–143
Elle **1**: 89
– Muskelanheftungen **1**: 117
Ellenbeuge, Arterien **1**: 140–141
– Hautnerven **1**: 151
– Nerven **1**: 140–141
– subkutane Nerven **1**: 151
– – Venen **1**: 151
Embryo **2**: 242
Eminentia arcuata **2**: 10, 291, 396
– collateralis ventriculi lateralis **2**: 322–323, 348, 350
– frontalis → Tuber frontale
– iliopubica **1**: 174, 176, 178
– intercondylaris tibiae **1**: 182–183, 197–198
– maxillaris → Tuber maxillare
– medialis fossae rhomboideae **2**: 309–311
– pyramidalis **2**: 396, 398
Enamelum **2**: 40
Encephalon **1**: 22; **2**: 303–306, 336–362
End phalanx of middle finger **1**: 167
Endolymphräume des Labyrinths **2**: 405
Endometrium → Tunica mucosa uteri
Endphalanx des Mittelfingers **1**: 167
Entwicklung, Handwurzelknochen **1**: 8
– Hüftbein **1**: 176
– Hüftgelenk **1**: 177
Ependyma **2**: 326
Epicardium **2**: 104, 123, 132–134, 142–143, 154
Epicondylus lateralis femoris **1**: 179–181
– – humeri **1**: 86, 90, 98, 105, 108–109, 116, 118–119, 135, 142–143, 148, 150, 154
– medialis femoris **1**: 179–181
– – humeri **1**: 86, 90–91, 98, 105–107, 114–116, 136, 140–141, 148, 152–153
Epidermis **1**: 11
Epididymis **1**: 17, 67; **2**: 170, 232, 270–274
– head **2**: 271
Epigastrium → Regio epigastrica
Epiglottis **2**: 46, 50–53, 65, 68–69, 101
Epimysium **1**: 10
Epiorchium **2**: 270–271, 274
Epipharynx **2**: 53
Epiphysis **1**: 8
Epithelium anterius corneae **2**: 375
– lentis **2**: 375
– pigmentosum iridis **2**: 375
– posterius corneae **2**: 375
Eponychium **1**: 11
ERCP → Cholangiopancreatography, retrograde, endoscopic
Erregungsleitungsbündel des Herzens **2**: 132–134
Erregungsleitungssystem des Herzens **2**: 140
Esophagus **2**: 105–107
– narrowings **2**: 106
Euryprosopia **2**: 2
Euryprosopie **2**: 2
Eurysomer **1**: 6
Excavatio disci nervi optici **2**: 374, 377
– rectouterina **1**: 17; **2**: 233, 238–239, 242–243, 256–257, 260
– rectovesicalis **2**: 170, 193, 206, 232, 247, 258
– vesicouterina **1**: 17; **2**: 233, 238, 243, 260
Excretion urogram **2**: 217
Expansion, digital, dorsal **1**: 124–125
– pelvic **1**: 178
Expiration **1**: 40–41; **2**: 153
Exspiration **1**: 40–41; **2**: 153
Extension **1**: 5
Extremitas acromialis claviculae **1**: 84, 95, 142–143
– anterior splenica **2**: 189
– inferior glandulae thyroideae **2**: 71
– – renis **2**: 208, 215, 217–218
– – testis **2**: 270
– posterior splenica **2**: 189
– sternalis claviculae **1**: 84
– superior glandulae thyroideae **2**: 71
– – renis **2**: 208, 215, 218, 222
– – testis **2**: 270
– tubaria ovarii **2**: 239–240
– uterina ovarii **2**: 239–240
Extremität, obere **1**: 82–83
– – Arterien **1**: 138–139
– – Dermatome **1**: 132–134, 244
– – Hautinnervation **1**: 132–133
– – segmentale Innervation **1**: 132–134, 244
– untere **1**: 170–171
– – Arterien **1**: 250–251, 269
– – Dermatome **1**: 134, 242–244
– – Hautinnervation **1**: 242–243
– – segmentale Innervation **1**: 134, 242–244
Eye **2**: 364–366
– extrinsic muscles **2**: 370–371, 379
– – – common tendinous ring **2**: 379
Eyeball **2**: 62, 362, 370–371, 374–375, 384–386
– arteries **2**: 372–373
– autonomic innervation **2**: 382
– parasympathetic innervation **2**: 382
– sympathetic innervation **2**: 382
– veins **2**: 372–373

F

Face **2**: 44
– arteries **2**: 86–90
– broad **2**: 2
– deep muscles **2**: 30
– – veins **2**: 80
– muscles **2**: 30, 37
– narrow **2**: 2
– nerves **2**: 85–90
– superficial arteries **2**: 85
– – muscles **2**: 28–29, 366
– – veins **2**: 80, 85
– veins **2**: 80
Facies anterior corneae **2**: 374
– – corporis maxillae **2**: 24
– – lateralis corporis humeri **1**: 86
– – lentis **2**: 374, 377
– – medialis corporis humeri **1**: 86
– – palpebralis **2**: 366
– – partis petrosae ossis temporalis **2**: 20
– – patellae **1**: 181
– – prostatae **2**: 237
– – radii **1**: 88
Facies anterior
– – renis **2**: 208, 218
– – scapulae → Facies costalis scapulae
– – ulnae **1**: 89
– anterolateralis cartilaginis arytaenoideae **2**: 64
– – corporis humeri → Facies anterior lateralis corporis humeri
– anteromedialis corporis humeri → Facies anterior medialis corporis humeri
– articularis acromialis claviculae **1**: 84–85
– – anterior dentis axis **1**: 30
– – arytaenoidea cartilaginis cricoideae **2**: 64
– – calcanea anterior tali **1**: 187, 207
– – – media tali **1**: 187, 207
– – – posterior tali **1**: 187, 207
– – capitis costae **1**: 44
– – – fibulae **1**: 183
– – carpalis radii **1**: 88, 94, 99
– – cuboidea calcanei **1**: 187
– – fibularis tibiae **1**: 183
– – incudis **2**: 397
– – inferior atlantis **1**: 30
– – – tibiae **1**: 183
– – ligamenti calcaneonavicularis plantaris tali → Facies articularis pro ligamento calcaneonaviculari
– – mallei **2**: 397
– – malleoli lateralis fibulae **1**: 182–183, 214
– – – medialis tibiae **1**: 182–183, 205, 214
– – navicularis tali **1**: 187, 207
– – patellae **1**: 181, 196, 198, 200
– – posterior dentis axis **1**: 30, 50
– – pro ligamento calcaneonaviculari **1**: 187, 207
– – stapedis **2**: 397
– – sternalis claviculae **1**: 84, 110; **2**: 96
– – superior atlantis **1**: 30, 50
– – – tibiae **1**: 182
– – talaris anterior calcanei **1**: 187, 207
– – – media calcanei **1**: 187, 207
– – – ossis navicularis **1**: 207
– – – posterior calcanei **1**: 187, 207
– – thyroidea cartilaginis cricoideae **2**: 64, 67
– – tuberculi costae **1**: 44
– – ulnae **1**: 89
– auricularis ossis ilii **1**: 176
– – – sacri **1**: 34–35
– buccalis dentis **2**: 40
– cerebralis alae majoris ossis sphenoidalis **2**: 19
– – – minoris ossis sphenoidalis **2**: 19
– – partis squamosae ossis temporalis **2**: 20
– colica splenica **2**: 189
– costalis pulmonis dextri **2**: 112, 151
– – – sinistri **2**: 112
– – scapulae **1**: 85
– diaphragmatica hepatis **2**: 180, 185, 191, 204
– – pulmonis dextri **2**: 111–112, 114
– – – sinistri **2**: 111–112
– – splenica **2**: 189
– distalis dentis **2**: 39–40
– dorsalis ossis sacri **1**: 35, 175
– facialis dentis → Facies vestibularis dentis
– gastrica splenica **2**: 189
– glutaealis ossis ilii **1**: 176, 191, 221, 225
– inferior hemisphaerii cerebri **2**: 306, 315, 317
– – linguae **2**: 38, 47
– infratemporalis alae majoris ossis sphenoidalis **2**: 19
– – corporis maxillae **2**: 24
– intervertebralis corporis vertebrae **1**: 30–33
– intestinalis uteri **2**: 242
– labialis dentis **2**: 40
– lateralis fibulae **1**: 182–183, 205
– – ossis zygomatici **2**: 26, 58, 367
– – radii **1**: 88
– – testis **2**: 270

Facies lateralis
– – tibiae **1**: 182–183, 205
– lingualis dentis **2**: 39–40
– lunata acetabuli **1**: 172, 176, 190–191, 193
– malleolaris lateralis tali **1**: 187–188
– – medialis tali **1**: 187
– maxillaris alae majoris ossis sphenoidalis **2**: 19
– medialis cartilaginis arytaenoideae **2**: 64
– – fibulae **1**: 182–183
– – hemisphaerii cerebri **2**: 317
– – ovarii **2**: 239–240
– – testis **2**: 270
– – tibiae **1**: 182, 230–231, 241, 270
– – ulnae **1**: 89
– mediastinalis pulmonis dextri **2**: 111, 151
– – – sinistri **2**: 111, 151
– mesialis dentis **2**: 39
– nasalis corporis maxillae **2**: 24
– occlusalis dentis **2**: 39
– orbitalis alae majoris ossis sphenoidalis **2**: 19, 58, 368
– – – minoris ossis sphenoidalis **2**: 19, 58
– – corporis maxillae **2**: 24, 58, 62, 367, 369
– – ossis frontalis **2**: 21, 58, 357, 367–369
– – – zygomatici **2**: 58, 367–368
– palatinalis dentis **2**: 40
– patellaris femoris **1**: 179, 181, 196, 198
– pelvica ossis sacri **1**: 35, 172
– poplitea femoris **1**: 180, 195, 199, 224, 233–234
– posterior corneae **2**: 374
– – fibulae **1**: 182–183
– – humeri **1**: 86
– – lentis **2**: 376–377
– – palpebralis **2**: 366
– – partis petrosae ossis temporalis **2**: 20, 405
– – prostatae **2**: 236–237
– – radii **1**: 88
– – renis **2**: 208, 218
– – scapulae **1**: 84
– – tibiae **1**: 182–183, 234
– – ulnae **1**: 89
– renalis splenica **2**: 189
– sacropelvica ossis ilii **1**: 176
– superior tali **1**: 187–188
– superolateralis hemisphaerii cerebri **2**: 314, 316
– symphysialis ossis pubis **1**: 40, 176, 191; **2**: 276
– temporalis alae majoris ossis sphenoidalis **2**: 19, 58
– – ossis frontalis **2**: 21
– vesicalis uteri **2**: 242
– vestibularis dentis **2**: 39–40
– visceralis hepatis **2**: 191, 204
– – splenica **2**: 189
Falx cerebelli **2**: 294
– cerebri **2**: 58, 289, 291, 294–295, 300, 336, 339, 343, 347, 354–358, 360–361
– inguinalis **1**: 66, 77
– septi → Valvula foraminis ovalis
Fascia antebrachii **1**: 106–109, 114–115, 119–120, 135–136, 141, 151–152, 160
– brachii **1**: 136
– cervical **2**: 33
– – deep (prevertebral) layer **2**: 33
– – middle (pretracheal) layer **2**: 33
– – superficial (investing) layer **2**: 33
– cervicalis **1**: 64; **2**: 29, 33, 36, 44–45, 91–92
– clavipectoralis **1**: 64
– cremasterica **1**: 67; **2**: 270–271, 274
– cribriform **1**: 218, 249
– cribrosa **1**: 218, 249
– cruris **1**: 218, 230, 233–234, 241, 246, 254, 266–268, 278–279
– deltoidea **1**: 53, 136
Fascia
– diaphragmatis pelvis inferior **2**: 250–251
– – – superior **2**: 250–251
– – urogenitalis inferior **2**: 250–251, 267
– dorsalis manus **1**: 164
– – pedis **1**: 230, 278
– genus anterior **1**: 200
– glutaea **1**: 53–54, 218, 226
– inferior diaphragmatis pelvis → Fascia diaphragmatis pelvis inferior
– – – urogenitalis → Fascia diaphragmatis urogenitalis inferior
– infraspinata **1**: 53–54, 108
– lata **1**: 53–54, 65, 218, 222–223, 230, 246, 248, 254, 256, 266
– masseterica **2**: 29, 33, 36, 91
– nuchae **2**: 33, 75, 91
– obturatoria **2**: 250–251, 264–266, 279–280
– palmaris superficialis **1**: 120, 160
– parotidea **2**: 29, 33, 36, 44, 91
– pectoralis **1**: 70, 74, 136
– penis profunda **2**: 274–276
– – superficialis **2**: 275–276
– perinei superficialis **2**: 250–251, 265–266, 279
– pharyngobasilaris **2**: 47, 49, 100
– poplitea **1**: 218, 230
– spermatica externa **1**: 67; **2**: 270–271, 274
– – interna **1**: 67; **2**: 270–271, 274
– superior diaphragmatis pelvis → Fascia diaphragmatis pelvis superior
– temporalis **2**: 30, 58
– thoracolumbalis **1**: 53–55, 222, 226
– transversalis **1**: 65–67
Fasciculus(-i) atrioventricularis **2**: 135, 140
– cerebrales **2**: 333
– cuneatus **2**: 288, 310
– gracilis **2**: 288, 310
– interfascicularis **2**: 288
– lateralis plexus brachialis **1**: 134, 140–141; **2**: 94–95, 166
– lenticularis **2**: 340
– longitudinalis(-es) aponeurosis palmaris **1**: 120
– – – plantaris **1**: 237
– – inferior cerebri **2**: 333
– – ligamenti cruciformis atlantis **1**: 50
– – superior cerebri **2**: 333
– mamillotegmentalis **2**: 328
– mamillothalamicus **2**: 324, 328, 334, 344, 350
– medialis plexus brachialis **1**: 134, 140–141; **2**: 94–95, 166
– posterior plexus brachialis **1**: 134, 140–141; **2**: 93–95, 166
– proprii medullae spinalis **2**: 288
– pyramidalis **2**: 340
– sulcomarginalis **2**: 288
– thalamicus **2**: 340
– transversi aponeurosis palmaris **1**: 120, 160
– – – plantaris **1**: 237
– uncinatus cerebri **2**: 333
Faser(n), somatomotorische **1**: 23
– somatosensible **1**: 23, 79
– viszeromotorische **1**: 23, 79
– viszerosensible **1**: 23, 79
Fasersysteme des Gehirns **2**: 333
Fastigium **2**: 307, 320
Faszie, Fussrücken **1**: 230
– Hals **2**: 33
– Muskel **1**: 10
– Unterschenkel **1**: 230
Faszienlogen des Halses **2**: 33
Fat tissue, subepicardial **2**: 135
Fazialiskanal **2**: 89, 393, 396, 399
Felsenbein **2**: 393, 395–396, 399–400, 402, 406–409
Femur **1**: 2, 7, 10, 170–171, 173, 177, 179–181, 194–200, 202–204, 216–217, 220–222, 224–226, 233–234, 247–249, 252–265, 269, 277; **2**: 209, 228, 253–255, 257–259
– Muskelanheftungen **1**: 229
Femurkopf **1**: 258
– Knochenkern **1**: 177
Femurschaft **1**: 258
Fenestra cochleae **2**: 391, 393, 396, 398, 401–402, 405
– vestibuli **2**: 393, 396, 401–402
Fenster, ovales **2**: 396
Fersenbein **1**: 187
Fettgewebe, subepikardiales **2**: 135
Fetus **2**: 243
Fibra(-ae) arcuatae cerebri **2**: 333
– corticorubrales **2**: 334
– corticospinales **2**: 328
– corticotectales **2**: 334
– dentatorubrales **2**: 309, 328, 334, 340
– frontopontinae **2**: 328
– intercrurales anuli inguinalis superficialis **1**: 66
– lentis **2**: 377
– obliquae tunicae muscularis gastricae **2**: 173
– parietotemporopontinae **2**: 328
– pontis longitudinales **2**: 340, 361
– – transversae **2**: 340
– praerectales musculi levatoris ani **2**: 236, 266
– thalamoparietales **2**: 332
– zonulares lentis oculi **2**: 374–375, 377
Fibre(s), afferent, somatic **1**: 23, 79
– – visceral **1**: 23, 79
– efferent, somatic **1**: 23
– – visceral **1**: 23, 79
Fibula **1**: 170–171, 182–183, 195–197, 205, 208, 210, 214, 269, 273–277
– muscle attachments **1**: 235
Filum(-a) radicularia **2**: 284–285, 288, 311
– – anteriora **1**: 23
– – posteriora **1**: 23
– spinale → Filum terminale
– terminale **1**: 22; **2**: 284
– – durale → Filum terminale externum
– – externum **2**: 284
Fimbria(-ae) hippocampi **2**: 324–326, 340, 351
– ovarica **2**: 239
– tubae uterinae **2**: 17, 233, 239–240
Finger, Sehnenscheiden **1**: 130–131
Finger(s), arteries **1**: 165
– middle **1**: 104, 124–125, 167
– nerves **1**: 165
– synovial sheaths **1**: 130–131
Fingerarterien **1**: 165
Fingernagel **1**: 11
Fingernail **1**: 11
Fingernerven **1**: 165
Fingerrücken, Muskeln **1**: 125
Fissura(-ae) antitragohelicina **2**: 390
– cerebelli **2**: 308
– choroidea ventriculi lateralis **2**: 322, 326
– horizontalis cerebelli **2**: 307–308, 340
– – pulmonis dextri **2**: 111–112, 150, 154–155
– intrabiventeris cerebelli **2**: 308–309
– ligamenti teretis **2**: 181
– – venosi **2**: 181
– longitudinalis cerebralis **2**: 58, 296, 306, 314–315, 336–343, 347–350, 354–358, 360–361
– mediana anterior medullae oblongatae **2**: 306, 340
– – – – spinalis **2**: 285, 288
– obliqua pulmonis dextri **2**: 111–112, 150, 154–155
– – – sinistri **2**: 111–112, 150, 155
– orbitalis inferior **2**: 4, 8, 23, 58, 367–369, 379

Fissura(-ae) orbitalis
– – superior **2**: 4, 9–10, 19, 23, 58–59, 367–369, 379
– petrooccipitalis **2**: 10, 12
– petrosquamosa **2**: 20
– petrotympanica **2**: 20
– posterolateralis cerebelli **2**: 308
– praebiventeris cerebelli **2**: 308–309
– prima cerebelli **2**: 307–308
– pterygomaxillaris **2**: 368
– secunda cerebelli **2**: 307–308
– tympanomastoidea **2**: 20
Fixio centralis funiculi umbilicalis **2**: 243
Flexion **1**: 5
Flexura coli dextra **2**: 154–155, 172, 191–192, 194, 205, 223
– – hepatica → Flexura coli dextra
– – sinistra **2**: 154–155, 172, 200, 205, 248
– – splenica → Flexura coli sinistra
– duodeni inferior **2**: 188
– – superior **2**: 191
– duodenojejunalis **1**: 16; **2**: 154–155, 188, 193–194, 199, 202, 204, 206
– perinealis recti **2**: 232
– sacralis recti **2**: 232
Flocculus **2**: 308–309, 340
Floor of oral cavity **2**: 39
– – – – muscles **2**: 36
– pelvic **2**: 264–266
– – muscles **2**: 264–266
Flügelgaumengrube **2**: 88–89
Fluid, cerebrospinal **2**: 321
– – circulation **2**: 321
Flumen(-ina) pilorum **1**: 11
Foetus → Fetus
Folium(-ia) cerebelli **2**: 308
– vermis cerebelli **2**: 307–308
Follicle, ruptured **2**: 245
Folliculus(-i) linguales **2**: 46, 101
– ovaricus(-i) atreticus **2**: 245
– – primarii **2**: 241
– – tertianii **2**: 245
– – vesiculosi → Folliculi ovarici tertianii
Follikel, gesprungener **2**: 245
Fonticulus anterior **2**: 17
– anterolateralis → Fonticulus sphenoidalis
– major → Fonticulus anterior
– mastoideus **2**: 17
– minor → Fonticulus posterior
– posterior **2**: 17
– posterolateralis → Fonticulus mastoideus
– sphenoidalis **2**: 17
Foot, arteries **1**: 271, 281
– back → Foot, dorsum
– bones **1**: 206, 208–211
– dermatomes **1**: 134, 244
– dorsum, arteries **1**: 278
– – fascia **1**: 230
– – muscles **1**: 231–232, 236
– – nerves **1**: 278
– – subcutaneous nerves **1**: 278
– – – veins **1**: 278
– – veins **1**: 278
– interosseous muscles **1**: 239
– joints **1**: 206–213, 215
– ligaments **1**: 206–208, 212–213, 215
– muscle attachments **1**: 240
– nerves **1**: 271
– skeleton **1**: 184–189, 209–211
– sole, arteries **1**: 279–280
– – muscles **1**: 237–238
– – nerves **1**: 279–280
– subcutaneous nerves **1**: 270
– – veins **1**: 270
– synovial sheaths **1**: 241
Foramen(-ina) alveolaria corporis maxillae **2**: 24
– apicis dentis **2**: 40
– axillare laterale **1**: 110–111, 143
– – mediale **1**: 106, 110–111, 140, 143
– caecum linguae **2**: 46, 50, 101
– – medullae oblongatae **2**: 306, 340
– – ossis frontalis **2**: 10, 21
– costotransversarium **1**: 46
– epiploicum → Foramen omentale
– ethmoidale anterius **2**: 21, 367–368
– – posterius **2**: 21, 367–368
– frontale → Incisura frontalis
– incisivum(-a) **2**: 24
– infraorbitale **2**: 4, 6, 24, 367–368
– infrapiriforme **1**: 223–225, 257; **2**: 264, 281
– interventriculare **2**: 300, 307, 320–322, 324, 326, 355
– intervertebrale(-ia) **1**: 28, 36, 38, 48, 51, 55
– ischiadicum majus **1**: 190–191, 193
– – minus **1**: 190–193, 225, 257
– jugulare **2**: 10, 27, 399–400
– lacerum **2**: 8–10, 17, 399–400
– magnum **1**: 22, 40, 50; **2**: 8–10, 12, 18, 72, 290–291, 394–395
– mandibulae **2**: 25, 27
– mastoideum **2**: 8, 20
– mentale **2**: 4, 6, 25, 41
– obturatum **1**: 172–173, 175–176, 190, 192–193, 220
– omentale **2**: 170–171, 191
– ovale **2**: 8–10, 19, 301, 395, 399–400
– – cordis **1**: 13
– palatinum(-a) majus **2**: 24
– – minora **2**: 24, 27
– parietale **2**: 3, 12, 22
– rotundum **2**: 10, 19, 301, 368, 400
– sacralia anteriora **1**: 34, 38, 172, 174
– – pelvica → Foramina sacralia anteriora
– – posteriora **1**: 34–35; **2**: 284
– sciaticum majus/minus → Foramen ischiadicum majus/minus
– singulare **2**: 401, 403
– sphenopalatinum **2**: 55–56, 368
– spinosum **2**: 8–10, 19, 395, 399–400
– stylomastoideum **2**: 8, 20, 84
– supraorbitale → Incisura supraorbitalis
– suprapiriforme **1**: 223–225, 257; **2**: 281
– supratrochleare humeri **1**: 86
– thyrohyoideum **2**: 46–48, 65–67, 70
– transversarium vertebrae cervicalis **1**: 30–31, 49–50; **2**: 289
– venae basivertebralis **1**: 39, 47–48
– – cavae **1**: 69
– vertebrale **1**: 30–33, 47
– zygomaticofaciale **2**: 4, 6, 26, 367–368
Forceps frontalis **2**: 345–346, 348
– major → Forceps occipitalis
– minor → Forceps frontalis
– occipitalis **2**: 345–346
Forearm **1**: 156–159
– arteries **1**: 152–155
– muscle attachments **1**: 117
– muscles **1**: 114–116, 118–119
– nerves **1**: 152–154
– subcutaneous nerves **1**: 150–151
– – veins **1**: 150–151
Forebrain, choroid plexus(es) **2**: 326
– nuclei **2**: 329–332
Formatio reticularis spinalis **2**: 288
Fornix **2**: 299–300, 304–305, 307, 317, 322, 324–326, 328, 334, 338, 340–341, 344, 346, 349–350, 355–356, 361
– conjunctivae inferior **2**: 364–365, 384
– – superior **2**: 365, 384
Fornix
– gastricus → Fundus gastricus
– sacci lacrimalis **2**: 365
– vaginae **1**: 17; **2**: 233, 240, 256–257
– ventricularis → Fundus gastricus
Fossa acetabuli **1**: 172–173, 176, 190–191, 193–194; **2**: 195, 250, 257–259
– axillaris **1**: 149
– canina **2**: 24
– cerebellaris **2**: 18
– cerebralis **2**: 18
– condylaris **2**: 18
– coronoidea **1**: 86
– cranii anterior **2**: 10, 56, 60, 291, 362, 369, 394
– – media **2**: 10–11, 60, 291, 369, 380, 382, 394, 399, 406–407
– – posterior **2**: 5, 10–11, 291, 394, 399, 407
– cubitalis **1**: 106
– digastrica **2**: 25
– glandulae lacrimalis **2**: 21, 367–368
– hypophysialis **2**: 7, 10, 12, 19, 52, 55, 90, 290, 294, 302
– iliaca **1**: 172, 176, 191
– incisiva **2**: 8, 24
– infraclavicularis **1**: 4
– infraspinata **1**: 84–85, 87, 95
– infratemporalis **2**: 88
– inguinalis lateralis **1**: 67, 77; **2**: 238, 244, 272
– – medialis **1**: 67, 77; **2**: 238, 272–273
– intercondylaris **1**: 180–181, 199
– interpeduncularis **2**: 306–307, 309, 344, 351
– ischioanalis **1**: 194, 261; **2**: 250–251, 253, 255–259, 261, 265–266, 278
– jugularis **2**: 8–9, 11–12, 20, 393, 395, 400, 402
– lateralis cerebralis **2**: 316, 323, 331, 337, 345
– malleoli lateralis **1**: 182
– mandibularis **2**: 8, 11, 20, 26, 395
– navicularis urethrae **1**: 17; **2**: 232, 275, 277–278
– olecrani **1**: 86, 91, 98, 148
– ovalis **1**: 12; **2**: 132, 140
– paravesicalis **2**: 238
– poplitea **1**: 217, 222, 261, 266–268
– popliteal **1**: 245
– – arteries **1**: 255–256, 267–268
– – lymph nodes **1**: 245
– – – vessels **1**: 245
– – nerves **1**: 255–256, 267–268
– – subcutaneous nerves **1**: 254, 266
– – – veins **1**: 254, 266
– – veins **1**: 255–256, 267–268
– pterygoidea **2**: 19, 23, 368
– pterygopalatina **2**: 88, 368
– radialis **1**: 86
– rhomboidea **2**: 284, 289, 309–311, 342, 409
– sacci lacrimalis **2**: 4, 367–369
– scaphoidea **2**: 19
– subarcuata **2**: 20
– subscapularis **1**: 85
– supraclavicularis major **1**: 63; **2**: 36
– – minor **1**: 63
– supraspinata **1**: 84–85, 87, 95
– supravesicalis **1**: 67, 77; **2**: 238
– temporalis **2**: 9, 11
– triangularis auriculae **2**: 390
– trochanterica **1**: 180, 194
– vesicae biliaris **2**: 181
– – felleae → Fossa vesicae biliaris
– vestibuli vaginae **2**: 269
Fossula fenestrae cochleae **2**: 396, 398
– – vestibuli **2**: 396
– petrosa **2**: 20
Fovea articularis capitis radii **1**: 88, 99
– capitis femoris **1**: 179–180
– centralis **2**: 374, 377

Fovea
– costalis **1**: 32
– – inferior **1**: 32
– – processus transversi **1**: 32, 46
– – superior **1**: 32, 46–47
– dentis atlantis **1**: 30
– inferior sulci limitantis **2**: 310
– oblonga cartilaginis arytaenoideae **2**: 64
– pterygoidea **2**: 25
– sublingualis **2**: 25
– submandibularis **2**: 25
– superior sulci limitantis **2**: 310
– triangularis cartilaginis arytaenoideae **2**: 64
– trochlearis **2**: 367–368
Foveola(-ae) **2**: 377
– ethmoidales **2**: 21
– gastricae **2**: 174
– granulares **2**: 3
– radialis **1**: 105, 125
Frenulum clitoridis **2**: 269
– labii inferioris oris **2**: 38
– labiorum pudendi **2**: 269
– linguae **2**: 38–39
– praeputii **2**: 274–276
– valvae ilealis **2**: 177–178
– veli medullaris superioris **2**: 310–311
Frill of iris **2**: 376
Frontalebenen **1**: 5
Frühschwangerschaft **2**: 242
Fundus gastricus **1**: 43; **2**: 128, 130, 155, 173–176, 182, 191
– uteri **1**: 17; **2**: 172, 233, 238–240, 242, 245, 250–251, 260
– ventricularis → Fundus gastricus
– vesicae biliaris **2**: 186, 190–191
– – felleae → Fundus vesicae biliaris
– – urinariae **2**: 235, 258
Funiculus anterior medullae oblongatae **2**: 340
– – – spinalis **2**: 288
– lateralis medullae oblongatae **2**: 310, 340
– – – spinalis **2**: 288
– – posterior medullae spinalis **2**: 288
– separans **2**: 310
– spermaticus **1**: 63–66, 216; **2**: 234, 261, 276
– umbilicalis **1**: 13; **2**: 242–243
Fuss, Gewölbekonstruktion **1**: 189
– Hautnerven **1**: 270
– Längsgewölbe **1**: 189
– Quergewölbe **1**: 189
– Sehnenscheiden **1**: 241
– Zwischenknochenmuskeln **1**: 239
Fussarterien **1**: 271, 281
Fussbänder **1**: 206–208, 212–213, 215
Fussgelenke **1**: 206–213, 215
Fussgewölbe **1**: 189
Fussknochen **1**: 206, 208–211
Fussknochenreihe, laterale **1**: 189
– mediale **1**: 189
Fussnerven **1**: 271
– subkutane **1**: 270, 278
Fussohlenarterien **1**: 279–280
Fussohlenmuskeln **1**: 237–238
Fussohlennerven **1**: 279–280
Fussrücken, Hautnerven **1**: 270, 278
Fussrückenarterien **1**: 278
Fussrückenfaszie **1**: 230
Fussrückenmuskeln **1**: 231–232, 236
Fussrückennerven **1**: 278
– subkutane **1**: 270, 278
Fussrückenvenen **1**: 278
– subkutane **1**: 278
Fussskelett **1**: 184–189, 209–211
– Muskelanheftungen **1**: 240
Fussvenen, subkutane **1**: 270
Fusswurzel, Sehnenscheiden **1**: 241
Fusswurzelbänder **1**: 208, 213
Fusswurzelgelenke **1**: 207–211, 213
Fusswurzelknochen **1**: 208–211

G

Galea aponeurotica **1**: 11, 53; **2**: 28–30, 85, 290, 347, 349, 352–362, 409
Gallbladder **2**: 186
Gallenblase **2**: 186
Gallengänge, Leber **2**: 183
Gallengangsystem **2**: 183
Gallenwege, extrahepatische **2**: 186–187
– intrahepatische **2**: 183
Ganglion(-ia) aorticorenale **2**: 167
– cardiacum(-a) **2**: 167
– cervicale medium **1**: 25; **2**: 95–96, 100, 149, 166–167
– – superius **1**: 25; **2**: 88, 95–96, 100, 166
– cervicothoracic, puncture **2**: 95
– cervicothoracicum **1**: 25; **2**: 95–96, 166–167
– – Punktion **2**: 95
– ciliare **2**: 381–382
– cochleare **2**: 404–405
– coeliacum(-a) **1**: 24–25; **2**: 167, 248–249
– geniculi **2**: 84, 87–89, 399
– impar **1**: 25
– inferius nervi vagi **2**: 100
– lumbalia **1**: 25; **2**: 248–249
– lumbaria → Ganglia lumbalia
– mesentericum inferius **1**: 24–25; **2**: 248
– – superius **1**: 24–25; **2**: 167, 248
– oticum **2**: 89
– pterygopalatinum **2**: 88–89
– renalia **2**: 248
– sacralia **1**: 25
– spinale **1**: 23, 48, 79; **2**: 95, 166, 284–287, 289
– spirale cochleae → Ganglion cochleare
– stellatum → Ganglion cervicothoracicum
– submandibulare **2**: 87
– superius nervi vagi **2**: 100
– thoracica **1**: 25; **2**: 96, 166–167
– trigeminale **1**: 22; **2**: 84, 86, 89, 293, 295, 306, 380
– trunci sympathetici **1**: 23, 25, 79; **2**: 285
– vertebrale **2**: 96, 167
– vestibulare **2**: 405
Gaster **1**: 16, 18–19, 21, 79; **2**: 106–107, 113, 155, 167, 170–176, 182, 186, 189–191, 194, 196, 200, 208, 223, 248
Gaumen **2**: 88
– weicher, Muskeln **2**: 395
Gaumenschleimhaut, Innervation **2**: 88
Gebärmutter **2**: 238, 240–242
– hochschwangere **2**: 243
Gebärmutterarterien **2**: 241
Gebärmutterkörper **2**: 242
Gebiss, definitives **2**: 39
– – Zahnanlagen **2**: 43
Gehirn **2**: 295–296, 304–306, 336–362
– Fasersysteme **2**: 333
– Seitenventrikel **2**: 322–323
– Ventrikel **2**: 320
Gehirnarterien **2**: 297–302
– oberflächliche **2**: 297
Gehirnvenen **2**: 297, 300
– oberflächliche **2**: 297
Gehörgang, äusserer **2**: 391–392, 406
– – knorpeliger Teil **2**: 392
– – nichtknorpeliger Teil **2**: 392
– innerer **2**: 399, 403, 406
Gehörknöchelchen **2**: 397
– Gelenke **2**: 397
Gehörorgan **2**: 391, 404
Gelenke **1**: 10
– Beckengürtel **1**: 190–191
– Fuss **1**: 206, 208–213
– Gehörknöchelchen **2**: 397
– Hand **1**: 102, 104
– Kehlkopf **2**: 65
– radioulnare **1**: 99, 101
– Unterschenkelknochen **1**: 205
Geniculum nervi facialis **2**: 84, 89
Genitalia, external, female **2**: 269
– – male **2**: 274, 276
– internal, female **2**: 239–243, 245
– – male **2**: 244
Genu **1**: 248, 252
– capsulae internae **2**: 332–333, 348, 355–356
– corporis callosi **2**: 307, 317, 322, 334, 345–350, 354–355
– occipitale radiationis opticae **2**: 335
– temporale radiationis opticae **2**: 335
Gesäss, Hautnerven **1**: 254
Gesässarterien **1**: 255–257
Gesässmuskeln **1**: 257
Gesässnerven **1**: 254–257
Gesässregion **1**: 257
Gesässvenen **1**: 254–257
– subkutane **1**: 254
Geschlechtsapparat, männlicher **2**: 232
– weiblicher **2**: 233
Geschlechtsorgane, äussere, männliche **2**: 274, 276
– – weibliche **2**: 269
– innere, männliche **2**: 232, 234
– – weibliche **2**: 233, 239–243
Geschlechtssystem **1**: 17
Gesicht **2**: 44
– Breitgesicht **2**: 2
– oberflächliche Muskeln **2**: 366
– Schmalgesicht **2**: 2
Gesichtsarterien **2**: 85–90
– oberflächliche **2**: 85
Gesichtsmuskeln, mimische **2**: 28–29
– oberflächliche **2**: 28–29
– tiefe **2**: 30
Gesichtsnerven **2**: 85–90
Gesichtsschädel **2**: 58, 369
Gesichtsvenen **2**: 80
– oberflächliche **2**: 80, 85
– tiefe **2**: 80
Gewölbekonstruktion, Fuss **1**: 189
Gingiva **2**: 40
Glabella **2**: 21
Gland, lacrimal, innervation **2**: 88
– parotid **2**: 44–45, 98
– – innervation **2**: 89
– sublingual **2**: 45
– – innervation **2**: 87
– submandibular **2**: 44–45
– – innervation **2**: 87
– suprarenal **2**: 216
– – fetal **2**: 210
– thyroid **2**: 70–71
– – arteries **2**: 101
– – nerves **2**: 101
– – scan **2**: 71
Glandula(-ae) adrenalis → Glandula suprarenalis
– areolares **1**: 70
– bulbourethralis **1**: 17; **2**: 250, 267, 277
– ceruminosae meatus acustici externi **2**: 392
– ciliares **2**: 366
– endocrinae **1**: 15
– lacrimalis **2**: 80, 365, 379–383, 385–386
– – accessoria **2**: 366
– – Innervation **2**: 88
– lingualis anterior **2**: 38, 47
– – apicalis → Glandula lingualis anterior

Glandula(-ae)
– mammaria **1**: 70
– parathyroidea inferior **1**: 15; **2**: 49–50, 101
– – superior **1**: 15; **2**: 49, 100–101
– parotidea **2**: 28, 36, 44–45, 49–50, 58, 63, 72–73, 85, 89, 92–93, 98, 100, 408
– – accessoria **2**: 44
– – Innervation **2**: 89
– pharyngeales **2**: 49
– pinealis → Corpus pineale
– pituitaria → Hypophysis
– prostatica → Prostata
– sebacea(-ae) meatus acustici externi **2**: 392
– – palpebrae **2**: 366
– sublingualis **2**: 38, 45, 47, 87, 99
– – Innervation **2**: 87
– submandibularis **2**: 33–34, 36, 44–45, 49–50, 85, 87, 98–99
– – Innervation **2**: 87
– suprarenalis(-es) **1**: 14–15, 19; **2**: 171, 182, 199, 210, 215–216, 218, 220–222, 248
– tarsales **2**: 366, 384
– thyroidea **1**: 15, 18, 20–21; **2**: 35–36, 47, 49–50, 52, 68, 70–71, 75, 96–98, 100–101, 149–151, 164–165, 167, 172
– – Arterien **2**: 101
– – Nerven **2**: 101
– tracheales **2**: 67, 109
– vestibularis major **2**: 251, 269, 280
Glans clitoridis **2**: 268–269, 281
– penis **1**: 17; **2**: 170, 232, 274–278
Gliederung, renkuläre, der Niere **2**: 210
– des zentralen Nervensystems **2**: 303
Gliedmasse, obere **1**: 82–83
– – Arterien **1**: 138–139
– – Dermatome **1**: 132–134, 244
– – Hautinnervation **1**: 132–133
– – segmentale Innervation **1**: 132–134, 244
– untere **1**: 170–171
– – Arterien **1**: 250–251, 269
– – Dermatome **1**: 134, 242–244
– – Hautinnervation **1**: 242–243
– – segmentale Innervation **1**: 134, 242–244
Globus pallidus lateralis **2**: 329–330, 332, 334–335, 338–339, 344, 348, 350, 356
– – medialis **2**: 329–330, 332, 334–335, 338–339, 344, 348, 350, 356
Glomus choroideum **2**: 322–323
Glottis **2**: 68
Granulatio(-ones) arachnoideae **2**: 295–297, 321
Grosshirn **2**: 314–319, 324–325, 327, 336–362
– Hinterhauptlappen **2**: 312–313
– limbischer Lappen **2**: 312–313
– Scheitellappen **2**: 312–313, 408
– Schläfenlappen **2**: 312–313, 409
– Stirnlappen **2**: 312–313, 336, 360
Grosshirnhemisphären **2**: 314–317
Grosshirnrinde **2**: 299, 318–319, 327
Grosshirnstiele **2**: 340–341
Gubernaculum testis **2**: 272
Gyrus(-i) ambiens **2**: 315, 317, 324, 334, 338
– angularis **2**: 316
– breves insulae **2**: 323, 344–345
– cerebri **2**: 316–317
– cinguli **2**: 304, 315, 317, 325–326, 337–338, 346–348, 350, 354–355, 357, 360–361
– dentatus **2**: 317, 324–326
– fasciolaris **2**: 317, 325
– frontalis inferior **2**: 314–316, 337–338, 348
– – medialis **2**: 337, 348
– – medius **2**: 314, 316, 337–338, 348
– – superior **2**: 314, 316–317, 337–338, 348
– hippocampi **2**: 304, 315, 317, 325–326, 331, 333–334, 338, 340, 348, 351, 361
– insulae **2**: 337, 348, 360

Gyrus(-i)
– intralimbicus **2**: 317, 325, 334
– lingualis **2**: 315, 317
– longus insulae **2**: 323, 344
– occipitales **2**: 314
– occipitotemporalis lateralis **2**: 304, 315, 317
– – medialis **2**: 315, 317, 333, 338
– orbitales **2**: 315, 335, 337
– parahippocampalis → Gyrus hippocampi
– paraterminalis **2**: 304, 317, 334, 337
– postcentralis **2**: 314, 316, 348, 362
– praecentralis **2**: 314, 316, 348, 362
– rectus **2**: 317, 335, 337
– semilunaris **2**: 317, 324, 334, 338
– supramarginalis **2**: 314, 316
– temporalis(-es) **2**: 331
– – inferior **2**: 315–316, 338
– – medius **2**: 316, 338
– – superior **2**: 316, 323, 337–338, 348
– – transversi **2**: 323, 338, 345
– uncinatus **2**: 317, 325

H

Haarströme **1**: 11
Haarzellen, Innenohr **2**: 404
Habenula **2**: 307, 325
Hair cells, inner ear **2**: 404
– tracts **1**: 11
Hallux **1**: 184–185, 188, 216–217
Hals **2**: 33, 47, 72–75, 91, 164–165
– autonomes Nervensystem **2**: 95–96
– Dermatome **2**: 82–83
– Hautinnervation **2**: 82–83, 91–92
– Lymphgefässe **2**: 81
– Lymphknoten **2**: 81
– oberflächliche Arterien **2**: 92
– – Venen **2**: 91–92
– segmentale Innervation **2**: 82–83
– subkutane Venen **2**: 91–92
– sympathisches Nervensystem **2**: 95–96, 166
Halsarterien **2**: 78–79, 92–93, 97–99, 118, 149, 165
Halsfaszie **2**: 33
– Lamina praetrachealis **2**: 33
– – praevertebralis **2**: 33
– – superficialis **2**: 33
– mittleres Blatt **2**: 33
– oberflächliches Blatt **2**: 33
– tiefes Blatt **2**: 33
Halsgeflecht **2**: 94
Halsmark **2**: 288
Halsmuskeln **1**: 55; **2**: 34–37
Halsnerven **2**: 91–93, 97–99, 149
Halsvenen **2**: 92–93, 97–99, 149, 164
Halswirbel **1**: 29–31, 60
Halswirbelsäule **1**: 31, 36–37, 49–51
Hammer **2**: 397
Hamstrings of thigh **1**: 221–224
Hamulus laminae spiralis **2**: 402–403
– ossis hamati **1**: 93–94, 102–103, 123–124
– pterygoideus **2**: 17, 19, 23–24, 27, 30, 48, 51, 56, 395
Hand **1**: 166–167
– arteries **1**: 160–163, 165
– dorsal interosseous muscles **1**: 126–127, 129
– dorsale Zwischenknochenmuskeln **1**: 126–127, 129
– Hautnerven **1**: 164
– interosseous muscles **1**: 126–127, 129
– joints **1**: 102, 104
– lumbrical muscles **1**: 127
– Lumbrikalmuskeln **1**: 127
– muscle attachments **1**: 128

Hand
– muscles **1**: 121–125
– nerves **1**: 160–162, 165
– palmar interosseous muscles **1**: 126–127
– Palmaraponeurose **1**: 120
– palmare Zwischenknochenmuskeln **1**: 126–127
– Sehnenscheiden **1**: 130–131
– skeleton **1**: 92–94
– Spulmuskeln **1**: 127
– subcutaneous nerves **1**: 164
– – veins **1**: 164
– subkutane Nerven **1**: 164
– – Venen **1**: 164
– Zwischenknochenmuskeln **1**: 126–127, 129
Handarterien **1**: 160–163, 165
Handgelenke **1**: 102, 104
Handmuskeln **1**: 121–125
Handnerven **1**: 160–162, 165
Handrücken **1**: 165
– Hautnerven **1**: 164
– subkutane Nerven **1**: 164
– – Venen **1**: 164
Handrückenarterien **1**: 165
Handrückenmuskeln **1**: 125
Handrückennerven **1**: 165
Handskelett **1**: 92–94
– Muskelanheftungen **1**: 128
Handwurzel **1**: 103–104, 123
– Sehnenscheiden **1**: 130–131
Handwurzelknochen **1**: 92–94
Harnapparat, männlicher **2**: 232
– weiblicher **2**: 233
Harnblase **2**: 235–237
Harnblasenboden **2**: 235
Harnblasendreieck **2**: 235
Harnblasenmuskulatur **2**: 235
Harnblasenwand **2**: 235
Harnleiter **2**: 213–214, 217, 235
– intramurale Strecke **2**: 235
Harnorgane **2**: 218
Harnröhre, männliche **2**: 235, 277–278
Harnröhrenöffnung, äussere **2**: 277
– innere **2**: 277
Harnsystem **1**: 17
Haustrum(-a) coli **2**: 177–179
Haut **1**: 11
– Spaltlinien **1**: 11
Hautinnervation **1**: 61
– Arm **1**: 132–133
– Bauchwand **1**: 78
– Bein **1**: 242–243
– Brustwand **1**: 78
Hautnerven **1**: 61
– Arm **1**: 132–133
– Bauchwand **1**: 78
– Bein **1**: 242–243, 248
– Brustwand **1**: 78
– Ellenbeuge **1**: 151
– Fuss **1**: 270
– Fussrücken **1**: 278
– Gesäss **1**: 254
– Hals **2**: 91–92
– Hand **1**: 164
– Handrücken **1**: 164
– Kniekehle **1**: 254, 266
– Oberarm **1**: 135–136
– Oberschenkel **1**: 248, 254
– Schulter **1**: 135–136
– Unterarm **1**: 150
– Unterschenkel **1**: 266, 270
Head **2**: 13
– arteries **2**: 78–79, 86–90
– cutaneous innervation **2**: 82–83
– dermatomes **2**: 82–83
– epididymis **2**: 271

Head
– femoral 1: 258
– – ligament 1: 193
– – ossification center 1: 177
– lymph nodes 2: 81
– – vessels 2: 81
– nerves 2: 84–90
– segmental innervation 2: 82–83
– superficial arteries 2: 85
– – veins 2: 85
Head's zones 1: 79
Head-Zonen 1: 79
Heart 2: 124–125, 128–140
– arteries 2: 142–143, 147–148
– cavities 2: 139
– conducting system 2: 140
– coronary arteries 2: 142–147
– – – origin 2: 135
– – veins 2: 142–143
– fibrous base 2: 135
– left atrium 2: 133
– – ventricle 2: 133–134, 140
– right atrium 2: 140
– – ventricle 2: 119, 132, 134, 140
– silhouette 2: 122
– veins 2: 142–143
Heel bone 1: 187
Helicotrema 2: 403
Helix 2: 390
Hemisphaerium cerebelli 2: 306, 308, 342–343, 350, 352–353, 358–359, 361–362, 400, 409
– cerebri 2: 314–317
Hemisphären des Grosshirns 2: 314–317
Hemispheres, cerebral 2: 314–317
Hepar 1: 12–13, 16, 18–21, 79, 148; 2: 128–131, 154–155, 160, 164, 170–172, 180–186, 190–191, 199–200, 204, 208, 213, 218, 222–224, 248
Hernia, femoral 1: 249
– femoralis 1: 249
– inguinal, congenital, complete 2: 273
– – lateral, indirect 1: 67; 2: 273
– – medial, direct 1: 67; 2: 273
– inguinalis 1: 67; 2: 273
– umbilical 1: 63
– umbilicalis 1: 63
Herz 2: 124–125, 128–140
– Binnenräume 2: 139
– Erregungsleitungssystem 2: 140
– linke Kammer 2: 133–134, 140
– linker Vorhof 2: 133
– rechte Kammer 2: 119, 132, 134, 140
– rechter Vorhof 2: 140
Herzarterien 2: 142–148
Herzbeutel 2: 107, 122–123
Herzkranzarterien 2: 142–148
– Ursprung 2: 135
– Variationen 2: 144–145
Herzkranzvenen 2: 142–143
Herzmuskulatur 2: 141
Herzschatten, röntgenologischer 2: 153
Herzsilhouette 2: 122, 153
Herzskelett 2: 135
Herzvenen 2: 142–143
Hiatus adductorius 1: 220
– analis 2: 264–265
– aorticus 1: 69; 2: 199
– basilicus fasciae brachii 1: 136, 151
– canalis nervi petrosi majoris 2: 396, 398, 400
– – – – minoris 2: 393, 396, 398–400
– maxillaris 2: 24, 55–57, 369
– oesophageus 1: 25, 69; 2: 104, 155, 162, 199, 218
– sacralis 1: 34–35, 175; 2: 265
– saphenus 1: 63–65, 218, 248–249
Hiatus saphenus
– – Lymphknoten 1: 248
– – Venenstern 1: 248
– semilunaris 2: 55–57
– tendineus → Hiatus adductorius
– urethralis 2: 266
– urogenitalis 2: 264–265
Hilum nuclei dentati 2: 309
– ovarii 2: 239, 241
– pulmonis 2: 126
– renale 2: 208, 210, 220–221
– splenicum 2: 189
Hinterhauptbein 2: 18
Hinterhauptlappen, Grosshirn 2: 312–313
Hinterhorn, Seitenventrikel 2: 342–343, 346–349, 354–356
Hip bone 1: 40, 176
– – development 1: 176
– – infantile 1: 176
– – muscle attachments 1: 227
– joint 1: 173, 192–194
– – development 1: 177
– – socket 1: 193
– muscles 1: 222–226
Hippocampus 2: 323–326, 333, 340–341, 348, 350–351, 356–358, 361–362, 408
Hirnhautarterien 2: 290
Hirnhäute 2: 285–286, 294–295
– harte 2: 289–292
– weiche 2: 295–296
Hirnhautvenen 2: 290
Hirnnerven 1: 22; 2: 84–90, 306
Hirnstamm 2: 289, 294, 307, 310–311, 324
Hirnventrikel 2: 320
Hoden 2: 270–273
– vorgeburtlicher Abstieg 2: 272
Hodenhüllen 2: 270
Hodensack 1: 67; 2: 271–274
– Inhalt 2: 271–272, 274
Hohlhandarterien 1: 160–162
Hohlhandnerven 1: 160–162
Horn, anterior, lateral ventricle 2: 337, 346–349, 354–355, 360
– inferior, lateral ventricle 2: 362
– posterior, lateral ventricle 2: 342–343, 346–349, 354–356
– temporal, lateral ventricle 2: 326
Hörrinde 2: 319
Horseshoe kidney 2: 211
Hufeisenniere 2: 211
Hüftbein 1: 40, 176
– Entwicklung 1: 176
– kindliches 1: 176
– Muskelanheftungen 1: 227
Hüftgelenk 1: 173, 192–194
– Entwicklung 1: 177
Hüftgelenkpfanne 1: 193
Hüftkopfband 1: 193
Hüftmuskeln 1: 222–226, 228–229
Hüllen des Hodens 2: 270
– – Rückenmarks 2: 286
Humeroulnargelenk 1: 101
Humerus 1: 82–83, 86–87, 90–91, 95, 98, 100, 110–111, 116, 137, 139, 144–149
– muscle attachments 1: 113
Hymen 2: 269
Hypochondrium → Regio hypochondriaca
Hypogastrium → Regio pubica
Hyponychium 1: 11
Hypopharynx 2: 53
Hypophyse 2: 293, 307
– Arterien 2: 301
Hypophysis 1: 15; 2: 13, 293, 295, 298, 301, 304–307, 321, 360
– arteries 2: 301
Hypothalamus 1: 15; 2: 307, 325, 328, 350–351, 358
– Kerne 2: 328
– nuclei 2: 328
Hypothenar 1: 105, 114–115, 120, 160
Hysterosalpingographie 2: 240
Hysterosalpingography 2: 240

I

Ileum 1: 16, 18, 20–21; 2: 170, 172, 177–178, 190, 192–196, 200, 202, 204–205, 220–221, 225, 232, 238, 243, 252, 256–261
Ilium → Os ilii
Impressio(-ones) cardiaca pulmonis dextri 2: 111, 114
– – – sinistri 2: 111
– colica hepatis 2: 181
– digitatae 2: 10
– duodenalis hepatis 2: 181
– gastrica hepatis 2: 181
– gyrorum → Impressiones digitatae
– ligamenti costoclavicularis 1: 84
– oesophagea hepatis 2: 181
– renalis hepatis 2: 181
– trigeminalis 2: 20, 399–400
Inca bone 2: 3
Incisura(-ae) acetabuli 1: 172, 176
– angularis gastrica 2: 173–176, 194
– anterior auriculae 2: 390
– apicis cordis 2: 125, 132, 141
– cardiaca gastrica 2: 173, 175
– – pulmonis sinistri 2: 111–112
– clavicularis 1: 41, 45
– costales 1: 45
– ethmoidalis 2: 21
– fibularis 1: 183
– frontalis 2: 4, 21, 367
– interarytaenoidea 2: 46, 50, 67, 101
– intertragica 2: 390
– ischiadica major 1: 172, 175–176
– – minor 1: 175–176
– jugularis ossis occipitalis 2: 18
– – – temporalis 2: 20
– – sterni 1: 45
– lacrimalis 2: 24, 367
– ligamenti teretis hepatis 2: 181, 190
– mandibulae 2: 25
– mastoidea 2: 20, 27
– nasalis 2: 24
– pancreatis 2: 188
– parietalis 2: 20
– parietooccipitalis 2: 316
– praeoccipitalis 2: 315–316
– pterygoidea 2: 19, 23
– radialis 1: 89, 99–100
– scapulae 1: 84–85, 95
– sphenopalatina 2: 57
– supraorbitalis 2: 4, 21, 367
– tentorii 2: 291
– terminalis auricularis 2: 390
– thyroidea inferior 2: 64, 70
– – superior 2: 64–65, 70
– trochlearis 1: 89–91, 99
– ulnaris 1: 88, 92
– vertebralis inferior 1: 32–33
– – superior 1: 32–33, 35
Inclinatio pelvis 1: 178
Inclination, pelvic 1: 178
Incus 2: 391, 397–398, 400, 405–406
Index, acetabular 1: 177
Index finger 1: 92–93, 105
Indusium griseum 2: 317, 324–326, 334–335, 338, 340, 345
Inferior 1: 5

Infundibulum hypophysis **2**: 291–293, 304, 307, 315, 330–331, 334–335, 358, 378, 385
– tubae uterinae **2**: 239–240
Inion **2**: 8, 18
Injection, intragluteal **1**: 257
Injektion, intraglutäale **1**: 257
Inkabein **2**: 3
Inlet, pelvic **1**: 178
Innenohr **2**: 391, 403
Innenrotation **1**: 5
Innervation, autonomic, eyeball **2**: 382
– cutaneous, dorsal body wall **1**: 61
– – head **2**: 82–83
– – lower limb **1**: 242–243
– – neck **2**: 82–83
– – upper limb **1**: 132–133
– – ventral body wall **1**: 78
– Gaumenschleimhaut **2**: 88
– lacrimal gland **2**: 88
– mucous membrane of nose **2**: 88
– – – – palate **2**: 88
– Nasenschleimhaut **2**: 88
– Ohrspeicheldrüse **2**: 89
– parotid gland **2**: 89
– segmental **1**: 134, 244
– – dorsal body wall **1**: 61
– – head **2**: 82–83
– – lower limb **1**: 242–243
– – neck **2**: 82–83
– – upper limb **1**: 132–133
– – ventral body wall **1**: 78
– segmentale **1**: 134, 244
– – Arm **1**: 132–133
– – Bein **1**: 242–243
– – Hals **2**: 82–83
– – hintere Rumpfwand **1**: 61
– – Kopf **2**: 82–83
– – vordere Rumpfwand **1**: 78
– sublingual gland **2**: 87
– submandibular gland **2**: 87
– teeth **2**: 86
– Tränendrüse **2**: 88
– Unterkieferspeicheldrüse **2**: 87
– Unterzungenspeicheldrüse **2**: 87
– vegetative, Augapfel **2**: 382
– Zähne **2**: 86
Insertio musculi **1**: 10
Inspiration **1**: 40–41; **2**: 153
Insula(-ae) **2**: 323, 338–340, 344–346, 350–351, 355, 361
– pancreaticae **1**: 15
Integumentum commune **1**: 11
Interneuron **1**: 23, 79
Intersectio(-ones) tendinea(-ae) musculi recti abdominis **1**: 62–65
Intestine, large **2**: 202–205
– – arteries **2**: 202
– small **2**: 176, 194–195, 203
Intestinum crassum **1**: 79
– tenue **1**: 16, 79; **2**: 170–171, 176, 192–196, 200, 203, 225, 232
Introitus labyrinthi ethmoidalis **2**: 22
Intumescentia cervicalis **1**: 22; **2**: 284
– lumbosacralis **1**: 22; **2**: 284
Iris **2**: 364, 370–371, 373–376, 384
Iriskrause **2**: 376
Ischium → Os ischii
Isthmus cartilaginis auricularis **2**: 390
– glandulae thyroideae **2**: 36, 52, 70–71
– gyri cinguli **2**: 304, 315, 317
– prostatae **2**: 237
– tubae uterinae **2**: 239–240
– uteri **2**: 233, 240, 256–257, 260

J

Jejunum **1**: 16, 18, 20–21; **2**: 154–155, 170, 172, 175–176, 192–194, 199, 208, 223–224
Joint(s) **1**: 9–10
– atlantoaxial **1**: 49–50
– atlantooccipital **1**: 49–51
– auditory ossicles **2**: 397
– cartilaginous **1**: 9
– costovertebral **1**: 46–47
– elbow **1**: 90–91, 98, 100–101, 147, 156
– fibrous **1**: 9
– foot **1**: 206, 208–213, 215
– hand **1**: 102, 104
– hip **1**: 173, 177, 192–194
– humeroulnar **1**: 101
– interphalangeal, distal **1**: 125
– – proximal **1**: 125
– intertarsal **1**: 207
– knee **1**: 10, 195–204, 265
– larynx **2**: 65
– osseous **1**: 9
– pelvis **1**: 190–191
– radiocarpal **1**: 159
– radioulnar, distal **1**: 99
– – proximal **1**: 99, 101
– sacroiliac **1**: 194
– shoulder **1**: 95–97
– spheroidal **1**: 10
– sternoclavicular **1**: 45
– sternocostal **1**: 45
– subtalar **1**: 188, 206–215
– talocalcaneonavicular **1**: 188, 206–207, 209–213, 215
– talocrural **1**: 188, 206, 208–214
– tarsometatarsal **1**: 207
– temporomandibular **2**: 26–27
– – capsule **2**: 26
– – ligaments **2**: 26–27
– tibiofibular **1**: 205
Jugum(-a) alveolare(-ia) mandibulae **2**: 25
– – maxillae **2**: 4, 24
– sphenoidale **2**: 10, 19, 291

K

Kammer, linke, Herz **2**: 133–134, 140
– rechte, Herz **2**: 132, 134, 140
Kapsel, innere, Gehirn **2**: 332, 355–356
Karpalkanal **1**: 103, 122–123
Karpaltunnel **1**: 103, 122–123
Kehldeckelknorpel **2**: 64–65
Kehlkopf **2**: 68–70
– Spiegelbilder **2**: 69
Kehlkopfarterien **2**: 101
Kehlkopfbänder **2**: 66
Kehlkopfgelenke **2**: 65
Kehlkopfknorpel **2**: 64–65
Kehlkopfmuskeln **2**: 47, 66–67
Kehlkopfnerven **2**: 101
Kehlkopfskelett **2**: 65–66
Keilbein **2**: 19
Keilbeinhöhlen **2**: 23, 61, 293, 369
Kerne, Hypothalamus **2**: 328
– Mittelhirn **2**: 330–331
– Vorderhirn **2**: 329–332
Kidney(s) **2**: 208–210, 212–215
– area of hyperalgesia **2**: 209
– arteries **2**: 214–215
– fetal **2**: 210
– lobulated **2**: 210
– lower polar artery **2**: 214–215
– multiple arteries **2**: 214
– upper polar artery **2**: 214
– zone of hyperalgesia **2**: 209
Kiefergelenk **2**: 26–27
Kiefergelenkbänder **2**: 26–27
Kiefergelenkkapsel **2**: 26
Kieferhöhlen **2**: 62–63, 383
Kleinhirn **2**: 308–309, 342–343, 350–353, 358–359, 362, 409
Kleinhirn-Brücken-Winkel **2**: 340
Kleinhirnstiele **2**: 308, 361
Kleinhirnwurm **2**: 357–358
Klitoris **2**: 268
– Schwellkörper **2**: 268
Knee, arteries **1**: 252
– cap **1**: 181
– joint **1**: 10, 195–204, 265
– – collateral ligaments **1**: 200
– – cruciate ligaments **1**: 200–201
– – menisci **1**: 201
– nerves **1**: 252
– veins **1**: 252
Kniearterien **1**: 252
Kniegelenk **1**: 10, 195–204, 265
– Kollateralbänder **1**: 200
– Kreuzbänder **1**: 200–201
– Menisci **1**: 201
Kniekehle **1**: 245
– Hautnerven **1**: 254, 266
– Lymphgefässe **1**: 245
– Lymphknoten **1**: 245
Kniekehlenarterien **1**: 255–256, 267–268
Kniekehlennerven **1**: 255–256, 267–268
– subkutane **1**: 254, 266
Kniekehlenvenen **1**: 255–256, 267–268
– subkutane **1**: 254, 266
Knienerven **1**: 252
Kniescheibe **1**: 181, 196, 198
Knievenen **1**: 252
Knöchel, Sehnenscheiden **1**: 241
Knochen, Fuss **1**: 208–211
– Handwurzel **1**: 92–94
– Unterschenkel **1**: 182–183
Knochenentwicklung **1**: 8
Knochenhaft **1**: 9
Knochenkern des Femurkopfes **1**: 177
Knochenkompakta **1**: 7
Knochenspongiosa **1**: 7
Knochenszintigramm **1**: 3
Knochenverbindungen **1**: 9
Knorpel, Kehlkopf **2**: 64–65
Knorpelhaft **1**: 9
Knotenpunkt der Achsen des Bulbus **2**: 374
Kollateralbänder des Kniegelenks **1**: 200
Kollodiaphysenwinkel des Oberschenkelknochens **1**: 179
Kommissur, vordere, Gehirn **2**: 338–339, 350
Kompakta **1**: 7
Konstitutionstypen **1**: 6
Konussehne **2**: 135
Kopf **2**: 13
– Dermatome **2**: 82–83
– Hautinnervation **2**: 82–83
– Lymphgefässe **2**: 81
– Lymphknoten **2**: 81
– segmentale Innervation **2**: 82–83
Kopfarterien **2**: 76–79, 85–90
Kopfmuskeln **2**: 37
Kopfnerven **2**: 84–90
Kopfschlagader, innere **2**: 292, 302
Kopfvenen, oberflächliche **2**: 85
Koronarangiographie **2**: 146–147
Koronararterien, Herz **2**: 142–148
– – Ursprung **2**: 135
– – Variationen **2**: 144–145
Koronarvenen, Herz **2**: 142–143

Körperachsen **1**: 5
Körperbau, athletischer **2**: 153
– leptosomer **2**: 153
– pyknischer **2**: 153
Körperebenen **1**: 5
Körperfühlsphäre **2**: 319
Körperlinien **1**: 4
Körperregionen **1**: 4
Körperskelett **1**: 2–3
Körperteile **1**: 2
Kreuzband, Atlas **1**: 50
Kreuzbänder, Kniegelenk **1**: 200–201
Kreuzbein **1**: 29, 34–35, 38–39
Kreuzbeingeflecht **2**: 226–227
Kugelgelenk **1**: 10
Kyphosis sacralis **1**: 28
– thoracica **1**: 28

L

Labium externum cristae iliacae **1**: 174, 176
– inferius oris **2**: 44–45, 47, 52
– internum cristae iliacae **1**: 174, 176
– laterale lineae asperae **1**: 180
– limbi vestibulare **2**: 404
– majus pudendi **1**: 17, 259; **2**: 233, 250, 252, 269, 281
– mediale lineae asperae **1**: 180, 223
– minus pudendi **1**: 17; **2**: 233, 240, 243, 250, 252, 269, 281
– superius oris **2**: 44–45, 52, 55
Labrum acetabulare **1**: 193
– acetabuli → Labrum acetabulare
– glenoidale scapulae **1**: 96–97
Labyrinth, bony **2**: 400–402
– – of newborn child **2**: 401
– häutiges **2**: 405
– knöchernes **2**: 400–402
– – eines Neugeborenen **2**: 401
– membranous **2**: 405
Labyrinthorgan **2**: 391
Labyrinthus ethmoidalis → Sinus ethmoidales
– membranaceus **2**: 405
– osseus **2**: 400–402
Lacertus musculi recti lateralis bulbi **2**: 370–371, 379
Lacuna(-ae) laterales sinus sagittalis superioris **2**: 295, 297
– urethrales **2**: 277
– vasorum retroinguinalis **1**: 219
Lage, extraperitoneale **2**: 171
– intraperitoneale **2**: 171
– retroperitoneale (sekundäre) **2**: 171
Laimer-Dreieck **2**: 49
Laimer's triangle **2**: 49
Lamina(-ae) affixa **2**: 300, 322, 326
– albae corporis medullaris cerebelli **2**: 309
– arcus vertebrae **1**: 33
– basilaris ductus cochlearis **2**: 404
– cartilaginis cricoideae **2**: 51–52, 64–68, 101, 109
– – thyroideae **2**: 64–66, 68, 70, 149
– chorionica **2**: 243
– choroidocapillaris choroideae **2**: 373
– cribrosa ossis ethmoidalis **2**: 7, 10, 12, 22–23, 55–56, 60, 88, 90, 291
– – sclerae **2**: 374
– episcleralis **2**: 374–375
– externa calvariae **2**: 3
– horizontalis ossis palatini **2**: 8, 12, 17, 24, 57–58
– interna calvariae **2**: 3
– lateralis processus pterygoidei ossis sphenoidalis **2**: 19, 23–24, 48, 368, 395
Lamina(-ae)
– medialis processus pterygoidei ossis sphenoidalis **2**: 19, 23–24, 30, 89, 395
– medullaris externa thalami **2**: 327, 340, 344
– – interna thalami **2**: 327, 340, 344
– – lateralis corporis striati **2**: 335, 338, 344, 348, 350
– – medialis corporis striati **2**: 335, 338, 344, 348, 350
– membranacea tubae auditoriae **2**: 395
– modioli cochleae **2**: 402
– orbitalis ossis ethmoidalis **2**: 7, 22, 59, 62–63, 368–369, 378, 385
– parietalis pericardii serosi **2**: 104, 123, 132, 142–143, 154–155
– – tunicae vaginalis testis **2**: 270–271, 274
– perpendicularis ossis ethmoidalis **2**: 12–13, 22–23, 58
– – – palatini **2**: 24, 55–57
– praetrachealis fasciae cervicalis **1**: 64; **2**: 33, 36
– praevertebralis fasciae cervicalis **2**: 33
– profunda fasciae cruris **1**: 233–234
– – – temporalis **2**: 30
– – – thoracolumbalis **1**: 54–55
– septi pellucidi **2**: 322
– spiralis ossea **2**: 402–404
– – secundaria **2**: 402–403
– superficialis fasciae cervicalis **2**: 29, 33, 44–45, 91–92
– – – temporalis **2**: 30
– – – thoracolumbalis **1**: 53–55
– tectalis **2**: 304–305, 315, 348
– tecti → Lamina tectalis
– terminalis **2**: 307
– tragi **2**: 390
– vasculosa choroideae **2**: 373
– visceralis pericardii serosi **2**: 104, 123–125, 132–134, 142–143, 154
– – tunicae vaginalis testis **2**: 270–271, 274
Längsachsen **1**: 5
Langschädel **2**: 2
Längsgewölbe, Fuss **1**: 189
Lappen, Grosshirn **2**: 312–313
– limbischer, Grosshirn **2**: 312–313
Laryngopharynx **2**: 53
Laryngoscopy **2**: 69
Larynx **1**: 16; **2**: 47, 66–70, 74, 101, 106
– Arterien **2**: 101
– arteries **2**: 101
– Bänder **2**: 66
– cartilages **2**: 64–65
– Gelenke **2**: 65
– joints **2**: 65
– Knorpel **2**: 64–65
– ligaments **2**: 66
– muscles **2**: 47, 66–67
– Muskeln **2**: 47, 66–67
– Nerven **2**: 101
– nerves **2**: 101
– skeleton **2**: 65–66
– Skelett **2**: 65–66
Lateralis **1**: 5
Leber **2**: 180–185
– Gallengänge **2**: 183
– eines Neugeborenen **2**: 181
– Pfortader **2**: 183
– Segmentgliederung **2**: 184–185
Leberarterien **2**: 183
Lebervenen **2**: 185
Leg **1**: 272–277
– arteries **1**: 267–268, 271
– dermatomes **1**: 134, 242–244
– fascia **1**: 230
– lower, bones **1**: 182–183
Leg
– muscle attachments **1**: 235
– muscles **1**: 231–234
– nerves **1**: 267–268, 271
– subcutaneous nerves **1**: 266, 270
– – veins **1**: 266, 270
– veins **1**: 267–268
Leistenbruch, lateraler, indirekter **1**: 67; **2**: 273
– – – angeborener **2**: 273
– medialer, direkter **1**: 67; **2**: 273
Leistenhernie → Leistenbruch
Leistenkanal **1**: 66–67; **2**: 272–273
Leistenregion **1**: 63, 65–67, 77; **2**: 272–273
– Lymphgefässe **2**: 163
– Lymphknoten **2**: 163
Leistenring, innerer **2**: 273
Lemniscus medialis **2**: 328
Lendengeflecht **2**: 226–227
Lendenkreuzbeingeflecht **2**: 226–227
Lendenmark **2**: 288
Lendenmuskeln **1**: 55
Lendenwirbel **1**: 29, 33, 48, 55
Lendenwirbelsäule **1**: 38–39, 48
– Muskelanheftungen **1**: 228
Lens bulbi oculi **2**: 61–62, 353, 359, 373–375, 377, 384–385
– crystalline, eyeball **2**: 377
– star, eyeball **2**: 377
Leptomeninx(-inges) **2**: 295–296
Leptoprosopia **2**: 2
Leptoprosopie **2**: 2
Leptosomer **1**: 6
Lid(er), Auge **2**: 364–366
Lid(s), eye **2**: 364–366
Lien → Splen
Ligament(s), annular, radius **1**: 99
– broad, uterus **2**: 239
– collateral, knee joint **1**: 200
– cruciate, knee joint **1**: 200–201
– cruciform, atlas **1**: 50
– femoral head **1**: 193
– foot **1**: 206, 208, 212–213, 215
– hepatoduodenal **2**: 183
– larynx **2**: 66
– pelvis **1**: 190–191
– temporomandibular joint **2**: 26–27
Ligamentum(-a) acromioclaviculare **1**: 9, 95–96, 108
– alare(-ia) **1**: 50
– anococcygeum **2**: 230, 265–266, 279
– anulare(-ia) radii **1**: 98–99
– – tracheae **2**: 47, 50, 65–66, 70, 108–109
– apicis dentis **1**: 50–51; **2**: 294
– arcuatum dorsale **1**: 102
– – laterale **1**: 69; **2**: 105
– – mediale **1**: 69; **2**: 105
– – medianum **1**: 69; **2**: 105
– – pubis **1**: 190, 193
– arteriosum **1**: 12; **2**: 107, 122–124, 142–143
– bifurcatum **1**: 206–207
– calcaneocuboideum dorsale **1**: 206–207
– – ligamenti bifurcati **1**: 206
– – plantare **1**: 211–213
– calcaneofibulare **1**: 206, 212, 214
– calcaneonaviculare ligamenti bifurcati **1**: 206
– – plantare **1**: 207, 212–213
– capitis costae radiatum **1**: 46–47
– – femoris **1**: 193; **2**: 256
– – fibulae anterius **1**: 195–196, 205
– – – posterius **1**: 195
– cardinale uteri **2**: 239, 246, 251, 256
– carpi radiatum **1**: 102
– carpometacarpalia dorsalia **1**: 102
– – palmaria **1**: 102

Ligamentum(-a)
– collaterale(-ia) articulationis interphalangealis manus **1**: 104
– – – – pedis **1**: 206
– – – metacarpophalangealis **1**: 104
– – – metatarsophalangealis **1**: 206
– – carpi radiale **1**: 102, 104
– – – ulnare **1**: 102, 104
– – fibulare **1**: 195–196, 199–200, 220, 272
– – mediale articulationis talocruralis → Ligamentum mediale articulationis talocruralis
– – radiale **1**: 98–99
– – tibiale **1**: 195–196, 199–200, 220–221, 272
– – ulnare **1**: 98–99
– conoideum **1**: 95–96
– coracoacromiale **1**: 9, 95–96
– coracoclaviculare **1**: 9, 95–96, 106–107, 110
– coracohumerale **1**: 95–96
– corniculopharyngeum **2**: 65
– coronarium hepatis **2**: 170, 180–181, 206–207
– costoclaviculare **1**: 45
– costotransversarium **1**: 47
– – laterale **1**: 47
– – superius **1**: 46
– costoxiphoideum(-a) **1**: 45, 63
– cricoarytaenoideum **2**: 66
– cricothyroideum medianum **2**: 48, 65–68, 70
– cricotracheale **2**: 65–66
– cruciatum anterius **1**: 196, 198–204, 265
– – posterius **1**: 196, 198–204, 272
– cruciforme atlantis **1**: 50
– cuboideonaviculare dorsale **1**: 206–207
– cuneonavicularia plantaria **1**: 213
– deltoideum → Ligamentum mediale articulationis talocruralis
– denticulatum **2**: 284–286, 289
– epididymidis inferius **2**: 270
– – superius **2**: 270
– falciforme hepatis **2**: 171, 180, 190–191, 204, 206, 222
– flavum(-a) **1**: 37, 39, 47, 49; **2**: 75, 285–286
– fundiforme penis **1**: 66
– gastrocolicum **2**: 190–191, 196, 204
– gastrolienale → Ligamentum gastrosplenicum
– gastrophrenicum **2**: 206
– gastrosplenicum **2**: 171, 191
– glenohumerale inferius **1**: 95–96
– – medium **1**: 95–96
– – superius **1**: 95–96
– hepatoduodenale **2**: 171, 183, 191, 204, 206
– hepatogastricum **2**: 170, 183, 191
– hepatorenale **2**: 170, 207
– hyoepiglotticum **2**: 66
– iliofemorale **1**: 192–193, 225
– iliolumbale **1**: 58–59, 190
– incudis posterius **2**: 398
– – superius **2**: 391, 398
– inguinale **1**: 14, 64–66, 75, 77, 190, 218–219, 248–249, 252; **2**: 209, 227–228
– intercarpalia dorsalia **1**: 102
– – interossea **1**: 104
– interclaviculare **1**: 45
– interfoveolare **1**: 66, 77
– interspinale(-ia) **1**: 39, 48
– intertransversarium(-a) **1**: 46–47, 59
– ischiofemorale **1**: 192–193
– lacunare **1**: 77, 190, 219, 249
– laterale articulationis temporomandibularis **2**: 26, 31
– latum uteri **2**: 239–241, 246
– lienorenale → Ligamentum splenorenale
– longitudinale anterius **1**: 37, 39, 46–49, 51, 190
– – posterius **1**: 37, 39, 47–48; **2**: 285
– mallei superius **2**: 391, 398

Ligamentum(-a)
– mediale articulationis talocruralis **1**: 206, 212–214
– meniscofemorale anterius **1**: 200
– – posterius **1**: 199–200
– metacarpale(-ia) transversum profundum **1**: 102, 124, 162
– – – superficiale **1**: 160
– – dorsalia **1**: 102
– – interossea **1**: 104
– – palmaria **1**: 102
– metatarsale(-ia) transversum profundum **1**: 213
– – dorsalia **1**: 206–207
– – interossea **1**: 215
– nuchae **1**: 60; **2**: 33, 73–74
– ovarii proprium **1**: 17; **2**: 238–241
– palpebrale laterale **2**: 365, 383
– – mediale **2**: 28, 365–366, 383
– patellae **1**: 10, 195–196, 198, 203–205, 218–220, 230–232, 235, 265, 272
– pectinatum sclerae → Reticulum trabeculare sclerae
– pectineale **1**: 192–193, 219
– pectineum → Ligamentum pectineale
– phrenicocolicum **2**: 191, 207
– pisohamatum **1**: 102, 124
– pisometacarpale **1**: 102
– plantare(-ia) longum **1**: 212–214, 238
– – articulationum metatarsophalangealium **1**: 213
– popliteum arcuatum **1**: 195, 233–234
– – obliquum **1**: 195, 221, 233–234
– pterygospinale **2**: 27
– pubicum superius **1**: 190, 193
– pubofemorale **1**: 192–193
– pubovesicale **2**: 246–247
– pulmonale **2**: 111
– quadratum **1**: 99
– radiocarpale dorsale **1**: 102
– – palmare **1**: 102
– rectouterinum **2**: 239, 246–247
– rectovesicale **2**: 246–247
– reflexum **1**: 66
– sacrococcygeum posterius profundum **1**: 190
– – – superficiale **1**: 190
– sacroiliaca anteriora **1**: 190–191
– – interossea **1**: 190
– – posteriora **1**: 190–191
– sacrospinale **1**: 190–193, 225–226, 257; **2**: 230, 265, 279, 281
– sacrotuberale **1**: 190–193, 223–226, 256–257; **2**: 230, 265, 279–281
– sphenomandibulare **2**: 27, 86
– spirale ductus cochlearis → Crista spiralis ductus cochlearis
– splenorenale **2**: 206–207
– sternoclaviculare anterius **1**: 45
– sternocostale intraarticulare **1**: 45
– – radiatum **1**: 45
– sternopericardiaca **2**: 122
– stylohyoideum **2**: 27, 35, 51
– stylomandibulare **2**: 26–27
– supraspinale **1**: 39, 46–48, 190
– suspensorium(-a) mammaria **1**: 70–71, 74
– – ovarii **2**: 233, 238–241, 251
– – penis **1**: 63–64, 66
– talocalcaneum interosseum **1**: 188, 206–208, 210–211, 214–215
– – laterale **1**: 206–207
– – mediale **1**: 207, 212
– – posterius **1**: 212
– talofibulare anterius **1**: 206
– – posterius **1**: 212, 214
– talonaviculare **1**: 212

Ligamentum(-a)
– tarsi dorsalia **1**: 206–207, 212
– – interossea **1**: 215
– tarsometatarsalia dorsalia **1**: 206–207, 212
– – plantaria **1**: 213
– teres hepatis **1**: 12; **2**: 180–181, 185, 190–191, 196, 200, 204, 248
– – uteri **1**: 17; **2**: 163, 172, 233, 238–240, 245–246, 250–251
– thyroepiglotticum **2**: 66
– thyrohyoideum laterale **2**: 65–67, 101
– – medianum **2**: 47–48, 65–66, 68, 70
– tibiofibulare anterius **1**: 205–206
– – posterius **1**: 206, 212
– trachealia → Ligamenta anularia tracheae
– transversum acetabuli **1**: 190, 193
– – atlantis **1**: 50–51; **2**: 294
– – genus **1**: 196
– – perinei **2**: 266–267
– – scapulae inferius **1**: 95, 143
– – – superius **1**: 95, 140, 143
– trapezoideum **1**: 95–96, 107
– triangulare dextrum hepatis **2**: 180, 207
– – sinistrum hepatis **2**: 180, 206–207
– tuberculi costae **1**: 47
– ulnocarpale palmare **1**: 102
– umbilicale mediale **1**: 12, 67, 77; **2**: 234, 245
– – medianum **1**: 67; **2**: 235, 245
– venae cavae **2**: 181
– venosum **1**: 12; **2**: 181
– vesicouterinum **2**: 246–247
– vestibulare **2**: 66, 68
– vocale **2**: 66
Limb, lower **1**: 170–171
– – arteries **1**: 250–251, 269
– – cutaneous innervation **1**: 242–243
– – deep veins **1**: 246
– – dermatomes **1**: 134, 242–244
– – lymph nodes **1**: 245
– – – vessels **1**: 245
– – perforating veins **1**: 246
– – segmental innervation **1**: 134, 242–244
– – superficial veins **1**: 246
– – surface relief **1**: 216–217
– – veins **1**: 246–247
– upper **1**: 82–83
– – arteries **1**: 138–139
– – cutaneous innervation **1**: 132–133
– – dermatomes **1**: 132–134, 244
– – segmental innervation **1**: 132–134, 244
Limbus acetabuli **1**: 172, 175–176, 193
– corneae **2**: 374–375
– fossae ovalis **2**: 132
– Giacomini **2**: 317, 325, 334
– laminae spiralis osseae **2**: 404
– palpebralis anterior **2**: 364, 366
– – posterior **2**: 364, 366
Limen insulae **2**: 323
– nasi **2**: 52, 55, 57
Linea(-ae) alba **1**: 63–66, 68, 77; **2**: 232
– anocutanea **2**: 179
– anorectalis **2**: 179
– arcuata ossis ilii **1**: 176; **2**: 244, 264
– – vaginae musculi recti abdominis **1**: 65
– aspera corporis femoris **1**: 180, 223–224
– axillaris anterior **1**: 4
– – posterior **1**: 4
– condylopatellaris lateralis **1**: 198
– – medialis **1**: 181
– corporis humani **1**: 4
– epiphysialis **1**: 8
– – femoris **1**: 202
– – radii **1**: 93
– – tibiae **1**: 202, 208, 214
– glutaealis anterior **1**: 176

Linea(-ae) glutaealis
– – inferior **1**: 176
– – posterior **1**: 176
– innominata **2**: 59, 367
– intercondylaris **1**: 180
– intermedia cristae iliacae **1**: 174, 176
– intertrochanterica **1**: 179, 192
– mediana anterior **1**: 4
– – posterior **1**: 4
– medioclavicularis **1**: 4
– musculi solei **1**: 182
– mylohyoidea **2**: 25
– nuchalis inferior **1**: 55; **2**: 8, 18
– – superior **1**: 55; **2**: 8, 18
– – suprema **2**: 18
– obliqua cartilaginis thyroideae **2**: 64, 66
– – mandibulae **2**: 25
– parasternalis **1**: 4
– paravertebralis **1**: 4
– pectinea femoris **1**: 180
– scapularis **1**: 4
– semilunaris **1**: 62, 65
– sternalis **1**: 4
– supracondylaris lateralis **1**: 180
– – medialis **1**: 180
– temporalis inferior ossis parietalis **2**: 6, 22
– – ossis frontalis **2**: 21
– – – temporalis **2**: 20
– – superior ossis parietalis **2**: 6, 22, 31
– terminalis pelvis **1**: 34, 172–174, 191; **2**: 195, 217
– transversae ossis sacri **1**: 9, 34
– trapezoidea **1**: 84
Lingua **2**: 13, 46–47, 51–52, 58, 68, 87, 336
Lingula cerebelli **2**: 307–309, 351
– mandibulae **2**: 25, 27
– pulmonis sinistri **2**: 111–112, 150, 160
– sphenoidalis **2**: 10, 19
Linse, Augapfel **2**: 377
Linsenstern, Augapfel **2**: 377
Liquor cerebrospinalis **2**: 75, 287, 294, 321
– – Zirkulation **2**: 321
Liver **2**: 180–185
– arteries **2**: 183
– bile ducts **2**: 183
– of newborn child **2**: 181
– portal vein **2**: 183
– segmentation **2**: 184–185
– veins **2**: 185
Lobe(s), cerebral **2**: 312–313
– – frontal **2**: 312–313, 336, 360
– – limbic **2**: 312–313
– – occipital **2**: 312–313
– – parietal **2**: 312–313, 408
– – temporal **2**: 312–313, 409
– insular **2**: 323
– pyramidal, thyroid gland **2**: 70
Lobulation, fetal, kidney **2**: 210
Lobulus(-i) auricularis **2**: 390
– biventer cerebelli **2**: 307–309
– centralis cerebelli **2**: 307–308
– glandulae mammariae **1**: 70
– gracilis cerebelli **2**: 308–309
– paracentralis cerebri **2**: 314, 317
– parietalis inferior cerebri **2**: 314, 316
– – superior cerebri **2**: 314, 316–317
– quadrangularis cerebelli **2**: 308
– semilunaris inferior cerebelli **2**: 307–309
– – superior cerebelli **2**: 308
– simplex cerebelli → Pars inferoposterior lobuli quadrangularis cerebelli
– testis **2**: 270–271
– thymi **2**: 104
Lobus(-i) caudatus hepatis **2**: 180–181, 183–184, 186, 191, 196, 204, 222, 248
Lobus(-i)
– cerebrales **2**: 312–313
– cerebri → Lobi cerebrales
– frontalis cerebri **2**: 305, 312–313, 329, 336, 339, 345, 347, 349, 354–358, 360, 362, 384, 386–387
– glandulae mammariae **1**: 70–71
– – thyroideae **2**: 35, 47, 49–50, 68, 70–71, 75, 97–98, 100–101, 149–151, 164–165, 167
– hepatis dexter **2**: 154–155, 172, 180–184, 191, 204, 220–224
– – sinister **2**: 154, 172, 180–184, 190–191, 196, 200, 204, 222, 248
– inferior pulmonis dextri **2**: 111–114, 116, 149–150, 155, 160, 165
– – – sinistri **2**: 111–112, 114, 117, 149–150, 154–156, 160
– insularis → Insula
– limbicus cerebri **2**: 312–313
– medius pulmonis dextri **2**: 111–114, 116, 149–150, 155, 160, 165
– occipitalis cerebri **2**: 305, 312–313, 329, 345, 347, 349, 354–359, 362
– parietalis cerebri **2**: 305, 312–313, 329, 341–343, 347, 354, 361–362, 408
– prostatae dexter **2**: 237
– – medius **2**: 237
– – posterior **2**: 237
– – sinister **2**: 237
– pyramidalis glandulae thyroideae **2**: 70, 149
– quadratus hepatis **2**: 181, 183–184, 191, 200, 223, 248
– renales **2**: 210
– superior pulmonis dextri **2**: 111–114, 116, 150, 154–155, 165
– – – sinistri **2**: 111–112, 114, 117, 149–150, 154–156, 160
– temporalis cerebri **2**: 61–63, 312–313, 329, 339, 341–343, 345, 349, 352–353, 356–362, 385, 400, 408–409
– thymi **2**: 104
Locus caeruleus **2**: 310
Loops, intestinal **2**: 242, 273
Lordosis cervicis **1**: 28
– colli → Lordosis cervicis
– lumbalis **1**: 28
Luftröhre **2**: 105–110
– Lymphgefässe **2**: 110
– Lymphknoten **2**: 110
Lung(s) **2**: 153
– borders **2**: 152
– left **2**: 111–114, 116–117
– right **2**: 111–114, 116–117
Lunge **2**: 153
– linke **2**: 111–114, 116–117
– rechte **2**: 111–114, 116–117
Lungenarterien **2**: 119
Lungengrenzen **2**: 152
Lungenlappengrenzen **2**: 152
Lungensegmente **2**: 115–117
Lunula capitis radii **1**: 86, 88
– unguis **1**: 11
Lymph nodes, abdomen **2**: 163, 198, 203
– – axilla **1**: 72–74
– – bronchi **2**: 110
– – chest wall **1**: 72–74
– – head **2**: 81
– – inguinal region **2**: 163
– – lower limb **1**: 245
– – mesenteries **2**: 203
– – neck **2**: 81
– – pelvis **2**: 163
– – thigh **1**: 248
– – thorax **2**: 163
Lymph nodes
– – trachea **2**: 110
– vessels, abdomen **2**: 163, 198, 203
– – axilla **1**: 72–74
– – bronchi **2**: 110
– – chest wall **1**: 72–74
– – head **2**: 81
– – inguinal region **2**: 163
– – lower limb **1**: 245
– – mesenteries **2**: 203
– – neck **2**: 81
– – pelvis **2**: 163
– – thorax **1**: 76; **2**: 163
– – trachea **2**: 110
Lymphgefässe, Axilla **1**: 72–74
– Bauch **2**: 163, 198, 203
– Becken **2**: 163
– Bein **1**: 245
– Bronchien **2**: 110
– Brusthöhle **2**: 163
– Brustkorb **1**: 76
– Brustwand **1**: 72–74
– Hals **2**: 81
– Kopf **2**: 81
– Leistenregion **2**: 163
– Luftröhre **2**: 110
– Mesenterien **2**: 203
– Oberbauch **2**: 198, 203
– Thorax **2**: 163
Lymphknoten, Axilla **1**: 72–74
– Bauch **2**: 163, 198, 203
– Becken **2**: 163
– Bein **1**: 245
– Bronchien **2**: 110
– Brusthöhle **2**: 163
– Brustwand **1**: 72–74
– Hals **2**: 81
– Hiatus saphenus **1**: 248
– Kopf **2**: 81
– Leistenregion **2**: 163
– Luftröhre **2**: 110
– Mesenterien **2**: 203
– Oberbauch **2**: 198, 203
– Oberschenkel **1**: 248
– Thorax **2**: 163

M

Macula cribrosa superior **2**: 402
– lutea **2**: 374, 377
Magen **2**: 173–176, 194
– atonischer **2**: 174
– hypertonischer **2**: 174
– hypotonischer **2**: 174
– orthotonischer **2**: 174
Malleolus lateralis **1**: 182–183, 205, 208, 212, 214, 216–217, 230–233, 236, 241, 276–278
– medialis **1**: 182–183, 205, 208–209, 212, 214, 216–217, 230–231, 233–234, 236–238, 241, 270, 276, 278–279
Malleus **2**: 391–392, 397–398, 400, 405–406
Mamma **1**: 70–74; **2**: 126
Mandibula **1**: 36, 51; **2**: 5, 7, 14–15, 17, 25, 40–41, 44–47, 73, 85, 87
Mandibularkanal **2**: 86
Mantelkante des Grosshirns **2**: 316–317
Manubrium mallei **2**: 391–392, 397–400
– sterni **1**: 40–41, 45; **2**: 118, 154
Manus **1**: 2, 105, 166–167
Margo alveolaris maxillae **2**: 24
– anterior fibulae **1**: 182–183
– – pulmonis dextri **2**: 111–112
– – – sinistri **2**: 111–112
– – radii **1**: 88
– – testis **2**: 270

Margo anterior
– – tibiae **1**: 182–183, 216, 230
– – ulnae **1**: 89
– ciliaris iridis **2**: 376
– falciformis hiatus sapheni **1**: 65, 218, 248
– fibularis pedis → Margo lateralis pedis
– frontalis alae majoris ossis sphenoidalis **2**: 19
– – – minoris ossis sphenoidalis **2**: 19
– – ossis parietalis **2**: 22
– gingivalis **2**: 40
– inferior cerebri **2**: 337
– – hepatis **2**: 180
– – pulmonis **2**: 153
– – – dextri **2**: 111–112, 154
– – – sinistri **2**: 111–112
– – splenica **2**: 189
– inferolateralis hemisphaerii cerebri → Margo inferior cerebri
– inferomedialis hemisphaerii cerebri → Margo medialis cerebri
– infraorbitalis corporis maxillae **2**: 24, 367–368
– – orbitae **2**: 4–5, 367, 371
– interosseus fibulae **1**: 182–183
– – radii **1**: 88
– – tibiae **1**: 182–183
– – ulnae **1**: 89
– lacrimalis corporis maxillae **2**: 24
– lambdoideus ossis occipitalis **2**: 18
– lateralis humeri **1**: 86, 98
– – pedis **1**: 216–217
– – renis **2**: 208
– – scapulae **1**: 84–85
– liber ovarii **2**: 239–240
– – unguis **1**: 11
– linguae **2**: 46
– mastoideus ossis occipitalis **2**: 18
– medialis cerebri **2**: 337
– – humeri **1**: 86, 98
– – pedis **1**: 216–217
– – renis **2**: 208
– – scapulae **1**: 84–85, 111, 142
– – tibiae **1**: 182
– mesovaricus ovarii **2**: 239–240
– nasalis ossis frontalis **2**: 21
– occipitalis ossis parietalis **2**: 22
– – – temporalis **2**: 20
– parietalis alae majoris ossis sphenoidalis **2**: 19
– – ossis frontalis **2**: 21
– – partis squamosae ossis temporalis **2**: 20
– posterior fibulae **1**: 182–183
– – radii **1**: 88
– – testis **2**: 270
– – ulnae **1**: 89, 118
– pupillaris iridis **2**: 376
– sagittalis ossis parietalis **2**: 22
– sphenoidalis partis squamosae ossis temporalis **2**: 20
– squamosus alae majoris ossis sphenoidalis **2**: 19
– – ossis parietalis **2**: 22
– superior cerebri **2**: 314, 316–317, 337–338
– – partis petrosae ossis temporalis **2**: 20
– – scapulae **1**: 84–85
– – splenica **2**: 189
– superomedialis cerebri → Margo superior cerebri
– supraorbitalis orbitae **2**: 4–5, 21, 367
– – ossis frontalis **2**: 21, 367
– zygomaticus alae majoris ossis sphenoidalis **2**: 19
Markhöhle **1**: 7
Massa lateralis atlantis **1**: 30–31, 36, 51; **2**: 72–73, 408
Masse des weiblichen Beckens **1**: 178
Masseter muscle **2**: 30–31
Mastdarm **2**: 179, 231
Mastdarmarterien **2**: 230–231
Mastdarmnerven **2**: 231
Mastoidzellen **2**: 393–394, 399, 407
Matrix unguis **1**: 11
Maxilla **2**: 4, 6, 8, 12, 14–17, 23–24, 54–59, 367–370
Measurements, pelvic **1**: 178
Meatus, acoustic, external **2**: 392, 406
– – – cartilaginous part **2**: 392
– – – non-cartilaginous part **2**: 392
– – internal **2**: 399, 403, 406
– acusticus externus **2**: 9, 11, 302, 361, 390–392, 394–395, 400, 406, 408
– – – cartilagineus **2**: 391–392
– – – osseus **2**: 391–392
– – internus **2**: 20, 395, 399, 403, 406–407, 409
– nasalis communis **2**: 22–23, 50, 58–59, 61–62
– – inferior **2**: 23, 52, 55–58, 365, 367
– – medius **2**: 23, 52, 55–58, 365
– – superior **2**: 55–56, 365
– nasopharyngeus **2**: 55, 57
Medialis **1**: 5
Medianebene **1**: 5
Mediastinum **2**: 104, 164–165, 191
– anterius **2**: 104
– Arterien **2**: 149, 165
– arteries **2**: 149, 165
– Gliederung **2**: 104
– inferius **2**: 104
– medium **2**: 104
– Nerven **2**: 149
– nerves **2**: 149
– posterius **2**: 104
– subdivision **2**: 104
– superius **2**: 104, 153
– testis **2**: 270–271
– veins **2**: 149, 164
– Venen **2**: 149, 164
Medulla glandulae suprarenalis **2**: 216
– oblongata **1**: 22, 25, 37; **2**: 13, 95, 166, 294, 303–305, 321, 342–343, 361, 408
– ossium flava **1**: 10
– – rubra **1**: 10
– ovarii **2**: 241
– renalis **2**: 208, 210, 212
– spinalis **1**: 22–23, 37, 39, 60; **2**: 73–75, 95, 158, 160–161, 166, 284–288, 292, 294, 303, 321, 324
Membrana atlantoaxialis anterior **1**: 49, 51
– atlantooccipitalis anterior **1**: 49, 51; **2**: 294
– – posterior **1**: 49, 55, 60
– cricovocalis → Conus elasticus
– fibrosa **1**: 10–11
– intercostalis interna **1**: 47
– interossea antebrachii **1**: 99, 102, 116–117, 122, 153–154, 157–158
– – – Muskelanheftungen **1**: 117
– – cruris **1**: 195–196, 205–206, 214, 271, 278
– – – Muskelanheftungen **1**: 235
– obturatoria **1**: 77, 190–191, 193, 249; **2**: 250–251, 256–257
– perinei **2**: 267, 279
– quadrangularis laryngis **2**: 67
– reticularis organi spiralis **2**: 404
– stapedialis **2**: 398
– synovialis **1**: 10–11
– tectoria articulationis atlantoaxialis medianae **1**: 49, 51
– – ductus cochlearis **2**: 404
– thyrohyoidea **2**: 44, 46–48, 65–68, 70, 74, 97, 101
– tympanica **2**: 17, 89, 391–392, 398–400, 405
– – secundaria **2**: 391, 398, 405
– vastoadductoria **1**: 219, 252–253
Membrana
– vestibularis → Paries vestibularis ductus cochlearis
Membrane, interosseous, of forearm, muscle attachments **1**: 117
– – – leg, muscle attachments **1**: 235
– tympanic **2**: 392
Membrum inferius **1**: 170–171
– superius **1**: 82–83
– – embryonis **2**: 242
Ménard-Shenton-Linie **1**: 177
Meninx(-inges) **2**: 285–286, 294–296
Meniscus(-i) articularis **1**: 10
– – knee joint **1**: 201
– – Kniegelenk **1**: 201
– lateralis **1**: 10, 196, 198–203
– medialis **1**: 196, 198–203
Mentum **2**: 44
Mesencephalon **2**: 303, 305, 315, 328, 334, 350–351
Mesenterialwurzeln **2**: 206
Mesenterien **2**: 204
– Arterien **2**: 203
– Lymphgefässe **2**: 203
– Lymphknoten **2**: 203
Mesenteries **2**: 204
– arteries **2**: 203
– attachment lines **2**: 206
– lymph nodes **2**: 203
– – vessels **2**: 203
Mesenterium **2**: 170–171, 176, 193, 224–225, 252, 260
Mesoappendix **2**: 177, 204
Mesocolon sigmoideum **2**: 193, 196, 230
– transversum **2**: 154, 170, 179, 191–193, 204
Mesometrium **2**: 239–240
Mesopharynx **2**: 53
Mesosalpinx **2**: 239–241
Mesotendineum **1**: 10
Mesovarium **2**: 239–241
Metacarpalia → Ossa metacarpi
Metacarpus **1**: 166–167
Metatarsalia → Ossa metatarsi
Metencephalon **2**: 303
Midbrain **2**: 309, 328, 351
– nuclei **2**: 330–331
Middle finger **1**: 165
Milchgebiss **2**: 42–43
Milchzähne, Durchbruch **2**: 42
Milkteeth, eruption **2**: 42
Milz **2**: 189, 199
Mittelfinger **1**: 104, 124–125, 165, 167
Mittelfussgelenke **1**: 207
Mittelhand **1**: 166–167
Mittelhirn **2**: 309, 315, 328, 334, 350–351
– Kerne **2**: 330–331
Mittelohr **2**: 391, 393–394
Mittelphalanx des Mittelfingers **1**: 167
Modiolus cochleae **2**: 403
Mons pubis **2**: 269, 281
Motion, directions **1**: 5
Mouth **2**: 38
– floor **2**: 39, 47
Multipara **2**: 268
Mund **2**: 38
Mundboden **2**: 39, 47
Mundbodenmuskeln **2**: 36
Mundhöhle **2**: 38–39
Mundhöhlendach **2**: 39
Muscle(s) **1**: 10
– abdomen **1**: 62–66
– attachments to base of skull **2**: 37
– – – fibula **1**: 235
– – – hip bone **1**: 227
– – – humerus **1**: 113

Muscle(s) attachments to
– – – hyoid bone **2**: 34–35
– – – interosseous membrane of forearm **1**: 117
– – – – – – leg **1**: 235
– – – lumbar spine **1**: 228
– – – pelvis **1**: 228
– – – radius **1**: 117
– – – shoulder girdle **1**: 112
– – – skeleton of foot **1**: 240
– – – – – hand **1**: 128
– – – skull **2**: 32
– – – thigh bone **1**: 229
– – – thorax **1**: 68
– – – tibia **1**: 235
– – – ulna **1**: 117
– back **1**: 52–55, 60; **2**: 37
– cremasteric **2**: 274
– deep, face **2**: 30
– dorsum of foot **1**: 231–232, 236
– – – hand **1**: 125
– extrinsic, eye **2**: 370–371, 379
– face **2**: 28–30, 37
– femoral, anterior **1**: 219
– floor of oral cavity **2**: 36
– forearm **1**: 114–116, 118–119
– genuine, back **1**: 56–59
– – thorax **1**: 68
– hand **1**: 121–125
– hip **1**: 222–226
– inner, larynx **2**: 67
– interosseous, dorsal, foot **1**: 239
– – – hand **1**: 126–127, 129
– – foot **1**: 239
– – hand **1**: 126–127, 129
– – palmar, hand **1**: 126–127
– – plantar, foot **1**: 239
– larynx **2**: 47, 66
– leg **1**: 231–234
– lumbar region **1**: 55
– lumbrical, hand **1**: 127
– neck **1**: 55, 60; **2**: 34–37
– pelvic floor **2**: 264–266
– perineum **2**: 265–266
– pharynx **2**: 47–49
– pterygoid **2**: 30–31
– scalp **2**: 28–30
– shoulder **1**: 10, 106, 108–111
– soft palate **2**: 395
– – – attachment to cranial base **2**: 395
– sole of foot **1**: 237–238
– sternocleidomastoid **2**: 35
– superficial, face **2**: 366
– suprahyoid **2**: 36
– temporalis **2**: 31
– thigh **1**: 219–222, 224, 226
– tongue **2**: 38, 46–47
– upper arm **1**: 106–111
– urinary bladder **2**: 235
– ventral, torso **1**: 62–65, 68
– vertebral, anterior **2**: 37
– – lateral **2**: 37
Musculus(-i) **1**: 10
– abdominis **1**: 55, 62–66, 68, 228
– abductor digiti minimi manus **1**: 103, 121–125, 127–128, 130, 165–166
– – – – pedis **1**: 208, 211, 214, 232, 236–238, 240, 279–280
– – hallucis **1**: 208, 214–215, 231, 236–238, 240–241, 277, 279–280
– – pollicis brevis **1**: 103, 116, 121–122, 124, 128, 130, 152, 162, 166
– – – longus **1**: 114–119, 121–122, 124, 128, 130, 152–154, 157–159, 162, 165
– adductor(-ores) brevis **1**: 220, 227–229, 253, 259–260, 262

Musculus(-i) adductor(-ores)
– – femoris **1**: 216–217; **2**: 234, 252, 254, 258, 278
– – hallucis **1**: 238, 240, 280
– – longus **1**: 219–221, 227–229, 252–253, 258–260, 262–263; **2**: 228
– – magnus **1**: 195, 219–224, 227–229, 233, 252, 255–256, 258–263; **2**: 253, 255
– – minimus **1**: 220, 223–224, 227–228
– – pollicis **1**: 121–124, 126, 128, 130, 162, 166–167
– anconaeus **1**: 101, 105, 109, 113, 117–119, 142–143, 154, 156
– antebrachii **1**: 114–119
– articularis genus **1**: 195, 229
– aryepiglotticus **2**: 67
– arytaenoideus(-i) **2**: 52
– – obliquus **2**: 51, 67, 101
– – transversus **2**: 51, 67–68
– auricularis anterior **2**: 28–29
– – posterior **2**: 29–30, 91–92
– – superior **2**: 28–29, 366
– biceps brachii **1**: 62, 64, 95–99, 101, 105–112, 114–117, 140–149, 151–153, 156; **2**: 93
– – femoris **1**: 195, 199–200, 204, 217, 221–224, 226–227, 229, 232–235, 255–256, 258, 261–265, 267–268, 271
– brachialis **1**: 101, 105–109, 111, 113–119, 140–143, 146–149, 152–154, 156
– brachii **1**: 105–111, 113
– brachioradialis **1**: 101, 105–109, 113–119, 140–143, 147–149, 151–154, 156
– buccinator **2**: 28, 30–32, 44, 48, 58, 80, 85–87
– bulbospongiosus **2**: 232, 234, 265–268, 274–276, 279–280
– caninus **2**: 32, 37
– capitis **2**: 32, 37
– cervicis → Musculi colli
– chondroglossus **2**: 46
– ciliaris **2**: 374–375, 377
– cinguli membri superioris **1**: 112
– coccygeus **1**: 219, 227–228; **2**: 230, 234, 244–245, 264–266, 279
– colli **2**: 32, 34–37
– constrictor(-ores) pharyngis **2**: 101
– – – inferior **2**: 33, 35, 46–49, 68, 74, 99–100
– – – medius **2**: 34–35, 46–49, 100
– – – superior **2**: 46–49, 100
– coracobrachialis **1**: 64, 106–107, 110, 112–113, 140–141, 144–145, 148–149; **2**: 93
– corrugator supercilii **2**: 32
– coxae **1**: 222–228
– cremaster **1**: 63–67; **2**: 270–272, 274
– cricoarytaenoideus lateralis **2**: 67
– – posterior **2**: 51, 66–68, 101
– cricothyroideus **2**: 35, 47–48, 66, 68, 70, 97, 149
– cruris **1**: 10, 231–235
– – vaginae synoviales tendinum **1**: 241
– deltoideus **1**: 52–53, 62–64, 96–97, 105–113, 140–145, 148–149; **2**: 92
– depressor anguli oris **2**: 28–29, 32, 44, 48, 99
– – labii inferioris **2**: 28–29, 32, 99
– – supercilii **2**: 28, 366
– detrusor vesicae **2**: 235
– digastricus **2**: 30, 32, 34, 36–37, 44–45, 47–51, 87, 97–100
– dorsi **1**: 52–59; **2**: 32, 37
– epicranius **2**: 28–30
– erector spinae **1**: 52, 54–59, 194, 227; **2**: 213, 222–225, 287
– extensor carpi radialis brevis **1**: 101, 113–114, 118–119, 123, 125, 128, 140, 152–154, 156–157, 159, 165

Musculus(-i) extensor carpi radialis
– – – – longus **1**: 101, 105–109, 113–116, 118–119, 123, 125, 128, 140, 142–143, 147–149, 152–154, 156–159, 165
– – – ulnaris **1**: 105, 113, 117–119, 123, 125, 128, 130, 154, 157–159, 165
– – digiti minimi **1**: 118–119, 123, 125, 130, 154, 157–159, 165
– – digitorum **1**: 103, 105, 109, 113, 118–119, 123–125, 128, 130, 154, 157–159, 165
– – – brevis **1**: 215, 231–232, 236, 239–241, 278
– – – longus **1**: 216, 231–232, 235–236, 239–241, 271, 273–278
– – hallucis brevis **1**: 231–232, 236, 240–241, 278
– – – longus **1**: 210–211, 231–232, 235–236, 240, 271, 274–278
– – indicis **1**: 105, 117, 119, 125, 154, 157–159, 165
– – pollicis brevis **1**: 115, 117–119, 125–126, 128, 130, 152, 154, 157–159, 162, 165
– – – longus **1**: 105, 117–119, 123, 125–126, 128, 130, 154, 157–158, 165
– externi bulbi oculi **2**: 59, 370–371, 379
– faciales **2**: 29–30
– faciei → Musculi faciales
– femoris **1**: 219–224, 226, 228–229
– fibularis brevis/longus/tertius → Musculus peronaeus brevis/longus/tertius
– flexor accessorius → Musculus quadratus plantae
– – carpi radialis **1**: 101, 105–106, 113–116, 121–124, 128, 130, 151–153, 156–159, 161–162
– – – ulnaris **1**: 101–102, 105, 113–117, 121–124, 128, 130, 152–153, 156–158, 161–162
– – digiti minimi brevis manus **1**: 121–124, 128, 130, 162, 166
– – – – – pedis **1**: 237–238, 240
– – digitorum brevis **1**: 208, 211, 214, 237–238, 240, 279–280
– – – longus **1**: 208, 214–215, 231, 233–235, 238, 240–241, 267–268, 273–277, 280
– – – profundus **1**: 101, 116–117, 121–124, 128, 153, 156–159, 162, 166–167
– – – superficialis **1**: 101, 113–117, 121–124, 128, 130, 152–153, 156–159, 162, 166–167
– – hallucis brevis **1**: 237–238, 240, 279
– – – longus **1**: 208, 211, 213–215, 233–235, 237–238, 240–241, 267–268, 273–277, 280
– – pollicis brevis **1**: 103, 121, 123–124, 128, 130, 162, 166–167
– – – longus **1**: 114–117, 121–124, 128, 130, 152–153, 157–159, 162, 166–167
– gastrocnemius **1**: 181, 195, 200, 203–204, 217, 221–222, 224, 226, 229, 231–234, 252, 255–256, 265, 267–268, 272–274, 276–277
– gemellus inferior **1**: 223–224, 226–227, 229, 256; **2**: 253, 279
– – superior **1**: 223–224, 226–227, 229, 256; **2**: 253, 279
– genioglossus **2**: 30, 32, 35–36, 38, 45–48, 97, 99
– geniohyoideus **2**: 30, 32, 34–36, 45–48, 87, 97–99
– glutaeus maximus **1**: 52–54, 194, 217, 222–224, 226–227, 229, 255–258, 261–262; **2**: 225, 230, 253, 255–259, 265–266, 278, 280
– – medius **1**: 52–54, 194, 217, 219, 223–224, 226–227, 229, 256–258; **2**: 225, 230, 252–259

Musculus(-i) glutaeus
– – minimus **1**: 194, 224, 227, 229, 257–258; **2**: 225, 252–259
– gracilis **1**: 200, 219–224, 227–228, 231, 233, 235, 252–253, 255–256, 259–265, 267, 272
– hyoglossus **2**: 34–35, 45–48, 97–99
– hyopharyngeus **2**: 35
– iliacus **1**: 69, 77, 194, 219–220, 227–228, 252–253, 258; **2**: 211, 218, 225, 228, 230, 252, 254
– iliococcygeus **2**: 264–266
– iliocostalis cervicis **1**: 56–58; **2**: 74
– – lumborum **1**: 54, 56–58
– – thoracis **1**: 56–58
– iliopsoas **1**: 194, 219–220, 224, 229, 249, 258; **2**: 256–259
– infrahyoidei **2**: 74
– infraspinatus **1**: 95, 97, 109, 111–113, 142–144, 148
– intercostales externi **1**: 54–55, 57–59, 64–65; **2**: 37, 93, 164
– – interni **1**: 64–65, 68; **2**: 164
– interossei dorsales manus **1**: 121–122, 124–130, 165–167
– – – pedis **1**: 231, 236–240
– – manus **1**: 126–127, 129
– – palmares manus **1**: 124, 126–128, 162, 167
– – pedis **1**: 239
– – plantares pedis **1**: 237–240
– interspinales cervicis **1**: 55
– intertransversarii laterales lumborum **1**: 58–59
– – posteriores cervicis **1**: 55, 59
– – thoracis **1**: 59
– ischiocavernosus **1**: 227; **2**: 234, 261, 265–268, 275, 279–280
– ischiocrurales **2**: 253, 255
– laryngis **2**: 47
– laryngopharyngeus **2**: 35
– latissimus dorsi **1**: 52–55, 63–65, 72, 105–107, 110, 113, 140–141, 148–149, 226–227; **2**: 93
– levator(-ores) anguli oris **2**: 28
– – ani **1**: 194, 227–228; **2**: 179, 231, 234, 236, 244, 250–251, 253–259, 261, 264–266, 276, 279–281
– – costarum breves **1**: 56, 58–59
– – – longi **1**: 56, 58–59
– – labii superioris **2**: 28–29, 32, 48, 366
– – – – alaeque nasi **2**: 28–30, 32, 85, 366
– – palpebrae superioris **2**: 58, 360, 365–366, 370–371, 379–384, 386–387
– – scapulae **1**: 54, 60, 110–112, 142; **2**: 30, 33–36, 74–75, 81, 92–93, 98, 149
– – veli palatini **2**: 37, 48, 51, 73, 395
– linguae **2**: 46–47
– longissimus capitis **1**: 56–58, 60; **2**: 32, 37, 73–74
– – cervicis **1**: 56–58; **2**: 74
– – thoracis **1**: 54, 56–58
– longitudinalis inferior linguae **2**: 38, 46–47
– – superior linguae **2**: 46–47
– longus capitis **2**: 35, 37, 63, 73, 294
– – colli **2**: 33, 35, 37, 71, 74–75
– lumbricales manus **1**: 120–124, 127, 130, 160, 162, 165, 167
– – pedis **1**: 237–238
– manus **1**: 121–125, 128
– masseter **2**: 28, 30–32, 34–37, 44, 49, 51, 58, 63, 73, 81, 85–86, 92, 98–99
– membri inferioris **1**: 216–217
– mentalis **2**: 28–30, 32
– multifidus **1**: 54–56, 58–59; **2**: 33, 74–75
– mylohyoideus **2**: 30, 32, 34–36, 44–48, 86–87, 97–99
– nasalis **2**: 28–30, 32, 366
– obliquus capitis inferior **1**: 55, 58–60

Musculus(-i) obliquus capitis
– – – superior **1**: 55, 58–60; **2**: 32, 37, 73
– – externus abdominis **1**: 53–55, 57, 63–68, 194, 226–228; **2**: 196, 220–221, 225, 228, 274
– – inferior bulbi **2**: 370–371, 384, 386
– – internus abdominis **1**: 53–55, 57, 64–67, 194, 226–228; **2**: 196, 220–221, 225, 228
– – superior bulbi **2**: 58, 360, 370–371, 379–383, 386–387
– obturator externus **1**: 220, 224, 227–229, 258–260; **2**: 250–252, 254, 256–257, 259
– – internus **1**: 194, 223–227, 229, 256; **2**: 230, 244, 250–259, 264–266, 279–281
– occipitofrontalis **1**: 53–54, 60; **2**: 28–30, 32–33, 85, 91–92, 366
– omohyoideus **1**: 106–107, 110, 112, 140, 143; **2**: 33–36, 44, 81, 92–93, 97–99
– opponens digiti minimi manus **1**: 103, 121–124, 128, 130, 162, 166
– – – – pedis **1**: 232, 238, 240
– – pollicis **1**: 103, 121–124, 128, 130, 166
– orbicularis oculi **2**: 28–30, 32, 48, 85, 366, 384–385
– – oris **2**: 28–30, 32
– palatoglossus **2**: 46–47
– palatopharyngeus **2**: 45, 51
– palmaris brevis **1**: 116, 120, 152, 160, 166
– – longus **1**: 101, 105, 114–116, 120, 123, 151–152, 160, 162
– papillaris(-es) ventriculi dextri **2**: 140
– – – sinistri **2**: 137, 140
– – anterior ventriculi dextri **2**: 132
– – – – sinistri **2**: 133–134, 140, 145
– – posterior ventriculi dextri **2**: 132
– – – – sinistri **2**: 133–134, 140, 145
– – septalis ventriculi dextri **2**: 132, 140
– pectinati atrii dextri cordis **2**: 132
– – – sinistri cordis **2**: 133
– pectineus **1**: 219–220, 227–229, 249, 252–253, 258–260; **2**: 227–228, 252, 257–259
– pectoralis major **1**: 62–65, 68, 70–74, 97, 105–107, 110–113, 140–141, 144; **2**: 92–93
– – minor **1**: 64, 68, 72, 97, 106–107, 110, 112, 140–141, 144; **2**: 93, 149
– pedis **1**: 231–232, 236–238, 240
– peronaeus brevis **1**: 206–208, 213–215, 231–236, 240–241, 267–268, 271, 273–277
– – longus **1**: 206, 208, 213–215, 231–235, 238, 240–241, 267–268, 271, 273–277
– – tertius **1**: 231–232, 236, 240–241, 278
– pharyngis **2**: 47
– piriformis **1**: 219, 223–226, 228–229, 256–258; **2**: 230, 234, 264–265, 281
– plantaris **1**: 195, 200, 222, 224, 229, 233–234, 267–268
– popliteus **1**: 181, 195, 199–200, 229, 233–235, 265, 267–268, 272, 276–277
– procerus **2**: 28–30, 366
– pronator quadratus **1**: 114–117, 122, 129, 153, 158, 162
– – teres **1**: 101, 106–107, 113–117, 140–141, 147, 152–154, 156
– psoas major **1**: 14, 54–55, 69, 194, 219, 228, 252–253; **2**: 171, 182, 209, 211, 213, 217–218, 220–221, 223–225, 227–228, 230, 233, 244–245, 248, 250–252, 254
– – minor **1**: 69, 219; **2**: 228
– pterygoideus(-i) **2**: 30–31
– – lateralis **2**: 26, 30–32, 37, 49–50, 63, 73, 80, 86–87
– – medialis **2**: 30–32, 36–37, 49–51, 63, 73, 80, 86–87, 89, 100
– pubococcygeus **2**: 261, 264–266
– puborectalis **2**: 179, 236, 264–266

Musculus(-i)
– pubovesicalis **2**: 246–247
– pyramidalis **1**: 64–65, 227–228; **2**: 276
– quadratus femoris **1**: 220, 223–224, 226–229, 256; **2**: 253, 255–256
– – lumborum **1**: 54–55, 58–59, 69, 219, 223, 227–228; **2**: 171, 209, 211, 218, 223–224, 227–228
– – plantae **1**: 208, 211, 214, 238, 240, 279–280
– quadriceps femoris **1**: 10, 195–196, 202, 219–220, 226, 231, 235, 252–253, 258–264; **2**: 252, 254–255
– rectouterinus **2**: 239, 246–247
– rectovesicalis **2**: 247
– rectus abdominis **1**: 62, 64–65, 67–69, 75–77, 194, 227–228; **2**: 193, 202, 225, 245, 256–261, 276
– – capitis anterior **2**: 37, 73
– – – lateralis **1**: 55; **2**: 37
– – – posterior major **1**: 55, 58–60; **2**: 32, 37, 73
– – – – minor **1**: 55, 58; **2**: 32, 37, 73
– – femoris **1**: 193–194, 216, 219–220, 226–228, 252–253, 258–259, 262–264; **2**: 228, 257–259
– – inferior bulbi **2**: 58, 63, 360, 362, 370–371, 379, 382, 384–387
– – lateralis bulbi **2**: 58, 352–353, 360, 370–371, 374, 379–382, 385, 387
– – medialis bulbi **2**: 58, 352–353, 360, 370–373, 379, 382, 385–387
– – superior bulbi **2**: 58, 360, 362, 370–371, 379–382, 384, 386–387
– rhomboideus major **1**: 53–54, 60, 110–112, 142
– – minor **1**: 54, 110–112, 142
– risorius **2**: 28–29, 99
– rotatores thoracis **1**: 56, 59
– salpingopharyngeus **2**: 51
– sartorius **1**: 194, 199–200, 204, 216, 219, 221–222, 224, 226–228, 231, 233, 235, 252–253, 255, 258–265, 267, 272, 277; **2**: 228, 256–259
– scalenus anterior **1**: 68; **2**: 33–37, 72, 75, 92–93, 96, 98, 149, 165
– – medius **1**: 68; **2**: 33–37, 72, 75, 81, 92–93, 96, 165
– – posterior **2**: 33–35, 37, 75, 92, 96
– semimembranosus **1**: 195, 199–200, 217, 221–222, 224, 226–227, 232–235, 255–256, 258, 261–265, 267–268, 272
– semispinalis capitis **1**: 53–54, 56–60; **2**: 32–33, 37, 73–75
– – cervicis **1**: 56, 58–59; **2**: 33, 74
– – thoracis **1**: 56–59
– semitendinosus **1**: 200, 219, 221–224, 227, 231, 233, 235, 255–256, 258, 261–265, 267, 272
– serratus anterior **1**: 63–65, 68, 72, 110, 112, 144, 148; **2**: 35
– – posterior inferior **1**: 54–55, 57
– – – superior **1**: 54, 57, 60
– soleus **1**: 211, 217, 221, 231–235, 241, 255, 267–268, 271, 273–274, 276–277
– sphincter ani externus **2**: 179, 230–233, 243, 251, 260–261, 265–266, 276, 279
– – – internus **2**: 179, 243, 251
– – pupillae **2**: 375
– – pyloricus **2**: 174
– – urethrae externus **2**: 235, 256, 267
– – vesicae **2**: 235
– spinalis **1**: 56–58
– splenius capitis **1**: 53–54, 56–57, 60; **2**: 30, 32–35, 37, 73–75, 92–93
– – cervicis **1**: 54, 56–57, 60; **2**: 33, 74
– stapedius **2**: 398

Musculus(-i)
– sternocleidomastoideus **1**: 53–54, 60, 62–64, 68, 112; **2**: 30, 32–37, 44, 68, 71–75, 81, 85, 92–93, 95, 98–99
– sternohyoideus **1**: 68, 112; **2**: 33–36, 44–45, 68, 71, 75, 93, 97–99
– sternothyroideus **1**: 68; **2**: 33–36, 68, 71, 75, 93, 97, 99, 149
– styloglossus **2**: 32, 34–35, 37, 46–49, 51, 87, 97, 99
– stylohyoideus **2**: 32, 34, 36–37, 44, 47, 49, 51, 86–87, 97–98, 100
– stylopharyngeus **2**: 35, 37, 46–51, 100
– subclavius **1**: 64, 68, 106–107, 112, 140; **2**: 93, 149
– suboccipitales **1**: 55
– subscapularis **1**: 64, 95, 97, 106–107, 110, 112–113, 140–141, 144, 149
– supinator **1**: 101, 113, 116–117, 119, 140, 152–154, 156
– suprahyoidei **2**: 36
– supraspinatus **1**: 95, 97, 106, 108–113, 142–143, 148–149
– tarsalis superior **2**: 366
– temporalis **2**: 31–32, 37, 58, 62–63, 73, 86–87, 352–353, 358–360, 362, 380, 387, 391
– temporoparietalis **2**: 28–29
– tensor fasciae latae **1**: 194, 216, 219, 226–227, 253, 259; **2**: 228, 256–259
– – tympani **2**: 391, 398–399
– – veli palatini **2**: 37, 48, 51, 73, 89, 395
– teres major **1**: 52–54, 97, 105–113, 140–144, 148–149
– – minor **1**: 95, 97, 109–113, 142–144, 148
– thoracis **1**: 62–64, 68
– thyroarytaenoideus **2**: 67–68
– thyrohyoideus **2**: 34–35, 44–48, 66, 93, 97, 99, 149
– tibialis anterior **1**: 198, 207, 211–213, 216, 231–232, 235, 240, 271, 273–278
– – posterior **1**: 208, 212–215, 231, 233–235, 238, 240–241, 267–268, 273–277
– trachealis **2**: 66
– transversus abdominis **1**: 54, 58, 65–69, 76–77, 194, 227–228; **2**: 196, 211, 220–221, 225, 228
– – linguae **2**: 46–47
– – perinei profundus **1**: 227; **2**: 232, 234, 250–251, 265–267
– – – superficialis **1**: 227; **2**: 265–267, 279–280
– – thoracis **1**: 68, 76
– trapezius **1**: 52–54, 60, 62–64, 97, 105, 111–112, 148–149; **2**: 30, 32–37, 73–75, 81, 92–93, 98
– triangularis → Musculus depressor anguli oris
– triceps brachii **1**: 52, 62, 95, 101, 105–119, 141–149, 152–154
– – surae **1**: 203, 211, 216, 221, 232–234, 240–241, 255, 267–268, 276–277
– uvulae **2**: 51, 395
– vastus intermedius **1**: 220, 229, 253, 258–264
– – lateralis **1**: 216–217, 219, 223–224, 226, 229, 231–232, 252–253, 255–256, 259–264; **2**: 228
– – medialis **1**: 216, 219, 221, 229, 231, 252–253, 258–260, 262–264; **2**: 228
– verticalis linguae **2**: 47
– vocalis **2**: 68
– zygomaticus **2**: 48
– – major **2**: 28–30, 32, 85, 98, 366
– – minor **2**: 28–30, 32, 366
Muskel(n) **1**: 10
– äussere, Auge **2**: 370–371, 379
– autochthone, Rücken **1**: 56–59
– Bauchwand **1**: 62–66
Muskel(n)
– Beckenboden **2**: 264–266
– Brustwand **1**: 62–64
– Damm **2**: 265–266
– Fingerrücken **1**: 125
– Fussohle **1**: 237–238
– Fussrücken **1**: 231–232, 236
– Hals **1**: 55; **2**: 34–37
– Hand **1**: 121–125
– Handrücken **1**: 125
– Harnblase **2**: 235
– Herz **2**: 141
– Hüfte **1**: 222–226
– ischiocrurale, Oberschenkel **1**: 221–224
– Kehlkopf **2**: 47, 66–67
– Kopf **2**: 37
– Larynx **2**: 47, 66–67
– Lendenregion **1**: 55
– mimische, Gesicht **2**: 28–29, 366
– Mundboden **2**: 36
– Nacken **1**: 55, 60
– Oberarm **1**: 106–111
– oberflächliche, Gesicht **2**: 28–29, 366
– Oberschenkel **1**: 219–224, 226
– Pharynx **2**: 47–49
– Rücken **1**: 52–60; **2**: 37
– Schlund **2**: 47–49
– Schulter **1**: 106–111
– tiefe, Gesicht **2**: 30
– – Hals **2**: 37
– Unterarm **1**: 114–116, 118–119
– Unterschenkel **1**: 231–234
– ventrale, Rumpf **1**: 62–65, 68
– weicher Gaumen **2**: 395
– Zunge **2**: 38, 46–47
Muskelanheftungen, Becken **1**: 228
– Brustkorb **1**: 68
– Elle **1**: 117
– Fussskelett **1**: 240
– Handskelett **1**: 128
– Hüftbein **1**: 227
– Lendenwirbelsäule **1**: 228
– Membrana interossea antebrachii **1**: 117
– – – cruris **1**: 235
– Oberarmknochen **1**: 113
– Oberschenkelknochen **1**: 229
– Schädel **2**: 32
– Schädelbasis **2**: 37, 395
– Schienbein **1**: 235
– Schultergürtel **1**: 112
– Speiche **1**: 117
– Wadenbein **1**: 235
– Zungenbein **2**: 34–35
Muskelfaszie
Mutterkuchen **2**: 243
Muttermund, äusserer **2**: 268
Myelencephalon **2**: 303
Myocardium **2**: 131, 133–135, 138, 141, 145
Myometrium → Tunica muscularis uteri

N

Nabelbruch **1**: 63
Nabelhernie → Nabelbruch
Nabelschnur **2**: 243
Nackenmuskeln **1**: 55, 60
Nagel **1**: 11
Nahtknochen, Schädeldach **2**: 3
Nail **1**: 11
Naris(-es) **2**: 44, 54
Nase, äussere, Skelett **2**: 54
– Schleimhautrelief **2**: 55–57
Nasenhöhle **2**: 58–63, 336, 365, 369, 386–387
– Seitenwand **2**: 54–57, 88–90
– – Innervation **2**: 88
Nasennebenhöhlen **2**: 59–63, 336, 369, 386–387
Nasenöffnungen, hintere **2**: 23
Nasenscheidewand **2**: 54, 90
– knöcherne **2**: 23
Nasenschleimhaut, Innervation **2**: 88–90
Nasenskelett **2**: 23
Nasenvorhof **2**: 55–57
Nasenwand, seitliche, knöcherne **2**: 55–57
Nasopharynx **2**: 53
Nasus externus **2**: 54, 63, 358–359
Nebenhoden **2**: 270–273
Nebenniere **2**: 216
– fetale **2**: 210
Neck **2**: 33, 47, 72–75, 164–165
– arteries **2**: 78–79, 93, 97–99, 118, 149, 165
– autonomic nervous system **2**: 95–96
– cutaneous innervation **2**: 82–83
– dermatomes **2**: 82–83
– fascia **2**: 33–35
– lymph nodes **2**: 81
– – vessels **2**: 81
– muscles **1**: 55, 60; **2**: 34–37
– nerves **2**: 92–93, 97–99, 149
– segmental innervation **2**: 82–83
– subcutaneous nerves **2**: 91–92
– – veins **2**: 91–92
– superficial arteries **2**: 92
– sympathetic nervous system **2**: 95–96, 166
– veins **2**: 93, 97–99, 149, 164
Neck-shaft angle of thigh bone **1**: 179
Nephros → Ren
Nerve(s), axilla **2**: 93, 149
– back of foot **1**: 278
– – – hand **1**: 165
– cranial **1**: 22; **2**: 306
– dorsum of foot **1**: 278
– – – hand **1**: 165
– elbow region **1**: 140–143
– face **2**: 86–90
– facial **2**: 84, 399
– fascicles of cerebrum **2**: 333
– female lesser pelvis **2**: 245
– – perineum **2**: 280–281
– fingers **1**: 165
– foot **1**: 271
– forearm **1**: 152–154
– gluteal region **1**: 255–256
– head **2**: 84–90
– knee **1**: 252
– larynx **2**: 101
– leg **1**: 267–268, 271
– lesser pelvis **2**: 234, 245
– male lesser pelvis **2**: 234
– – perineum **2**: 279
– mediastinum **2**: 149
– neck **2**: 92–93, 97–99, 149
– optic **2**: 352–353, 359, 378, 384–385
– palm of hand **1**: 160–162
– perineum **2**: 279–281
– peripharyngeal space **2**: 100
– popliteal fossa **1**: 255–256, 267–268
– posterior abdominal wall **2**: 228
– pudendal **2**: 281
– rectum **2**: 231
– shoulder **1**: 140, 142–143
– sole of foot **1**: 279–280
– spinal **1**: 22–23, 48; **2**: 286
– subcutaneous, back of foot **1**: 278
– – – – hand **1**: 164
– – dorsum of foot **1**: 278
– – – – hand **1**: 164
– – elbow region **1**: 151
– – foot **1**: 270
– – forearm **1**: 150–151

Nerve(s), subcutaneous
– – gluteal region **1**: 254
– – leg **1**: 266, 270
– – neck **2**: 91
– – popliteal fossa **1**: 254, 266
– – shoulder **1**: 135–136
– – thigh **1**: 248, 254
– – upper arm **1**: 135–136
– thigh **1**: 252, 255–256; **2**: 228
– thorax **2**: 93, 97
– thyroid gland **2**: 101
– tongue **2**: 97
– trigeminal **2**: 84, 86
– upper arm **1**: 140–143
– vestibulocochlear **2**: 399, 405, 408–409
Nerven, Achselhöhle **2**: 93, 149
– Brustwand **2**: 93
– Ellbogenregion **1**: 142–143
– Ellenbeuge **1**: 140–141
– Finger **1**: 165
– Fuss **1**: 271
– Fussohle **1**: 279–280
– Fussrücken **1**: 278
– Gesäss **1**: 255–256
– Gesicht **2**: 86–90
– Hals **2**: 92–93, 97–99, 149
– Handrücken **1**: 165
– hintere Bauchwand **2**: 228
– Hohlhand **1**: 160–162
– Kehlkopf **2**: 101
– kleines Becken **2**: 234, 245
– Knie **1**: 252
– Kniekehle **1**: 255–256, 267–268
– Kopf **2**: 84–90
– männliches kleines Becken **2**: 234
– Mastdarm **2**: 231
– Mediastinum **2**: 149
– Oberarm **1**: 140–143
– oberer Brustraum **2**: 97
– Oberschenkel **1**: 252, 255–256; **2**: 228
– Rachenwand **2**: 100
– Schilddrüse **2**: 101
– Schulter **1**: 140, 142–143
– Spatium peripharyngeum **2**: 100
– subkutane, Ellenbeuge **1**: 151
– – Fuss **1**: 270
– – Fussrücken **1**: 278
– – Gesäss **1**: 254
– – Handrücken **1**: 164
– – Kniekehle **1**: 254, 266
– – Oberarm **1**: 135–136
– – Oberschenkel **1**: 248, 254
– – Schulter **1**: 135–136
– – Unterarm **1**: 150–151
– – Unterschenkel **1**: 266, 270
– Unterarm **1**: 152–154
– Unterschenkel **1**: 267–268, 271
– weibliches kleines Becken **2**: 245
– Zunge **2**: 97
Nervensystem, autonomes **1**: 24–25
– – Brustraum **2**: 167
– – Hals **2**: 95–96, 166
– – Oberbauch **2**: 167
– – oberer Brustraum **2**: 95–96, 166
– – Retroperitonealraum **2**: 248–249
– – Thorax **2**: 95–96, 166–167
– parasympathisches **1**: 24; **2**: 231
– – Retroperitonealraum **2**: 248–249
– peripheres **1**: 22
– sympathisches **1**: 24; **2**: 231
– – Brustraum **2**: 167
– – Hals **2**: 95–96, 166
– – Oberbauch **2**: 167
– – oberer Brustraum **2**: 95–96, 166
– – Retroperitonealraum **2**: 248–249
Nervensystem
– vegetatives **1**: 24–25
– zentrales **1**: 22
– – Gliederung **2**: 303
Nervus(-i) abducens [VI] **1**: 22; **2**: 291–295, 298, 306, 309, 311, 378–382
– accessorius [XI] **1**: 22, 51, 60, 68, 112; **2**: 32, 37, 92–93, 96, 98–99, 149, 284, 289, 291–292, 294–295, 306, 311, 340
– alveolaris(-es) inferior **2**: 84, 86–89
– – superiores **2**: 84, 86
– ampullaris anterior **2**: 405
– – lateralis **2**: 405
– – posterior **2**: 405
– anales inferiores → Nervi rectales inferiores
– anococcygei **2**: 279
– auricularis magnus **1**: 60–61, 78; **2**: 82–83, 85, 91–94, 96, 98
– – posterior **2**: 85–86, 91–92, 99
– auriculotemporalis **2**: 82–87, 89, 98–99
– axillaris **1**: 61, 78, 112–113, 132–134, 140–141, 143; **2**: 94
– buccalis **2**: 82–87
– canalis pterygoidei **2**: 88–89
– capitis **2**: 84
– cardiacus cervicalis inferior **2**: 95–96, 166–167
– – – medius **2**: 95, 100, 166–167
– – – superior **2**: 95–96, 100, 166
– caroticus internus **2**: 100
– cervicales **1**: 22, 51; **2**: 32, 37, 83, 94–96, 166, 284, 286, 288–289, 291, 294, 306, 311
– ciliares breves **2**: 381–382
– – longi **2**: 372, 382
– clunium inferiores **1**: 61, 243, 254–255; **2**: 279–280
– – medii **1**: 61, 243, 254–255
– – superiores **1**: 61, 243, 254–255
– coccygeus **1**: 22; **2**: 226, 284
– cochlearis **2**: 399, 405
– craniales **1**: 22; **2**: 306
– cutaneus(-i) abdominis **1**: 78
– – antebrachii lateralis **1**: 132–133, 136, 140–143, 151–153
– – – medialis **1**: 132–136, 140–141, 150–151, 160; **2**: 93–94, 149
– – – posterior **1**: 133, 135, 142–143, 150, 154
– – brachii lateralis inferior **1**: 132–133, 135–136, 142–143, 150
– – – – superior **1**: 61, 78, 132–133, 135, 142–143
– – – medialis **1**: 61, 78, 132–136, 140–142, 150–151; **2**: 94, 149
– – – posterior **1**: 61, 133, 135, 142–143, 150–151
– – capitis **2**: 82–83
– – colli **2**: 82–83
– – dorsalis intermedius (nervi fibularis superficialis) **1**: 270, 278
– – – lateralis (nervi suralis) **1**: 242, 266–267, 278
– – – medialis (nervi fibularis superficialis) **1**: 270, 278
– – dorsi **1**: 61
– – femoralis posterior **1**: 61, 243–244, 254–257, 266; **2**: 226, 279–280
– – femoris lateralis **1**: 61, 78, 242–244, 248–249, 252, 254–255, 266; **2**: 209, 226–228, 248
– – femoris posterior → Nervus cutaneus femoralis posterior
– – membri inferioris **1**: 242–243
– – – superioris **1**: 132–133
– – surae lateralis **1**: 242–243, 248, 254–256, 266–268
– – – medialis **1**: 255–256, 266–268
Nervus(-i) cutaneus(-i)
– – thoracis **1**: 78
– digitales dorsales manus **1**: 133, 161, 164–165, 167
– – – pedis **1**: 270–271, 278
– – palmares communes **1**: 132, 161–162
– – – proprii **1**: 132–133, 160–162, 164–165, 167
– – plantares communes **1**: 279–280
– – – proprii **1**: 279–280
– dorsalis clitoridis **2**: 231, 280–281
– – penis **1**: 78; **2**: 231, 267, 279
– – scapulae **1**: 60, 112, 134; **2**: 93–94, 98, 149
– erigentes → Nervi splanchnici pelvici
– ethmoidalis anterior **2**: 82–86, 88, 90, 381–383
– – posterior **2**: 88, 381–382
– facialis [VII] **1**: 22, 25, 51; **2**: 32, 34, 37, 84–85, 87–89, 91–93, 98–100, 284, 289, 291–292, 294–295, 306, 308–309, 311, 340, 383, 391, 398–399, 408–409
– femoralis **1**: 78, 227–229, 235, 242, 244, 248–249, 252, 270; **2**: 226–228, 245, 248, 256–258
– fibularis communis **1**: 229, 235, 242–243, 245, 255–256, 263–265, 267–268, 271–272
– – profundus **1**: 235, 240, 242, 270–271, 274–275, 278
– – superficialis **1**: 235, 240, 242, 270–271, 278
– frontalis **2**: 58, 84, 379–382, 384
– genitofemoralis **1**: 78, 242, 244, 248; **2**: 226–228, 274
– glossopharyngeus [IX] **1**: 22, 25, 51; **2**: 34–35, 37, 89, 96–97, 100, 284, 289, 291–292, 294–295, 306, 311, 340
– glutaeus inferior **1**: 227, 229, 244, 256–257; **2**: 226, 279
– – superior **1**: 227, 229, 244, 257; **2**: 226, 228
– hallucis plantaris medialis **1**: 279
– hypogastricus **1**: 25; **2**: 231, 248–249
– hypoglossus [XII] **1**: 22, 51; **2**: 32, 34, 37, 47, 81, 87, 94, 96–100, 284, 291–292, 294–295, 306, 311
– iliohypogastricus **1**: 61, 78, 242–244, 248, 254–255; **2**: 209, 226–228, 248
– ilioinguinalis **1**: 78, 244, 248; **2**: 209, 226–228, 248, 274
– infraorbitalis **2**: 58, 82–87, 379, 383–384
– infratrochlearis **2**: 82–83, 85–87, 381–383
– intercostales **1**: 25, 61, 68, 78, 132–133, 136, 227–228; **2**: 93, 96, 166–167
– intercostobrachialis **1**: 72, 132, 136; **2**: 93, 149
– intermediofacialis → Nervus facialis [VII]
– intermedius **1**: 22; **2**: 295, 306, 308–309, 311, 340, 399
– interosseus antebrachii anterior (nervi mediani) **1**: 153
– – – posterior (nervi radialis) **1**: 154
– ischiadicus **1**: 223, 227, 229, 235, 240, 244, 256–257, 261–262; **2**: 226, 230, 253, 256–259, 279–280
– jugularis **2**: 96, 100
– labiales anteriores **2**: 280
– – posteriores **2**: 280–281
– lacrimalis **2**: 58, 82–86, 88, 379–382
– laryngealis inferior/recurrens/superior → Nervus laryngeus inferior/recurrens/superior
– laryngeus inferior **2**: 101
– – recurrens **1**: 25; **2**: 96–97, 101, 149, 167
– – superior **2**: 67, 70, 96–101, 167
– lingualis **2**: 45, 47, 84, 86–89, 97, 99
– lumbales **1**: 22, 228; **2**: 226, 284
– lumbares → Nervi lumbales
– mandibularis [V/3] **1**: 22; **2**: 82–84, 86–87, 89, 293, 295, 306, 380, 395
– maxillaris [V/2] **1**: 22; **2**: 82–84, 86, 88–89, 293, 295, 306, 380

Nervus(-i)
– medianus **1**: 113, 117, 123, 128, 132–133, 140–141, 145–147, 152–153, 156–159, 161–162; **2**: 93–94, 149
– mentalis **2**: 82–86
– musculocutaneus **1**: 112–113, 117, 132–134, 140–141, 145–146; **2**: 93–94, 149
– mylohyoideus **2**: 84, 87, 89, 98
– nasociliaris **2**: 84, 379–382
– nasopalatinus **2**: 88, 90
– obturatorius **1**: 227–229, 235, 242–244, 248, 252, 254–255, 266; **2**: 226–228, 234, 244–245, 248, 256
– occipitalis major **1**: 60; **2**: 82–83, 85, 91–93
– – minor **1**: 60–61; **2**: 82–83, 85, 91–94, 96, 98
– – tertius **1**: 60; **2**: 82–83
– oculomotorius [III] **1**: 22, 25; **2**: 88, 291–295, 298, 306, 315, 344, 358, 379–382
– olfactorii [I] **1**: 22; **2**: 88, 90, 298, 306, 315
– ophthalmicus [V/1] **1**: 22; **2**: 82–84, 86, 89, 293, 306, 379–382
– opticus [II] **1**: 22; **2**: 58, 88–89, 291–295, 298, 301, 304, 306, 315, 324, 335–336, 338, 352–353, 358–360, 362, 370–374, 378–382, 384–385, 387
– palatinus(-i) major **2**: 84, 88–89
– – minores **2**: 84, 89
– pectoralis lateralis **1**: 68, 112–113, 134; **2**: 93–94, 149
– – medialis **1**: 68, 112–113, 134; **2**: 93–94, 149
– pelvici splanchnici → Nervi splanchnici pelvici
– pelvis **2**: 234, 245
– perineales **1**: 255; **2**: 267, 279–281
– peronaeus communis/profundus/superficialis → Nervus fibularis communis/profundus/superficialis
– petrosus major **2**: 84, 88–89, 398–399
– – minor **2**: 89, 393, 398
– – profundus **2**: 88–89
– phrenicus **2**: 33, 93–94, 96, 149
– plantaris lateralis **1**: 214, 240, 243, 268, 279–280
– – medialis **1**: 214, 240, 243, 268, 279–280
– praesacralis → Plexus hypogastricus superior
– pudendus **1**: 61, 227, 243–244, 254–257; **2**: 226, 231, 244, 251, 256, 267, 279–281
– radialis **1**: 61, 112–113, 117, 128, 132–134, 140–143, 145–147, 150–154, 160–161, 164–165; **2**: 93–94
– rectales inferiores **1**: 254–255; **2**: 231, 279–281
– saccularis **2**: 405
– sacrales **1**: 22; **2**: 232, 284
– saphenus **1**: 242–243, 248, 252, 263, 266, 270, 278
– sciaticus → Nervus ischiadicus
– scrotales anteriores **1**: 78
– – posteriores **2**: 279
– spinalis(-es) **1**: 22–23, 48, 61, 79, 227; **2**: 285–286
– spinosus → Ramus meningeus nervi mandibularis
– splanchnicus(-i) imus **2**: 167
– – lumbales **1**: 25; **2**: 248–249
– – lumbares → Nervi splanchnici lumbales
– – major **1**: 25; **2**: 167, 248–249
– – minor **1**: 25; **2**: 167
– – pelvici **1**: 25; **2**: 226, 231, 248–249
– – sacrales **1**: 25; **2**: 231
– subclavius **1**: 112, 134; **2**: 93–94, 98, 149
– subcostalis **2**: 209, 226–227, 248
– suboccipitalis **1**: 60; **2**: 289
– subscapulares **1**: 112–113, 134, 140; **2**: 94
– supraclaviculares **1**: 61, 78, 132–133, 142; **2**: 94, 98
– – intermedii **1**: 136; **2**: 82–83, 91–92

Nervus(-i) supraclaviculares
– – laterales **1**: 135–136; **2**: 82–83, 91–93
– – mediales **2**: 82–83, 91–92
– – posteriores → Nervi supraclaviculares laterales
– supraorbitalis **2**: 82–83, 85–87, 380–381, 383
– suprascapularis **1**: 112–113, 134, 140, 143; **2**: 93–94, 149
– supratrochlearis **2**: 82–83, 85–87, 380–381, 383
– suralis **1**: 242–243, 266–267, 278
– temporales profundi **2**: 87, 89
– thoracicus(-i) **1**: 22, 133; **2**: 94, 166, 284
– – longus **1**: 68, 72, 112, 134; **2**: 93–94, 98, 149
– thoracodorsalis **1**: 72, 112–113, 134, 140–141, 227; **2**: 93–94, 149
– tibialis **1**: 227, 229, 235, 240, 242–243, 245, 255–256, 263–265, 267–268, 270, 272–275, 277, 279
– transversus colli **1**: 78; **2**: 82–83, 91–94, 98–99
– trigeminus [V] **1**: 22; **2**: 32, 34, 37, 82–84, 86–89, 284, 289, 291–295, 306, 309, 311, 341, 344, 352, 359, 380, 382
– trochlearis [IV] **1**: 22; **2**: 88, 284, 289, 291–295, 306, 310–311, 379–382
– tympanicus **2**: 89, 393
– ulnaris **1**: 113, 117, 123, 128, 132–134, 140–143, 145–147, 152–153, 156–162, 165; **2**: 93–94, 149
– utricularis **2**: 405
– utriculoampullaris **2**: 405
– vagus [X] **1**: 22, 25, 51; **2**: 33–35, 37, 75, 82, 93, 96–97, 100, 149, 167, 248–249, 284, 289, 291–292, 294–295, 306, 311, 340
– vestibularis **2**: 399, 405
– vestibulocochlearis [VIII] **1**: 22, 51; **2**: 284, 289, 291–292, 294–295, 306, 308–309, 311, 340, 399, 405, 408–409
– zygomaticus **2**: 82–86, 379
Netzhaut **2**: 377
Neurofibrae cerebrales **2**: 333
Neurohypophysis **2**: 293, 307, 328
Niere **2**: 2, 208–210, 212–213, 215
– fetale **2**: 210
– gefurchte **2**: 210
– obere Polarterie **2**: 214
– untere Polarterie **2**: 214–215
– Zone der Schmerzausstrahlung **2**: 209
Nierenarterien **2**: 214–215
– multiple **2**: 214
– obere Polarterie **2**: 214
– untere Polarterie **2**: 214–215
– Variationen **2**: 214
Nierenbecken **2**: 212, 216–217
– ampulläres **2**: 216
– dendritisches **2**: 216
Nierenkelche **2**: 212, 217
Nipple **1**: 70
Nodulus(-i) lymphatici tonsillae lingualis → Folliculi linguales
– valvularum semilunarium valvae aortae **2**: 133, 135
– – – – trunci pulmonalis **2**: 132, 135
– vermis **2**: 307–309
Nodus(-i) atrioventricularis cordis **2**: 140
– jugulodigastricus → Nodus lymphaticus jugulodigastricus
– juguloomohyoideus → Nodus lymphaticus juguloomohyoideus
– lacunaris medialis → Nodus lymphaticus lacunaris medialis
– lymphaticus(-i) abdominis **1**: 15; **2**: 163, 198, 203
– – appendicularis **2**: 203

Nodus(-i) lymphaticus(-i)
– – axillares **1**: 15, 72–74; **2**: 163
– – – apicales **1**: 72–73
– – – brachiales **1**: 74
– – – centrales **1**: 72–74
– – – interpectorales **1**: 73–74
– – – laterales **1**: 72
– – – pectorales **1**: 72–73
– – – subscapulares **1**: 72
– – – superficiales **1**: 74
– – bronchopulmonales **2**: 110–111, 113
– – buccinatorius **2**: 81
– – capitis **2**: 81
– – cervicales **1**: 15
– – – laterales profundi **2**: 81, 163
– – – superficiales **2**: 81, 91
– – coeliaci **2**: 198, 203
– – colici dextri **2**: 203
– – – medii **2**: 203
– – colli **2**: 81
– – cubitales **1**: 74
– – cysticus **2**: 198
– – faciales **2**: 81
– – foraminalis **2**: 198
– – gastrici dextri **2**: 198
– – – sinistri **2**: 198
– – gastroepiploici dextri/sinistri → Nodi lymphatici gastroomentales dextri/sinistri
– – gastroomentales dextri **2**: 198
– – – sinistri **2**: 198
– – hepatici **2**: 183, 198
– – ileocolici **2**: 203
– – iliaci communes **2**: 163
– – – externi **2**: 163
– – infraauriculares **2**: 91
– – infrahyoideus **2**: 81
– – inguinales **1**: 15
– – – profundi **2**: 163
– – – superficiales **1**: 248; **2**: 163
– – – – inferiores **1**: 245
– – – – superolaterales **1**: 245
– – – – superomediales **1**: 245
– – intercostales **2**: 163
– – interpectorales → Nodi lymphatici axillares interpectorales
– – jugulodigastricus **2**: 81
– – juguloomohyoideus **2**: 81
– – lacunaris medialis **1**: 249
– – lienales → Nodi lymphatici splenici
– – lumbalis(-es) intermedii **2**: 198
– – – dexter praecavalis **2**: 198
– – lumbares → Nodi lymphatici lumbales
– – mandibularis **2**: 44, 81
– – mastoidei **2**: 81
– – mesenterici **2**: 203
– – – juxtaintestinales **2**: 203
– – – superiores **2**: 203
– – occipitales **2**: 81
– – pancreatici inferiores **2**: 203
– – – superiores **2**: 198, 203
– – pancreaticoduodenales **2**: 198, 203
– – paramammarii **1**: 74
– – parasternales **1**: 73, 76
– – paratracheales **2**: 110, 113
– – parotidei **2**: 81
– – – profundi **2**: 81
– – – superficiales **2**: 81
– – pelvis **1**: 15; **2**: 163
– – phrenici inferiores **2**: 113
– – – superiores **1**: 76; **2**: 113
– – popliteales profundi **1**: 245
– – – superficiales **1**: 245
– – praeaortici → Nodi lymphatici coeliaci
– – praelaryngealis **2**: 81, 110
– – praetracheales **2**: 81, 110, 163

Nodus(-i) lymphaticus(-i)
– – praevertebrales **2**: 113
– – retroauriculares **2**: 81
– – retropylorici **2**: 198
– – sacrales **2**: 163
– – splenici **2**: 198, 203
– – submammarii **1**: 73
– – submandibulares **2**: 44, 81
– – submentales **2**: 44, 81
– – subpylorici **2**: 198
– – subscapulares → Nodi lymphatici axillares subscapulares
– – supraclaviculares **2**: 81
– – suprapyloricus **2**: 198
– – thoracis **1**: 15; **2**: 163
– – thoracoepigastrici **1**: 72
– – tracheobronchiales **2**: 157
– – – inferiores **2**: 104, 110, 113
– – – superiores **2**: 110, 113
– lymphoideus(-i) → Nodus(-i) lymphaticus(-i)
– sinuatrialis cordis **2**: 140
Norma basalis cranii **2**: 8, 16, 17
– facialis cranii → Norma frontalis cranii
– frontalis cranii **2**: 2, 4, 14
– lateralis cranii **2**: 6, 15, 17
– verticalis cranii **2**: 2–3, 17
Nose, mucous membrane, innervation **2**: 88–90
– skeleton **2**: 23, 54
Nucleus(-i) anteriores thalami **2**: 325–327, 331, 340, 344
– caudatus **2**: 300, 322–323, 326, 329–334, 338–341, 344, 346–351, 354–356, 360–361
– centromedianus thalami **2**: 327, 340
– corporis geniculati lateralis **2**: 327
– – – medialis **2**: 327
– – mamillaris **2**: 328
– dentatus **2**: 309, 334
– dorsalis(-es) posterior thalami **2**: 327
– – thalami **2**: 326–327, 330, 340, 344
– emboliformis **2**: 309
– fastigii **2**: 309
– forebrain **2**: 329–332
– globosus **2**: 309
– habenulae **2**: 331, 348
– hypothalamicus anterior **2**: 328
– – dorsomedialis **2**: 328
– – posterior **2**: 328
– – ventromedialis **2**: 328
– interpeduncularis **2**: 325
– intralaminares thalami **2**: 327
– lentiformis **2**: 300, 329, 332
– lentis **2**: 375
– mediales thalami **2**: 326–327, 331, 340, 344
– mediani thalami **2**: 327
– mesencephali **2**: 330–331
– midbrain **2**: 330–331
– motorii laterales medullae spinalis **2**: 288
– – mediales medullae spinalis **2**: 288
– olivaris inferior **2**: 309
– parafascicularis thalami **2**: 327
– paraventricularis anterior thalami **2**: 331
– – hypothalami **2**: 328
– pontis **2**: 340
– praeoptici **2**: 328
– prosencephali **2**: 329–332
– pulposus disci intervertebralis **1**: 39, 48
– pulvinares **2**: 327
– reticularis thalami **2**: 326–327, 330–331, 340, 344
– ruber **2**: 309, 315, 328, 331, 334–335, 340–341, 350, 357
– subthalamicus **2**: 330, 340, 344, 350
– suprachiasmaticus **2**: 328
– supraopticus **2**: 328
– tegmentales **2**: 328
Nucleus(-i)
– thalami **2**: 237, 340
– tuberales **2**: 328
– ventralis(-es) anterior thalami **2**: 327
– – intermedius thalami **2**: 327
– – posteriores thalami **2**: 340
– – posterolateralis thalami **2**: 327
– – posteromedialis thalami **2**: 327
– – thalami **2**: 326–327, 331, 344
Nullipara **2**: 268

O

Oberarm **1**: 144–149
– Hautnerven **1**: 135–136
– subkutane Nerven **1**: 135–136
– – Venen **1**: 135–137
Oberarmarterien **1**: 140–143
Oberarmknochen **1**: 86–87
– Muskelanheftungen **1**: 113
Oberarmmuskeln **1**: 106–111
Oberarmnerven **1**: 140–143
Oberbauch **2**: 182, 191, 199, 208, 222–223
– Arterien der retroperitonealen Organe **2**: 203
– autonomes Nervensystem **2**: 167
– Lymphgefässe **2**: 198
– – der retroperitonealen Organe **2**: 203
– Lymphknoten **2**: 198
– – der retroperitonealen Organe **2**: 203
– retroperitoneale Organe **2**: 199, 203
– sympathisches Nervensystem **2**: 167
– Venen der retroperitonealen Organe **2**: 203
Oberbaucharterien **2**: 196–198
Oberbauchorgane **2**: 154–155
Oberflächenrelief, Arm **1**: 105
– Bauch **1**: 62
– Bein **1**: 216–217
– Brust **1**: 62
– Rücken **1**: 52
Oberkiefer **2**: 24
Oberschenkel **1**: 258–265; **2**: 228
– Adduktoren **1**: 220–221
– Hautnerven **1**: 248, 254
– ischiocrurale Muskeln **1**: 221
– Lymphknoten **1**: 248
Oberschenkelarterien **1**: 251–253, 255–256; **2**: 228
Oberschenkelfaszie **1**: 218
Oberschenkelknochen **1**: 173, 177, 179–181, 258
– Muskelanheftungen **1**: 229
Oberschenkelmuskeln **1**: 219–224, 226, 228–229
– dorsale **1**: 221–224
– ischiocrurale **1**: 222–224
– mediale **1**: 220–221
– ventrale **1**: 219
Oberschenkelnerven **1**: 252, 255–256; **2**: 228
– subkutane **1**: 248, 254
Oberschenkelvenen **1**: 252, 255–256; **2**: 228
– subkutane **1**: 248, 254
Obex **2**: 310–311
Oculus **2**: 364
Odontoblast layer **2**: 40
Odontoblastenschicht **2**: 40
Odontogenesis **2**: 42–43
Oesophagus **1**: 14, 16, 19, 79; **2**: 47–53, 75, 100–101, 104–107, 113, 155–161, 164–165, 173–175, 206
Oesophagusenge, mittlere **2**: 106
– obere **2**: 53, 106
– untere **2**: 106–107
Oesophaguserweiterung, epiphrenische **2**: 107
Oesophagusmund **2**: 106
Ohr, äusseres **2**: 390–391
– inneres **2**: 391, 403
Ohrenspiegel **2**: 392
Ohrknorpel **2**: 390
Ohrmuschel **2**: 390
Ohrspeicheldrüse **2**: 44–45, 98
– Innervation **2**: 89
Ohrtrompete **2**: 393–395, 406
Olecranon **1**: 89–91, 98–101, 105, 108–109, 118–119, 135, 142, 149–150
Oliva **2**: 311, 340
Omentum majus **1**: 63; **2**: 170, 173, 190–193, 196, 202, 204
– minus **2**: 170–171, 173, 191
Opening, saphenous **1**: 248–249
– – lymph nodes **1**: 248
– – superficial veins **1**: 248
Operculum frontale **2**: 316, 323, 338
– frontoparietale **2**: 323
– parietale → Operculum frontoparietale
– temporale **2**: 323, 338
Ophthalmoskop **2**: 377
Ora serrata retinae **2**: 373–374, 376
Orbiculus ciliaris **2**: 374, 376
Orbit **2**: 5, 11, 58–63, 336, 360, 362, 365, 367–371, 378–382, 384–387
Orbita **2**: 5, 11, 58–63, 336, 360, 362, 365, 367–371, 378–382, 384–387
– Dach **2**: 357
– Venen **2**: 383
Orchis → Testis
Organ(s), abdominal **2**: 170–172, 190–193
– endocrine **1**: 15
– female lesser pelvis **2**: 245–247
– hearing **2**: 391
– male lesser pelvis **2**: 244, 246–247
– of Corti **2**: 404
– pelvic **2**: 232–234, 238
– – female **2**: 233, 238–243, 245
– – male **2**: 232, 234, 236–237, 244, 246–247
– retroperitoneal, upper abdomen **2**: 199, 203
– – – – arteries **2**: 203
– – – – lymph nodes **2**: 203
– – – – – vessels **2**: 203
– – – – veins **2**: 203
– thoracic **2**: 126–127, 150–151, 154–161, 172
– upper abdomen **2**: 154–155
– urinary **2**: 218
Organe, endokrine **1**: 15
– kleines Becken **2**: 244–247
– männliches kleines Becken **2**: 244, 246–247
– Oberbauch **2**: 154–155
– retroperitoneale, Oberbauch **2**: 199, 203
– – – Arterien **2**: 203
– – – Lymphgefässe **2**: 203
– – – Lymphknoten **2**: 203
– – – Venen **2**: 203
– weibliches kleines Becken **2**: 245–247
Organum(-a) extraperitoneale **2**: 171
– genitalia feminina externa **2**: 269
– – – interna **2**: 239–243, 245
– – masculina externa **2**: 274, 276
– – – interna **2**: 244
– intraperitoneale **2**: 171
– pelvis **2**: 238, 244–245
– retroperitoneale **2**: 171
– spirale **2**: 404
– vestibulocochleare **2**: 391
Orifice(s), hernial, inner **1**: 249
– – outer **1**: 249
– ureteral **2**: 235
– urethral, external **2**: 277
– – internal **2**: 277
Origo musculi **1**: 10
Oropharynx **2**: 53
Os uteri, external **2**: 268

Os (Ossa) basilare **2**: 19
– capitatum **1**: 8, 92–94, 102–104, 123, 129
– carpi/carpalia **1**: 8, 82–83, 92–93
– coccygis **1**: 28, 34–35, 38–40, 170–171, 173–175, 225; **2**: 232, 243, 256–259, 265–266, 280, 284
– corporis humani **1**: 2–3
– costalia **1**: 44
– coxae **1**: 40, 170, 172, 176, 220–221, 225; **2**: 194–195, 217, 250
– cruris **1**: 182
– cuboideum **1**: 184–186, 188–189, 209–211, 215
– cuneiforme(-ia) **1**: 186, 209
– – intermedium **1**: 184–186, 188–189, 209–211, 215
– – laterale **1**: 184–186, 188–189, 209, 211, 215
– – mediale **1**: 184–186, 188–189, 209–210, 215
– digitorum manus **1**: 82–83, 92–94
– – pedis **1**: 170–171, 184–186, 188
– ethmoidale **2**: 10, 12–17, 22–23, 55–56, 58, 60, 63, 88, 90, 291, 367–369
– femoris → Femur
– frontale **1**: 87; **2**: 3–4, 6–7, 10, 12–17, 21, 23, 55–56, 58, 336, 347, 349, 356–357, 360, 367–369, 384
– hamatum **1**: 8, 92–94, 102–104, 123–124, 129
– hyoideum **1**: 18, 20–21; **2**: 27, 34–36, 44–49, 53, 65–68, 70, 78–79, 93, 98–101
– ilii **1**: 172–173, 176–177, 194, 258; **2**: 205, 220–221, 230, 251, 253–257, 264
– ilium → Os ilii
– incisivum **2**: 17, 23–24
– interparietale **2**: 3
– ischii **1**: 172–173, 176–177, 194, 261; **2**: 253, 255, 257, 268, 281
– lacrimale **2**: 6, 14–15, 17, 55–57, 367–368
– lunatum **1**: 8, 92–94, 102, 104, 129, 159
– manus **1**: 92–94
– metacarpi/metacarpalia **1**: 8, 82–83, 92–94, 102–104, 124–127, 129, 165–167
– metatarsi/metatarsalia **1**: 170–171, 184–186, 188, 209–211, 215
– nasale **1**: 87; **2**: 4, 6, 12, 14–15, 17, 23, 54–56, 62, 90, 368–369
– naviculare **1**: 184–186, 188, 207, 209–211, 215
– occipitale **1**: 37, 49, 51, 60, 87; **2**: 3, 6–8, 10, 12–13, 15–19, 49, 51, 72, 290, 294, 321, 347, 349, 352–353, 355–359, 362, 409
– palatinum **2**: 8, 12, 15–17, 23–24, 55–58, 368
– parietale **1**: 87; **2**: 3–4, 6–7, 12–17, 22, 290, 295, 368
– pedis **1**: 184–186, 188–189, 208–211
– pelvicum → Os coxae
– pisiforme **1**: 8, 92–94, 102–104, 121–124, 130, 152–153, 161–162
– pubis **1**: 77, 172, 176–177, 194, 259; **2**: 234, 244, 252, 254, 256–261, 267–268, 275–276, 278, 281
– sacrum **1**: 28, 34–35, 38–39, 170–175, 177, 191, 194, 225; **2**: 225, 232–233, 243–244, 253, 255, 258–261, 264–265, 281, 284
– scaphoideum **1**: 8, 92–94, 102, 104, 123, 129–130, 159
– sesamoidea manus **1**: 92–94, 102
– – pedis **1**: 184–186, 188
– sphenoidale **1**: 51; **2**: 4, 6, 8, 10, 12, 14–17, 19, 23, 55–56, 58, 60, 63, 367–369, 380
– suturalia **2**: 3
– tarsi/tarsalia **1**: 170–171, 184–185, 188
– temporale **2**: 4–6, 8, 10–12, 14–17, 20, 26–27, 34–35, 47–48, 51, 58, 100, 302, 352–353, 359, 361–362, 368, 391–395, 399, 405–409
– trapezium **1**: 8, 92–94, 102–104, 123, 129
– trapezoideum **1**: 8, 92–94, 102–104, 129
Os
– triquetrum **1**: 8, 92–94, 102, 104, 123, 129, 159
– zygomaticum **2**: 4, 6, 14–17, 26, 48, 58–59, 62–63, 367–368
Ossicles, auditory **2**: 397
– – joints **2**: 397
Ossiculum(-a) auditoria **2**: 361, 397
– auditus → Ossicula auditoria
Ossification center of femoral head **1**: 177
Osteogenesis **1**: 8
Ostium abdominale tubae uterinae **2**: 239–240
– aortae **2**: 141
– appendicis vermiformis **2**: 177–178
– arteriae coronariae dextrae **2**: 133–134, 140
– – – sinistrae **2**: 133–134
– atrioventriculare dextrum **2**: 135, 139, 141
– – sinistrum **2**: 133, 135, 139, 141
– cardiacum **2**: 174, 204
– ileale → Valva ileocaecalis
– maxillare accessorium **2**: 56
– pharyngeum tubae auditivae → Ostium pharyngeum tubae auditoriae
– – – auditoriae **2**: 45, 52, 55, 57, 88, 90
– pyloricum **2**: 174–176, 194
– trunci pulmonalis **2**: 132, 141
– tympanicum tubae auditivae → Ostium tympanicum tubae auditoriae
– – – auditoriae **2**: 398
– ureteris **1**: 17; **2**: 232–233, 235, 237, 246, 250
– urethrae externum **1**: 17; **2**: 232–233, 240, 265, 268–269, 276–277, 280
– – internum **1**: 17; **2**: 232–233, 235, 246, 258
– uteri **2**: 233, 239–240, 268
– – externum **2**: 243
– – internum **2**: 243
– uterinum tubae uterinae **2**: 240
– vaginae **2**: 265, 268–269, 280
– valvae ilealis **2**: 177–178
– venae cavae inferioris **2**: 132, 134
– – – superioris **2**: 132
Otoscope **2**: 392
Outlet, pelvic **1**: 178
Ovar **2**: 238–239, 241, 245
– Arterien **2**: 241
Ovarium(-ia) **1**: 15, 17; **2**: 163, 233, 238–241, 245, 251–252
Ovary(-ies) **2**: 238–239, 241, 245
– arteries **2**: 241

P

Palate **2**: 88
– mucous membrane, innervation **2**: 88
– soft, muscles **2**: 395
Palatum durum **1**: 51; **2**: 5, 7, 13, 38–39, 41, 45, 52, 62, 73, 336, 386–387, 395
– molle **2**: 7, 38–39, 45, 50, 52, 55, 57, 90, 395
– osseum **2**: 45, 55–57, 365
Pallium → Cortex cerebri
Palm of hand, arteries **1**: 160–162
– – – nerves **1**: 160–162
Palma manus **1**: 105, 160–162
Palmaraponeurose **1**: 120
Palpebra inferior **2**: 362, 364, 379, 384
– superior **2**: 362, 364, 366, 379, 384
Pancreas **1**: 16, 18–19; **2**: 154–155, 170–172, 186, 188, 191, 196, 198–200, 203–204, 206–208, 223–224, 248
Papilla(-ae) ductus parotidei **2**: 45
– duodeni major **2**: 187–188
– – minor **2**: 188
– filiformes **2**: 39
– foliatae **2**: 39, 46
Papilla(-ae)
– fungiformes **2**: 46
– gingivales **2**: 38
– ilealis **2**: 177
– incisiva **2**: 38–39
– interdentales → Papillae gingivales
– lacrimalis inferior **2**: 364–365
– – superior **2**: 364–365
– mammaria **1**: 70–71, 73–74
– renalis **2**: 212
– vallatae **2**: 39, 46, 101
Paraflocculus **2**: 309
Parenchyma testis **2**: 271
Paries anterior abdominis **1**: 75, 77; **2**: 238
– – pharyngis **2**: 101
– – thoracis **1**: 75
– – vaginae **2**: 240, 245
– externus ductus cochlearis **2**: 404
– inferior orbitae **2**: 58–59
– labyrinthicus cavitatis tympanicae **2**: 396, 398
– lateralis orbitae **2**: 11, 58, 371
– medialis orbitae **2**: 11, 58–59, 367, 371
– membranaceus cavitatis tympanicae **2**: 398
– – trachealis **2**: 49–50, 65–66, 109
– posterior gastricus **2**: 191
– – vaginae **2**: 239–240
– superior orbitae **2**: 58–59, 336, 360, 386–387
– vestibularis ductus cochlearis **2**: 404
Pars(-tes) abdominalis aortae **1**: 12–14; **2**: 105, 107, 155, 162, 171, 182, 186, 189, 196, 199, 202–203, 206–208, 211, 213–215, 218–224, 228–230, 234, 243
– – musculi pectoralis majoris **1**: 63–65
– – oesophagi **2**: 105, 174–175, 206–207
– – ureteris **2**: 216
– acromialis musculi deltoidei **1**: 108
– alaris musculi nasalis **2**: 30
– alveolaris mandibulae **2**: 25, 41
– anterior fornicis vaginae **2**: 240
– – lobuli quadrangularis cerebelli **2**: 308
– – pedunculi cerebri **2**: 309, 358, 385
– anularis vaginae fibrosae digitorum manus **1**: 121
– – – – hallucis **1**: 238
– – – – indicis **1**: 121
– ascendens aortae **1**: 12, 14; **2**: 78, 96, 104, 107, 122, 124, 128–130, 132, 134–135, 137, 140, 142, 146–147, 149, 157–158, 161–162, 164–165, 167
– – duodeni **2**: 188, 194, 206
– atlantica arteriae vertebralis **1**: 60; **2**: 73, 79, 292, 302
– basilaris ossis occipitalis **1**: 37, 49; **2**: 8, 18, 49, 51, 290, 294
– cardiaca gastrica **2**: 106–107, 113, 173
– cartilaginea septi nasi **2**: 54
– – tubae auditoriae **2**: 51
– cavernosa arteriae carotidis internae **2**: 292–293, 301–302, 359–360, 378
– centralis ventriculi lateralis **2**: 300, 320, 322–323, 326, 340–341, 346–347, 349, 354–356, 361, 408
– ceratopharyngea musculi constrictoris pharyngis medii **2**: 34, 46–47
– cerebralis arteriae carotidis internae **2**: 77, 301, 360, 378
– cervicalis arteriae carotidis internae **2**: 77, 292, 301–302
– – – vertebralis → Pars transversaria arteriae vertebralis
– – columnae vertebralis **1**: 28, 31, 36–37, 49–51
– – oesophagi **2**: 53, 105–106
– chondropharyngea musculi constrictoris pharyngis medii **2**: 34–35, 46–47
– ciliaris retinae **2**: 374–376, 384

Pars(-tes)
– clavicularis musculi deltoidei **1**: 108
– – – pectoralis majoris **1**: 63–64
– – – sternocleidomastoidei **2**: 35–36
– compacta substantiae nigrae **2**: 328
– convoluta corticis renalis **2**: 212
– corporis humani **1**: 2
– costalis diaphragmatis **1**: 68–69; **2**: 222
– – pleurae parietalis → Pleura costalis
– cricopharyngea musculi constrictoris pharyngis inferioris **2**: 33, 47
– cruciformis vaginae fibrosae digitorum manus **1**: 121
– – – – hallucis **1**: 238
– – – – indicis **1**: 121
– descendens aortae **1**: 12, 14; **2**: 78, 104–107, 126, 157–163, 165, 167
– – duodeni **2**: 155, 187–188, 191, 194, 199–200, 223, 248
– diaphragmatica pleurae parietalis → Pleura diaphragmatica
– flaccida membranae tympanicae **2**: 392
– glossopharyngea musculi constrictoris pharyngis superioris **2**: 46–47
– horizontalis duodeni **2**: 170, 175, 188, 194, 199, 202, 206, 224
– inferior duodeni → Pars horizontalis duodeni
– – nervi vestibularis **2**: 405
– inferoposterior lobuli quadrangularis cerebelli **2**: 308
– infraclavicularis plexus brachialis **1**: 134, 140–141
– insularis arteriae cerebri mediae **2**: 298, 300
– interarytaenoidea rimae glottidis **2**: 69
– intercartilaginea rimae glottidis **2**: 69
– intermedia bulborum vestibuli **2**: 268
– intermembranacea rimae glottidis **2**: 69
– intracanalicularis nervi optici **2**: 378
– intracranialis arteriae vertebralis **2**: 79, 292, 302
– – nervi optici **2**: 358, 370, 378
– intramuralis ureteris **2**: 235
– iridica retinae **2**: 374–375, 384
– labialis musculi orbicularis oris **2**: 28–30
– lacrimalis musculi orbicularis oculi **2**: 32
– laryngea pharyngis **1**: 16; **2**: 50, 52–53, 68, 106, 165
– lateralis cruris dextri partis lumbalis diaphragmatis **2**: 105
– – – sinistri partis lumbalis diaphragmatis **2**: 105
– – fornicis vaginae **2**: 240
– – ossis occipitalis **1**: 49; **2**: 17–18
– – – sacri **1**: 34–35
– libera columnae fornicis **2**: 324, 328
– lumbalis columnae vertebralis **1**: 28, 38–39, 48
– – diaphragmatis **1**: 69; **2**: 105–106, 155, 182, 189, 196, 200, 208, 220–223
– mastoidea ossis temporalis **2**: 20
– medialis cruris dextri partis lumbalis diaphragmatis **2**: 105
– – – sinistri partis lumbalis diaphragmatis **2**: 105
– mediastinalis pleurae parietalis → Pleura mediastinalis
– membranacea septi interventricularis **2**: 133
– – urethrae masculinae **2**: 235–237, 266, 277
– muscularis septi interventricularis **2**: 134, 145
– nasalis pharyngis **1**: 16; **2**: 45, 50, 52–53, 55, 63, 73
– obliqua musculi cricothyroidei **2**: 66, 70
– opercularis gyri frontalis inferioris → Operculum frontale
– optica retinae **2**: 374, 376, 384
– oralis pharyngis **1**: 16; **2**: 13, 38, 45, 52–53
– orbitalis glandulae lacrimalis **2**: 365
– – gyri frontalis inferioris **2**: 316

Pars(-tes) orbitalis
– – musculi orbicularis oculi **2**: 28–30, 32, 366
– – nervi optici **2**: 336, 352–353, 359–360, 362, 370, 378, 384–385, 387
– – ossis frontalis **2**: 4, 12, 21, 360, 384
– ossea septi nasi **2**: 54
– palpebralis glandulae lacrimalis **2**: 365
– – musculi orbicularis oculi **2**: 28–30, 366
– patens arteriae umbilicalis **2**: 230
– petropharyngea musculi constrictoris pharyngis superioris **2**: 49
– petrosa arteriae carotidis internae **2**: 292, 301–302, 378
– – ossis temporalis **2**: 5, 10, 17, 302, 353, 359, 362, 399, 405–409
– pigmentosa retinae **2**: 374
– postcommunicalis arteriae cerebri anterioris **2**: 292, 298, 301–302
– – – – posterioris **2**: 292, 298
– posterior fornicis vaginae **1**: 17; **2**: 233
– – pedunculi cerebri **2**: 309
– postsulcalis dorsi linguae **2**: 51, 101
– praecommunicalis arteriae cerebri anterioris **2**: 292, 298, 301
– – – – posterioris **2**: 292, 298
– praesulcalis dorsi linguae **2**: 101
– praevertebralis arteriae vertebralis **1**: 60; **2**: 79, 118
– profunda musculi masseteris **2**: 30, 98
– prostatica urethrae masculinae **2**: 235–237, 258, 277
– pylorica gastrica **2**: 173–176, 191, 194, 200, 223
– radiata corticis renalis **2**: 212
– recta musculi cricothyroidei **2**: 66, 70
– reticularis substantiae nigrae **2**: 328
– retrolentiformis capsulae internae **2**: 332–333
– sphenoidalis arteriae cerebri mediae **2**: 298
– spinalis musculi deltoidei **1**: 108
– spongiosa urethrae masculinae **2**: 237, 266, 277
– squamosa ossis temporalis **2**: 4, 6, 10, 12, 17, 20, 26, 368, 391–392, 399
– sternalis diaphragmatis **1**: 68–69
– – musculi sternocleidomastoidei **2**: 34–36
– sternocostalis musculi pectoralis majoris **1**: 63–65
– sublentiformis capsulae internae **2**: 332–333, 344
– superficialis musculi masseteris **2**: 30, 98–99
– superior duodeni **2**: 171, 173–176, 188, 194, 199, 204, 206, 223
– – nervi vestibularis **2**: 405
– supraclavicularis plexus brachialis **1**: 134
– supratarsalis palpebrae superioris **2**: 364
– talonavicularis articulationis talocalcaneonavicularis **1**: 209, 215
– tarsalis palpebrae superioris **2**: 364
– tecta columnae fornicis **2**: 324, 328, 334
– tensa membranae tympanicae **2**: 392
– terminalis arteriae cerebri posterioris **2**: 298
– – ilei **2**: 177–178, 190, 194–195, 200, 202, 204, 225, 238
– thalamolentiformis capsulae internae **2**: 333
– thoracica aortae **1**: 14, 76; **2**: 105–107, 162–163
– – columnae vertebralis **1**: 28, 46–47
– – oesophagi **2**: 105–106
– thyropharyngea musculi constrictoris pharyngis inferioris **2**: 46–47, 68
– tibiocalcanea ligamenti medialis articulationis talocruralis **1**: 212–214
– tibionavicularis ligamenti medialis articulationis talocruralis **1**: 206, 212

Pars(-tes)
– tibiotalaris anterior ligamenti medialis articulationis talocruralis **1**: 212
– – posterior ligamenti medialis articulationis talocruralis **1**: 212
– transversa musculi nasalis **2**: 28–30
– transversaria arteriae vertebralis **1**: 60; **2**: 72, 74, 79, 118, 292, 302
– triangularis gyri frontalis inferioris **2**: 316
– tympanica ossis temporalis **2**: 20, 391–392
– uterina tubaria **2**: 240
Part, cartilaginous, external acoustic meatus **2**: 392
– central, lateral ventricle **2**: 326, 354–355, 361
– non-cartilaginous, external acoustic meatus **2**: 392
– vaginal, cervix uteri **2**: 268
Patella **1**: 10, 170, 181, 195–198, 200, 202–204, 216, 219–221, 226, 230–231, 247, 259, 265, 270–271
Pathway, visual **2**: 335
Paukenhöhle **2**: 391, 393–394, 396, 398–400, 406
– laterale Wand **2**: 398
– mediale Wand **2**: 396, 398
Pecten ossis pubis **1**: 172, 174, 176
Pectoralis, pocket **1**: 110
Pediculus arcus vertebrae **1**: 32–33, 47
Peduncles, cerebellar **2**: 308, 361
– cerebral **2**: 340–341
Pedunculus cerebellaris inferior **2**: 308, 310–311
– – medius **2**: 306, 308–311, 340, 343, 361
– – superior **2**: 308–311, 334, 343
– cerebralis → Pedunculus cerebri
– cerebri **2**: 306, 309–311, 315, 328, 332–335, 340–341, 344, 351, 358, 385
– floccularis **2**: 309
– tonsillae cerebelli **2**: 309
Pektoralistasche **1**: 110
Pelvis **1**: 2, 178, 190–191; **2**: 217, 232–233, 246–247, 250–261
– arteries **2**: 234
– female **2**: 233, 243
– joints **1**: 190–191
– lesser **2**: 235, 244, 246–247, 276
– – female **2**: 238–243, 245–247, 250–253, 256–257, 260
– – – arteries **2**: 245
– – – connective tissue **2**: 246–247
– – – nerves **2**: 245
– – – organs **2**: 245–247
– – male **2**: 236–237, 244, 246–247, 250–251, 254–255, 258–259, 261, 276
– – – arteries **2**: 244
– – – connective tissue **2**: 246–247
– – – organs **2**: 246–247
– – veins **2**: 244
– ligaments **1**: 190–191
– lymph nodes **2**: 163
– – vessels **2**: 163
– male **2**: 232, 234, 244
– muscle attachments **1**: 228
– nerves **2**: 234
– renal **2**: 212, 216–217
– – ampullary **2**: 216
– – dendritic **2**: 216
– renalis **1**: 17; **2**: 208, 211–212, 214, 216–217, 287
– skeletal **1**: 172–175, 178
– – female **1**: 172, 174–175, 178
– – infantile **1**: 177
– – male **1**: 172, 174
– veins **2**: 234
Penis **1**: 17; **2**: 170, 232, 234, 250, 254, 259, 261, 267, 274–278
– arteries **2**: 277

Penis
– body **2**: 275, 278
– bulb **2**: 267
– erectile bodies **2**: 275
– oberflächliche Faszie **2**: 276
– root **2**: 278
– shaft **2**: 274, 276
– superficial fascia **2**: 276
Penisarterien **2**: 277
Penisschaft **2**: 274–276, 278
Penisschwellkörper **2**: 275
Penisvenen, dorsale, oberflächliche **2**: 276
– – tiefe **2**: 276
Peniswurzel **2**: 278
Pericardium **2**: 107, 123–125, 142, 172
– fibrosum **2**: 107, 122–125, 149, 151, 164–165
– serosum **2**: 104, 123–125, 134, 142–143, 154–155
Perichondrium **1**: 9
Perikard **2**: 122–125, 172
– fibröses **2**: 123–125
– seröses **2**: 123–125
Perilymphräume des Labyrinths **2**: 405
Perimetrium → Tunica serosa uteri
Perineum **1**: 17; **2**: 265–266, 279–281
– female **2**: 265, 280–281
– – arteries **2**: 280
– – nerves **2**: 280
– – veins **2**: 280
– male **2**: 266, 279
– – arteries **2**: 279
– – nerves **2**: 279
– – veins **2**: 279
– muscles **2**: 265–267
Periodontium **2**: 40
Periorbita **2**: 58, 379, 381, 384
Periorchium **2**: 270–271, 274
Periosteum **1**: 9–10, 200; **2**: 285–286
Peritendineum **1**: 10
Peritonealhöhle **2**: 170–171
Peritoneum **1**: 249; **2**: 179, 193, 241, 250–251
– parietale **1**: 63, 67, 77; **2**: 171, 196, 200, 204, 218, 222, 230, 232–233, 247, 273, 276
– reflexion lines **2**: 207
– Umschlaglinien **2**: 207
– viscerale **2**: 171, 176, 232
Pes **1**: 2, 216–217, 270–271
– anserinus **1**: 219, 221, 231, 272
– hippocampi **2**: 320, 323–325, 344
Petiolus epiglottidis **2**: 64–65
Pfannendach **1**: 177
Pfannendachwinkel **1**: 177
Pfortader → Vena portae hepatis
Phalanx(-anges) distalis digiti manus **1**: 92–94, 104, 167
– – – pedis **1**: 184–186, 188
– – hallucis **1**: 184–186, 188
– – pollicis **1**: 92–93, 123
– manus/pedis → Ossa digitorum manus/pedis
– media digiti manus **1**: 11, 92–94, 104, 167
– – – pedis **1**: 184–186, 188
– proximalis **1**: 8
– – digiti manus **1**: 11, 92–94, 104, 129, 167
– – – pedis **1**: 184–186, 188
– – hallucis **1**: 184–186, 188
– – pollicis **1**: 92–93, 123
Pharynx **1**: 16; **2**: 13, 38, 45–53, 63, 68, 73, 100–101, 106, 165
– cavity **2**: 50–51
– diverticulum in Laimer's triangle **2**: 49
– Höhle **2**: 50–51
– laryngeal part **2**: 52–53
– muscles **2**: 47–49
– Muskeln **2**: 47–49
– nasal part **2**: 52–53
Pharynx
– oral part **2**: 52–53
– Pars laryngea **2**: 52–53
– – nasalis **2**: 52–53
– – oralis **2**: 52
– wall **2**: 45–51
– Wand **2**: 45–51
Phase, expiratory **2**: 153
– exspiratorische **2**: 153
– inspiratorische **2**: 153
– inspiratory **2**: 153
Philtrum **2**: 38, 44
Pia mater encephali → Pia mater cranialis
– – cranialis **2**: 295, 326, 405
– – spinalis **2**: 285–286, 295
Pilus(-i) **1**: 11
Placenta **1**: 13; **2**: 243
– delivered **2**: 243
Plane(s) **1**: 5
– coronal **1**: 5
– frontal **1**: 5
– horizontal **1**: 5
– median **1**: 5
– paramedian **1**: 5
– sagittal **1**: 5
– transverse **1**: 5
Planta pedis **1**: 217, 279–281
Plantaraponeurose **1**: 237
Planum nuchale **2**: 8, 18
– occipitale **2**: 18
– temporale **2**: 6, 22
Platysma **2**: 28–30, 32–33, 45, 49, 74, 85, 91–93, 98–99
Plazenta **2**: 243
– geborene **2**: 243
Pleura **2**: 111, 152
– borders **2**: 152
– costalis **2**: 150–151, 154–157
– diaphragmatica **2**: 150
– mediastinalis **2**: 150–151
– parietalis **1**: 74; **2**: 153, 209
– pulmonalis **2**: 115, 151, 154–157
– visceralis → Pleura pulmonalis
Pleuragrenzen **2**: 152
Pleuraraum **2**: 153
Plexus aorticus abdominalis **1**: 25; **2**: 248–249
– – thoracicus **2**: 167
– basilaris **1**: 51; **2**: 291, 293–294
– brachialis **1**: 68, 72, 134, 140–141, 144; **2**: 35, 72, 75, 81, 92–96, 98, 165–167
– cardiacus **2**: 96, 149, 166–167
– caroticus communis **1**: 25; **2**: 100
– – externus **1**: 25; **2**: 87, 89, 96
– – internus **1**: 25; **2**: 87–88, 95–96, 166, 382
– cervicalis **1**: 68; **2**: 34–35, 37, 93–94
– choroideus(-i) prosencephali **2**: 326
– – ventriculi lateralis **2**: 300, 322–323, 326, 340, 344, 346–349, 351, 354–356, 361–362
– – – quarti **2**: 306–309, 321, 340
– – – tertii **2**: 300, 307, 320–321, 326, 340, 344
– coccygeus **2**: 227
– coeliacus **1**: 25; **2**: 248–249
– dentalis inferior **2**: 84, 86
– – superior **2**: 84, 86
– entericus **2**: 248–249
– gastricus **1**: 25
– hepaticus **1**: 25; **2**: 248–249
– hypogastricus inferior **1**: 24–25; **2**: 231, 248–249
– – superior **1**: 25; **2**: 231, 248–249
– iliacus **1**: 25; **2**: 248
– intermesentericus **2**: 248
– intraparotideus **2**: 84–85, 98–99
– lienalis → Plexus splenicus
– lumbalis **1**: 227–229, 244; **2**: 209, 226
Plexus
– lumbaris → Plexus lumbalis
– lumbosacralis **1**: 244; **2**: 226–227
– lymphaticus axillaris **1**: 15; **2**: 163
– mesentericus inferior **1**: 25; **2**: 248–249
– – superior **1**: 25; **2**: 248–249
– nervosus pharyngeus → Plexus pharyngeus nervi vagi
– oesophageus **1**: 167; **2**: 25
– ovaricus **1**: 25; **2**: 248–249
– pampiniformis **1**: 67; **2**: 270–272, 274, 278
– pelvicus → Plexus hypogastricus inferior
– pharyngealis nervi vagi → Plexus pharyngeus nervi vagi
– – venae jugularis internae→ Plexus pharyngeus venae jugularis internae
– pharyngeus nervi vagi **2**: 96, 100
– – venae jugularis internae **2**: 100
– pterygoideus **2**: 80, 383
– pulmonalis **1**: 25; **2**: 96, 167
– rectalis inferior **2**: 231
– – medius **2**: 231, 248–249
– – superior **2**: 248
– renalis **1**: 25; **2**: 248–249
– sacralis **1**: 227–229, 244; **2**: 226–227, 231, 234, 244–245, 248
– splenicus **2**: 248–249
– suprarenalis **2**: 248
– testicularis **2**: 248–249
– thyroideus impar **1**: 14; **2**: 149
– tympanicus **2**: 89, 398
– uretericus **2**: 248–249
– venosus areolaris **1**: 75
– – basilaris → Plexus basilaris
– – pharyngeus → Plexus pharyngeus venae jugularis internae
– – prostaticus **2**: 250, 254, 276
– – rectalis **2**: 251
– – sacralis **2**: 287
– – vaginalis **2**: 251
– – vertebralis internus **2**: 287
– – – – anterior **2**: 285
– – – – posterior **2**: 285–286
– – vesicalis **2**: 250, 254, 276
Plica(-ae) alaris **1**: 198, 200
– aryepiglottica **2**: 50, 52, 101
– axillaris anterior **1**: 62
– chordae tympani **2**: 398
– ciliares **2**: 376
– circulares duodeni **2**: 174, 188, 194
– – jejuni **2**: 176
– duodenalis inferior **2**: 206
– – superior **2**: 206
– duodenomesocolica → Plica duodenalis inferior
– fimbriata linguae **2**: 38
– gastricae **2**: 174–175
– glossoepiglottica lateralis **2**: 46
– – mediana **2**: 46
– ileocaecalis **2**: 177
– interureterica **2**: 235
– iridis **2**: 376
– lacrimalis **2**: 365
– mallearis anterior **2**: 391–392, 398
– – posterior **2**: 392, 398
– mucosae urethrae masculinae **2**: 277
– – vesicae biliaris **2**: 186
– nervi laryngei superioris **2**: 50
– palatinae transversae **2**: 38–39, 45
– palmatae canalis cervicis uteri **2**: 233
– pancreaticocolica **2**: 154
– petroclinoidea anterior **2**: 291, 293
– – posterior **2**: 291, 293
– pharyngoepiglottica **2**: 50–51
– rectouterina **2**: 239, 241

Plica(-ae)
– salpingopharyngea **2**: 45, 50, 90
– semilunaris(-es) coli **2**: 177–178
– – conjunctivae **2**: 364–365
– sublingualis **2**: 38
– synovialis **1**: 10–11
– – infrapatellaris **1**: 198, 200
– – tibiofibularis **1**: 214
– transversa(-ae) recti **2**: 179, 232, 243
– – – media **2**: 179
– tubariae **2**: 241
– umbilicalis lateralis **1**: 77; **2**: 196, 200, 238, 246
– – medialis **1**: 77; **2**: 196, 200, 238, 246
– – mediana **1**: 77; **2**: 196, 200, 238, 246
– vesicalis transversa peritonei **1**: 77
– vestibularis **2**: 52, 68
– vocalis **2**: 52, 68
Pocket of pectoralis **1**: 110
Point, nodal, of axes of bulb **2**: 374
Polarterie, obere, Niere **2**: 214
– untere, Niere **2**: 214–215
Pollex **1**: 92–93, 105, 123
Polus anterior bulbi oculi **2**: 374
– – lentis **2**: 374, 377
– frontalis **2**: 296, 306, 314–316, 348
– occipitalis **2**: 296, 306, 314–316, 334, 348
– posterior bulbi oculi **2**: 374
– – lentis **2**: 377
– temporalis **2**: 306, 315–316, 323
Pons **1**: 22; **2**: 13, 303–306, 309, 311, 321, 324, 333, 341, 344, 352–353, 359, 361, 384, 408–409
Porta arteriarum cordis **2**: 123
– hepatis **2**: 181
Portio supravaginalis cervicis uteri **2**: 233, 239–240
– vaginalis cervicis uteri **1**: 17; **2**: 233, 239–240, 251, 268
Porus acusticus externus **2**: 6–8, 20, 26, 47–48, 395
– – internus **2**: 10, 12, 20, 27, 400, 403
Position, extraperitoneal **2**: 171
– intraperitoneal **2**: 171
– retroperitoneal (secondary) **2**: 171
Posterior **1**: 5
Praecuneus **2**: 317
Praeputium clitoridis **2**: 269, 281
– penis **1**: 17; **2**: 232, 274–276
Pregnancy, early **2**: 242
Processus accessorius vertebrae lumbalis **1**: 33
– alveolaris maxillae **2**: 24, 58–59, 369
– anterior mallei **2**: 397
– articularis inferior axis **1**: 30
– – – vertebrae **1**: 30–33, 36, 38, 46, 55; **2**: 224, 286
– – superior atlantis **1**: 31
– – – axis **1**: 30; **2**: 72
– – – ossis sacri **1**: 34–35, 38, 172
– – – vertebrae **1**: 30–33, 36, 38, 42, 46–48, 55; **2**: 224, 286
– – vertebrae **1**: 29
– calcaneus ossis cuboidei **1**: 186
– caudatus lobi caudati hepatis **2**: 181
– ciliares **2**: 374–376
– clinoideus anterior **2**: 7, 10, 19, 293, 301, 368, 378, 380
– – medius **2**: 10
– – posterior **2**: 10, 19, 291, 301
– cochleariformis **2**: 396
– condylaris mandibulae **2**: 6, 9, 25–26
– coracoideus scapulae **1**: 84–85, 87, 95–96, 106–107, 110, 140–141, 148–149
– coronoideus mandibulae **2**: 6, 25–26, 41, 63, 73
– – ulnae **1**: 89–91, 99–100, 116
Processus
– costalis vertebrae lumbalis **1**: 33, 38, 41, 48; **2**: 284
– ethmoidalis conchae nasalis inferioris **2**: 56–57
– falciformis ligamenti sacrotuberalis **1**: 190
– frontalis maxillae **2**: 4, 12, 24, 54, 57, 367–369
– – ossis zygomatici **2**: 367
– intrajugularis ossis occipitalis **2**: 18
– – – temporalis **2**: 20
– jugularis ossis occipitalis **2**: 18
– lateralis mallei **2**: 392, 397
– – tali **1**: 187
– – tuberis calcanei **1**: 186–187
– lenticularis incudis **2**: 397–398
– mammillaris vertebrae lumbalis **1**: 33
– mastoideus **1**: 55, 59–60; **2**: 4, 6, 8, 12, 20, 26–27, 35, 49–50, 73, 92, 100, 393, 395, 399
– maxillaris ossis zygomatici **2**: 58
– medialis tuberis calcanei **1**: 186–187
– muscularis cartilaginis arytaenoideae **2**: 64–65
– orbitalis ossis palatini **2**: 57, 368
– palatinus maxillae **2**: 8, 12, 17, 23–24, 55–56, 59, 88, 90, 369
– posterior tali **1**: 184, 187–188
– pterygoideus ossis sphenoidalis **2**: 8, 12, 17, 19, 23–24, 27, 30, 48, 55–56, 63, 368, 395
– pyramidalis ossis palatini **2**: 24, 57, 368
– retroarticularis ossis temporalis **2**: 20
– sphenoidalis ossis palatini **2**: 57
– spinosus axis **1**: 30–31, 36–37, 49–51, 55; **2**: 294
– – vertebrae **1**: 29–33, 36–40, 42, 46–48, 55, 60, 190; **2**: 75, 287, 289
– – – prominentis **1**: 31
– styloideus ossis metacarpalis III **1**: 92, 94
– – – temporalis **2**: 6, 8, 12, 20, 27, 34–35, 47–49, 51, 100, 290, 395, 399, 402
– – radii **1**: 88, 92–94, 99, 102, 105, 116, 159
– – ulnae **1**: 89, 92, 94, 99, 102, 105
– supracondylaris humeri **1**: 86
– transversus atlantis **1**: 30–31, 49, 55, 59–60; **2**: 33, 37, 72, 289
– – axis **1**: 30
– – vertebrae **1**: 29–32, 35–36, 42, 47, 51; **2**: 33, 37, 72, 74–75, 284
– uncinatus ossis ethmoidalis **2**: 22, 55–56
– – pancreatis **2**: 188, 199, 202
– vaginalis apertus testis **2**: 272–273
– – offener, persistierender **2**: 273
– – ossis sphenoidalis **2**: 19, 23
– – peritonei **2**: 272–273
– – persistent **2**: 273
– vocalis cartilaginis arytaenoideae **2**: 64, 68
– xiphoideus **1**: 40–41, 45, 62, 64, 68; **2**: 150
– zygomaticus maxillae **2**: 8, 24
– – ossis frontalis **2**: 4, 21, 58, 367
– – – temporalis **2**: 20, 26, 399
Prolapse of disc **1**: 48
Proliferationsphase **2**: 242
Prominentia canalis facialis **2**: 396
– – semicircularis lateralis **2**: 396
– laryngea **2**: 33–34, 36, 44, 64–65, 70, 97, 99, 149, 164
– mallearis **2**: 392
– spiralis ductus cochlearis **2**: 404
Promontorium cavitatis tympanicae **2**: 391, 393, 396, 399, 402
– ossis sacri **1**: 14, 28, 35, 38–40, 69, 172, 178, 190; **2**: 202, 220–221, 228, 233–234, 238, 244–245, 248, 260–261, 264
– tympani → Promontorium cavitatis tympanicae
Pronation **1**: 5
Prosencephalon **2**: 303
Prostata **1**: 17, 20–21, 260; **2**: 170, 232, 234–237, 244, 247, 250, 254, 258–259, 261, 266, 277
Protuberantia mentalis **2**: 4, 25
– occipitalis externa **1**: 53, 55, 57–58; **2**: 8, 12, 17–18
– – interna **2**: 10–11, 18, 409
Proximalis **1**: 5
Pubis → Os pubis
Pudendum femininum **2**: 163
Pudendusanästhesie **2**: 281
Pulmo(-ones) **1**: 13, 18–19; **2**: 152
– dexter **1**: 16, 20; **2**: 96, 111–114, 116–117, 119, 128–131, 149–150, 154–161, 165, 172, 191, 220, 222
– sinister **1**: 12, 16, 21; **2**: 111–114, 116–117, 119, 128–131, 149–150, 154–161, 172, 191, 209, 220, 222
Pulpa dentis **2**: 40
Pulsionsdivertikel im Laimer-Dreieck **2**: 49
– pharyngoösophageales **2**: 49
Pulvinar acetabuli **1**: 193
– thalami **2**: 310–311, 329–332, 335
Punctum lacrimale **2**: 364–365
Puncture, suboccipital **2**: 321
Pupilla **2**: 364, 371–372, 376, 384
Putamen **2**: 323, 329–335, 337–341, 344, 346–351, 355–357, 360–362
Pylorus **2**: 154, 173
Pyramis medullae oblongatae **2**: 306, 309
– renalis **2**: 212
– vermis **2**: 307–309

Q

Querachsen **1**: 5
Quergewölbe des Fusses **1**: 189
Querkolon **2**: 179, 205

R

Rachen → Schlund
Rachenring, lymphatischer **2**: 45
Rachenwand, Arterien **2**: 100
– Nerven **2**: 100
– Venen **2**: 100
Radialiskanal **1**: 143
Radiatio(-ones) acustica **2**: 332, 350
– corporis callosi **2**: 337, 342, 345–346, 348
– optica **2**: 332, 335, 348, 350, 355–357
– thalamicae anteriores **2**: 332–333
Radii lentis bulbi oculi **2**: 377
Radiokarpalgelenk **1**: 159
Radioulnargelenk, distales **1**: 99
– proximales **1**: 99, 101
Radius **1**: 8, 82–83, 88, 90–94, 98–100, 102–104, 116, 118–119, 122, 125, 129–130, 139, 154–159
– annular ligament **1**: 99
– muscle attachments **1**: 117
– Muskelanheftungen **1**: 117
Radix(-ices) anterior ansae cervicalis **2**: 87, 93–94, 97–100
– – nervi cervicalis **2**: 95, 166, 288–289, 306, 311
– – – spinalis **1**: 23; **2**: 285, 288
– – – thoracici **2**: 286
– clinica dentis **2**: 40
– craniales nervi accessorii **1**: 22; **2**: 289, 292, 294, 306, 311
– dentis **2**: 40
– lateralis nervi mediani **1**: 140–141; **2**: 93, 149
– – tractus optici **2**: 335
– linguae **2**: 46
– medialis nervi mediani **1**: 140–141; **2**: 93, 149
– – tractus optici **2**: 335
– mesenterii **2**: 171, 199–200, 202, 204, 206–207
– mesocoli sigmoidei **2**: 206–207
– – transversi **2**: 200, 206

Radix(-ices)
– motoria nervi cervicalis/spinalis/thoracici → Radix anterior nervi cervicalis/spinalis/thoracici
– – – trigemini **2**: 89, 295, 306, 380
– nasociliaris ganglii ciliaris → Radix sensoria ganglii ciliaris
– nervi cervicalis **1**: 51
– – sacralis **2**: 232
– – spinalis **2**: 285
– oculomotoria ganglii ciliaris → Radix parasympathetica ganglii ciliaris
– parasympathetica ganglii ciliaris **2**: 381–382
– penis **2**: 276
– posterior ansae cervicalis **2**: 93–94, 97–98
– – nervi cervicalis **2**: 95, 166, 284, 288–289, 311
– – – spinalis **1**: 23; **2**: 285, 288
– – – thoracici **2**: 284, 286
– sensoria ganglii ciliaris **2**: 381–382
– – nervi cervicalis/spinalis/thoracici → Radix posterior nervi cervicalis/spinalis/thoracici
– – – trigemini **2**: 84, 89, 306, 380
– spinales nervi accessorii **1**: 22; **2**: 289, 292, 294–295, 306, 311
– sympathetica ganglii ciliaris **2**: 382
– unguis **1**: 11
Ramus(-i) acetabularis arteriae obturatoriae **1**: 193
– acromialis arteriae thoracoacromialis **2**: 93, 98
– alveolares superiores posteriores nervi maxillaris **2**: 86–87
– anastomoticus arteriae meningeae mediae cum arteria lacrimali **2**: 77
– anterior arteriae obturatoriae **1**: 193, 253
– – – pancreaticoduodenalis inferioris **2**: 198
– – – recurrentis ulnaris **1**: 153
– – – renalis **2**: 218
– – nervi auricularis magni **2**: 82, 91–92
– – – cervicalis **2**: 94, 284, 286
– – – cutanei antebrachii medialis **1**: 132, 136, 151
– – – lumbalis **2**: 226
– – – obturatorii **1**: 252
– – – sacralis **2**: 284
– – – spinalis **1**: 23, 79; **2**: 285
– – – thoracici **2**: 94
– – sulci lateralis cerebri **2**: 316
– – ventriculi dextri arteriae coronariae dextrae **2**: 124, 135, 142, 144, 146
– apicalis rami lobi superioris arteriae pulmonalis dextrae **2**: 113
– articularis arteriae descendentis genicularis **1**: 248, 269
– – nervi ischiadici **1**: 256
– ascendens arteriae circumflexae femoris lateralis **1**: 251–253
– – – – – medialis **1**: 253
– – sulci lateralis cerebri **2**: 316
– atriales arteriae coronariae dextrae **2**: 144
– – – – sinistrae **2**: 144
– auricularis(-es) anteriores arteriae temporalis superficialis **2**: 98
– – – venae temporalis superficialis **2**: 98
– – arteriae auricularis posterioris **2**: 85, 91–92
– – – occipitalis **2**: 76
– – nervi vagi **2**: 82
– bronchiales nervi vagi **1**: 25
– buccales nervi facialis **2**: 84–85, 92, 98–99
– calcanei arteriae tibialis posterioris **1**: 267, 279–281
– – laterales nervi suralis **1**: 266–267
– – mediales nervi tibialis **1**: 242, 267–268, 270, 279
– calcarinus arteriae occipitalis medialis **2**: 298–299, 302
Ramus(-i)
– capsularis arteriae renalis **2**: 162
– cardiacus(-i) cervicales inferiores nervi vagi **1**: 25; **2**: 96, 149
– – – superiores nervi vagi **1**: 25; **2**: 100, 149
– – cervicalis medius nervi vagi **2**: 149
– – thoracici gangliorum thoracicorum **2**: 166–167
– – – nervi vagi **2**: 149
– carpalis dorsalis arteriae radialis **1**: 155, 163, 165
– – – – ulnaris **1**: 138, 155, 161–162, 165
– – palmaris arteriae radialis **1**: 152–153, 161–162
– – – – ulnaris **1**: 162
– cervicalis nervi facialis → Ramus colli nervi facialis
– choroidei posteriores laterales arteriae cerebri posterioris **2**: 300
– – – mediales arteriae cerebri posterioris **2**: 300
– circumflexus arteriae coronariae sinistrae **2**: 107, 128–130, 135, 141–144, 147–148
– clivales partis cerebralis arteriae carotidis internae **2**: 301
– clunium → Nervi clunium
– coeliaci trunci vagalis posterioris **2**: 167, 248
– collateralis arteriae intercostalis posterioris **1**: 76
– colli nervi facialis **2**: 84–85, 91–93, 98–99
– communicans(-antes) albus nervi spinalis **1**: 23, 79; **2**: 96, 166–167, 285
– – cochlearis nervi vestibularis **2**: 405
– – cum nervo ulnari **1**: 161–162
– – – – vago **2**: 398
– – – – zygomatico **2**: 88, 379–380
– – – ramo laryngeo recurrente **2**: 101
– – fibularis **1**: 266
– – griseus nervi spinalis **1**: 23, 79; **2**: 95–96, 166–167, 285
– – nervi spinalis **1**: 23, 25; **2**: 248
– – ulnaris nervi radialis **1**: 133, 164
– coni arteriosi arteriae coronariae dextrae **2**: 135, 142
– – – – – sinistrae **2**: 135
– cricothyroideus arteriae thyroideae superioris **2**: 98
– cutaneus(-i) anterior(-iores) nervi femoralis **1**: 78, 242, 248, 270; **2**: 227–228
– – – – iliohypogastrici **1**: 242
– – – nervorum intercostalium **1**: 23, 78, 132; **2**: 93
– – arteriae cervicalis superficialis **2**: 91
– – – thoracoacromialis **2**: 91
– – cruris mediales nervi sapheni **1**: 242–243, 266, 270, 278
– – lateralis(-es) arteriae intercostalis posterioris **1**: 76
– – – nervi iliohypogastrici **1**: 242–243, 248, 254–255
– – – – intercostalis (rami anterioris nervi thoracici) **1**: 23; **2**: 93
– – – nervorum intercostalium **1**: 61, 78, 132–133, 136
– – nervi buccalis **2**: 85
– – – obturatorii **1**: 242–243, 248, 252, 254–255, 266; **2**: 228
– deltoideus arteriae profundae brachii **1**: 143
– – – thoracoacromialis **2**: 93
– descendens arteriae circumflexae femoris lateralis **1**: 251–253
– – – occipitalis **2**: 76
– dexter arteriae hepaticae propriae **2**: 183–184, 197
Ramus(-i) dexter
– – venae portae hepatis **1**: 12–13
– digastricus nervi facialis **2**: 84, 87
– dorsalis arteriae intercostalis posterioris **1**: 76
– – nervi ulnaris **1**: 133, 150, 153–154, 162, 164
– epiploici → Rami omentales
– externus nervi laryngei superioris **2**: 96–98, 100–101
– – trunci nervi accessorii **2**: 100
– femoralis nervi genitofemoralis **1**: 78, 242, 248; **2**: 227
– frontalis(-es) anteromedialis arteriae callosomarginalis **2**: 299
– – arteriae callosomarginalis **2**: 297
– – – meningeae mediae **2**: 77, 290, 380
– – – temporalis superficialis **2**: 76, 85–86
– – intermediomedialis arteriae callosomarginalis **2**: 299
– – posteromedialis arteriae callosomarginalis **2**: 299
– ganglionaris trigeminalis arteriae carotidis internae **2**: 301
– gastrici anteriores trunci vagalis anterioris **2**: 167, 248
– – posteriores trunci vagalis posterioris **2**: 167
– genitalis nervi genitofemoralis **1**: 78, 242; **2**: 227, 274
– glandulares arteriae facialis **2**: 76
– hepatici trunci vagalis anterioris **2**: 248–249
– iliacus arteriae iliolumbalis **2**: 228
– inferior(-iores) nervi oculomotorii **2**: 379, 381–382
– – – transversi colli **2**: 91–92
– – ossis pubis **1**: 172–173, 175–176, 221; **2**: 250, 268, 281
– infrahyoideus arteriae thyroideae superioris **2**: 98
– infrapatellaris nervi sapheni **1**: 242, 248, 252, 270
– intercostales anteriores arteriae thoracicae internae **1**: 75–76
– interganglionares trunci sympathetici **1**: 23, 25; **2**: 167
– internus trunci nervi accessorii **2**: 100
– – nervi laryngei superioris **2**: 65, 67, 70, 96–101
– intersegmentales venae pulmonalis **2**: 115
– interventricularis(-es) anterior arteriae coronariae sinistrae **2**: 124, 129, 132, 135, 141–142, 144, 147–148
– – posterior arteriae coronariae dextrae **2**: 135, 143–144, 146, 148
– – septales arteriae coronariae dextrae **2**: 144
– – – – – sinistrae **2**: 144
– labialis(-es) posterior arteriae perinealis **2**: 280
– – superiores arteriae infraorbitalis **2**: 87
– – – nervi infraorbitalis **2**: 85
– laryngopharyngei ganglii cervicalis superioris **2**: 96, 100
– lateralis arteriae coronariae sinistrae **2**: 144, 147–148
– – nervi supraorbitalis **2**: 82–83, 85–87, 380–381, 383
– lobi inferioris arteriae pulmonalis dextrae **2**: 113, 119
– – – – – sinistrae **2**: 119
– – medii arteriae pulmonalis dextrae **2**: 119
– – superioris arteriae pulmonalis dextrae **2**: 113, 119
– – – – – sinistrae **2**: 119
– malleolares laterales arteriae fibularis **1**: 250, 266–268
– – mediales arteriae tibialis posterioris **1**: 250, 266–268

Ramus(-i)
– mammarii laterales arteriae thoracicae lateralis **1**: 72
– – – rami cutanei lateralis arteriae intercostalis posterioris **1**: 76
– – mediales ramorum perforantium arteriae thoracicae internae **1**: 76
– mandibulae **2**: 4–7, 25–26, 41, 49, 51, 58, 73, 100
– marginalis dexter arteriae coronariae dextrae **2**: 124, 135, 142, 144, 146, 148
– – mandibulae nervi facialis **2**: 84–85, 92, 98–99
– – sinister arteriae coronariae sinistrae **2**: 135, 143, 147
– mastoideus arteriae occipitalis **2**: 76–77
– medialis nervi supraorbitalis **2**: 82–83, 85–87, 380–381, 383
– meningeus anterior arteriae ethmoidalis anterioris **2**: 90
– – arteriae carotidis internae **2**: 301
– – – vertebralis **2**: 77
– – nervi mandibularis **2**: 380
– – – maxillaris **2**: 380
– – – ophthalmici → Ramus tentorii nervi ophthalmici
– – – spinalis **1**: 23; **2**: 285–286
– – – vagi **2**: 100
– mentalis arteriae alveolaris inferioris **2**: 76, 85–86
– mylohyoideus arteriae alveolaris inferioris **2**: 76–77, 87, 89
– nasalis(-es) anteriores laterales arteriae ethmoidalis anterioris **2**: 77, 88
– – externus nervi ethmoidalis anterioris **2**: 82–83, 85–86, 90, 383
– – interni laterales nervi ethmoidalis anterioris **2**: 88
– – – mediales nervi ethmoidalis anterioris **2**: 90
– – posteriores inferiores nervi palatini majoris **2**: 88–89
– – – superiores laterales nervi maxillaris **2**: 88–89
– – – – mediales nervi maxillaris **2**: 90
– nodi atrioventricularis arteriae coronariae dextrae **2**: 135, 144
– – sinuatrialis arteriae coronariae dextrae **2**: 144
– obturatorius arteriae epigastricae inferioris **2**: 244
– – venae epigastricae inferioris **2**: 244
– occipitalis arteriae auricularis posterioris **2**: 91–92
– – nervi facialis **2**: 84
– oesophagealis(-es) nervi laryngei recurrentis **2**: 101
– – partis thoracicae aortae **2**: 162
– – trunci thyrocervicalis **2**: 100
– omentales arteriae gastroomentalis dextrae **2**: 196, 202
– orbitalis arteriae meningeae mediae **2**: 77, 380
– ossis ischii **1**: 172, 176, 220–221; **2**: 268, 275, 279
– ovaricus arteriae uterinae **2**: 241, 245
– palmaris nervi mediani **1**: 132, 151–153, 160–161
– – – ulnaris **1**: 132, 151–153, 160
– – profundus arteriae ulnaris **1**: 153, 161–163
– – superficialis arteriae radialis **1**: 152–153, 155, 160–163
– palpebralis(-es) inferiores nervi infraorbitalis **2**: 85, 87
– – nervi lacrimalis **2**: 82–83, 85–86
– pancreatici arteriae splenicae **2**: 197

Ramus(-i)
– parietalis(-es) arteriae meningeae mediae **2**: 77, 290, 380
– – – occipitalis medialis **2**: 299
– – – temporalis superficialis **2**: 76, 85–86
– parietooccipitalis arteriae occipitalis medialis **2**: 302
– pectorales arteriae thoracoacromialis **2**: 93
– pedunculares arteriae cerebri posterioris **2**: 299
– perforans(-antes) arteriae fibularis **1**: 250, 271, 278
– – – thoracicae internae **1**: 76
– pericardiacus nervi phrenici **2**: 149
– – partis thoracicae aortae **2**: 162
– perineales nervi cutanei femoralis posterioris **2**: 279–280
– petrosus arteriae meningeae mediae **2**: 380
– pharyngealis nervi vagi → Ramus pharyngeus nervi vagi
– pharyngeus arteriae palatinae descendentis **2**: 88
– – nervi vagi **2**: 100
– posterior(-iores) arteriae obturatoriae **1**: 193
– – – pancreaticoduodenalis inferioris **2**: 198
– – – recurrentis ulnaris **1**: 153
– – nervi(-orum) auricularis magni **2**: 82–83, 91–92
– – – cervicalium **2**: 32, 37, 83, 284, 286
– – – cutanei antebrachii medialis **1**: 132, 135–136, 150–151
– – – lumbalium **2**: 284
– – – obturatorii **1**: 252
– – – sacralis **2**: 284
– – – spinalis(-ium) **1**: 23, 60–61, 227; **2**: 285
– – – thoracicorum **1**: 133
– – sulci lateralis cerebri **2**: 316
– – ventriculi sinistri arteriae coronariae sinistrae **2**: 135, 143–144, 147–148
– posterolateralis dexter arteriae coronariae dextrae **2**: 135, 143, 146
– profundus arteriae circumflexae femoris medialis **1**: 253
– – – plantaris medialis **1**: 279–281
– – – transversae cervicis **1**: 60
– – nervi plantaris lateralis **1**: 279–280
– – – radialis **1**: 113, 117, 140, 152–154, 157
– – – ulnaris **1**: 152–153, 161–162
– pubicus arteriae obturatoriae **2**: 244
– – venae obturatoriae **2**: 244
– radiculares arteriae vertebralis → Rami spinales arteriae vertebralis
– saphenus arteriae descendentis genicularis **1**: 252, 269
– scrotales anteriores nervi ilioinguinalis **1**: 248
– – posteriores arteriae perinealis **2**: 279
– segmentalis arteriae pulmonalis **2**: 115
– septalis anterior arteriae ethmoidalis anterioris **2**: 90
– – posterior arteriae sphenopalatinae **2**: 88, 90
– sinister arteriae hepaticae propriae **2**: 183–184, 196–197
– – venae portae hepatis **1**: 12–13; **2**: 183
– spinalis(-es) arteriae intercostalis posterioris **1**: 76; **2**: 285–286
– – – sacralis lateralis **2**: 244
– – – vertebralis **2**: 286
– – venae intercostalis posterioris **2**: 285–286
– – – vertebralis **2**: 286
– splenici arteriae splenicae **2**: 197, 219
– sternales arteriae thoracicae internae **1**: 76
– sternocleidomastoidei arteriae occipitalis **2**: 76
– superficialis arteriae circumflexae femoris medialis **1**: 252–253
– – – plantaris medialis **1**: 279–281

Ramus(-i) superficialis
– – nervi plantaris lateralis **1**: 279–280
– – – radialis **1**: 132–133, 140, 150–154, 157, 160–161, 164
– – – ulnaris **1**: 152–153, 161–162
– superior(-iores) nervi oculomotorii **2**: 379, 381
– – – transversi colli **2**: 91–92
– – ossis pubis **1**: 172–176, 221; **2**: 268
– suprahyoideus arteriae lingualis **2**: 98
– temporales nervi facialis **2**: 84–85, 98
– – superficiales nervi auriculotemporalis **2**: 86–87
– tentorii nervi ophthalmici **2**: 293, 380
– thyrohyoideus ansae cervicalis **2**: 94, 98–99
– tonsillaris arteriae facialis **2**: 76
– tracheales trunci thyrocervicalis **2**: 100
– transversus arteriae circumflexae femoris lateralis **1**: 251
– tubarius arteriae ovaricae **2**: 241
– – – uterinae **2**: 241, 245
– vaginalis arteriae uterinae **2**: 241
– zygomatici nervi facialis **2**: 84–85, 98–99, 383
– zygomaticofacialis nervi zygomatici **2**: 82–83, 85–86, 379
– zygomaticotemporalis nervi zygomatici **2**: 82–83, 85, 379
Raphe palati **2**: 38–39
– palpebralis lateralis **2**: 364
– penis **2**: 274–275
– perinealis **2**: 269
– perinei → Raphe perinealis
– pharyngis **2**: 49, 100
– pterygomandibularis **2**: 48
– scroti **2**: 274, 276
Rascetta **1**: 105
Recessus axillaris articulationis humeri **1**: 95–97, 109
– cochlearis **2**: 402
– costodiaphragmaticus **1**: 43; **2**: 126, 153–155, 182, 209, 222
– costomediastinalis **2**: 157–159
– duodenalis inferior **2**: 193, 206
– – superior **2**: 193
– ellipticus **2**: 401–402
– epitympanicus **2**: 391, 398, 407
– hepatorenalis recessus subhepatici **2**: 171
– hypotympanicus **2**: 391
– ileocaecalis inferior **2**: 177
– – superior **2**: 177
– inferior bursae omentalis **2**: 170
– infundibuli **2**: 293, 307, 320, 338
– intersigmoideus **2**: 204
– lateralis ventriculi quarti **2**: 308–311, 320
– membranae tympanicae anterior **2**: 398
– – – posterior **2**: 398
– opticus **2**: 301, 307, 320, 351
– pharyngeus **2**: 50–51, 57
– pinealis **2**: 307, 320, 324
– piriformis **2**: 46, 50, 53, 68, 74, 101
– sacciformis articulationis radioulnaris distalis **1**: 102, 104
– sphaericus **2**: 401–402
– sphenoethmoidalis **2**: 9, 11, 55, 62, 359
– subhepaticus **2**: 170
– subphrenicus **2**: 154, 170, 190
– subpopliteus **1**: 10, 200
– superior bursae omentalis **2**: 170, 206–207
– suprapinealis **2**: 307, 310–311, 320
– triangularis **2**: 307, 320
Rectum **1**: 16–21, 194, 261; **2**: 170, 179, 202, 204–206, 218, 220, 230–234, 238, 245–248, 251, 256–261, 272, 276
– arteries **2**: 230–231
– nerves **2**: 231
Regenbogenhaut **2**: 376

Regio(-ones) brachialis anterior **1**: 4
– cervicalis anterior **1**: 4
– – lateralis **1**: 4
– – posterior **1**: 4
– corporis humani **1**: 4
– deltoidea **1**: 4, 135–137
– dorsalis pedis → Dorsum pedis
– epigastrica **1**: 4
– facialis **2**: 44
– – profunda **2**: 86–90
– – superficialis **2**: 85
– femoralis anterior **1**: 4
– – posterior **1**: 4
– femoris anterior/posterior → Regio femoralis anterior/posterior
– glutaealis **1**: 4
– hypochondriaca **1**: 4
– infrascapularis **1**: 4
– inguinalis **1**: 4, 64–67, 77, 216, 249; **2**: 272–273
– lateralis abdominis **1**: 4
– lumbalis **1**: 4
– lumbaris → Regio lumbalis
– nasalis **2**: 44
– nuchalis → Regio cervicalis posterior
– occipitalis **1**: 4
– olfactoria lateralis nasi **2**: 90
– – septalis nasi **2**: 90
– orbitalis **2**: 366, 383
– parietalis **1**: 4
– pectoralis **1**: 4
– perinealis **2**: 279–281
– plantaris pedis → Planta pedis
– praesternalis **1**: 4
– pubica **1**: 4
– respiratoria nasi **2**: 90
– scapularis **1**: 4
– sternocleidomastoidea **1**: 4
– sublingualis **2**: 47
– suprascapularis **1**: 4
– umbilicalis **1**: 4
– vertebralis **1**: 4
Region, axillary, arteries **2**: 149
– – nerves **2**: 149
– – veins **2**: 149
– gluteal **1**: 257
– – arteries **1**: 255–257
– – muscles **1**: 257
– – nerves **1**: 255–257
– – subcutaneous nerves **1**: 254
– – – veins **1**: 254
– – veins **1**: 255–257
– inguinal **1**: 63, 65–67, 77, 249; **2**: 272–273
– – lymph nodes **2**: 163
– – – vessels **2**: 163
– olfactory **2**: 90
– orbital **2**: 366, 383
– perineal, female **2**: 280
– – male **2**: 279
– sublingual **2**: 47
Rektum, Arterien **2**: 230–231
– Nerven **2**: 231
Rektumpfeiler **2**: 246
Ren **2**: 171, 182, 186, 199, 208–215, 217–218, 220–224, 248
– arcuatus **2**: 211
– fetalis **2**: 210
– lobatus **2**: 210
Renculus(-i) **2**: 210
Restricta **1**: 105
Rete acromiale arteriae thoracoacromialis **1**: 135, 142; **2**: 91–92
– articulare cubiti **1**: 142–143, 154
– – genus **1**: 250, 252
– calcaneum **1**: 266–268, 279–280
Rete
– carpale dorsale **1**: 138, 154, 165
– – palmare **1**: 162
– malleolare laterale **1**: 271, 278
– – mediale **1**: 278
– patellare **1**: 248, 252, 271
– testis **2**: 270
– venosum dorsale manus **1**: 105, 164
– – – pedis **1**: 278
Reticulum trabeculare sclerae **2**: 375
Retina **2**: 373–377, 384–385
Retinaculum(-a), cervical **2**: 246
– cutis **1**: 11
– extensorum → Retinaculum musculorum extensorum manus
– flexorum → Retinaculum musculorum flexorum manus
– musculorum extensorum inferius pedis **1**: 230–232, 236, 241, 271
– – – manus **1**: 118–119, 125, 130, 150, 154, 164–165
– – – superius pedis **1**: 230–232, 236, 241
– – fibularium inferius/superius → Retinaculum musculorum peronaeorum inferius/superius
– – flexorum manus **1**: 102–103, 121–124, 130, 152–153, 161–162, 166
– – – pedis **1**: 233–234, 238, 241, 267–268, 279–280
– – peronaeorum inferius **1**: 206, 232, 241
– – – superius **1**: 232–234, 241
– patellae laterale **1**: 195, 200, 265
– – mediale **1**: 195, 200, 265
– rectal **2**: 246
– vesical **2**: 246
Retrokardialraum **2**: 107, 127
Retroperitonealraum, autonomes Nervensystem **2**: 248–249
Retrosternalraum **2**: 127
Retroversion **1**: 5
Rhombencephalon **2**: 303
Ribs **1**: 29, 41–42, 44
Riechregion **2**: 90
Rima ani **2**: 256–259
– glottidis **2**: 68–69
– oris **2**: 44
– palpebrarum **2**: 364, 384
Rinde, motorische **2**: 319
Rindenfelder, funktionelle, primäre, der Grosshirnrinde **2**: 319
– – sekundäre, der Grosshirnrinde **2**: 319
– Grosshirnrinde **2**: 327
Ring, hernial **1**: 63; **2**: 273
– inguinal, deep **2**: 273
– lymphatic, of pharynx **2**: 45
– tendinous, common, of extrinsic eye muscles **2**: 379
Ringknorpel **2**: 64–65
Rippen **1**: 29, 40–42, 44
Rippenwirbelgelenke **1**: 46–47
Roof, acetabulum **1**: 177
– oral cavity **2**: 39
– orbital cavity **2**: 357
Root(s), penis **2**: 278
– spinal **2**: 285
– teeth **2**: 41
Rostrum corporis callosi **2**: 299, 307, 317, 325, 337
– sphenoidale **2**: 19
Rotation, lateral **1**: 5
– medial **1**: 5
Rücken, Oberflächenrelief **1**: 52
Rückenmark **1**: 23; **2**: 284–286, 288
Rückenmarkarterien **2**: 286
Rückenmarkhäute **2**: 284–286
Rückenmarkhüllen **2**: 286
Rückenmarkvenen **2**: 286–287
Rückenmuskeln **1**: 52–60; **2**: 37
– autochthone **1**: 56–59
Ruga(-ae) vaginales **2**: 233, 240, 269
Rumpfmuskeln, ventrale **1**: 62–65, 68
Rumpfwand, dorsale, Arterien **2**: 162
– – Venen **2**: 162

S

Sac, hernial **1**: 63, 67, 249; **2**: 273
– pericardial **2**: 122–123
– preputial **2**: 276
Sacculus vestibularis **2**: 405
Saccus conjunctivalis **2**: 385
– endolymphaticus **2**: 400, 405
– lacrimalis **2**: 365, 383
Sacrum **1**: 29, 34–35, 38–39
Sagittalachsen **1**: 5
Sakralmark **2**: 288
Sakroiliakalgelenk **1**: 194
Salpinx → Tuba uterina
Samenbläschen **2**: 236–237
Samenleiter **2**: 236–237
Samenstrang **2**: 274
– Inhalt **2**: 274
Scala tympani **2**: 391, 402–405
– vestibuli **2**: 391, 402–405
Scalene gap **2**: 34–35, 37
Scalp, muscles **2**: 28–30
Scan of bones **1**: 3
– – thyroid gland **2**: 71
Scapha **2**: 390
Scapula **1**: 20, 43, 73, 82–85, 87, 97, 108, 111–112, 137, 142, 144, 148–149; **2**: 156, 161
Schädel **2**: 2, 4–7, 12–15, 58
– Breitschädel **2**: 2
– eines Neugeborenen **2**: 17
– Langschädel **2**: 2
– Muskelanheftungen **2**: 32
Schädelbasis **2**: 16, 395
– äussere **2**: 8–9, 395
– innere **2**: 10–11, 293, 400
– Muskelanheftungen **2**: 37, 395
Schädeldach **2**: 2–3, 295
Schädelform **2**: 2
Schädelhöhle **2**: 294
Scheitelbein **2**: 22
Scheitellappen, Grosshirn **2**: 312–313, 408
Schenkelbruch **1**: 249
Schenkelhernie → Schenkelbruch
Schienbein **1**: 182–183
– Muskelanheftungen **1**: 235
Schilddrüse **2**: 70–71
– Szintigramm **2**: 71
Schilddrüsenarterien **2**: 101
Schildknorpel **2**: 64–65
Schläfenbein **2**: 20, 393–395, 399–400, 406–409
Schläfenlappen, Grosshirn **2**: 312–313, 409
Schlund **2**: 45–52
Schlundhöhle **2**: 50–51
Schlundmuskeln **2**: 47–49
Schlundwand **2**: 45–47, 50–51
Schlüsselbein **1**: 84, 112
Schlüsselbeingelenk, mediales **1**: 45
Schmalgesicht **2**: 2
Schnecke **2**: 399–400, 402–404, 406
Schulter **1**: 144
– Hautnerven **1**: 135–136
– subkutane Nerven **1**: 135–136
– – Venen **1**: 135–137
Schulterarterien **1**: 140, 142–143
Schulterblatt **1**: 84–85, 112
Schultergelenk **1**: 95–97

Schultergürtel **1**: 84, 87
– Muskelanheftungen **1**: 112
Schultermuskeln **1**: 106–111
Schulternerven **1**: 140, 142–143
Schwellkörper, Klitoris **2**: 268
– Penis **2**: 275
Sclera **2**: 372–376, 384
Scrotum **1**: 17, 67; **2**: 170, 232, 234, 261, 271–274, 276
– contents **2**: 271–272, 274
– Inhalt **2**: 271–272, 274
Segment(s), bronchopulmonary **2**: 115–117
Segmentation, liver **2**: 184–185
– lung **2**: 115–117
Segmentgliederung, Leber **2**: 184–185
– Lunge **2**: 115–117
Segmentum(-a) anterius inferius hepatis **2**: 184
– – pulmonis dextri **2**: 116
– – – sinistri **2**: 117
– – superius hepatis **2**: 184
– apicale pulmonis dextri **2**: 116
– apicoposterius pulmonis sinistri **2**: 117
– basale anterius pulmonis dextri **2**: 116
– – – – sinistri **2**: 117
– – laterale pulmonis dextri **2**: 116
– – – – sinistri **2**: 117
– – mediale pulmonis dextri **2**: 116
– – posterius pulmonis dextri **2**: 116
– – – – sinistri **2**: 117
– bronchopulmonalia **2**: 115–117
– cardiacum pulmonis dextri → Segmentum basale mediale pulmonis dextri
– hepatis **2**: 184–185
– laterale inferius hepatis **2**: 184
– – pulmonis dextri **2**: 116
– – superius hepatis **2**: 184
– lingulare inferius pulmonis sinistri **2**: 117
– – superius pulmonis sinistri **2**: 117
– mediale inferius hepatis **2**: 184
– – pulmonis dextri **2**: 116
– – superius hepatis **2**: 184
– posterius inferius hepatis **2**: 184
– – pulmonis dextri **2**: 116
– – superius hepatis
– pulmonis **2**: 115–117
– superius pulmonis dextri **2**: 116
– – – sinistri **2**: 117
Sehbahn **2**: 335
Sehne **1**: 10
Sehnenring, gemeinsamer, der äusseren Augenmuskeln **2**: 379
Sehnenscheiden, Finger **1**: 130–131
– Fussknöchel **1**: 241
– Handwurzel **1**: 130–131
Sehnerv **2**: 352–353, 359, 378, 384–385
Sehnervenkreuzung **2**: 358, 360
Sehrinde **2**: 319, 322, 335
Sehstrahlung **2**: 335
Seitenventrikel, Gehirn **2**: 300, 322–323
– Hinterhorn **2**: 342–343, 346–349, 354–356
– Unterhorn **2**: 326, 362
– Vorderhorn **2**: 337, 346–349, 354–355, 360
– Zentralteil **2**: 326, 354–355, 361
Sekretionsphase **2**: 242
Sella turcica **2**: 23, 57
Semicanalis musculi tensoris tympani **2**: 393, 396, 399
– tubae auditivae → Semicanalis tubae auditoriae
– – auditoriae **2**: 393, 396, 399
Septulum(-a) testis **2**: 270–271
Septum(-a) canalis musculotubarii **2**: 393, 396, 398
– femorale **1**: 249
– glandis penis **2**: 275
Septum(-a)
– interalveolare(-ia) mandibulae **2**: 25
– – maxillae **2**: 24
– interatriale **2**: 131–132, 140
– intermusculare brachii laterale **1**: 108–109, 118–119, 135, 142–143, 150, 154
– – – mediale **1**: 106–107, 114–116, 140–141, 152–153
– – cruris anterius **1**: 271
– – femoris mediale **1**: 220
– interradiculare(-ia) mandibulae **2**: 25
– – maxillae **2**: 24
– – medullae spinalis **2**: 285–286
– interventriculare **2**: 132–134, 137, 139, 145, 159
– leptomeningicum **2**: 285–286
– linguae → Septum linguale
– linguale **2**: 46–47
– lucidum → Septum pellucidum
– nasal **2**: 54, 90
– – bony **2**: 23
– nasi **1**: 51; **2**: 5, 9, 11, 23, 41, 50–51, 54, 59–63, 336, 352, 359–360, 367, 369, 385–387
– – osseum **2**: 4, 23, 58–59, 61, 365
– orbitale **2**: 365, 383–384
– pellucidum **2**: 304–305, 307, 317, 323–324, 332, 337–340, 347–350, 354–355, 360
– penis **2**: 275
– scroti **2**: 232, 271, 274
– sinuum frontalium **2**: 59, 61
– – sphenoidalium **2**: 61, 293
Shaft of penis **2**: 274, 276
Sheaths, synovial, on ankle of foot **1**: 241
– – – fingers **1**: 130–131
– – – wrist **1**: 130–131
Shenton's line **1**: 177
Shin bone **1**: 182–183
Shoulder **1**: 144
– arteries **1**: 140, 142–143
– girdle **1**: 84, 87
– – muscle attachments **1**: 112
– joint **1**: 95–97
– muscles **1**: 106–111
– nerves **1**: 140, 142–143
– subcutaneous nerves **1**: 135–136
– – veins **1**: 135–137
Siebbein **2**: 22
Siebbeinhöhlen **2**: 57, 61–62, 369
Siebbeinzellen → Siebbeinhöhlen, Sinus ethmoidales
Silhouette des Herzens **2**: 122
– of heart **2**: 122
Sinus anales **2**: 179
– aortae **2**: 133
– caroticus **2**: 76, 78–79, 292
– cavernosus **2**: 62, 80, 293, 352–353, 359–360, 378, 380, 383
– coronarius **2**: 125, 135, 139, 141, 143
– durae matris **2**: 291
– epididymidis **2**: 270–271
– ethmoidales **2**: 5, 7, 9, 11, 22, 58–62, 305, 352–353, 360, 367, 395
– – anteriores **2**: 60, 62, 336, 358–359, 365, 369, 380, 385–386
– – medii **2**: 57, 60, 62, 359, 369, 385
– – posteriores **2**: 13, 57, 60, 62–63, 359, 369, 385, 387
– frontalis **2**: 5, 7, 12–13, 23, 52, 55–62, 88, 90, 291, 294, 305, 347, 357, 362, 367, 369–370, 380–382, 386
– intercavernosus **2**: 293–294
– lactiferi **1**: 70–71
– maxillaris **2**: 5, 7, 24, 41, 58–63, 73, 80, 336, 360, 362, 365, 367–370, 379, 383–384, 386–387
Sinus
– obliquus pericardii **2**: 123, 143
– occipitalis **2**: 291, 294
– paranasales **2**: 59–63
– petrosus inferior **2**: 291, 294
– – superior **2**: 80, 291
– prostaticus **2**: 235, 277
– rectus **2**: 52, 289, 291, 294, 321, 354–359
– renalis **2**: 208, 210, 212–213, 218, 223
– sagittalis inferior **2**: 291, 294, 300, 321, 326
– – superior **2**: 289–291, 294–295, 297, 300, 321, 336, 339, 341–343, 347, 349, 354–361, 408
– sigmoideus **2**: 49, 100, 291, 294, 352–353, 395, 400, 409
– sphenoidalis **2**: 7, 9, 11–13, 23, 27, 52, 55–57, 60–62, 88–90, 293–294, 302, 305, 352–353, 359–360, 369, 378, 385, 406–407, 409
– sphenoparietalis **2**: 291, 294
– tarsi **1**: 184, 188, 209–211, 215
– transversus **2**: 289, 291, 294, 359, 362, 409
– – pericardii **2**: 104, 123, 130, 143
– trunci pulmonalis **2**: 139
– tympani **2**: 396
– venosus sclerae **2**: 372–375
Sinus(es), cavernous **2**: 293, 359, 380
– ethmoidal **2**: 57, 61–62, 369
– frontal **2**: 62
– maxillary **2**: 62–63, 383
– paranasal **2**: 59–63, 336, 369, 386–387
– sphenoidal **2**: 23, 61, 293, 369
– venous, dural **2**: 291
Skalenuslücke **2**: 34–35, 37
Skeleton **1**: 2–3
– axiale **1**: 40–41
– foot **1**: 184–189, 209–211
– – muscle attachments **1**: 240
– hand **1**: 92–94
– – muscle attachments **1**: 128
– larynx **2**: 65–66
– nose **2**: 23, 54
Skelett **1**: 2–3
– Fuss **1**: 184–189, 209–211
– Hand **1**: 92–94
– – Muskelanheftungen **1**: 128
– Kehlkopf **2**: 65–66
– Nase **2**: 23, 54
Skin **1**: 11
Skull **2**: 2, 4–7, 12–15, 58
– arteries **2**: 76–77
– base **2**: 8–11, 16, 395, 400
– brachycephalic **2**: 2
– broad **2**: 2
– cap **2**: 2–3
– dolichocephalic **2**: 2
– long **2**: 2
– muscle attachments **2**: 32
– of newborn child **2**: 17
– shape differences **2**: 2
Socket of hip joint **1**: 193
Sole of foot, arteries **1**: 279–280
– – – muscles **1**: 237–238
– – – nerves **1**: 279–280
Space(s), endolymphatic, of labyrinth **2**: 405
– infracolic **2**: 170–171
– parietocolic **2**: 171
– perilymphatic, of labyrinth **2**: 405
– peripharyngeal, arteries **2**: 100
– – nerves **2**: 100
– – veins **2**: 100
– retrocardial **2**: 107, 127
– retroperitoneal, autonomic nervous system **2**: 248–249
– retrosternal **2**: 127

Space(s)
– subarachnoid **2**: 321
– suprasternal **2**: 33
Spalt, mesenteriokolischer **2**: 170–171
– parietokolischer **2**: 171
Spaltlinien, Haut **1**: 11
Spatium(-ia) anguli iridocornealis **2**: 375
– epidurale **1**: 39; **2**: 73, 232, 285–287
– episclerale **2**: 384
– extradurale → Spatium epidurale
– extraperitoneale **2**: 171
– intercostale(-ia) **1**: 41
– intervaginale subarachnoidale nervi optici **2**: 374, 378
– peridurale → Spatium epidurale
– perilymphaticum **2**: 405
– perinei profundum **1**: 17; **2**: 233, 237, 243, 247, 276
– – superficiale **2**: 250–251
– peripharyngeum, Arterien **2**: 100
– – Nerven **2**: 100
– – Venen **2**: 100
– profundum perinei → Spatium perinei profundum
– retroperitoneale **2**: 248–249
– retropubicum **2**: 232–233, 237, 246, 252, 260–261
– subarachnoideum **2**: 73, 75, 232, 285–287, 289, 294–295, 321, 326, 352–353, 405
– subdurale **2**: 285–286, 289, 295
– subperitoneale **2**: 250–251
– superficiale perinei → Spatium perinei superficiale
– suprasternale **2**: 33
– zonularia lentis oculi **2**: 374–375, 377
Speech area, motor **2**: 319
– – sensory **2**: 319
Speiche **1**: 88
– Muskelanheftungen **1**: 117
Speiseröhre **2**: 105–107
– Engstellen **2**: 106
Spiegelbild, Augenhintergrund **2**: 377
– Kehlkopf **2**: 69
– Trommelfell **2**: 392
Spina(-ae) helicis auriculae **2**: 390
– iliaca anterior inferior **1**: 172–174, 176, 191–192, 249
– – – superior **1**: 62–65, 172–176, 191–193, 216, 218–220, 226, 248–249, 252–253, 257; **2**: 228
– – posterior inferior **1**: 175–176
– – – superior **1**: 52, 55, 175–176, 191, 218, 222, 254–255, 257
– ischiadica **1**: 172, 174–176, 190–191, 220–221, 224–226; **2**: 244, 264, 281
– mentalis **2**: 25
– nasalis anterior corporis maxillae **2**: 4, 6, 12, 23–24, 56–57
– – ossis frontalis **2**: 21
– – posterior laminae horizontalis ossis palatini **2**: 8, 12, 23–24, 30, 57
– ossis sphenoidalis **2**: 19
– palatinae processus palatini maxillae **2**: 24
– scapulae **1**: 53, 84–85, 87, 95–96, 105, 108–109, 111, 142, 148; **2**: 92
Spinalnerven **1**: 22–23, 48; **2**: 286
Spinalnervenwurzeln **2**: 285
Spine, cervical **1**: 31, 36–37, 49–51; **2**: 288
– lumbar **1**: 38–39, 48; **2**: 288
– – muscle attachments **1**: 228
– sacral **2**: 288
– thoracic **1**: 42, 46–47; **2**: 288
Splen **1**: 15–16, 18–19, 21; **2**: 155, 171–172, 182, 186, 189, 191, 198–199, 204, 208, 220–222, 248
Splenium corporis callosi **2**: 307, 315, 317, 322, 324–325, 335, 342–343, 345–347, 349, 354–356, 361
Splenographie **2**: 197
Splenography **2**: 197
Spongiosa **1**: 7
Sprachzentrum, motorisches **2**: 319
– sensorisches **2**: 319
Sprungbein **1**: 187
Sprunggelenk, oberes **1**: 188, 206, 208–212, 214
– unteres **1**: 188, 206–215
Squama frontalis **1**: 87; **2**: 3–4, 6–7, 10, 12, 21, 23, 368
– occipitalis **1**: 51, 87; **2**: 3, 6–7, 10, 12, 17–18, 290, 294, 321, 352
Stage, proliferative **2**: 242
– secretory **2**: 242
Stammskelett **1**: 40–41
Stapes **2**: 391, 397–399, 405
Steigbügel **2**: 397
Steissbein **1**: 34–35, 38–39
Stellknorpel **2**: 64–65
Sternoklavikulargelenk **1**: 45
Sternokostalgelenk **1**: 45
Sternum **1**: 40–41, 45, 68, 76, 82, 87; **2**: 104, 107, 127, 150, 156–157, 172
Stiele, Grosshirn **2**: 340–341
Stillingsche Schere **2**: 334
Stirnbein **2**: 21
Stirnhöhlen **2**: 62
Stirnlappen, Grosshirn **2**: 312–313, 336, 360
Stirnnaht, persistierende **2**: 3
Stomach **2**: 173–176, 194
– atonic **2**: 174
– hypertonic **2**: 174
– hypotonic **2**: 174
– orthotonic **2**: 174
Stratum(-a) alba colliculi superioris **2**: 350
– circulare tunicae muscularis gastricae **2**: 173
– – – – oesophagi **2**: 49
– corneum epidermidis **1**: 11
– fibrosum vaginae tendinis **1**: 10
– germinativum epidermidis **1**: 11
– grisea colliculi superioris **2**: 350
– longitudinale tunicae muscularis gastricae **2**: 173
– – – – oesophagi **2**: 47, 49, 105–106
– subependymale **2**: 338
– synoviale capsulae articularis **1**: 10
– – – articulationis genus **1**: 198, 200
Stria(-ae) diagonalis **2**: 315, 317, 335
– longitudinalis lateralis **2**: 322, 324, 334–335, 338, 345
– – medialis **2**: 322, 324, 334–335, 338, 345
– mallearis **2**: 392
– medullaris(-es) thalami **2**: 300, 307, 324–326, 340
– – ventriculi quarti **2**: 310–311, 320
– olfactoria lateralis **2**: 306, 335
– – medialis **2**: 306, 335
– terminalis **2**: 300, 322, 325–326, 331, 340, 344, 346
– vascularis ductus cochlearis **2**: 404
Stroma iridis **2**: 375
– ovarii **2**: 241
Structures, investing, of spinal cord **2**: 286
Subarachnoidalraum **2**: 321
Subarachnoidalzisternen **2**: 321
Subiculum **2**: 326
– promontorii tympani **2**: 396
Subject, athletic **2**: 153
– leptosome **2**: 153
– pyknic **2**: 153
Subokzipitalpunktion **2**: 321
Substantia alba **2**: 289, 338
– – medullae spinalis **2**: 285
– compacta **1**: 7, 10
– grisea **2**: 289, 338
– – centralis mesencephali **2**: 289, 315, 328, 335, 340
– – medullae spinalis **2**: 285
– nigra **2**: 309, 315, 328, 330–331, 335, 340, 344, 350–351
– perforata anterior **2**: 306, 315, 324–325, 331, 335
– – interpeduncularis **2**: 306, 335
– – posterior → Substantia perforata interpeduncularis
– propria corneae **2**: 375
– – sclerae **2**: 375
– spongiosa **1**: 7, 10
– trabecularis → Substantia spongiosa
Sulcus(-i) anterolateralis medullae oblongatae **2**: 311, 340
– – – spinalis **1**: 23; **2**: 285, 288, 306
– arteriae meningeae mediae **2**: 19, 22, 400
– – occipitalis **2**: 20
– – subclaviae **1**: 44
– – temporalis mediae **2**: 20
– – vertebralis **1**: 30–31, 49, 59; **2**: 289
– arteriales **2**: 3, 12
– arteriosi → Sulci arteriales
– basilaris **2**: 309, 344
– calcanei **1**: 187
– calcarinus **2**: 304–305, 315, 317, 319, 322, 335, 346, 348, 355
– caroticus **2**: 10, 19
– carpi **1**: 94
– – distalis **1**: 105
– – medius **1**: 105
– – proximalis **1**: 105
– centralis cerebri **2**: 304–305, 314, 316–317, 319, 323, 348, 362
– – insulae **2**: 323
– cerebral, lateral **2**: 362
– cerebri **2**: 316–317
– cinguli **2**: 304–305, 317, 319, 337–338, 348
– circularis insulae **2**: 323, 337, 344–345, 350
– collateralis cerebri **2**: 315, 317, 326, 340
– coronarius cordis **2**: 124–125
– corporis callosi **2**: 304, 317, 319, 337
– costae **1**: 44
– deltoideopectoralis **1**: 63, 136
– frontalis inferior **2**: 314, 316
– – medius **2**: 316
– – superior **2**: 314, 316
– glutaealis **1**: 218
– glutaeus → Sulcus glutaealis
– hippocampalis → Sulcus hippocampi
– hippocampi **2**: 315, 317, 326, 334
– hypothalamicus **2**: 307, 324, 328
– infraorbitalis **2**: 4, 24, 58, 367–368
– intermedius posterior medullae spinalis **2**: 284, 288
– – – – oblongatae **2**: 310–311
– intertubercularis **1**: 86
– interventricularis anterior **2**: 124, 132, 134, 141, 145
– – posterior **2**: 125, 141, 145
– intraparietalis **2**: 314, 316
– lacrimalis maxillae **2**: 24
– lateralis cerebri **2**: 306, 315–317, 319, 338, 344, 348, 351, 355, 361–362
– limitans fossae rhomboideae **2**: 310–311
– lunatus cerebri **2**: 316
– malleolaris fibulae **1**: 182–183
– – tibiae **1**: 182, 212
– matricis unguis **1**: 11

Sulcus(-i)
– medianus fossae rhomboideae **2**: 284, 310–311, 320
– – linguae **2**: 39, 46
– – posterior medullae oblongatae → Sulcus medianus fossae rhomboideae
– – – – spinalis **2**: 284, 286, 288, 310–311
– mentolabialis **2**: 44
– musculi subclavii **1**: 84
– mylohyoideus **2**: 25, 27
– nasolabialis **2**: 44
– nervi petrosi majoris **2**: 10, 396, 400, 402
– – – minoris **2**: 396, 400
– – radialis **1**: 86
– – spinalis **1**: 30–31
– – ulnaris **1**: 86
– obturatorius **1**: 176
– occipitalis(-es) **2**: 316
– – transversus **2**: 314–315
– occipitotemporalis **2**: 315, 317
– olfactorius **2**: 317, 335
– orbitales **2**: 315, 335
– palatinus(-i) major maxillae **2**: 24
– – maxillae **2**: 24
– palatovaginalis **2**: 19
– palpebralis superior **2**: 364
– palpebronasalis **2**: 364
– paraolfactorius anterior **2**: 304, 317
– – posterior **2**: 317
– parietooccipitalis **2**: 304–305, 314–317
– popliteus **1**: 180–181
– postcentralis **2**: 314, 316
– posterolateralis medullae spinalis **1**: 23; **2**: 284, 286, 288
– – – oblongatae **2**: 310–311
– praecentralis **2**: 314, 316
– praechiasmaticus **2**: 19, 291
– promontorii tympani **2**: 396
– rectus **2**: 315
– rhinalis **2**: 304, 315, 317, 333–334
– sclerae **2**: 374
– sinus petrosi inferioris **2**: 10, 18, 20
– – – superioris **2**: 10, 20
– – sagittalis superioris **2**: 3, 10, 18, 21–22, 295
– – sigmoidei **2**: 10–11, 18, 20, 22, 400, 402, 406–407
– – transversi **2**: 10, 12, 18
– spiralis externus **2**: 404
– – internus **2**: 404
– subparietalis **2**: 304, 317
– tali **1**: 187
– temporalis inferior **2**: 316
– – superior **2**: 316
– tendinis musculi fibularis longi → Sulcus tendinis musculi peronaei longi
– – – flexoris digitorum longi **1**: 213
– – – – hallucis longi **1**: 187, 212–213
– – – peronaei longi **1**: 186–187
– terminalis cordis **2**: 125
– – linguae **2**: 46, 101
– tubae auditivae → Sulcus tubae auditoriae
– – auditoriae **2**: 8, 19
– tympanicus **2**: 396
– vaginarum tendinum radii **1**: 88
– venae cavae **2**: 181
– vomerovaginalis **2**: 19
Supercilium **2**: 364
Superior **1**: 5
Supination **1**: 5
Surface, internal, cranial base **2**: 293, 400
– relief, abdomen **1**: 62
– – arm **1**: 105
– – back **1**: 52
– – lower limb **1**: 216–217
– – thorax **1**: 62
Sustentaculum tali **1**: 184, 186–188, 210
Sutura coronalis **2**: 3–4, 6, 10, 12, 17
– frontalis **2**: 17
– – persistens **2**: 3
– frontoethmoidalis **2**: 10, 12
– frontolacrimalis **2**: 4
– frontomaxillaris **2**: 4, 365, 367
– frontonasalis **2**: 4
– frontozygomatica **2**: 4, 6, 368
– infraorbitalis **2**: 367
– intermaxillaris **2**: 4
– internasalis **2**: 4
– lacrimomaxillaris **2**: 6, 368
– lambdoidea **2**: 3, 5–8, 10, 12
– mendosa **2**: 17
– nasomaxillaris **2**: 4, 6
– occipitomastoidea **2**: 6, 8, 10, 12
– palatina mediana **2**: 8, 24
– – transversa **2**: 8, 24
– parietomastoidea **2**: 6, 10
– sagittalis **2**: 3, 5, 17, 295
– sphenoethmoidalis **2**: 10
– sphenofrontalis **2**: 4, 6, 10, 12, 368
– sphenoparietalis **2**: 4, 6, 10, 12
– sphenosquamosa **2**: 6, 8, 10, 12
– sphenozygomatica **2**: 4, 6, 368
– squamosa **2**: 6, 12
– temporozygomatica **2**: 6, 8
– zygomaticofrontalis → Sutura frontozygomatica
– zygomaticomaxillaris **2**: 4, 6, 8, 367–368
– zygomaticosphenoidalis → Sutura sphenozygomatica
Suture, frontal, persisting **2**: 3
Symmetrieebene des Körpers **1**: 5
Symphysis manubriosternalis **1**: 40–41, 45
– pubica **1**: 17, 170–175, 177–178, 190, 221; **2**: 232–233, 237, 243, 245, 247, 252, 256–261, 264–265, 268
– xiphosternalis **1**: 45
Synchondrosis **1**: 9
– manubriosternalis → Symphysis manubriosternalis
– sphenooccipitalis **2**: 10, 17
– sternocostalis costae primae **1**: 45
– xiphosternalis → Symphysis xiphosternalis
Syndesmosis **1**: 9
– tibiofibularis **1**: 205–206, 208–209, 212, 214, 275
Synostosis **1**: 9
System, alimentary **2**: 52–53
– circulatory **1**: 12–14
– conducting **2**: 132–134
– digestive **1**: 16; **2**: 52–53
– extrapyramidal **2**: 288
– extrapyramidales **2**: 288
– limbic **2**: 319, 325
– limbisches **2**: 319, 325
– lymphatic **1**: 15
– lymphatisches **1**: 15
– nervous, autonomic **1**: 24–25
– – – neck **2**: 95–96
– – – retroperitoneal space **2**: 248–249
– – – thorax **2**: 95–96, 166–167
– – central **1**: 22
– – – division **2**: 303
– – parasympathetic **1**: 24; **2**: 231, 248–249
– – peripheral **1**: 22
– – sympathetic **1**: 24; **2**: 231, 248–249
– – – neck **2**: 95–96, 166
– – – thorax **2**: 95–96, 166–167
– – – upper abdomen **2**: 167
– respiratory **1**: 16; **2**: 52–53
– urogenital **1**: 17
– ventricular, brain **2**: 320
Systema alimentarium **1**: 16
– cardiovasculare **1**: 12–14
– conducens cordis **2**: 140
– digestorium → Systema alimentarium
– lymphaticum **1**: 15
– lymphoideum → Systema lymphaticum
– nervosum **1**: 22
– – autonomicum **1**: 24–25
– – centrale **1**: 22; **2**: 303
– – periphericum **1**: 22
– respiratorium **1**: 16
– skeletale **1**: 2–3
– urogenitale **1**: 17
Systole **2**: 138
Szintigramm, Knochen **1**: 3
– Schilddrüse **2**: 71

T

Taenia choroidea **2**: 322, 326
– – ventriculi quarti **2**: 308–311
– fimbriae hippocampi **2**: 326
– fornicis **2**: 300, 322, 324, 326
– habenulae **2**: 326
– libera coli **2**: 177, 179, 192–193, 202, 204
– mesocolica coli **2**: 177
– omentalis coli **2**: 177
– thalami **2**: 310–311, 324, 326
Talus **1**: 184–185, 187–188, 207–210, 214, 277
Tapetum **2**: 333–334
Tarsalia → Ossa tarsi
Tarsalkanal **1**: 209
Tarsus inferior **2**: 365, 383–384
– superior **2**: 365–366, 383–384
Tectum mesencephali → Tectum mesencephalicum
– mesencephalicum **2**: 289, 328, 334, 350–351, 357–358, 361
Teeth, eruption **2**: 39, 42
– innervation **2**: 86
– permanent **2**: 40–41
– – precursors **2**: 43
– – surfaces **2**: 40
Tegmen tympani **2**: 391, 396
Tegmentum mesencephali → Tegmentum mesencephalicum
– mesencephalicum **2**: 304, 310, 324, 328, 335, 350–353, 358–359
Teil, knorpeliger, des äusseren Gehörgangs **2**: 392
– nichtknorpeliger, des äusseren Gehörgangs **2**: 392
Tela adiposa subepicardialis **2**: 133–134
– choroidea ventriculi quarti **2**: 304, 307, 309, 324
– – – tertii **2**: 299–300, 307, 326, 344, 346, 348
– subcutanea **1**: 11, 63, 65, 67
– subserosa peritonei parietalis **2**: 171, 230, 232
– – – visceralis **2**: 171
Telencephalon **2**: 303
Tendo **1**: 10
– calcaneus **1**: 206–207, 210–213, 215, 217, 230, 232–234, 241, 267–268, 275–277, 279–280
– conjunctivus → Falx inguinalis
– infundibuli cordis **2**: 135
– musculi **1**: 10
– – bicipitis brachii **1**: 140, 152–153
Tendon **1**: 10
Tentorium cerebelli **2**: 289, 291–294, 343, 353, 356–359, 361–362
Testis **1**: 15, 17, 67, 79; **2**: 170, 232, 258, 261, 270–273, 278
– coverings **2**: 270
– prenatal descent **2**: 272

Thalamus **2**: 300, 304–305, 307, 317, 324, 326–330, 332, 340–341, 344, 346–349, 354–356, 360–361, 408
Thalamuskerne **2**: 327
– Projektionsfelder **2**: 327
Thenar **1**: 105, 114–115, 120, 160
Thigh **1**: 258–265; **2**: 228
– adductor muscles **1**: 220–221
– anterior muscles **1**: 219
– arteries **1**: 251–253, 255–256; **2**: 228
– bone **1**: 173, 177, 179–181
– – muscle attachments **1**: 229
– dermatomes **1**: 134, 244
– fascia **1**: 218
– hamstrings **1**: 221–224
– lymph nodes **1**: 248
– muscles **1**: 219–222, 224, 226
– nerves **1**: 252, 255–256; **2**: 228
– subcutaneous nerves **1**: 248, 254
– – veins **1**: 248, 254
– veins **1**: 252, 255–256; **2**: 228
Thorax **1**: 2, 40–43, 75–76, 82–83; **2**: 93, 96, 118, 126–131, 149–151, 153–162, 164–167, 172
– Arterien **1**: 76; **2**: 93, 97, 118
– arteries **1**: 76; **2**: 93, 97, 118
– autonomic nervous system **2**: 95–96, 166–167
– Dermatome **1**: 134, 244
– dermatomes **1**: 134, 244
– genuine muscles **1**: 68
– lymph nodes **2**: 163
– – vessels **2**: 163
– Lymphgefässe **2**: 163
– Lymphknoten **2**: 163
– muscle attachments **1**: 68
– nerves **2**: 93, 97
– phase of expiration **1**: 40–41
– – – inspiration **1**: 40–41
– segmental innervation **1**: 134, 244
– segmentale Innervation **1**: 134, 244
– surface relief **1**: 62
– sympathetic nervous system **2**: 95–96, 166–167
– veins **1**: 76; **2**: 93, 97
Thymus **1**: 15; **2**: 104, 151
– infantile **2**: 104
– kindlicher **2**: 104
Thymusfettkörper **2**: 104
Tibia **1**: 10, 170–171, 182–183, 188, 195–203, 205, 208–212, 214, 220–221, 226, 230–232, 234, 241, 247, 269–270, 272–277
– muscle attachments **1**: 235
Tissue, connective, of female lesser pelvis **2**: 246–247
– – – male lesser pelvis **2**: 246–247
Tongue **2**: 38, 46–47
– arteries **2**: 97
– dorsum **2**: 46
– muscles **2**: 38, 46–47
– nerves **2**: 97
– veins **2**: 97
Tonsil(s), lingual **2**: 45
– palatine **2**: 45
– pharyngeal **2**: 45
Tonsilla adenoidea → Tonsilla pharyngealis
– cerebelli **2**: 294, 307–309
– lingualis **2**: 45–46, 50, 101
– palatina **2**: 38–39, 45–46, 50–52, 101
– pharyngea → Tonsilla pharyngealis
– pharyngealis **2**: 13, 45, 50–51, 57, 63
Torso, ventral muscles **1**: 62–65, 68
Torus levatorius **2**: 57
– tubarius **2**: 45, 50, 52, 57
Trabecula(-ae) carneae cordis **2**: 132–134, 140
Trachea **1**: 14–16, 18, 43; **2**: 36, 48–50, 52–53, 65, 68, 71, 75, 100, 104–110, 113–115, 119–121, 126–127, 130–131, 150–151, 155–156, 161, 165, 172
– lymph nodes **2**: 110
– – vessels **2**: 110
Tractus corticobulbaris → Tractus corticonuclearis
– corticonuclearis **2**: 332
– corticospinalis **2**: 332
– – anterior **2**: 288
– – lateralis **2**: 288
– frontopontinus **2**: 332
– iliotibialis **1**: 65, 195, 218–219, 222–224, 226, 230–232, 255–256, 259–260, 271; **2**: 228
– intermedius aponeurosis dorsalis digiti manus **1**: 125
– lateralis aponeurosis dorsalis digiti manus **1**: 125
– olfactorius **1**: 22; **2**: 298, 306, 315, 324–325, 333, 335, 337, 387
– opticus **1**: 22; **2**: 293, 315, 333–335, 338–339, 344, 351
– pyramidalis → Tractus corticospinalis
– rubrospinalis **2**: 288
– spinocerebellaris anterior **2**: 288
– – posterior **2**: 288
– spinoreticularis **2**: 288
– spinotectalis **2**: 288
– spinothalamicus anterior **2**: 288
– – lateralis **2**: 288
– spiralis foraminosus **2**: 401–403
– temporopontinus **2**: 332
Tragus **2**: 390
Trajectories **1**: 7
– stress, compressive **1**: 7
– – tensile **1**: 7
Trajektorien **1**: 7
Tränenapparat **2**: 365
Tränendrüse, Innervation **2**: 88
Tränennasengang **2**: 365
Tränenröhrchen **2**: 365
Transversalebenen **1**: 5
Tree, bronchial **2**: 114–115, 119–121
– – subdivision **2**: 114
Triad, portal **2**: 183, 185
Triangle, femoral **1**: 249
– pleura-free, lower **2**: 150
– – upper **2**: 150
Trias hepatica **2**: 183, 185
Trigeminusganglion **2**: 293
Trigone of urinary bladder **2**: 235
Trigonum cervicale anterius → Regio cervicalis anterior
– – posterius → Regio cervicalis lateralis
– clavipectorale **1**: 63
– collaterale **2**: 322–323, 354, 356, 362
– colli laterale **1**: 62
– femorale **1**: 4, 249
– fibrosum dextrum cordis **2**: 135, 141
– – sinistrum cordis **2**: 135, 141
– habenulae **2**: 310–311, 348
– habenulare → Trigonum habenulae
– lemnisci lateralis **2**: 311
– lumbale **1**: 53–55, 57, 226
– lumbare → Trigonum lumbale
– lumbocostale **1**: 69
– mentale **2**: 4
– nervi hypoglossi **2**: 310–311
– – vagi **2**: 310–311
– olfactorium **2**: 306, 315, 325, 333, 338
– omoclaviculare **1**: 63; **2**: 36
– omotrapezoideum **2**: 36
– retromolare mandibulae **2**: 25
Trigonum
– sternocostale **1**: 68–69
– vesicae urinariae **2**: 235, 237
Trochanter major **1**: 173, 177, 179–180, 192, 194, 216–217, 220–221, 223–225, 254–255, 257–258; **2**: 209, 253–254, 257, 259
– minor **1**: 173, 177, 179–180, 192–193, 220–221, 224–225, 258; **2**: 209, 256–257
– tertius **1**: 180
Trochlea **2**: 370–371, 381–383
– fibularis → Trochlea peronaealis
– humeri **1**: 86, 90–91, 98, 100–101, 149
– peronaealis **1**: 187–188
– phalangis manus/pedis → Caput phalangis manus/pedis
– tali **1**: 184, 187–188, 210–211
Trommelfell **2**: 392
– Quadrantengliederung **2**: 392
– Spiegelbild **2**: 392
Truncus(-i) anterior arteriae iliacae internae **1**: 251
– brachiocephalicus **2**: 78–79, 96–97, 100, 104–107, 118, 122–125, 130, 139, 142–143, 149, 151, 156, 161–163, 165, 167
– bronchomediastinalis **1**: 15
– coeliacus **1**: 12, 14; **2**: 105, 162, 196–200, 203, 211, 218–219
– corporis callosi **2**: 300, 307, 317, 325, 337–341, 344–345, 360
– costocervicalis **1**: 139; **2**: 79, 118
– embryonis **2**: 242
– encephali **2**: 289, 303, 306–307, 310–311
– fasciculi atrioventricularis **2**: 140
– inferior plexus brachialis **1**: 134; **2**: 94–95, 166
– intestinalis **1**: 15; **2**: 163
– jugularis **1**: 15; **2**: 81, 149, 163
– lumbalis(-es) **1**: 15; **2**: 163
– lumbaris(-es) → Truncus(-i) lumbalis(-es)
– lumbosacralis **2**: 226, 231, 234, 248
– lymphaticus jugularis **1**: 72, 76
– – subclavius **1**: 76
– medius plexus brachialis **1**: 134; **2**: 94–95, 166
– nervi accessorii **2**: 100
– – cervicalis **2**: 286
– – spinalis **1**: 23; **2**: 95, 166, 285–286
– plexus brachialis **1**: 134
– posterior arteriae iliacae internae **1**: 251
– pulmonalis **1**: 12–13, 43; **2**: 107, 115, 119, 122, 124, 126–130, 132–135, 139, 142, 144, 149, 154, 157–158, 164
– subclavius **1**: 15; **2**: 149, 163
– superior plexus brachialis **1**: 134; **2**: 94–95, 166
– sympatheticus **1**: 23; **2**: 33, 95–96, 100, 166–167, 231, 248–249
– thyrocervicalis **1**: 139; **2**: 78–79, 93, 100, 118
– tibiofibularis **1**: 269
– vagalis anterior **1**: 25; **2**: 167, 248–249
– – posterior **1**: 25; **2**: 167, 248–249
Tuba auditiva → Tuba auditoria
– auditoria **2**: 51, 393–395, 398, 405–406
– uterina **1**: 17; **2**: 163, 172, 233, 238–241, 245, 250–252
– – Arterien **2**: 241
Tube(s), auditory **2**: 393–395, 406
– uterine **2**: 239–241
– – arteries **2**: 241
Tuber calcanei **1**: 184–188, 212–213, 217, 232–234, 237–238
– cinereum **2**: 301, 306–307, 315, 330–331, 334–335
– frontale **2**: 17, 21
– ischiadicum **1**: 175–176, 190–191, 193, 221, 223–226, 249, 256, 258; **2**: 230, 251, 253, 255–259, 265–267, 278–279, 281
– maxillare **2**: 24, 368

Tuber
– omentale hepatis **2**: 181
– – pancreatis **2**: 188
– parietale **2**: 17, 22
– vermis **2**: 307
Tuberculum adductorium femoris **1**: 179–180
– anomale dentis (Carabelli) **2**: 39
– anterius atlantis **1**: 30, 50; **2**: 37
– – vertebrae cervicalis **1**: 30–31; **2**: 33
– articulare ossis temporalis **2**: 20, 26
– auriculare **2**: 390
– buccale anterius dentis **2**: 39
– – posterius dentis **2**: 39
– calcanei **1**: 186–187
– caroticum **1**: 31; **2**: 37
– conoideum **1**: 84
– corniculatum **2**: 50–52, 68, 101
– costae **1**: 41, 44; **2**: 75
– cuneatum **2**: 310–311
– cuneiforme **2**: 50–52, 68, 101
– dorsale radii **1**: 88, 92, 102, 116, 125, 130
– epiglotticum **2**: 68
– gracile **2**: 310–311
– iliacum **1**: 256
– infraglenoidale **1**: 84–85, 95–96
– intercondylare laterale tibiae **1**: 182, 197, 204–205
– – mediale tibiae **1**: 182, 197, 204–205
– jugulare **2**: 18
– labii superioris oris **2**: 44
– laterale tali **1**: 187
– majus humeri **1**: 86–87, 95, 97, 109, 111
– mediale tali **1**: 187–188
– mentale **2**: 25
– minus humeri **1**: 86–87, 97
– musculi scaleni anterioris **1**: 44
– ossis metacarpalis V **1**: 94
– – scaphoidei **1**: 94, 102, 123, 130
– – trapezii **1**: 94, 102–103
– palatinale anterius dentis **2**: 39
– – posterius dentis **2**: 39
– pharyngeum **2**: 18, 47, 395
– posterius atlantis **1**: 30–31, 49–50, 55, 59–60; **2**: 302
– – vertebrae cervicalis **1**: 30–31; **2**: 33
– pubicum **1**: 66, 172, 174, 176, 219; **2**: 228, 245
– sellae **2**: 10
– supraglenoidale **1**: 85
– supratragicum **2**: 390
– thyroideum inferius **2**: 64
– – superius **2**: 64
– trigeminale **2**: 310–311
Tuberositas deltoidea **1**: 86
– glutaea → Tuberositas glutaealis
– glutaealis **1**: 180
– iliaca **1**: 174, 176
– ligamenti coracoclavicularis **1**: 84
– masseterica **2**: 25, 31
– musculi serrati anterioris **1**: 44
– ossis cuboidei **1**: 186
– – metatarsalis I **1**: 186, 188
– – – V **1**: 186, 188, 232
– – navicularis **1**: 186, 188, 279
– – sacri → Tuberositas sacralis
– phalangis distalis manus **1**: 94
– – – pedis **1**: 186
– radii **1**: 88, 90–91, 100
– sacralis **1**: 34–35
– tibiae **1**: 10, 181–183, 195–197, 205, 216, 218–219, 230–232, 252, 270–271
– ulnae **1**: 89, 99
Tunica albuginea corporis spongiosi **2**: 275
– – corporum cavernosorum **2**: 275, 278
– – testis **2**: 270–271
Tunica
– conjunctiva bulbaris **2**: 364–365, 370–371, 374–375
– – palpebralis **2**: 364–365
– dartos **1**: 67; **2**: 271, 274
– fibrosa bulbi **2**: 372
– mucosa canalis analis **2**: 265–267
– – cavitatis oris **2**: 45
– – linguae **2**: 46
– – nasi **2**: 90
– – palati **2**: 39, 45, 58
– – pharyngis **2**: 50
– – tracheae **2**: 68
– – uteri **2**: 233, 242
– – vaginae **2**: 233
– – vesicae urinariae **2**: 235, 277
– muscularis gastrica **2**: 173
– – oesophagi **2**: 47, 106
– – pharyngis **2**: 48–49
– – tubae uterinae **2**: 241
– – uteri **2**: 233, 242–243
– – vaginae **2**: 233
– – vesicae urinariae **2**: 235
– serosa jejuni **2**: 176
– – omenti majoris **2**: 173
– – – minoris **2**: 173
– – peritonei parietalis **2**: 171
– – – visceralis **2**: 171
– – uteri **2**: 233, 243
– – vesicae urinariae **2**: 235, 245
– vaginalis testis **1**: 67; **2**: 170, 270–274
Tunnel, carpal **1**: 103, 122–123

U

Ulna **1**: 82–83, 89–94, 98–100, 102–105, 116, 118–119, 122, 125, 129–130, 139, 149, 154–158
– muscle attachments **1**: 117
– Muskelanheftungen **1**: 117
Umbilicus **1**: 12–13, 63–65, 75; **2**: 172, 196, 202, 209, 243
Umbo membranae tympanicae **2**: 392, 398
Uncus corporis vertebrae cervicalis **1**: 30–31
– gyri hippocampi **2**: 315, 324–325, 333–334
Unguis **1**: 11, 167
Unterarm **1**: 156–159
– Hautnerven **1**: 150
– Muskelanheftungen **1**: 117
– subkutane Nerven **1**: 150–151
– – Venen **1**: 150–151
Unterarmarterien **1**: 152–155
Unterarmmuskeln **1**: 114–116, 118–119
Unterarmnerven **1**: 152–154
Unterbauch **2**: 192–193, 225
Unterbaucharterien **2**: 202
Unterhorn, Seitenventrikel **2**: 326, 362
Unterkieferspeicheldrüse **2**: 44–45
Unterschenkel **1**: 272–277
– Hautnerven **1**: 266, 270
– Muskelanheftungen **1**: 235
Unterschenkelarterien **1**: 267–268, 271
Unterschenkelfaszie **1**: 230
Unterschenkelgelenke **1**: 205
Unterschenkelknochen **1**: 182–183
Unterschenkelmuskeln **1**: 231–234
Unterschenkelnerven **1**: 267–268, 271
– subkutane **1**: 266, 270
Unterschenkelvenen **1**: 267–268
– subkutane **1**: 266, 270
Unterschläfengrube **2**: 88–89
Unterzungenregion **2**: 47
Unterzungenspeicheldrüse **2**: 45
Urachus **1**: 13; **2**: 246
Ureter **1**: 14, 17–20, 77; **2**: 162, 199, 208–214, 216–218, 223, 230, 234–236, 238–239, 241, 244–246, 248, 251, 261, 276
– intramural part **2**: 235
Ureterostien **2**: 235
Urethra **1**: 20–21
– feminina **1**: 17; **2**: 233, 243, 250, 256–257, 260
– masculina **1**: 17; **2**: 170, 232, 235–237, 250, 258–259, 266–267, 275, 277–278
Urogenitalsystem **1**: 17
Uterus **1**: 17; **2**: 163, 172, 233, 238–243, 246–247, 250–252, 256–257, 260
– Arterien **2**: 241
– arteries **2**: 241
– pregnant **2**: 243
– schwangerer **2**: 243
Utriculus prostaticus **2**: 235, 237
– vestibularis **2**: 405
Uvula palatina **2**: 13, 38–39, 45, 50–52
– vermis cerebelli **2**: 307–309
– vesicae urinariae **2**: 235

V

Vagina(-ae) **1**: 17; **2**: 233, 239, 243, 245, 251, 256–257, 260, 268–269
– bulbi oculi **2**: 384
– carotica **2**: 33
– communis tendinum musculorum fibularium → Vagina tendinum musculorum peronaeorum communis
– – – – flexorum manus **1**: 123, 130
– externa nervi optici **2**: 371–372, 374, 378, 384
– fibrosa(-ae) tendinis hallucis **1**: 238
– – – indicis **1**: 121
– – tendinum digitorum manus **1**: 102, 121–122, 124, 130, 162
– – – – pedis **1**: 237
– interna nervi optici **2**: 372, 374, 378, 384
– musculi recti abdominis **1**: 63–65, 68, 75, 77
– processus styloidei **2**: 20
– synovialis(-es) tendinis(-inum) **1**: 10
– – – digitorum manus **1**: 121–122, 130–131
– – – – pedis **1**: 238
– – – musculorum cruris **1**: 241
– tendinis(-inum) **1**: 10
– – intertubercularis **1**: 95–96, 106
– – manus **1**: 130–131
– – musculi(-orum) abductoris longi et extensoris pollicis brevis **1**: 130
– – – extensoris(-orum) carpi radialium **1**: 130
– – – – – ulnaris **1**: 130
– – – – digiti minimi **1**: 130
– – – – digitorum et extensoris indicis **1**: 130
– – – – digitorum longi **1**: 241
– – – – hallucis longi **1**: 241
– – – – pollicis longi **1**: 130
– – – fibularis longi plantaris → Vagina tendinis musculi peronaei longi plantaris
– – – fibularium communis → Vagina tendinum musculorum peronaeorum communis
– – – flexoris carpi radialis **1**: 130
– – – – digitorum longi **1**: 214–215, 241
– – – – hallucis longi **1**: 214–215, 241
– – – – pollicis longi **1**: 121, 130–131
– – – obliqui superioris bulbi **2**: 370–371
– – – peronaei longi plantaris **1**: 238
– – – peronaeorum communis **1**: 214–215, 241
– – – tibialis anterioris **1**: 241
– – – – posterioris **1**: 214–215, 241
Vallecula cerebelli **2**: 308
– epiglottica **2**: 46, 53, 68
Vallum unguis **1**: 11

Valva aortae **2**: 104, 128–130, 133–138, 141
– atrioventricularis dextra **2**: 104, 132, 134–135, 140, 159–160
– – sinistra **2**: 104, 133–138, 140
– ilealis → Valva ileocaecalis
– ileocaecalis **2**: 177, 194–195, 205
– mitralis → Valva atrioventricularis sinistra
– tricuspidalis → Valva atrioventricularis dextra
– trunci pulmonalis **2**: 132, 134–135, 137, 141
Valve, aortic **2**: 133
– ileocecal **2**: 177, 195
Valvula(-ae) anales **2**: 179
– foraminis ovalis **2**: 133
– semilunaris(-es) anterior valvae trunci pulmonalis **2**: 132, 135
– – dextra valvae aortae **2**: 133–135, 137
– – – – trunci pulmonalis **2**: 132, 135
– – posterior valvae aortae **2**: 133–135, 137, 141
– – sinistra valvae aortae **2**: 133–135, 137
– – – – trunci pulmonalis **2**: 132, 135, 141
– – valvae aortae **2**: 129–130, 138
– – – trunci pulmonalis **2**: 134, 137
– sinus coronarii **2**: 132, 140
– venae cavae inferioris **2**: 132, 140
– venosa(-ae) **1**: 73, 137, 246–247
Vas (Vasa) hepatica **2**: 183
– lymphatica mammae **1**: 72–74
– – membri inferioris **1**: 245
– – peribronchialia **2**: 113
– – subpleuralia **2**: 113
– – thoracis **1**: 76
– spirale **2**: 404
Vault, cerebral **2**: 295
Vein(s), anterior abdominal wall **1**: 77
– axilla **2**: 93, 149
– brain **2**: 300
– coronary, heart **2**: 142–143
– deep, lower limb **1**: 246
– dorsal body wall **2**: 162
– – deep, penis **2**: 276
– – superficial, penis **2**: 276
– dorsum of foot **1**: 278
– eyeball **2**: 372–373
– facial **2**: 80
– – deep **2**: 80
– – superficial **2**: 80
– female lesser pelvis **2**: 244
– gluteal region **1**: 255–256
– heart **2**: 142–143
– knee **1**: 252
– leg **1**: 267–268
– lesser pelvis **2**: 234, 244
– liver **2**: 185
– lower limb **1**: 246–247
– male lesser pelvis **2**: 234
– mediastinum **2**: 149, 164
– meningeal **2**: 290
– mesenterial, superior **2**: 200–201
– neck **2**: 93, 97–99, 149, 164
– orbital **2**: 383
– perforating, Boyd's **1**: 246
– – Cockett's **1**: 246
– – Dodd's **1**: 246
– – lower limb **1**: 246
– peripharyngeal space **2**: 100
– popliteal fossa **1**: 255–256, 267–268
– portal **2**: 183
– posterior abdominal wall **2**: 228
– retroperitoneal organs of upper abdomen **2**: 203
– spinal cord **2**: 286–287
– subcutaneous, back of hand **1**: 164
– – dorsum of foot **1**: 278
– – elbow region **1**: 151
Vein(s), subcutaneous
– – foot **1**: 270
– – forearm **1**: 150–151
– – gluteal region **1**: 254
– – leg **1**: 266, 270
– – lower limb **1**: 246
– – neck **2**: 91–92
– – popliteal fossa **1**: 254, 266
– – shoulder **1**: 135–137
– – thigh **1**: 248, 254
– – upper arm **1**: 135–137
– superficial, brain **2**: 297
– – face **2**: 85
– – head **2**: 85
– thigh **1**: 252, 255–256; **2**: 228
– thorax **1**: 76; **2**: 93, 97
– tongue **2**: 97
– ventral body wall **1**: 75
– vertebral column **2**: 286–287
Velum medullare inferius **2**: 307, 309
– – superius **2**: 304, 307–311
– palatinum → Palatum molle
Vena(-ae) abdominis **2**: 203
– adrenalis → Vena suprarenalis
– alveolaris inferior **2**: 80
– anastomotica superior **2**: 296–297
– angularis **2**: 80, 383
– anterior(-iores) cerebri **2**: 300
– – septi pellucidi **2**: 300, 322, 337
– appendicularis **2**: 202
– auricularis posterior **2**: 80, 85
– axillaris **1**: 73, 137, 148; **2**: 149, 163
– azygos **1**: 14; **2**: 97, 111, 122, 130, 132, 155–159, 161–165, 167, 222
– basilica **1**: 105, 136–137, 140, 146–147, 150–151, 164
– basivertebrales **2**: 287
– brachialis(-es) **1**: 137, 140, 145–147, 156
– brachiocephalica **1**: 12, 14, 76, 137; **2**: 96–97, 104, 107, 122, 128–130, 149–151, 156, 162–164
– buccalis **2**: 80
– bulbi oculi **2**: 372–373
– capitis **2**: 80
– cardiaca magna → Vena coronaria sinistra
– – parva → Vena coronaria dextra
– cava inferior **1**: 12–15; **2**: 123, 125–127, 131, 139–141, 143, 154–155, 159–160, 162, 164, 167, 171, 180–183, 185–186, 196, 198–199, 203, 206–208, 211, 213–214, 218, 220, 222–224, 228, 248
– – superior **1**: 12–15; **2**: 97, 107, 111, 122–125, 129–130, 132–133, 139–143, 157, 161–165, 167, 172
– centralis(-es) glandulae suprarenalis **2**: 216
– – retinae **2**: 372–374, 378, 384
– cephalica **1**: 72–73, 75, 105, 135–137, 140, 145–147, 150–151, 156–157, 164; **2**: 98, 149
– – accessoria **1**: 150
– cerebri **2**: 300
– cervicalis superficialis **2**: 98
– choroidea(-ae) superior **2**: 300
– – oculi → Venae vorticosae
– ciliaris anterior **2**: 373
– circumflexa femoris lateralis **1**: 247
– – iliaca profunda **2**: 244
– – – superficialis **1**: 75, 248
– – scapulae **1**: 135
– colica dextra **2**: 201–202
– – media **2**: 200, 202
– – sinistra **2**: 202
– comitans nervi hypoglossi **2**: 99
– conjunctivalis anterior **2**: 373
– cordis magna → Vena coronaria sinistra
– – parva → Vena coronaria dextra
Vena(-ae)
– coronaria(-ae) **2**: 142–143
– – dextra **2**: 129, 135, 140, 143
– – sinistra **2**: 107, 129–130, 133, 135, 142–143
– cystica **2**: 196, 200
– digitales dorsales pedis **1**: 278
– diploicae **2**: 295
– directae laterales **2**: 300
– dorsalis(-es) profunda penis **2**: 267, 275–277
– – superficiales penis **2**: 275–277
– emissaria occipitalis **2**: 49
– – parietalis **2**: 295
– epigastrica inferior **1**: 67, 75, 77, 252; **2**: 196, 200, 244, 246, 272–273
– – superficialis **1**: 75, 245, 248, 252; **2**: 276
– – superior **1**: 75
– episcleralis **2**: 373
– ethmoidalis anterior **2**: 381
– – posterior **2**: 381
– facialis **2**: 80–81, 85, 92, 98–99, 383
– femoralis **1**: 14, 64–65, 75, 218–219, 246–249, 259–260, 262–263; **2**: 163, 228, 256–259
– fibularis **1**: 247, 273–274
– frontales **2**: 297
– gastrica dextra **2**: 188, 196, 203
– – sinistra **2**: 196, 200
– gastroepiploica dextra/sinistra → Vena gastroomentalis dextra/sinistra
– gastroomentalis dextra **2**: 196, 200
– – sinistra **2**: 196
– glutaea inferior/superior → Vena glutaealis inferior/superior
– glutaealis inferior **1**: 257
– – superior **1**: 257
– hemiazygos **1**: 14; **2**: 162–163, 167
– – accessoria **1**: 14; **2**: 163
– hepaticae **1**: 12–14; **2**: 131, 160, 162, 167, 180, 182, 185, 199, 206–207, 218, 221–222
– – dextrae **2**: 185
– – intermediae **2**: 185
– – sinistrae **2**: 185
– ileales **2**: 188, 200–201, 204
– ileocolica **2**: 201–202
– iliaca communis **1**: 12, 14, 194; **2**: 162–163, 218, 225, 228, 234, 243–245, 252, 254, 260–261
– – externa **1**: 14, 75, 77, 247, 252; **2**: 163, 220–221, 228, 230, 233–234, 238, 244–245, 250, 252, 256, 258–259
– – interna **1**: 14; **2**: 163, 220–221, 228, 234, 244–245, 253, 255
– inferiores cerebelli **2**: 295
– – cerebri **2**: 291, 295, 300
– infraorbitalis **2**: 80
– intercapitales manus/pedis → Venae intercapitulares manus/pedis
– intercapitulares manus **1**: 164
– – pedis **1**: 278
– intercostales posteriores **1**: 14, 76; **2**: 162–163, 167, 285–286
– internae cerebri **2**: 299–300, 326, 344, 355, 361
– interossea anterior **1**: 157
– interventricularis anterior **2**: 132, 142
– – posterior **2**: 135, 143
– intervertebralis **2**: 287
– iridis **2**: 373
– jejunales **2**: 188, 200–201, 204
– jugularis anterior **2**: 33, 74–75, 91, 97–98, 122, 149
– – externa **2**: 33, 80, 85, 91–93, 97–99, 122, 149, 162, 164
– – interna **1**: 12, 14; **2**: 33, 36, 68, 71, 73–75, 80–81, 92–93, 96–100, 122, 149, 162–164, 383, 395

Vena(-ae)
– labialis inferior **2**: 80, 383
– – superior **2**: 80, 383
– lienalis → Vena splenica
– lumbalis(-es) **1**: 14; **2**: 162
– – ascendens **2**: 287
– magna cerebri **2**: 291, 299–300, 307, 342–343, 346
– marginalis dextra cordis **2**: 135, 142
– – lateralis pedis **1**: 278
– – sinistra cordis **2**: 143
– maxillaris **2**: 80, 99, 383
– media profunda cerebri **2**: 300
– – superficialis cerebri **2**: 291, 295–297, 300
– medialis atrii ventriculi lateralis **2**: 300
– mediana antebrachii **1**: 105, 136, 151
– – cubiti **1**: 105, 136–137, 151
– – – profunda **1**: 136, 140, 151
– mediastinalis(-es) **2**: 122
– membri inferioris **1**: 246–247
– meningeae mediae **2**: 80, 290
– mentalis **2**: 80
– mesenterica inferior **2**: 188, 200, 202–203
– – superior **2**: 188, 199–203, 207, 223–224
– metacarpales dorsales **1**: 164
– metatarsales dorsales **1**: 278
– nasales externae **2**: 80
– nasofrontalis **2**: 80
– nuclei caudati **2**: 300
– obliqua atrii sinistri cordis **2**: 143
– obturatoria **2**: 244, 252, 254, 256
– occipitalis(-es) **1**: 60; **2**: 80, 85, 297
– ophthalmica inferior **2**: 58, 80, 379, 383–384
– – superior **2**: 58, 80, 379, 382–384
– ovarica **1**: 14; **2**: 162–163, 238–239
– pancreaticae **2**: 188
– pancreaticoduodenales **2**: 188
– paraumbilicales **1**: 75
– parietales **2**: 297, 300
– pelvis **2**: 234, 244
– perforantes **1**: 246, 262–263, 270
– – epifasciales **1**: 246
– – subfasciales **1**: 246
– perinealis(-es) **2**: 276, 279
– plantaris lateralis **1**: 214
– – medialis **1**: 214
– poplitea **1**: 245, 247, 255–256, 261, 264–265, 267–268, 272
– portae hepatis **1**: 12–13; **2**: 181, 183, 185, 188, 196, 198–199, 201, 203, 213, 248
– posterior septi pellucidi **2**: 300
– profunda brachii **1**: 145–146
– – facialis → Vena profunda faciei
– – faciei **2**: 80, 383
– – femoris **1**: 260, 262
– pudenda(-ae) externae **1**: 248
– – interna **1**: 257; **2**: 251, 255–256, 276, 279–280
– pulmonales **1**: 12–13; **2**: 96, 115, 155
– – dextrae **2**: 111, 113, 123, 125, 131–133, 139, 141, 143, 158, 164–165, 167
– – sinistrae **2**: 107, 111, 123–125, 131, 133, 139–141, 143, 157–158, 164
– radialis **1**: 140, 164
– rectalis inferior **2**: 276
– – media **2**: 276
– renalis **1**: 14; **2**: 162, 208, 210–212
– – dextra **2**: 213, 218
– – sinistra **2**: 199, 203, 213–214, 218, 223
– retromandibularis **2**: 80, 98–99, 383
– sacralis lateralis **2**: 244, 287
– – mediana **1**: 14; **2**: 162, 218, 234, 244, 255
– saphena accessoria **1**: 245, 248, 254–255, 266; **2**: 228

Vena(-ae) saphena
– – magna **1**: 75, 245–249, 252, 254, 262–267, 270, 272, 274, 278; **2**: 163, 228, 256–259
– – parva **1**: 245, 247, 254–256, 266–268, 272, 274–275, 277–278
– sigmoidea **2**: 202
– spinales anteriores **2**: 285–286
– – posteriores **2**: 285–286
– splenica **2**: 188–189, 198–200, 203, 207, 221
– sternocleidomastoidea **2**: 99
– subclavia **1**: 12, 14, 64, 72–73, 75, 137; **2**: 93, 97, 122, 149, 162–164
– sublingualis **2**: 47, 99
– submentalis **2**: 80, 383
– superficiales cerebri **2**: 297
– – membri inferioris **1**: 246
– superiores cerebelli **2**: 295
– – cerebri **2**: 294, 296–297
– supraorbitalis **2**: 80
– suprarenalis **1**: 14; **2**: 162
– suprascapularis **2**: 122
– supratrochlearis **2**: 80, 85, 383
– temporalis media **2**: 85
– – superficialis **2**: 80–81, 85, 98–99, 383
– terminalis → Vena thalamostriata superior
– testicularis **1**: 77; **2**: 162, 218, 228, 270–272, 274, 278
– thalamostriata superior **2**: 300, 322, 326, 344, 346, 348
– thoracica interna **1**: 75–76; **2**: 97, 122, 149, 151, 154, 162
– – lateralis **2**: 93
– thoracis **1**: 75–76
– thoracoacromialis **1**: 73, 137
– thoracoepigastrica **1**: 75; **2**: 149
– thyroidea inferior **2**: 97, 122, 149, 151, 162, 164
– – media **2**: 97, 149
– – superior **2**: 80, 98–99, 149
– tibialis anterior **1**: 247, 273–275
– – posterior **1**: 246–247, 267, 273–275, 277
– transversa cervicis **2**: 149, 164
– trunci corporis **1**: 14
– ulnaris **1**: 164
– umbilicalis **1**: 13; **2**: 243
– uterina **2**: 251
– vaginae nervi optici **2**: 373
– ventriculi dextri cordis anterior **2**: 135, 142
– – sinistri cordis posterior **2**: 135, 143
– vertebralis **2**: 68, 73, 75, 165, 286
– vesicalis inferior **2**: 276
– vorticosae **2**: 372–373, 381
Venen, Achselhöhle **2**: 93, 149
– Augapfel **2**: 372–373
– Augenhöhle **2**: 383
– Bein **1**: 246–247
– Boyd-Venen **1**: 246
– Brusthöhle **2**: 97
– Brustkorb **1**: 76
– Brustwand **1**: 75; **2**: 93
– Cockett-Venen **1**: 246
– Dodd-Venen **1**: 246
– Fussrücken **1**: 278
– Gehirn **2**: 300
– Gesäss **1**: 255–256
– Gesicht **2**: 80
– Hals **2**: 93, 97–99, 149, 164
– Herz **2**: 142–143
– hintere Rumpfwand **2**: 162
– – Bauchwand **2**: 228
– Hirnhäute **2**: 290
– kleines Becken **2**: 234, 244
– Knie **1**: 252
– Kniekehle **1**: 255–256, 267–268
– männliches kleines Becken **2**: 234

Venen
– Mediastinum **2**: 149, 164
– oberflächliche, Bein **1**: 246
– – Ellenbeuge **1**: 151
– – Fuss **1**: 270
– – Fussrücken **1**: 278
– – Gehirn **2**: 297
– – Gesäss **1**: 254
– – Gesicht **2**: 80, 85
– – Hals **2**: 91–92
– – Hand **1**: 164
– – Handrücken **1**: 164
– – Kniekehle **1**: 254, 266
– – Kopf **2**: 85
– – Oberarm **1**: 135–137
– – Oberschenkel **1**: 248, 254
– – Schulter **1**: 135–137
– – Unterarm **1**: 150–151
– – Unterschenkel **1**: 266, 270
– Oberschenkel **1**: 252, 255–256; **2**: 228
– Orbita **2**: 383
– perforierende, Bein **1**: 246
– Rachenwand **2**: 100
– retroperitoneale Organe des Oberbauchs **2**: 203
– Rückenmark **2**: 286–287
– Spatium peripharyngeum **2**: 100
– subkutane → Venen, oberflächliche
– tiefe, Bein **1**: 246
– – Gesicht **2**: 80
– Unterschenkel **1**: 267–268
– vordere Bauchwand **1**: 75, 77
– weibliches kleines Becken **2**: 244
– Wirbelsäule **2**: 286–287
– Zunge **2**: 97
Venenstern im Hiatus saphenus **1**: 248
Venter anterior musculi digastrici **2**: 32, 34, 36, 44–45, 47–48, 98–99
– frontalis musculi occipitofrontalis **2**: 28–30, 85, 366
– inferior musculi omohyoidei **1**: 106–107, 110, 140, 143; **2**: 34–36, 92, 98
– musculi **1**: 10
– occipitalis musculi occipitofrontalis **1**: 53–54, 60; **2**: 29–30, 32–33, 85, 91–92
– posterior musculi digastrici **2**: 30, 34, 37, 44, 49–51, 87, 98–100
– superior musculi omohyoidei **2**: 34, 36, 81, 98–99
Ventralis **1**: 5
Ventricle(s), brain **2**: 320
– fourth, brain **2**: 307, 310–311, 342–343, 359
– lateral, anterior horn **2**: 337, 346–349, 354–355, 360
– – brain **2**: 300, 322–323, 326
– – central part **2**: 326, 354–355, 361
– – inferior horn **2**: 362
– – posterior horn **2**: 342–343, 346–349, 354–356
– – temporal horn **2**: 326
– left, heart **2**: 133–134, 140
– right, heart **2**: 119, 132, 134, 140
– third, brain **2**: 307, 326, 348–349, 355–356, 361
Ventriculus(-i) → Gaster
– dexter cordis **1**: 12–13; **2**: 104, 119, 124–125, 128–129, 132, 134, 136–137, 139, 141, 142–143, 145, 154, 158–159, 160, 166, 167
– encephali **2**: 320
– laryngis **2**: 52, 68
– lateralis(-es) **2**: 289, 300, 302, 320, 322–323, 326, 332–335, 337–344, 346–350, 354–358, 360–362, 385, 408
– quartus **2**: 304–305, 307–311, 320–321, 324, 342–343, 359, 409

Ventriculus(-i)
– sinister cordis **1**: 12–13; **2**: 107, 122, 124–126, 128–134, 137–139, 141–143, 145, 154, 159–160, 166–167
– tertius **2**: 299, 301, 304, 307, 310, 320–321, 326–327, 329, 332, 338–341, 344, 348–351, 355–357, 361, 408
Ventrikel, dritter, Gehirn **2**: 307, 326, 348–349, 355–356, 361
– Gehirn **2**: 320
– vierter, Gehirn **2**: 307, 310–311, 342–343, 359
Venula macularis inferior **2**: 377
– – superior **2**: 377
– medialis retinae **2**: 377
– nasalis retinae inferior **2**: 377
– – – superior **2**: 377
– temporalis retinae inferior **2**: 377
– – – superior **2**: 377
Verbindungen, zerebellorubrale **2**: 334
Verdauungssystem **1**: 16; **2**: 52–53
Vermis cerebelli **2**: 306, 308–309, 322, 334, 342–343, 345, 348, 350–353, 356–359, 409
Vertebra(-ae) **1**: 29
– cervicalis(-es) **1**: 28–31, 36–37, 40, 49, 51, 60, 82–83; **2**: 7, 33, 37, 53, 68, 71–75, 79, 118, 284, 286, 289, 294, 302, 321, 408
– – I → Atlas
– – II → Axis
– coccygeae → Os coccygis
– homologous parts **1**: 29
– lumbales **1**: 28–29, 33, 38–40, 48, 55, 82–83, 170–171, 173, 190–191, 219; **2**: 105, 182, 194, 203, 205, 213–214, 217, 220–221, 223–224, 243–245, 253, 264, 284, 287
– lumbares → Vertebrae lumbales
– prominens **1**: 28, 31, 37, 40, 60
– sacrales **1**: 35, 191, 194; **2**: 225, 232, 243–244, 260–261, 264
– thoracicae **1**: 28–29, 32, 36, 38, 42–43, 47, 60; **2**: 37, 72, 75, 104, 107, 127, 156–161, 182, 209, 214–215, 217, 222, 284
Vertex corneae **2**: 370
– cranii **2**: 6
Vesica biliaris **1**: 16, 18, 20, 79; **2**: 154, 172, 180–181, 183, 186–187, 190–191, 196, 198, 200, 204, 208, 223, 248
– fellea → Vesica biliaris
– urinaria **1**: 12–13, 16–18, 20–21, 77, 79, 194; **2**: 163, 170, 172, 193, 202, 204, 206, 211, 217–218, 220–221, 232–238, 242–244, 246–247, 250, 252, 254, 256–261, 276–278
Vesicula seminalis **1**: 17, 77, 260; **2**: 234, 236–237, 244, 247, 255, 258–259, 261
Vestibulum labyrinthi **2**: 400–402, 406
– laryngis **2**: 68, 74
– nasi **2**: 52, 55–57
– oris **1**: 16; **2**: 45, 52
– vaginae **2**: 233, 269
Vinculum(-a) tendinum digitorum manus **1**: 124
Viscera, abdominal **1**: 18–21; **2**: 154–155, 170–172, 190–193, 199
– colli **2**: 97
– pelvic **2**: 232–234, 238
– – female **2**: 233, 238–243, 245
– – male **2**: 232, 234, 236–237, 244
– retroperitoneal, of upper abdomen **2**: 199
– thoracic **1**: 18–21; **2**: 126–127, 150–151, 154–161, 172
Viscerocranium **2**: 369
Vomer **2**: 8, 12–17, 23, 58, 395
Vorderhirn, Adergeflechte **2**: 326
– Kerne **2**: 329–332
– Plexus choroidei **2**: 326
Vorderhorn, Seitenventrikel **2**: 337, 346–349, 354–355, 360
Vorhaut, Penis **2**: 276
Vorhof, linker, Herz **2**: 133
– rechter, Herz **2**: 132, 140
Vorsteherdrüse **2**: 235–237
V-phlegmona **1**: 131
V-Phlegmone **1**: 131

W

Wadenbein **1**: 182–183
– Muskelanheftungen **1**: 235
Waldeyer-Rachenring **2**: 45
Waldeyer's ring of lymphoid tissue **2**: 45
Wall, abdominal, anterior **1**: 75, 77, 79; **2**: 238, 272–273
– – – arteries **1**: 77
– – – veins **1**: 77
– – posterior **2**: 206–207, 228
– – – arteries **2**: 228
– – – nerves **2**: 228
– – – veins **2**: 228
– lateral, membranous, of tympanic cavity **2**: 398
– medial, labyrinthine, of tympanic cavity **2**: 396, 398
Wand, laterale, der Paukenhöhle **2**: 398
– mediale, der Paukenhöhle **2**: 396, 398
Ward-Dreieck des Oberschenkelknochens **1**: 7
Ward's triangle of femur **1**: 7
Wirbel, homologe Abschnitte **1**: 29
Wirbelkanal **2**: 284–285, 289
Wirbelsäule **1**: 28, 30–39, 42–43, 46–51
Wirbelsäulenarterien **2**: 286
Wirbelsäulenvenen **2**: 286–287
Wirbelschlagader **2**: 292, 302
Woman, multiparous **2**: 268
– nulliparous **2**: 268
Wrist, synovial sheaths **1**: 130–131
Wurmfortsatz **2**: 177–178, 195
– verschiedene Lagen **2**: 178

Z

Zahnbogen **2**: 24–25
Zähne, bleibende **2**: 39–41
– Durchbruch **2**: 39, 42
– Innervation **2**: 86
– Wurzeln **2**: 41
Zenker-Divertikel **2**: 49
Zenker's diverticulum **2**: 49
Zentralteil des Seitenventrikels **2**: 326, 354–355, 361
Zervixpfeiler **2**: 246
Ziliarkörper **2**: 376
Zirbeldrüse **2**: 348
Zirkulation des Liquor cerebrospinalis **2**: 321
Zöliakographie **2**: 197
Zona cutanea canalis analis **2**: 179
– fasciculata glandulae suprarenalis **2**: 216
– glomerulosa glandulae suprarenalis **2**: 216
– haemorrhoidalis canalis analis **2**: 179
– incerta **2**: 340
– orbicularis articulationis coxae **1**: 192
– reticularis glandulae suprarenalis **2**: 216
Zonula ciliaris **2**: 373–377
Zugspannungstrajektorien **1**: 7
Zunge **2**: 38, 46–47
Zungenarterien **2**: 97
Zungenbein **2**: 27, 35, 47
– Muskelanheftungen **2**: 34–35
Zungenbeinmuskeln, obere **2**: 36
Zungenmuskeln **2**: 38, 46–47
Zungennerven **2**: 97
Zungenrücken **2**: 46
Zungenvenen **2**: 97
Zwerchfell **1**: 69
Zwischenhirn **2**: 334
Zwischenknochenmuskeln, dorsale, Fuss **1**: 239
– – Hand **1**: 126–127, 129
– Fuss **1**: 239
– Hand **1**: 126–127, 129
– palmare, Hand **1**: 126–127
– plantare, Fuss **1**: 239
Zwischenwirbelscheibe **1**: 31, 48
Zygapophysis → Processus articularis inferior/superior vertebrae
Zysten, persistierende, im Samenstrang **2**: 273